T0251557

HEAT TRANSFER ENHANCEMENT AND ENERGY CONSERVATION

HEAT TRANSFER ENHANCEMENT AND ENERGY CONSERVATION

Editor in Chief

Song-Jiu Deng
Research Institute of Chemical Engineering
South China University of Technology, Guangzhou, PRC

Editors

T. N. Veziroğlu
Clean Energy Research Institute
University of Miami
Coral Gables, Florida, USA

Ying-Ke Tan
Lie-Qiang Chen
Research Institute of Chemical Engineering
South China University of Technology, Guangzhou, PRC

CRC Press
Taylor & Francis Group
Boca Raton London New York

CRC Press is an imprint of the
Taylor & Francis Group, an **informa** business

Contents

v

CONDENSATION

BOILING AND EVAPORATION

FURNACE AND REACTORS

ANALYSIS OF INDUSTRIAL PROCESS

MISCELLANEOUS

Preface

The International Symposium on Heat Transfer Enhancement and Energy Conservation (ISHTEEC) was held at the South China University of Technology on August 2–5, 1988. The symposium was sponsored by the Research Institute of Chemical Engineering of the South China University of Technology, the Clean Energy Research Institute of the University of Miami, and the Guangzhou Institute of Energy Conversion of the Chinese Academy of Sciences. Also, the symposium was supported by the K. C. Wong Education Foundation.

Since 1973, when the world-wide energy crisis stimulated progress in the technology of heat transfer onhancement and energy conservation, many brilliant achievements have been attained in these two areas of research. However, the shortage of energy and the low efficiency of energy utilization are two remaining obstacles to the improvement of the standard of living for all mankind, especially for those in developing countries with large populations and limited energy resources. Spreading current knowledge about energy conservation more efficiently and developing more useful energy conservation techniques will certainly aid in over-coming these obstacles.

Heat transfer enhancement is usually closely connected with the better utilization of thermal energy. ISHTEEC (Guangzhou 1988) provided a place for the international exchange of the latest information concentrating on the status of heat transfer enhancement and energy conservation research, development, applications, and the relation between these two scopes.

The Symposium admitted scientific papers and accepted scholars from countries and areas including Bahrain, Canada, China, France, the Federal Republic of Germany, Hong Kong, Italy, India, Japan, New Zealand, Poland, the United Kingdom, the United States of America, and Yugoslavia. One hundred twelve papers in full text and 14 papers in the form of abstracts are published in this book. It is expected that this book will be a valuable reference for people who are interested in heat transfer enhancement and energy conservation.

It is a great pleasure for me to take this opportunity to express our sincere appreciation to members of the Organizing Committee of ISHTEEC, to the authors of papers, invited keynote speakers, to

Professor T. N. Veziroğlu for his help in printing and distributing the symposium announcement, to K. C. Wang for his financial support, and to all the persons who made contributions to the Symposium and the Proceedings.

Song-Jiu Deng

Organizing Committee

Aung, Win, *National Science Foundation, U.S.A.*
Bergles, A. E., *Rensselaer Polytechnic Institute*
Chen, J. C., *Xian Jiaotong University*
Durst, F., *Friedrich Alexander University*
Fujii, Tetsu, *Kyushu University*
Hijikata, Kunio, *Tokyo Institute of Technology*
Holland, F. A., *Salford University*
Kakaç, S., *University of Miami*
Lin, Jifan, *Dalian University of Technology*
Liu, Zhenqun, *South China University of Technology*
Mayinger, F., *Technische University Munchen*
Nakayama, Wateru, *Hitachi, Ltd.*
Ogino, Fumimaru, *Kyoto University*
Tan, Yinke, *South China University of Technology*
Tanasawa, Ichiro, *University of Tokyo*
Tien, C. L., *University of California*
Tseung, A. C. C., *The City University, London*
Veziroğlu, T. N., *University of Miami (Co-Chairman)*
Wang, Buxuan, *Guangzhou Institute of Conversion (Co-Chairman)*
Wu, Wen, *Guangzhou Institute of Energy Conversion (Co-Chairman)*
Wu, Zhijian, *Guangzhou Institute of Energy Conversion*

General Secretary

Chen, Lieqiang, *South China University of Technology*

KEYNOTE REPORTS

Energy Problems in Chemical and Allied Industries in China and Corresponding Research Work in South China University of Technology

SONG-JIU DENG
Chemical Engineering Research Institute
South China University of Technology
Wushan, Guangzhou 510641, PRC

Abstract
   The importance of energy conservation and heat transfer enhancement research to the development of industries in China is discussed. In respect to the energy problems in the chemical and allied industries, a review of the research work in the field of heat transfer enhancement and energy conservation in South China University of Technology is presented.

Introduction
   China is the third largest energy consumer in the world, yet in per capita terms the figure is in the level of developing countries [1]. According to the national economic developing program, the annual energy production is expected to have 2-fold increase during the years of 1980-2000, in the same period, the industrial and agricultural annual production is required to have 4-fold increase. Industries share more than 60% of the total energy consumption. The efficiency of energy utilization in our industries is quite low as compared to those of the developed countries at the level of 1980. Many authors have predicated that energy saving of approximately 30% might be realized through the application of present day energy conservation technologies to the industries in U.S.A.[2]. It is necessary and also possible for us to fulfil the national industrial developing program with the application of proper energy conservation technologies to our industries.
   The chemical and allied industries (including chemical, petroleum, food and paper etc.) consumed about 20% of the total energy, as an institue of chemical engineering, it is very interesting for us to concern about the energy conservation problems in these industries.

Current Situation of the Chemical and Allied Industries in China

   In China, most of the small and/or middle size chemical plants were built in the decade of 1950-1960, which were designed without the consideration of adequate utilization of the thermal energy, most of the cold streams are heated to the desired temperature by the steam directly coming from a low pressure boiler, and the hot streams are cooled by cooling water. In recent years some improvements have been adopted in many plants but energy consumption is still 50-100% higher than those in the advanced countries, only a few of the plants are matching to the advanced level. For example, the energy consumption of the atmospheric and vacum distillation unit for crude petroleum has been decreased from 1.1 to 0.6 billions Joule per ton oil through the improvement in heat exchanger network and other arrangements. A few of new systems are designed for ever

3

lower energy consumption, which are close to the advanced level (about 0.42 billions Joule per ton oil). One of the barriers limiting further improvements is the low heat transfer coefficient of the heat exchangers [4].

In the decade of 1970-1980, many large scale chemical plants were imported from the developed countries. The energy utilization systems are much better than the old ones in China, the energy consumption per unit weight of product is close to the advanced level in 1980's. But there are still many places which can be improved. For example, after a detail analysis of an imported organic chemical assembly, we found that the energy consumption per unit weight of product can be down to 60% of the original value with an additional investment which can be paybacked in about one year[5].

We should learn the advanced energy conservation techniques from all other countries, and make full use of these techniques to improve the energy utilization efficiency of the industries in our country.

Research Work on Heat Transfer Enhancement and Energy Conservation in South China University of Technology

Better use of the thermal energy in the chemical and allied industries is one of the key respects to save energy. According to the second law of thermodynamics, the higher the efficiency of thermal energy utilization, the lower the heat transfer temperature drop which must be adopted for the heat exchangers used in the heat recovery system[3]. A good heat utilization system must have high thermal energy efficiency, proper equipment investment and low pumping power requirement, so heat transfer enhancement research is important for the optimal design of a recovery system.

1. Heat Transfer Enhancement

1.1 Spiral Ridged Tube(S.R.Tube)
S.R. tube is a simple and effective means in heat transfer enhancement, studies on this kind of tube have been reviewed in reference [6]. About the unclear mechanism of heat transfer enhancement in S.R.tube, Deng[7] suggested that there are two kinds of flow patterns inside the S.R. tube; the spiral ridges with large interception angle with the tube axis will produce more separation wake behind the ridges and the spiral ridges with small interception angle will produce more spiral flow. Theoretical analysis showed that the former kind of flow pattern will give better heat transfer enhancement with less fluid power. Li[8] proved this predication by using hydrogen bubble flow visualization technique and found S.R. tube with the largest interception angle has the highest heat transfer enhancement possibility as compared with the smooth tube at the same pumping power consumption using water as tested fluid. Comprehensive correlations for heat transfer coefficient and friction factor have been provided on the basis of large amount of tested tubes and experiment data. Ye[9] and Xiao[10] did similar experiments with air and high viscosity glycol solutions as tested fluids, determined the velocity profile inside S.R. tube by a hot-wire anemometer and made a modification to Li's correlation. All the above correlations are only suitable to single start S.R. tube. Deng[11] made the theoretical analysis of the thermal resistances of the laminar, buffle and turbulent core regions in the smooth tube. By applying this knowledge, a generalization heat transfer correlation for both single and multiple start S.R. tube has been obtained[12].

4

Flow boiling inside S.R. tubes was also studied with F-113 as boiling
fluid, 50-100% heat transfer enhancement result has been observed[13].
Addition to the laboratory work, S.R. tube also has been successfully
used in the tube and shell heat exchangers in many places. For the case
of corrosive fluids, S.R. tube had been used in the stainless steel waste
heat boiler of formalin hot gas mixture from the chemical reactor[14],
in the titanium heater for corrosive solutions in electro-plating[15].
For the case of scale forming fluids, S.R. tube had been used as raw sugar
juice heater in sugar mills and crude oil heater in petroleum plants,
they showed much better antifouling ability than smooth tubes. It   also
showed good performances in steam condenser and other applications[16].
These experiences showed that S.R. tube can be successfully used in
chemical and allied industries.

1.2. Tube Inserts
Tube inserts have been extensively investigated by many authors for
a long time. Huang[17] studied the performance of different kinds of tube
inserts for the augmentation of convective heat transfer of gas at low
Reynolds number (500-10,000), comparing with common inserts such as twisted taps, coils and static-mixers etc., he found that the narrow ribbon
coil showed higher heat transfer enhancement   under the same perssure
drop.Xu[18] studied the tube inserts for heat transfer augmentation of
high viscosity fluids at Re from 300 to 3,500, a new kind of tube insert
(cross interconnected ribbon) has been developed,which showed better performance than other kinds of inserts. Liu[19] studied the mechanism of
vapor condensation inside a tube with different kinds of tube inserts.
Many of the above tube inserts have been adopted by chemical industries.
Topics mentioned above mainly deal with the heat transfer of single-
phase fluids, more have been done for the heat transfer enhancement of
vapor condensation, sublimation and liquid boiling. Professor Tan of our
university will give more detail description about this subject, only
brief introduction will be presented here.

1.3 Three kinds of tubes have been studied for the heat transfer enhancement of vapor condensation, especially for organic compounds, such as freons, alkanes, olefines etc.. They are saw-teeth fin tube, radially-ridged-
fin tube and longitudianly fluted tube. The heat transfer mechanism of
saw-teeth tube was studied[20]. Based on this study, a new kind of low
fin tube, the radially-ridged fin tube has been developed with the condensing heat transfer coefficient 10-30% higher than the saw-teeth tube.
The circumferential temperature distribution has been investigated and
a more detail discussion of the mechanism heat transfer augmentation is
presented[21]. Vertically placed longitudinally fluted tube with 1-3
meters in length has· been studied with isobutane condensed on the fluted
outside surface. The condensation heat transfer coefficient was found
around 4,000-5,000(w/m^2.C) which is 4-6 times larger than smooth tube[22].

1.4 Three kinds of tubes have been studied for the heat transfer enhancement for liquid boiling, especially for organic compounds and liquid solutions. They are mechanically-made porous surface tube and its improved
type, and T-shape tunnel tube. The heat transfer enhancement mechanism
of porous surface tube was studied[23,24] and a new type of mechanically-
made porous surface tube with finned tunnel has been developed. Under
a low and medium heat load 1-4X10^{4}w/m^2 the superheat for bubble generation
of the new porous tube may be lowered to 0.6-2°C, it also has a much higher critical heat flux than the common porous surface tube[25]. The effect
on the antifouling and the mechanism of the porous surface tube for water

solutions of NaCl and CaSO$_4$ have been studied, it was found that porous tube with suitable structure parameters have much better antifouling ability than smooth tube[26].

1.5 The tube side flow liquid boiling and flow vapor condensations have been studied inside internal spiral fin tubes, an increase of heat transfer coefficient from 60% to 220% has been obtained, compared with common smooth tubes[27,28].

1.6 The mechansim of simultaneous heat and mass transfer in the partial condensation and desublimation process of phthalic anhydride vapor from gas mixture containing large amount of noncondensible gas has been studied. A kind of zigzag segmented fin tube with optimized structure parameters was recommended for the desublimation process with high desbulimating rate[29]. The longitudinally fluted tube with small spiral angle was recommended for the vertical partial condenser of phthalic anhydride vapor gas mixture of higher concentration[30]. Heat exchangers composed of these kinds of tubes are now operated successfully in phthalic anhydride plants, compared to the old heat exchangers, much lower heat transfer area and much better capture of phthalic anhydride vapor have been achieved with new equipment. Their heat transfer performance can be comparable with the advanced level in the developed countries.

2. Energy Conservation

We have emphasized on the significance of this topic particlarly in the chemical and allied industries in our country in recent many years [3,4] The subjects that we have been investigating in this field will be summerized as follows.

2.1 Thermodynamic Analysis and Systematic Optimization of Energy Utilization in Process Systems.

This subject has been investigated by many authors in the world for many years. Cooperating with SINOPEC (The Petroleum Corporation of China), Hua has studied and developed a systematic method for energy and exergy analysis of process systems based on the investigation of the practice in refineries. "Three Links Model", including the energy conversion, energy utilization and energy recovery links in the whole plant, has been applied to different extent in many SINOPEC's refineries [31, 32]. Incorporating with the Linnhoff's pinch technology[33], Wang [34] extended the application of "Three Links Method" to an imported modern organic chemical plant. The approach consisted of the procedure and principles of energy synthessis, the allied usage of pinch technology and an evaluation strategy based on the exergy-economic evaluation, which harmonizes the optimization of the total system. Their reports showed that the energy consumption per unit weight of one kind of main organic products can be decreased at least 40% with an additional investment of about ten million RMB which can be paybacked by energy saving about one year. Furthermore, based on the thermodynamic basic and rigorous first and second laws, a benzene hydrogenation process has been systematically analyzed by Chen[35]. A new concept, energy degradation has been employed and optimum reaction temperature of the benzene hydrogenation reactor has been found out, and various thermodynamic efficiencies of the process have been meaninffully defined.

6

2.2 Incorporation Between the Optimization of Both the Energy Utilization System and Heat Transfer Equipment

In the past, optimization of energy utilization system was composed of heat exchangers with constant given heat transfer coefficients. Higher thermal energy utilization efficient could be obtained at the expense of more heat transfer area. If optimization of heat transfer equipment design (including the adoption of suitable heat augmentation technique, choice of optimization operating conditions) can be proceeded incorporating with the optimization of energy utilization system, more efficient energy utilization system can be obtained with less equipment investment and fluid pumping power consumption. For these purposes, we have been developing some softwares in order to get optimization use of the heat transfer enhancing techniques in practical industrial design application. Computer programs for optimizing the characteristic parameters of many kinds of enhancing tubes have been complied. In a recent work, a mode-opting program has been developed, which involves three types of enhancing technique based on the national standard series of shell-and-tube exchanger and can be used for practical design[36].

2.3 Investigation of Energy Recovery Equipment

In our university, an absorption heat pump with water-lithium bromide as working medium has been investigated. Lo [37] developed a refrigerator with hot water as heat source. Three kinds of enhanced tubes have been used in the heat exchangers in the absorption system, so that its total heat transfer area is much less than the same type of refrigerator with smooth tube. Lin [38] and Li [39] investigated the computer simulation of the use of absorption heat transformer for the recovery of the waste heat from water vapour at 80°C in sugar mills, computed graphs showed optimum design and operating conditions for this kind of heat pump. If heat transfer augmantation techniques are adopted, a heat transformer of 1--2 Billions Joule/hr capacity with a temperature rise of 30°C may be pay-backed within one year by energy saving. For the heat transfer augmentation process of lithium bromide solution, Ma[40] and Yuen [41] have studied the mechanism of the heat and mass transport process in the liquid layer and developed some kinds of tubes which showed an increase in heat transfer coefficient more than 100% as compared with smooth tube.

Sun [42] has studied another type of heat pump, metal hydride heat pump, with a detail study of the heat and mass transfer process in the metal hydride bed. A computer program for the optimum design of metal hydride container was finished.

The topics I mentioned above is a part of our research work in the recent ten years. Much more work need be done in the future and we may learn a lot of latest technologies, techniques and information at this symposium from the attendents.

References
1. Smil, V. and Knowland, W.E., "Energy in the Developing world, the Real Energy Crisis" Oxford University Press, 1980
2. Chiogioji, M.H.,"Industrial Energy Conservation" Marcel Dekker Inc. 1979
3. Deng, S.J., Heat Transfer and Energy Conservation, Chemical Engineering, No.6, p.1, 1981.[*]
4. Hua, B., Energy Conservation and Update of Heat Transfer Equipment

in Chinese Refineries, Petroleum refinning No. 3, p.28, 1984[*]
5. Wu, K.D., " The Improvement and Application of Optimization Method for Energy Use in Chemical Process" M.S. Thesis in SCUT, 1988[*]
6. Li, H.M., Ye, K.S., and Deng, S.J., The Investigation and Application of Spiral Ridged Tubes, A Review, Journal of Chemical Industrial and Engineering, No.4, p. 259, 1982[*]

7. Deng, S.J. et al., Investigation of the Heat Transfer and Fluid Friction in Spiral Ridged Tubes, Chemical Engineering, No. 6 p.1, 1980[*]
8. Li, H.M., Ye, K.S., Tan, Y.K. and Deng, S.J., Investigation on the Tube-side Flow Visualization, Friction Factor and Heat Transfer Characteristics of Helical-Redging Tubes, Proceedings of the Seventh International Heat Transfer Conference, Vol.3. No.2, p.57, 1982[*]
9. Ye, Q.Y., Tan, Y.K., Deng, S.J., Investigation of Fluid Friction and Heat Transfer Characteristics of Spiral Ridged Tubes, Journal of Chemical Engineering of Chinese Universities, Vol.2. No.2, p.57, 1987[*]
10. Xiao, J.W., "Investigation of Heat Transfer Performance of High Viscosity Liquids in Spiral Ridged Tubes" M.S. Thesis in SCUT, 1986[*]
11. Deng, X.H., and Deng, S.J., An Approximate Analytical Solution for the Nesselt Number Expression in Turbulent Pipe Flow with Constant Physical Properties, Journal of Chemical Industry and Engineering No.4, p.494, 1987[*]
12. Deng, X.H., Tan, Y.K. and Deng, S.J., Investigation of the Correlation of Heat Transfer in Single and Multiple Start Spiral Tubes with Single Phase Turbulent Flow, Advances in Heat Transfer Enhancement and Energy Conservation, Hemisphere Pub. 1989
13. Wei, H.L. and Tan, Y.K., Heat Transfer and Pressure Drop Characteristics of Refrigerants Evaporation in Horizontal Spiral-Corrugated Tubes, Proceedings of 4th Miami Int. Symp. on Multiphase Transport Phenomena, 1986
14. Luo, Y.L., Design of Waste Heat Boiler for Formalin Production, Paper presented in 3rd Chinese National Conference of Chemical Engineering, Guangzhou, 1981 [*]
15. Ye, G.X., Spiral Ridged Tube Heat Exchanger Used in Electroplating plant, Paper presented in Guangdong Advanced Technique Conference, 1984 [*]
16. Qian, S.W. et al,. "The Application of Redged Tube Heat Exchanger for the Heat Exchange of Crude Oil and Residual Oil" Internal Report in Technical Appraisement by Nanking Petroleum Refinery Works. 1985[*]
17. Huang, G.H., Cui N.Y., Lu, Y.S., Tan, Y.K. and Deng, S.J., The Investigation of Heat Transfer Augmentation of Single Phase Flow by Tube Inserts. Journal of Chemical Industry and Engineering, No.1, p. 23, 1983 [*]
18. Xu, T.H., Cui, N.Y. and Tan Y.K., Heat Transfer Augmentation of High Viscosity Liquids by Tube Inserts, Guangdong Chemical Technology, No. 4, p. 1, 1986[*]
19. Liu, Z.Q., Tan, Y.K. and Cui, N.Y., The Enhancement of R-113 Condensation Inside Horizontal Tubes with Novel Tube Inserts, " Advances in Phase Change Heat Transfer" proceedings of ISPCHT, Chongqing, China, 1988
20. Wang, S.P., Tan, Y.K., and Dend S.J., Investigation of the Condensation Heat Transfer on Saw Teeth Tubes, Journal of Engineering Thermophysics, No.4, p.374, 1984[*]
21. Wang, S.P., Yin, Q.H., Tan, Y.K. and Deng, S.J., Investigation of Condensation Heat Transfer Enhancement Mechanisims of Particularly Shaped Fin Tube with Numerical Method, Advances in Heat Transfer Enhancement and Energy Conservation, Hemisphere Pub. 1989
22. Tan, Y.K. and Deng, S.J. A Novel and High Efficient Condenser Used

8

in A Geothermal Electric Power Station, Proceedings of 8th Miami International Conference on Alternative Energy Sources, 1987

23. Cai, B.S., Zhuang, L.X., Tan Y.K., and Deng, S.J.. Investigation of Boiling Heat Transfer Mechanism and Performance of Mechanically Fabricated Porous Surface', proceedings of Joint Meet of CIESC and AICHE" Vol. 2, p.664, 1982

24. Fang, Q.Y. and Deng, S.J., Investigation of Boil Heat Transfer Mechanisim of Mechanially Fabricated Porous Surface Tube Symposium of Heat and Mass Transfer of Chinese Institute of Thermophysics, 1986[*]

25. Zhang, Li, Boiling Heat Transfer Characteristics of Mechanical Fabricated Porous Surface Tubes with Ridgs on the Surface of Tunnels. M.S. Thesis of SCUT, 1989[*]

26. Zheng, K.M. and Deng, S.J., Antifouling Characteristics of Mechanically Fabraicated Porous Surface Tubes, Proceedings of the Symposium of Thermophysics in Chinese Universities, Oct.1986, Paper No.862024[*]

27. Lu, Y.S., Zhuang, L.X. and Ruan, Z.Q., Internal Spiral Fin Tube for the Heat Transfer Enhancement of Flow Boiling and Condensation, Journal of Refrigeration, No.1, p.49, 1986[*]

28. Yang, W.N., Tan, Y.K., Lu, Y.S., Wei, H.L. and Ruan,Z.Q., Investigaof Heat Transfer Enhancement of Flow Boiling of R-12 inside Internal Spiral Fin Tubes, Chemical Engineering, No.1,p.19, 1988[*]

29. Yang, X.X., Ye, G.X., Cai, J.D., Tan, Y.K. and Deng,S.J., Investigation of Desublimation Process of Phthalic Anhydride Vapour Mixture on Single Finned Tube, Advances in Heat Transfer Enhancement and Energy Conservation, Hemisphere Pub. 1989

30. Zhou, H.S., Ye, G.X., Yang, X.X., Cai, J.D., Deng, S.J., Condensation Process of Dilute Phthalic Anhydride Vapour Mixture on a Vertical Surface, Advances in Heat Transfer Enhancement and Energuy Conservation, Hemisphere Pub. 1989

31. Hua, B., The Analysis and Synthesis of Process Energy Systems, Hydrocarbon Processing Press, 1988[*]

32. Hua, B., A Systemic Methodology for Analysis and Synthesis of Process Energy Systems, ASME Proceedings, AES. Vol.2-1, p.57, 1986

33. Linhoff, B. and Ahmad, S., Optimum Synthesis of Energy Management Systems, ASME Proceedings, AES. Vol.2-1. p. 1, 1986

34. Wang, W.J., Investigation of the Synthesis of Energy Systems of a Complex Chemical Process, M.S. Thesis of SCUT, 1988[*]

35. Chen, L.Q. and Ruan, F.C., Energy Degradation Analysis of a Banzene Hydriogenation System, Advances in Heat Transfer Enhancement and Energy Conservation, Hemisphere Pub. 1989

36. Hua, B. and Xu, T.H., The Exergy-Economic Evaluation and Optimization of Shell and Tube Heat Exchangers and its Heat Transfer Enhancement, Advances in Heat Transfer Enhancement and Energy Conservation, Hemisphere, Pub. 1989

37. Luo, Y.L., Xu, Q.C., Chang, X.X., Development and Operation of the Hot-water Type Lithium Bromide Absorption Refrigerator, Guangdong Energy Conservation, No. 1, p. 5, 1985[*]

38. Lin, W.T., The Research and Development of Absorption Thermal Energy Transformer, M.S. Thesis of SCUT, 1987[*]

39. Li, H., The Study of the Performance and Economics of Type-II Lithium Bromide Absorption Heat Pump, M.S. Thesis of SCUT, 1988[*]

40. Ma, S.P., Chen, L.Q., Luo, Y.L., and Deng, S.J., Flow Analysis of Falling Film Absorption Process Enhanced by New-type Finned Tubes in Libr Absorption Refrigerator. Advances in Heat Transfer Enhancement and Energy Conservation, Hemisphere Pub. 1989

41. Ruan, F.C., Chen, L.Q., Lo, Y.L.and Deng, S.J., The Investigation

of Heat Mass Transfer in Absorption of Low Pressure Water Vapour by a solution of Lithium Bromide in Falling Film Flow, Advances in Heat Transfer Enhancement and Energy Conservation, Hemisphere Pub. 1989

42. Sun, D.W., Li, Z.X. and Deng, S.J., Heat and Mass Transfer Analysis of Metal Hydride Beds, Advances in Heat Transfer Enhancement and Energy Conservation, Hemisphere Pub. 1989

Note. The papers with the remark [*] at the end have been published in Chinese.

The Current Status of Heat Transfer Enhancement

A. E. BERGLES
Rensselaer Polytechnic Institute
Troy, New York 12180-3590, USA

ABSTRACT

During the past twenty-five years, heat transfer enhancement has grown at a rapid rate to the point where it can be regarded as a major field of endeavor, a second generation heat transfer technology. After some historical background, mention of the driving trends, and a review of the various convective enhancement techniques, four areas of major contemporary interest are discussed: structured surfaces for shellside boiling, rough surfaces in tubes, offset strip fins, and microfin tubes for refrigerant evaporators and condensers. The review concludes with developments in the major areas of application.

INTRODUCTION

The enhancement of heat transfer has concerned researchers and practitioners since the the earliest documented studied of heat transfer. In his pioneering paper directed toward development of a temperature scale, Newton(1701) suggested an effective way of increasing convective heat transfer"...not in a calm air, but in a wind that blew uniformly upon it..."

Joule(1861) reported significant improvement in the "conductivity" or overall heat transfer coefficient for in-tube condensation of steam when a wire, spiralled around the condenser tube, was inserted in the cooling water jacket. Whitham(1896) reported increases up to 18% in fire-tube boiler efficiency when "retarders" or twisted-tape inserts were inserted in the tube; it was suggested that the inserts should be used only when "the boiler plant is pushed and the draught is strong."Enhanced surfaces for boiling were part of one of the first systematic studies of nucleate pool boiling by Jakob and Fritz(1931). The U.S. Patent literature dealing with enhanced heat exchangers dates back to the 1920s (shell-side fins, Lea, 1921) as does the manufacturer's literature(corrugated tubes, Alberger Heater Company, 1921). The latter example of commercialization, albeit brief, is shown in Fig.1; an increase of the hot water heating capacity of 50% was claimed as a result of enhancing both the tube-side single-phase flow and the shell-side condensing of steam.

In spite of these early efforts, this aspect of heat transfer received relatively little attention until about 30 years ago, as evidenced by the small amount of technical and commercial literature. In the mid 1950s the field began to develop in response to the need for more efficient power and process heat exchangers, the advent of commercial nuclear power

FIGURE 1. Steam-heated water heater using corrugated tubes(Alberger, 1921).

and the demands of space flight systems. Another sharp increase in activity was associated with heat recovery and alternate energy systems stimulated by the 1973 oil embargo. The exponential increase in world technical literature is evident from Fig.2 and a similar trend with U.S. Patents is noted in Fig.3. It is now estimated that each year over 500 papers and reports are published on the subject. Over 10% of the papers at the quadrennial International Heat Transfer Conference concern this subject. Chinese researchers are increasingly important contributors to the literature. In view of the level of activity and the continued growth, enhanced heat transfer has truly become a second generation heat transfer technology.

There have been a number of recent survey articles and handbook sections prepared that deal with the field in general(e.g., Bergles, 1978 1983, 1985; Nakayama, 1982) or with specific aspects(e.g., finned tubes: Webb, 1980; nucleate boiling: Webb, 1981; electrohydrodynamics: Kulacki, 1981; condensation: Marto, 1986). The purpose of the present review is not to repeat the extensive citations given in such reviews, but rather to single out recent examples of progress in understanding and predicting enhancement.

Some general observations can be made as to the evolution of enhancement technology during the past ten years. The widely scattered literature has been collected, classified, entered into a computerized data base, and published in several reports. Sophisticated experiments have been conducted to determine local heat transfer coefficients with complex geometries. Numerical simulations of increasingly complicated configurations are being attempted. Of greatest significance is the extent to which the more effective and feasible enhancement techniques have graduated from the laboratory to full-scale industrial equipment. This is documented by the over 200 manufacturers who offer products ranging from enhanced tubes to entire thermal systems incorporating enhancement technology(Bergles et al., 1984). Research and development at the present time is driven primarily by applications rather than by curiosity.

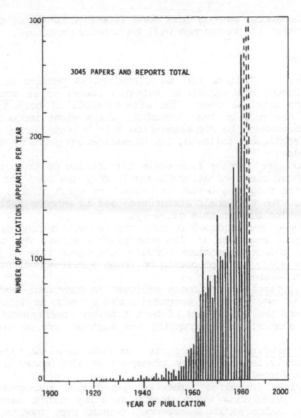

FIGURE 2. Annual publications in heat transfer enhancement (Bergles et al., 1983).

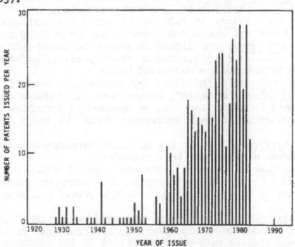

FIGURE 3. U.S. Patents issued annually in heat transfer enhancement (Webb et al., 1983).

13

Before discussing some of the more recent progress in understanding and applications, the techniques will be briefly described.

THE TECHNIQUES

Enhancement techniques can be classfied as <u>passive</u> methods, which require no direct application of external power, or as <u>active</u> schemes, which require external power. The effectiveness of both types depends strongly on the mode of heat transfer, which might range from single-phase free convection to dispersed-flow film boiling. Brief descriptions of passive techniques, following the classfication given by Bergles(1985) are given below.

<u>Treated surfaces</u> involve fine-scale alternation of the surface finish or coating (continuous or discontinuous). They are used for boiling and condensing; the roughness height is below that which affects single-phase heat transfer. The fine-scale structures used to enhance nucleate boiling are generally included in this category.

<u>Rough surfaces</u> are produced in many configurations ranging from random sand-grain-type roughness to discrete protuberances. The configuration is generally chosen to promote turbulence ranther than to increase the heat transfer area. The application of rough surfaces is directed primarily toward single-phase flow.

<u>Extended surfaces</u> are routinely employed in many heat exchangers. The development of new types of extended surface, such as intergral inner-fin tubing, and the improvement of heat transfer coefficients on extended surface, by shaping or interrupting the surface, are of particular interest.

<u>Displaced enhancement devices</u> are inserted into the flow channel so as to indirectly improve energy transport at the heated surface. They are used with forced flow.

<u>Swirl devices</u> include a number of geometrical arrangements or tube inserts for forced flow which create rotating and/or secondary flow: coiled tubes, inlet vortex generators, twisted-type inserts, and axial-core inserts with a screw-type winding.

<u>Surface-tension devices</u> consist of wicking or grooved surfaces to direct the flow of liquid in boiling and condensing.

<u>Additives for liquids</u> include solid particles and gas bubbles in single-phase flows and liquid trace additives for boiling systems.

<u>Additives for gases</u> are liquid droplets or solid particles, either dilute phase (gas-solid suspensions) or dense phase (fluidized beds).

The active techniques are described below.

<u>Mechanical aids</u> stir the fluid by mechanical means or by rotating the surface. Surface "scraping", widely used for viscous liquids in the chemical process industry, can also be applied to duct flow of gases. Equipment with rotating heat exchanger ducts in found in commercial practice.

<u>Surface vibration</u> at either low or high frequency has been used primarily to improve single-phase heat transfer.

<u>Fluid vibration</u> is the most practical type of vibration enhancement, given the mass of most heat exchangers. The vibration range from pulsations of about 1 Hz to ultrasound. Single-phase fluids are of primary concern.

<u>Electrostatic fields</u> (dc or ac) are applied in many different ways to dielectric fluids. Generally speaking, electrostatic fields can be directed to cause greater bulk mixing of fluid in the vicinity of the heat transfer surface. An electrical field and a magnetic field may be combined to provide a forced convection via electromagnetic pumping.

14

Injection involves supplying gas to flowing liquid through a porous heat transfer surface or injecting similar fluid upstream of the heat transfer section. Surface degassing of liquids can produce enhancement similar to gas injection. Only single-phase flow is of interest.

Suction involves either vapor removal through a porous heated surface in nucleate or film boiling, or fluid withdrawal through a porous heated surface in single-phase flow.

Two or more of the above techniques may be utilized simultaneously to produce an enhancement larger than that produced by only one technique. The simultaneous use is termed compound enhancement.

It should be emphasized that one reason for studying enhanced heat transfer is to assess the effect of an inherent condition on heat transfer. Some practical examples include roughness produced by standard manufacturing, degassing of liquids with high gas content, surface vibration resulting from rotating machinery or flow oscillations, fluid vibration resulting from pumping pulsation, and electric fields present in electrical equipment.

To give an idea of the greatest activity, Table 1 breaks down the citations represented in Fig.2 according to technique and mode. It is clearly impossible in a brief space to give an overview of progress in all areas. Instead, just four commercially important passive techniques will be discussed. These techniques are typical of many in that they have been the object of sophisticated experiments and analysis.

TABLE 1. Classification of augmentation techniques and number of references in each category (Bergles et al., 1983)

	Single-Phase Natural Convection	Single-Phase Forced Convection	Pool Boiling	Forced-Convection Boiling	Condensation	Mass Transfer
Passive Techniques (No external power required)						
Treated surfaces	NA	NA	149	17	53	NA
Rough surfaces	7	418	42	65	65	29
Extended surfaces	23	416	75	53	175	33
Displaced enhancement devices	NA	59	4	17	6	15
Swirl flow devices	NA	140	NA	83	17	10
Coiled tubes	NA	142	NA	50	6	9
Surface tension devices	NA	NA	12	1	NA	NA
Additives for liquids	3	22	61	37	NA	6
Additives for gases	NA	211	NA	NA	5	NA
Active Techniques (External power required)						
Mechanical aids	16	60	30	7	23	18
Surface vibration	52	30	11	2	9	11
Fluid vibration	44	127	15	5	2	39
Electric or magnetic fields	50	53	37	10	22	22
Injection or suction	6	25	7	1	6	2
Jet impingement	NA	17	2	1	NA	2
Compound enhancement (Two or more techniques)	2	50	4	4	4	2

NA = not applicable

As noted in the preceding sections, this classification refers to fine-scale alteration of the surface finish, i.e., the tube surface appears rather smooth. A coating may be applied to the plain tube or the surface may be deformed to produce subsurface channels or pores.

Nucleate and transition pool boiling are usually quite strongly dependent on the surface condition as characterized by the material, the surface finish, and the surface chemistry. Certain types of roughness, fouling, and oxidation have been shown to reduce wall superheats, increase peak critical heat flux(CHF), and destabilize film boiling; however, these naturally occurring conditions are too unpredictable to permit their exploitation in commercial equipment. The action of nonwetting coatings, such as Teflon, for promoting boiling of water or aqueous solutions is well documented. These insulating coatings of Teflon may also be used to promote the rate of cooling of hot materials through reduction of the surface temperature to the level where transition or nucleate boiling occurs rather than film boiling. While the mechanisms of both situations are well understood, there have been no practical applications.

With highly-wetting liquids (refrigerants, fluorochemicals, other organics, cryogens, alkali liquid metals), doubly reenstrant cavities are required to ensure vapor trapping so that nucleation sites are increased by selective machining, forming, or coating the surface structure with a porous material. Furthermore, the subsurface structures represent large surface areas that are conducive to high rates of vapor generation. The liquid flows via selected paths or channels to the interior where it is vaporized; the vapor is then ejected through other paths by a sort of "bubbling."

Boiling data for three of the most widely used structured surface are shown in Fig.4. Note that the heat flux is based on the area of the equivalent smooth tube for a particular outside diameter. The shifts in the nucleate boiling curve to lower superheat are representative of the excellent performance of such surfaces. It must be emphasized, however, that the performance is very sensitive to surface configuration and the working fluid. The gap width, pore size, or particle size is tailored to the fluid and the pressure for optimum performance using empirical experience or guidelines inferred from models.

Porous Metallic Matrix Coating

Mechanistic models of the liquid-vapor exchange process have been proposed for the major commercial structured surfaces. Consider first the porous coating consisting of sintered particles or fibers, particles bonded by electroplating, metal sprayed powder, or electroplated polyurethane foam. An example of sintered particles is shown in Fig.5. Webb (1983) examined a wide variety of data for coatings of nearly spherical particles and concluded that particle diameter has very little effect on performance but that the preferred coating thickness is three to four layers of particles.

O'Neill et al.(1972) postulated a nucleate boiling model assuming an idealized matrix of uniform diameter spherical particles in a known packing arrangement with pores of uniform size each containing vapor bubbles. Thin liquid films exist on the surfaces of the particles. Heat flows from the prime surface through the particles and vaporizes the liquid film. The pores are assumed to be interconnected so that the vapor can exit to the surrounding liquid and liquid can be supplied to the

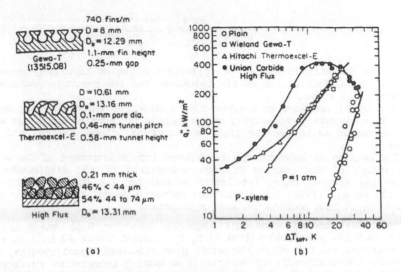

FIGURE 4. Pool boiling from smooth and structured surface on the same apparatus. (a) Cross-sectional sketches of surface. (b) Boiling curves. (Yilmaz et al., 1980).

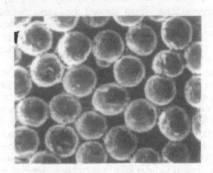

FIGURE 5. Plan view of sintered surface, 50x (Kim and Bergles, 1985).

interior of the matrix. Webb(1983) has clarified a number of the assumptions employed in this model.

The heat flux based on the projected area of the base surface, A_p, is simply

$$q'' = \frac{q}{A_p} = k_L \frac{S}{A_p} \frac{(T_W - T_L)}{t} \tag{1}$$

where K_L is the liquid thermal conductivity, S is the total surface area occupied by the liquid film, t is the average liquid film thickness, T_W is the particle surface temperature (assumed to be the base temperature), and T_L is the liquid film temperature at the surface of the bubble. Introducing the Gibbs equation, the wall superheat for a pore or bubble radius R is given by

17

$$T_W - T_S = \frac{\beta q'' R_b^2}{k_L} + \frac{2\sigma v_{fg} T_S}{h_{fg} R_b} \qquad (2)$$

where the geometry factor, β , a collection of terms involving geometry, can be calculated if the matrix thickness and the particle packing arrangement are known.

This model provides an optimum pore radius for which the wall superheat is a minimum. A relatively coarse matrix is required for water while a fine matrix is better for light hydrocarbons, fluorocarbons, and cryogens.

The packing arrangement may be deduced from measurement of the weight and volume of the coating or from measurement of the displacement of a wetting liquid. Using the latter method for the High Flux surface, O'Neill et al. (1972) found that the porosity ranged from 0.50 to 0.65; hence, an in-line packing was most appropriate.

Webb(1983) tested equation(2) against data for R-11 and R-113 and found that the predicted values of $T_W - T_S$ ranged from 0.62 to 2.05 times the observed values. While the model gave reasonably good results, considering the difficulty of accurately measuring temperature differences of the order of 1 K, the model has some shortcomings. As noted later by Czikk and O'Nell (1979), commercial coatings utilize powders that are not spherical and have a range of particle sizes. Also, since not all pores physically present are functionally active, it was necessary to determine the active sites from actual boiling data. With these modifications, the predictions were in rather good agreement with the data. Of course, by requiring the actual boiling data, the model becomes an interpretive rather than a predictive tool. Its success suggests that the physics is correct.

On the other hand, Webb(1983) argues that even with this realism, the basic concept of a static model is incorrect and that proper modeling must take into account time-dependent two-phase flow within the matrix. No suggestions are offered as to how this might be accomplished.

While the established nucleate boiling performance of porous metallic matrix surfaces is excellent, there are concerns about the initiation of boiling. The typical fluids used with these surfaces are highly wetting; that is, they have contact angles approaching zero on all engineering surfaces. As a result, doubly reentrant cavities are required if pure vapor nuclei are to be retained during subcooled liquid conditions. Such cavities are possible with spherical particles; however, Kim and Bergles(1985) have shown that even the fine particles normally employed are too large to allow reentrant cavities small enough to permit a significant subcooling without flooding the cavities. The burden of initiating boiling is thus placed on naturally occurring cavities on the base surface or on the particles. Since these are expected to be small, the wall superheat at incipient boiling is expected to be high, as is observed(Bergles and Chyu,1982; Kim and Bergles,1985). As shown in Fig.6, the incipient boiling superheats and subsequent temperature excursions are repeatable and large for a plain surface and several sintered surfaces. The results in the well-known boiling curve hysteresis. The significant reduction in superheat with the enhanced surfaces, from 30-40 K to 20-30 K, is attributed to the greater probability of finding reentrant cavities in the large contorted surfaces within the matrix.

Given the impossibility of controlling incipient boiling superheat through geometry, other means must be found to insure the onset of boiling, particularly in systems that are ΔT controlled. Introduction of

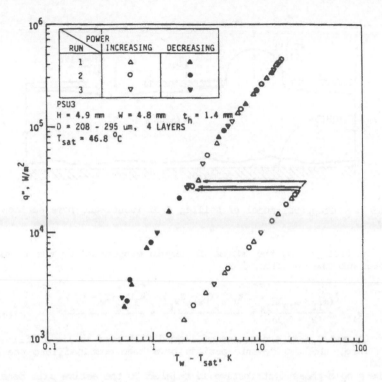

FIGURE 6. Boiling data for a sintered surface illustrating the large wall superheat required to initiate boiling (Bergles and Kim, 1985).

vapor or foreign gas is effective (Kim and Bergles, 1985), but the theory of its action is not well understood.

Tunnel-and-Pore Forming

The second type of structured surface is the Hitachi Thermoexcel-E surface characterized by well-defined pores that expose the subsurface channels. Nakayama et al.(1980) proposed a dynamic model based on the escape of vapor from some pores while liquid is being drawn into other pores. A substantial portion of the surface is subject to single-phase convection. The heat flux is based on the projected area corresponding to the outside diameter and is given by

$$q'' = q''_1 + q''_{ex}$$ (3)

where q''_1 is the latent heat flux and q''_{ex} is the enhanced free convection contribution. This is shown in Fig.7.

The latent heat term was basically considered to involve three periods: a dormant period where the thin liquid film in the tunnel evaporates causing the pressure to build up, a bubbling period where some pores are active which leads to a reduction in pressure within the tunnel, and a short liquid intake period where the pressure is low enough to permit liquid to flow into the tunnel. The core of the analysis is to

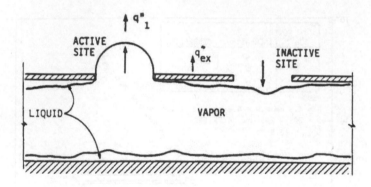

FIGURE 7. Conceptual model of boiling from tunnel-and-pore surface (Ayub, 1986).

find for each period the amount of liquid evaporated or the total heat transfer and the duration, i.e.,

$$q_1'' = \frac{(m_{11} + m_{12})h_{fg}}{(\theta_1 + \theta_2)A} - \frac{k_1 C_{t_1}(T_W - T_{V_1})}{A} + \frac{k_1 C_{t_2}(T_W - T_{V_2})}{A} \qquad (4)$$

where C_{t1} and C_{t2} are the ratios of liquid film area to liquid film thickness. Here, the two dynamic periods have been combined into the second period.

The single-phase contribution is related to the active site density:

$$q_{ex}'' = \left(\frac{\Delta T}{C_q}\right)^{1/y}\left(\frac{N_A}{A}\right)^{-x/y} \qquad (5)$$

The analysis is very complex and in places difficult to interpret(Ayub, 1986). Much empirical information is required to establish the basic characteristics of the heat transfer for a given surface. In particular, at an arbitrary reference state, say T=1K, q" is recorded and average values of the frequency of bubble emission, bubble departure diameter, and number of active sites are measured. This establishes the values of C_{t1} and C_{t2} along with two other key empirical parameters, the constant in the bubble departure equation and a constant characterizing the liquid curvature during the liquid intake.

Equation(5) was established from data for a variety of regular surfaces and several of the structured surfaces. It was necessary to estimate the natural convection component by subtracting the latent component from the total heat flux:

$$q_{ex}'' = q'' - \frac{\pi}{12} d_b^3 f_b h_{fg} \rho_v \frac{N_A}{A} \qquad (6)$$

The exponents x and y were presumed to apply to all surfaces and fluids, but C_q was strongly dependent on the fluid.

For a given fluid and surface, once the five constants are determined, the boiling curve can be predicted with fair accuracy beyond the reference point. It is evident that this model is also interpretive and of no use in engineering calculations. Potentially important parameters such as the best pore size cannot be determined from the model. The general physics of the process does seem to be correct, but the details cannot really be verified because of the large number of empirical constants. In spite of these shortcomings, the work of Nakayama et al.(1980) stands out as an example of original thinking and physical insight. They have shown that dynamic modelling is necessary for the structured surfaces and that it is possible to formulate the equations and obtain solutions.

The problem of incipient boiling with the Thermoexcel-E surface is similar to that described earlier. While such behavior is not well documented, Torii et al.(1978), among others, report considerable boiling curve hysteresis.

T-shaped Fin Forming

Turning now to the final surface shown in Fig. 4, the GEWA-T surface has also been studied in detail in an attempt to obtain a first-order model of the even more complex liquid-vapor exchange mechanism. Here, the liquid and vapor are not constrained to flow through certain openings Although it has been suggested that bubbles flow in the channels around the circumference and are ejected near the top(Stephan and Mitrovic, 1981), Marto and Hernandez(1983) observed that these bubbles are ejected at other locations, even at the bottom of the tube. Ayub and Bergles (1987), on the basis of tests with simulators and actual tubes, confirmed that liquid flows in and vapor is ejected around the entire tube circumference when boiling at high flux. Their model shown in Fig. 8 is similar to that for the Thermoexcel-E surface except that the liquid and vapor are free to cross the tunnel opening at any point around the circumference.

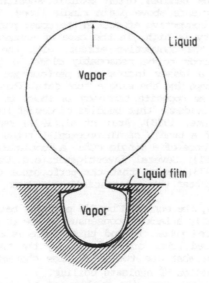

FIGURE 8. Presumed configuration for boiling in a Gewa-T channel (Ayub and Bergles, 1987).

Considering only time-averaged tube behavior, the total heat flux
was formulated simply as

$$q'' = k_1 \frac{C_\ell \Delta T}{A} + \left(\frac{\Delta T}{C_q}\right)^{1/y} \left(\frac{N_A}{A}\right)^{-x/y} \tag{7}$$

where a single average vaporization term is considered. The values of
C_q were essentially the same as those used by Nakayama et al.(1980).
The value of N_A was measured and correlated as a function of ΔT. The
film parameter C_ℓ was then obtained from a $q''-\Delta T$ reference point. The
predicted values of other points along the boiling curve were found to
be in good agreement with the data. In the case of water, the latent
heat component dominated, whereas with R-113, the natural convection component was most important. This model ignores the dynamic behavior necessary for liquid-vapor exchange, but even with the simplifications,
the resulting equation can't be used for engineering predictions.

The Gewa-T tubes exhibit boiling curve hysteresis with wetting liquids
but the effect is generally less than that observed with the Thermoexcel-
E and High Flux surfaces(Ayub and Bergles, 1988). This is attributed
to enhancement of the heat transfer prior to incipient boiling, through
pumping of hot liquid from the channels or even by formation and trapping of vapor within the channels.

Tube Bundle Performance

The performance of enchanced tubes in horizontal bundles is receiving
increased attention because this is the major application for these tubes
As in most other application of enhancement technology, the understanding of enhanced tube behavior depends on the mechanism of the normal
situation. Plain tube bundles often exhibit substantial increases in
heat transfer coefficients above plain single tubes. This bundle factor
is attributed to the convective effect in the upper portion of the bundle
when the vapor quality is high. In the case of enhanced tubes, nucleate
boiling overshadows the convective effect, with the result that the
bundle performance tends to be reasonably close to that of the single
tube. The result is a lesser increase in performance than would be expected from simply comparing the single tube data. This behavior is shown
in Fig.9. Although the opposite tendency is shown in Fig.9 at low heat
flux, there is some evidence that bundles of Gewa-T tubes perform better
than single tubes(Menze, 1988). Arai et al.(1977) reported that the a-
verage performance of a bundle of Thermoexcel-E tubes was generally superior to the performance of a single tube, a conclusion that is disputed
by Yilmaz et al.(1981). Several investigators(e.g.,Czikk et al.,1970;
Lewis and Sather, 1978) report that the performance of High Flux tubes
in a bundle may be better than, essentially the same, or less than that
of a single tube.

While in all cases, the enhanced bundle performs better than the plain
bundle, there is clearly a lack of consensus on how to predict the bundle
performance of enhanced tubes. Beyond this, there is a lack of knowledge
as to how the enhanced tubes are effected by the two-phase crossflow.
It is not even known what the two-phase flow characteristics are when
there is intense promotion of nucleate boiling.

Bundles are not immune from the problem of boiling curve hysteresis
with highly wetting fluids. As shown in Fig. 10, an ammonia flooded
evaporator with enhanced tubes, intended for closed-cycle ocean thermal
energy conversion, came up to full performance only after three days!

Boiling Superheat, ΔT , C

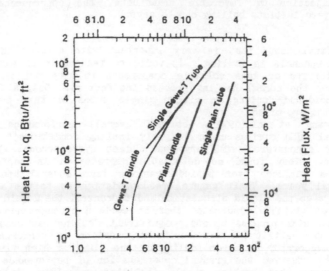

FIGURE 9. Tube bundle boiling curves with plain tubes and Gewa-T Tubes (Yilmaz et al., 1981).

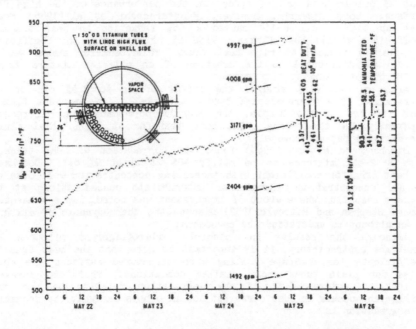

FIGURE 10. Time variation of overall heat transfer coefficient in a High Flux flooded bundle ammonia evaporator, due to gradual spreading of boiling (Lewis and Sather, 1987).

While this problem does not occur in all systems because of normal vapor injection or favorable transients, the consequences of the inability to initiate boiling can be severe.

Fouling
The final issue that is very important with surface having small interior channels is fouling. It would be reasonable to expect solids to precipitate or less volatile components to hide out in the pores and render the normal boiling process ineffective. Only a few studies have considered fouling on either single tubes or tube bundles, and the results are mixed.

Gottzmann et al.(1973) obtained overall performance data for industrial tube bundles under "lightly fouling conditions," that which might be encountered with cryogens, light hydrocarbons, and aqueous solutions. They found no detectable degradation in performance of High Flux surfaces over periods ranging from several hundred hours to several years. Their general recommendation is to use a pore size that keeps up the liquid circulation and to control the liquid chemistry. O'Neill et al.(1980) indicated that in certain circumstances typically associated with process upsets, significant fouling can occur. Lorenz et al.(1981) attributed the severe degradation of performance in an ammonia evaporator to prior fouling of the aluminum high flux coating. Chyu(1984) observed the gradual deterioration of performance of a high flux surface when boiling singly distilled water(Fig. 11); this was also attributed to some preexisting contaminants on the surface.

Compressor lubricating oil is a possible contaminant in flooded refrigeration evaporators. The three surfaces under discussion have been tested with refrigerant-oil mixtures. Czikk et al.(1970) found that 2% oil in R-11 had no effect on the performance of the high flux surface. More recently, however, Wanniarachchi et al.(1986) found that up to 3% oil R-114 caused a 35% reduction in heat transfer coefficient at all heat fluxes. With 6% or more oil, the coefficient was sharply reduced at high heat flux. This behavior summarized in Fig.12 was attributed to the creation of an oil-rich mixture within the porous structure.

Arai et al.(1977) studied the effect of up to 3.4% oil on the performance of a Thermoexcel-E bundle with R-12. At low heat fluxes, near the top of the bundle, the overall performance was increased, apparently due to foaming. The heat transfer at higher heat fluxes was degraded but not severely.

Stephen and Mitrovic(1981) found the degradation of heat transfer with R-12-oil mixtures to be nearly 40% at about 9% oil. Decreases in heat transfer coefficient with increasing concentration were observed at all concentrations except for intermediate concentrations at the highest heat flux where about 5% improvement was noted. In a subsequent paper, Stephen and Mitrovic(1982) discuss the thermodynamic of mixtures in an attempt to understand the phenomena.

Although the results for miscible oil-refrigerant mixtures are sometimes contradictory, it is important to note that the heat transfer coefficients for nucleate boiling with structured surfaces are above those for plain tubes under similar conditions. There is, however, a lack of understanding of why degradation relative to the pure refrigerant occurs in some cases but not in others and what the mechanism of degradation is.

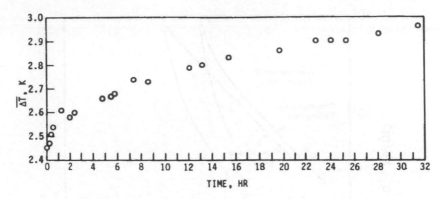

FIGURE 11. Deterioration of boiling performance of a sintered surface due to boiling (Chyu, 1984).

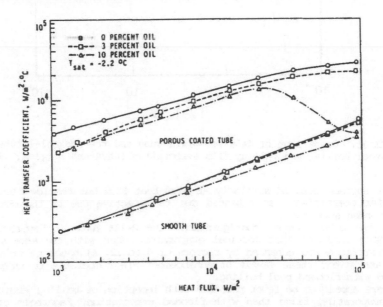

FIGURE 12. Influence of oil on heat transfer performance of a smooth tube and a sintered surface tube (Wanniarachchi et al., (1986).

Falling Film Evaporation

The use of structured surfaces in falling film(spray film) evaporation is another possibility. The same surfaces depicted in Fig.4 were studied with single horizontal tubes subject to a feed of water (Chyu and Bergles 1985). Both the High Flux and the Thermoexcel-E tubes exhibited low flux, nonboiling behavior similar to plain tubes. At high fluxes, nucleate boiling appeared to dominate and the data corresponded essentially to the extrapolation of pool boiling for the same surface with the same conditioning. Data for Thermoexcel-E surface are shown in Fig.13. The

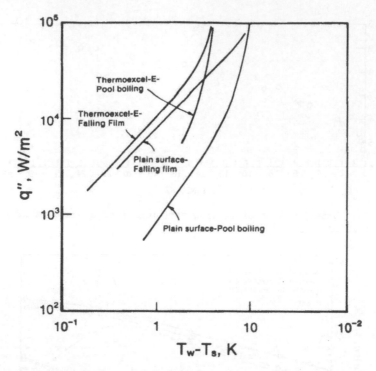

FIGURE 13. Comparison of data for plain tube and thermoexcel-E tube in both pool boiling and falling film evaporation (Chyu and Bergles, 1985).

Gewa-T surface behaved similarly at high heat flux but the low flux heat transfer coefficient was enhanced due to effective use of the channels as extended surfaces.

In the other common configuration, the R-114 data of Fagerholm et al.(1985) indicate that vertical evaporator tubes with the same structured surfaces are enhanced by factors of 6 to 12. Although no reference data are given, these levels of enhancement are comparable to those observed with flooded pool boiling.

There appear to be fewer problems with inception of boiling with falling evaporating films than with flooded evaporation. Fagerholm et al. (1985) report only a slight thermal excursion with the High Flux tube. This is expected because the structure is full of vapor prior to system startup. Some erratic nucleation behavior has been observed by Hillis et al.(1979) with horizontal bundles. When ammonia flowed over the unheated tubes, the vapor was removed from the High Flux surface and boiling was initiated only very slowly.

There do not appear to be any investigations of bundle effects with structured surface tubes. Although the film flows are likely to be quite disturbed, because of dripping and vapor shear, the convective effects will probably be overshadowed if there is nucleate boiling. This is likely to be the case because the structured surfaces exhibit boiling at very low superheat.

Fouling of enhanced tubes in falling film evaporation is expected to be similar to that for flooded boiling, with single or multiple tubes.

There is, however, the additional complication of dryout with a long vertical tube evaporator.

ROUGH SURFACES IN TUBESIDE SINGLE-PHASE FLOWS

An enormous variety of surface roughnesses have been tested in both single- and two-phase flows under both natural and forced circulation. The roughness may be applied to any of the usual prime or extended heat exchange surface, e.g., tubes or fins. Both two- and three-dimensional integral roughness elements have been produced by the traditional processes of machining, forming, casting, or welding. Various inserts or wrap-around structures can also provide surface protuberances.

Attention is focused here on the helical repeated rib roughness that is readily manufactured and results in good heat transfer performance (up to 300% increases over the smooth tube) in single-phase turbulent flow without severe pressure drop penalty. Sketches of typical surface configurations are shown in Fig.14. The main effect is that the ribs cause flow separation and reattachment that results in higher average heat transfer coefficients. Analytical models, such as those of Lewis (1975) and Wassel and Mills(1979) have not been particularly successful because of the need for many empirical constants describing the heat transfer and flow friction or form drag. Furthermore, they have been applied only to transverse repeated ribs. Numerical simulations have thus far been applied only to single ribs(Huang et al.,1987). Accordingly the fundamental work on this type of roughness has tended to focus on the development of semi-empirical correlations.

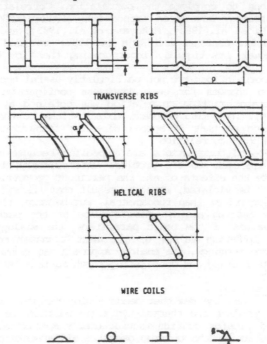

TRANSVERSE RIBS

HELICAL RIBS

WIRE COILS

PROFILE SHAPES

FIGURE 14. Examples of repeated rib roughness (Ravigururajan and Bergles, 1985).

Analogy-Based Correlations

Following the lead of Dipprey and Sabersky(1963), Webb et al.(1971) used the heat transfer-momentum transfer analogy to correlate data for tubes with transverse repeated rib roughness. The procedure involves interpreting the friction factor data in terms of Nikuradse's similarity function, U_c^+, and the roughness Reynolds number, e^+. Applying the analogy, the general form of the heat transfer correlation is

$$St = \frac{f/2}{1 + (f/2)^{0.5}(\bar{g}F - U_c^+)} \tag{8}$$

The functions $\bar{g}$, F, and U_c^+, were obtained empirically for carefully constructed repeated rib roughnesses. As pointed out by Webb(1979), the correlation does not allow easy physical interpretation because of use of unconventional variables, e.g., roughness Reynolds number($e^+=(e/d)Re\sqrt{f/2}$) instead of pipe Reynolds number. Originally, these variables were a barrier to understanding and accepting the correlations, but this has changed within the past ten years.

Strictly speaking, the analogy is valid for a single p/e, i.e., geometrically similar roughness. Since this would severely limit the application of the correlations, pitch is included in the friction correlation ($U_c^+=f(e^+,p/e)$)and the heat transfer function($\bar{g}=f(e^+,p/e)$). Webb et al. (1971) found that for sufficiently high e^+, U_c^+ depends only on p/e and $\bar{g}$ is dependent only on e^+.

Because most indenting or swaging processes lead to spiralled ribs, the helix angle is an additional variable that alters the geometrical similarity. An angle correction factor is usually added to both the U_c^+ and $\bar{g}$ expressions to complete the correlation. Correlations following this general strategy have been proposed by Withers(1980a,1980b), Gee and Webb(1980),Li et al.(1982), Nakayama et al.(1983), and Sethumadhavan and Raja Rao(1986). Ganeshan and Raja Rao(1982) extended the method to non-Newtonian power law liquids, incorporating the flow behavior index into U_c^+.

The Withers correlations are not particularly useful because the friction correlation changes for each roughness configuration; however, a reasonably accurate friction correlation was obtained by Ravigururajan and Bergles(1986). They also proposed modifications to the Gee and Webb (1980) correlation so that it agrees with the transverse rib limit of the Webb et al.(1971) correlation.

The analogy-based correlations are appealing because they represent a fundamental approach to the problem. Unfortunately, because of the need to correlate the effects of all the pertinent geometrical parameters, similarity must be violated, with the result that the extended correlations are more empirical than fundamental. Furthermore, the correlations have had little testing against data outside of the particular study. For identical values of the basic parameters, the analogy correlations predict widely differing values of the heat transfer coefficient. In spite of these shortcomings, the analogy approach has emerged as a powerful tool for interpreting and correlating rough surface behavior.

Statistical Correlation

It is evident that the designer needs a wide-ranging correlation that can be used to predict the thermal-hydraulic performance of tubes with helical repeated ribs(or confirm manufacturer's specifications for such tubes) or to determine the optimum geometrical parameters for a particular application. Given the large and growing data base that is now available for this type of roughness, it is now feasible to do this statistically using a large-scale computer.

Ravigururajan and Bergles(1985) developed such correlations from large data bases for friction factor and heat transfer. The 1807 heat transfer points covered the following ranges: Re from 6000 to 440000, Pr from 0.66 to 37.6, e/d from 0.01 to 0.22, p/d from 0.1 to 17.8, and $\alpha/90$ from 0.3 to 1.0. Most ribs were of semi-circular cross section, but many data points were included for circular, rectangular, and triangular cross sections.

The heat transfer correlation is given by

$$Nu_a/Nu_s = \left\{ 1 + \left[2.64\ Re^{0.036}(e/d)^{0.212}(p/d)^{-0.21}(\alpha/90)^{0.29}(Pr)^{-0.024} \right]^7 \right\}^{1/7}$$

(9)

where

$$Nu_s = 0.5\ f\ RePr/(1 + 12.7(0.5\ f)^{0.5}(Pr^{0.667} - 1)$$

(10)

with this equation, 69% of the data were correlated to within ±20%.

The friction factor correlation is more complex because it was necessary to include a shape factor characterizing the profile by the number of sharp corners facing the flow and the contact angle:

$$f_a/f_s = \left\{ 1 + \left[29.1\ Re^{(0.67 - 0.06\ p/d - 0.49\ \alpha/90)} \right. \right.$$
$$\times (e/d)^{(1.37 - 0.157\ p/d)}$$
$$\times (p/d)^{(-1.66 \times 10^{-6}\ Re - 0.33\ \alpha/90)}$$
$$\times (\alpha/90)^{(4.59 + 4.11 \times 10^{-6}\ Re - 0.15\ p/d)}$$
$$\left. \left. \times \left(1 + \frac{2.94}{n} \right) \sin \beta \right]^{15/16} \right\}^{16/15}$$

(11)

where

$$f_s = (1.58 \ln Re - 3.28)^{-2}$$

(12)

In this case, 64% of the data were correlated to within ±20%. It should be noted that it is common to find data sets that have widely varying friction factors for nearly identical geometries.

The analogy correlations, extended if necessary, were applied to the same data base. It was concluded that Eqs.(9) to (12) displayed all of the right trends and were more accurate than any of the other correlations. The success of this approach suggests that statistical methods are an effective, if not elegent, way to correlate data for rough surfaces.

OFFSET STRIP FINS

One of the most popular geometries for compact heat exchangers is the offset strip fin depicted in Fig. 15. Substantial enhancement over uninterrupted rectangular channels results from periodic development of laminar boundary layers and their at least partial dissipation in the fin wakes. As documented by Joshi and Webb(1987), a broad assault has been underway for over forty years to establish the flow friction and heat transfer characteristics as well as the basic transport mechanisms for this deceptively simple enchanced extended surface. The approaches include obtaining heat transfer and flow friction data for actual cores or scaled-up models, empirical correlations of such data, mass transfer data for scaled-up models, flow visualization, analytical models, and numerical solutions.

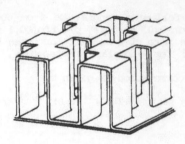

FIGURE 15. Offset strip fins used in plate-fin heat exchangers.

Empirical Data and Correlations

Although much data for prototype cores reside in company files, relatively few results are available in the open literature. Assuming many fins in the flow direction, the geometry should be described by fin height h, length L, transverse spacing s, and thickness t. The offset is usually uniform and equal to half the fin spacing; if not, an additional geometrical variable is introduced. The actual cores embody the usual manufacturing irregularities such as burred or scarfed fin edges, bonding imperfections, and separating plate roughness. The relatively old data presented in Kays and London(1984) are commonly referred to for design.

From such data, an example of the enhanced performance of the offset strip fin can be given. In Fig. 16, f and j data are presented for an offset strip fin and a plain fin scaled to provide the same hydraulic diameter. At Re_h =1000, the j-factor of the offset strip fin is 150% higher than that for the plain fin; however, the friction factor increase is 200%. The j/f ratio(at constant Reynolds number) is thus only 0.83; however, the benefits of the enhancement are better represented by one of the many thermal-hydraulic performance evaluation criteria. For example, if the flow rate and friction power are constrained to be constant, the offset strip fin will provide the same hA as the plain fin with 55% less surface area while requiring only 10% increase in flow frontal area(Webb and Bergles, 1983).

Until recently, the only broad-based correlations were those of Wieting(1975). He presented power-law curve fits of f and j data for 22 geometries. The correlations were broken down into low and high Reynolds number regions. For Re_h < 1000, both f and j were functions of L/Dh,s/h, and Re_h. For Reh 2000, f and j were found to be functions of L/Dh, t/Dh, and Re. Prandtl number does not appear to because the tests were for the intended application of air.

Joshi and Webb(1987) reexamined some of the older data for 21 geometries and proposed a more accurate correlation that has a more refined limit on the laminar and turbulent regions:

For $Re_h \leq Re_h^*$

$$f = 8.12(Re_h)^{-0.74}(L/D_h)^{-0.41}(s/h)^{-0.02} \qquad (13)$$

$$j = 0.53(Re_h)^{-0.50}(L/D_h)^{-0.15}(s/h)^{-0.14} \qquad (14)$$

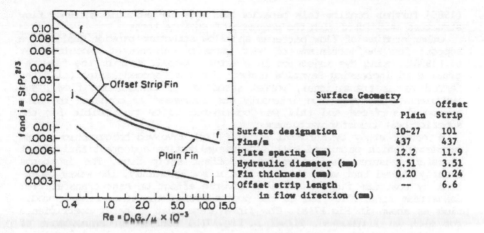

FIGURE 16. Typical data for offset strip fins (Webb and Bergles, 1983).

For $Re_h \geq Re_h^* + 1000$

$$f = 1.12(Re_h)^{-0.36}(L/D_h)^{-0.65}(t/D_h)^{0.17} \qquad (15)$$

$$j = 0.21(Re_h)^{-0.40}(L/D_h)^{-0.24}(t/D_h)^{0.02} \qquad (16)$$

where

$$Re_h^* = \frac{257(L/s)^{1.23}(t/L)^{0.58}D_h}{t + 1.328 \, L/(Re_L)^{0.5}} \qquad (17)$$

Overall, 82% of the f data and 91% of the j data are correlated to within ± 15%. The problem with these correlations as well as those if Wieting (1975) is the gap of 1000 in Reynolds numbers between laminar and turbulent. In any event, the Joshi and Webb(1987) work is a good example of the use of flow visualization to determine flow transition behavior.

Qualitative Observations

Many visual studies, usually of scaled-up arrays, have aided the interpretation of the experimental data. As the Reynolds number is increased, the wake exhibits time-dependent velocity behavior. Further increases in Reynolds number lead to vortex shedding. The oscillations thus created act as freestream turbulence for the downstream fin. Even though the boundary layers are basically laminar, heat transfer and momentum transfer are increased. Loehrke and Lane(1982) and Joshi and Webb(1987), among others, found that first disturbances corresponded approximately to the departure from the log-linear behavior of f and j.(See Fig.16.)In their study of in-line arrays, Loehrke and Lane(1982) included acoustic measurements of the flow noise. At the departure from laminar friction behavior, there was a 10-20 db increase in sound pressure level clearly audible in the 1-2 kHz range. In a recent paper, Mullisen and Loehrke

31

(1986) further confirm this behavior and show the periodic unsteady flow structure associated with the sequence of audible tones.

Other studies of flow patterns and flow structure provide qualitative support for the mechanisms of heat transfer enhancement. Mochizuki et al.(1988), using dye injection in a water channel, confirm the flow regimes with increasing Reynolds number: laminar, second laminar(also referred to as transitional, vortex shedding, or oscillating flow), and turbulent. The turbulence intensity was increased as compared to uninterrupted surfaces and this was considered to be responsible for the superior heat transfer performance.

A recent study by Kurosaki et al.(1988) involved holographic interferometry which permitted both flow visualization and quantitative heat transfer measurements for scaled-up offset strip fins. The isotherms clearly showed that when the fins are offset uniformly, the wakes generated by upstream fins can have an adverse effect on heat transfer from downstream fins. This led to the proposal that the staggering be modified as shown in Fig.17(a). The fin-average heat transfer coefficients for such an arrangement shown in Fig.17(b) confirm an improvement of about 10% over the usual configuration. This seems to be a promising strategy for the development of higher performance surfaces.

Analytical Approach

The analytical approach of Kays(1972) modeled the fins as a series of short plates on which laminar boundary layers develop. Complete dissipation of the boundary layers is assumed in the wake region. Both f and j are calculated from the usual laminar boundary layer equations except that the former includes a term to account for profile drag. The rather poor agreement with data is attributed to neglect of any effect of the wake on the boundary layer. The analysis of Joshi and Webb(1987) is more refined in that they include heat transfer from fin ends and parting sheets(top and bottom walls) as well as the fin sides. The end heat transfer was assumed the same as the sides, and existing analytical solutions were used to model heat transfer from the other surfaces. The wall and side friction was handled the same way and the effect of the

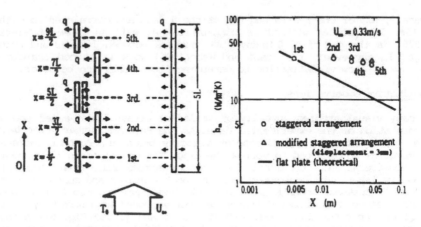

FIGURE 17. Comparison of data for offset strip fins with various staggerings (Kurosaki, 1988).

ends was represented as a form drag. In the turbulent region, standard correlations were used to obtain f and j for the walls assuming that fully developed channel flow was applicable. The f and j for the fin sides were then backed-out of the data for 21 surfaces and a correlation obtained. The agreement over the entire range of Reynolds number was about ±20%, lending strong support to the methodology. Because the additive equations are somewhat cumbersome, the average f and j empirical correlations given by equations(13) to (17) were suggested as a more useful alternative.

Numerical Studies

The few numerical studies that have been attempted suffer from restrictions on the geometrical parameters and the assumption of a stable laminar wake. The numerical prediction of vortex shedding, through unsteady flow equations, has yet been undertaken. The study of Kelkar(1985) is more advanced in that the two-dimensional assumption(large h/s) is relaxed; however, zero fin thickness is assumed. The computations were carried out for the entrance region and continued until the flow exhibited periodically fully developed behavior, which occured after 5-10 fins or modules. Fig.18 represents the module average heat transfer coefficients for various values of L/s. For L/s = 0.002, the fin and wall results are separated to show that the coefficient for the fin is substantially higher than that for the wall because the wall flow is not being interrupted.

The computed fully developed Nusselt numbers are generally higher than the data of Wieting(1975), while the computed friction factors are below the data. These discrepancies can be explained at least in part by neglect of the temperature gradient in the fin and by geometrical irregularities such as burrs. While the numerical formulations are not yet accurate simulations, the calculations yield local flow and thermal fields rather than the gross averages deduced from experiments with actual or scaled-up cores. The only experimental confirmation could come from the interferometric measurements noted earlier or from local measurements of naphthalene sublimation. However, in the only cases where this has been done, Kurosaki et al.(1988) and Sparrow and Hajiloo(1980), respectively, only average coefficients for each fin were reported.

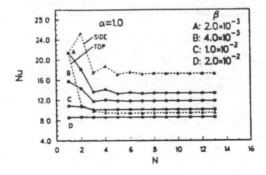

FIGURE 18. Module-averaged heat transfer coefficient predictions for offset strip fins (Kelkar and Patankar, 1985).

33

The final detailed discussion concerns a new type of inner-fin tubing that is receiving much attention around the world. The original configurations for enhancement of in-tube evaporation of refrigerants were offset strip fin inserts soldered to the copper tubes or aluminum star-shaped inserts secured by drawing the tube over the insert. For reasons of cost and high pressure drop, these composite tubes are being replaced by tubes with integral inner fins of moderate number and height. The current trend is toward tubes with more numerous and very short fins that have good thermal-hydraulic performance and are cost-effective.

Performance with Pure Refrigerants

Cross-sectional views of typical "micro-fin" tubes are shown in Fig.19. These copper tubes are the popular 3/8 in.(9.5 mm) O.D. and have 60 to 70 spiral fins ranging from 0.10 to 0.19 mm in height. Representative data for evaporation of R-22 in these tubes are shown in Fig.20. The heat transfer coefficients, based on area of an equivalent smooth tube, are increased 30 to 80% above the smooth tube values depending on the vapor quality and fin profile. The pressure drop penalties are in the same percentage range for this series of tests; however, some investigators report greater increases in average heat transfer coefficients than inceases in pressure drop(Schlager at al.,1987). Similar favorable performance has been observed in condensation, but in this case the heat transfer coefficients(not the pressure drops) are more sensitive to the geometry(Khanpara et al.,1985). Regarding geometry, the higher fins performed better, flat valleys are preferred, but peak geometry is not critical.

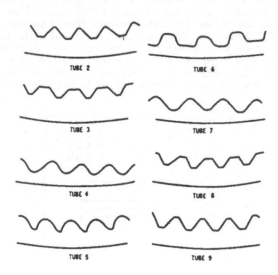

TUBE 2   TUBE 6

TUBE 3   TUBE 7

TUBE 4   TUBE 8

TUBE 5   TUBE 9

FIGURE 19. Fin profiles for micro-fin tubes used in refrigerant evaporators and condensers (Khanpara et al., 1985).

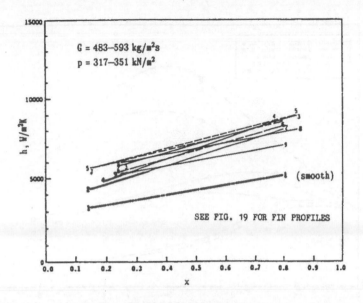

FIGURE 20. Average heat transfer coefficients for vaporization of R-22 in micro-fin tubes (Khanpara et al., 1986).

The outstanding thermal-hydraulic performance is related to increased surface area, increase in the turbulence level of the annular liquid film, and alteration of the flow pattern. The latter is particularly important in extending the dryout to higher quality. These tubes are at a stage that is often found in enhancement technology: They can be manufactured by several patented processes and are incorporated into commercial units, but relatively few experiments have been carried out to determine the performance, and no correlation are available. There is not even clear consensus on the combination of geometrical parameters that yields optimum performance. Ongoing flow pattern studies and mechanistic model development are under way.

Influence of Oil

The usual contaminant in refrigeration systems is compressor lubricating oil, which may be present in concentrations up to 10%. Recent experiments have defined the effects of oil on the performance of a typical micro-fin tube. It is of interest to compare the average heat transfer coefficient with smooth tube and low-fin tube behavior. In Fig. 21 it is seen that oil generally degrades the performance of the low-fin tube while a small amount of oil slightly enhances the micro-fin tube performance at about 2% concentration. In contrast, the plain tube heat transfer coefficient is enhanced by oil throughout the test range, with a particularly sharp enhancement occurring at 3% concentration. With condensation on the other hand, all three tubes exhibit a reduction in heat transfer coefficient with oil. While the inner-fin tubes always enhance heat transfer relative to the smooth tube, it is evident that the absolute performance and the enhancement are strongly dependent on the oil concentration. Correlations of these data are being developed.

35

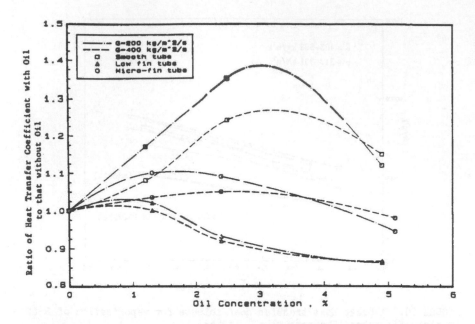

FIGURE 21. Influence of oil on average evaporation heat transfer coefficients for plain and internally finned tubes (Schlager et al., 1988).

MAJOR AREAS OF APPLICATION

To demonstrate the trends of applications of enhancement technology, seven industrial areas will be considered. Each of these areas has particular needs for enhancement and particular constraints, such as fouling or cost, that limit the techniques.

Heating, Ventilating, and Air Conditioning
This industry is well known for it ready adoption of enhancement for both air and refrigerant heat transfer. In the common evaporator or condenser coil, air-side heat transfer is routinely improved with fins that are louvered, corrugated, or serrated, as discussed in the section on offset strip fins. The intent is to both extend the area and increase the heat transfer coefficient. With the reduction in the air-side thermal resistance, the refrigerant-side resistance becomes relatively more important. As discussed in the previous section, the micro-fin tube is one answer to double enhancement.

A satisfactory micro-fin configuration can usually be found that will work well in either evaporation of condensation. This is particularly important in heat pump air-conditioning systems where the role of evaporator and condenser are periodically reversed. As another practical matter, the best flat tip profiles are preferable because mechanical expansion of the tube to secure the fins produces flat fins.

Noting the significant effect of oil on the heat transfer coefficient with enhanced tubes, methods have been developed to determine the oil concentration in refrigeration systems without sample removal(Baustian et al.,1988).

36

The structured boiling surfaces are being adopted more frequently for flooded refrigerant evaporators. Bundle simulator tests for this particular application, using High Flux tubes with R-11, are reported by Tatara and Payvar(1985). Their data confirm some previous observations that bundle convective enhancement is negligible with these tubes. This conclusion, although not validated in general, is justification for continuation of single-tube studies such as that of McManus et al.(1986). They included in the test program a new tube, the new Turbo-B tube, which has a structured boiling surface on the outside and helical repeated ribs on the inside. The exterior boiling enhancement is made by rolling low integral fins, cutting diagonally across the fins, and rolling to compress the fins and give a uniform outside diameter. Re-entrant passageways are formed with a cross-hatched pattern. The wall superheat with pure R-114 was reduced as much as a factor of 6.4. The maximum enhancement was reduced to a factor of 4.9 with 6% oil.

Low, integral fins have long been used for enhancement of shellside condensation of refrigerants. The recent trend has been to carefully configure the fins(Webb et al.,1985; Rudy and Webb, 1985; Marto, 1986) and allow for drainage. Sauer et al.(1980) found that oil does not degrade the condensing performance of low fin tubes. As surveyed by Webb (1984) and Bergles(1985),three-dimensional extended surfaces yield higher performance than the low-fin tubes. For condensing as well as boiling, tubes are now provided with tube-side enhancements of the spiral repeated rib type. These doubly enhanced tubes are indicative of the tendency to provide the highest possible overall heat transfer coefficients.

Automotive

Enhanced air-side surfaces have contributed to the dramatic reduction in size and weight of automotive radiators as well as the other heat exchangers found in modern vehicles. Mori and Nakayama(1980) report this reduction to be about 60% during the period 1957 to 1980. Additional gains have been made in the last seven years through the use of thin (down to 0.025 mm) copper fin stock that is louvered and tightly packed (up to 2000 fins/m).

There has been renewed interest in enhancement of the viscous flows that occur in oil coolers. Many types of convoluted channels and mixing inserts have been utilized; however, there is little in the way of analysis or correlations to guide the design. The one exception is twisted-tape inserts,which have received considerable attention.

Twisted-tape inserts have been tested with liquids of high viscosity under both heating and cooling conditions. The temperature difference has a strong influence on the heat transfer coefficient; however, the data are quite well correlated with a wall-to-bulk viscosity ratio correction similar to that used for smooth tubes. Manglik and Bergles(1987) developed the first correlation for uniform temperature tubes with twisted-tape inserts.

Power

High-performance air-side surfaces have been developed for dry cooling towers used for fossil power plant heat rejection. Current designs utilize structured boiling surfaces in the intermediate heat exchanger that evaporates the transfer fluid, ammonia.

Internally finned or rifled boiler tubes have been studied in more detail, particularly under dryout and post-dryout conditions. These tubes delay dryout to higher steam qualities and reduce wall temperature in dispersed-flow film boiling(Kitto and Albrecht, 1988).

Although new nuclear plant construction in the U.S.is nearing a standstill, there is a considerable business in reload fuel. The new bundles have specially designed rod spacers so that the dryout power is increased, thereby allowing an increase of plant power. Some rather sophisticated experiments have been run to define the spacer configurations; however, the data are generally proprietary(Groeneveld and Yousef,1980).

Ocean thermal energy conversion is still in the planning stage, but enhanced boiling and condensing surfaces are vital to efficient designs. Many of the surfaces were developed for desalination, an area that has not fulfilled its promise relative to potential use of enhancement technology. The design of condensing surfaces is particularly well advanced because of recent analytical studies of surface-tension-driven condensate flow. The enhanced surfaces thus developed are being used for evaporation and condensation of a variety of organic fluids being considered for organic Rankine cycles, both fired and driven by geothermal brine.

The potential degradation of heat transfer due to fouling has inhibited the use of tube-side enhancements in steam condensers. To the contrary, a rather optimistic report emerged from the retubing of a large power condenser for the Tennessee Valley Authority. Sartor(1982) found that the corrugated tubes(helical repeated rib roughness) did not have fouling rates greater than plain tubes. Standard cleaning methods were effective in restoring performance. In this application, the objective was to improve plant thermal efficiency through a lower turbine exhaust temperature. The payback period on initial investment was estimated to be less than one year.

Process

The chemical process industry has adopted enhancement technology sparingly because of concerns about fouling. Plant engineers do not wish to risk shutdown of an entire process facility because of degradation of a heat eachanger that is costwise a small part of the facility. The Cal Gavin Heatex wire loop inserts and Vapor Sphere Matrix fluted spheres(tubeside), or solid spheres(shell-side, not only improve heat transfer but reduce fouling with typical process fluids(Mascone, 1986). Once again, commercial installation has preceded research on these inserts.

Some attention has been given recently to the enhancement of heat transfer to laminar in-tube flow of non-Newtonian liquids. Percentage improvements in heat transfer coefficients with spirally corrugated tubes and twisted-tape inserts are comparable to those observed with Newtonian liquids.

The High Flux surface has been noted in connection with refrigeration applications, but its main application is in process plant reboilers. Numerous theoretical papers and economic analyses have been directed toward this application.

Since most process reboilers and evaporators involve mixtures, many of the structured boiling surface are being tested with mixtures of two or more components(Thome, 1988). Fig.22 compares the boiling curves for the new Gewa-TX tube with those of a plain tube for a five-component hydrocarbon mixture. This tube is similar to the Gewa-T tube shown in Fig.4; however, the interior channels have small notches around the circumference. The performance gains are similar to those shown in Fig.4 for a pure fluid.

Industrial Heat Recovery

Advanced surfaces for high-temperature heat recovery are being designed so that both convection and radiation are enhanced. Of particular in-

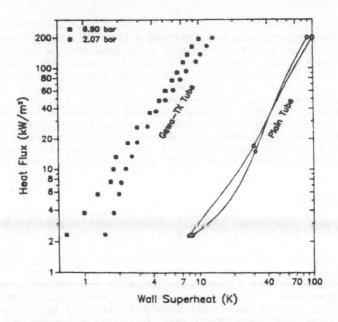

FIGURE 22. Boiling of a five-component hydrocarbon mixture on a Gewa-TX tube (Thome, 1988).

terest recently are ceramic tubes that are enhanced externally and/or internally. These tubes show great promise for the recovery of heat from waste streams in excess of 850°C(Bergles,1987).

Fouling and corrosion must be minimized if high-temperature heat recovery is to be practical. While most studies of fouling are with plain surfaces, serrated finned tubes and offset-strip-fin heat exchangers have been studied. These surfaces may remain effective in moderate fouling conditions; however, it is possible that nearly total plugging occurs with certain streams, e.g.,diesel exhaust and glass furnace exhaust (Marner and Webb, 1983).

Electronics Cooling

Enhanced extended surfaces have long been used for air cooling of electronic devices ranging from radar tubes to transistors. The recent emphasis has been on the development of cooling schemes for micro-electronic chips used in computers. Structures are provided to conduct the heat generated by an array of chips out to the air or water coolant. The final thermal coupling is through finned arrays that may be quite complex for air cooling or simple for liquid cooling. At present, much attention is being given to direct immersion of the chips in an inert, dielectric liquid. Saturated or subcooled boiling occurs due to the high chip powers. With the dual objective of reducing the wall superheat and increasing the burnout heat flux, enhanced boiling surfaces are attached to individual chips(Park and Bergles, 1986). There is much work yet to be done to establish configurations that achieve both objectives.

Jet impingement of dielectric liquid has recently been explored as a means of accommodating very high chip heat dissipations, over 2×10^6 W/m² (Ma and Bergles,1983).

Aerospace

Gas turbine blade cooling has challenged the heat transfer community for the past decade. Several enhancement techniques are now being utilized to increase the heat transfer coefficient from blade wall to internal coolant, thereby reducing the wall temperature. Transverse repeated ribs and pin fins can be cast into the blade,or the air supply can be arranged to direct jets against the leading-edge wall(Simoneau, 1985).

Difficult heat transfer problems are being encountered in connection with the aerospace plane proposed for transatmospheric flight. In addition to protection of components in the propulsion system, active cooling of the vehicle will likely be required to reduce material temperature during hypersonic flight. Enhancement of heat transfer will probably be required.

CONCLUDING REMARKS

Heat transfer enhancement today is characterized by vigorous activity in both research and industrial practice. Due to extensive research and much successful field experience during the post twenty-five years, enhancement techniques are routinely considered for improving convective heat transfer in a wide variety of equipment. Current research is driven primarily by applications rather than curiosity and typically involves multiple approaches. In many cases, however, analytical or numerical predications of heat transfer and flow friction behavior are not yet possible and empirical correlations must be relied on for design. Modelling should be encouraged, however, because the resulting understanding of the phenomena can result in better predictions and improved surface or insert device configurations.

The growing maturity of this second generation heat transfer is evidenced by its incorporation into undergraduate textbooks(e.g.,White,1984) and it is likely that the subject will be featured in many of the new editions. At least two graduate courses(Penn State and RPI) are devoted entirely to enhanced heat transfer. The greatest challenge in the coming years, perhaps, will be to collect the literature, evaluate it, and communicate it in concise form by way of formal courses, short courses, and review articles.

It is quite apparent that in China, heat transfer enhancement will be a strong component of research, practice,and technical communications. With the large effort that is going on worldwide, there is every reason to expect the development of much more advanced enhancement techniques that could become a third generation heat transfer technology.

ACKNOWLEDGMENT

Major portions of this manuscript are taken from an article appearing in a special 1988 issue of the Journal of Heat Transfer.

NOMENCLATURE

A, A_p = projected or nominal surface area

C_q = constant in equation(5)

C_t = ratio of liquid film area to thickness

C_p = specific heat at constant pressure

d	=	tube diameter
d_b	=	bubble departure diameter
e	=	protuberance height
e^+	=	roughness Reynolds number, $(e/d)Re(f/2)^{0.5}$
F	=	function in equation(8)
f	=	Fanning friction factor
f_b	=	frequency of bubble emission
G	=	mass flux
$\bar{g}$	=	function in equation(8)
h	=	heat transfer coefficient; height of offset strip fin
h_{fg}	=	enthalpy of vaporization
j	=	Colburn factor, $StPr^{2/3}$
k	=	fluid thermal conductivity
k_1, k_L	=	liquid thermal conductivity
L	=	length of offset strip fin
m_1	=	mass vaporized
N_A	=	number of nucleation sites
N_u	=	Nusselt number, hd/k
n	=	number of sharp corners facing flow
P	=	spacing between roughness elements
Pr	=	Prandtl number, $c_p\mu/k$
q	=	rate of heat transfer
q''	=	heat flux
q_1''	=	superficial latent heat flux
q_{ex}''	=	superficial heat flux resulting from free convection
R_b	=	bubble radius
R_e	=	Reynolds number, Gd/μ
R_e^*	=	Reynolds number parameter defined by equation(17)
S	=	total surface area occupied by evaporating liquid film
S_t	=	Stanton number, h/Gc_p
s	=	spacing of offset strip fin
t	=	average liquid film thickness; thickness of offset strip fin plates
T_L	=	liquid film temperature at bubble interface
T_S	=	saturation temperature
T_V	=	vapor temperature
T_W	=	wall temperature
ΔT	=	temperature difference(wall superheat)
U_e^+	=	Nikuradse similarity function, $(2/f)^{0.5}+2.5\ln 2e/d+3.75$

v_{fg} = specific volume change during vaporization

x,y = exponents

Greek Symbols

α = angle from axis of helical rib roughness

β = geometry factor in equation(2); rib contact angle in equation(11)

θ = time

μ = dynamic viscosity

σ = surface tension

ρ = vapor density

Subscripts

1 = first period pertaining to Thermoexcel-E model

2 = second period pertaining to Thermoexcel-E model

a = pertains to enhanced tube

h = based on hydraulic diameter

L = based on fin element length

s = pertains to plain tube

REFERENCES

Alberger Heater Company, 1921, Catalog No.3, Buffalo, NY.

Arai,N., Fukushima, T., Arai,A., Nakayama,T., Fujie, K., and Nakayama, Y., 1977, "Heat Transfer Enhancing Boiling and Condensation in Heat Exchanger of a Refrigerating Machine," ASHRAE Transactions, Vol.83,pt.2, pp.58-70.

Ayub,Z.H.., 1986, "Pool Boiling from Gewa Surfaces in Water and R-113," Ph.D. Dissertation, Iowa State University, Ames, IA.

Ayub,Z.H., and Bergles,A.E., 1987, "Pool Boiling from GEWA Surfaces in Water and R-113, "Wärme-und Stoffübertragung, Vol.21, pp.209-219.

Ayub,Z.H. and Bergles, A.E., 1988, "Nucleate Pool Boiling Curve Hysteresis for GEWA-T Surfaces in Saturated R-113," presented at National Heat Transfer Conference, Houston, TX.

Baustian,J.T., Pate,M.B., and Bergles,A.E., 1988, "Measuring the Concentration of a Flowing Oil-Refrigerant Mixture: Instrument Test Facility and Initial Results," ASHRAE Transactions, Vol.94,Pt.1.

Bergles,A.E., 1978, "Enhancement of Heat Transfer, "Heat Transfer 1987, Proceedings, 6th International Heat Transfer Conference, Hemisphere Publishing Corp., Washington, DC, Vol.6,pp.89-108.

Bergles,A.E., 1983, "Augmentation of Heat Transfer, "Heat Exchanger Design Handbook, Vol.2, Hemisphere Publishing Corp., Washington, DC, pp.2.5.11-1-12.

Bergles,A.E., 1985, "Techniques to Augment Heat Transfer, "Handbook of Heat Transfer Applications, McGraw-Hill Book Co., New York, NY, pp. 3-1-80.

Bergles,A.E., 1987, "Heat Transfer Enhancement-Application to High-Temperature Heat Exchangers "Heat Transfer in High Technology and Power Engineering, Hemisphere Publishing Corp., Washington, DC and Springer-Verlag, Berlin, F.R. Germany,pp.333-355.

Bergles,A.E., Nirmalan,V., Junkhan,G.H., and Webb,R.L., 1983, "Bibliography on Augmentation of Convective Heat and Mass Transfer-II," HTL-31, ISU-ERI-Ames-8422, DE-84018484, Iowa State University, Ames, IA.

Bergles,A.E., Nelson,R.M., Junkhan,G.H., and Webb,R.L. 1984, "Assessment, Development and Coordination of Technology Base Studies in Enhanced Heat Transfer," HTL-33,ISU-ERI-Ames-85024, Iowa State Univ., Ames, IA.

Bergles,A.E., and Chyu,M.C., 1982, "Characteristics of Nucleate Boiling from Porous Metallic Coatings, "Journal of Heat Transfer, Vol.104,pp.279-285.

Chyu,M.C., 1984, "Falling Film Evaporation on Horizontal Tubes with Smooth and Structured Surfaces, "Ph.D. Dissertation, Iowa State Univ., Ames, IA.

Chyu,M.C. and Bergles, A.E., 1985, "Enhancement of Horizontal Tube Spray Film Evaporators by Structured Surfaces, "Advances in Enhanced Heat Transfer-1985, HTD-Vol.43, ASME, New York, NY, pp.39-48.

Czikk,A.M., Gottzmann,C.F., Ragi,E.G., Withers,J.G. and Habdas,E.P., 1970, "Performance of Enhanced Heat Transfer Tubes in Refrigerant-Flooded Liquid Coolers," ASHRAE Transactions, Vol.1, Pt.1, pp.96-109.

Czikk,A.M. and O'Neill, P.S., 1979, "Correlations of Nucleate Boiling from Porous Metal Films," Advances in Enhanced Heat Transfer, ASME, New York, NY, pp.53-60.

Dipprey,D.G. and Sabersky,R.H., 1963, "Heat and Momentum Transfer in Smooth and Rough Tubes at Various Prandtl Numbers, "International Journal of Heat and Mass Transfer, Vol.6,pp.329-353.

Fagerholm, N.-E., Kivioja,K., Ghazanfari,A.-R., Jarvinen,E., 1985, "Using Structured Surfaces to Enhance Heat Transfer in Falling Film Flow," I.I.F.-I.I.R.-Commission E2 Trondheim(Norway), pp.187-192.

Ganeshan,S. and Raja Rao,M., 1982, "Studies on Thermohydraulics of Single and Multistart Spirally Corrugated Tubes for Water and Time-Independent Power Law Fluids," International Journal of Heat and Mass Transfer, Vol.25,pp.1013-1022.

Dee,D.L. and Webb,R.L., 1980, "Forced Convection Heat Transfer in Helically Rib-Roughened Tubes, "International Journal of Heat and Mass Transfer, Vol.23,pp.1127-1136.

Gottzmann,C.F., O'Neill, and Minton,P.E. 1973, "High Efficiency Heat Exchangers," Chemical Engineering Progress, Vol.69, No.7, pp.69-75.7

Groeneveld,D.C.,and Yousef,W.W.,1980, "Spacing Devices for Nuclear Fuel Bundles: A Survey of Their Effect on CHF, Post-CHF Heat Transfer and Pressure Drop,Proceedings of the ANS/ASME/NRC International Topical Meeting on Nuclear Reactor Thermal-Hydraulics, NUREGI/CP 0014, Vol.2, pp. 1111-1130.

Hillis,D.L., Lorenz,J.J., Yung,D.T., and Sater, N.F.,1979,"OTEC Performance Tests of the Union Carbide Sprayed-Bundle Evaporator," AN1/OTEC-PS-3, Argonne National Lab., Argonne, IL.

Huang,Y.H., Liou,T.M., and Syang,Y.C., "Heat Transfer Enhancement of Turbulent Flow in Pipes with an Internal Circular Rib,"Advances in Enhanced Heat Transfer 1987, HTD-Vol.68, pp.55-63.

Jakob,M., and Fritz,W., 1931, "Versuche über den Verdampfungsvorgang," Forschung auf dem Gebiete des Ingenieurwesens, Vol.2. pp.435-447.

Joshi, H.M., and Webb,R.L.,1987,"Heat Transfer and Friction in the Offset Strip-Fin Heat Exchanger,"International Journal of Heat and Mass Transfer, Vol.30, pp.69-84.

Joule,J.P.,"On the Surface-condensation of Steam," 1861, Philosphical Transactions of the Royal Society of London, Vol.151, pp.133-160.

Kays,W.M., 1972,"Compact Heat Exchanger," AGARD Lecture Series No.57, Heat Exchangers, AGARD-LS-57-72.

Kays,W.M., and London,A.L.,1984, Compact Heat Exchangers, 3rd Ed., Mc-Graw-Hill Book Co., New York, NY.

Kelkar,K.M.,and Patankar, S.V., 1985, "Numerical Prediction of Heat Transfer and Fluid Flow in Retangular Offset-Fin Arrays," Augmentation of Heat Transfer in Energy Systems,HTD-Vol.52, ASME, New York, pp.21-28.

Khanpara,J.C., Bergles,A.E.,and Pate, M.B., 1985,"Augmentation of R-113 In-Tube Condensation with Micro-Fin Tubes," Heat Transfer in Air Conditioning and Refrigeration Equipment, HTD-Vol.65,ASME, New York, pp.21-32.

Khanpara,J.C., Bergles,A.E., and Pate,M.B.,1986, "Augmentation of R-113 In-Tube Evaporation with Micro-Fin Tubes," ASHRAE Transactions, Vol.92, Part 2B, pp. 506-524.

Kim,C.J., and Bergles,A.E.,1985, "Structured Surfaces for Enhanced Nucleate Boiling,"HTL-36,ISU-ERI-Ames-86220, Iowa State Univ., Ames, IA.

Kitto,J.B., and Albrecht,M.J.,1988,"Elements of Two-phase Flow in Fossil Boilers,"Two-phase Flow Exchangers; Thermal Hydraulic Fundamentals and Design,Kluwer Academic Publishers, The Netherlands, pp.495-551.

Kulacki, F.A.,1981,"Electrohydrodynamic Enhancement of Convective Heat and Mass Transfer," Advances in Transport Processes, Vol.II, Halstead Press, New York, NY, pp.105-147.

Kurosaki,Y., Kashiwagi,T.,Kobayashi,H., Uzuhashi,H, and Tang,S.C.,1988, "Experimental Study on Heat Transfer from Parallel Louvered Fins by Laser Holographic Interferometry,"Experimental Thermal and Fluid Sci. ,Vol.1. pp. 59-67.

Lea,R.B., 1921, "Oil Cooler." U.S. Patent No.1,367,881.

Lewis,M.J.,1975,"An Elementary Analysis for Predicting the Momentum and Heat-Transfer Characteristics of a Hydraulically Rough Surface,"Journal of Heat Transfer, Vol.97, pp.249-254.

Li,H.M., YE, K.S., Tan,Y.K., and Deng, S.J.,1982,"Investigations on Tube-Side Flow Visualization, Friction Factors and Heat Transfer Characteristics of Helical-Ridging Tubes," Heat Transfer 1982, Proceedings,7th International Heat Transfer Conference,Hemisphere Publishing Corp.,Washington, DC. Vol.3.

Loehrke,R.I. and Lane,J.C.,1982,"Heat Flow Through Array of Interrupted, Parallel Plates," Heat Transfer 1982,Proceedings,7th International Heat Transfer Conference, Hemisphere Publishing Corp.,Washington DC,Vol.3, pp.81-86.

Lorenz,J.J., Yung,D.,Howard,P.A.,Panchel,C.B., and Poucher,F.W., 1981,"OTEC-1 Power System Test Program: Performance of One-Megawatt Heat Exchangers," ANL/OTEC-PS-10, Argonne National Lab., Argonne, IL.

Lewis,L.G. and Sather,N.F.,1978,"OTEC Performance Tests of the Union Carbide Flooded-Bundle Evaporator,"ANL-OTEC-PS-1,Argonne National Lab., Argonne, IL.

Mascone,C.F.,1986,"CPI Strive to Improve Heat Transfer In Tubes,"Chemical Engineering , February, pp. 22-25.

Ma C.F. and Bergles,A.E.,and "Boiling Jet Impingement Cooling of Simulated Microelectronic Chips," Heat Transfer in Electronic Equipment,HTD-Vol. 28, ASME, New York, NY, pp.5-12.

Manglik,R.K. and Bergles,A.E.,1987,"A Correlation for Laminar Flow Enhanced Heat Transfer in Uniform Wall Temperature Circular Tubes with Twisted-Tape Inserts," Advances in Enhanced Heat Transfer-1987, HTD-Vol. 68, ASME, New York, NY, pp.35-45.

Marner,W.J. and Webb,R.L.,1983,"A Bibliography on Gas-Side Fouling," Proceedings of the ASME-JSME Thermal Eng. Joint Conference,Vol.1,pp.559-570.

Marto, P.J.,1986,"Recent Progress in Enhancing Film Condensation Heat Transfer on Horizontal Tubes,"Heat Transfer 1986,Proceedings,8th International Heat Transfer Conference,Vol.1,Hemisphere Publishing Corp., Washington,DC, pp.161-170.

Marto,P.J. and Hernandez,B.,1983,"Nucleate Pool Boiling Characteristics of a GEWA-T Surface in Freon-113,"AIChE Symposium Series, No.225 Vol.79, pp.1-10.

McManus,S.M.,Marto,P.J. and Wanniarachchi,A.S.,1986,"An Evaluation of Enhanced Heat Transfer Tubing for Use in R-114 Water Chillers," Heat Transfer in Air Conditioning and Refrigeration Equipment, HTD-Vol.65, ASME, New York, NY, pp.11-19.

Menze,K.,1988, Wieland-Werke,Ulm, F.R. Germany,Personal Communication, February 4.

Mochizuki,S.,Yagi,Y., and Yang W.J.,1988,"Flow Pattern and Turbulence Intensity in Stacks of Interrupted Parallel-Plate Surfaces,"Experimental Thermal and Fluid Science, Vol.1, pp.51-57.

Mori,Y. and Nakayama,W., 1980,"Recent Advanced in Compact Heat Exchangers in Japan," Compact Heat Exchangers-History, Technological Advancement and Mechanical Design Problems,HTD-Vol.10, ASME, New York,NY, pp.5-16.

Mullisen,R.S.and Loehrke,R.I.,1986,"A Study of The Flow Mechanisms Responsible for Heat Transfer Enhancement in Interrupted-Plate Heat Exchangers," Journal of Heat Transfer, Vol.108, pp.377-385.

Nakayama,W., Daikoku,T.,Kuwhara,H., and Nakajima,T., 1980 "Dynamic Model of Enhanced Boiling Heat Transfer on Porous Surfaces, Part II: Analytical Modeling," Journal of Heat Transfer, Vol.102, pp.451-456.

Nakayama,W., 1982,"Enhancement of Heat Transfer," Heat Transfer 1982, Proceedings,7th International Heat Transfer Conference, Hemisphere Publishing Corp., Washington,DC, Vol.1, pp.223-240.

Nakayama,W., Takahashi, K., and Daikoku, T.,1983,"Spiral Ribbing to Enhance Single-Phase Heat Transfer Inside Tubes," ASME-JSME Thermal Engineering Joint Conference Proceedings,Vol.1, pp.365-372.

Newton,I.,(Anon.),1701,"Scale Graduum Caloris," The Philosophical Transactions of the Royal Society of London, Vol.22,1701, pp.824-829. Translated from the Latin in The Philosophical Transactions of Royal Society of London, Abridged, Vol.IV(1694-1702), London,1809,pp.572-575.

O'Neill,P.S.,Gottzmann,C.F., and Turbot,J.W., 1972,"Novel Heat Exchanger Increases Cascade Cycle Efficiency for Natural Gas Liquefaction,"Advances in Cryogenic Engineering, Vol.17,pp.420-437.

O'Neill,P.S.,King,R.C.,and Ragi,E.C.,1980,"Application of High Performance Evaporator Tubing in Refrigeration Systems of Large Olefins Plants,"AIChE Symposium Series, No.199, Vol.76, pp.289-300.

Park,K.A. and Bergles,A.E.,1986,"Boiling Heat Transfer Characteristics of Simulated Microelectronic Chips with Detachable Heat Sinks"Heat Transfer 1986, Proceedings,8th International Heat Transfer Conference, Hemisphere Publishing Corp.,Washington,DC,Vol.4, pp.2099-2104.

Ravigururajan,T.S. and Bergles, A.E., 1985, "General Correlations for Pressure Drop and Heat Transfer for Single-Phase Turbulent Flow in Internally Ribbed Tubes,"Augmentation of Heat Transfer in Energy Systems, HTD-Vol.52, ASME, New York, NY, pp.9-20.

Rudy,T.M. and Webb,1985,"An Aanlytical Model to Predict Condensate Retention on Horizontal Integral-Fin Tubes,"Journal of Heat Transfer,Vol.107 pp.361-368.

Sauer,H.S., Davidson,G.W., and Chungrungreong,S.,1980,"Nucleate Boiling of Refrigerant-Oil Mixtures from Finned Tubing,"ASME Paper No.80-HT-111.

Sator,W.E.,1982,"Extended and Enhanced Tube Surfaces to Improve Heat Transfer,"Proceedings of 2nd Symposium on Shell and Tube Heat Exchangers, Houston,TX, pp.411-418.

Sethumadhavan,R. and Raja Rao,M.,1986"Turbulent Flow Friction and Heat Transfer Characteristics of Single- and Multistart Spirally Enhanced Tubes," Journal of Heat Transfer, Vol.108, pp.55-61.

Schlager,L.M.,Pate,M.B., and Bergles, A.E., 1988, "Evaporation and Conden - sation of Refrigerant-Oil Mixture in a Low-Fin Tube, "ASHRAE Transactions, DC, and Springer-Verlag, Berlin, F.R. Germany, pp.285-319.

Schlager,L.M.,Pate,M.B., and Bergles, A.E., 1988, "Evaporation and Condensa- tion of Refrigerant-Oil Mixture in a Low-Fin Tube," ASHRAE Transactions, Vol.94, part 1.

Simoneau,R.J.,1987,"Heat Transfer in Aeropropulsion Systems,"Heat Trans- fer in High Tech.and Power Eng.Hemisphere Publishing Corp. Washington,DC, and Springer-Verlag,Berlin, F.R. Germany, pp.285-319.

Sparrow,E.M. and Haliloo,A.,1980,"Measurements of Heat Transfer and Pres- sure Drop for an Array of Staggered Plate Aligned to an Air Flow,"Journal of Heat Transfer, Vol.102, pp.426-432.

Stephan,K.and Mitrovic,J.,1982,"Heat Transfer in Natural Convective Boil- ing of Refrigerants and Refrigerant-Oil-Mixtures in Bundles of T-shaped Finned Tubes,"Advances in Enhanced Heat Transfer-1981,HTD-Vol.18,ASME, New York, NY, pp.131-146.

Stephan,K.and Mitrovic,J.,1982,"Heat Transfer in Natural Convective Boil- ing of Refrigerant-Oil Mixtures,"Heat Transfer 1982,Proceedings,7th Inter- national Heat Transfer Conference, Hemisphere Publishing Corp.,Washington DC, Vol.4, pp.73-87.

Tatara,R.A. and Payvar,P.,1986,"Pressure Drop and Heat Transfer Measure- ments of Boiling Refrigerant in Normal Flow Through a Porous Coated Tube Bundle,"Heat Transfer in Air Conditioning and Refrigeration Equipment, HTD-Vol.65,ASME,New York,NY,pp.1-9.

Thome,J.R.,1988,"Application of Enhanced Boiling Tubes to Reboilers," Two-Phase Flow Heat Exchangers: Thermal-Hydraulic Fundamentals and Design, Kluwer Academic Publishers, The Netherlands, pp.747-778.

Torii,T.,Hirasawa,S.,Kuwahara,H.,Yanagida,T., and Fujie,K.,1978,"The Use of Heat Exchangers with Thermoexcel's Tubing in Ocean Thermal Energy Power Plants," ASME Paper No.78-WA?HT-65.

Wanniarachchi,A.S.,Marto,P.J., and Reilly,J.T.,1986,"The Effect of Oil Contamination on The Nucleate Pool-Boiling Performance of R-114 from a Porous Coated Surface,"ASHRAE Transactions, Vol.92,Pt.2.

Wassel,A.T. and Mills,A.F.,1979,"Calculation of Variable Property Turbu- lent Friction and Heat Transfer in Rough Pipes,"Journal of Heat Transfer, Vol.101,pp.469-474.

Webb,R.L.,Rudy,T.M.,and Kedzierski,M.A.,1985,"Prediction of the Condensa- tion Coefficient on Horizontal Integral-Fin Tubes,"Journal of Heat Trans- fer, Vol.107, pp. 369-376.

Webb,R.L.,1984,"Shell-side Condensation in Refrigerant Condensers," ASHRAE Transactions, Vol.90,Pt.1, pp.5-24.

Webb,R.L. 1979,"Toward a Common Understanding of the Performance and Selection of Roughness for Forced Convection,"Studies in Heat Transfer, Hemisphere Publishing Corp. Washington,DC. pp.257-272.

Webb,R.L. and Bergles,A.E.,1983, "Performance Evaluation Criteria for Selection of Heat Transfer Surface Geometries Used in Low Reynolds Number Heat Exchangers,"Low Reynolds Number Heat Exchangers,Hemisphere Publishing Corp.,Washington,DC and Springer-Verlag,Berlin, F.R.Germany,pp.735-752

Webb,R.L.,Bergles,A.E., and Junkhan,G.H., "Bibliography of the U.S. Patents on Augmentation of Convective Heat and Mass Transfer-II," HTL-32,ISU-ERI-Ames-84257,Iowa State University,Ames,IA.

Webb,R.L.,1980,"Air-Side Heat Transfer in Finned Tube Heat Exchangers," Heat Transfer Engineering, Vol.1,No.3,pp.33-49.

Webb,R.L. 1981,"The Evolution of Enhanced Surface Geometries for Nucleate Boiling,"Heat Transfer Engineering,Vol.2,Nos.3-4,pp.46-69.

Webb,R.L,1983,"Nucleate Boiling in Porous Coated Surface,"Heat Transfer Engineering, Vol.4, Nos.3-4, pp.71-82.

Webb,R.L., Eckert,E.R.G.,and Goldstein,R.J.,1971,"Heat Transfer and Friction in Tubes with Repeated Rib Roughness,"International Journal of Heat and Mass Transfer, Vol.14,pp.601-618.

White,F.,1984,Heat Transfer,Addison-Wesley Publishing Company,Reading,MA.

Whitham,J.M.,1896,"The Effect of Retarders in Fin Tubes of Steam Boilers," Street Railway Journal, Vol.12, p.374.

Wieting,A.R.,1975,"Empirical Correlations for Heat Transfer and Flow Friction Characteristics of Rectangular Offset-Fin Plate-Fin Heat Exchangers," Journal of Heat Transfer, Vol.97, pp.488-490.

Withers,J.A.,1980a,"Tube-side Heat Transfer and Pressure Drop for Tubes Having Helical Ridging with Turbulent/Transitional Flow of Single-Phase Fluid,Part 1.Single-Helix Ridging,"Heat Transfer Engineering, Vol.2. No.1, pp.48-61.

Withers,J.A.,1980b, "Tube-side Heat Transfer and Pressure Drop for Tubes Having Helical Internal Ridging with Turbulent/Transitional Flow of Single-Phase Fluid. Part 2. Multiple-Helix Ridging,"Heat Transfer Engineering, Vol.2, No. 2, pp.43-50.

Yilmaz,S., Palen,J.W., and Taborek,J., 1981,"Enhanced Boiling Surfaces as Single Tubes and Bundles,"Advances in Enhanced Heat Transfer-1981, HTD-Vol.18, ASME, New York, NY, pp.123-129.

Yilmaz,S., Hwalck,J.J., and Westwater,J.W., 1980,"Pool Boiling Heat Transfer Performance for Commercial Enhanced Tube Surfaces,"ASME Paper No. 80-HT-41.

Condensation of NARBs inside a Horizontal Tube

TETSU FUJII
Institute of Advanced Material Study
Kyushu University
Japan

1. INTRODUCTION

Recently, research and development of heat pump and refrigeration systems using nonazeotropic refrigerant blends (NARBs) as the working fluid is being successfully performed. The aim of using NARBs is to drastically improve the coefficient of performance (COP) by realizing the Lorenz cycle, to choose the most suitable working fluid under various conditions of air conditioning and industrial heating and cooling systems and to provide substitutes of stratospherically safe refrigerants.

In the present paper, future problems concerning condensation heat transfer research on horizontal in-tube condensers will be explained through presentation of some theoretical and experimental results on the relation between the actual Lorenz cycle and overall heat transfer coefficient, the relation between mass fraction of NARBs and condensation heat transfer coefficient in a horizontal tube , and heat transfer enhancement by increasing vapor velocity.

2. EFFECT OF HEAT TRANSFER UPON COP OF ACTUAL HEAT PUMP SYSTEM[1]

We consider a simple compression heat pump system, in which water of flow rate of 0.0155kg/s is heated from 40°C to 70.8°C , and heat source water of 30°C is available. In order to realize a Lorenz cycle , we assume that the condenser and evaporator are of counterflow double tube type, which have the same values of the product of heat transfer area F and overall heat transfer coefficient K. When a 25wt%R22+75wt%R11 blend is used as the working fluid, the temperature difference between the refrigerant and high temperature water is kept constant along the tube axis in the condenser, where the temperature difference is about 10°C for FK = 200W/K and the flow rate of the refrigerant W_r is 0.0106kg/s. For the evaporator, the flow rate of low temperature water is chosen so that the temperature increase of the working fluid is equal to the temperature decrease of the low temperature water.

Figure 1 shows a schematic diagram of axial temperature distributions of heat transfer water and working fluid, which are respectively expressed by broken lines and solid lines with arrows denoting flow directions, for the system. The cycle for the system is called here " the actual cycle". As the FK value becomes large, the solid lines approach the broken lines in Fig.1. The limit case is called "the ideal cycle". In the T-s diagram of Fig.2 the actual and ideal cycles are shown by solid and two-dot-dashed lines, respectively. It can be easily understood that the COP of the actual cycle is much smaller than that of the ideal cycle.

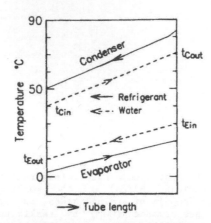

FIGURE 1. Axial temperature distributions of heat transfer water and refrigerant. [25wt%R22+75wt%R11, t_{Cin}=40°C, t_{Cout}=70.8°C, t_{Ein}=30°C, W_C= 0.0115kg/s, W_E=0.0186kg/s, W_r= 0.0106kg/s, FK=200W/K, Q_C=2kW, Q_E=1.6kW]

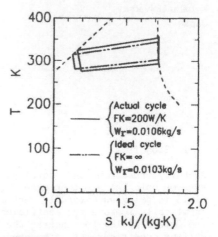

FIGURE 2. Comparison between actual and ideal cycles for 25wt%R22 +75wt%R11 in the T-s diagram. [Given conditions are the same as those in Fig.1 except for FK]

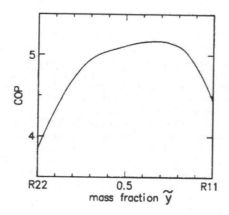

FIGURE 3. Relation between COP and mass fraction $\tilde{y}$ for R22+R11. [Given conditions are the same as those in Fig.1 except for mass fraction]

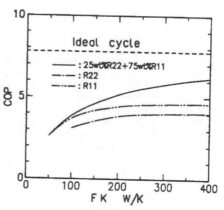

FIGURE 4. Relation between COP and FK for R22+R11. [Given conditions are the same as those in Fig.1 except for FK]

Figure 3 shows the relation between COP and mass fraction of R11 $\tilde{y}$ of the R22+R11 blend, which was calculated under the same conditions as those in Fig.1 except for mass fraction. The axial temperature distributions of water and refrigerant are not parallel except for $\tilde{y}$ = 0.75, at which the maximum value of COP takes places. The $(COP)_{max}$ is larger than the COPs of pure R22 and pure R11 by about 35% and 16%, respectively.

Figure 4 shows the relation between COP and FK, which is obtained under the same conditions as those in Fig.1. In the figure the COP of the actual NARB cycle is much smaller than that of the ideal NARB cycle, and the difference in COP between actual NARB and pure refrigerant cycles is small when FK is small, whereas the superiority of the NARB cycle becomes considerable with the increase of FK. It should be reminded that usual comparison in COP is made between ideal NARB cycles and pure refrigerant cycles. Conclusively, the study of heat transfer is necessary for both the prediction and improvement of COP for the NARB cycle.

3. CONDENSATION HEAT TRANSFER COEFFICIENT OF R22+R114 BLEND INSIDE A HORIZONTAL TUBE[2]

3.1 Axial Distribution of Heat Transfer Coefficient

Figure 5 shows some examples of the axial distribution of heat transfer coefficient for the condensation of the NARB R22+R114 inside a horizontal tube. The tested tube has a length of 4.8m, a mean inner diameter of 8.32mm and sixty spiral grooves of 0.15mm in height inside the tube. The local heat transfer coefficient α is defined by

$$\alpha = \frac{q}{T_s - T_w} \tag{1}$$

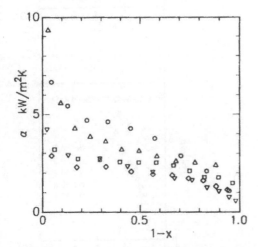

Symbol	$\frac{R114}{mol\%}$	$\frac{G_r}{kg/m^2 s}$
○	0	199
□	25	214
◇	50	193
▽	75	204
△	100	200

FIGURE 5. Axial distribution of heat transfer coefficient for condensation of R22+R114 inside a horizontal tube with spiral grooves.

51

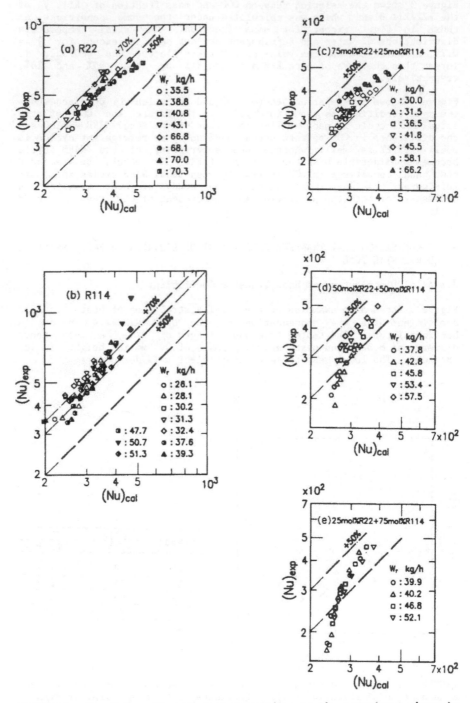

FIGURE 6. Local Nusselt number corresponding to the experiment shown in Fig.5.

where q is the local heat flux, T_w is the temperature of the inner tube surface and T_s is the refrigerant equilibrium temperature which is obtained from the heat balance equation assuming uniform temperature in the cross section. The quality x in the abscissa varies from 1 to 0 towards the refrigerant flow direction.

In Fig.5 the heat transfer coefficients of all refrigerants decrease towards downstream, and those of the mixtures are lower than those of the pure vapors. Precisely speaking, the data of R22 are higher than those of R114, and those of 50mol%R22+50mol%R114 and 25mol%R22+75mol%R114 are the lowest.

3.2 Local Nusselt Number

Figures 6(a)~(e) show the local Nusselt number $(Nu)_{exp} (= \alpha d/\lambda_L)$ corresponding to the aforementioned experiments for a spirally grooved tube compared with the local Nusselt number $(Nu)_{cal}$, which is estimated from a semiempirical equation applicable to the condensation of pure refrigerants inside a horizontal smooth tube[3]. In the figure, only the data for the vapor of equilibrium temperature are plotted.

$(Nu)_{exp}$ is higher than $(Nu)_{cal}$ by 40~70% for R22 and by 50~80% for R114. The increment is caused by heat transfer enhancement in the condensate side by the spiral grooves . The degree of the increment for the mixtures is smaller than that for pure refrigerants. Furthermore, the axial variation of $(Nu)_{exp}$ is affected by physical properties, bulk molar fraction and flow rate of refrigerants, especially for the mixtures.

3.3 Average Nusselt Number

The average Nusselt number $\overline{Nu}$ for condensation area from x=1 to x=0 is defined by

$$\overline{Nu} = \frac{\overline{q}\ell}{(\overline{T}_s - \overline{T}_w)\lambda_L} = \frac{\overline{\alpha}\ell}{\lambda_L} \tag{2}$$

where $\overline{q}$ is heat flux averaged over the condensation area, $\overline{T}_s$ is the arithmetic mean value of equilibrium temperatures at x=1 and x=0, and $\overline{T}_w$ is wall temperature averaged over the condensation area. $\overline{Nu}$ in Fig.7 is calculated from the experimental results shown in Fig.6, and the ordinate in Fig.7 is referred from the previous results[4] for pure vapors except for ξ. All experimental data can be correlated well by the following equation:

$$\overline{Nu} = 0.53\xi(d/\ell)^{0.4}Ph^{-0.6}(Re_\ell Pr_L/R)^{0.8} \tag{3}$$

where

$$\xi = 1 - 0.73y + 0.37y^2 + 0.36y^3 \tag{4}$$

$Ph = Cp_L(\overline{T}_s - \overline{T}_w)/(h_{vs,x=1} - h_{Ls,x=0})$ is phase change number, Cp_L is isobaric specific heat capacity, h is specific enthalpy, $Re_\ell = \rho_L U_v \ell/\mu_L$ is two-phase Reynolds number, ρ is density, U_v is mean vapor velocity at the tube inlet (x=1), μ is dynamic viscosity, Pr is Prandtl number,

53

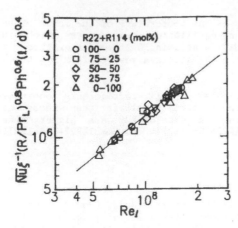

FIGURE 7. Average Nusselt number corresponding to the data in Fig.6.

$R=(\rho_L\mu_L/\rho_V\mu_V)^{1/2}$ is $\rho\mu$-ratio, and y is molar fraction of the less volatile vapor at the tube inlet.

The fact that the scattered data in local Nusselt number are correlated well in average Nusselt number is still questionable.

4. EFFECT OF VAPOR VELOCITY UPON NARBS HEAT TRANSFER COEFFICIENT

There is a lot of uncertainty in heat transfer characteristics for condensation of NARBs inside a horizontal tube as aforementioned. Therefore, by using a similarity solution of the two phase boundary layer equations for laminar film condensation on a vertical surface, the mechanism of the decrease in heat transfer coefficient of NARBs is explained.

Figure 8 shows an example of the relation between mass fraction $\tilde{y}$ in the main vapor stream and average heat transfer coefficient $\bar{\alpha}$, which is defined by

$$\bar{\alpha} = \frac{\bar{q}}{T_V-T_W} \tag{5}$$

where $\bar{q}$ is heat flux averaged over the heat transfer length ℓ, T_V is main stream vapor temperature and T_W is cooling surface temperature. Figures 8 (a) and (b) correspond to forced convection condensation and free convection condensation, respectively, and solid lines represent results of the similarity solution while broken lines represent values calculated on the assumption that vapor-condensate interface temperature T_i is equal to T_V like an azeotropic mixture.

The difference between solid and broken lines in $\bar{\alpha}$ corresponds to the decrease in heat transfer coefficient due to the temperature decrease through the vapor boundary layer, the reason of which is explained as below. Since the condensation mass flux of more volatile component is smaller than that of less volatile component, the mass fraction of the former component vapor increases at the vapor-condensate interface and

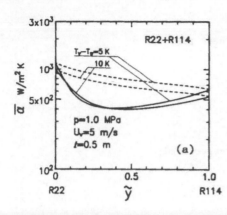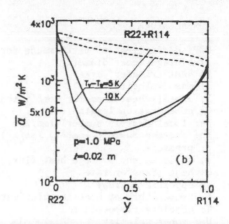

FIGURE 8. Relation between average heat transfer coefficient and mass fraction of vapor for laminar film condensation of R22+R114 on a vertical flat surface.
(a)forced convection condensation, (b)free convection condensation.

the interface temperature decreases as seen in the characteristics of condensation of vapor containing noncondensable gas. In each figure the solid curves are convex downwards, which is in contrast with the characteristics of the COP vs. $\tilde{y}$ relation shown in Fig.3. Furthermore, the heat transfer coefficient decreases as the temperature difference (T_v-T_w) decreases.

A comparison between Figs.8 (a) and (b) clarifies that the decrease in heat transfer coefficient for forced convection condensation is smaller than that for free convection condensation. This fact suggests that the increase of vapor velocity prevents the decrease in heat transfer coefficient in condensation inside a horizontal tube as well. However, free convection condensation inevitably takes place near the exit of condenser tubes. In this portion any device for heat transfer enhancement in the vapor phase is necessary.

5. CONCLUSION

1. High COP of actual Lorenz cycle can be realized only when condensers and evaporators with high K values are developed.

2. Minimum values of condensation heat transfer coefficient appear near a mass fraction corresponding to maximum COP for a given NARB cycle.

3. The decrease in condensation heat transfer coefficient of NARBs is mainly caused by the increase in thermal resistance in the concentration boundary layer. Therefore, any heat transfer enhancement for the vapor phase should be devised.

4. Detail study on the mechanism of condensation inside a horizontal tube is necessary so as to be able to predict heat transfer characteristics by using only the data of thermophysical properties of NARBs.

NOMENCLATURE

COP: coefficient of performance for heat pump cycle
d: average inner diameter
F: heat transfer area
G: mass velocity
K: overall heat transfer coefficient
ℓ: condensation length
Nu: local Nusselt number, $(\alpha d/\lambda_L)$
$\overline{Nu}$: average Nusselt number, $(\overline{\alpha}\ell/\lambda_L)$
p: pressure
q, $\overline{q}$: local and average heat flux, respectively
Q: heat transfer rate
s: specific entropy
t: temperature of heat transfer water
T: absolute temperature
U_v: vapor velocity
W: mass flow rate
x: quality
y: molar fraction of less volatile component
$\overline{y}$: mass fraction of less volatile component
α, $\overline{\alpha}$: local and average heat transfer coefficient, respectively
λ: thermal conductivity

Subscripts

c: heat transfer water
cal: calculation
C: condenser
exp: experiment
E: evaporator
in: inlet

L: liquid
out: exit
r: refrigerant
s: saturation or eqilibrium
v: vapor
w: wall

REFERENCES

1. Fujii, T., Koyama, Sh. and Miyara, A., Theoretical Consideration on the Characteristics and the Performance Evaluation for a Heat Pump Cycle of Nonazeotropic Refrigerant Mixtures (in Japanese), Trans. Japanese Association of Refrigeration, Vol.4, no.1, pp.27-34, 1987.

2. Fujii, T., Koyama, Sh., Miyara, A. and Takamatsu, H., Condensation and Evaporation of Nonazeotropic Binary Mixture of Refrigerants R22 and R114 inside a Horizontal Tube (in Japanese), Report on Research and Survey of Heat Pump Technology, Japanese Association of Refrigeration, pp.125-144, 1987.

3. Fujii, T., Honda, H. and Nozu, Sh., Condensation of Fluorocarbon refrigerants inside a Horizontal Tube—Proposals of Semi-empirical Expressions for the Local Heat Transfer Coefficient and the Interfacial Friction Factor—(in Japanese), Refrigeration, Vol.55, no.627, pp.3-20, 1980.

4. Fujii, T. and Nagata, T., Condensation of Vapors in a Horizontal Tube (in Japanese), Report of Research Institute of Industrial Science, Kyushu University, no.57, pp.35-50, 1973.

CONVECTIVE HEAT TRANSFER

Fundamental Study on Enhancement of Heat Transfer in a Stagnant Liquid Layer

KUNIO KATAOKA, YASUNORI MORIKAWA, YASUYUKI OKADA,
and TAKAHIKO MIYAKE
Department of Chemical Engineering
Kobe University
Rokkodai, Nada, Kobe 657, Japan

Abstract --- The enhancement of heat transfer due to the vortex motion
induced in a stagnant liquid layer was studied experimentally by using
an electrochemical technique under the assumption of analogy between
heat and mass transfer. Alternately rolling-up motion was produced
periodically by reciprocating an array of parallel rods (vortex
generator) near the bottom of the stagnant liquid layer. The mass
transfer was enhanced effectively by means of the vortex generator
even at sufficiently low frequencies. It has been confirmed that the
surface renewal motion occurs at the same frequency as the rolling-up
vortex motion.

INTRODUCTION

The interest in the use of vortical motions for the enhancement and
control of local heat/mass transfer has increased enormously in recent
years. Our interest is directed to the surface renewal effect of
unsteady vortical motion. As the first phase, our previous paper [1]
reported on the enhancement of mass transfer across a free gas-liquid
interface due to periodical vortex motions. It has been found that
the liquid-phase mass transfer is in proportion to the square root of
the velocity amplitude divided by the wavelength of the vortex motion.
The objective of the present work is to examine the effect of the same
periodical vortex motion on the heat/mass transfer to a liquid-solid
interface of stagnant liquid layer from the viewpoint of the surface
renewal.

EXPERIMENT

An electrochemical technique [2] was made use of under the assumption
of analogy between heat and mass transfer. The experiment was
designed to observe local, time-dependent velocity gradient, s, on a
mass transfer surface as well as local mass transfer coefficient, k.

The experimental setup is shown in Figure 1. A stagnant liquid layer
was formed between two large horizontal plates placed in the middle of
the apparatus. The distance between them was changed by raising or
lowering the upper plate. The lower plate had a large anode made of
nickel plate. The upper plate had a main cathode, made of nickel
plate, confronting the anode. As shown in Figure 2, small test
cathodes of two kinds were embedded in the wall of the main cathode
which served as the mass transfer surface. The working fluid used was

59

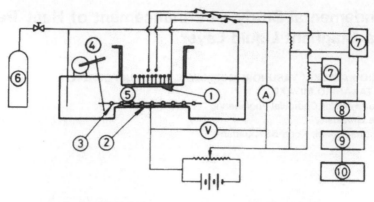

FIGURE 1. Experimental apparatus and electrode circuit.

1	MAIN CATHODE WITH ISOLATED TEST CATHODES	6	NITROGEN VESSEL
2	ANODE	7	DC AMPLIFIER
3	VORTEX GENERATOR	8	DATA RECORDER
4	CRANK SHAFT	9	AD CONVERTER
5	ELECTROLYTIC SOLUTION	10	DIGITAL COMPUTER

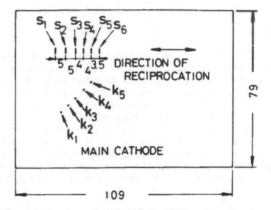

$S_1, ---, S_6$: RECTANGULAR TEST CATHODES FOR VELOCITY GRADIENT MEASUREMENT
$k_1, ---, k_5$: POINT TEST CATHODES FOR MASS TRANSFER MEASUREMENT

FIGURE 2. Layout of isolated test cathodes embedded in main cathode. Dimensions given are in mm.

an electrolytic aqueous solution of about 0.05 kmol/m³ $K_3Fe(CN)_6$, 0.05 kmol/m³ $K_4Fe(CN)_6$ and 1 kmol/m³ NaOH. Sodium silicate (the so-called water glass) was added into the electrolytic solution for varying widely its viscosity and Schmidt number. The experiment was conducted isothermally in the range of temperature 14.2 ∿ 21 C.

60

Table 1 Physical properties of the electrolytic solution

Property		
Density, ρ kg/m^3	1.044×10^3	$\sim 1.165 \times 10^3$
Viscosity, μ kg/m s	1.21×10^{-3}	$\sim 2.95 \times 10^{-3}$
Diffusivity, D m^2/s	2.52×10^{-10}	$\sim 6.28 \times 10^{-10}$
Schmidt number, Sc	1.85×10^3	$\sim 1.01 \times 10^4$

The physical properties of the working fluid are listed in Table 1.
The electrochemical measurement was performed using a diffusion-
controlled cathode reaction of ferricyanide ions. Measurement of
local velocity gradients was made using rectangular test cathodes (s_1
through s_6) with the main cathode inactive. Measurement of local mass
transfer coefficients was made using both the circular and rectangular
test cathodes under the condition of keeping the main cathode active
at the same potential.

A parallel rod array was used as the vortex generator, as shown in
Figure 3. The 2 mm-dia. circular rods of PVC plastics were arrayed
and fixed in parallel at intervals of $\lambda = 20$ mm. The vortex
generator, positioned 3 mm above the anode surface, was reciprocated
smoothly in parallel with the anode and cathode surfaces by means of a

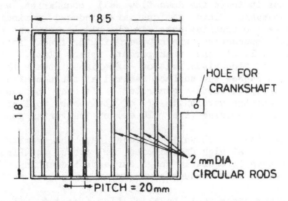

FIGURE 3. Vortex generator. Dimensions given are in mm.

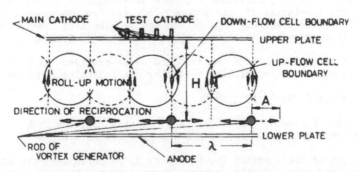

FIGURE 4. Schematic picture of an array of rolling-up vortex rolls.

variable-speed motor with a crankshaft. The stroke, 2 A, and frequency, f_0, of reciprocation were freely varied. A flow-visualization experiment was made suspending fine aluminum platelets (nominal size = $40 \mu m$) in the vortex flow. It was confirmed that an array of well- organized vortex rolls was produced and decayed periodically in phase with the reciprocation of the vortex generator, as shown schematically in Figure 4.

The experimental data of fluctuating velocity gradients on the surface of the upper flat plate were obtained by means of a data recorder through a two-channel DC amplifier. The time traces of the fluctuating velocity gradients were digitized at 300 Hz by an AD converter. The spectral analysis was made by a digital computer. The sampling time-interval was 50 ms and the number of data points sampled from the time series digitized was 1024.

EXPERIMENTAL RESULTS AND DISCUSSION

Figure 5 shows three sets of photographs of the vortex rolls visualized with aluminum platelets suspended in the stagnant liquid layer. A transparent plate of acrylic resin was used as the upper plate to take the top-view photographs. In Figure 5a, the dark, thick straight lines indicate the down-flow cell boundaries where separation-like downward fluid motions are induced periodically.
The up-flow cell boundaries exist in the middle between the adjacent down-flow cell boundaries, where upward fluid streams impinge periodically on the wall of the transparent upper plate. It has been confirmed in the flow-visualization experiment that the periodical roll-up motion causes the surface renewal at the same frequency as the reciprocation of the vortex generator. It can be seen from the photographs that the wavelength, λ, of the vortex roll array is equal to the rod-to-rod spacing (i.e. 20 mm) of the vortex generator.

At low frequencies, as shown in Figure 5a, the vortex rolls have smooth surfaces. At high frequencies, as shown in Figure 5c, the vortex rolls have irregularly rough surfaces indicating the superposition of irregular disturbances on the main periodical roll-up motion.

This flow system does not have a time-averaged velocity as the macroscopic velocity scale because of no mean motion. The following two characteristic flow parameters should, therefore, be considered as the scales: velocity amplitude A f_0 and wavelength λ. The velocity amplitude, consisting of the product of the length-amplitude (half stroke), A, and frequency, f_0, of the vortex generator reciprocation, can be interpreted as the intensity of velocity fluctuations.

As shown in Figure 6, the space- and time-averaged mass transfer coefficients, k, were correlated well by the following equation

$$Sh = 0.88 \, Re^{1/2} \, Sc^{1/3} \, (H/\lambda)^{-0.6}$$

where the Sherwood, Reynolds, and Schmidt numbers are defined respectively as $Sh = k \, \lambda/D$, $Re = A \, f_0 \lambda \, \rho/\mu$, and $Sc = \mu/\rho \, D$.

The height of the stagnant liquid layer, H, is defined as the distance from the rod-axes of the vortex generator to the mass transfer surface. From the viewpoint of the penetration depth of reciprocating

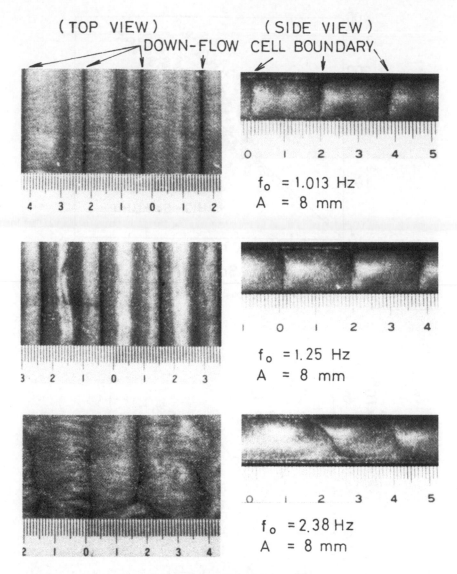

FIGURE 5. Photographs of rolling-up vortex rolls.

motions, it is reasonable that Sh is inversely proportional to $(H/\lambda)^{0.6}$. It has been confirmed that the mass transfer in a stagnant liquid layer is enhanced effectively by the vortices periodically rolling-up.

Figure 7 shows lateral distribution of the time-averaged mass transfer coefficients in the direction of reciprocation. As can be expected, the time-averaged mass transfer varies corresponding to the array of vortex rolls. The minimum point of k appearing at $\bar{x}/\lambda = 0.7$ suggests that one of the down-flow cell boundaries is located there.

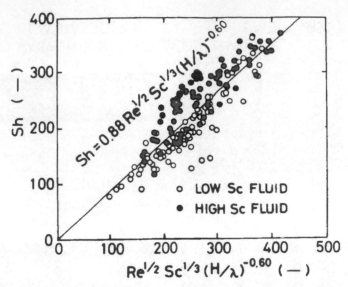

FIGURE 6. Correlation of the space- and time-averaged mass transfer coefficients.

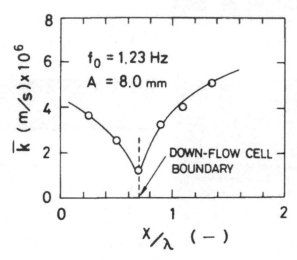

FIGURE 7. Lateral distribution of the time-averaged mass transfer.

Figure 8 shows a power spectrum of the local velocity-gradient fluctuations at the surface of mass transfer. It is clear that there appears a single peak at the same frequency as the reciprocation of the vortex generator and that the second peak implies the harmonic component. This flow state corresponds to that of Figure 5a. Figure 9 shows a power spectrum of the mass transfer fluctuations. These results suggest that the mass transfer surface undergoes the surface

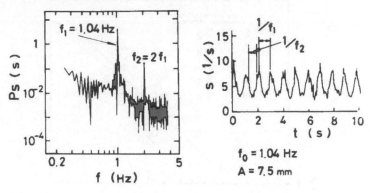

FIGURE 8. Time traces of local velocity gradient at wall and the power spectrum.

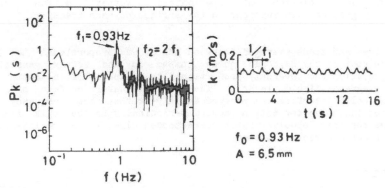

FIGURE 9. Time traces of local mass transfer coefficient and the power spectrum.

renewal action periodically at the fundamental frequency, f_1, which is exactly equal to f_0.

It is worth while examining the relation between the macroscopic vortex motion in the bulk and the near-wall fluid motion on the mass transfer surface. A straight line with slope of 45 degree could be drawn through the scattered data points, as shown in Figure 10. This implies that the RMS values of the velocity gradient fluctuations on the mass transfer surface correspond directly to the macroscopic Reynolds number, i.e. A f_0. That is, the reciprocating bulk fluid motion induces the near-wall turbulences controlling the mass transfer enhancement.

CONCLUSION

The mass transfer in a stagnant liquid layer was enhanced effectively by reciprocating the vortex generator at sufficiently low frequencies.

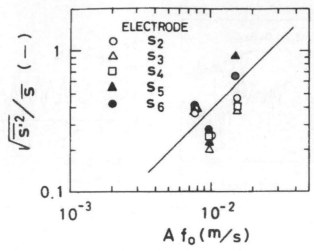

FIGURE 10. Relation of periodical vortex motion with velocity-gradient fluctuations on the surface of mass transfer.

The time- and space-averaged Sherwood number was proportional to the square root of the Reynolds number based on the vortex generator reciprocation and inversely proportional to 0.6 power of the depth of the stagnant liquid layer. The mass transfer surface of the liquid-solid interface was renewed exactly at the same frequency as the rolling-up vortex motion induced periodically in the bulk fluid. The macroscopic Reynolds number corresponded directly to the intensity of the surface renewal motion.

REFERENCE

1. Kataoka, K., Okada, Y., Morikawa, Y. and Iwata, A., Mass Transfer across an Interface of Stagnant Liquid Layer with Periodical Vortex Motion, J. Chem. Eng. Japan, vol.20, pp.363-368, 1987.

2. Mizushina, T., The Electrochemical Method in Transport Phenomena, in Advances in Heat Transfer, Academic Press, vol.7, pp.87-161, NY, 1971.

Natural Convection Heat Transfer Enhancement
by an External Magnetic Field in Horizontal Annulus

JEN-SHIH CHANG
Department of Physics Engineering
McMaster University
Hamilton, Ontario, Canada L8S 4M1

ABSTRACT

A numerical study of natural convection of mercury under the presence of a constant magnetic field being applied parallel to the gravity vector is carried out as part of this work. The magnetic field has a large influence on flow, overall heat transfer and local heat transfer at both the inner and outer cylinder. Larger enhancement of heat transfer is observed around the outer surface.

1. INTRODUCTION

Flow and heat transfer enhancement due to an electromagnetic field (EMF) and pulses (EMP) on the single and two phase fluid flow in a heat exchanging system has increased in importance as a result of recent developments in liquid-metal cooling type nuclear power plants[1]. The boiling phenomena becomes important to estimate heat flow in the nuclear fuel bundle surface when the reactor or turbine is being tripped, since the heat transport system are under EMF or EMP, and quite often "lightning" strikes or other types of EMP are cause of these trips. In a nuclear fusion reactor primary heat transport system, EMP and EMF become a more common problem, since relatively large magnetic field and pulse currents must be used to confine or generate plasmas.[2] However, the effects of the EMF or EMP on the quantities of engineering interest such as the critical heat flux, heat transfer rate and pressure drops are poorly understood at the present stage.

Most of the recent impetus toward the establishment of a firm theorectical basis for this discipline has, in fact, been due to the requirements of modern fission and fusion reactor design. Natural convection of fluids under an electromagnetic field has also received increasing attention. Many experimental and analytical studies have been made on the natural convection of electrically conducting fluids in a magnetic field.[3-9] However the magnetic field is applied perpendicular to the gravity vector in most of these studies. When the direction of the magnetic field is perpendicular to the gravity vector, the flow induced by the buoyant force crosses it. In that case, a term for the electromagnetic retarding force appears in the momentum equation for the vertical velocity component together with the buoyancy forces. In the case that the magnetic field is parallel to the gravity vector, it interacts with the the velocity component that is perpendicular to the gravity vector. The momentum equations for the velocities in both parallel and perpendicular to the gravity must be solved. Because of this coupling of governed equation, only a few works have been done. Seki et al.[9] carried out an experimental and numerical study on the natural convection of mercury in a rectangular container heated from a vertical wall with a magnetic field being applied parallel to the gravity vector and to the heated wall. Their experimental results showed that the magnetic field decreased the Nusselt number considerably in the low region of the Rayleigh number. The effect of the parallel field was found to be less than that for a field normal to the gravity vector but they are not negligible. Numerical results on the Nusselt number were found to predict approximately experimental ones. Calculated velocity profiles also displayed noticeable changes due to the application of the magnetic field.

In the present work, some numerical studies were conducted on the natural convection of mercury in a magnetic field parallel to the gravity vector in a horizontal cylindrical annulus with a radius ratio of 2 over a range of Hartman number, Ha = 0.0 to 8.0 Rayleigh number, Ra = 8000.

2. GOVERNING EQUATIONS AND BOUNDARY CONDITIONS

In the present model, laminar flow is assumed since this has been found experimentally for the range of parameters being used. The flow is also assumed to be invariant along the axis of the cylinders which lead to two-dimensional approach. Utilizing cylindrical coordinates and letting the angular coordinate ϕ measured from the upward vertical as $\phi = 0$. For steady state conditions, the flow is assumed to be symmetric about a vertical plane through axes of the cylinder. Even if secondary flow exists, they must necessarily appear as counter rotating eddy pairs and therefore be symmetrical. Thus, attention is restricted to $0 < \phi < \pi$. The governing equations in cylindrical coordinates can be written as follows:[10]

1) **Mass Conservation Equation**

$$\frac{\partial \rho}{\partial t} + \nabla \cdot (\rho \overline{U}) = 0 \tag{1}$$

2) **Momentum Equation**

The linear momentum equation, which is basically the Navier Stokes equation with the addition of the electrohydrodynamic force, is as follows:

$$\rho \frac{d\overline{U}}{dt} + \rho \overline{U} \cdot \nabla \overline{U} = -\rho g \beta (T - T_i) - \nabla P + f(\overline{B}) + \mu \nabla^2 \overline{U} \tag{2}$$

The first term in the RHS is the gravity force. In this term, we assume that fluid density varies linearly with the temperature, which is previously described as Boussinesq's approximation.

3) **The Energy Equation**

We assumed that our system operates at constant density and the viscous dissipation term can be neglected. In this case, the energy equation reduces to

$$\frac{\partial T}{\partial t} + \overline{U} \cdot \nabla T = \frac{k}{\rho C_p} \nabla^2 T \tag{3}$$

where k is the thermal conductivity and is assumed to be constant, c_p is the specific heat.

If we introduced non-dimensional variables as follows:

$$Ra = \frac{d^3 g \beta (T_o - T_i)}{\alpha v} ; \qquad Pr = \frac{v}{\alpha} ; \qquad \text{Velocity } u = \frac{U}{U_s} ;$$

$$\text{Time } \tau = \frac{t u_s}{d} ; \qquad \text{Grad operator } \nabla = d\nabla ; \qquad \text{Pressure } p = \frac{P}{P_s} .$$

68

where U_s is the themal reference velocity (α/d), T_o, and T_i are the inner and outer temperature, respectively, and $P_s = \rho_n U_R{}^2$ is the reference pressure.

We would like to transform these equations in terms of vorticity and stream function as defined by:

$$u_r = \frac{1}{r}\frac{\partial \psi}{\partial \phi} \text{ and } u_\Phi = -\frac{\partial \psi}{\partial r} \; ; \quad \nabla^2 \psi = -\omega \tag{4}$$

$$\frac{\partial \omega}{\partial \tau} = \text{Ra Pr}\left[\sin\phi\,\frac{\partial T}{\partial r} + \frac{\cos\phi}{r}\frac{\partial T}{\partial \phi}\right] - \frac{1}{r}\left[\frac{\partial \rho}{\partial \phi}\frac{\partial \omega}{\partial r} - \frac{\partial \rho}{\partial r}\frac{\partial \omega}{\partial \phi}\right] + f'(B) + \text{Pr}\nabla^2\omega \tag{5}$$

$$\frac{\partial \theta}{\partial \tau} = \frac{1}{\text{Pr}}\nabla^2\theta - \frac{1}{r}\left[\frac{\partial \psi}{\partial \phi}\frac{\partial \theta}{\partial r} + \frac{\partial \psi}{\partial r}\frac{\partial \theta}{\partial \phi}\right] \tag{6}$$

The boundary conditions can be expressed as follows. At the inner cylinder surfaces $(r = 1)$,

$$\theta(r=1) = 1 \text{ and } \psi = \frac{\partial \psi}{\partial r}\bigg|_{r=1} = 0 \; ; \quad \omega = -\frac{\partial^2 \psi}{\partial r^2}\bigg|_{r=1}$$

At the outer cylinder surface $(r = r_0)$,

$$\theta(r=0)=0 \text{ and } \psi = \frac{\partial \psi}{\partial r} = 0 \; ; \quad \omega = -\frac{\partial^2 \psi}{\partial r^2}$$

The local Nusselt number which indicates the heat transfer rate at the inner and outer surfaces is determined as follows:

$$Nu_{\ell i} = \left(u_r - \frac{\partial \theta}{\partial r}\right)\ell n\, r_0 \quad (\text{at } r=1) \quad \text{or} \quad (\text{at } r=r_0)$$

Then the total Nusselt number at the inner and outer cylinder is calculated through the form.

$$Nu_t = \frac{1}{\pi}\int_{\phi=0}^{\phi=\pi} Nu_{\ell i}(\phi)\, d\phi$$

The block diagram of the calculation procedure is shown in Figure 2. In this diagram, the vorticity, stream function and energy equations are solved by self-consistent iterative method untild sufficient convergence is attained. The decision is made by requiring

$$\sum \left| n_{ij}^{k+1} - n_{ij}^k \right| \leq 10^{-4}$$

where k represents the present iteration step. When sufficient convergence is reached within one time step, the procedure is repeated for the next time step until steady state is attained. The decision on steady state convergence is made by the following criterion.

$$\left| Nu_{ai}^{k+1} - Nu_{ai}^k \right| + \left| Nu_{ao}^{k+1} - Nu_{ao}^k \right| \leq 10^{-4}$$

where Nu_{ai} and Nu_{ao} are the total Nusselt number at the inner and outer cylinder, respectively, and are determined by Simpson's rule.

For Natural convection of liquid metal under the influence of a constant magnetic field parallel to the direction of gravity with an assumption that the induced electric current

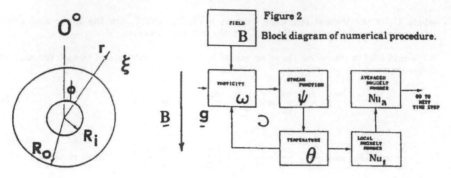

Figure 2

B Block diagram of numerical procedure.

Fig. 1   Schematics of cylindrical annulus and field directions.

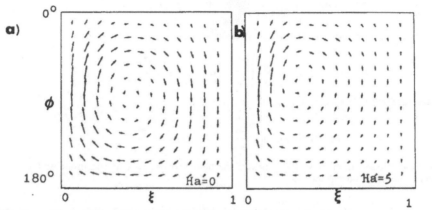

Fig. 3   Flow pattern of mercury natural convection with and without magnetic field for Pr = 0.023, Ra = 8×10³, R = 2.0, a) Ha = o, and b) Ha = 5.0.

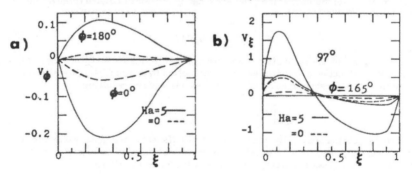

Fig. 4   Velocity profiles with and without magnetic fields for Pr = 0.023, Ra = 8×10³, R = 2.0, a) angular velocity and, b) radial velocity profiles.

does not distort appreciably the applied magnetic field, the magnetic force expression is obtained in a similar fasion with Seki et al.[9] and is shown in dimensionless form as:

$$f'(B) = -Ha^2 \overline{\nabla \omega}$$

where ω is the vorticity, Ha is the Hartman number, $(\sigma/\eta)^{1/2} Bd$, B is the magnetic field, σ is the conductivity, and η is the viscosity.

3.   NUMERICAL RESULTS FOR LIQUID METAL CASE

The effect of a constant magnetic field parallel to gravity vector is investigated in this section by finding the influence of different strength magnetic fields on the flow pattern, velocity profiles, local Nusselt number and averaged Nusselt number, and temperature profiles. Figure 3 shows that the natural convection flow pattern of mercury, $Pr = 0.23$, $Ra = 8 \times 10^3$, for $Ha = 0.0$ and $Ha = 5.0$. Figure 3 shows that the center of rotation is shifted toward the inner cylinder and also toward $\phi = 100°$, compared to 0 to 90° when $Ha = 5.0$ instead of $Ha = 0.0$.

Figure 4 shows that the angular and radial velocity are significantly enhanced by the applied magnetic field, even though this magnetic field is still small. The point of reversing polarity (center of rotation) can be observed to shift toward the inner cylinder surface and this could be due to the influence of the magnetic force created by the presence of the applied magnetic field. It should also be noted here that both velocity components are affected by the applied magnetic field because of the geomtery of the system. This observed enhancement in local velocity leads to an enhancement of local Nusselt number as shown in Figure 5. The enhancement of local Nusselt number evaluated at the inner cylinder surface is observed at the top half of the cylinder while a slight dehancement in $Nu_{\ell i}$ is observed for the bottom half. Since the center of rotation is moved upward with the presence of magnetic field, a wider stagnant area should exist at the bottom half, this lack of fluid movement in this region leads to the dehancement of the local Nusselt number. The enhancement of heat transfer at the outer cylinder $Nu_{\ell u}$ is also observed in Figure 5, under the presence of a magnetic field. Temperature profiles are shown to be slightly influenced by the magnetic field as shown in Figure 6 for $Ha = 0.0$ and 5.0 respectively. An attempt to obtain solution at a higher Hartman number failed because of numerical instability. An upwind difference technique should be used, at least in the vorticity equation (especially in the convective term), so a more stable solution could be obtained. Even though there is only sufficient information here, it

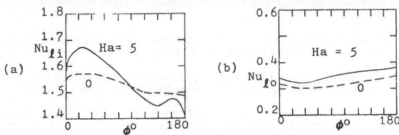

(a)   (b)

Fig 5   Local Nusselt number profiles at the inner cylinder (a) and outer cylinder (b) with and without magnetic field for $Pr = 0.023$, $Ra = 8 \times 10^3$, and $R = 2.0$.

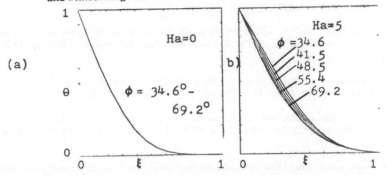

(a)   b)

Fig. 6   Temperature profiles with and without magnetic field for $Pr = 0.023$, $Ra = 8 \times 10^3$, and $R = 2.0$

can generally be seen that the average Nusselt number increases for increasing Hartman number within the present range of parameters. This result is at first contrary to Seka et al.'s results for a rectangular channel in which Nusselt decreases with increasing magnetic field. These differences may arise as expected since two different geometries are used.

4. CONCLUSIONS

In this study, natural convection of mercury in a horizonatal cylinder annulus under the influence of a constant magnetic field being applied parallel to the gravity vector is investigated. The results show that the magnetic field enhances both the angular and radial velocity components. The center of rotation is shifted toward the inner surface and also upward in the top half of the cylinder. This leads to an enhancement of local Nusselt number in the top half and a dehancement in the bottom region.

ACKNOWLEDGEMENT

The author wishes to express his appreciation to F.B.P. Tran for technical assistance. This work was supported by the Natural Sciences and Engineering Research Council of Canada.

REFERENCES

1.   N. Sheriff, and N.W. Davies, "Review of Liquid Metal Natural Convection Relevant to Fast Rector Conditions, " TRG-2959, 1977.

2.   J S Chang, "Electromagnetic Hydrodynamics -The Role of Electromagnetic Field on Fluid Flow", Proceedings of the Electrostatic Society of Japand, EHD Symposium, pp. 1-40, (1986).

3.   E.M. Sparrow and R.D. Cess, "The Effect of a Magnetic Field on Free Convection Heat Transfer, Int. J. Heat Mass Transfer, Vol. 3, 1961, pp 267-274

4.   P.S. Lykoudis, "Natural Convection of an Electrically Conducting Fluid in the Presence of a Magnetic Field", Int. J Heat Mass Transfer, Vol. 5, 1962, pp. 23-34.

5.   A.S. Gupta, "Steady and Transient Free Convection of an Electrically Conducting Fluid from a Vertical Plate in the Presence of a Magnetic Field", Appl. Sci. Res. Section A, Vol. 9, 1960-1961, pp. 319-333.

6.   G. Poots, "Laminar Natural Convection Flow in Magneto-Hydrodynamics", Int. J. Heat Transfer, Vol. 3, 1961, pp. 1-25.

7.   D.D. Papaliou and P.S. Lykoudis,"Magneto-Fluid-Mechanic Laminar Natural Convection - An Experiment", Int. J. Heat Mass Transfer, Vol. 11, 1968, pp. 1385-1391.

8.   D.D. Papailiou and P.S. Lykoudis, "Magneto-Fluid-Mechanics Free Convection Turbulent Flow", Int. J Heat Mass Transfer, Vol. 17, 1974, pp. 1181-1189.

9.   M. Seki, H. Kawamura, and K. Sanokawa, "Natural Convection of Mercury in a Magnetic Field Parallel to the Gravity"J, ASME J. Heat Transfer, Vol. 101, 1979, pp. 227-232.

10.   J.S. Chang and F. Tran,"Numerical Analysis of Transient Natural Convection of Gas in Horizontal Annulus Under DC Electric Field", Natural Convection, I. Cattan and R.N. Smith, Eds., ASME Pub.., HTD, Vol. 16, pp. 51-59 (1980).

Laminar Flow and Heat Transfer in Sine-Shaped Convergent-Divergent Round Pipes

ZHENG ZHANG
Beijing Institute of Chemical Technology
Beijing, PRC

ABSTRACT

This paper shows a great deal for a sine–shaped convergent–divergent round pipes to enhance heat tranfer in low Reynolds number laminar flow largely when friction stress increases not as much, especially when Pr is large. Nu$\propto$Pr$^{1/3}$ law is also found as if Pr is large enough, which is quite similar to the results in ⟨1⟩ for a sine–shaped convergent–divergent plate channel.

These results all come from numerical computation with a nonorthogonal coordinates 2D–periodical elliptical program Zhang developed at Univ. of Minnesota ⟨2⟩ .

1. INTRODUCTION

More and more people pay great attention to study low Re laminar flow and heat transfer in a compact heat exchanger ⟨3–6⟩ recently. Augmentation of heat transfer is usually an important thing to consider for this kind of flow since heat coefficient is small especially for high Pr fluid. Zhang ⟨1⟩ studies laminar flow and heat transfer in 2D sine–shaped conv.–div. channel by numerical method. This paper tends to extend the research work to an axisymmetric waved round pipe with same shape. The longitudinal section of the pipe is shown on fig.1. The calculating domain is the shaded area of the section for its periodic feature. As it is done in ⟨1⟩ , nonorthogonal body fitted coordinates are also adapted here, which are better to fit the geometry we are interested in. The grids of new coordinates are shown on fig.2.

2. BASIC FORMULA

2.1 Control Differential Equations

The control differential equations are same as usual ⟨1⟩ , i.e.

$$div \vec{J} = S \tag{1}$$

in which, the total flux $\vec{J}$ of the general variable Φ is $\vec{J} = \rho \vec{v} \Phi - \Gamma \, grad\Phi$, ρ is density, Γ is diffusion coefficient of Φ, $\vec{v}$ is the vector of velocity , and S represents source term of Φ. The difference between eq. (1) and its counterpart on ⟨1⟩ is that gradients and divergence presentation

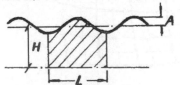

Fig.1 Longitudinal Section
of the Pipe

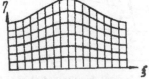

Fig.2 Grid Distribution
in Physical Domain

should be taken in axisymmetric form here and in (x,y) form for ⟨1⟩ .

2.2 Discretization Equations in (ξ,η) Coordinates

A typical control volume in the new coordinates is shown on fig.3. The total flux for this control volume is

$$J_e - J_w + J_n - J_s = b \tag{2}$$

in which, $J_e = J_{P,e} - J_{S,e}$ \tag{3}

the primary flux

$$J_{P,e} = (m\Phi)_e - (\Gamma\alpha_\xi \frac{\partial\Phi}{h_\xi \partial\xi})_e \tag{4}$$

the secondary flux (additional flux for nonorthogonal coordinates) :

$$J_{S,e} = (-\Gamma\beta_\xi \frac{\partial\Phi}{h_\eta \partial\eta})_e \tag{5}$$

The primary area for the flux is $\alpha_\xi = r\, h_\eta / Sin\delta$, and the secondary area for the flux (also introduced from non-orthogonal coordinates) is $\beta_\xi = \alpha_\xi Cos\delta$. h_ξ and h_η represents the length of ξ & η coordinates respectively (fig.3) . δ is the angle between ξ & η coordinates. m is the mass flux through the control volume surface. Of course, J_w, J_n & J_s can be written in same manner.

$$b_s = \int_{\Delta\xi}\int_{\Delta\eta} r\,S\,h_\xi h_\eta\,Sin\delta d\xi d\eta = J_a S \tag{6}$$

in which, Jacob factor J_a can be appoximately calculated as $r\,h_\xi h_\eta Sin\delta$. The formulas above are almost same as the counterparts in ⟨1⟩ except all of the area terms, which need to be multiplied by distance r between point P & symmetric axis (fig.3) . By using power law scheme ⟨8⟩ & a proper mean method ⟨2⟩ to calculate $(\frac{\partial\Phi}{h_\eta \partial\eta})$, $J_{P,e}$ can be written as

$$J_{P,e} = a_E (\Phi_P - \Phi_E) + m_e \Phi_P \tag{7}$$

the final discretization eqs. are

$$a_{P,\Phi}\Phi_P = \sum a_{nb,\Phi}\Phi_{nb} + b_{,\Phi} + b_{No,\Phi} \tag{8}$$

Here additional source term $b_{No,\Phi}$ introduced from non-orthogonal coordinates is

$$b_{No,\Phi} = J_{S,e} - J_{S,w} + J_{S,n} - J_{S,s} \tag{9}$$

In all of the formulas above, the general variable Φ can represent T,u,v etc., but the discretization eq.for the covarient velocity components $u_\xi = \vec{V}\cdot\vec{e}_\xi$ and $u_\eta = \vec{V}\cdot\vec{e}_\eta$ should be slightly different

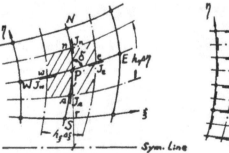
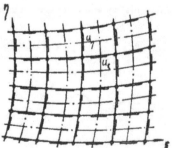

Fig.3 Control Volume in (ξ,η)　　Fig.4 Staggered locations for u_ξ and u_η

from above.First, their locations should be staggered from the loctions where the other variables are.Second,there are some extra source terms needed to be added into u_ξ & u_η discretization equation to suit direction changing from location to locations in (x,r) coordinates .The staggered locations of the covarient velocity component u_ξ & u_η are shown on fig.4. Now we may pay our attention to care about making u_ξ discretization equation.A staggered control volume for the u_ξ is shown on fig.5.Since ξ direction would be changed,normally a discretization of u_ξ at P could be written as

$$a_{P,\xi} u_{\xi,P} = \sum a_{nb,\xi} u'_{\xi,nb} + b_\xi \tag{10}$$

in which,

$$u'_{\xi,nb} = \vec{v}_{nb} \cdot \vec{e}_{\xi,P} \ \& \ u_{\xi,nb} = \vec{v}_{nb} \cdot \vec{e}_{\xi,nb}.$$

By using real covarient neighbouring velocity component $u_{\xi,nb}$ instead of $u'_{\xi,nb}$,equation (10) can be changed as

$$a_{P,\xi} u_{\xi,P} = \sum a_{nb,\xi} u_{\xi,nb} + b_\xi + b_{\xi,cw}. \tag{11}$$

in which,$b_{\xi,cur} = \sum a_{nb,\xi}(u'_{\xi,nb} - u_{\xi,nb})$is an additional curvature source term from coordinates direction changing.Similarly, we may have

$$a_{P,\eta} u_{\eta,P} = \sum a_{nb,\eta} u_{\eta,nb} + b_{\eta,cur} + b_{\eta,cur} \tag{12}$$

in which ,$b_{\eta,cur} = \sum a_{nb,\eta}(u'_{\eta,nb} - u_{\eta,nb})$, $u'_{\eta,nb} = \vec{v}_{nb} \cdot \vec{e}_{\eta,P}$ & $u_{\eta,nb} = \vec{v}_{nb} \cdot \vec{e}_{\eta,nb}$, $\vec{e}_\xi$ & $\vec{e}_\eta$ are unit vectors on direction of ξ & η respectively.

Similarly,discretized pressure p eq. & presure correction p' eq. can be written in new coordinates of axisymmetric system from its continuity differential eq..They are:

$$a_{P,p} p_P = \sum a_{nb,p} p_{nb} + b_{S,p} + b_{No,p} \tag{13}$$
&
$$a_{P,p} p'_P = \sum a_{nb,p} p'_{nb} + b_{S,p'} + b_{No,p'}. \tag{14}$$

2.3 Solution of the Velocity Field

Similar to <1>,the basic periodic conditions for the flow field are:
$$u_\xi(\xi,\eta) = u_\xi(\xi + L,\eta) = u_\xi(\xi + 2L,\eta) = \cdots \cdots \tag{15}$$
$$u_\eta(\xi,\eta) = u_\eta(\xi + L,\eta) = u_\eta(\xi + 2L,\eta) = \cdots \cdots \tag{16}$$
$$p(\xi,\eta) - p(\xi + L,\eta) = p(\xi + L,\eta) - p(\xi + 2L,\eta) = \cdots \cdots \tag{17}$$

in which,L is the periodic length.By defining a new variable—pressure function $P(\xi,\eta) = p(\xi,\eta) + (p(\xi,\eta) - p(\xi + L,\eta)) \cdot \xi / L$,eq. of P & its correction P'could be derived as follows,

$$a_{P,P} P_P = \sum a_{nb,P} P_{nb} + b_{S,P} + b_{No,P} \tag{18}$$

$$a_{P,P} \cdot P' = \sum a_{nb,P'} \cdot P'_{nb} + b_{S,P'} + b_{No,P'}. \tag{19}$$

in which,$P(\xi,\eta) = P(\xi+L,\eta) = P(\xi+2L,\eta) = \cdots \cdots$,and P'itself is also a periodic function as P. considering with the proper boundary conditions of velocity on the wall & symmetric axis,eqs. (11),(12),(18),&(19) can be solved in same manner as it in <1>.

2.4 Solution of Temperature Field

Two typical thermal boundary conditions(q_w = const & T_w = const) are considered in this paper as usual.Their solution procedures & periodical condition treatment are also same as in <1>.

75

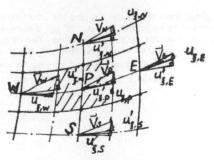

Fig.5 Control Volume of $U_{\xi,P}$
& its neighbours

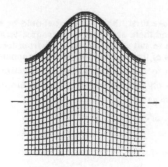

Fig.6 Grid Distribution

3. CALCULATION DETAILS

3.1 Main Parameters

The main parameters adapted are: L / H,A / H,Re,Pr,fRe & Nu.The definitions of Re,f & Nu are:

$$Re = \rho \bar{u}_{max} D_H / \mu, \quad f = \frac{P(\xi,\eta) - P(\xi + L,\eta)}{L} \frac{D_H}{\frac{1}{2}\rho u_{max}^2}, and \quad Nu = \alpha D_H / k_f$$

in which,hydraulic diameter D_H of the convergent—divergent round pipe is the average value of whole calculation domain,i.e.

$$D_H = \frac{4\iint_\Omega r\, h_\eta h_\xi Sin\delta\, d\xi\, d\eta}{\int_{\Omega_L} r\, h_\xi d\xi} \tag{20}$$

in which, Ω represents the whole calculating domain,Ω_L is the boundary area of the pipe wall. r refers to the distance of the point to symmetric axis.

Average velocity $\bar{u}_{max}$ refers to the value at the minimum cross section of the calculation domain.Heat transfer coefficient α,which equals to $q_w / (T_w - T_b)$,refers to a local cross section of ξ coordinate & T_b means average bulk temperature of the same cross section.

After local Nusselt numbers are obtained,average Nusselt number of the domain may be calculated by the discretization form of

$$\overline{Nu} = \frac{\int_{\Omega_L} Nu\, r\, h_\xi d\xi}{\int_{\Omega_L} r\, h_\xi d\xi} \tag{21}$$

3.2 Grid Distribution of the Calculation Domain

36×36 grid points are used in all of the calculating cases. The grid distribution of the calculation domain is shown on fig.6. The grids arc equal—control volume divided in ξ direction.High density of grid lines in ξ direction must be put in near wall boundary ,for the boundary effect of the temperature distribution, especially in the high Pr case .Our calculation domain is divided into 20 uniform parts in η direction .The first 19 parts from the symmetric axis are refered to 19 control volumes .The last one is further divided into 15 control volumes.The size of each control volume of 15 is about half size of the last control volume in η direction .In this manner, the size of the control

76

volume by the wall is only about 1.5×10^{-6} of total size in η direction of the domain, which is small enough to cover very thin thermal boundary layer in high Pr case. Calculation shows that the results of temperature distribution near the wall would be still reliable until Pr reachs 10^{14} which is much higher than Pr of any practical fluids.

3.3 Parameter Ranges of the Calculation

$$A / H = 0 - 0.25, \quad L / H = 1 - 10, \quad Re = 2 - 1000, \quad Pr = 0.01 - 10^{10}$$

The whole calculation was running on the super—computer of cray—2 at the University of Minnesota. The relaxation factor in the axisymmetric case is still an important factor to effect convergence of the calculation. It must be treated very carefully.

4. COMPUTATIONAL RESULTS & DISCUSSION

4.1 Flow Field

The results show that the influnce of the geometry of the pipe & Re on flow field is about the same as the similar case in the 2D convergent—divergent channel 〈1〉 .Fig.7 shows the stream line at the small A / H case (A / H = 0.2),in which recirculation is obvious in both higher Re (Fig.8(a),Re = 520), & lower Re(Fig.8(b),Re = 52). Anyway, much stronger recirculation appears at higher Re.

4.2 Influence of A / H on fRe & Nu

Fig.9 shows influence of A / H on fRe & Nu (when Pr = 1 & L / H = 1). Nu_H & Nu_T represent the values of Nu of constant wall heat flux & constant wall temperature along the wall respectively.

In smaller Re case (Re = 62), along with increase of A / H, both Nu(including Nu_T & Nu_H) & fRe decrease simultaneously. Decline rate of Nu is larger than the rate of fRe. Nu_H & Nu_T decreases about 45% & 42% respectively, along with 34% decrease of fRe when A / H increases from 0 to 0.2 . These results are somewhat like the ones of convergent divergent plate channel.

In larger Re case(Re = 567), Nu & fRe tend to increase along with increase of A / H ,and the increase rate of Nu & fRe tend to increase when A / H become larger .This shows a little difference from the case in 2D convergent—divergent plate channel 〈1〉 .When A / H increases from 0 to 0.2 ,Nu_H,Nu_T & fRe increase about 34%, 41% & 66% respectively in the case of convergent—divergent round pipe, comparing with almost unchange or even slightly decrease of these parameters in the case of 2D convergent—divergent plate channel 〈1〉 .

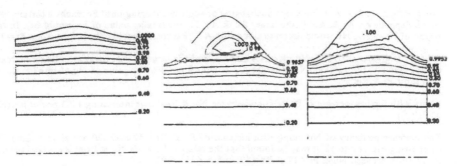

Fig.7 Stream Lines at L / H = 1,　　　(a)　Re = 520　　　　　(b)　Re = 52
A / H = 0.05,Re = 52 & 520　　　Fig.8 Stream Lines at L / H = 1,A / H = 0.2

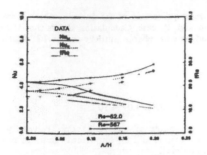 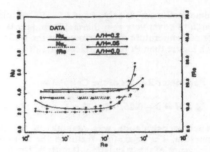

Fig.9 Effect of wave Amplitued on Nu &   Fig.10 Effect of Renolds No.on Nu &
fRe   Pr = 1 Waved Pipe L / H = 1   fRe Pr = 1 Waved Pipe L / H = 1

4.3 Influence of Re on Nu & fRe

Fig.10 shows influence of Re on Nu & fRe at 3 different geometries(A / H = 0, 0.05, & 0.2).Obviously,our calculating results at A / H = 0 are exactly same as the exact solution for a fully—developed flow & heat transfer in round pipes($Nu_H = 4.36$, $Nu_T = 3.65$ & $fRe = 16$) .Along with increase of A / H ,Nu & fRe would decrease gradually when Re is small (Re < 100—200) ,since the area of wall surfaces increases. For a small A / H ($<$0.05),Nu_H, Nu_T& fRe keeps almost unchanged until Re = 200.

The values of these parameters only be slightly smaller than the values at A / H = 0 respectively. After Re goes over 200, these parameters start to increase slightly. The values of these parameters at Re = 1000 are almost 20% higher than the values at low Re separately. For a larger A / H(A / H = 0.2), these parameters start to increase earlier(about Re = 100), and increase rate is much higher. Nu_H, Nu_T& fRe at Re = 630 is about 2.7, 2.05 & 2.5 times of each value at Re = 100 respectively.

The effect due to Re can be also explained by the complex effects of appearance of recirculation in the concave surface region and boundary layer thining near the convex surface.

4.4 Influence of Pr on Nu

All of the results above are based on Pr = 1. The results of different Pr show great influence as it was found in ⟨1⟩ . Fig.11 & 12 with different A / H as a parameter show influence of Pr on Nu at Re = 52 & 520 respectively(for cheaper computing, only Nu_H is mainly calculated. Nu represents Nu_H in the figures).

The main effect can be summarized as follows.

For a certain Re & A / H, Nu keeps unchanging or slightly changing until Pr reachs a certain Pr_C, which depends on Re & A / H(the larger Re & A / H, the smaller value of Pr_Cwould be). In tne region of Pr < Pr_C, Nu would decrease along with A / H increase just like the results when Pr = 1;

After Pr is greater than Pr_C, Nu increases quickly, and the larger A / H makes the larger increase of Nu. Quickly, the value of Nu at larger A / H would go over the value of Nu at smaller A / H with same Pr & Re;

Along with further increase of Pr, relation between Nu & Pr shows interesting 1 / 3 power law(Nu $\propto Pr^{1/3}$);

The increase tendency of Nu along with increase of Pr at Re = 52 and 520 are almost same. By compairing Fig.11 and 12, it may be found that the relation between Nu & Pr at Re = 520 is about the relation between Nu & Pr / 17 at Re = 52 with A / H = 0.05;

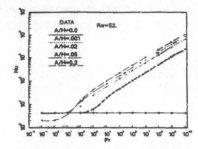
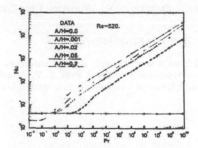

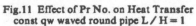

Fig.11 Effect of Pr No. on Heat Transfer
const qw waved round pipe L / H = 1

Fig.12 Effect of Pr No. on Heat Transfer
const qw waved round pipe L / H = 1

Influence of Pr on Nu may be explained as that the thickness of thermal boundary layer terns thinner when Pr gets larger.

5. CONCLUSION REMARKS

A nonorthogonal coordinates finite difference method based on Patankar's control volume method is introduced to analyze a periodic laminar flow and heat transfer problem in a sine–shaped convergent–divergent round pipe. Influence of Re & A / H on fluid field and fRe, and influence of Re, Pr & A / H on Nu are tested. The results are similar to the results of a sine–shaped con.–div. plate channel. It also shows that Pr has important inpluencce on Nu, especially for high Pr. In this case, Nu would increase as 1 / 3 power of Pr. This result tells us that heat transfer coefficient for high Pr fluid in a waved channel would be largely greater than the coefficient in a straight channel. As an example, a sine–shaped con.–div. round pipe with small wave(A / H = 0.05) may cause Nu more than 10 times increase comparing with a straight round pipe if fluid's Pr is 10^4 or over. On the ether hand, pressure drag is almost unchanged.

REFERENCES

〈1〉 Zheng Zhang, "2D Laminar Flow & Heat Transfer in a sine–shaped convergent divergent plat channel" to be presented on 2nd International symposium on heat Transfer Aug.8–12, 1988, Tsinghua University, Beijing, P. R. China

〈2〉 Zheng Zhang: "Numerical method of periodical flow & heat transfer in non–orthognal coordinates system" 1987, Internal report

〈3〉 Uttawar S. B. & Rao M. K., "Augmentation of Laminar Flow Heat Transfer in Tubes by means of wire Coil Inserts" ASME Journal of Heat Transfer, Vol.107, No.4, pp.930–935, 1985

〈4〉 Geiger et al, "Heat transfer & friction factors of high Pr laminar flow through an annulus with circumferential fins", Proceeding of sixth international heat conference(Torroto) 1978, pp.607–611

〈5〉 Bergles A. G. & Jashi S. D., "Augmentation techniques for low Reynolds No. in tube flow", Low–Renolds No. Flow Heat Exchangers, edited by S. Kakac, R. K. Shah & A. E. Bergles, Hemisphere / McGraw Hill, Washington D. C., 1983

〈6〉 Shah R. K., "Research needs in low.Reynolds No. flow Heat exchangers", Low–Reynolds No. Flow Heat Exchangers, edited by S. Kakac, R. K. Shah & A. E. Bergles, pp.983–1000, Hemisphere / McGraw Hill, Washington D. C., 1983

〈7〉 Kailash C.K., "A Calculation procedure for viscous flow at all speeds in complex geometries" Ph.D. Thesis, University of Minnesota, June, 1986

⟨8⟩   Patankar, S. V.: "Numerical Heat Transfer & Fluid Flow" 1980, Hemisphere Pubishing Co.

ACKNOWLEDGEMENT

This work was originally done at university of Minnesota, U.S.A. The author would like to take this opportunity to thank Prof. S. V. Patankar for his kind support. Author would also like to thank University super-computer center, University of Minnesota, U.S.A. for their help.

Natural Convection Heat Transfer along a Vertical Flat Plate Disturbed by a Circular Cylinder

EISUKE MARUMO
Department of Mechanical Engineering
Akashi Technological College
Akashi, Hyogo 674, Japan

KENJIRO SUZUKI
Department of Mechanical Engineering
Kyoto University
Kyoto 606, Japan

1. Introduction and Purpose

Natural convection along a heated vertical wall is related to many practical applications. In some cases, the heated wall is not a smooth flat plate and, in other cases, obstacles located near the wall disturbes the incurred flow. Therefore, investigations of the natural convective heat transfer in such complicated situations are required. Earlier studies on this subject include the one of Vliet and Liu [1], on a vertical flat plate with an attachment of a circular cylinder, and Fujii et. al. [2], on a vertical cylinder with surface roughness. Both of these are concerned with the laminar-to-turbulent transition of natural convection boundary layer. Former study reported that the obstacle promoted the turbulent transition but the latter reported negative on this. the present study treats a case such that the natural convection boundary layer along the vertical heated flat plate is disturbed by an unheated circular cylinder located detached from the wall, and discusses both of the heat transfer and laminar-to-turbulent transition characteristics. Recently, Ichimiya and Miyazawa [3] did a similar experiment as the present one. However, the spatial resolution of their experiment on the Nusselt number distribution is not high enough. Therefore, the present study supplements their data and makes it possible to discuss the phenomenon in detail.

2. Experimental Apparatus and Procedure

The heated flat plate consisted of a settling chamber with other three walls made of plywood plate of 5 mm in thickness as shown in Fig.1. The chamber was 2.5 m high and had 3mx1m horizontal cross section. The bottom of the chamber was located 100 mm above the floor. To prevent thermal stratification of air in the chamber, two 100 mm wide slit openings were arranged at the ceiling of the chamber and at the wall facing the test surface. This makes it possible for the heated buoyant air to flow out from the ceiling slit just above the heated surface and for the same amount of air to flow in through another slit.

The vertical heated plate was 2.4 m high and 0.9 m wide. The test surface was covered by three vertical strips of 304 stainless steel foil. Each strip was 2 m long, 0.3 m wide, and 30 μm thick. The three strips were glued to 15 mm thick plywood base plate with a 1 mm wide gap between them, leaving the bottom part of 0.4 m long uncovered. The strips were connected in series and heated electrically using a-c current. The back of the plywood base plate was covered by another plywood plate of 9 mm thickness with a provision of an intermediate space of 4 mm thick air layer, in order to reduce the conductive heat loss through the base plate.

Surface temperatures T_w on the test wall were measured with forty-

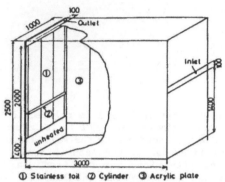

Fig.1 Experimental apparatus

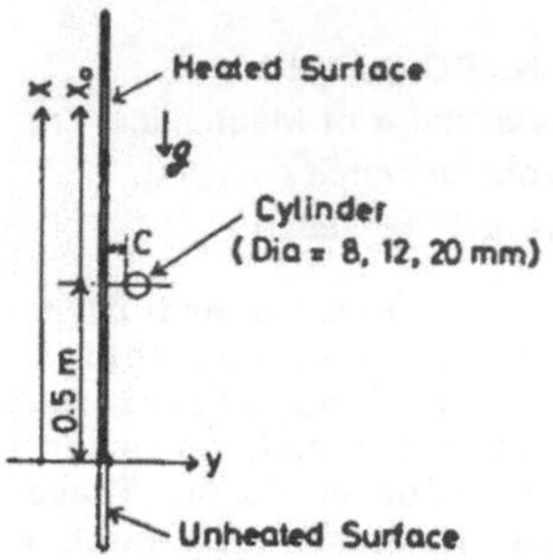

① Stainless foil ② Cylinder ③ Acrylic plate

Fig.2 Flat plate-cylinder set up
and coordinate system

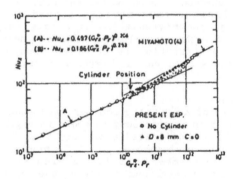

Fig.3 Plots of Nu_x against Ra_x^*
for an undisturbed case and
a disturbed case (D=8mm, C=0)

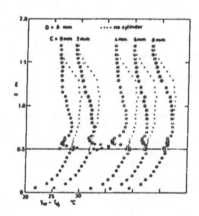

Fig.4(a) Wall temperature
distributions for D=8 mm

two Almel-Cromel thermocouples of 0.1 mm diameter, which were placed in contact with the stainless steel foil along the centerline of the test plate. Other six thermocouples were located on both sides of the centerline at three different positions from the bottom leading edge in order to check the spanwise uniformity of surface temperature. Some other thermocouples were attached on to the back surface of the base plate in order to evaluate the conduction heat loss through the base plate.

Figure 2 shows a schematic view of the cylinder-flat plate system. Three circular cylinders of different diameter were used to disturb the natural convection boundary layer. The diameters chosen were D=8,12,and 20 mm. Each cylinder was placed horizontally and parallel to the test surface at a position of 0.5 m above the lower edge of the heated part of the plate. The clearance C between the cylinder and the test surface was changed in five steps: C=0,2,4,6,and 8 mm. In the following, X and Xo denote the vertical distance measured upward, respectively, from the lower edge of the heating metal foil and from the cylinder.

Local heat transfer coefficients were calculated by using the following equation

$$h_{\lambda} = \frac{q_{cv}}{T_w - T_a} = \frac{q_t - (q_r + q_{cd})}{T_w - T_a}$$

where

q_{cv}=natural convection heat flux
q_t =electric power dissipated
q_r =radiation heat loss from the plate to the surrounding
q_{cd}=conduction heat loss through the base plate

The q_r and q_{cd} were evaluated by assuming the emissivity of the stainless steel foil to be 0.13 and the heat conductivity of the plywood plate to be 0.16 W/mK respectively. The resulting net convection heat flux was about 100 W/m^2. The ambient air temperature T_a was measured with a thermocouple placed 1.3 m above the bottom of the box and 0.3 m apart from the test surface. Temperature distributions in the natural convection boundary layer were measured with another Almel-Cromel thermocouple of 0.1 mm diameter. The all measurements were carried out in a steady state conditions established about five hours after the initiation of heating.

3. Experimental Results and Discussion

The present heat transfer results for an undisturbed case and a disturbed case of C=0 mm when D=8 mm are shown in Fig.3 in terms of local Nusselt number Nu_x plotted against the modified local Rayleigh number $Ra_x^* = Gr_x^* Pr$. The fluid properties used in the definitions of Nu_x and Ra_x^* were evaluated at the film temperature relevant for each test except for the coefficient of volumetric expansion which was evaluated at the ambient temperature. For the undisturbed case, the present results agree very well with the experimental results of Miyamoto and Fukumura [4] both in the laminar and turbulent regimes. This demonstrates that the provision of unheated part at the bottom of the plate affects the heat transfer results little. Turbulent transition is found to occur in the range of $10^{11} < Gr_x^* Pr < 10^{12}$. This agrees with the results of Vliet and Ross [5]. For the disturbed case, turbulent transition is found to be tripped by the cylinder. This confirms the results of Vliet and Liu [1].

Figs.4(a),(b),and (c) show the vertical distributions of wall excess temperature, $T_w - T_a$, for the cases of three cylinder diameters respectively. Each figure includes the results of five cases of different clearance C between the cylinder and the test surface. In each case, small dotts show the results obtained in the undisturbed case. Upward

83

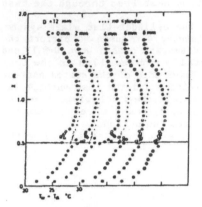

Fig.4(b) Wall temperature
distributions for D=12 mm

Fig.4(c) Wall temperature
distributions for D=20 mm

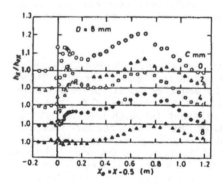

Fig.5(a) Degree of local heat
transfer augmentation (D=8 mm)

Fig.5(b) Degree of local heat
transfer augmentation (D=12 mm)

decrease of wall tempereture indicates the occurrence of laminar-to-turbulent transition of the boundary layer. Except in two cases of C=0 and 2 mm when D=12 mm, turbulent transition is found to be well tripped by cylinder. In all cases of different cylinder diameters, laminar-to-turbulent transition seems to be promoted by an insertion of cylinder most effectively when C=4 mm and least effectively when C=8 mm. From the viewpoint of heat transfer augumentation, the decrease of wall temperature is desirable. The present results suggest that the clearance C affects somewhat significantly the heat transfer augmentation.

Figs.5(a),(b),and (c) show the degree of heat transfer augmentation obtained by an insertion of cylinder. Here, h_x/ho_x is the ratio of the local heat transfer coefficients between the disturbed case and the undisturbed case. Mild peak of (h_x/ho_x) distribution is found around Xo=0.7 m in every case except for one case of C=0 mm and D=12 mm. This peak resulted from the promotion of turbulent transition caused by an insertion of cylinder as discussed in the above. The value of the ratio (h_x/ho_x) approaches unity at positions of large Xo in every cases. This results simply from the fact that the fully turbulent boundary layer is attained there in the undisturbed case, therefore, that the value of ho_x is high enough there.

The shape of (h_x/ho_x) distribution in the region of small Xo is also interesting from the viewpoint of heat transfer augmentation. It also changes with the value of clearance. When the cylinder is attached to the wall (C=0), the large peak value of (h_x/ho_x) appears at a position of small distance from the cylinder in every cases of different cylinder diameters. Figure 6 shows the variation of the peak value $(h_x/ho_x)m$ and the peak position Xom with a changr of the clearance C. For all the three cases of different cylinder diameters, the peak value decreases once when C=2 mm, recovers when C=4 mm, and then decreases again to disappear when C=8 mm. This again indicates the clearance C is an important parameter. There is found a region where the heat transfer deteriorates just above the cylinder. This heat transfer deterioration is most noticiable in the case of C=2 mm, and it is less noticiable at C=0 and 4 mm. This is now discussed briefly. In case of forced convective boundary layer disturbed by a cylinder [6,7], heat transfer deterioration occurring when C=0 in the same region is well recovered if the cylinder is a little detached from the wall. This has been concluded to result from a combination of a few different heat transfer mechanisms but basically from the inflow of fresh fluid through a space between the cylinder and the plate [6]. In the present case of natural convection, at least in the range of Rayleigh number presently studied, high fluid velocity cannot be expected in the space between the plate and the cylinder. Thus, at a small clearance like C=2 mm, the amount of inflow through the space is not still large enough to effectively disturb the stagnant flow region behind the cylinder. Therefore, detaching the cylinder a little from the plate is rather equivalent to use a cylinder of larger size. Therefore, the heat transfer deterioration cannot be improved and the size of the region of deteriorated heat transfer is rather enlarged. When C=4 mm, fluid can flow through the space more easily. Thrrefore, the heat transfer peak near the cylinder becomes higher and the depth of the dip in the (h_x/ho_x) distribution found just behind the cylinder decreases, from those in the case of C=2 mm.

Now, attention is turned to the peak distance Xom shown in Fig.6. For the case of C=0, the peak distance is almost the same for three cases of diffrent cylinder diameter. Its non-dimensional value normalized with the cylinder diameter D is 7.5, 5.4, and 3.5 for each case of the cylinder diameter D=8,12, and 20 mm respectively. In the case of forced convective boundary layer disturbed by a cylinder, a peak of (h_x/ho_x) was

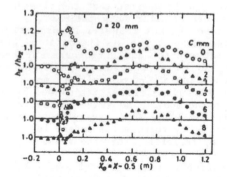

Fig.5(c) Degree of local heat
transfer augmentation (D=20 mm)

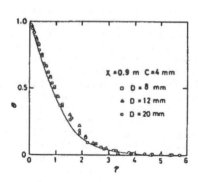

Fig.6 Plots of $(h_x/ho_x)m$ and Xom
against clearance C

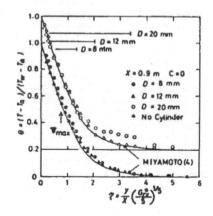

Fig.7(a) Temperature profiles in
boundary layer (X=0.9 m, C=0)

Fig.7(b) Temperature profiles in
boundary later (X=0.9 m, C=4 mm)

also obserbed after the region of deteriorated heat transfer. The peak
position was concluded to correspond to the position of intermittent
reattachment of separating shear layer starting from the far side of
cylinder surface from the wall. However, the above values of (Xom/D) are
different from 10 for the same value found in the case of forced
convective boundary layer disturbed by a cylinder. This indicates that
the flow behavior in the near wake of cylinder is quite different between
the cases of the present natural convection and the previous forced
convection. Entrainment of ambient fluid into the boundary layer caused
by buoyancy governs the flow in the rear of cylinder and may suppress the
flow separation up to a position of much larger angle measured from the
front stagnation point of cylinder.

Figs.7(a) and (b) show the temperature profiles in the boundary
layer at the section of X=0.9 m in the cases of C=0 and C=4 mm,
respectively. The temperature profiles of the disturbed cases are found
to have slightly higher values than the one of the undisturbed case. In
the experiments of Fujii et. al. [8], a similar temperature profile was
observed in an initial stage of laminar-to-turbulent transition of
natural convection boundary layer along a vertical cylinder surface. They
found that ,in this stage, a vortex flow is dominated and heat transfer
rate increases, though the temperature profile is laminar-like.
Therefore, in the present experiments, the disturbed flow at the positon
of X=0.9 m is considered to correspond to the initial stage of the
transition. Detailed flow visualization must be made in a future work to
clarify both the flow around the cylinder and the developement of the
disturbance caused by the cylinder.

4. Conclusions

Natural convective heat transfer along a vertical flat plate has
been studied experimentally in cases that a cylinder of different
diameter was located near the plate. For three kinds of cylinder
diameters examined, the clearance of C=4 mm is most promising to enhance
the natural convective heat transfer. The measured heat transfer
characteristics have been discussed in comparison with the previously
reported measurement for the cases of the forced convection. In the
present case, small gap between the cylinder and the plate (C=2 mm) does
not lead to a noticiable improvement in the heat transfer in the region
just downstream of the cylinder. Further downstream, heat transfer is
augmented by the promotion of laminar-to-turbulent transition of the flow
due to the flow disturbance caused by the cylinder.

References
[1] Vliet G.C. and Liu C.K. An experimental study of turbulent natural
convection boundary layers. Trans. ASME, J. Heat Transfer, 1969, 91,
pp.517-531
[2] Fujii T. et.al. Natural convection heat transfer from a vertical
cylinder surface with small roughness to liquid fluid. Trans. JSME,
1970, 36(286), pp.994-999
[3] Ichimiya K. and Miyazawa T. Effect of a rod in the free convection
boundary layer to the local heat transfer (Upfacing horizontal and
vertical heated plate). Flow Visualization, 1986, 6(21) pp.49-59 (in
Japanese)
[4] Miyamoto M. and Fukumura H. Natural convection heat transfer along a
vertical heated flat plate with constant heat flux (From laminar region
to turbulent region). 14th National Heat Transfer Symposium of Japan,
1977, pp.415-417
[5] Vliet G.C. and Ross D.C. Turbulent natural convection on upward and
downward facing inclined constant heat flux surfaces. Trans. ASME, J.

Heat Transfer, 1975, 97, pp.549–555

[6] Marumo E., Suzuki K., and Sato T. Turbulent heat transfer in a flat plate boundary layer disturbed by a cylinder. Int.J.Heat & Fluid Flow, 1985, 6(4), pp.241–248

[7] Fujita H. et. al. The forced convective heat transfer on a plate with a cylinder inserted in the boundary layer. (2nd report, Effects of cylinder diameter and comparison with a square bar). Trans. JSME, 1981, 47(414), 317

[8] Fujii T. et. al. On flow properties and temperature profiles in a natural convection boundary layer. Trans. JSME, 1970, 36(288), pp.1349–1357

Regression Analysis with Unknown Operational Parameter Variables, Optimal Design of Sequential Experiment and Its Application to Optimization of High Efficiency Heat Transfer Tubes

ZHAO-QIANG HUANG and SHI-QIONG YANG
South China University of Technology
Wushan, Guangzhou 510641, PRC

ABSTRACT   This paper offers the theory and method for the regression analysis with operational parameter variables on the fundamental of mathematical statistic. This is the kind of two-step fitting regression model. The second part of this paper in which we allow experimental step numbers as variable parameters, the optimization is combined with the recursive identification algorithm to deduce sequential experiment optimum design with operational parameter variables. Finally the methods of this paper is applied to the study of intensified heat transfer.

INTRODUCTION   Regression coefficients are always constants about the problem of regression analysis in the past. This paper considers that the regression coefficients are not constants but the functions of some operational parameter variables. It has really significant in the chemical thermal transfer process. For example, the design parameters consider not only the structral variables of tubes but also the operational parameter variables for operative flow fluid in optimal design of high efficiency heat transfer tubes.

REGRESSION ANALYSIS OF QUADRATIC MODEL WITH THE OPERATIONAL PARAMETER VARIABLES

2.1 Description of general mathematical form

Let $X=[x_1, \ldots ,x_n]^T$ be n-dimentional vector denoting n structral variables, $Z=[z_1, \ldots ,z_m]^T$ be m-dimensional vector denoting m operational parameter variables, and Y (quantitative index) be the quadratic form having the operational parameter variables, that is,

$$Y=\beta_0(z_1,\ldots,z_m)+\sum_{i=1}^{n}\beta_i(z_1,\ldots,z_m)x_i+\sum_{i=1}^{n}\beta_{ii}(z_1,\ldots,z_m)x_i^2$$

$$+\sum_{i<j}^{n}\beta_{ij}(z_1,\ldots,z_m)x_ix_j+ e \qquad (2.1.1)$$

where e is the radom error.
To fit the equation (2.1.1), we do the following experiments.

The project Supported by National Natural Science Foundation of China

Table 1. Experimental Scheme

No. of experiment	Levels of structral variables $x_1 \quad x_n$	Observing value y — Levels of operational parameter variables $(z_{11},\ldots,z_{m1})$	...	$(z_{1m},\ldots,z_{mM})$
1	$x_{11} \cdots x_{n,1}$	$y_1(1,\ldots,1)$	...	$y_1(M,\ldots,M)$
.	.	.	.	.
N-1	$x_{1,N-1} \cdots x_{n,N-1}$	$y_{N-1}(1,\ldots,1)$	...	$y_{N-1}(M,\ldots,M)$
N	$x_{1,N} \cdots x_{n,N}$	$y_N(1,\ldots,1)$	...	$y_N(M,\ldots,M)$

Hence, we obtain,

$$y_u(v)=\beta_0(v)+\sum_{i=1}^{n}\beta_i(v)x_{iu}+\sum_{i=1}^{n}\beta_{ii}(v)x_{iu}^2+\sum_{i<j}^{n}\beta_{ij}(v)x_{iu}x_{ju}+e_u(v);$$

$$u=1,\ldots,N; \quad v=1,\ldots,M. \qquad (2.1.2)$$

where, the random errors $(e_u(v),u=1,\ldots,N; \quad v=1,\ldots,M)$ satisfy the four assumptions: (1) independence; (2) unbiasedness; (3) equality in variance: $Var[e(v)]=0$, $u=1,\ldots,N; V=1,\ldots,M;$ (4) normality: $N(0, \sigma^2)$.

2.2 First step fitting models —— the regression equation for the structral variables and the quantitative variable

2.2.1. The least square estimation

According to the least square identification, we find the least square estimators $\bar{B}(v)$ of the coefficients $B(v)$ in the equation (2.1.1). Therefore we obtain the first step fitting models which have M regression equations. Where $\bar{B}(V)$ and $B(v)$ are denoted by

$$B(v)=[\beta_0(v),\beta_1(v),\ldots,\beta_n(v),\beta_{11}(v),\ldots,\beta_{nn}(v),\beta_{12}(v),\ldots,$$

$$\beta_{n-1,n}(v)]^T$$

$$\bar{B}(v)=[b_0(v),b_1(v),\ldots,b_n(v),b_{11}(v),\ldots,b_{nn}(v),b_{12}(v),\ldots,$$

$$b_{n-1,n}(v)]^T$$

the symbol "т" represents transpose.

2.2.2 The analysis of variance

For any fixed v,the variance sum of squares is expressed as follows,

$$\sum_{u=1}^{N}[y_u(v)-\bar{y}(v)]^2=\sum_{u=1}^{N}[y_u(v)-\hat{y}_u(v)]^2+\sum_{u=1}^{N}[\hat{y}_u(v)-\bar{y}(v)]^2, \text{ where}$$

$$y(v)=(1/N)\sum_{i=1}^{N}y_u(v)$$

$$\hat{Y}_u(v) = \beta_0(v) + \sum_{i=1}^{n} \beta_i(v)x_{iu} + \sum_{i=1}^{n} \beta_{ii}(v)x_{iu}^2 + \sum_{i<j}^{n} \beta_{ij}x_{iu}x_{ju}$$

We may give the table of analysis of variance as usually, here the F-ratio is only given as following,

$$F_{N-[2n+\frac{n(n-1)}{2}]-1}^{2n+\frac{n(n-1)}{2}} = \frac{N-[2n+\frac{n(n-1)}{2}]-1}{2n+\frac{n(n-1)}{2}} \cdot \frac{\sum_{u=1}^{N}[\hat{Y}_u(v)-\bar{y}(v)]^2}{\sum_{u=1}^{N}[y_u(v)-\hat{Y}_u(v)]^2}$$

2.2.3. Analysis of regression

Since the estimators of regression coefficients $b_0(v)$, $b_i(v)$, $b_{ii}(v)$, $b_{ij}(v)$ $(i,j=1,2,\ldots,n; i<j)$ obey the normal distributions $N(\beta_0(v), C_{00}\sigma^2)$, $N(\beta_i(v), C_{ii}\sigma^2)$, $N(\beta_{ii}(v)$, $C_{iiii}\sigma^2)$, $N(\beta_{ij}(v), C_{ij\ ij}\sigma^2)$, respectively, and $S_1(v)/\sigma^2$ follows X^2—distribution with degree of freedom $N-2n-n(n-1)/2-1$, where $S(v) = \sum_{u=1}^{N}[y_u(v)-\hat{Y}_u(v)]^2$, Therefore

$$\frac{(b_i(v)-\beta_i(v))/\sqrt{C_{ii\ ii}\ \sigma^2}}{(1/\sqrt{N-2n-\frac{n(n-1)}{2}-1}\ ) \cdot \sqrt{\frac{S_1(v)}{\sigma^2}}} = t_{N-2n-\frac{n(n-1)}{2}-1}$$

follows "Student" distribution with d.f. $N-2n-n(n-1)/2-1$, and we can test the following hypothesis,
$$\beta_i=0, \quad i=1,2,\ldots,n.$$

We can also get the same conclusions about $\beta_0(v)$, $\beta_{ii}(v)$, $\beta_{ij}(v)$, if we change the note (i) to (0),(ii),or(ij).

2.3 Second step fitting models —— the fitting regression coefficients with the operational parameter variables respecting to the operational parameter variables

The coefficients $\beta_0(z_1,\ldots,z_m)$, $\beta_i(z_1,\ldots,z_m)$, $\beta_{ii}(z_1,\ldots,z_m)$, $\beta_{ij}(z_1,\ldots,z_m)$,with the operational parameter variables $z_1,\ldots,z_m$ in the equation (2.1.1) can be expended with Talor series. We use a quadratic form to denote its relation between $\beta_i(z_1,\ldots,z_m)$ and $(z_1,\ldots,z_m)$ as follows,

$$\beta_i = \alpha_0^{(i)} + \sum_{\bullet=1}^{m} \alpha_\bullet^{(i)}z_\bullet + \sum_{\bullet=1}^{m} \alpha_{\bullet\bullet}^{(i)}z_\bullet^2 + \sum_{\bullet<\bullet}^{} \alpha_{\bullet\bullet}^{(i)}z_\bullet z_\bullet + e^{(i)}$$

$$b_i(v) = a_0^{(i)} + \sum_{\bullet=1}^{m} a_\bullet^{(i)}z_{\bullet v} + \sum_{\bullet=1}^{m} a_{\bullet\bullet}^{(i)}z_{\bullet v}^2 + \sum_{\bullet<\bullet}^{} a_{\bullet\bullet}^{(i)}z_{\bullet v}z_{\bullet v} + e_v^{(i)}$$

$$v=1,2,\ldots,M; \quad i=1,2,\ldots,n \qquad (2.3.1)$$

From the assumptions in 2.1 section, $(e_v^{(i)}, v=1,2,\ldots,m)$ satisfy the four conclusions: (i) independence; (ii)

unbiasedness; (iii) Equality invariance: $Var[e_1^{(i)}]=C_{ii}\sigma^2$,....,
$Var[e_M^{(i)}]=C_{ii}\sigma^2$; (iv)Normality: $N(0,C_{ii}\sigma^2)$.

The least square estimators $\bar{A}^{(i)}$ of the coefficients
$\bar{A}^{(i)}$ in the equation (2.3.1) can be obtained through the same
procedures as 2.2 section, so do the analysis of variance and
the analysis regression. where,

$$\bar{A}^{(i)}=[\alpha_0^{(i)},\alpha_1^{(i)},\ldots,\alpha_m^{(i)},\alpha_{11}^{(i)},\ldots,\alpha_{mm}^{(i)},\alpha_{12}^{(i)},\ldots,\alpha_{m-1,m}^{(i)}]^T$$

$$\bar{A}^{(i)}=[a_0^{(i)},a_1^{(i)},\ldots,a_m^{(i)},a_{11}^{(i)},\ldots,a_{mm}^{(i)},a_{12}^{(i)},\ldots,a_{m-1,m}^{(i)}]^T$$

We can also get the same conclusions about $\bar{A}^{(0)}, \bar{A}^{(ii)}$,
$\bar{A}^{(ij)}$, if we change the superscript (i) to (0),(ii),(ij).
Now, we yield the total regression equation:

$$\hat{Y} = \hat{\beta}_0 + \sum_{i=1}^{n}\hat{\beta}_i x_i + \sum_{i=1}^{n}\hat{\beta}_{ii} x_i^2 + \sum_{i<j}^{n}\hat{\beta}_{ij} x_i x_j$$

$$=(a_0^{(0)} + \sum_{\bullet=1}^{m}a_\bullet^{(0)}z_\bullet + \sum_{\bullet=1}^{m}a_{\bullet\bullet}^{(0)}z_\bullet^2 + \sum_{\bullet<\bullet'}^{m}a_{\bullet\bullet'}^{(0)}z_\bullet z_\bullet)$$

$$+ \sum_{i=1}^{n}(a_0^{(i)} + \sum_{\bullet=1}^{m}a_\bullet^{(i)}z_\bullet + \sum_{\bullet=1}^{m}a_{\bullet\bullet}^{(i)}z_\bullet^2 + \sum_{\bullet<\bullet'}^{m}a_{\bullet\bullet'}^{(i)}z_\bullet z_\bullet)x_i$$

$$+ \sum_{i=1}^{n}(a_0^{(ii)} + \sum_{\bullet=1}^{m}a_\bullet^{(ii)}z_\bullet + \sum_{\bullet=1}^{m}a_{\bullet\bullet}^{(ii)}z_\bullet^2 + \sum_{\bullet<\bullet'}^{m}a_{\bullet\bullet'}^{(ii)}z_\bullet z_\bullet)x_i^2$$

$$+ \sum_{i=1}^{n}(a_0^{(ij)} + \sum_{\bullet=1}^{m}a_\bullet^{(ij)}z_\bullet + \sum_{\bullet=1}^{m}a_{\bullet\bullet}^{(ij)}z_\bullet^2 + \sum_{\bullet<\bullet'}^{m}a_{\bullet\bullet'}^{(ij)}z_\bullet z_\bullet)x_i x_j$$

$$(2.3.2)$$

2.4 Tatol variance analysis

Consider that the total variance sum of squares is

$$\sum_{v=1}^{M}\sum_{u=1}^{N}[y_u(v)-\bar{y}(v)]^2 = \sum_{v=1}^{M}\sum_{u=1}^{N}[y_u(v)-\hat{y}_u(v)]^2 + \sum_{v=1}^{M}\sum_{u=1}^{N}[\hat{y}_u(v)-\hat{\hat{y}}_u(v)]^2$$

$$+ \sum_{v=1}^{M}\sum_{u=1}^{N}[\hat{\hat{y}}_u(v)-\bar{y}_u(v)]^2 + 2\sum_{v=1}^{M}\sum_{u=1}^{N}[\hat{\hat{y}}_u(v)-\hat{y}_u(v)][\hat{y}_u(v)-\bar{y}(v)]$$

$$(2.4.1)$$

Besides, the variance sum of squares in the equation
(2.2.1) can be written as follows,

$$\sum_{v=1}^{M}\sum_{u=1}^{N}[y_u(v)-\bar{y}(v)]^2 = \sum_{v=1}^{M}\sum_{u=1}^{N}[y_u(v)-\hat{y}_u(v)]^2 + \sum_{v=1}^{M}\sum_{u=1}^{N}[\hat{y}_u(v)-\bar{y}(v)]^2$$

$$(2.4.2)$$

For comparison with two equations, the sum of 3rd and
4th terms in the right side of the equation (2.4.1) is less
than the sum of 2nd terms in the right side of the equation
(2.4.2), and we desire that they are approaching nearly.
Hence we have

$$\sum_{v=1}^{M}\sum_{u=1}^{N} [\hat{\Upsilon}_u(v)-\hat{\hat{\Upsilon}}_u(v)]^2 < \varepsilon \ (>0), \ \varepsilon \text{ is given}$$

When having $\sum_{v=1}^{M}\sum_{u=1}^{N} [y_u(v)-\hat{\Upsilon}_u(v)]^2$ as a measurement, we express the relative error form as follows:

$$\frac{\sum_{v=1}^{M}\sum_{u=1}^{N} \{[\hat{\Upsilon}_u(v)-\hat{\hat{\Upsilon}}_u(v)] / \hat{\Upsilon}_u(v)\}^2}{\sum_{v=1}^{M}\sum_{u=1}^{N} \{[y_u(v)-\hat{\Upsilon}_u(v)]/y_u(v)\}^2} < \delta_1 \ (>0), \ \delta_1 \text{ is given}$$

(2.4.3)

If the equation (2.4.3) does not satisfy, then we should increase such order of the regression equation in the second step fitting procedure that the equation (2.4.3) satisfied.

Till now, we finish the fit of initial models through the original experiments and statistical analysis. The initial models determine the structrue of the process models.

AN ALGORITHM OF OPTIMAL DESIGN WITH SEQUENTIAL EXPERIMENT

To form the quantitative function $Q=f(X,Z)$, we use the total regression equation and we want to search such optimal treatment combination $(\hat{X},\hat{Z})$ that Q is got by extremum. This paper offers the algorithm of optimal design with sequential experiment to search the optimal treatment combination.

The recursive model mentioned above is different from the general recursive identification algorithm, because it has the factors of the operational parameter variables Z . Here we consider to yield the total recursive model by using two steps.

Firstly to find the recursive form of the equation (2.1.2), suppose N is the variable parameter, we obtain

$$\hat{\mathbf{B}}_{N+1}(v) = \hat{\mathbf{B}}_N(v) + \bar{X}_{N+1}[y_{N+1}(v)-\bar{\Psi}_{N+1}^T \hat{\mathbf{B}}_N(v)]$$

$$\bar{X}_{N+1} = \bar{R}_N \bar{\Psi}_{N+1}^T (1+\bar{\Psi}_{N+1} \bar{R}_N \bar{\Psi}_{N+1}^T)^{-1}$$

$$\bar{R}_{N+1} = \bar{R}_N - \bar{R}_N \bar{\Psi}_{N+1}^T (1+\bar{\Psi}_{N+1} \bar{R}_N \bar{\Psi}_{N+1}^T)^{-1} \bar{\Psi}_{N+1} \bar{R}_N$$

$$v=1,2,\ldots,M$$

$$\hat{\mathbf{B}}_{N+1}(M+1) = (\bar{X}_{N+1}^T \bar{X}_{N+1})^{-1} \bar{X}_{N+1}^T \bar{Y}_{N+1}(M+1)$$

(3.1.1)

where the matrix X_{N+1} is consisted of the observing values of variables x according to the equation (2.1.2).

Secondly to find the recursive form of the equation (2.3.1), and let M be the variable parameter, we obtain

$$\hat{\bar{A}}_{M+1,N+1}^{(i)} = \hat{\bar{A}}_{M,N+1}^{(i)} + \bar{Z}_{M+1}[\hat{\mathbf{b}}_{i,N+1}(M+1)-\bar{A}_{M+1}^T \hat{\bar{A}}_{M,N+1}^{(i)}]$$

$$\bar{Z}_{M+1} = \bar{\mathcal{T}}_M \bar{A}_{M+1}^T [1+\bar{A}_{M+1} \bar{\mathcal{T}}_M \bar{A}_{M+1}^T]^{-1}$$

$$\bar{\mathcal{T}}_{M+1} = \bar{\mathcal{T}}_M - \bar{\mathcal{T}}_M \bar{A}_{M+1}^T (1+\bar{A}_{M+1}^T \bar{\mathcal{T}}_M \bar{A}_{M+1})^{-1} \bar{A}_{M+1} \bar{\mathcal{T}}_M$$

(3.1.2)

where $\hat{\mathbf{b}}_{i,N+1}(M+1)$, is the element in $\hat{\mathbf{B}}_{N+1}(M+1)$, and $\bar{A}_{M,N+1}$

, can be expressed as follows,

$$\bar{A}_{M,N+1}^{(i)} = (\bar{Z}_M^T \bar{Z}_M)^{-1} \bar{Z}_M^T \bar{b}_{M,N+1}^{(i)}$$

the elements of $\bar{b}_{M,N+1}^{(i)}$ will get the elements in

$$\bar{B}_{N+1}(1), \ldots, \bar{B}_{N+1}(M).$$

The matrix Z_M is consisted of the observing values of variables z according to the equation (2.3.1).
We can also obtain the others by changing the superscript (i) to(0),(ii),(ij).
Therefore, we can correct the total regression equation (2.3.2) with the operational parameter variables in the light of the above recursive equation.

SIMULATING EXAMPLE AND RESULTS

We use the method in this paper to discuss a problem of the optimization of high efficiency of heat transfer structural variables and operational parameter variables influencing the optimization algorithm,we obtain the optimum design parameters.
We deal with experimental data and go on simulation for the paper[5].
The structural parameter variables of spirally fluted tube are $x_1 = e/D$, $x_2 = H/e$ .The operational parameter variable is $Z_1 = Re$ (Reynolds number).We expect to obtain the models of X_1, X_2, Z_1 about friction factor f and Stanton number St,i.e.

$$St = g_1(X_1, X_2, Z_1) \qquad (4.1)$$
$$f = g_2(X_1, X_2, Z_1) \qquad (4.2)$$

We take the objective function is

$$A = \sqrt{C_1 \frac{1}{P} C_2} \qquad (4.3)$$

where

A——heat transfer area ; P——heat transfer power loss; C_1——A factor is determined by concrete exchange heat mission; C_2—— quantity is determined by heat transfer and friction characteristics, i.e. $C_2 = f/Pr^3 St^3$
where
Pr——Prandtl number
The above formula 4.3 can be written to

$$A \cdot \sqrt{P} \propto C_2$$

C_2 is called effect factor. A is minimum,when C_2 is minimum for choosing P.We look for the optimum of heat transfer effect,hence,we consider that the effect factor is the objective function,i.e.

$$Q = f/Pr^3 St^3$$

94

where Pr is Prandtl number of gas, Pr=0.7

Fitting models of two steps are obtained as follows,

St----model:

$$St=\beta_{st}^{(0)}+\beta_{st}^{(1)}x_1+\beta_{st}^{(2)}x_2+\beta_{st}^{(3)}x_1^2+\beta_{st}^{(4)}x_2^2+\beta_{st}^{(5)}x_1x_2$$

$$=(-6.65-1.9\times10^{-3}z_1+2.2\times10^{-7}z_1^2-7.9\times10^{-12}z_1^3$$
$$+1.2\times10^{-16}z_1^4-6.5\times10^{-22}z_1^5)$$

$$+(486.1+7.1\times10^{-2}z_1-7.5\times10^{-6}z_1^2+2.7\times10^{-10}z_1^3$$
$$-4.1\times10^{-15}z_1^4+2.2\times10^{-20}z_1^5)x_1$$

$$+(0.63+1.3\times10^{-4}z_1-1.5\times10^{-8}z_1^2+5.6\times10^{-13}z_1^3$$
$$-8.6\times10^{-18}z_1^4+4.6\times10^{-23}z_1^5)x_2$$

$$+(-3506.1-0.492z_1+5.3\times10^{-5}z_1^2-2.1\times10^{-9}z_1^3$$
$$+3.1\times10^{-14}z_1^4-1.7\times10^{-19}z_1^5)x_1^2$$

$$+(-7.2\times10^{-3}-1.7\times10^{-6}z_1+2.1\times10^{-10}z_1^2-8.0*10^{-15}z_1^3$$
$$+1.2\times10^{-19}z_1^4-6.7\times10^{-25}z_1^5)x_2^2$$

$$+(-16.35-2.5\times10^{-3}z_1+2.7\times10^{-7}z_1^2-9.9\times10^{-12}z_1^3$$
$$+1.5\times10^{-16}z_1^4-7.9\times10^{-22}z_1^5)x_1x_2$$

f----model:

$$f=\beta_f^{(0)}+\beta_f^{(1)}x_1+\beta_f^{(2)}x_2+\beta_f^{(3)}x_1^2+\beta_f^{(4)}x_2^2+\beta_f^{(5)}x_1x_2$$

$$=(-17.50-7.0\times10^{-3}z_1)$$
$$+2437.5x_1$$
$$+(5.3\times10^{-4}z_1-4.0\times10^{-8}z_1^2+1.2\times10^{-12}z_1^3)x_2$$
$$+(12668.1)x_1^2$$
$$+(5.9\times10^{-2})x_2^2$$
$$+(-95.47)x_1x_2$$

According to form (2.4.3), we find the relative deviation for St, f can be written to:

St: $5.634511\times10^{-16}<\delta_1$, δ_1 is given 10^{-2};

f : $8.2517138\times10^{-18}<\delta_2$, δ_2 is given 10^{-2}.

respectively.

Objective function $Q=f/Pr^3St^3$, Pr=0.7

$$\min_{x_1,x_2,z_1} Q=Q(X_1^*,X_2^*,Z_1^*)$$

X_1, X_2, and Z_1 are subject to

$$0.011 < X_1 < 0.070$$

$$8 < X_2 < 25$$

$$5000 < Z_1 < 10^5.$$

According to optimisation of nonlinear function, we find that optimal structure parameters and optimal operation parameter are

$$\frac{e}{D} = 0.0497, \frac{H}{e} = 8, Re = 1.4 \times 10^4$$

respectively.

References

1. Wilks, s.s., Mathematical Statistics, John Wiley & Sons. Inc. 1962

2. Setsuo Sagara, et al., (in Japanese), System Identification. The Institute of Measurement and Automatic Control. (1981)

3. Applied Central Station of Chemical Engineering Research and Computer, Department of Chemistry and Industry, (in Chinese), A collection of Translated Text for The Methods of Design of Sequential Experiment, The Publishing House of Chemistry and Industry, (1983)

4. Goodwin, G.C., and Payne, R.L., Dynamic System Identification, (Experiment Design and Data Analysis), Academic Press, 1977

5. Ye, Q.Y., Investigation on tube-side friction factor and heat transfer Characteristics of spirally fluted tube, Thesis for Master, South China University of Technology, 1984

Investigation of the Correlations of Heat Transfer in Single and Multiple Start Spiral Tubes with Single-Phase Turbulent Flow for Fluid of Constant Physical Properties

XIAN-HE DENG, YIN-KE TAN, and SONG-JIU DENG
South China University of Technology
Wushan, Guangzhou 510641, PRC

ABSTRACT

A heat transfer correlation based on the analysis of the distribution of heat transfer resistance in circular tubes for fluid of constant physical properties had been given out, which can be fitted with the experimental data for both single and multiple start spiral tubes and adapted to a wide range of the Pr number (.7 to 35) in the single phase turbulent flow.

INTRODUCTION

Since the reduction of the size and cost of heat exchangers is an important factor in improving the economy of industry thermal systems, many efforts have been made to investigate different methods for the enhancement of heat transfer. As spiral tubes display the simplification in application and the effectiveness in the enhancement of heat transfer, studies on spiral tubes have emergied continually in recent years[1 to 4]. But many studies gave out correlations of heat transfer based on the data of their own experiments. Although further investigations have been made, such as Xiao Jie- Wen[3] and T.S. Ravigururajan[4], the solution for finding a general equation of heat transfer for spiral tubes is still imperfect . According to Deng Xian-He etc[5], the distribution of heat transfer resistance in circular tubes had been theoretically derived, and Deng Xian-He etc[6], applying this knowledge, had obtained a general heat transfer correlation for both single and multiple start spiral tubes in the range of the Pr number .7 to 35. In this paper, a comparison had been made to show the deviations about the calculation results of correlations [4,6] to the heat transfer experimental data[1 to 3].

HEAT TRANSFER CORRELATIONS

According to the analysis of Deng Xian-He etc[6], the heat transfer correlation of spiral tubes could be written as,

$$St = \frac{F}{Pr^{2/3} + .78} \qquad (1)$$

where,

$$F = 0.7(e/d)^{.4388}(H/e)^{-.1182}(\alpha/90)^{1.799\sqrt{\alpha/90}-.114} Re^n \qquad (2)$$

$$n = -.2216 - 1.304 \times 10^{-4}(1.1 - 1/Pr^{1/3})(e/d)Re^{.8} - 1.08(\alpha/90)/\sqrt{Re} \qquad (3)$$

Eqs.(1-3) can be used in the range, $e/d = .01-.07, H/e = 7-40$, $\alpha/90 = .46-.97, Pr = .7-35, Re = 8 \times 10^3 - 1 \times 10^5$, the start number $N = 1-4$.

The oblique angle of curves of the St number versus the Re number in the heat transfer experimental data[1-3] depends mainly on the variables of $e/d, Pr, \alpha/90$ and Re. Through the analysis of these heat transfer experimental data, Eqs.(1-3) were formulated according to the following ideals:
a)   As the fact that when the Re increases, with the same tube parameters, the St number of fluid with higher Pr number will decrease more quickly than that of fluid with lower Pr number, it can be assumed that the oblique angle of curves of St versus Re is influenced by the ratio of thickness of thermal sublayer δ_t and viscous sublayer δ, which is in proportion to $1/Pr^{1/3}$. The term $(1.1 - 1/Pr^{1/3})$ in equation (3) is equivalent to the difference between the value of $1/Pr^{1/3}$ of air and other tested liquids, or the value of $\delta_{t_g}/\delta_g - \delta_{t_l}/\delta_l$. Since $e/d > \delta_t/d$ was most of the case in the experiments in reference[1-3], the increase of $(e/d)Re^{.8}$ (approximately proportion to the value of e^+) will have less effect to increase the heat transfer coefficient of fluid of higher Pr.Since the index n on Re represents the influence of Re to heat transfer coefficient, the term $(1.1 - 1/Pr^{1/3})(e/d)Re^{.8}$ included in equation (3) represents the degree of reduction of the influence of Re to heat transfer coefficient with air as datum.
b)   The angle α of spiral tubes is a second important factor in the index of the Re number in Eq.(2). The greater the angle, the smaller the index. Becaues the greater the angle is ,the greater the friction factor will be, and the e^+ will get bigger. According to the same reason mentioned above,the term $1.08(\alpha/90)/\sqrt{Re}$ also represents the degree of the reduction of the influence of Re to heat transfer coefficient. However,the influence of spiral angle on the index of the Re is slight. Especially, with the rise of the Re number, the spiral angle will give less and less effect on the index of the Re.

COMPARISON

There have been several correlations of heat transfer for spiral tubes applied to single phase turbulent flow ; however,the range of application is usually limited to comparatively narrow areas.

(1) Li X.M. etc[1]

$$St= \frac{\sqrt{f/2}}{3.417\ln[(d/e)/2]-4.636 +G_1} \qquad (4)$$

where,

$$G_1 =.4775(e/d)^{-.621}(\alpha/90)^{-.869}Pr^{.57}(e^+)^{.641+.105\ln(e/d)} \qquad (5)$$

$$\sqrt{2/f} = 3.417\ln[(d/e)/2] -4.6361 +1.249(e/d)^{-0.0566}(H/e)^{.499}x$$
$$(\alpha/90)^{1.14}\exp[(\ln Re-9.62)^2/(H/e)^{-1.378}/1000] \qquad (6)$$

available range, 1, single start spiral tube;
2, Re=1x10^4 to 7 2x10^4, water;
3, e/d=.0099 to   .069 , H/e=7.69 to 39.7.

(2) Ye Qiao-Yan[2],

$$St= \frac{\sqrt{f/2}}{2.5\ln[(d/e)/2]-3.75+ G_2} \qquad (7)$$

where,

$$G_2 =7.629(e/d)^{-.0349}(H/e)^{-.1159}(e^+)^{.202}Pr^{.612} \qquad (8)$$

$$\sqrt{2/f} =2.5\ln[(d/e)/2]- 3.75 + 1.03(e/d)^{-.1426}(H/e)^{.5729}x$$
$$\exp[(\ln Re-9.91)^2/19.11/(e/d)^{-.0466}] \qquad (9)$$

available range, 1, single start spiral tube;
2, Re= 6x10^3 to 1x10^5, water and air;
3, e/d= .01 to .053 , H/e= 8 to 25.

(3) Xiao Jie-Wen[3]

$$St= \frac{\sqrt{f/2}}{2.5\ln[(d/e)/2]-3.75 + G_3} \qquad (10)$$

where, $G_3 = 5.7(e/d)^{-.014}(H/e)^{.029}(e^+)^{.10}Pr^{.56}$ $\qquad (11)$

$$\sqrt{2/f} =2.5\ln[(d/e)/2]- 3.75 + 1.16(e/d)^{-.183}(H/e)^{.429}x$$
$$\exp[(\ln Re-9.67)^2/252/(e/d)^{.342}] \qquad (12)$$

available range, 1, single start spiral tube;
2, Re= 2.35x10^3 to 1x10^5 , Pr= .7 to 35;
3, e/d=.01 to .056 , H/e= 7.9 to 39.

(4) T.S. Ravigururajan[4],

$$Nu_a/Nu_s = \{1+ [\ 2.64Re^{.036}(e/d)^{.212}(H/d)^{-.21}(\alpha/90)^{.19}x$$

$$Pr^{-.024}]^7\}^{1/7} \tag{13}$$

where, Nu_a , Nusselt number of spiral tube,

Nu_s , Nusselt number of smooth tube.

$$Nu_s = \frac{(f_s/2)RePr}{1 + 12.7\sqrt{f_s/2}\ (Pr^{2/3} - 1)} \tag{14}$$

where, $f_s = (1.58 lnRe - 3.28\ )^{-2}$ (15)

available range, 1, Re= 5000 to 250,000, Pr= .66 to 37.6;
2, e/d= .01 to .2 , H/d= .1 to 7.0 ,
α/90= .3 to 1.0.

Correlations[4-12] can't extend to the area of multiple start spiral tubes, so these correlations can't fit the heat transfer experimental data of multiple start spiral tubes[1-3].

Figs.(1-2) showed the results of the comparisons between the calculated results by Eq.(1) and Eq.13) with the heat transfer experimental data [1-3]. The total experimental data [1-3] is 335 points.The maximal deviation of Eq.(1), except 4 points, is within ±15% ;however,for Eq.(13), only 39% points are within ±15% deviation but 61% points within 15-100% deviation.

Figs.(3-5) plotted the heat transfer experimental data [1-3] with Nu versus Re. The calculation results of Eqs.(1-3) ,plotted in curves, fit these data well.

CONCLUSION

1. Although there have been several correlation methods for the heat transfer coefficient of spiral tubes, a comparative-ly simple and effective one has been provided [6],which gave out the correlation of heat transfer for both single and mul-tiple start spiral tubes and a wide range of the Pr number. Eqs.(1-3) can fit the heat transfer experimental data[1-3] much better than Eqs.(13-15).
2, The oblique angle of St versus Re depends mainly on the values of e/d, Pr, α/90 and Re, which can be discovered not only in reference[1-3] but also in reference[7].In this paper ,the rule of the influence of e/d, Pr, α/90 and Re on the ob-lique angle,or the index n in Eq.(2), had been revealed. The rule of the influence of tube roughness parameters, Pr and Re on the oblique angle of St versus Re revealed in this paper can be predicted applicable to other form of roughness of tubes besides spiral tubes.

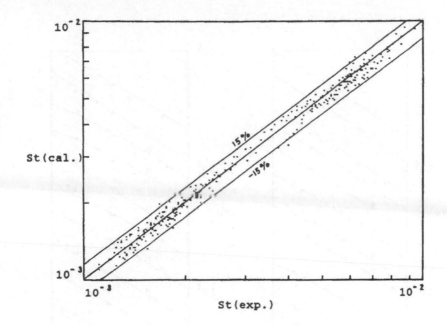

FIGURE 1   Comparison of the calculated St by Eq.(1) with the experimental data of the St in reference[1-3].

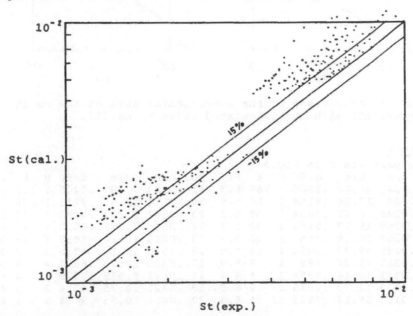

FIGURE 2   Comparison of the calculated St by Eq.(13) with the experimental data of the St in reference[1-3].

101

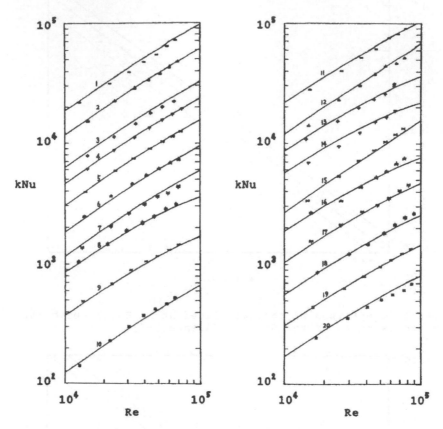

FIGURE 3 Comparison of the experimental data of the Nu in reference[1] with the calculated value by Eq.(1).

TABLE 1
Tube parameters in Fig.3

No.	e/d	H/e	α/90	N	k	Pr	No.	e/d	H/e	α/90	N	k	Pr
1	.0247	28.28	.8606	1	140	6.3	11	.0383	7.692	.9405	1	110	5.5
2	.024	17.54	.9154	1	80	5.6	12	.0114	12.5	.9711	1	100	5.5
3	.0246	28.41	.8608	1	50	5.5	13	.0547	8.333	.9083	1	40	5.6
4	.0269	15.63	.9154	1	30	5.5	14	.069	7.787	.8922	1	25	5.4
5	.0261	16.16	.915	1	20	5.7	15	.0099	17.14	.9659	1	25	5.4
6	.0245	39.73	.8091	1	16	5.9	16	.0575	15.5	.4596	4	18	5.2
7	.0261	21.24	.889	1	8	6.0	17	.0445	10.23	.6654	4	8	5.2
8	.0531	10.44	.8888	1	4	6.3	18	.0471	9.569	.7389	3	4	4.5
9	.0433	12.79	.8889	1	2	6.0	19	.0421	10.75	.8214	2	2	4.9
10	.019	29.12	.8889	1	1	6.3	20	.0411	10.96	.9094	1	1	4.6

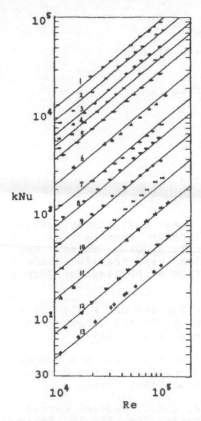

kNu

Re

kNu

Re

FIGURE 4 Comparison of the
experimental data of the Nu
in reference[2] with the
calculated value by Eq.(1)

FIGURE 5 Comparison of the
experimental data of the Nu
in reference[3] with the
calculated value by Eq.(1).

TABLE 2
Tube parameters in Fig.4

No.	e/d	H/e	d/90	N	k	Pr
1	.0511	11.1	.8867	1	230	.7
2	.0217	12.05	.947	1	230	.7
3	.0241	11.9	.9419	1	160	.7
4	.0276	12.42	.9307	1	120	.7
5	.0338	16.75	.8867	1	90	.7
6	.0528	10.79	.6039	4	60	.7
7	.0518	10.91	.6866	3	30	.7
8	.044	12.87	.7802	2	20	.7
9	.030	22.64	.8651	1	14	.7
10	.0115	17.5	.96	1	10	.7
11	.0455	9.03	.9196	1	3	.7
12	.0377	9.23	.9311	1	1.5	.7
13	.0301	13.5	.9198	1	1	.7

TABLE 3
Tube parameters in Fig.5

No.	e/d	H/e	d/90	N	k	Pr
1	.0538	10.75	.884	1	100	36
2	.0341	16.94	.884	1	50	34
3	.0538	10.75	.884	1	25	20
4	.0341	16.94	.884	1	8	20
5	.0345	11.71	.919	1	4	20
6	.0538	10.75	.884	1	2	11
7	.0341	16.94	.884	1	1	11

NOMENCLATURE

D,d   diameter of tube [M],
e   height of rib of spiral tube [M],
e^+ nondimensional height of rib [=(e/d)Re$\sqrt{\phi}$].
f   friction factor,
H   pitch between ribs [M],
N   start number of rib of spiral tube,
Nu   Nusselt number,
Pr   Prandtl number,
Re   Reynolds number,
St   Stanton number,
α spiral angle.
ϕ friction factor [=f/2].

REFERENCE

1.　Li,X.M., Ye,K.S., Tan,Y.K.,and Deng, S.J.,"Investigation on Tube-Side Flow Visualization , Friction Factors and Heat Transfer Characteristics of Helical Ridging Tubes," Heat Transfer 1982 , Proc. 7th Internation Heat Transfer Conference, Vol.3, PP.75-80 (1982) , Hemisphere Publishing Corp., Washington,D.C..

2.　Ye Qiao-Yan, "Study on Friction Factors of Fluid Flow and Heat Transfer Characteristics of Spiral Tubes," M.Sc. thesis, South China University of Technology, (1984).

3.　Xiao Jie-Wen,"The Influence of the Fluid Prandtl Number on Heat Transfer Characteristics of Spiral Tubes," M.Sc. thesis, South China University of Technology,(1986).

4.　Ravigururajan, T.S. and Bergles, A.E., "General Correlations for Pressure Drop and Heat Transfer for Single Phase Turbulent Flow in Internal Ribbed Tubes," Augmentation of Heat Transfer in Energy Systems, New York: ASME, 52:9-20, (1986).

5.　Deng Xian-He, Deng Song-Jiu,"An Approximate Analytical Solution for the Nusselt Number Expression in Turbulent Pipe Flow with Constant Physical Properties," J. of Chem. Industry and Eng.,(in Chinese) Vol.4, PP.494, (1987).

6.　Deng Xian-He,Tan Yin-Ke and Deng Song-Jiu,"Investigation of the Single and Multiple Start Spiral Tubes with Single Phase Turbulent Flow," to be published by J. of Chem. Industry and Eng.(in Chinese).

7.　B.A.Kader and A.M.Yaglom,"Turbulent Heat and Mass Transfer from a Wall with Parallel Roughness Ridges",Int. J. Heat and Mass Transfer Vol.20 PP.345-357,(1977).

Heat Transfer of Gas Flow inside Spirally Corrugated Tubes

QIANG-TAI ZHOU, ZI-XIN ZHANG, SAI-YIN-YE, YONG-FU ZHANG, and
ZHI-MING LIANG
Nanjing Institute of Technology
Nanjing, PRC

ABSTRACT

In the present paper experimental investigations have been carried out on
tube-side friction factor and heat transfer of single-start spirally corrugated tubes of 55-56.5 mm ID, having corrugation parameters, h/D_i and
p/h, of 0.0146-0.0467 and 11.6-69, for turbulent air flow with Reynolds
number ranging from 8×10^3 to 10^5.

The experimental data have been correlated with exponential type equat-
-ions. Thermal performance comparison of these tubes has been made with
smooth tube. It was found that the tubes with corrugation parameters of
h/D_i=0.025-0.046 and p/h=11.6-16 were shown to be more efficient.

INTRODUCTION

Over the past 20 years, many types of tube-side heat transfer enhancement
technique have been developed for improving heat exchanger performance.
The application of these enhancement devices to heat exchangers can result in the reduction of pumping power or heat transfer area or metal
temperature of the wall, or the increase in the heat transfer rate. Among
these devices, spirally corrugated tubes have been shown to have several
advantages over other rough surfaces such as: (1) easy fabrication, (2)
limited fouling and (3) higher enhancement in the heat transfer rate
compared with the same increase in the friction factor.

There has been a considerable body of literature on the investigation of
heat transfer and pressure drop inside the spirally corrugated tubes[1-9],
but most of the experimental data was obtained from small diameter test
sections with liquid being working fluid, e.g. water flowing inside tubes
with diameter of 15-25 mm. There was a lack of experimental data for air
flowing inside tubes with larger diameter. This paper presented the experimental data of pressure drop and heat transfer to air flow inside the
spirally corrugated tubes with inside diameter of 55-56.5 mm.

EXPERIMENTAL APPARATUS

The experimental apparatus was an air to air heat exchanger whose schematic drawing was shown in Fig. 1. Heated air by electrical heating device
5 flowed through the tube side of the test section 6 from air compressor
1, which was followed by an air tank 2 to insure a stable flow. The flow

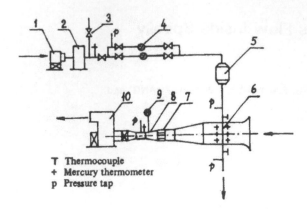

FIGURE 1. Experimental apparatus

T   Thermocouple
+   Mercury thermometer
p   Pressure tap

rate was controlled by bypass valve 3. Cooling air flowed outside the
test section with cross flow pattern by a variable speed, damper controlled and centrifugal fan 10. Tube-side flow rate was measured by two
turbine flow meters 4 in parallel. The flow rate of cooling air was
measured by Pitot tube 9, in front of which in the flow direction a
special-curve nozzle 8 and a flow straightener 7 were used to obtain a
flat velocity profile. The temperature of heated air was measured by calibrated Chromel-Constantan thermocouples which were installed at the inlet and the exit of the test section and connected to 7V 07 Programmable
Scanner with 100-point recorder. The temperature of cooling air was measured with precision mercury thermometers having 0.1°C minimum graduat-
ions. Two ring-type static pressure taps were located at either ends of
the test section and connected to the U-tube or inclined manometer, depending on the value of the water column, for measurement of the inlet
pressure and test section pressure drop.

The test section were single-start spirally corrugated tubes made of carbon steel, having 1000 mm length with 55-56.5 mm ID, whose dimensions
were listed in Tables 1 and 2. The dimensionless corrugation, h/D_i, varied from 0.0146 to 0.0467, and the corrugation pitch to height ratio, p/h,
ranged from 11.6 to 69. In order to establish the flow pattern and prevent the outlet effect, a steady-flow entrance zone of 1000 mm having the
same internal configuration was provided upstream of the test section
which was followed again by a 350 mm steady-flow section. Therefore, the
total length of the test section was 2350 mm.

Heat transfer studies were carried out with the heated air flowing inside
the tubes being the heating medium of the heat exchanger in the range of
Re from 8×10^3 to 10^5. The inlet temperature of the test sections was
maintained at the value of about 200°C. Data acquisition was conducted
under steady-state conditions. Heat flow rate Q was determined by the enthalpy drop of the heated air which generally agreed with the value calculated from the enthalpy rise of the cooling air except for the conditions when the temperature rise of the cooling air was less than 1°C. In
order to measure the wall temperature, for some spirally corrugated tubes,
twelve Copper-Constantan thermocouples were embedded on the wall of the
test section at four positions with three thermocouples at each position

TABLE 1. Tube Dimensions for Heat Transfer Test

Tube no.	D_o mm	D_i mm	p mm	h mm	h/D_i	p/h	Symbol
1	63.26	55.12	20.00	1.38	0.0250	14.49	●
2	62.70	55.14	20.00	1.65	0.0299	12.12	○
3	63.45	56.31	29.90	1.71	0.0304	17.48	+
4	63.46	56.30	29.90	2.58	0.0458	11.59	●
5	63.46	56.26	30.00	1.96	0.0348	15.31	■
6	63.50	56.02	30.00	2.58	0.0461	11.63	□
7	63.52	56.22	31.85	2.12	0.0377	15.02	▲
8	63.70	56.34	40.00	2.38	0.0422	16.81	⊞
9	63.48	56.26	43.90	2.63	0.0467	16.69	⊠
10	63.48	56.06	44.00	2.37	0.0423	18.56	◇
11	63.51	56.19	53.50	2.09	0.0372	25.60	◆
12	63.75	55.47	55.90	0.81	0.0146	69.01	▲
13	63.64	55.88	56.00	1.18	0.0211	47.06	⊙
Smooth	63.50	56.50	-	-	-	-	○

TABLE 2. Tube Dimensions for Friction Factor Test

Tube no.	D_i mm	p mm	h mm	h/D_i	p/h	Symbol
1	55.66	20.00	1.25	0.0224	16.00	+
2	53.72	20.38	1.45	0.0270	14.00	●
3	55.93	29.97	2.76	0.0493	10.86	◇
4	56.09	30.00	2.35	0.0419	12.76	■
5	56.37	31.85	2.54	0.0451	12.54	×
6	56.33	39.97	2.22	0.0394	18.00	♀
7	56.02	43.92	2.87	0.0512	15.30	●
8	55.98	43.97	2.51	0.0448	17.52	▲
9	56.55	56.85	2.72	0.0481	20.90	◆
10	56.20	58.18	2.36	0.0420	24.65	■
Smooth	56.34	-	-	-	-	○

placed in circumference by equal interval. The isothermal pressure drop experiments were conducted at ambient temperature.

EXPERIMENTAL RESULTS

Experimental procedures were described in (10) in detail. Smooth tube tests for pressure drop and heat transfer were conducted first. The results were compared with the existed equations, which was used to examine the experimental apparatus. And, then, the spirally corrugated tube experiments were carried out. The isothermal pressure drop data were presented in terms of friction factor, f, versus Reynolds number, Re, defined as

$$f = \Delta P / \left(\frac{L}{D_i} \frac{\rho u^2}{2} \right) \tag{1}$$

and

107

$$Re = \rho u D_i / \mu \tag{2}$$

The friction factor for smooth tube was 10 percent higher than that calculated by Knudsen-Katz equation

$$f = 0.184 \, Re^{-0.2} \tag{3}$$

which was ascribed to the fact that the inside surface of the carbon-steel tubes for our friction factor experiment was covered with rust in part. The friction factor for all tubes listed in Table 2 was shown in Fig. 2. It can be seen from Fig. 2 that for the spirally corrugated tubes tested and Reynolds number ranging from 10^4 to 10^5, the completely rough flow did not appear.

The overall heat transfer coefficient U_i can be obtained from the heat transfer experiment and calculated by

$$U_i = Q / (\pi D_i L \, \Delta T_m) \tag{4}$$

In order to obtain the tube-side heat transfer coefficient α, a technique of Wilson plot was used for seperating α from U_i. For some spirally corrugated tubes, the tube-side heat transfer coefficient obtained with Wilson plot technique was compared with that calculated from outside wall temperature measurement. Fair agreement was found between the two values in general.

Tube-side heat transfer data obtained for fourteen tubes listed in Table 1 were plotted in Fig. 3 in terms of Nu versus Re. The heat transfer data for smooth tube was found to be in good agreement with Dittus-Boelter type equation

$$Nu = 0.021 \, Re^{0.8} \, Pr^{0.4} \tag{5}$$

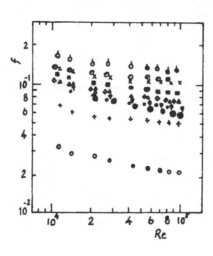

FIGURE 2. Friction factor vs. Reynolds number

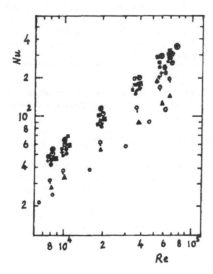

FIGURE 3. Nusselt number vs. Reynolds number

It can be seen from Fig. 2 and Fig. 3 that for the spirally corrugated tubes with h/D_i=0.025-0.046 and p/h=11.6-16, the heat transfer rate went up by 70-200 percent while the pressure drop went up by 150-400 percent in comparison with smooth tube.

Having analysed Figs. 2 and 3, we used Blasius type equation for friction factor and Dittus-Boelter type equation for heat transfer to correlate our experimental data, with tube geometry parameters being allowed for. The results were plotted in Figs. 4 and 5, for friction factor and heat transfer, respectively. Equations (6) and (7) fitted the data well with maximum deviation of 9.9 and 12.5 percent, respectively:

$$f=29.3 \ Re^{-0.14}(h/D_i)^{0.825}(p/h)^{-0.6} \tag{6}$$

$$Nu=0.107 \ Re^{0.837} \ Pr^{0.4}(h/D_i)^{0.15}(p/h)^{-0.295} \tag{7}$$

PERFORMANCE EVALUATION AND APPLICATION

Bergles et al.[11], and Webb and Eckert[12] have proposed several performance criteria for enhanced tubes based on tube-side heat transfer coefficient, keeping various parameters constant for a particular criterion. We calculated three criteria: (1) the reduction of heat transfer area for equal pumping power and heat duty (P/P_s=1 and Q/Q_s=1), (2) the increase in heat transfer rate for equal pumping power and heat transfer area (P/P_s=1 and A/A_s=1), and (3) the reduction of pumping power for equal heat duty and heat transfer area (Q/Q_s=1 and A/A_s=1). The calculation was performed by assuming that the resistance of the outside film and the metal wall was zero. By rule of thumb, the more the heat transfer area reduced, the more the heat transfer rate increased and the more the pumping power saved. Therefore, the criterion for the reduction of heat transfer area was presented here.

The performance comparison showed that for the single-start spirally corrugated tubes with h/D_i=0.025-0.047 the enhancement was limited if p/h>25. The less improvement of performance was found for Tubes 11, 12 and 13.

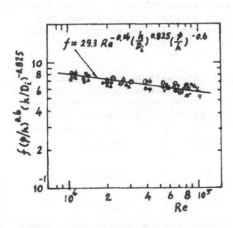

FIGURE 4. Correlation of friction factor data

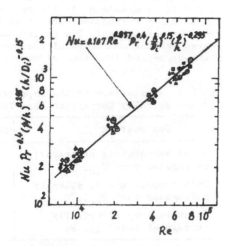

FIGURE 5. Correlation of heat transfer data

Table 3. Values of A/A_s for Spirally Corrugated Tubes

Re_s	Tube no. in Table 1									
	1	2	3	4	5	6	7	8	9	10
10^4	0.611	0.605	0.647	0.636	0.643	0.631	0.648	0.670	0.677	0.68
2×10^4	0.600	0.596	0.635	0.626	0.632	0.626	0.637	0.659	0.666	0.67
3×10^4	0.593	0.589	0.628	0.617	0.625	0.618	0.628	0.651	0.658	0.66
5×10^4	0.584	0.580	0.618	0.609	0.615	0.610	0.619	0.640	0.647	0.65
8×10^4	0.576	0.572	0.608	0.600	0.607	0.601	0.611	0.632	0.638	0.64

And Tube 2 was identified to be most efficient, producing a maximum saving of 40-43 percent in heat transfer area. Tube 1 was similar to Tube 2 with 39-42 percent saving in heat transfer area. The results of performance evaluation on the reduction of heat transfer area for Tubes 1-10 were listed in Table 3 in terms of A/A_s versus Re_s. It is seen from Tables 1 and 3 that for the tubes with $h/D_i = 0.025-0.047$, dimensionless corrugation pitch, p/h, may be the most important parameter for heat transfer enhancement, and $p/h = 11-14$ will gain more benefits than those with higher value of p/h.

It was interesting to compare the present result with Sethumadhavan data[4]. In (4) the best tube was Tube 4, having corrugation parameter of $h/D_i = 0.023$ and $p/h = 12.9$, producing a reduction of 35-45 percent in heat transfer area, which was quite similar to Tubes 1 and 2 in the present study.

A spirally corrugated tube with similar characteristic parameters to Tube 5 in Table 1 was used to enhance convection heat transfer inside the fire tubes in a coal-fire packaged boiler with 1.11 kg/s (4 t/h) of steam output which was manufactured in Nanjing Boiler Factory with the heat transfer area of the fire tube having been reduced by 35 percent, and was in operation in 1985.

The experimental data of the spirally corrugated fire tube from the packaged boiler tests were shown in Table 4. And the design values were listed for comparison. It can be seen from the experimental values that the spirally corrugated tube is an advantageous device for tube-side heat transfer enhancement. The boiler has been working for more than two years. Another character of the boiler is that it needs less time for start-up period than that of smooth fire-tube boiler.

TABLE 4. Comparison of the Experimental Data with Design Values for the Spirally Corrugated Fire Tube in the Packaged Boiler

Designation	Design value	Experimental data
Inlet temperature of spirally corrugated tube T_{in}, °C	914.3	897-927
Exit temperature of spirally corrugated tube T_{ex}, °C	303	250-260
Pressure drop of spirally corrugated tube Δp, Pa	700	550-560

CONCLUSION

1. The single-start spirally corrugated tubes with different values of parameter and 55-56.5 mm ID had been investigated experimentally in the air flow. The tubes with h/D_i=0.025-0.046 and p/h=11.6-16 showed an enhancement of tube-side heat transfer ranging from 70 to 200 percent and an increase in friction factor of 150-400 percent in comparison with the smooth tube.

2. The performance comparison showed that the tubes with corrugation parameters of h/D_i=0.025-0.046 and p/h=11.6-16 would be more efficient for single-start spirally corrugated tube, which yielded a reduction in heat transfer area by 35-43 percent, and that p/h=11-14 and h/D_i=0.025-0.030 would be the best values.

3. A spirally corrugated tube with similar corrugation parameters to Tube 5 in Table 1 was used in coal-fire packaged boiler to enhance fire-tube heat transfer. The heat transfer area of the fire tube and the boiler drum metal have been reduced by 30-35 percent. The boiler has been in operation quite well for more than two years.

NOMENCLATURE

A	heat transfer area, m^2	c_p	specific heat, kJ/kg K
D_i	tube inside diameter, m	D_o	tube outside diameter, m
f	Darcy friction factor, $f=\Delta p(D_i/L)(2/\rho u^2)$	h	corrugation height, m
k	thermal conductivity, W/m K	L	length, m
Nu	Nusselt number, Nu= $\alpha D_i/k$	p	corrugation pitch, m
P	pumping power, W	Δp	pressure drop, Pa
Pr	Prandtl number, Pr= $\mu c_p/k$	Q	heat flow rate, W
Re	Reynolds number, Re= $\rho u D_i/\mu$	St	Stanton number, St= $\alpha/\rho u c_p$
T	temperature, °C	ΔT_m	mean temperature difference for cross flow, °C
u	axial mean velocity, m/s	U	overall heat transfer coefficient, W/m^2 K
α	tube-side heat transfer coefficient, W/m^2 K	μ	dynamic viscosity, Pa s
ρ	fluid density, kg/m^3	τ_o	wall shear stress, Pa

Subscripts

i	based on inside diameter	s	for smooth tube

REFERENCES

1. Withers, J.G., Tube-Side Heat Transfer and Pressure Drop for Tubes Having Helical Internal Ridging with Turbulent/Transitional Flow of Single-Phase Fluid. Part 1. Single-Helix Ridging. Heat Transfer Engineering, Vol. 2, no. 1, pp. 48-58, 1980.

2. Withers, J.G., Tube-Side Heat Transfer and Pressure Drop for Tubes Having Helical Internal Ridging with Turbulent/Transitional Flow of Single-Phase Fluid. Part 2. Multiple-Helix Ridging. Heat Transfer

Engineering, Vol. 2, no. 2, pp. 43-50, 1980.

3. Li, H.M., Ye, K.S., Tan, Y.K. and Deng, S.J., Investigation on Tube-Side Flow Visualisation, Friction Factors and Heat Transfer Characteristics of Helical-Ridging Tubes. Proc. of 7th Int. Heat Transfer Conference, Vol. 3, pp. 75-80, 1982.

4. Sethumadhavan, R. and Rao, M.R., Turbulent Flow Friction and Heat Transfer Characteristics of Single- and Multi-Start Spirally Enhanced Tubes. J. of Heat Transfer, Vol. 108, pp. 55-61, 1986.

5. Cunningham, J. and Milne, H.K., The Effect of Helix Angle on the Performance of Roped Tubes. Proc. of 6th Int. Heat Transfer Conference, Vol. 4, pp. 601-605, 1978.

6. Nakayama, W., Takahashi, T. and Daikoku, T., Spirally Ribbing to Enhance Single-Phase Heat Transfer Inside Tubes. Proc. of ASME-JSME Thermal Engineering Joint Conference, Vol. 1, pp. 365-372, 1983.

7. Bogolyubov, Yu.N., Lifshitz, M.N. and Grigor'ev, G.V., Investigation and Industrial Application of Spirally Corrugated Tubes. Teploenergetika, no. 7, pp. 48-50, 1981 (in Russian).

8. Brodov, Yu. M., Effectiveness of Application of Spirally Corrugated Tubes to Heat Exchangers in Turbine Equipment. Teploenergetika, no. 12, pp. 36-40, 1982 (in Russian).

9. Chizeskaya, E.M., Brodov, Yu.M., Savel'ev, R.Z. and Men, P.G., Heat Transfer in Spirally Corrugated Tubes for Single-Phase Flow. Energetika, no. 5, pp. 109-112, 1984 (in Russian).

10. Ye, S.Y., Investigation of Heat Transfer Enhancement Inside and Outside Spirally Corrugated Tubes for Single-Phase Flow. M.Sc. Thesis, Nanjing Institute of Technology, 1987 (in Chinese).

11. Bergles, A.E., Blumenkrantz, A.R. and Taborek, J., Performance Evaluation Criteria for Enhanced Heat Transfer Surfaces. Proc. of 5th Int. Heat Transfer Conference, Vol. 2, pp. 239-243, 1974.

12. Webb, R.L. and Eckert, E.R.G., Application of Rough Surface to Heat Exchanger Design. Int. J. Heat Mass Transfer, Vol. 15, pp. 1647-1658, 1972.

Augmentation of Heat Transfer for Single-Phase Thermopositive-Oil SD280 in the Tube with Inserts

LUSHENG XIE
Department of Environmental Engineering
Qingdao Architectural Engineering College
Shandong, PRC

RUIYING GU
Department of Environmental Engineering
Xian College of Metallurgical and Constructional Engineering
Shaanxi, PRC

XUEQUN WANG
Department of Heat and Ventilating Engineering
Cyangsha Ferrous Metal Mine Design and Research Institute
Hunan, PRC

ABSTRACT

The method of electric heating was utilized to simulate the actual industrial conditions, the heat transfer and friction characteristics of single-phase thermopositive-oil SD280 were examined when the fluid was forced to flow in a horizontal tube with and without insert. Four different types of inserts were tested separately. The performance of augmentatiom of heat transfer for each type of inserts was evaluated by means of equivalent pumping power comparison.

INTRODUCTION

SD280, the thermopositive-oil, is characterized by its stable heat properties, its higher boiling point and its lower solid ifying point. It is one of ideal intermediates for recovery of waste heat, but it usually flows in the transition region due to its higher viscosity, and its heat transfer coefficient is low. The purpose of the paper is to make investigation of the heat transfer and friction characteristics of single-phase thermo-positive-oil SD280 as the oil is forced to flow in the tube with and without various types of inserts, and to compare the perform-ance of the augmented tube and the empty tube at the equal pump-ing power.

TEST DEVICE AND INSERTS

XIE LUSHENG, Prof.,Department of Environmental Engineering, Qingdao Architectural Engineering College, Shandong,China.
GU RUIYING, Assoc. Prof., Department of Environmental Engineer-ing, Xian College of Metallurgical & Constructional Engineering Shaanxi, China.
WANG XUEQUN, Engineer, Department of Heat and Ventilating Engineering, Changsha Ferrous Metal Mine Design & Research Institute, Hunan, China.

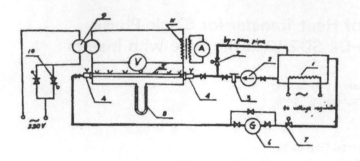

FIGURE 1. Test facility. 1)oil tank and U-type electric heater, 2)circulating pump, 3)oil filter, 4)temperature measuring device at entry and exit of the tube, 5)heated test section, 6)flow-meter, 7)valve for flow rate regulation, 8)U-tube manometer, 9)transformer, 10)silicon controlled voltage regulator, 11)current transformer.

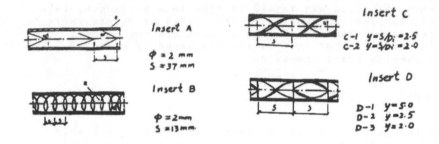

FIGURE 2.   Four types of inserts.

The test device is shown in figure 1, the test section (5) is a tube of stainless steel (1Cr18Ni9Ti), its inside diameter is 18 mm, wall thickness 1 mm respectively, the heated length is 1952 mm.   The tube was heated by a A-C current passed directly through the tube wall.   Four different shapes of the inserts are shown in figure 2.   The insert A is like a fork of tree branches. The insert B is a spiral wire and looks like a spring.   The insert C is twisted-tape, two kinds of pitch-to-diameter ratio y were tested, one was 2.5 (insert C-1), another was 2 (insert C-2). The insert D is static mixer, for which three kinds of pitch-to-diameter ratio, y=5 (insert D-1), 2.5 (insert D-2) and 2 (insert D-3) were tested. The average clearance between insert

114

and inside tube wall is about 1 mm. In order to assure good electric insulation, all of the inserts were coated with poly-amideimine paint before they were put into the tube.

HEAT TRANSFER AND FRICTION TEST FOR EMPTY TUBE

Test Range.

The temperature of the fluid at the entry of test section: 383—433 K.

Heat flux for test section: 4370—18310 w/m².

Reynolds number: 2300—22000.

Prandtl number: 40—88.

Test Results. Friction factor was measured at heating state. The test data obtained accord with the formula recommended in reference 8.

$$ f = \frac{0.3164}{Re_f^{0.25}} \left(\frac{Pr_w}{Pr_f} \right)^{\frac{1}{3}} \qquad \text{for } Re_f > 2300 \qquad (1) $$

The relative errors between data from test and from formula (1) are less than +4.0 %.

The deviations between the data from tests and from formula (2) are less then +6.0 %.

$$ Nu = 0.011 (Re_f^{0.87} - 280) Pr_f^{0.4} \left[1 + \left(\frac{Di}{L} \right)^{\frac{2}{3}} \right] \left(\frac{Pr_f}{Pr_w} \right)^{0.11} \qquad (2) $$

The difference between the formula given in reference 9 within transition region and formula (2) is that the coefficent of the former is 0.012, the later is 0.011. Even though 0.012 is adopted, the maximum error (−14.3%) for all test data measured is still not beyond +20% which is the error limit given in reference 9. Therefore, the test device is proved reliable, formula (1) and (2) will be the standard for calculating f and Nu of the empty tube.

HEAT TRANSFER AND FRICTION TEST FOR THE TUBE WITH INSERT

Test Range.

The temperature of the fluid at the entry of the test section: 427 K.

Heat flux for test section: 6130—16440 w/m².

115

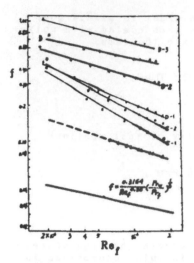

FIGURE 3. f versus Re_f.

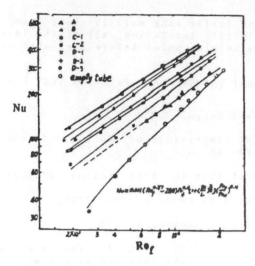

FIGURE 4. Nu versus Re_f.

TABLE 1. Regression equations for inserts.

No. of inserts	f		Nu_f	
A	$\dfrac{1\cdot 878}{Re_f^{0.381x}}\left(\dfrac{P_{rw}}{P_{rf}}\right)^{\frac13}$	$\pm 1.7\%$	$0.064\,Re_f^{0.7023}\,P_{rf}^{0.4}\left(\dfrac{P_{rf}}{P_{rw}}\right)^{0.11}$	$\pm 7.6\%$
B	$\dfrac{4\cdot 605}{Re_f^{0.342}}\left(\dfrac{P_{rw}}{P_{rf}}\right)^{\frac13}$	$\pm 2.2\%$	$0.1043\,Re_f^{0.711}\,P_{rf}^{0.4}\left(\dfrac{P_{rf}}{P_{rw}}\right)^{0.11}$	$\pm 2.2\%$
C - 1	$\dfrac{47.68}{Re_f^{0.6236}}\left(\dfrac{P_{rw}}{P_{rf}}\right)^{\frac13}$	$\pm 9.6\%$	$0.041\,Re_f^{0.77}\,P_{rf}^{0.4}\left(\dfrac{P_{rf}}{P_{rw}}\right)^{0.11}$	$\pm 10.5\%$
C - 2	$\dfrac{36.95}{Re_f^{0.3782}}\left(\dfrac{P_{rw}}{P_{rf}}\right)^{\frac13}$	$\pm 11.8\%$	$0.1089\,Re_f^{0.6779}\,P_{rf}^{0.4}\left(\dfrac{P_{rf}}{P_{rw}}\right)^{0.11}$	$\pm 7.1\%$
D	$f=(1.629y^2-8.351y+17.386)Re_f^{(0.0222y^2-0.2276y+0.1086)}\left(\dfrac{P_{rw}}{P_{rf}}\right)^{\frac13}$			$\pm 6.6\%$
	$Nu_f=0.158\,Re_f^{0.696}\,P_{rf}^{0.4}\,y^{-0.293}\left(\dfrac{P_{rf}}{P_{rw}}\right)^{0.11}$			$\pm 6.7\%$

Reynodls number: 1700--18000.

Prandtl number: 41.7.

Test results. Friction factor was measured at heating state.
The test data and regression equations are shown in figure 3
and 4 and table 1, where the percentage after the regression
equation represents the deviation between values from tests and
that calculated from regression equations. By way of trial-and-
error, and comparison of the test results (Nu) of SD280 with the

116

calculated results (Nu) from the formulas in previous works (reference 1 to 7), in this paper, 0.4 is taken as the exponent of Pr for Nu regression equations.

EVALUATION ON THE PERFORMANCES OF INSERTS TESTED

Figure 5 shows the curves of (ha/he)p versus Re_e. It indicates,
1. For $Re_e < 10^4$, all four kinds of inserts are able to augment heat transfer to varying degree, and (ha/he)p decreases gradually with increase of Re_e, until $Re_e > 3 \times 10^4$ the four kinds of inserts lost their effects of augmentation of heat transfer.

2. For $4000 < Re_e < 10^4$, insert D-2, the static mixer, has the best effect of augmentation as its pitch-to-diameter ratio y is 2.5. Insert C-1, the twisted-tape, is nearly ineffective as y is 2.5. And (ha/he)p is 1.64 and 1.11 respectively.

3. For twisted-tape inserts $Re_e < 2 \times 10^4$, (ha/he)p increases as y decreases from 2.5 to 2, this is different from the conclusion in reference 1 where the ratio y 2.5 is optimum for air.

4. Among the static mixers, insert D-2 of y=2.5 is better than insert D-3 of y=2 and D-1 of y=5. Therefore there must be an optimum pitch-to-diameter ratio that shall make the (ha/he)p maximum between y=2 and 5.

Because of the test arrangement, the optimum ratio y could not be found out directly by test, but it could be obtained by calculation from the formula given in table 1. Figure 6 shows the results calculated, the optimum ratio y is 3.5 approximately.

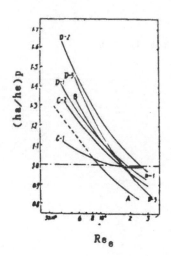

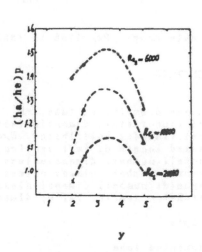

FIGURE 5. (ha/he)p versus Re . FIGURE 6. (ha/he)p versus y .

117

CONCLUSIONS

1. It is appropriate to use formula (1) and (2) to calculate the friction factor and Nusselt number for the oil SD280 within transition region when there is no insert in the tube.

2. When the insert is in the tube it is available to take 0.4 as the exponent of Pr number in the Nusselt number correlation within the test range.

3. Under the conditions of same pumping power, when $Re_e < 10^4$, all kinds of the tested inserts are able to improve heat transfer. When $Re_e > 3 \times 10^4$, all the inserts have no effect on augmentation of heat transfer. The static mixer produces the best effect.

4. Under the conditions of same pumping power, for twisted-tape inserts, the pitch-to-diameter ratio 2.5 is not optimum to augment heat transfer, while, for static mixer, y=2.5 is the best among all inserts in the test. The calculations indicate that the optimum pitch-to-diameter ratio is 3.5 approximately.

5. If static mixer is used to augment heat transfer for the oil SD280, the heat transfer coefficient and friction factor can be calculated by the following correlations:

$$f = (1.629\,y^2 - 8.351\,y + 17.386)\,Re_f^{(0.1086 - 0.2276\,y + 0.222\,y^2)}\left(\frac{Pr_w}{Pr_f}\right)^{\frac{1}{3}}$$

$$Nu = 0.158\,Re_f^{0.696}\,Pr_f^{0.4}\,y^{-0.293}\left(\frac{Pr_f}{Pr_w}\right)^{0.11}$$

Applicable range, Re_f =1700 to 18000 and y=2.0 to 5.0.

NOMENCLATURE

Di   inside diameter of tube, mm
f   friction factor, dimensionless
h   heat transfer coefficient, $W/m^2\,K$
L   heated length of test section, mm
Nu   Nusselt number, dimensionless
Pr   Prandtl number, dimensionless
Re   Reynolds number, dimensionless
y   Pitch-to-diameter ratio, dimensionless

Subscripts

a   augmented tube
e   empty tube

f   bulk fluid temperature condition
p   pumping power
w   wall temperature condition

REFERENCES

1. A. E. Bergles, Survey and Evaluation of Techniques to
    Augment Convective Heat Transfer, Progress in Heat and Mass
    Transfer, Vol. 1, Pergmon Press, New York, pp. 331–424, 1969.

2. T. C. Carnaves,Heat Exchangers:Design and Theory Sourcebook,
    Scripta Book Company, Washington, D. C., pp. 441–489, 1974.

3. Date, Flow in Tubes Containing Twisted Tapes, Heat and
    Ventilating Engineering, Vol. 46, pp.240–249, 1978.

4. E. Smithberg, F. Landis, Friction and Forced Convection Heat
    Transfer Characteristics in Tubes with Twisted Tape Swirl
    Generators, J. Heat Transfer, Vol. 86, pp.39–49, 1964.

5. R. F. Lopina, A. E. Bergles, Heat Transfer and Pressure Drop
    in Tape Generated Swirl Flow of single-phase Water, J. Heat
    Transfer, Vol.91, pp. 434–442, 1969.

6. C. D. Grace, Static Mixing and Heat Transfer, Chemical and
    Process Engineering pp.57–59, July 1971.

7. S. T. Lin, L. T. Fan and N. Z. Azer, Augmentation of Single
    Phase Convective Heat Transfer with In-Line Static Mixers,
    Proceedings of the 1978 Heat Transfer and Fluid Mechanics
    Institute, Stanford University Press, Stanford, California,
    pp. 117–130, 1978.

8. M. A. Mihaiyefu, Fundamentals of Heat Transfer, Advanced
    Education Publishing House, Beijing, China, 1954.

9. Tianjin University, Heat Transfer, Constructional Indusrty
    Publishing House, Beijing, China, 1980.

Effect of Vortex Flow on Heat Transfer Enhancement in Heat Exchanges

TIAN YE
Department of Power Machinery Engineering
Xian Jiaotong University
Xian, PRC

ABSTRACT

Effect of vortex flow on heat transfer is studied via the naphthalene sublimation technique, local and average heat transfer in three typical heat exchangers is determined to ascertain the effect of vortex flow, heat/mass transfer visualization is made to affirm the measurements. From the results a qualitative prediction of the vortex flow's behavior is proposed.

INTRODUCTION

The performance of a heat exchanger is greatly dependent on the flow patterns of fluid flow. Although the heat transfer can be effectively increased by extending heat transfer area, we are more interested in enhancing the heat transfer coefficient. One of the main factors that influence the coefficient is the flow pattern. Vortex flow, which is usually considered very amenable to heat transfer, is a common pattern encounted in many cases, its effect has been discussed by many authors, [1-7] However, because of its complicated structure and formation, it may behave positively or negtively to heat transfer in different situations. So far it is still difficult to predict the behavior on the basis of theoritical analysis, therefore an empirical prediction, either a quantitative evaluation or a qualitative statement is urgently required to meet the need of practical applications.

In present paper we studied three typical heat exchanger models of popular applications, local mass/heat transfer was carried out to detect the vortex structure and its effect on heat transfer, a qualitative rule was extracted out from the results to judge the behavior of vortex flow. The intensity and scale of vortex is expressed with heat transfer since our concern is on heat/mass transfer, then the mechanism of vortex flow is discussed.

EXPERIMENTAL SYSTEM

The experiments were carried out with naphthalene sublimation technique.
The technique has been discussed in many papers [5-8], so it will not
be expounded here. The experimental system is illustrated in the figure
1, a sucking wind tunnel is used to keep the air from contamination
of naphthalene during tests. The test elements which were cast in a
aluminum mould were assembled in the test section, the flow rate is
measured with the flowmeter. The exchanger models are presented in the
figure 2, in Fig.2 (a) the tubes are not of naphthalene because its
area is much smaller than that of the fins, since the structure of
the exchanger is symmetric, only half of the surfaces is tested as
expressed with the hatched area in the figure. In (b) the four side-
walls are of naphthalene. The parameters are given in the figure 2.

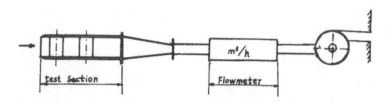

Fig.1 Experimental system

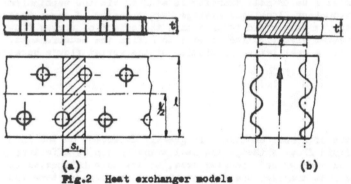

(a)   (b)

Fig.2 Heat exchanger models
(a) Plate fin and tube heat exchanger
(b) Rectangular wavy channel heat exchanger

The local mass transfer is measured by detecting the depth of sublimation
point by point at certain locations on the test elements, the procedure
was explained in reference [8], from the measured depth $s(x,y)$ the local
Sherwood number $Sh(x,y)$ may be calculated

$$Sh(x,y) = s(x,y) \cdot De \cdot \rho_n / D \cdot (\rho_n - \rho(y)) \cdot \tau \qquad (1)$$

where De is the characteristic diameter, τ is the test time and D is
the mass diffusivity which is determined as

$$D = \nu / Sc , \qquad Sc = 2.5 \qquad (2)$$

Local Nusselt number Nu(x,y) can be obtained according to the analogy between heat transfer and mass transfer

$$Nu(x,y) = Sh(x,y) \cdot (Pr/Sc)^n \qquad (3)$$

where Pr =0.7 for air, furthermore, the spanwise average heat transfer at a certain cross-section on the test element may be defined as

$$Nu_y(y) = \frac{1}{Lx} \int_0^{Lx} Nu(x,y)dx \qquad (4)$$

where the spanwise width is Lx.

RESULTS AND DISCUSSION

(a) Vortex flow in plate fin and tube heat exchanger

The local mass transfer on the fin of a plate fin and tube heat exchanger is shown in the figure 3. The local Sherwood number is very intricate at each cross-section from the inlet y/L = 0.03 to the exit y/L = 0.9, although it is difficult to analyse it theoritically, the impressive variation is regular in someway. In the region around the tubes the mass transfer is much intense than that in other regions, this phenomenon has been discussed by E.M.Sparrow and he attributed it to a vortex flow caused by flow separation in front of the tubes. Our results testify this fact.

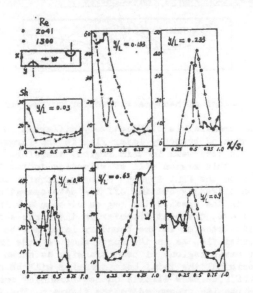

Fig.3   Local Sherwood number on the fin
of a plate fin and tube heat exchanger

There exist a horse shoe shape region around the tube, the horse shoe which is defined on heat transfer intensity corresponds to a vortex flow pattern. It is intuitively demonstrated in the visualisation of mass transfer in the figure 4 that the horse shoe is of multi-layer in structure and it dissipates with flowing down stream. Obviously, the mass transfer in the region behind the tubes is very weak comparing with that in the horse shoe even though it is also in vortex flow, this means that the effects of vortex flow on heat transfer will be different with different formation.

From above discussion we suggest a devision of the fin according to the heat/mass transfer on it, the fin is devided in to three regions, as shown in the figure 5.

Fig.4 Visualization of local mass transfer
and the horse shoe on the fin

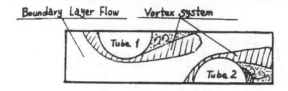

Fig.5 Regions of heat/mass transfer on the fin

The local heat/mass transfer is distinctively different in the regions, the area of the region of boundary layer flow is the largest, the mass transfer, however, is very weak in this region, therefore, to enhance the heat transfer in this region will be effective to the improvement of total heat transfer on the fin. Augment of heat transfer in boundary layer flow may be achieved by cutting off the development of boundary layer, or by reducing the area of the region of boundary layer flow. As we have discussed above, the heat transfer in the horse shoe is quite intense, so the heat transfer in the boundary layer may be effectively improved if the vortex flow as in the horse shoe is made in the boundary layer flow. If the basic regions of heat transfer as shown in the figure 5 are maintained original the total heat transfer will be changed slightly even by means of some hard efforts on heat transfer enhancement, this is proved with the results presented in the figure 6. The figure shows this variation of spanwise average heat/mass transfer at the cross-sections caused by rearrangement of the tubes. Obviously, the similar profiles of the Shy represent that the rearrangement of the tubes does not change the heat transfer perminently, the regions are kept unchanged eventhough their locations are moved with the rearrangement. Furthermore, the nadir is found to be always the mass/heat transfer in the region behind the tubes, the highest points of heat transfer are in front of the tubes, clearly, corresponding to the different formations, the vortex flows may be even opposite in their effects on heat transfer.

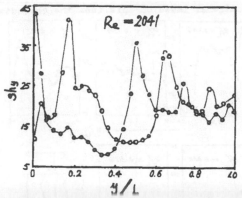

Fig.6 Spanwise average Sherwood numbers
under two arrangements of tubes

(b) Vortex flow in rectangular wavy channel

Local mass/heat transfer in a rectangular wavy channel is presented in
the figure 7, the figure shows the influence of the wavy side-walls of
the channel to the local mass transfer on the other two flat walls. The
wavy walls are arranged as a symmetric and a asymmetric configurations,
therefore the flow patterns corresponding to the two arrangements are
different from each other. The local Sherwood numbers shows that the
mass transfer in the two situations is similar not only to each other
but also to that in the plate fin and tube heat exchanger in their
mechanism, the wavy side walls function as the tube in the plate fin
and tube heat exchanger, vortex flows formed in the separation of flow
on the waves result in a rapid increase of heat/mass transfer around
the waves, the horse shoes are made with the vortex flows.

From above discussion we have affirmed that in some cases the vortex flow
flows can be very helpful to heat transfer. In spite of the complicated
structure and formation of the vortex flow, we notice that one of the
important features of the vortex flows is their axes' directions. The
results of local mass/heat transfer demonstrate that the vortex flows
will enhance the heat transfer if their axes are parallel to the surfaces,
and the vortex flows with their axes vertical to the surface may aggrevate
the heat transfer, the vortex flow of which the axes are parallel to the
surface can be formed in the stagnation flows ahead a obstacle normal
to the surface, the vortex flows in the separation flows behind the
obstacle are usually vertical to the surface.

(c) Intensity and scale of vortex flow

It is usually difficult to evaluate the features of vortex flows, the
intensity and scale of a vortex flow are the most important characteristics which are used to define the vortex flow quantitatively, however,
they can not be calculated on the theory of fluid mechanics in some cases
such as in turbulent flow. But our concern is often on heat transfer,
so we suggest that the intensity and scale of the vortex flow be

125

expressed with the heat/mass transfer, this expression has a direct
meaning of heat/mass transfer therefore it will be very convenient to
the practical applications. Based on this idea the characteristics are

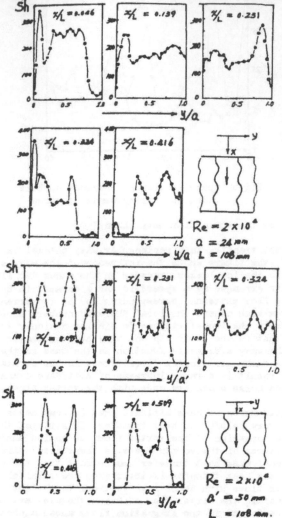

Fig.7 Local mass/heat transfer on the flat walls
of the rectangular wavy channels

presented quantitatively in the figure 8. At the center of the vortex
the intensity is the greatest, the two layer structure appears with two
peaks of Sherwood number at the two centers of the vortex. The scale
of the vortex is about 1/2 of the characteristic diameter of the obstacle.
The quantitative evaluation of the vortex's intensity and scale may be
set up on the further detail investigations on local mass/heat transfer
in the vortex flow.

126

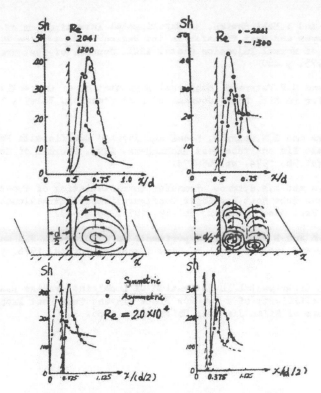

Fig.8 The intensity and scale of vortex
expressed with local mass transfer

SUMMARY

The vortex flows in three typical heat exchanger are studied, the local
mass/heat transfer in the vortex flows is measured and it proves that
the vortex flows parallel to the surface are positive to the heat/mass
transfer. The intensity and scale of the vortex flows are suggested to
be expressed with the heat/mass transfer and presented in Fig.8. Some
further investigations would be carried out to realise a quantitative
evaluation.

REFERENCES

1. W.M.Kays and A.L.London, Compact Heat Exchanger, McGraw-HillBook Co,
   2nd Ed., New York, 1958.

2. R.J.Goldstein, et al, Laminar Separation Reattarchment and Transition
   of Flow over a Down Stream Faciting Step, ASME Journal of Basic
   Engineering, Vol.92, 1970, pp.732-741

127

3. R.C.Foster and A.Hali-Sheikh, An Experimental Investigation of Boundary Layer and Heat Transfer in the Region of Separation Flow Downstream of Normal Injection Slots, ASME. Journal of Heat Transfer Vol. 97, 1975, p.260

4. J.K.Hagge and G.H.Peterson, Mechnical Augmentation of Convective Heat Transfer in Air, ASME. Journal of Heat Transfer, Vol.97, 1975, p.516

5. F.E.M.Saboya and E.M.Sparrow, Local and Average Coefficients For One-Row Plate Fin and Tube Heat Exchangers, ASME. Journal of Heat Transfer, Vol.96, 1974, pp.265-274.

6. F.E.M.Saboya and E.M.Sparrow, Transfer Characteristics of Two-Row Plate Fin and Tube Heat Exchanger Configurations, International Journal of Heat Mass Transfer, Vol.19, 1975, pp.41-49.

7. F.E.M.Saboya and E.M.Sparrow, Experiments on a Three-Row Fin and Tub Heat Exchanger, ASME. Journal of Heat Transfer, Vol.98, 1976, pp.520 522.

8. Tian Ye, An Experimental Investigation of Local and Average Heat Transfer Coefficients of a Two-Row Plate Fin and Tube Heat Exchanger Master Thesis of Xi'an Jiaotong University, 1986.

Heat Transfer from a Circular Cylinder with Tripping Wires in Cross Flow

HIDEOMI FUJITA and TADAO KAWAI
Department of Mechanical Engineering
Nagoya University
Furocho, Chikusaku, Nagoya 464, Japan

1. INTRODUCTION

The purpose of the present study is to clarify the effects of two-dimensional protuberances on the hydraulic drag and heat transfer of a cylinder in cross flow, using a cylinder to which two wires of the diameter d were attached at two symmetrical positions of an angle θ from the frontal stagnation point as shown in Fig. 1. In the preceding report [1], we showed that the pressure distribution around the cylinder changed remarkably with d and θ, the measured distributions were able to be classified into three patterns as illustrated in Fig. 4, and the drag coefficient reduced to about a half of that of the bare cylinder irrespective of the wire diameter when the distribution belonged to the second type of pattern II.
In this report the local heat transfer coefficient distributions around a cylinder measured under almost the same conditions as those used in the preceding experiment are shown, the effects of the tripping wires on the heat transfer characteristics are investigated and the relation between the heat transfer coefficient and drag coefficient is also discussed.
Similar experiments were conducted by Aiba, who attached the fine wires at θ = 65° [2], and by Johnson and Joubert, who attached the vortex generator of special shape at θ = 50° [3]. In both experiments the dimension and position of the protuberance were fixed. Igarashi and Iida examined also the influences of the Reynolds number and the height of the vortex generators installed at θ = 60° [4].

2. EXPERIMENTAL APPARATUS AND PROCEDURE

A test cylinder measuring D = 50 mm in diameter and 400 mm in length was made

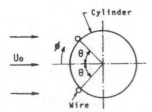

FIGURE 1. Position of tripping wires and notations.

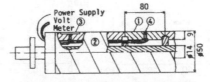

① Bakelite cylinder   ③ Electrode
② Stainless steel ribbon   ④ Thermocouple

FIGURE 2. Test cylinder to measure a heat transfer coefficient distribution.

129

by wrapping a bakelite cylinder with a stainless steel ribbon of 30 μm thick and 30 mm wide helically as shown in Fig. 2. In order to measure the surface temperature distribution around the cylinder, 73 thermocouples are installed in close contact with the back of the ribbon. The local heat transfer coefficient on the electrically heated cylinder was calculated from

$$h = q/(t_w - t_o) \tag{1}$$

under an assumption of constant heat flux, where t_w is the cylinder surface temperature, t_o is the free stream temperature and q denotes the heat flux which is obtained from the electric power supplied to the ribbon. We attempted to measure the heat loss through the bakelite cylinder by conduction at four points of $\phi = 0$, 90, 180 and 270°. The results showed that the heat transfer coefficient might be underestimated by about 2 % at $\phi = 0$ and 180°, and overestimated by about 7 % at 90 and 270°. In the present study, however, values obtained by Eq. (1) are used without any correction.

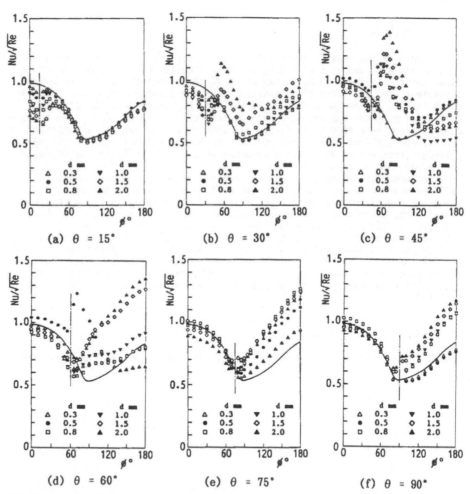

FIGURE 3. Local heat transfer coefficient distributions around the cylinders. A solid curve represents the distribution measured without tripping wires.

130

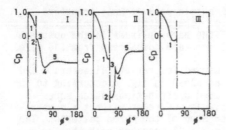

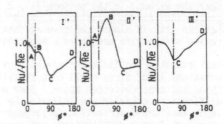

FIGURE 4. Typical patterns of
the pressure distribution.

FIGURE 5. Typical patterns of the heat
transfer coefficient distribution.

As the tripping wires, nylon ones of d = 0.3, 0.5, 0.8 and 1.0 mm and steel
ones covered with enamel of d = 1.5 and 2.0 mm were used. The wire positions
are θ = 15, 90° and every 15 degrees from 30 to 75° for d = 0.3 to 1.0 mm, and
every 15 degrees from 15 to 90° for d = 1.5 and 2.0 mm. The measurements were
made in a wind tunnel with a test section measuring 400 mm by 500 mm and at a
Reynolds number $Re = U_oD/\nu$ = 5 × 10⁴, where U_o is the free stream velocity
and ν is the kinematic viscosity, and at a heat flux q = 1730 W/m².

3. EXPERIMENTAL RESULTS AND DISCUSSION

3.1 Local Heat Transfer Coefficient Distributions

Some local heat transfer coefficient distributions around a cylinder with
tripping wires are shown in Fig. 3. In the ordinate Nu is the Nusselt number
defined as $Nu = hD/\lambda$, where λ is the thermal conductivity of air. In the
figures a chain line indicates the wire position, and a solid curve represents
the heat transfer coefficient distribution around the bare cylinder without
wires. The measured distribution changes remarkably in shape with the diameter
d and the position θ of the wire as with the pressure distribution shown in
the preceding report. The distributions of local heat transfer coefficient
were able to be classified into three typical patterns of I', II' and III'
illustrated schematically in Fig. 5, corresponding well to those of the
pressure distributions of I, II and III shown in Fig. 4, respectively.
The pattern I', which was observed when the wires were attached on the frontal
part of the cylinder, is similar on the whole to the distribution of the bare
cylinder with the exception of the local decrease in heat transfer coefficient

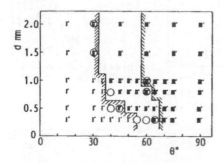

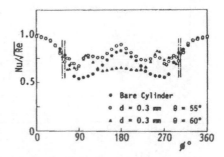

FIGURE 6. Classification of heat
transfer coefficient distribution.

FIGURE 7. Examples of peculiar
distributions.

near the wires.   In the pattern II', a remarkable peak of heat transfer
coefficient exists downstream from the wire.   The distribution belonging to the
pattern III' is almost the same as that of the bare cylinder in the upstream
region of the wire and shows a monotonical increase downstream from the wire.

Heat transfer coefficient distributions measured for various diameters d and
positions θ of wires could be classified as shown in Fig. 6 according to the
three patterns described above.   Some measured distributions which were
difficult to classify are indicated by a circle.   Circles without a number
imply that the distribution corresponded to none of the three patterns.   Two
examples of such distributions are shown in Fig. 7 together with the
distribution of the bare cylinder.   Such distributions were generally less
symmetrical.
Figure 6 is divided into three regions by hatching, which shows the regions of
pressure distribution classified into three based on the patterns shown in Fig.
4, corresponding to the pattern I, II and III from left to right.   This figure
shows clearly that there is a close relation between the fluid flow and heat
transfer.

As seen in Fig. 6, the pattern I' appeared in a wide range up to $\theta = 50°$ when
the wire diameter is as small as $d = 0.3$ mm, and its range becomes narrow as
the diameter increases.   In this pattern, as mentioned before, the heat
transfer coefficient decreases locally near the wires, which is thought due to
the existence of stable vortices formed just before and behind the wires and
becomes remarkable with increase of the wire diameter.   This situation is shown
more clearly in Fig. 8(a).   In Fig. 8 some typical distributions are plotted as
a ratio of the local Nusselt number to that measured at the same position on
the bare cylinder.
Pattern II', which has a large peak B behind the wire indicated in Fig. 5,
mainly appeared when the wire diameter was larger than the boundary layer
thickness of $0.2 \sim 0.3$ mm, and the θ range of this pattern was rather narrow
and shifted to a smaller θ value with increase of the wire diameter as seen
from Fig. 6.   The peak B is larger as the wire diameter is larger, and Fig.
8(b) shows that Nu/Nu_s at the peak was 1.5 to 2.3 for $d = 0.5$ to 2.0 mm.   It is
also noticed that the peak B exists at a position where the heat transfer
coefficient is rather low on the bare cylinder.   Therefore its contribution to

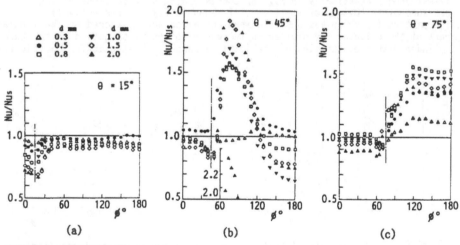

FIGURE 8.   Distributions of heat transfer coefficient non-dimensionalized by
that of the bare cylinder.

132

heat transfer enhancement is very effective, although in almost all cases Nu/Nu_* is less than unity in the rear part of the cylinder. Besides, the range of the pattern II' in Fig. 6 almost coincides with that of pattern II of the pressure distribution.

The preceding report [1] clarified that when the pressure distribution belongs to pattern II, the pressure drag of the cylinder considerably decreases. Therefore, it can be said in general that the condition in which the pressure distribution shows pattern II and the heat transfer coefficient distribution displays pattern II', is very preferable from the practical viewpoint of heat exchanger technology, because it brings about both a decrease in drag and an increase in heat transfer.

As seen from Fig. 6, pattern III' virtually corresponds to pattern III, which suggests that the boundary layer separates from the cylinder at or near the wire. In this pattern, the heat transfer is enhanced on the rear part of the cylinder as seen clearly in Fig. 8(c), and it increases with increase of the wire diameter.

From these results it can be said on the whole that pattern II' brings about a heat transfer enhancement on the sides of the cylinder and that pattern III' does so on the rear part of the cylinder.

3.2 Mean Heat Transfer Coefficient

The mean heat transfer coefficients obtained from the local heat transfer coefficent distributions such as shown in Fig. 3 are plotted against the wire position θ in Fig. 9. In the case of pattern I', the mean Nusselt number $\overline{Nu}$ is lower than that of the bare cylinder because of local decrease in Nu near the wires. In pattern II', $\overline{Nu}$ is generally higher than that of the bare cylinder because of large peaks existing downstream of the wires. However, in some cases of larger wire diameters $\overline{Nu}$ does not always increase because Nu decreases deeply on the rear part of the cylinder. In pattern III', on the other hand, heat transfer is enhanced in the wider range of the rear part of the cylinder, whereas little enhancement is found on the frontal part. Figure 9 suggests that installing wires at suitable positions brings about 20 % increase in heat transfer coefficient for both patterns of II' and III'.

In Fig. 10 the drag coefficients of the cylinder with tripping wires are shown by a ratio to that of the bare cylinder against the wire position of θ. The coefficient reduces to about a half of that of the bare cylinder when its

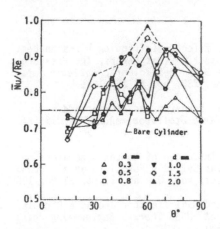

FIGURE 9. Mean Nusselt number.

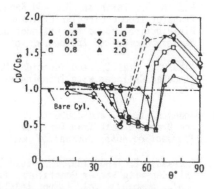

Figure 10. Drag coefficient.

133

pressure distribution belongs to pattern II. However, when it belongs to
pattern III, the drag coefficient exceeds that of the bare cylinder as shown in
Fig.10. Therefore, when we require both enhancement of the heat transfer and
reduction of the drag, it is necessary to select both d and θ which make for
the patterns II and II' in Fig. 6. As for this experiment, when the wires of d
= 0.5 mm are installed at a position between θ = 50 and 60°, the tripping
wires were most effective. Since the drag coefficient is sensitively varied
with the wire position as seen in Fig. 10, extreme care is necessary in
determining the wire position to secure a larger reduction in drag.

4. Conclusions

The heat transfer coefficient distribution around a cylinder with tripping
wires was measured in a cross flow at the Reynolds number of 5×10^4 for
various diameters and positions of the wires. The results are summarized as
follows.
(1) Heat transfer coefficient distributions, which changed remarkably with
both diameter and position of the wire, could be classified into the three
patterns of I' to III' shown in Fig. 5 like pressure distributions. When the
distributions belong to pattern I', II' and III', the pressure distributions
almost correspondingly showed pattern I, II and III.
(2) In pattern II' the heat transfer was enhanced on sides of the cylinder,
while in pattern III' it was so affected on the rear part. In both cases the
mean heat transfer coefficient increased to about 1.2 times as large as that of
the bare cylinder.
(3) When the diameter and position of the wires were selected so that
distributions of the pressure and the heat transfer coefficient may belong to
patterns II and II', respectively, the heat transfer was enhanced and the drag
was reduced. In this experiment the most effective conditions were found when
the wires of d = 0.5 mm (d/D = 0.01) were attached at a position θ between 50
and 60°.

The senior author would like to thank Professor Emeritus W. H. Giedt of the
University California at Davis for his valuable suggestions in setting up this
study. The authors are also grateful to Messrs. H. Kawamura, A. Fujishiro, S.
Nishimura and N. Shiraki of the Heat Engine Laboratory of Nagoya University for
their experimental assistance.

REFERENCES

1. Fujita, H., Takahama, H., and Kawai, T., Effects of Tripping Wires on Heat
   Transfer from a Circular Cylinder in Cross Flow (1st Report, The Pressure
   Distribution around the Cylinder and the Drag Coefficient), *Bulletin Japan
   Society Mechanical Engineers*, vol. 28, no. 235, pp. 80-87, 1985.

2. Aiba, S., Ota, T., and Tsuchida, H., Heat Transfer and Flow around a
   Circular Cylinder with Tripping-Wires, *Wärme-Stoffübertragung*, Bd. 12, Nr.
   3/4, pp. 221-231, 1979.

3. Johnson, T. R., and Joubert, P. N., The Influence of Vortex Generators on
   the Drag and Heat Transfer from a Circular Cylinder Normal to an Airstream,
   Transaction ASME, Journal of Heat Transfer, vol. 91, no. 1, pp. 91-99, 1969.

4. Igarashi, T., and Iida, Y., Fluid Flow and Heat Transfer around a Circular
   Cylinder with Vortex Generators, *Proc. ASME-JSME Thermal Engineering Joint
   Conf., Honolulu*, vol. 4, pp. 143-149, 1987.

Heat Transfer Enhancement in Narrow Gaps and Wedged Passage with Staggered Pin-Fin Arrays

WIEZAO GU, YUMING ZHANG, and CHANGCHUN LIU
Institute of Engineering Thermophysics
Chinese Academy of Sciences
P.O. Box 2706
Beijing, 100080, PRC

ABSTRACT

In this paper the test results of friction and heat transfer in the narrow gaps with staggered pin-fin arrays and in the pin-finned wedged passage with preset perforated plate and oval guide pillar rows are presented. Local heat transfer around pins in the pin-finned passages are very different. Heat transfer enhancement in the pin-finned wedged passage is higher than in the narrow gaps with pin-fin arrays and preset perforated plate can intensify heat transfer in the pin-finned wedged passage.

INTRODUCTION

In recent years extensive development has been seen in the field of research on and use of pin-finned chordwise wedged passage for enhanced cooling of gas turbine blades. Due to the effect of end wall, air flow in the pin-finned passage is three dimensional and is difficult to make analysis using numerical methods, and therefore friction and heat transfer in the blade cooling passage are determined mainly by scaled-up model tests. The geometric parameters of the cooling passage and pin-fin arrays strongly affect the flow resistance and heat transfer intensity. The pin-fin length in the blade cooling passage is determined by its width, which often varies along the chordwise direction. Pins in the channel cannot be made as slender as desired in lost wax casting. Generally, the ratio of pin length to its diameter L/d in the blade cooling passages is about 0.5-3.

In this paper experimental results are presented of friction and heat transfer in two pin-finned narrow gaps and one wedged passage with preset perforated plate and staggered pin-fin arrays.

DESCRIPTION OF TEST RIG

The test stand is shown in Fig.1. Air from the compressor passed through the cleaner, orifice plate, valve and stabilizer, then entered the test section and flowed out through the exhaust valve. Temperatures of the air stream and of the wall were measured by calibrated chromel-alumel thermocouples, the indicating instrument being a digital voltmeter of type PF15 with 1 μv resolving capability. Nickel-chromic alloy ribbon was wrapped around the test pieces as heater.

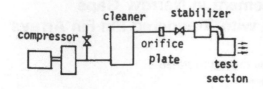

Fig.1 Sketch of test apparatus

TEST RESULTS OF PIN-FINNED NARROW GAPS

The width of narrow gaps is 2.5 mm and the diameter of pin-fin is 5 mm, so the ratio L/d equals 0.5 and the area of pin surface swept by air flow is equal to that portion of the gap walls which were covered by the pin ends.

The total area of heat transfer surface in the gap with pin-fin array is the same as the wall surface of the gap without pin-fins and the increase in heat transfer is inevitably attributed to the effect of turbulence promoters.

The pitch to diameter ratios of pin-fin rows are $S_1/d = 3.0$, $S_2/d = 5.2$ and $S_1/d = 5.2$, $S_2/d = 6.0$. Experimental data showed that the higher flow resistance and heat transfer coefficients occurred in the passage with smaller pitch to diameter ratios of the pin-fin arrays [1]. Test results are depicted in Figs.2 where ζ is the friction factor through one pin-fin row based on the velocity without pin-fins, and C is the coefficient of heat transfer correlation equation in pin-finned narrow gaps, namely

$$Nu_d = C\, Re_d^{0.8} \qquad (1)$$

Obviously, the maximum heat transfer appeared just behind the pin fins and is about 120% higher than in smooth channels and the average heat transfer can be expressed

$$\overline{Nu}_d = 0.041\, Re^{0.8}\, Pr^{0.4} \qquad (2)$$

PIN-FINNED WEDGED PASSAGE WITH PRESET PERFORATED PLATE

The geometric dimensions and thermocouple numbers of the test piece made of stainless steel are shown in Fig.3.

Air from the radial inlet channel passed through the perforated plate and was forced into the pin-fin wedged passage with a dip angle of 3°52' and initial width 18.6 mm. 9 mm away from the plate was a guide pillar row with the oval cross section and then were 8 pin-fin rows. A nickel-chromel electric heater with insulating material was attached to the side wall and inlet channel of the test piece. The top and bottom

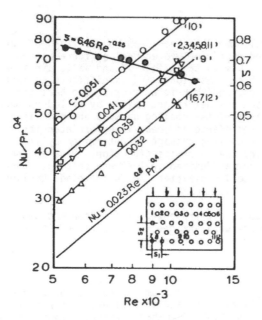

Fig.2 Test results of pin-finned passage $S_1/d = 3.0$, $S_2/d = 5.2$

136

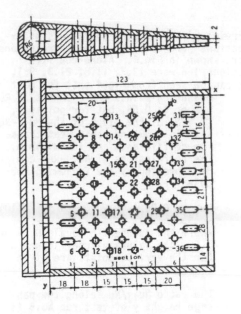

Fig.3 Geometric dimensions and thermocouple numbers of passage

of the passage were not heated as the gas turbine vanes.

The flow rate G of cooling air was measured with a calibrated double orifice plate. Heat received by air in the test section can be calculated

$$Q_t = GC_p(T_{out} - T_{in}) \qquad (3)$$

In all the 5 test runs Q_t value deviated from the electric heater power Q_e by less than 5% due to good insulation. The flow areas and wetted perimeters at six measuring sections were very different because of wedged cross section and different pin numbers, and they can be calculated by the following equations for section i:

$$A_i = L_i(H - n_id), \qquad (4)$$

$$U_i = 2(L_i + H - n_id + n_iL_i)$$

where H-passage height equal to 126 mm, n_i-pin number at section i, d-pin diameter and L_i-pin length, calculated by the expression

$$L_i = 18.6 - 2X\tan3^\circ52' \quad \text{mm} \qquad (5)$$

where X was distance from the perforated plate. The hydraulic diameter on section i of the passage was defined as $D_{hi}=4A_i/U_i$.

The average air temperature over section i T_{fi} was obtained from the ratio of local heating length L_{ti} and total heating length L_t, namely

$$\overline{T}_{fi} = T_{in} + (T_{out} - T_{in})\, L_{ti}/L_t \qquad (6)$$

The heat transfer coefficient and Nu_{ij} at point j of section i was determined by the following correlations

$$\alpha_{ij} = \overline{q} / (T_{wij} - \overline{T}_{fi}), \qquad Nu_{ij} = \alpha_{ij}\, D_{hi} / \lambda_i \qquad (7)$$

The average air velocity $\overline{w}_i$ and Re_i were defined as follows

$$\overline{w}_i = G / A_i\rho_i \qquad , \qquad Re_i = \rho_i\overline{w}_iD_{hi} / \mu_i \qquad (8)$$

The air flow resistance in the test passage was measured by the pressure difference from the pressure taps at the inlet and exit guide ducts. The general friction factor ζ can be determined by the following equation

$$\Delta P = \zeta\overline{\rho}\overline{w}^2 / 2 \qquad (9)$$

where $\overline{\rho}$ was the average air density and $\overline{w}$ was the mean value of air velocities at the 10 wedged passage sections with pin rows.

RESULTS AND ANALYSIS

5 test runs were conducted with varying air flow rates and 5 sets of experimental data for 6 measuring sections were obtained. The correlations $Nu_{ij}-Re_i$ for 36 measuring points are shown in Fig.4. Along the flow direction the values of D_{hi} at sections 1-6 were 12.7, 11.8, 21.3, 9.1, 14.3 and 9.4 mm, respectively. It can be seen from Fig.4 that for most measuring points the exponent of Re number is 0.59, but for points 1, 19, 25, 31 the slope of the $Nu_{ij}-Re_i$ line is 0.50 and their heat transfer coefficients are the minimum. The correlations of $Nu_{ij}-Re_i$ can be expressed by the following formula

$$Nu_{ij} = C\, Re_i^n \qquad (10)$$

where the exponent n equals 0.59 or 0.50, the values of coefficient C for the various points are given in Fig.4.

The ratio Nu_{ij}/Nu_0 along the passage height y of test run No.5 is shown in Fig.5, where Nu_0 is obtained by Dittus-Boelter's equation

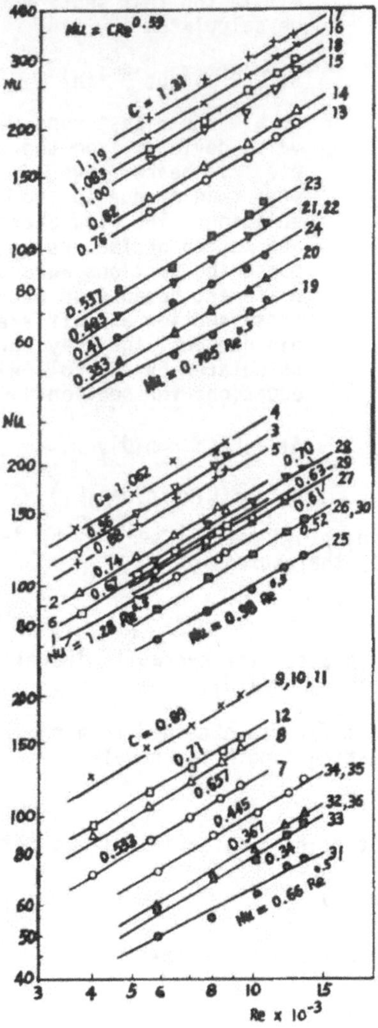

Fig.4 Test data of measuring points

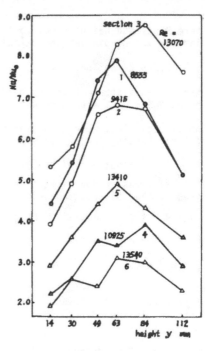

Fig.5 Ratio Nu/Nu_0 at measuring points

for air flow, namely

$$Nu_o = 0.02 \, Re^{0.8} \tag{11}$$

It must be remembered that the values of D_{hi} are very different for measuring sections, and therefore the significant difference between Nu_{ij} values for different sections does not mean the actual difference of heat transfer coefficients. In Fig.6 are shown the heat transfer coefficients α_{ij} at 36 points for 6 sections. The maximum heat transfer appears in the middle part of section 2 from point 9 to 11 and it does not coincide with the distribution of Nusselt's number shown in Fig.5 due to the influence of D_{hi} values.

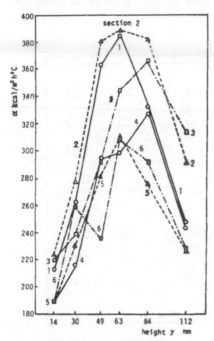

Fig.6 Distribution of heat transfer coefficients

In rectangular passages with staggered pin-fin arrays heat transfer in the first pin-fin row is the lowest and the maximum appears in 3rd to 5th pin-fin rows. Then, heat transfer slowly diminishes along with the increase in the number of pin-fin rows [2,3] . In wedged passages the length of first pin-fin row is larger than of others. This results in not only the larger heat transfer surface area, but also the minimum retarding effect of the end walls on the air flow turbulence. Therefore, the highest heat transfer intensity is obtained in the first pin-fin row region, and then heat transfer gradually diminishes along with the increase in pin-fin row numbers [4,5]. In this paper the first measuring section is in the region where the oval pins are located and the distrubance of air flow is weaker than at the second section. Thus the heat transfer at the second measuring section is higher than at other sections.

If the Nusselts number and Reynolds number are based on the pin diameter d,i.e.,

$$Re_{di} = \rho w_i d / \mu_i \, , \quad Nu_{dij} = \alpha_{ij} d / \lambda_i \tag{12}$$

the difference of Nu_{ij} values caused by D_{hi} can be avoided. For this reason the average heat transfer intensity in the test passage can be expressed

$$\overline{Nu}_{dij} = 0.53 \, \overline{Re}_{di}^{0.57} \tag{13}$$

The values $\overline{Nu}_d$ calculated by Eq.(13) are much higher than those from reference 2 due to the longer pin-fins in the wedged passage, and they are 16% higher than from reference 5 due to the jet flow caused by the perforated plate.

139

The flow friction factor can be correlated as follows

$$\zeta = 54.6\ \overline{Re}_d^{-0.16} \tag{14}$$

In Fig.7 experimental data of test run No.5 reduced as Nu_{dij}-R_{di} relation
are illustrated. The value of the mean velocity $\overline{w}_i$ is varied at different measuring sections and the values Re_{di} from section 1 through section
6 are 4040, 4770, 3680, 7180, 5620 and 8620, respectively. Nu_{dij} values
on the upper side of the passage are minimum. Along the height of the
passage Nu_{dij} gradually increases and reaches maximum at the middle part,
then gradually decreases. Due to the higher air velocity in the lower
part compared with that in the upper part from the flow visualization,
heat transfer in the lower part of the passage is lower than in other
parts, but the difference of Nu_{dij} values is not as large as shown in
Fig.5.

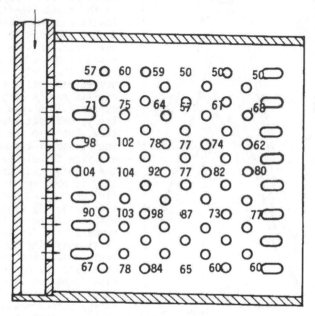

Fig.7 Experimental data of Nu_{dij}-Re_{di}

CONCLUSIONS

1. Pin-fins in flow gaps and wedged channels are very effective for the
enhancement of gas turbine blade cooling, and local heat transfer coefficients around pins in the pin-fin wedged passage or narrow gaps are very
different.

2. Heat transfer augmentation in the pin-fin wedged passage is higher
than in narrow gaps with pin-fin arrays, and preset perforated plate can
intensify heat transfer in the pin-finned wedged passage.

3. Flow resistance in the wedged pin-fin passage with preset perforate
plate is much higher than in pin-finned narrow gaps.

ACKNOWLEDGEMENT

This work was sponsored by the Chinese National Natural Science Foundation under grant No. 5860031. Their support is gratefully acknowledged.

REFERENCE

1. Gu Weizao, Xu Hongkun, Zhang Yuming, Heat transfer in narrow gaps and its augmentation with pillar rows, J. Engineering Thermophysics, Vol.4, No.2 (1983).

2. D.E. Metzger, R.A. Berry, J.P. Bronson, Developing heat transfer in rectangular ducts with staggered arrays of short pin fins, J. Heat Transfer, Vol.104, 700 (1982).

3. S.C. Arora, W.A. Messeh, Heat transfer experiments in high aspect ratio rectangular channel with epoxied short pin fins, ASME paper 83-GT-57 (1983).

4. F.E. Faulkner, Analytical investigation of chord size and cooling methods on turbine blade cooling requirements, NASA CR-120882-BK-1, 1971.

5. A. Brown, B. Mandijkas, J.M. Mudyiwa, Blade trailing edge heat transfer, ASME paper 80-GT-45 (1980).

An Experiment Study of the Heat Transfer
of the Horizontal Elliptical Tubes with the Fluidized Bed

JIDONG LU, JINBO WANG, RENZHANG QIAN, and WENDI HUANG
Department of Power Engineering
Huazhong University of Science and Technology
Wuhan, Hubei, PRC

ABSTRACT

The experiment results of the heat transfer of the immersed horizontal
elliptical tubes with the gas-solid fluidized bed are presented in this
paper. The ratio of two axes of the tubes are 3.26, 2.44, 1.75 and 1.00
respectively. The diameters of the sand particles are 0.50, 0.72.0.968
and 1.538 mm. An experiential relation is correlated from the experiment
results. It is shown in the results that the heat transfer coefficients
of the tubes with elliptical section are higher obviously than that of
circular tubes, which has some advantage influence on the optimum design
of the heat exchanger in the fluidized beds.

INTRODUCTION

It is well known that one of the most important characteristics of the
fluidized beds is the high heat transfer rate between the immersed surfaces and the beds. How to get further rate higher is very important for
the optimum design of the fluidized beds with heat exchanger, the improvement of the gas-particle two phase flow in the bed and decrease of
the equipment costs.

According to the previous measurements to the local heat transfer coefficient on the immersed horizontal circular tubes with the fluidized
beds [1], it is seemed, in generally, the heat transfer rate in the upstream and downstream regions are lower than these in the side regions.
It is determined by the characteristic of the gas-solid flow adjacent to
the immersed tubes. A lean phase bubble with few particles stays in the
upstream region and an unfluidized packet is formed in the downstream
region. The frequency of particle replacement in these regions with the
bed is rather low, especially near the minimum fluidization state. But
the side regions are touched with particle packets and gas bubbles in
higher frequency. The reason of the high heat transfer rate of immersed
tubes in the fluidized beds is just the continuous replacement of the
particle packets adjacent to the heat transfer tubes. Thus it can be concluded that the heat transfer can be further enhanced if the immersed
tube section is changed to increase the strong transfer part and decrease the weak part.

Kurochkin [2] investigated experimentally the influence of the tube profile in the total heat transfer coefficients between electrical heated horizontal tubes of circular, elliptical and lenticular section and flowing packed beds of quartz sand. As the tube profile varied from circular to lenticular, the value of h_w increased. But he only considered the situation of the major to minor axis ratio being 1.45 for the elliptical tube and the correlation is only valid for dry granular materials of mean size from 0.16 to 0.30 mm.

So a further investigation to the heat transfer between elliptical tubes with different ellipticity and the fluidized bed with different particle sizes is carried out and are relatively rational experiential formula is correlated for the design of the heat exchangers in the fluidized beds.

EXPERIMENT APPARATUS AND PROCEDURE

The experiment are carried out on a fluidized bed with the inner section of 307×215 mm at the ambient temperature and pressure. The fluidizing fluid is compressed by an air blower and adjusted by a gate valve. The gas distributor is a multiorifice plate with the open area fraction of 3.2%. The heat transfer test tube is set horizontally at the axis of the bed 310 mm from the distributor. The major axis of the tube section is in accordance to the direction of the fluid flow. The bed material are sand particles with the area averaged diameter of 0.5, 0.72, 0.986 and 1.538 mm. The static bed height is maintained at about 350 mm.

The geometrical dimensions of the heat transfer tube are shown in Tab. 1. Four copple-constantan thermoelectric couple are immersed in the tube surface to measure the surface temperature. The inner heater is made with the nicklechrome wire. The gap between the tube wall and heater is filled with magnesia to give the electric isolation and decrease the heat resistance. The two ends of the tube are wrapped with asbestos for the thermal isolation, so the end loss can be omitted.

The velocity of the fluidizing fluid is measured with a pitot in the front of the valve. A U-shape tube is used to measure the bed differential pressure. A thermoelectric couple for the bed temperature is set in the center of the bed 190 mm above the distributor. After the temperature of the tube surface, bed temperature, the electric current and voltage for the heater being measured in the experiment, the average heat transfer coefficient can be calculated with the following equation,

$$h_w = Q/[A_w(T_w - T_b)] \tag{1}$$

where T_w is the average temperature of the tube surface. The maximum error of the experiment data is not larger than 5%.

TABLE. 1 The Dimension of the Heat Transfer Tubes

Tube number	Symbol in figures	2a(mm)	2b(mm)	a/b
1	o	35.55	10.90	3.26
2	Δ	33.80	13.85	2.44
3	●	31.95	18.85	1.75
4	▲	25.00	25.00	1.00

EXPERIMENT RESULTS AND DISCUSSION

The experiment results of the average heat transfer coefficients are
shown in Figs. 1-8. Taking the ratio of major to minor axis of the heat
transfer tubes as the parameter, Fig. 1-4 compare the effects of the
various profile and the dimension of the four different kind of particles. It can be seen from the figures that the heat transfer coefficients
increase with the major to minor axis ratio in the experiment range
$(1 \leq a/b \leq 3.26)$. For instance, the heat transfer coefficient of the elliptical tube with $a/b=3.26$ is 25% higher than that of the circular tube,
which is very advantageous to the heat transfer enhancement. The reason
is the larger a/b is, the smaller the unfavorable affects of the lean
phase region and the un-fluidized region. Figs. 5-8 show the relation of
the heat transfer coefficients h_w with the superficial gas velocity for
each section, in which the particle dimension is taken as the parameter.
It is shown in figures that the heat transfer of the elliptical tubes
with the bed is decreased with the increasing particle dimension, which
is the same trend as the circular tubes. When the bed particle diameter
is relatively large, the variation of h_w with U is not obvious.

Many authors have given a series of experiential correlations to describe
the effects of various factors on the heat transfer coefficients of the
immersed tubes with the fluidized beds. From the view of heat transfer
enhancement we hope to obtain the correlation for the maximum heat transfer coefficients. It can be found in many calculating formulae for the
circular tubes given in [3],[4] that the correlations in the form of
dimensionless parameter Nu and Ar are very simple. But they are agreement
with the experiment data in a rather large region. One of the useful experiment experiential correlations for the horizontal circular tubes is
Zabrodsky's,

$$Nu_{pmax}=0.88Ar^{0.213} \tag{2}$$

The heat transfer of the horizontal elliptical tubes has some similari-
ties with circular tubes. So the relation of the dimensionless parameters
Nu_{pmax} and Ar used in this paper to correlate the experiment results.
The ratio of major to minor axis of the immersed tubes is also introduced
to consider the effect of the tube profile. Thus the assumed formula is,

$$Nu_{pmax}=nAr^{m}(a/b)^{n} \tag{3}$$

The dual linear regression is carried out for the test data. The experiment correlation for the heat transfer of the immersed horizontal elliptical tubes can be obtained as following,

$$Nu_{pmax}=0.278 \, Ar^{0.266} \, (a/b)^{0.19} \tag{4}$$

the suitable applied range of this formula is $1 \leq a/b \leq 3.26$ and $7000 < Ar <$
230,000. The comparison of the correlation curve and test results are
shown in Fig.9.

CONCLUSION

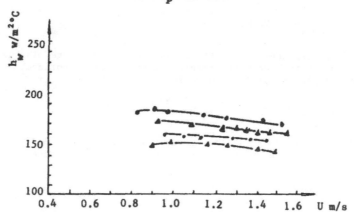

FIGURE. 1 The variation of h_w versus U
for d_p=0.5 mm.

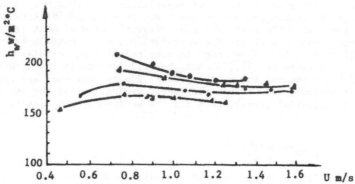

FIGURE. 2 The variation of h_w versus U
for d_p=0.27 mm.

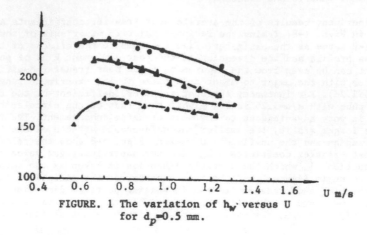

FIGURE.3 The variation of h_w versus U
for d_p=0.968 mm.

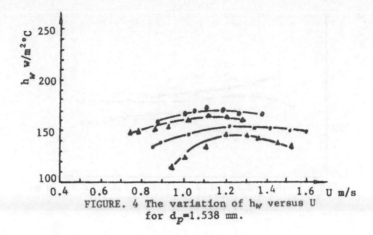

FIGURE. 4 The variation of h_w versus U
for d_p=1.538 mm.

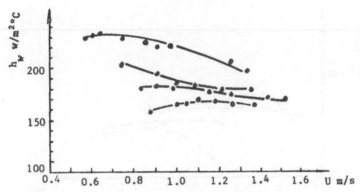

FIGURE. 5 The variation of h_w versus U
for a/b=3.26.

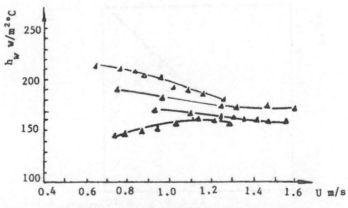

FIGURE. 6 The variation of h_w versus U
for a/b=2.44.

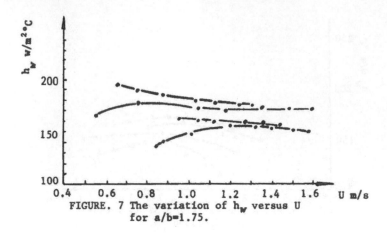

FIGURE. 7 The variation of h_w versus U
for a/b=1.75.

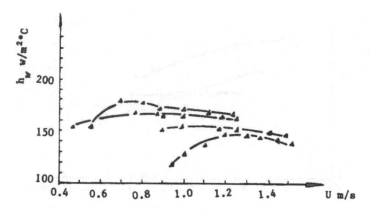

FIGURE. 8 The variation of h_w versus U
for a/b=1.00.

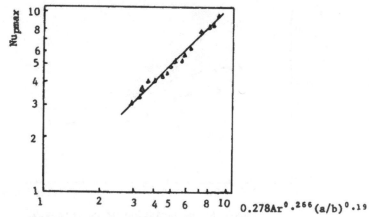

FIGURE. 9 The comparison of correlation
and experiment data.

148

1. The change of the profile of the heat transfer surface can improve the heat transfer in a certain extension. The larger the ratio of the two axes is, the larger the increase is.
2. In the normal fluidization state, the heat transfer coefficients don't change obviously with the fluidizing fluid velocity.
3. The correlation obtained from the experiment results can be used for the design of the heat exchangers with the elliptical tube in the fluidized beds.

NOMENCLATURE

a	half of the major axis length;
Ar	Archimedes number, $(\rho_f(\rho_s-\rho_f)gd_p^3/\mu_f^2)$;
b	half of the minor axis length;
h_w	average heat transfer coefficient;
k_f	heat conductivity of the fluid;
Nu	Nusselt number, $h_w d_p/k_f$;
U	superficial velocity;
μ_f	fluid dynamic viscosity;
ρ_f	fluid density;
ρ_s	solid density.

REFERENCES

1. Gelperin N.I. and Einstein V. G. Heat Transfer in Fluidized Beds, in Fluidization, Eds. J.F. Davidson and D. Harrison, Academic, New Yark, pp. 471-540, 1987.
2. Kurochkin, Yu.p. Heat Transfer between Tubes of Various Profiles and a Flow of Granular Material, J. Eng. Phys. V. 10, p.549, 1966.
3. Saxena, S. C. Grewal, N. S, Tabor J. D, Heat Transfer between a Gas Fluidized Bed and Immersed Tubes, Adv. Heat Transfer, V. 14, pp. 149-247, 1978.
4. Mohammed Shan M, Generalized Prediction of Maximum Heat Transfer to Single Cylinder and Spheres in Gas Fluidized Bed, Heat Transfer Eng. V.4, p.104, 1983.

Fluid Dynamic Features and Heat Transfer Enhancement of NTIC–DDB Shell and Tube Heat Exchangers

SONGWEN QIAN, WENMING ZEN, XIAOMING MA, and JUNSHENG ZHAO
Department of Chemical Engineering Machinery
South China University of Technology
Guangzhou 510641, PRC

ABSTRACT

This paper presents an experimental study on fluid dynamic properties and heat transfer for No Tube In Centre Disk-Doughnut Baffle(NTIC-DDB) heat exchanger. Results are compared with those of other types of exchangers. It is found that it has the properties of high heat transfer coefficient, low pressure drop, and good anti-vibration performance.

INTRODUCTION

According to flow path analysis, there exist dead-zone and leakage flow in cross flow heat exchangers, such as segmental baffle and disk-doughnut baffle exchangers, so that heat transfer area can not be made of fullest use. At baffle cut zone or flow turn region, tubes are often failed due to flow induced vibration. To improve performance of these exchangers, two newly developed shellside structures are introduced, i.e., No Tube In Window(NTIW), and NTIC tube banks [1] [3]. Experimental studies are carried out to guide practical design.

TEST ON HEAT TRANSFER AND PRESSURE DROP

To measure heat transfer coefficient and pressure drop, a test unit is designed(Fig.1). Results of shellside α_o and ΔP are related with Re in the following semi-empirical expressions:

$$\alpha_o = A \cdot 10^{-2} (\lambda/d_e) \, (P_r)^{0.3-0.4} \cdot (Re)^m \tag{1}$$

$$\Delta P = B \cdot N_b \cdot 10^{4} (Re)^n \, (\mu^7/(\rho d_e^2)) \tag{2}$$

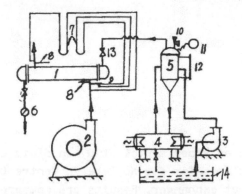

FIGURE 1(a)   A NTIC-DDB tube bank

FIGURE 1(b) Test flow chart (shellside fluid is air)
1-NTIC-DDB heat exchanger; 2-blower; 3-pump; 4-electronic
boiler; 5-steam trap; 7-U gauge(filled with water or mercury);
8-thermometer; 9-pitot tube; 10-relief valve; 11-barometer;
12-lever indicator; 13-valve; 14-pool.

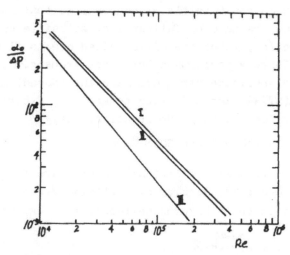

FIGURE 2. Comparison of the overall performance of three types
of heat exchangers(NTIC-DDB, RODbaffle and segmental baffle).

TABLE 1. Test results(shell side fluid is air)

test no.	1	2	3	4	5	6
$Re \times 10^{-4}$	2.145	2.680	3.284	3.719	4.339	4.498
$\alpha_{\circ}$ ($W/m^2 \cdot °C$)	67.06	83.46	92.55	102.5	121.1	118.2
ΔP (Pa)	3201	3867	6668	7467	10935	11069
$\alpha_{\circ}/\Delta P \times 10^2$ ($W/m^2 °C$)/(Pa)	2.095	2.158	1.388	1.373	1.107	1.068

TABLE 2. Part of data for a field NTIC-DDB exchanger

	shell side	tube side
mass flow rate(kg/hr.)	52106	59139
inlet temperature (°C)	155	415
outlet temperature (°C)	361	260
density (kg/m^3)	7.46	1.98
viscosity ($\mu \cdot 10^5$, Pa·s)	2.78	2.97
specific heat (kcal/kg·°C)	0.252	0.284
conductivity (kcal/m·hr·°C)	0.0381	0.0436
maximum allowed pressure drop (kgf/cm²)	0.075	0.135

where, A,B,m,n are constants depend on geometries and structures. They are discussed in detail in reference [4] The overall performance parameter ($\alpha/\Delta P \sim Re$) are compared with rodbaffle and segmental baffles exchangers. It is seen from Fig.2 that $\alpha/\Delta P$ of NTIC-DDB exchanger is very close to that of rodbaffle heat exchanger($\frac{\alpha}{\Delta P}$ of rodbaffle H.E. is 1.33 times that of segmental baffle H.E., and $\alpha/\Delta P$ of rodbaffle H.E. is 50% higher than that of single segmental baffle H.E..) and is superior to that of segmental baffle heat exchanger(SBHE). Though there is no tube location in the central region, shell diameter of NTIC-DDB H.E. may remain unchanged owing to shell side heat transfer enhancement, and at the same time, the anti-vibration property is improved greatly.

A NTIC-DDB H.E. used in the production of phosphonitrogen is selected to calculate shellside heat transfer coefficient and pressure drop. The processing data are listed in table(3) and the geometric parameters are as following: baffle spacing L is o.6m; baffle number N_b is 12; the totle tube number N_t is 630; tube outside diameter is 31.8mm. with tube wall thickness of 2 mm.; the inner and outer diameters of doughnut are 0.59 and 1.494 m.; the diameter of disk is 1.35m.; and the structure constants A,B,m,and n are 5.82 , 2.24, 0.74 and 1.05 respectively. The calculation results show that α_o =184.78kcal/(m^2·hr·℃) and the overall coefficient K=77.11 kcal/(m^2hr.℃)_, which agree with the designed values. The total shellside pressure drop is 0.0731, which is less than the allowed value 0.075 kgf/cm^2.

From this example, we may conclude that the shellside pressure drop is low and the heat transfer coefficient is high, so that the overall coefficient for gas-to-gas type heat transfer may be high, and if the enforced tubes (eg., screw-grooved tube or finned tubes) are used, the overall coefficient may even higher.

FLUID EXCITATION RESPONSE

The flow distribution in the channel may be taken as parabolical and approximated by sine function:

$$U_g(x) = V_g \cdot \sin(\pi x/1) \tag{3}$$

where, V_g is the maximum gap velocity in the bank.

With the consideration of fluid force, fluid damping force and fluidelastic force per unit length of tube, tube equation of motion is written as:

$$EJ\frac{\partial^2 y}{\partial x^4} + m\frac{\partial^2 y}{\partial t^2} + c\frac{\partial y}{\partial t} + K_f y = f(x,t) \tag{4}$$

For simply supported boundary condition, tube response takes the form:

$$y(x,t) = \sum_{i=1} \sin(i\pi x/l_e)\frac{KA_1 \rho V_g^2 \sin(\omega t - \psi_i)}{4mg\sqrt{(\omega_i^2 - \omega^2) + (2\zeta\omega_i\omega)^2}} \tag{5}$$

$$\psi_i = \arctan((2\zeta\omega_i\omega/(\omega_i^2 - \omega^2))$$

where, $V_g = (p/(p-d_0))\cdot(\pi/2)\overline{V}$, $\omega_i^2 = (i\pi/1)(EJ/m)$, $2\zeta\omega_i = c/m$ and $\omega_k^2 = K_f/m$. When the fluid excitation frequency coincides with the fundamental frequency of tube, maximum displacement occurs:

$$y_{1max} = (KA_1\rho V_g^2)/(8mg\zeta\omega_1^2) \tag{6}$$

Critical flow velocity is usually defined as that at which tube response amplitude increases abruptly. If the amplitude is confined within a coefficient of β, i.e., $y_{1max} = \beta A_1$, the above equation may be further reduced into

$$U_g/(f_n d_0) = K_c (4\pi \beta \zeta mg/(\rho d_0^2))^{\frac{1}{2}} \tag{7}$$

Note that for lightly damped fluid, $\delta = 2\pi\zeta$, therefore eq.(7) is in fact of the same form as Connors' formula.

TUBE RESPONSE TEST

A NTIC-DDB H.E. test apparatus is shown in Fig.(3). The shellside fluid is water with maximum flux of 80m³/hr. and the tube side fluid is still air or water. The tube bank contains 54 brass and copper tubes with outer diameter of 14 mm. and tube wall thickness of 1.5 mm.. Tubes are arranged in equilateral-triangular patten with tube pitch of 20 mm.. The diameter of disk baffle is 180 mm. and the outer and inner diameters of doughnut baffle are 196 and 70 mm.. Measurements are taken for tubes in different locations.

Parts of the results are shown in Fig.(4). It is clear that the response is concerned with tube location: curve (1) represents mid-span amplitude of the second row at the inlet region,

FIGURE 3. Test Unit

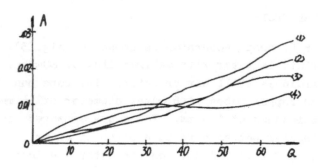

FIGURE 4. Tube response to liquid flow(A-mm., Q-m³/hr.).

and of the four curves, this is the largest; curve ⟨4⟩ indicates response of peripheral tubes in the DDB tube bank, and the amplitude is relatively small mainly for fluid velocity in this area is lower; curve ⟨3⟩ stands for the amplitude of central tubes and curve ⟨2⟩ displays the response in the outlet zone. All curves are wavy for tubes undergo flutter in unsteady flow field.

CONCLUSION

For NTIC-DDB heat exchangers, the pressure drop is low, and its heat transfer coefficient remains high in spite of the reduction in heat transfer area. Moreover, its anti-vibration function is satisfactorily good, so we suggest it should be used in the replacement of traditional segmental baffle shell and tube heat exchanger when possible.

NOMENCLATURE

A, A_i = constant, ith mode amplitude, (m.)
B = constant, dimensionless
c = damping constant, (N.s/m)
d = diameter, (m.) or (mm.)
E = modulus of elasticity of tube material, (Pa)
f = fluid distribution force, (N/m)
f_n = natural frequency of tube, (Hz)
g = gravitational acceleration constant, (=9.8m/s^2)
J = moment of inertia of tube cross section, (m^4)
K = overall heat transfer coefficient, (kcal/m^2·hr·℃)
l = distance between adjacent baffles, (m) or (mm)
m = constant; tube mass per unit length, (kg/m)
n = constant, dimensionless
N_b, N_t = total baffles and tube numbers, respectively
p = tube pitch, (m or mm)
ΔP = pressure drop, (Pa or kgf/m^2)
U, V = velocity, (m/s)
y, x = displacement of tube, coordinate, (m)
α_0 = shellside heat transfer coefficient, (see K)
ψ = phase lag, (degree)
ρ = fluid density, (kg/m^3)
λ = thermal conductivity, (kcal/m·hr·℃)

u = viscosity, (Pa·s)

δ = logarithmic decrement of damping, (dimensionless)

ω = modal natural frequency, (1/s)

REFERENCES

[1] Qian Songwen et al, Test on Heat Transfer and Flow Induced Vibration of NTIW Heat Exchanger, Petrochemical Equipment (in Chinese), pp17-20, Apr.1985.

[2] Zen Wenming et al, Experimental Study of the Shell Side Heat Transfer Enhancement of RODbaffle Heat Exchanger, Petroleum Refining and Chemical Machinery(in Chinese), vol.12, no.2, pp1-6, Feb.1983.

[3] Qian Songwen et.al.,Exploration and Experimentation on Newly Developed Anti-vibration and High Efficiency Heat Exchangers, Chemical Equipment Design(in Chinese), no.1, pp1-4, Mar.1984.

[4] Zen Wenming et. al., Design and Experimental report of NTIC-DDB Heat Exchanger, Part 1 and Part 2, South China University of Technology, July 1983(unpublished, in Chinese).

Enhanced Heat Transfer Mechanism Using Turbulence Promoters in Rectangular Duct

HISASHI MIYASHITA, KOJI FUKUSHIMA, MASAHIRO KOMETANI,
and SHINKICHI YAMAGUCHI
Department of Chemical Engineering
Toyama University
Gofuku, Toyama, 930, Japan

INTRODUCTION

In order to improve the heat transfer from a wall surface, it is well known to occur the turbulence by use of turbulence promoter having regular geometric roughness element on the transfer surface in a duct. The enhanced heat transfer is accompanied by the increase in the resistance to fluid flow. These problems on the best heat transfer performance for a given flow friction have been studied by many investigators[1,2,3,4,5]. The evaluation of heat transfer performance should be done in practical application, from the view point of efficient usage of energies and conservation resources. A few investigations on an enhanced heat transfer mechanism have presented. Mori et al.[1] and Fujita et al.[2] suggested that enhanced heat transfer depend mainly on turbulence intensity near the wall surface in down stream for single cylinder type turbulence promoter placing on transfer wall in rectangular duct. Miyashita et al.[6] pointed out that the enhancement depends on turbulence intensity near the wall surface in case of no clearance between the promoter and the wall, and depends not on turbulence intensity but also on shear stress at the wall in some clearances in rectanglar duct. In previous paper[7], wall shear stress and turbulence intensity were measured by the electrochamical methods and enhancement of heat transfer were explained by corresponding them with the flow patterns by visualization.

In this paper, in order to check the mechanism of enhanced heat transfer, the enhancement ratio of local and average mass transfer coefficients were measured by varying a kind of promoter, the pitch of the promoter and the clearance between the promoter and the wall surface, accompaning with flow behaviors by visualization. In order to examine the evaluation of heat transfer performance, the friction factors for flow resistance were also measured by the pressure drop in a rectangular duct. The heat transfer performance was obtained by the method of Sano et al.[8] considering same pumping power.

EXPERIMENTAL APPARATUS AND PROCEDERE

A schematic diagram of the experimental apparatus and a detail of the test section on the wall surface are shown in Fig.1 and Fig.2, respectively. The dimensions of a duct are height 50 mm, width 60 mm, length 4120 mm. The test section followed 61 hydraulic diameters in length calming section of hard vinyl chloride to allow the hydraulically fully developed flow at the mass transfer section. Following the entrance region, to obtain the fully developed condition of mass transfer, the cathode had a

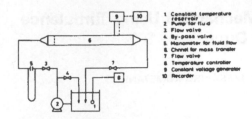

FIGURE 1. Schematic diagram of experimental apparatus.

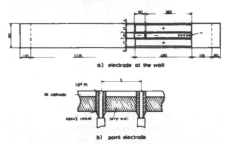

a) electrode at the wall

b) point electrode

FIGURE 2. Details of test section.

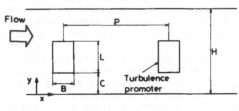

FIGURE 3. Coordinates and notations of test section.

mass transfer developing region of 10x90 mm. Next, cathode (10x360 mm) for measurment of average mass transfer coefficients was located on the transfer wall. Further, 1.0 mm platinium point electrodes (30 points) for the measurement of the local mass transfer coefficients, wall shear stress and mass transfer intensity were arranged at interval of 5 mm on the nickel cathode. Anode were located on the both side of cathode having area of (17x450) mm^2 x2. each electrodes were isolated electrolically by epoxy resin.

The fluid (electrolyte solution) was kept at a constant temperature by the controller in the reservoir and pumped to the test section through the entrance region, then returned to the constant temperature reservoir. The flow rate was measured by an orifice meter and adjusted by flow control valve.

Experiment was carried out varying the diameter of turbulence promoter, Dp (3,5,7 and 10 mm) for cylinder type turbulence promoter, L (5 and 10 mm) for rectangular type and flow Reynolds number Re (5.1x10^3 $\sim$ 1.52x10^4). In experiment for electrochemical method, 0.005 M potassium ferri/ferro cyanide and 2 M sodium hydrooxide used as electrolyte solution. It was controlled within 303 ± 0.5 K. The dencity and viscosity of the electrolyte were 1075 kg/m^3 and 0.0013 Pa·s (N·s/m^2) respectively. Diffusivity for ferricyanide ion was 5.776x10^{-10} m^2/s given by Michell and Hanratty's correlation[9], and the Schmidt number was equal to 2097. In experiment for visualization, water suspending alminum-powder used as fluid[7]. The experimental conditions were similar to them for the measurement of mass transfer. The coordinate & notation of test section were shown in Fig.3.

TRANSPORT PHYSICAL FACTORS

The mass transfer coefficients, shear stress and fluid velocity were measured by using the potassium ferri-ferrocyanide redox electrochemical reaction as reviewed in detail by Mizushina [10].

The mass transfer coefficient is given by Eq.(1)

$$k = i/(n_e \cdot F \cdot A \cdot C_b) \tag{1}$$

The wall shear stress can be calculated from following equation in the

160

case of circular surface.

$$\tau = 3.55 \times 10^{-15}(\mu \cdot i)/(D^2 \cdot C_b{}^2 \cdot d^5) \qquad (2)$$

Mass transfer intensity is defined by following equation.

$$I = 100\sqrt{\overline{k'^2}}/k_o \qquad (3)$$

k_o and k' is calculated from Eq.(1), where $\sqrt{\overline{k'^2}}$ is the root mean square value for the fluctuating component of mass transfer coefficient. k_o is time average mass transfer coefficient in smooth duct. Mass transfer intensity is a transport property to be obtained the information on turbulence in the vicinity of the wall surface.

EXPERIMENTAL RESULTS AND DISCUSSION

FLOW PATTERNS

It was classified in the previous paper[7] that the flow patterns between turbulence promoters in a duct were distinguished by P/L=5 and P/L=7 (P/Dp is used in the case of cylinder type promoters) according to the occurrence of the reattachment flow to a wall surface or not, and as to the effect of clearance, distinction was able to be made by NDC≤ 0.09 when Karman's vortex street did not arise and NDC=0.133 when the release of vortices was observed, but the reattachment to a wall surface did not occur. As the examples, the flow patterns with the rectangular turbulence promoters of P/L=10 and the cylinder type turbulence promoters of P/Dp=9 are shown in Fig.4. As shown in the figure, in the case of NDC=0, the flow separated at the upper part of the promoters reattached to the wall surfaces at the nearly same positions, that is, X/L=5 ∿ 6 for the rectangular type and X/Dp=4 ∿ 6 for the circular type. A part of those flowed reversely on the wall surface, and formed the circulating flow large scale. In the case of the rectangular type, just before the next promoter, a part of the flow which collided with a promoter strongly strikes the wall surface, thereby the flow becomes turbulent state, but in the case of the cylinder type, it was not able to be confirmed clearly.

In the case of NDC=0.044, because of existence of a clearance, jet flow acrose just below a promoter, and it joined with the separated flow from the upper part of the promoter, consequently, counterclockwise and clockwise vortices occurred at X/L=1 ∿ 2. This phenomenon has been confirmed also for cylinder type turbulence promoters. In both promoters, reattaching points were observed in the vicinity of X/L=4 ∿ 5, and shifted

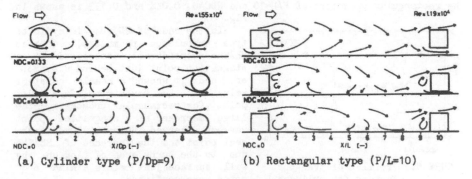

(a) Cylinder type (P/Dp=9)   (b) Rectangular type (P/L=10)

FIGURE 4. Flow pattern.

to upstream side than the case of NDC=0. It was observed that the separated flow from a hind promoter separated at sharp angle as compared with that in cylinder type.

In the case of NDC=0.089, in the reattachment of the separated flow from the upper part of a promoter to a wall surface, the reattachment region was wide due to the influence of the jet flow through just below the promoter, and was unclear.

In the case of NDC=0.133, for cylinder type promoters, a Karman's vortex street was formed the vicinity of X/L=2, whereas for rectangular type, because the negative pressure behind a promoter is high, a Karman's vortex street was formed from vicinity of X/L=1.0 ～ 1.5. Besides, the dissipation of a Karman's vortex street occurred at X/L=8.5 before the position of a hind promoter for rectangular promoters, in this way, it dissipated at considerably forward position.

DISTRIBUTION OF LOCAL MASS TRANSFER COEFFICIENTS

In the previous study[7], the increase of local mass transfer coefficients were quantitatively measured through shear stress profiles and mass transfer intensity profiles, using cylinder type promoters and it was attempted to elucidate the increasing mechanism. as a general tendency, the enhanced ratio Sh/Sho of transfer coefficients showed a peak just below a promoter, and this peak coincided with the position of the maximum value of the absolute values of shearing stress and the minimum value of mass transfer intensity. Therefore, it was found that the increase of transfer coefficients was induced by a thin laminar layer on a wall surface due to the jet flow just below a promoter. As an example, the distributions of Sh/Sho, shearing stress $|\tau/\tau_0|$ and mass transfer intensity I which is the measure of the turbulence near the wall surface in the case of P/Dp=9 and NDC=0.044 are shown in Fig.5.

The ratio of increase of transfer coefficients Sh/Sho showed a somewhat small value at X/Dp=4, but mostly evenly high values except just below a promoter. This distribution coincided well with the distribution of mass transfer intensity. From this fact, it was found that the turbulence on a wall surface due to reattachment was one important factor for the enhancement of transfer coefficients.

In this way, it is known that mass transfer (heat transfer) is controlled largely by the flow behavior, that is, jet flow due to a clearance, reattachment of flow to a wall surface, Karman's vortex street and so on.

Here, discussion was performed on local mass transfer coefficients while comparing those with flow patterns. The distribution in the case of the rectangular promoters of P/L=10 and NDC=0, 0.044 and 0.133 is shown in Fig. 6.

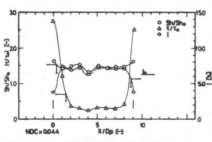

FIGURE 5. Profiles of transport factors for cylinder type promoters.

In the case of NDC=0, in cylinder type turbulence promoters, whereas Sh/Sho considerably decreased at a stagnation point just behind a promoter, and it showed a nearly constant value in the downstream from that point for rectangular promoters, the large decrease at a stagnation point was not observed. However, a reattachment point and the increase of Sh/Sho due to the turbulence of flow on the wall surface just before a hind promoter were confirmed.

In the case of NDC=0.044, for cylinder type promoters, Sh/Sho showed almost flat distribution, but for rectangular promoters, jet flow occurred by the effect of accelerating flow through a clearance, therefore, the increase of Sh/Sho was fairly conspicuous as compared with that for cylinder type. Besides, in the place corresponding to the vicinity of X/L=2, the decrease of Sh/Sho was observed. However in the downstream from that, flat distribution was shown similarly to the case of cylinder type promoters, and just before a hind promoter, Sh/Sho increased again. Similarly to the case of NDC=0, this corresponded to the degree of turbulence in the flow patterns just before promoters.

In the case of NDC=0.133, the distribution profiles of Sh/Sho were similar in both promoters. It was observed that Sh/Sho just below a promoter remarkably increased for rectangular promoters.

AVERAGE MASS TRANSFER COEFFICIENT

When the average mass transfer coefficient at the time of installing promoters was represented with the ratio to that of a smooth duct, Sh/Sho, and it was plotted in relation to Reynolds numer, it resulted in Fig.7. As seen in the figure, in turbulent region, Sh/Sho showed the tendency of decrease accompanying the increase of Re. From this fact, it was found that the increase of transfer coefficients by installing promoters is effective in the transient region from laminar flow to turbulent flow. In a fully developed turbulent flow region, it is considered that the flow between promoters stagnated by the influence of accelerating bulk flow, consequently, Sh/Sho lowered. As to this tendency, the same thing can be

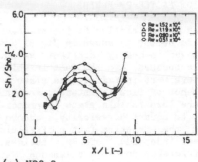

(a) NDC=0

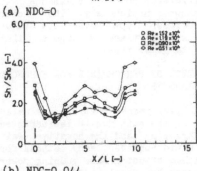

(b) NDC=0.044

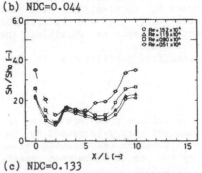

(c) NDC=0.133

FIGURE 6. Distribution of local mass transfer coefficients.

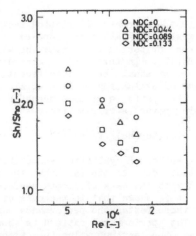

FIGURE 7. Average mass transfer coefficient for P/L=10.

said also about circular type promoters. Besides, the increase by about 10 % was observed in rectangular type as compared with cylinder type promoters. From the previous flow patterns, it is considered that this was caused by the intensity of the rolling-in of flow behind promoters, the acceleration intensity of jet flow and so on.

Next, the influence of a clearance is discussed. As a clearance increased, Sh/Sho decreased for both rectangular and cylinder type promoters. From the distribution of local mass transfer coefficients mentioned before, whereas the increase of Sh/Sho due to the reattachment to a wall surface in NDC=0 spread over a wide range, in the case of a clearance, the rate of the increase of Sh/Sho at a reattaching point decreased, accordingly, this is because the increase of Sh/Sho became only that just below a promoter.

FRICTION FACTOR OF FLUID

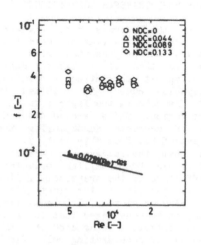

The fluid friction factor at each Reynolds number is shown in Fig.8, and it showed a nearly constant value. this is because, in the duct in which promoters were installed, the state of turbulence of flow is intense, and the pressure loss in this state is proportional to square of velocity, accordingly, the friction factor of fluid becomes independent of Re. Besides, the difference due to a clearance was not shown conspicuously. The factor for cylinder type showed the value about $2 \sim 2.5$ times as large as that for rectangular type.

FIGURE 8. Friction factor for P/L≈10.

EVALUATION OF PERFORMANCE FOR ENHANCED HEAT TRANSFER

The performance must be evaluated by taking the gain due to the enhanced heat transfer and the negative gain due to the increase of friction factor of fluid together. Many methods for its evaluation have been proposed. The method of using same pumping power as the criterion for evaluation was reported by Bergles[11], Webb and others[12]. In this study, according to the policy that the criteria for evaluation should be as significant, direct and simple as possible, the method of evaluation by Sano et al. [8] was adopted.

The ratio η of mass transfer (heat transfer) coefficients in same pumping power is represented by

$$\eta = (Sh/Sho)/(f/fo)^{0.291} \qquad (4)$$

This method of evaluation shows that when turbulence promoters were installed, due to the fact that the same consumed energy increased as compared with the case of smooth surfaces, mass transfer was promoted. Namely, it is shown that in the case of η>1, by the use of turbulence promoters, those better the performance, and are effective.

The performance ratio η is shown in Fig.9. As shown in figure, η showed the maximum value in a transition region in each clearance, and lowered accompanying the increase of Re. as mentioned before in the Chapters of average mass transfer coefficient and the friction factor, this is

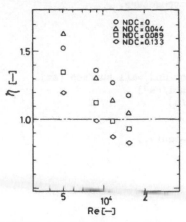

FIGURE 9. Evaluation of performance for P/L=10.

due to the multiplied effect of the decrease of Sh/Sho and the increase of f/fo with the increase of Re.

Concerning the influence of a clearance, the best performance can be obtained in rectangular promoters, in a transition region of flow, and in the case of NDC=0. It is considered that this is because the effect of jet flow just below a promoter and the effect of the reattachment to a wall surface of the flow separated from the upper part of the promoter acted most effectively. It is known that outside a transition region of flow, the performance lowered accompanying the increase of a clearance since the friction factor for circular type turbulence promoters, increased conspicuously as compared with a smooth duct without promoter, the installation of promoters was effective when flow was in a transition region and NDC=0, but it is known that generally the installation is not very desirable.

CONCLUSION

The experiment on the enhancement mechanism of mass transfer when rectangular turbulence promoters were installed in a rectangular duct was carried out, and by comparing with the case of installing circular type turbulence promoters, the following results were obtained.
1) according to the visualization for rectangular turbulence promoters, the negative pressure behind a promoter was stronger than the case of cylinder type, and the reattachment of separated flow to a wall surface and the position of forming a Karman's vortex street shifted forward. However, the behavior of both as a whole was similar.
2) In the previous study, in the case of cylinder type turbulence promoters, the increase of mass transfer coefficients between the promoters was explained based on flow patterns by the quantitative measurement of transport factors (shear stress, transfer intensity, etc.). In the case of rectangular type, the explanation was made only on the basis of flow patterns, standing on those facts. Namely, the decrease of boundary layer thickness due to jet flow just below a promoter, the turbulence in the vicinity of a wall surface at a reattaching point and just before a promoter contributed largely to the mass transfer enhancement.
3) The increase of average mass transfer coefficient arose effectively in the transient region from laminar flow to turbulent flow. Besides, by providing with a clearance, the degree of its increase due to the reattachment to a wall surface lowered, and only the increase just below a promoter contributed, therefore, it is not effective to provide a clearance for a promoter.
4) The friction factor was not affected remarkably by re and a clearance. Besides, cylinder type obtained the value about twice as large as that for rectangular type promoters.
5) As to the evaluation of performance, good results were shown in the transition region of flow in each clearance, but accompanying the increase of Re, the performance lowered.
As to the installation of promoters, that of rectangular type was

165

more effective than that of cylinder type promoters.

NOMENCLATURE

A   surface area of electrote [cm^2]
B   width of turbulence promoter [mm]
c   clearlance between turbulence promoter and wall surface [mm]
C_b bulk concentration of ferricyanide [mol/cm^3]
D   diffusivity of ferricyanide ion [cm^2/s]
Dp   diameter of turbulence promoter [mm]
d   diameter of point electrode [mm]
F   Faraday's constant (=9.652x10^4) [C/g-equiv.]
f   friction factor [-]
H   height of rectangular duct [mm]
I   mass transfer intensity [%]
i   electric current [A]
k   mass transfer coefficient [cm/s]
L   height of turbulence promoter [mm]
NDC   non-dimensional clearance [-]
n_e valence charge of an ion [-]
P   pitch of turbulence promoter [mm]
Re   Reynolds number [-]
Sh   Sherwood number [-]
x,y   coordinates of test section [mm]
η   performance ratio [-]
μ   viscosity [Pa·s]
τ   shear stress [Pa]

superscript
'   value at same pumping power

overline
-      time-smoothed

subscript
o   smoothed duct

REFERENCES

1. Mori,Y., and Daikoku,T., JSME, 38, 832 (1982).
2. Fujita,H., Takahama,H., and Yamashita, R., JSME, 42, 2828 (1976).
3. Rao,C.K., and Picot,J.J.C., 4th Int. Heat Transfer Conf., FC 8.4 (1970).
4. Han,J.C., Gricksman,L.R., and Rohsenow,W.M., Int. J. Heat Mass Transfer, 21, 1143 (1978).
5. Konno,H., Okuda,K., Sasabayashi,K., and Ohtani,S., Kagaku Kogaku, 31, 872 (1967).
6. Miyashita,H., Shiomi,Y., and Wakabayashi,K., Kagaku Kogaku Ronbunshu, 7, 349 (1981).
7. Miyashita,H., Shiomi,Y., and Wakabayashi,K., Proceedings of PACHEC, 38, vol.1, I-1, Seoul, Korea (1983).
8. Sano,Y., and Usui,H., Kagaku Kogaku Ronbunshu, 8, 516 (1982).
9. Reiss,L.P., and Hanratty,T.J., A.I.Ch.E. Journal, 8,245 (1962).
10. Mizushina,T., Advances in Heat Transfer, vol.7, p.87, Academic Press, N.Y. (1971).
11. Bergles,A.E., Blumenklrantz,A.R., and Taborek,J., Proceeding of 4th Int. Heat Transfer Conf.,vol.2, FC 6.3, (1974).
12. Webb,R.L., Int. J. Heat Mass Transfer vol.24, 715 (1981).

Heat Transfer in Irregular Passages with and without Pin-Fins

SHEN JIARUI, YUMING ZHANG, and WEIZAO GU
Institute of Engineering Thermophysics
Chinese Academy of Sciences
P.O. Box 2706
Beijing, PRC

INTRODUCTION

Enhancement of heat transfer is the subject of growing importance in myriad industrial applications. In the areas of new energy development and energy conservation, high performance heat exchangers play important roles. Effective cooling to protect high temperature members is one of the most significant technical subjects in nuclear, aerospace, electrical and electronic engineering.

It is noted that pin-fin rows located in the regular passages can substantially enhance heat transfer for straight air flow [1-4]. But to date only a limited number of published papers are available that deal with pin-finned irregular passages [5,6]. The present paper reports the author's measurements of heat transfer in the curved, rectangular, trapezoidal and diverging passages with and without pin-fins. The purpose of this work was to investigate the heat transfer augmentation in the pin-finned irregular passages, and to determine the effects on heat transfer of different passage geometries having the same pin-fin parameters and arrangement.

EXPERIMENTAL METHODS AND DATA REDUCTION

The experimental facility is shown in Fig.1. The air from the compressor passes through the cleaner, orifices and stabilizer, enters the test apparatus, and finally flows out through the exhaust valve. The test apparatuses, as shown in Fig.2, were made of stainless steel. Their typical geometric parameters are listed in Table 1. End walls of each apparatus had equal and uniform heat flux, which were provided by electrical tape heaters. As the air flowed through the apparatus, the end walls and pin-fins were cooled. Copper-constantan thermocouples, 0.12 mm dia., were laid on the pin-fin bases of the end walls for temperature measurements. The inlet and outlet air temperatures of the apparatus were likewise measured by thermocouples, while the air flow rate through the apparatus by orificemeter and U-tube manometer. With the air flow rate, inlet and outlet air temperatures, and end wall temperatures of the apparatus, heat transfer in the test apparatus can be obtained as follows.

$$N_x = Q \, Dh \, / \, \lambda \, F \, (T_{wx} - T_{fx}) \tag{1}$$

where $\quad Q = GC_p (T_f'' - T_f')$

$$T_{fx} = T_f' + (T_f'' - T_f') F_x / F$$

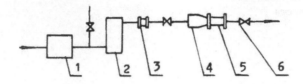

Fig.1 Experimental facility
1. compressor   2. cleaner   3. orifices
4. stabilizer   5. test apparatus 6. exhaust valve

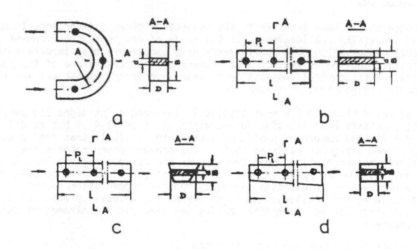

Fig.2 A shematic diagram of the test apparatuses
a. curved passages (I, II)   b. rectangular passages (I, II)
c. trapezoid passages (I, II)   d. diverging passage

As short pin-fin length to diameter ratio and pin-fin length change along the air flow direction, the end wall effects were considered.

In the test, the Reynolds numbers of the air supply to the apparatuses varied within a ragne of 5×10^3 to 4×10^4 based on the hydraulic diameter defined as

$$Dh = 4A \sum_1^n P_{Li} / \sum_1^n A_{fi} \tag{2}$$

Table 1. Typical Geometric Parameters of the Apparatuses

description		a curved		b rectangular		c trapezoid		d diverging
		I	II	I	II	I	II	
B	mm	23	19	6.5	6.6	10	7	12.5
D	mm	8.5	14	21	21	15.8	18.9	12.0
L	mm	50	50	150	150	150	150	165
d	mm	3	3	3	3	3	3	3
P_L	mm	15	15	15	15	15	15	15
n		3	3	9	9	9	9	11
D/R		2	1.3					

RESULTS AND DISCUSSION

Heat transfer data concerning curved, rectangular, trapezoidal and diverging passages with and without pin-fins are given in Fig.3 to Fig.6.
They can be correlated as shown in Table 2, with maximum deviation of
± 6%.

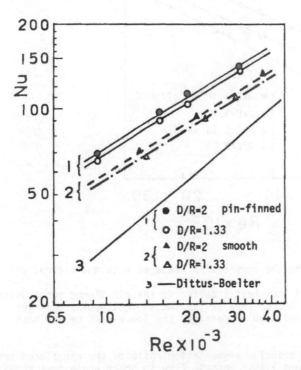

Fig.3 Heat transfer in the curved passages with and without pin-fins

169

It can be seen from Fig.3 that heat transfer in the curved passages is controlled by the ratio of bend width to curvature radius. The larger the ratio, the higher will be its heat transfer coefficient. Heat transfer in the curved passages with pin-fins is about 30% and 100% higher than in curved passages without pin-fins and straight smooth tube, respectively.

Heat transfer in curved passages is closely connected with the flow patterns and shear distribution. Flow visualizations in reference 7 give support to the author's test results. The flow patterns and shear on the convex side of the curved passage were characterized by a fairly uniform distribution, but the flow patterns on the concave side were less uniform, resulting in reduced shear. In reference 7, a rather strong crossflows from the convex to the concave sides were observed on the upper and lower boundaries over the entire length of the curved passage, and a large, slowly rotating vortex pair was formed along the center portion of the curved passage. Therefore, it is reasonable to say that heat transfer in curved passages is considerably higher than in straight passages. On the other hand, the center portion of the curved passages is weak in heat transfer. Obviously, the pin-fins located in the center portion can substantially enhance heat transfer in the curved passages.

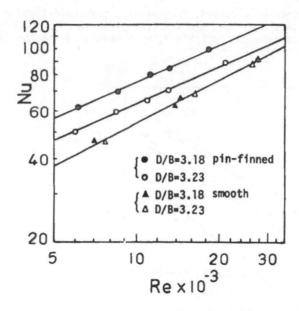

Fig.4 Heat transfer in the rectangular passages with and without pin-fins

It is found from Fig.4 that heat transfer in the pin-finned rectangular passages is about 35% higher than in the smooth rectangular passages. And the narrower the rectangular passage, the lower will be its heat transfer coefficient.

Fig.5 shows that heat transfer augmentation ratio of the pin-finned trapezoid passages is about 1.10. Because flow in sharp angle area appears in stagnant state 8, the smaller the acute angle of the trapezoid passage, the lower will be its heat transfer.

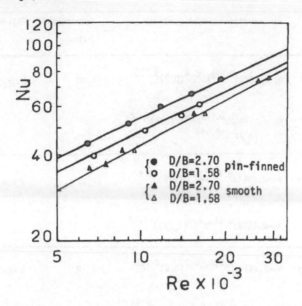

Fig.5 Heat transfer in the trapezoid passages with and without pin-fins

From Fig.6 it can be observed that enhancement ratio of the pin-finned diverging passages is about 1.40. In the diverging passage, while the cross-sectional area becomes larger and larger, the fluid moves slower and slower, thereby heat transfer worsens. In the presence of the pin-fins, as the flow is disturbed, heat transfer is substantially enhanced.

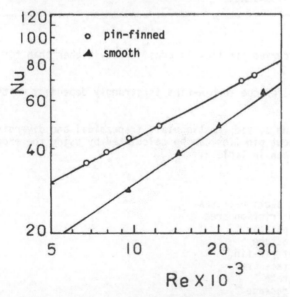

Fig.6 Heat transfer in the diverging passages with and without pin-fins

Table 2. Empirical Relations

passage geometry		empirical relation	enhancement ratio	Fig.No.
curved passages	with pin-fins	$Nu=0.239Re^{0.625}Pr^{0.4}(D/R)^{0.125}$	1.30	Fig.3
	without pin-fins	$Nu = 0.194Re^{0.625}Pr^{0.4}(D/R)^{0.125}$ $(T_w/T_f)^{-0.19}$		
rectangular passages	with pin-fins	$Nu=4.254Re^{0.432}Pr^{0.4}(D/B)^{-0.9}$	1.35	Fig. 4
	without pin-fins	$Nu=0.515Re^{0.52}Pr^{0.4}(T_w/T_f)^{-0.19}$		
trapezoid passages	with pin-fins	$Nu=0.743Re^{0.456}Pr^{0.4}(D/B)^{0.229}$	1.10	Fig.5
	without pin-fins	$Nu=0.510Re^{0.52}Pr^{0.4}(T_w/T_f)^{-0.19}$		
diverging passage	with pin-fins	$Nu=0.389Re^{0.528}Pr^{0.4}$	1.40	Fig.6
	without pin-fins	$Nu=0.066Re^{0.68}Pr^{0.4}$		

CONCLUSIONS

1. Heat transfer for curved air flow is considerably higher than for straight air flow.

2. Enhanced heat transfer due to pin-fins is strongly dependent on the passage geometries.

3. Heat transfer in the curved, rectangular, trapezoidal and diverging passages with and without pin-fins can be calculated by using the experimental correlations given in Table 2.

NOMENCLATURE

A	flow cross sectional area
A_f	total flow friction area
B	height of passage
D	width of passage
d	diameter of pin-fin
F	heated surface area
G	mass flow rate
L	length of passage
n	number of pin-fins

P_L pitch of pin-fins
Q heat flow rate
R curvature radius of bend
T_f, T_f inlet and outlet air temperatures
T_w end wall temperature

REFERENCES

1. Larson, E.G. and Sparrow, E.M., Performance Comparison among Geometrically Different Pin-Fin Arrays Situated in an Oncoming Longitudinal Flow, Int. J. Heat and Mass Transfer, Vol.25, P. 723-725, 1982.

2. Metzger, D.E., Berry, R.A. and Bronson, J.P., Developing Heat Transfer
   in Rectangular Ducts with Staggered Arrays of Short Pin-Fins, J. Heat
   Transfer, Vol.104, P. 700-706, 1982.

3. Van Fossen, G.J., Heat Transfer Coefficients for Staggered Arrays of
   Short Pin-Fins, J. Engineering for Power, Vol.104, P. 268 274, 1982.

4. Metzger, D.E., Fan, Z.X., and Shepard, W.B., Pressure Loss and Heat
   Transfer Through Multiple Rows of Short Pin-Fins, Heat Transfer 1982,
   Vol.3, P. 137-142, 1982.

5. Shen Jiarui, Zhang Yuming, Gu Weizao, and Liu Changchun, An Investigation of Heat Transfer Augmentation and Friction Loss Performance
   in a Continuously Sharp Turned Channel, J. Engineering Thermophysics,
   Vol.8, No.2, 1987.

6. A. Brown, B. Mandjikas and J.M. Mudyiwa, Blade Trailing Edge Heat
   Transfer, ASME. 80-GT-45, 1980.

7. Paul, F.B. and Robert, W.G., Flow and Heat Transfer in a Curved
   Channel, NASA TN D-8464, 1977.

8. Gu Weizao, Shen Jiarui and Zhang Yuming, Investigation on the Heat
   Transfer in the Acute Angle of Triangular Ducts, J. Engineering Thermophysics, Vol.7, No.2, 1986.

Analysis of Tubeside Laminar and Turbulent Flow Heat Transfer with Twisted Tape Inserts

B. DONEVSKI
Faculty of Engineering
University of Bitola
97000 Bitola, p. fah. 99, Yugoslavia

M. PLOCEK and J. KULESZA
Institute of Heat and Refrigeration
Engineering
Technical University of Lodz
90-924 Lodz, Poland

M. SASIC
Department of Mechanical Engineering
University of Belgrade
11000 Belgrade, Yugoslavia

1. INTRODUCTION

Numerous works on friction and heat transfer characteristics for twised tape flow have been published in the last two decades. Most of these experiments were conducted in the USA, USSR and Poland, in a wide range of fluid flow and heat transfer parameters with the presentation of various empirical correlations.

Several approaches have been made to solve this problem both theoretically and experimentally. Smithberg and Landis [1] were the first to attempt to determine the friction coefficient (at the turbulent and transient regimes) for twisted tape flow using a theoretical approach. They described a mathematical model for friction coefficient based on velocity profiles in the boundary layer.

Donevski and Kulesza [2,3,4] classified the twisted tape flow into regimes according to the predominance of three basically physical phenomena caused by viscous action, secondary flow and turbulence. Empirical correlations have been formulated to represent the friction coefficients for twisted tape flow in each of these regimes. This method of evaluation of the problem proved to be successful, and empirical correlations gave good prediction with the available experimental data by other authors [1,5,6,7].

However, it should be noticed that most of the above mentioned experimental investigations were conducted using water and air, i.e. for media having insignificant changes of physical properties with temperature. Therefore it strongly disagrees with experimental data obtained for heated and unheated flow channels. The theoretical investigations were based mainly on Colburn's analogy of heat and mass transfer, and they agree well with experimental results.

A survey carried out by Du Plesis [6] showed that the available experimental data were not extensive, and the correlations were applicable only for a certain range of values of the independent parameters.

The purpose of the present paper is primarily, to establish a general correlation for successful prediction of friction and heat transfer characteristics of laminar, transient and turbulent flow regimes, and secondly to establish the equations for criteria separating these types of flow regimes. It also proves that Colburn's analogy is appropriate for these cases, so it may be used, for example, for similar cases, such as: divided tapes, entry section of the tape, etc. and may represent a basic form for design of equipment without requiring detailed experimental investigations.

2. REGIMES OF PREDOMINANCE OF DIFFERENT PHENOMENA

The investigation of Donevski and Kulesza [2,3,4] showed that three kinds of forces: - viscous forces, - turbulent pulsation forces and - centrifugal forces are responsible for this type of flow A system of primary forces, generated by fluid flow combined with secondary forces derived by the velocity distribution, results in the formation of certain flow regimes, which can be described as follows:

i) The flow regime with predominant role of viscous forces. In this condition the fluid particles move in definite paths called streamlines, and there are no components of fluid velocity normal to the duct axis. The fluid appears to move by sliding laminae of infinitesimally small thickess relative to the adjacent layers. This flow regime is pure laminar twisted tape flow.

ii) The flow regime, when the cross-sectional secondary flow occurs due to the centrifugal forces. This flow is fully developed laminar flow region with secondary circulation.

iii) The flow regime when the turbulent fluctuations start to occur having no influence on the flow. During this condition the dominant centrifugal forces will make the turbulent pulsations smoother and will laminarize the flow. This will be the transient flow region.

(iv) The flow regime when the turbulent pulsation forces are strong and dominant over the viscous forces. They will obviously reduce the influence of centrifugal forces and the secondary flow will be generated by this phenomenon. This flow is fully developed turbulent flow. Because of the effect of this existing secondary flow circulation, the stream line of twisted turbulent flow is not identical with the twisted streamline marked by the tape.

3. PHYSICAL MODEL FOR THE CORRELATIONS

3.1 Evaluation of friction coefficient correlations

In the model of the investigated flow, it has been assumed, according to previously established facts, that three basic kinds of forces define the nature of the fluid flow: - viscous forces, - forces connected with turbulent fluctuations and - centrifugal forces. According to the predominance of the above physical phenomena, the twisted flow mode can be classifed into four regimes, depending on flow velocities as follows:

i) In the first lowest velocity flow regime, under the prevalence of viscous forces, centrifugal forces do not have any significant influence on the nature of the flow. In this case the fluid element moves along a path parallel to the surface of the tape. The flow is purely laminar without secondary circulation. This condition is denoted by region I in Figure 1, and may hydrodynamically be correlated by the relation:

$$f_I = f_\ell \, \xi_\ell \tag{1}$$

where: $f_\ell = 63/Re$ - is a friction coefficient for a pure laminar flow in a semi- circular cross section [3] $\tag{2}$

ξ_ℓ - is the ratio between twisted and straight lengths of the streamlines (L_s/L) $\tag{3}$

$$L_s = L \, [1 + 3.4 \, (D/H)^2]^{1/2} \tag{4}$$

$$\xi_\ell = L_s/L = [1 + 3.4 \, (D/H)^2]^{1/2} \tag{5}$$

176

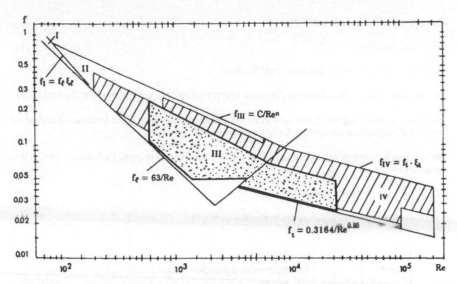

FIGURE 1.   Numerical values for friction factor for twisted tape flow for different H/D ratios.
                   ▨ –   Regimes investigated in present study.
                   ▨ –   Regimes investigated by other authors [1, 5, 6, 7]

$$f_I = (63/Re)[1 + 3.4\,(D/H)^2]^{1/2} \tag{6}$$

ii)   In the next lower velocity regime II the viscous and centrifugal forces are predominant and they determine the fluid flow character. These forces being secondary forces, result from the velocity profile and the geometry of the duct. The flow is still laminar but with secondary flow caused by centrifugal forces, and this flow will be fully developed laminar flow. According to Du Plessis [8,9] hydrodynamically, this flow region may be correlated by:

$$f_{II} = f_D(\varepsilon)\left\{1 + \left[\frac{Re_D}{(H/D)^{1.3}}\right]^{1.5}\right\}^{1/3} \tag{7}$$

where: The value of $f_D(\varepsilon)$ is derived for the effective flow concept for any tube and twisted combination. The value of Re_D is based on the internal diamater of the tube and is calculated for full flow cross-sectional area in the range of Reynolds number $50 \le Re_D \le 2000$.

The criteria for separating the regimes II and III takes place at the Reynolds number:

$$Re_{II\text{-}III} = 25 + 2175/[1 + 135\,(D/H)] \tag{8}$$

iii)   In higher velocity region III, the turbulent pulsations begin to appear, but the centrifugal forces still have greater influence on the nature of the flow regime. However, the appearance of the turbulent force pulsations (which have significant influence on the character of centrifugal forces) displaces the regime boundaries towards a higher Reynold's number. This flow regime is transient flow and hydrodynamically the flow may be correlated by

$$f_{III} = C/Re^n \tag{9}$$

where: $C = 63[1 - 0.936 \, (D/H)^{0.04}]$ (10)

$n = 1 - 0.75 \, (D/H)^{0.3}$ (11)

are functions of geometric parameters of the duct.

For these flow conditions changes of velocity and pressure drop are functions of the ratio H/D.

The characteristic region of pressure drop for the transition flow from laminar to turbulent flow in a straight tube disappears for low values of H/D.

The criteria for separating between regime III and the fully developed turbulent regime IV is indicated by:

$Re_{III-IV} = 3300 + 34000 \, (D/H)^{3/2}$ (12)

Equations for region boundaries are presented in Fig. 2.

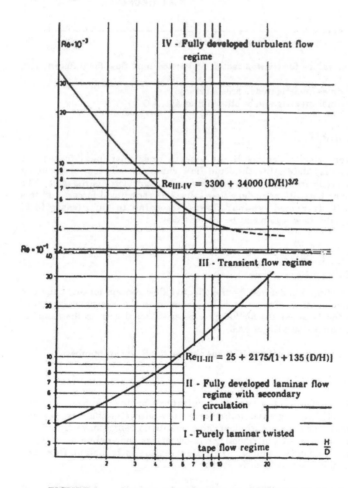

FIGURE 2.   Regimes of predominance of different phenomena.

178

iv)   In the highest velocity regime IV, the forces connected with turbulent pulsations are predominant. The flow is strongly turbulent (fully developed) and essentially without any influences from the secondary flow and with simultaneous additional changes in the specific path of the fluid. This kind of flow will be twisted turbulent flow and may hydrodynamically be correlated by:

$$f_{IV} = f_t \cdot \xi_t \tag{13}$$

where ξ_t is a coefficient taking into consideration the geometric changes of the flow path, and

$$\xi_t = 1 + 11.64 \, (D/H)^{1.61} \tag{14}$$

f_t - is a friction coefficient for a straight tube

$$f_t = 0.3164/Re_H^{0.25} \tag{15}$$

$$f_{IV} = (0.3164/Re_H^{0.25})[1 + 11.64 \, (D/H)^{1.61}] \tag{16}$$

An analysis of the experimental results and majority of available data regarding Reynold's number $Re < 150.000$ was also performed for region IV, [3,4].

Reference data used as a base in this analysis are present in Table 1.

3.2   Evaluation of the Heat Transfer Coefficient

An experimental investigation was also conducted to obtain the heat transfer coefficient but only for conditions of fluid flow regimes denoted by III and IV (transient and fully developed turbulent regime).

If used an analogy for momentum and heat transfer may be correlated by an expression for the average Nusselt number for the region III in the following form

$$Nu_{III} = C_{III} \, Re^n \, P_r^{0.4} \tag{17}$$

The values of C_{III} and the exponent n in eq. (17) of the twisted flow streamline on the tube wall may be correlated by the following as a function of geometric parameters of the duct.

TABLE 1. Reference data used as a base in this analysis

Author	Working medium	D [mm]	H/D	Methods of calculation
Smithberg, E., and Landis, F[1]	air water	35	3.62 + ~	$D_H \simeq (\pi D/\pi + 2)$
Koch, R. [5]	air	50	5 + 22	D
Migaj, V. [6]	air	20	3.5 + 20	$D_{ek} = D/2$
Thorsen, R., and Landis, F. [7]	air	15	3.16 + 8	$D_H \simeq (\pi D/\pi + 2)$

179

$$C_{III} = 0.023 + 0.967\ (D/H)^{0.768} \tag{18}$$

$$n = 0.8 - 0.35\ (D/H)^{0.25} \tag{19}$$

where the heat transfer coefficient is a function of the ratio H/D:

$$\tag{20}$$

$$Nu_{III} = [0.023 + 0.967\ (D/H)^{0.768}]\ Re^{(0.8\ -\ 0.35(D/H)^{0.25})}\ Pr^{0.4}$$

In the region IV, the expression for Nusselt number may be correlated in the following form:

$$Nu_{IV} = C_{IV}\ Re^{0.8}\ Pr^{0.4} \tag{21}$$

The value of C_{IV} in Eq. (21) is expressed by relation:

$$C_{IV} = 0.023 + 0.035\ (D/H) \tag{22}$$

Substituting eq. (22) to eq. (21)

$$Nu_{IV} = [0.023 + 0.035\ (D/H)]Re^{0.8}\ Pr^{0.4} \tag{23}$$

Figure 3 shows typical results of Nu vs. Re number. The Nusselt number increases with the twist ratio H/D of the tape. For a given H/D, the Nusselt number increases due to the presence of the twisted tape. The dotted line in Fig. 3 is the criteria separating regime III and the fully developed turbulent regime IV predicted by the following equation:

$$Re_{III-IV} = \{[1 + 42.04(D/H)^{0.768}]/[1 + 1.52(D/H)]\}^{[1/0.35(D/H)^{0.25})]} \tag{24}$$

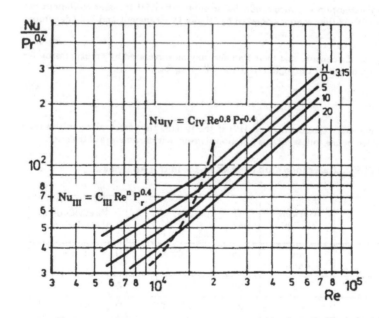

FIGURE 3.   Numerical values for Nusselt number for twisted tape flow for different H/D ratios.

4. CONCLUDING REMARKS

A comparison of results presented by several investigators [2-5, 12, 13] shows that the character of correlated relations both for the friction and heat transfer coefficient is analogous, and may be written in the following forms:

$$f_{IV} = f_t \cdot \xi_t \tag{25}$$

$$Nu_{IV} = Nu_t \cdot \xi_t \tag{26}$$

This is in agreement with the Colburn's analogy. However it should be noted that Donevski and Kulesza [2,7,8] in their investigations of friction factor, assumed that the criteria separating the flow regimes III and IV take place at the Reynolds number $Re_H = 3300$. In this case the hydraulic diameter of cross sectional flow area was the characteristic parameter, and the value was $D_H = 7.107$ mm. While, in the case of heat transfer Plocek and Kulesza [10,11], investigated fully developed turbulent flow for $Re_D > 10^4$, and used as the characteristic dimension the inside tube diameter. As a result the criterial line separating regime III and IV was shifted to a high value of Reynolds number in the latter case.

Agreement with these results may be found in investigations [1, 5, 6, 7] as shown on Fig. 4-7. However, a better expression based on the improved temperature profile is required. Here, the temperature profile for the media used in their experimental investigations is insignificant.

The regime III (transient) shows signficant differences in results obtained by some authors [7,8,9]. However, these differences are a consequence of different methods of the approach to this problem. Donevski and Kulesza [2,3,4] suggested that this condition of flow regime will

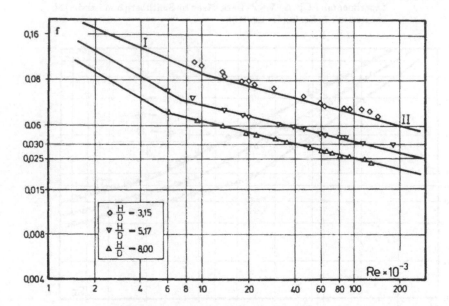

FIGURE 4.   Comparison between experimental data and the correlation formulae for friction factor.
Experimental: ◇ ▵ ▽ - Data given by Thorsen and Landis [7].
Theoretical: I - Eq. (9); II - Eq. (16).

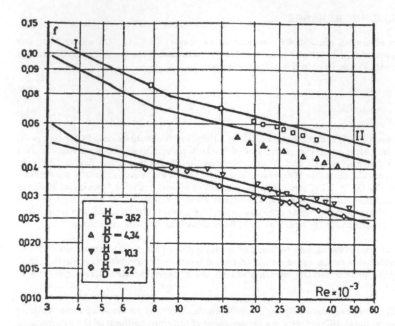

FIGURE 5.   Comparison between experimental data and the correlation formulae for
friction factor.
Experimental: □ △ ▽ ◇ - Data given by Smithberg and Landis [1].
Theoretical: I - Eq. (9); II - Eq. (16).

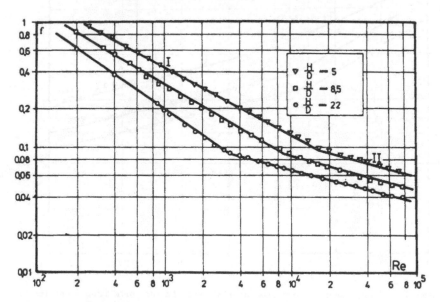

FIGURE 6.   Comparison between experimental data and the correlation formulae for
friction factor.
Experimental: ▽ □ O - Data given by Koch [5].
Theoretical: I - Eq. (9); II - Eq. (16).

182

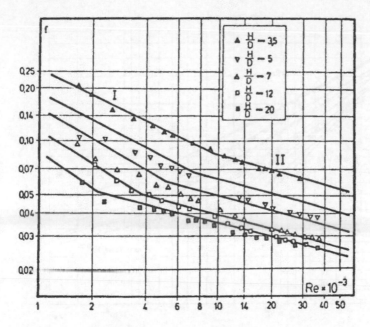

FIGURE 7.   Comparison between experimental data and the correlation formulae for
friction factor.
Experimental: ▲ ▽ △ ☐ ■ - Data given by Migaj [6].
Theoretical: I - Eq. (9); II - Eq. (16).

take place at $Re_H < 3300$ and they compared it with the laminar flow region. However, Plocek and Kulesza [10,11] classified this regime that will take place at $Re_D < 10^4$ and they compared their results with the regime of fully developed turbulent flow. Consequently, there are different correlated expression which do not modify the basic character of the phenomena. As the brief survey shows, the problem thus demands a formal unification of the analogy discussed. This is confirmed by Smithberg and Landis [1] expression, which theoretical expression predicts experimental results well only for regime III (transient region), Fig 5

The regime II, i.e. laminar flow with secondary circulation has been investigated by Donevski and Kulesza [2,3], Fig. 8, and Du Plessis and Kröger [8,9]. They obtained results which are of greater interest for future experiments and analysis, expecially to obtain completeness of experimental results over the whole flow regime.

The regime I is difficult to measure   However, these correlations were established at a time when combined nature and forced laminar flow have become important for a number of engineering applications.

ACKNOWLEDGEMENTS

This work was supported partly by the Macedonian Council of Scientific Research Activities, Skopje, Yugoslavia and the Institute of Heat and Refrigeration Engineering at the Technical University of Lodz, Lodz, Poland where the experiments were conducted

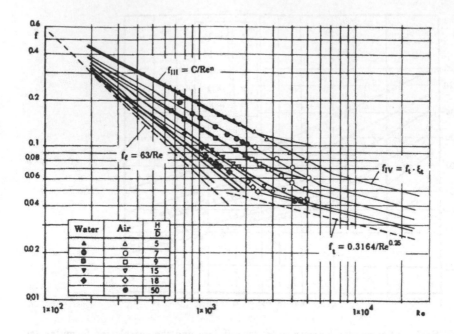

FIGURE 8.   Comparison between experimental data and the correlation formulae for friction factor in present study

NOMENCLATURE

C	— constant	[−]
D	— inside tube diameter	[m]
f	— friction coefficient for twisted flow	[−]
H	— twisted pitch (defined for 360 deg. rotation of twisted tape)	[m]
L	— length of tube, axial length of twisted tape	[m]
L_s	— average twisted length	[m]
n	— exponent	[−]
Nu	— Nusselt number	[−]
Pr	— Prandtl number	[−]
Re_H	— Reynolds number for twisted flow	[−]
y	— tape twist parameter, H/D	[−]
ξ	— coefficient, parameter, L_s/L	[−]

Subscripts

D	— based on diameter at tube
H	— based on hydroulic diameter
ℓ	— laminar
s	— for twisted flow
t	— turbulent
I,II,III,IV	— flow regions

REFERENCES

1. Smithberg, E., and Landis, F., Friction and Forced Convection Heat Transfer Characteristics in Tubes with Twisted Tape Swirl Generators, *J. Heat Transfer*, vol. 86, no. 2, pp. 39-49, 1964.

2. Donveski, B. and Kulesza, J., Resistance Coefficient for Laminar and Turbulent Flow in Swirling Ducts (in Polish), *Archives of Thermodynamics and Combustion*, vol. 9, no. 3, pp. 497-506, 1978.

3. Donevski, B., Investigations of Pressure Drop in One and Two-Phase Flow through Horizontal Tube with Loos Swirl Tapes (in Polish), Ph.D. Thesis, Warsaw Technical University, Warsaw, Poland, 1979.

4. Donevski, B., and Kulesza, J., Friction in Isothermal Turublent Flow in Tubes with Swirl Tape (in Polish), Scientific Bulletin of Lodz Technical University, *Mechanics*, no. 58, pp. 2-25, 1980.

5. Koch, R., Druckverlust und Warmeüberzgang bei Verwirbelter Stromung, *VDI-Fosschungsheft*, vol. B24, no. 469, pp. 1-44, 1958.

6. Migaj, V.K., Friction and Heat Transfer Characteristics in Swirl Flow Inside Tube (in Russian), *Izvestija AN USSR*, Energetika and Transport, No. 5, 1966.

7. Thorsen, R., and Landis, F., Friction and Heat Transfer Characteristics in Turbulent Swirl Flow Subjected to Large Transverse Temperature Gradients. *J. Heat Transfer*, vol. 90, no. 2, pp. 87-88, 1968.

8. Du Plessis, J.P., Laminar Flow and Heat Transfer in a Smooth Tube with Twisted Tape Insert, *Ph.D. Thesis*, University of Stellenbosch, South Africa, 1982.

9. Du Plessis, J.P., and Kröger, D.G., Friction Factor Prediction for Fully Developed Laminar Twisted Tape Flow, *Int. J. Heat Mass Transfer*, vol. 27, no. 11, pp. 2095-2100, 1984.

10. Plocek, M. and Kulesza, J., Heat Transfer in Tubes with Twisted Tapes (in Polish), *Proc. 6th Sym., on Heat and Mass Transfer*, Warsaw - Jablona, pp. 207-212, October 1986.

11. Plocek, M., Intensification of Heat Transfer in Tube by Short Twisted Tapes. *Ph. D. Thesis*, Lodz Technical University, Lodz, Poland, 1986.

12. Kamenishikov, F.T., Reshetov, V.A., and Rjabov, A.N., Introduction to Mechanics of Rotated Flows and Heat Transfer in Nuclear Power Equipment (in Russian), 1st ed., pp. 108-128, *Energoatomizdat*, Moscow, 1984.

13. Watanabe, K., Taira, T., and Mori, Y., Heat Transfer Augmentation in Tubular Flow by Twisted Tapes at Hgih Temperatures and Optimum Performance. Trans. Jap. Soc. Mech. Engr. Ser. B., vol. 49, no. 349, pp. 685-694, 1983.

Heat Transfer Augmentation in Entry Region of a Duct with Rib-Roughened Walls

J. S. PARK and B. B. KISLAN
Navy Ships Parts Control Center
Department of the Navy
Mechanicsburg, Pennsylvania, USA

J. C. HAN
Department of Mechanical Engineering
Texas A&M University
College Station, Texas, USA

ABSTRACT

The enhancement of the local Nusselt number for turbulent flow in the hydrodynamic and thermal entrance region of a square duct with two opposite rib-roughened walls was studied experimentally. Air was used as a working fluid in the ranges of Reynolds numbers between 7,000 and 81,000. The rib height-to-equivalent diameter ratio(e/D) was kept at a constant value of 0.063, the rib-pitch-to-height ratio(P/e) was varied from 10 to 20, and the rib angle-of-attack(A) was varied from 90 to 60 to 45 to 30. Local wall temperature distribution, local Nusselt number distribution, average Nusselt numbers between x/D=2.9 and 16.8, friction factor, and thermal performance comparison are presented.

1. INTRODUCTION

Enhancement of heat transfer through the use of roughness element schemes has been of interest since 1950. Many investigators have studied this field, both experimentally and analytically, in an attempt to develop a method for predicting the effect of a given roughness geometry or to identify the geometry which provides the best heat transfer performance for a given flow friction.

The majority of the previous studies of flow over rough surfaces have been concerned with the fully rough region where the friction factor is independent of the Reynolds number. Based on the previous studies, the effects of rib height-to-equivalent diameter ratio, e/D, rib pitch-to-height ratio, P/e, and rib angle-of-attack, A, on the average heat transfer coefficient and friction factor in the fully developed region have been established over a wide range of Reynolds numbers. Semi-empirical correlations in the fully developed region have been developed. It has been suggested that the best performance can be obtained with P/e=8 at a 33 degree rib-angle-of-attack in annular flow[1], with P/e=10 at a 45 degree rib-angle-of-attack in parallel-plate flow[2], and with p/e=10-15 at a 50-60 degree rib-angle-of-attack in circular tubes[3,4].

In advanced gas turbine blade cooling applications, repeated-ribs

187

have been selected as one of the most effective turbulence promoters used to enhance the internal cooling capability[5,6]. Based on thermal and mechanical considerations, most of the blade cavities have been designed in a rectangular cross section rather than a circular channel, and only two opposite walls in a channel may be required to be rib-roughened. A cutaway of a turbine blade with a convective internal cooling geometry and with a geometry of repeated-rib roughened wall is shown in Figure 1. The heat transfer and friction characteristics in channels of this kind are expected to be different from circular tubes, parallel-plates, or annuli.

In 1970, Burggraf[7] studied the square channel with two opposite rib-roughened walls only with a 90 degree angle-of-attack, P/e=10, and e/D=0.055. The results indicated that the augmentation of the Nusselt number on the ribbed side wall and on the smooth side wall was approximately 140 percent and 20 percent higher than the four sided smooth duct flow values; whereas, the friction factor was about 9 times higher. Han, Park, and Lei[8] further investigated the effect of rib angle-of-attack on the pressure drop and heat transfer coefficients for fully developed turbulent air flows in a square duct with two opposite rib-roughened walls.

The local heat transfer distribution on the smooth walls and between the ribs of rough walls is especially important for use in advanced blade cooling design. However, no local heat transfer measurements on a smooth wall or between the ribs of a rough wall have been studied extensively. Recently, Park[9] studied the local heat transfer distribution on a square duct with two opposite rib roughened walls and the effect of rib-angle-of-attack at the entrance and fully developed region.

The present study is to investigate the effects of the rib angle-of-attack, A, and rib pitch-to-height ratio, P/e, on the local and average heat transfer enhancements for the entrance regions in a duct with two opposite ribbed walls.

2. EXPERIMENT

2.1 Apparatus

Figure 2 and 3 show the rib geometries and a schematic diagram of the apparatus. A 5 HP blower forced air through a 10.16 cm (4 in) diameter pipe equipped with a 3.8 cm (1.5 in) diameter orifice plate to measure flow rate. A plexiglass plenum with a cross section of 38 cm by 38 cm (15 in by 15 in) and a length of 76 cm (30 in) was connected between the pipe and the test duct to ensure that the air entering the test duct was uniform and had a sudden contraction condition. The contraction ratio was 5:1. A round corner with a radius of 0.63 cm (0.25 in) was provided between the plenum and the test duct. The test section was designed to simulate the inlet condition of the cooling flow in the turbine blade. At the end of the test section, the air was exhausted into the atmosphere. The Reynolds number in the test duct was varied between 7,000 and 80,000.

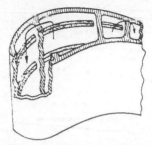

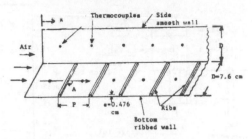

FIGURE 1.Two opposite rib-roughened
in an advanced multipass turbine
blade

FIGURE 2. Rib geometries of
test section

The test duct which consisted of four parallel stainless steel
plates was constructed with the upper and bottom walls rib-
roughened, but the two side walls were smooth. 80 thermocouples
were instrumented in the test section. The detailed apparatus
and procedures are described in a reference[10].

2.2 Procedure and data reduction

The low Reynolds number run was conducted first at the nominal
flow rate and at the desired heating rate, which maintained the
maximum wall temperature less than 90 degree C. Equation (1) was
used to determine the friction factor based on the adiabatic
conditions.

$$f = \Delta P/[4 \ (L/D)(G^2/2 \mathfrak{f} g_c)] \tag{1}$$

The maximum uncertainty of friction factor was estimated to be
less than 7 percentage for Reynolds number greater than 10,000.
The local Nusselt numbers were determined by:

$$Nu(S) = [q''(S)/(TW(S)-TB)]/(D/K) \tag{2}$$

$$Nu(R) = [q''(R)/(TW(R)-TB)]/(D/K) \tag{3}$$

$$Nu(AV) = [Nu(S) + Nu(R)]/2 \tag{4}$$

The net heat flux is the heat flux generated by the heater
substracting the heat loss to outside and the axial heat
conduction in the test duct, i.e., q"(net) = q"(heater)-
q"(loss) - q"(net heat conduction). The maximum heat loss from
the smooth side and the ribbed side wall was estimated to be less
than 7 % and 4 % for Reynolds number greater than 10,000; whereas
the axial conduction could be up to 9 % at the entrance region in
some operating conditions. The ribbed side heat flux, q"(R),
was based on only the projected heat transfer area by not
including the increased rib surface area. The maximum
uncertainty in the Nusselt number was estimated to be less than
15 % at x/D < 6 and less than 10 % at x/D > 6 by using the
uncertainty method[11].

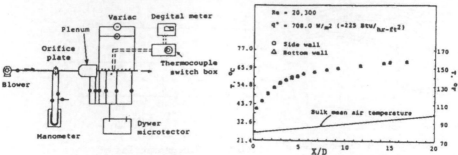

FIGURE 3. Schematic diagram for
experimental apparatus.

FIGURE 4. Temperature distribution for
four sided smooth walls.

3. RESULTS WITH FOUR SIDED SMOOTH WALLS

Before initiating experiments with ribbed walls, the pressure drop and heat transfer were calibrated for a four sided smooth fully developed region for uniform heat flux was determined by plotting the duct wall temperatures and bulk mean temperatures against the axial distance. The flow was assumed to be fully developed when the slope of the wall temperature was equal to the axial gradient of the bulk air temperature.

In the present study the thermal development was accomplished reasonably within a distance of 10 diameters as shown in Figure 4. Nusselt numbers calibrated at x/D=14.9 were compared with those predicted by the modified Dittus and Boelter correlation and showed good agreements within 10 %.

4. EXPERIMENTAL RESULTS WITH TWO OPPOSITE RIB-ROUGHENED WALLS

Table 1 shows a total of 9 rib geometries tested for the square duct with sudden contraction entrance. Only the representative results will be discussed.

TABLE 1. Rib geometries for present tests.

A	P/e = 10		P/e = 20	
	Friction	Heat Transfer	Friction	Heat Transfer
A = 90°	X	X	X	X
= 60°	X	X	X	X
= 45°	X	X	X	X
= 30°	X	X	X	X
= 15°	X			

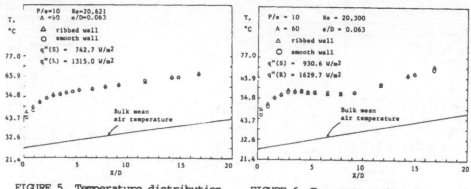

FIGURE 5. Temperature distribution for A=90

FIGURE 6. Temperature distribution for A=60

4.1 Local Wall Temperature Distribution.

Figures 5 and 6 show the typical temperature distributions of the ribbed side wall and the smooth side wall along the test section. It is noted that the wall temperature distribution for the rib angle-of-attack of A=90 is very similar to that of the conventional smooth duct turbulent flow, indicating the smooth transition from the entrance to the fully developed region. However, the wall temperature distribution for A=60 has "overshoot" at x/D about 3 then decreases to a minimum value at x/D about 9, then increases again along the test duct. This wall temperature "overshoot" characteristics was found for all tests of A=60, 45, 30, P/e=10, and Re=7,000 to 80,000.

4.2 Local Nusselt Number Distributiion.

Figures 7 and 8 show the typical local Nusselt number distribution along the centerline in both ribbed side and smooth side walls for different rib flow-attack-angle at different P/e and Re. It is noted that every thermocouple was located at the middle between two adjacent ribs along the centerline of the ribbed side wall. The local Nusselt number of the four sided smooth duct is included for comparison. It is seen that the local Nusselt number distribution for A=90 decreases smoothly with distance for all Reynolds numbers, which is similar to that of the smooth duct results except that the former has a larger values for both of ribbed side and smooth side walls. However, the Nusselt number for A=60 and 30 reaches a minimum at x/D=3 then increased to reach a maximum value at about x/D=9 and decreases with further increasing x/D. The minimum Nusselt number occurs at x/D=3, while the maximum Nusselt number occurs at x/D=9. This overshoot distribution is diminished a lot when the P/e change from 10 to 20 as shown in Figure 9, or when the test section is connected to a long duct entrance as shown in Figure 10. Figure 10 shows Nusselt number distributions for A=45 with the different entrance geometries. The unheated long duct entrance, which had the same cross-section and length as those of the test duct, was made of plexi-glass plates and was ribbed over its length on two opposite walls in the same way as the test

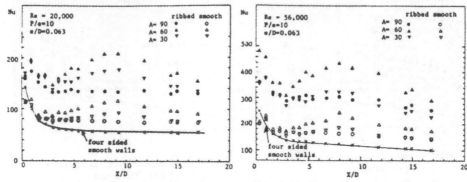

FIGURE 7. Nusselt number distribution for P/e=10, Re=20,000

FIGURE 8. Nusselt number distribution for P/e=10, Re=56,000

duct.   Between x/D=4 and 12, the local heat transfer values for
sudden contraction entrance are higher than those for long duct
entrance.

4.3   Average Nusselt Number Between x/D=2.85 and 16.81.

The data shown in Figures 11 and 12 were based on the average
value of the local Nusselt number as x/D=2.85 to 16.81. The data
show that the Nusselt number increases with increasing Reynolds
number as the conventional turbulent pipe flow. Nusselt number
of the ribbed side wall with A=90 is about 2.5 times higher than
the four sided smooth duct as shown in Figure 11 for P/e=10. The
Nusselt number with A=30 is about 20 % higher than A=90, while
the Nusselt number with A=45 and 60 is about 25 % and 33 %
higher, respectively. The Nusselt number of the smooth side wall
is also enhanced by 50-100 % due to the presence of the ribs on
the adjacent walls when the rib angle-of-attack, A, changes from
90-30-45-60. Figure 12 shows that the average Nusselt number for
the ribs with the oblique angle to the flow is higher than that
for those with A=90.   The results for P/e=20 have the similar
trends as P/e=10 except that the enhanced value is lower.   It is
noted that the heat transfer data for sudden contraction entrance
has about 5-15 % more enhancement when the ribs have an oblique
angle to the flow.

4.4   Average Friction Factor Between x/D=3.4 and 18.5.

Figures 13 shows the average friction factor vs Reynolds number
for flow in the region with x/D>3 and P/e=10. For A=90, 60, and
45, the friction factor approaches an approximately constant
value as the Reynolds number increases while the friction factor
decreases with Reynolds number for A=30 and 15. For the P/e=10,
the friction factor with A=90 is about 3-10 times higher than the
four sided smooth duct over the range of Reynolds number. The
fricition factor   with A=45 is slightly less than A=90,
however, it decreases by 30-40 % when the A changes from 90 to
30.   The friction factor with A=60 is about 50 % higher than
A=90. The results for p/e=20 generally agree with that for
P/e=10 except with relatively lower values.

192

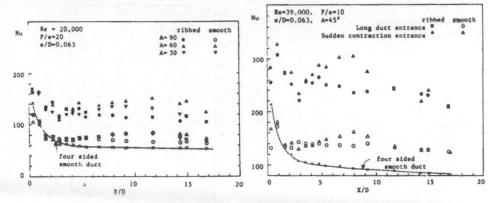

FIGURE 9. Nusselt number distribution for P/e=20, Re=20,000

FIGURE 10.Nusselt number distribution
for A=45 with different entrance geometry

4.5 Thermal Performance Comparison

Figure 14 shows the average friction factor between x/D=3.4 and
18.5 and average Stanton number between x/D=2.85 and 16.81 vs the
rib angle-of-attack. The data of the 0 degree of angle-of-attack
were obtained from the present smooth duct results. Both the
friction and heat transfer increase with decreasing A, and reach
a maximum value at A about 60, then decrease with further
decreasing A. It is noted that the amount of the friction factor
decrease is relatively larger than that of the Stanton number
when the rib angle-of-attack changes from 90 to 30. This
suggests that the best thermal performance may be obtained at the
A around 30.

5. CONCLUSION

(1) As expected, with two opposite rib-roughened walls, local
heat transfer distribution for A=90 decreases smoothly with
distance for all Reynolds numbers. This is similar to that of
the all smooth duct results except that the values were higher.

(2) For P/e=10 the heat transfer for all angles-of-attack, A=60,
45,30, did not decrease smoothly with distance, but reached a
minimum at x/D=3 then increased to reach a maximum value at x/D=9
and decreased with further increasing x/D. For example, at A=60,
the heat transfer increased between x/D of 3 and 9, reaching a
maximum that was 50 % greater than that for A=90.

(3) For the same geometrical condition with P/e=10 and A=45,
between x/D=4 and 12, the local heat transfer values for sudden
contraction entrance are higher than those for long duct
entrance.

(4) For P/e=10, the average Nusselt number of A=90 is about 2
times higher than that of the four sided smooth duct while the
average friction factor is about 4-6 times higher. The average

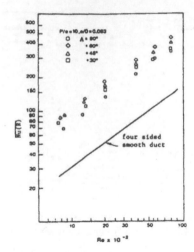

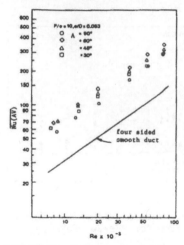

FIGURE 11.Average Nusselt number for ribbed side wall between X/D=2.9 & 16.8

FIGURE 12. Average Nusselt number for both of ribbed and smooth side walls between X/D=2.9 and 16.8

Nusselt number of A=30 is about 20 % higher than A=90 while the average friction factor is about 20-45 % lower. The ribs with an oblique angle to the flow for A=30,45,60 give a relatively higher heat transfer distribution, that is, overshoot distribution, along the duct after x/D > 3 than the ribs with a transverse angle to the flow for A=90. The overshoot distributio n is diminished a lot when P/e=20 or when the test section has a long duct entrance.

(5) The best thermal performance may be obtained at rib angle-of-attack around 30 degree.

ACKNOWLEDGEMENTS - The study reported herein has been supported by the U. S. Army Research and Technology Laboratories and NASA-Lewis Research Center under Grant NAG 3-311. Experiments were conducted at Turbomachinery Laboratory of Texas A&M University. The authors grately appreciate the review from W. Emery and the support from W. Lowry during preparation of the manuscript.

NOMENCLATURE

A = flow attack angle
D = side dimension or equivalent diameter of square duct
e = rib height
f = friction factor
g_c = conversion factor
G = mass flux
K = thermal conductivity of fluid
L = test section length for friction pressure drop
Nu(S) = local Nusselt number for smooth side wall
Nu(R) = local Nusselt number for ribbed side wall
$\overline{Nu(R)}$ = average Nusselt number for ribbed side wall
          between x/D=2.85 and 16.81

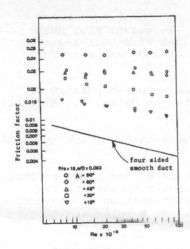

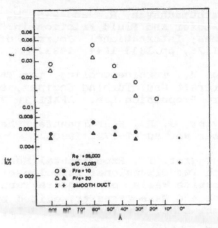

FIGURE 13. Friction factor between X/D=3.4 and 18.5

FIGURE 14. Comparison of friction factor and heat transfer with different rib angle-of-attacks, A

Nu(AV) = local average Nusselt number in a duct with two opposite ribbed walls
$\overline{Nu(AV)}$ = average Nusselt number between x/D=2.85 and 16.81
ΔP = pressure drop across the test section
P = rib pitch
q"(S) = net local heat flux for smooth side wall
q"(R) = net local heat flux for ribbed side wall
$\overline{St}$ = average Stanton number between x/D=2.85 and 16.81
TB = bulk mean air temperature
TW(S) = local wall temperature for smooth side wall
TW(R) = local wall temperature for ribbed side wall
X = axial distance for the heated test duct
ρ = average density of fluid

REFERENCE

1.   White, L., and Wilkie, D., The Heat Transfer and Pressure Loss Characteristics of some Multi-start Ribbed Surfaces, Augmentation of Convective Heat and Mass Transfer, edited by A.E Bergles and R. L. Webb, ASME, New York, 1970.

2.   Han, J. C., Glicksman, L. R., and Rohsenow, W. M., An Investigation of Heat Transfer and Friction for Rib-Roughened Surfaces, International Journal of Heat and Mass Transfer, Vol.21, pp. 1143-1156, 1978.

3.   Gee, D. L., and Webb, R. L., Forced Convection Heat Transfer in Helically Rib-Roughened Tubes, International Journal of Heat and Mass Transfer, Vol.23, pp. 1127-1136, 1980.

4. Sethumadhavan, R., and Rao, M. R., Turbulent Flow Heat
    Transfer and Fluid Friction in Helical-Wire-Coil-Inserted
    Tubes, International Journal of Heat and Mass Transfer,
    Vol.26, pp.1833-1844, 1983.

5. Suo, M., Turbine Cooling, in: The Aerothermaodynamics of
    Aircraft Gas Turbine Engines, edited by Oates, G., Air Force
    Aero Propulsion Lab., AFAPL TR 78-52, 1978.

6. Taylor, J. R., Heat Transfer Phenomena in Gas Turbine, ASME
    Paper No. 80-GT-172, 1980.

7. Burggraf, F., Experimental Heat Transfer and Pressure Drop
    with Two-Dimensional Turbulence Promoter Applied to Two
    Opposite Walls of a Square Tube, Augmentation of Convective
    Heat and Mass Transfer, edited by A. E. Bergles and R. L.
    Webb, ASME, New York, pp. 70-79, 1970.

8. Han, J. C., Park, J. S., and Lei, C. K., Heat Transfer
    Enhancement in Channels with Turbulence Promoters, ASME
    Journal of Engineering for Gas Turbines and Power, Vol. 107,
    pp. 628-635, 1985.

9. Park, J. S., Turbulent Heat Transfer Enhancement in
    Rectangular Channels with Two Opposite Rib-Roughened Walls,
    Ph.D Thesis, Department of Mechanical Engineering, Texas A&M
    University, 1986.
10. Han, J. C., Park, J. S., and Lei, C. K., Heat Transfer and
    Pressure Drop in Blade Cooling Channels with Turbulence
    Promoters, NASA Contractor Report 3837, Grant NAG 3-311, Nov.
    1984.

11. Kline, S. J., and McClintock, F. A., Describing Uncertainties
    in Single-Sample Experiments, Mechanical Engineering, Vol.
    75, pp. 3-8, Jan. 1953.

Heat Transfer Characteristics in a Circular Tube Downstream of Mixing Junctions

YOUSUKE KAWASHIMA, KAZUHIRO MURAI, and SHIGEYASU NAKANISHI
Department of Chemical Engineering
Himeji Institute of Technology
2167 Shosha, Himeji, Hyogo, 671-22 Japan

INTRODUCTION

The mixing junction is of great importance in many fields as a element of pipeline systems, and a large number of works have been done on energy losses for many types of mixing junctions[1]. However, only a few works has been done on heat transfer in mixing junctions, particularly in wyes.

The inflow of the branch stream to the main stream at the mixing junction causes high flow distortions and disturbances in the merged downstream flow from the point of mixing, depending on the cross-sectional area ratio of the main pipe to the branch pipe, the flow rate ratio of the main stream to the merged downstream and the crossing angle of the branch pipe to the main pipe (i.e. branch angle). Such complicated flow strongly affects heat transfer coefficients in the region situated downstream of mixing junction.

Sparrow and Kemink[5] studied heat transfer for turbulent air flow in a circular tube situated downstream of a mixing tee with cross-sectional area ratio of 1.0. They showed that the mixing of two pipe flows gave rise to a substantial augmentation of heat transfer compared with that in a conventional thermal entrance region.

In our previous papers[2], we discussed the effects of the confluent flow on heat transfer in a rectangular tube situated downstream of mixing wyes and tees, and also showed that the heat transfer coefficients were much affected by the cross-sectional area ratio and/or the branch angle.

In this study, we experimentally investigated heat transfer for turbulent air flow in the relatively short region of a circular tube situated just downstream of mixing wyes and tees under conditions of constant heat flux, and examined the effects of the cross-sectional area ratio, the flow rate ratio and the branch angle on the axial and circumferential distributions of heat transfer coefficient.

EXPERIMENTAL APPARATUS

The schematic diagram of the experimental apparatus used in this work is shown in figure 1. The main stream flows from the left to the right in the figure. The main stream consists of a circular plastic tube having an internal diameter of 51.7mm. Three plastic tubes having internal diameters of 51.7mm, 29.0mm and 19.8mm, respectively, were fitted for the branch stream. The mixing junctions were machined of plastic resin block.

197

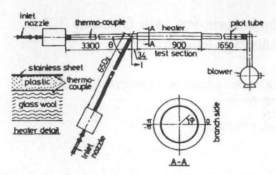

FIGURE 1. Schematic diagrams of experimental apparatus

The lengths of the upstream entrance region in the main and branch streams
are $63D_1$ and $65D_2$, respectively, and the downstream region from the point
of mixing is about $50D_1$ long. Calibrated inlet nozzles were attached to
the upstream ends of the main and branch pipes to measure the flow rates
of the main and branch streams. The total flow rate was checked by a
pitot tube located at the exit of the main pipe.

The length of the heated test section is 900mm, and the axial
distance from the downstream corner of the mixing junction to the start of
heating is 34mm, as shown in figure 1. The test section was heated by an
axial A-C electric current through 0.1mm thick stainless steel sheet which
were sticked to the inner wall surface of the test section tube which was
made of plastics (polyvinylchloride). The electric power supplied to the
test section is about 80Watt. The temperatures of the heat transfer
surface were measured with 120 thermo-couples attached to the stainless
steel sheet at 10 axial positions. At each position, twelve thermo-couples
were placed around the circumference with a uniform interval. The first
set of thermo-couples is at $l/D_1=1.0$. Several thermo-couples affixed to
the adiabatic side of the test section tube to check the heat loss through
the test section tube wall (15mm thick). The inlet and exit bulk
temperatures were measured with two thermo-couples inserted in the
upstream and downstream of the main stream. The pipeline system was
insulated with glass wool thermal insulator.

The experiments were carried out under conditions of constant heat
flux (about $550W/m^2$), and air is the working fluid. The cross-sectional
area ratios m employed in these experiments are 1.0, 3.18 and 6.82, and
the branch angles θ are 50°, 70°, 90°, 110° and 130°. The test section
Reynolds numbers Re are 30000 and 40000. The flow rate ratio Q_1/Q_3 was
varied over the range 0~1.0 (for the m=6.82 case, the range of the flow
rate ratio is from 0.2~1.0 and the test section Reynolds number is 30000
only, because of the capacity limitation of the blower).

RESULTS AND DISCUSSIONS

Local Nusselt number $Nu_x(\phi)$ is evaluated from

$$Nu_x(\phi)=D_1h_x(\phi)/\kappa_x , \quad h_x(\phi)=q/(t_{w,x}(\phi)-t_{b,x}) \tag{1}$$

where q is heat flux which is determined by subtracting the heat loss by
conduction across the test section tube wall from the heating power of the
heater, $t_{b,x}$ is the bulk temperature calculated from the heat flux q and
the total flow rate Q_3, κ_x is the thermal conductivity of air at $t_{b,x}$.

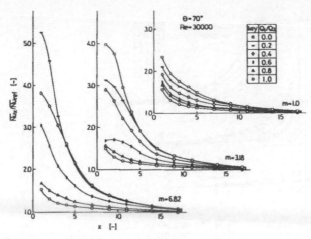

FIGURE 2. Representative results of the effects of m and Q_1/Q_3 on $\overline{Nu}_x/\overline{Nu}_{ref}$

Circumferentially and Axially Averaged Local Nusselt Numbers

In the downstream region from the point of mixing, there is a variety of complicated flow patterns depending on Q_1/Q_3, m and θ, and it is expected that these complicated flows affect the local and the average heat transfer characteristics very strongly. Thus, in this place, we examine the effects of Q_1/Q_3, m and θ on the circumferentially and axially averaged heat transfer characteristics.

Figure 2 shows the experimental results for $\theta=70°$ and Re=30000 as a representative example showing the effects of the cross-sectional area ratio m and the flow rate ratio Q_1/Q_3 on the circumferentially averaged local Nusselt number $\overline{Nu}_x$ normalized by $\overline{Nu}_{ref}$. Where, $\overline{Nu}_{ref}$ is the circumferentially averaged Nusselt number at x=17 (i.e. at the most downstream position in the present experiment) for $Q_1/Q_3=1.0$.

In the $Q_1/Q_3=1.0$ case, only the main stream flows straight through the main pipe with no inflow of the branch stream, and hence the stream is affected only by the aperture of the branch pipe in the wall at the junction. Thus, as shown in graphs, in the case of $Q_1/Q_3=1.0$, the $\overline{Nu}_x/\overline{Nu}_{ref}$ curves are almost invariant irrespective of m, and agree with those in a conventional thermal entrance region[4]. However, as Q_1/Q_3 becomes smaller, the velocity of the branch stream becomes faster, and the effect of the inflow of the branch stream on heat transfer becomes stronger in the downstream region. In the same way, the velocity of the branch stream also becomes faster with increasing m. Thus, on the whole, the value of $\overline{Nu}_x/\overline{Nu}_{ref}$ increases as the Q_1/Q_3 ratio decreases and as the m value increases.

Figure 3 shows representative results of the effect of the branch angle θ on the $\overline{Nu}_x/\overline{Nu}_{ref}$ ratio. In graph (a) for m=1.0 and in graph (b) for m=6.82. Each of these graphs is a composite of three graphs, respectively for $\theta=50°$, $\theta=90°$ and $\theta=130°$. In the m=1.0 case, the differences of the $\overline{Nu}_x/\overline{Nu}_{ref}$ value between the cases of $Q_1/Q_3=1.0$ and 0.0 becomes greater as the branch angle θ increases. However, in the m=6.82 case, the differences between the cases of $Q_1/Q_3=1.0$ and 0.2 becomes smaller as θ increases.

Figure 4 shows the relative value of the axially averaged Nusselt

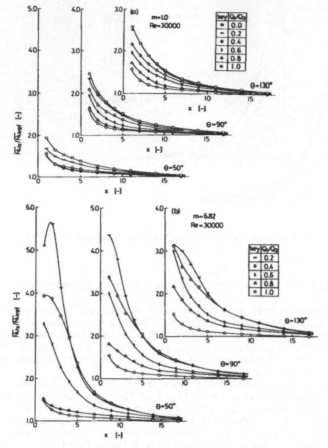

FIGURE 3. Representative results of the effect of θ on $\overline{Nu}_x/\overline{Nu}_{ref}$

number expressed in terms of the ratio λ_x, defined as

$$\lambda_x = \int_{x_s}^{x} Nu_x dx \bigg/ \int_{x_s}^{x} Nu_{x,1.0} dx \qquad (x_s = 1.0) \qquad (2)$$

Namely, λ_x is the Nusselt number axially averaged over the range $x = x_s \sim x$, normalized by that for the case of $Q_1/Q_3 = 1.0$. In graph (a) for $m=1.0$, in graph (b) for $m=3.16$ and in graph (c) for $m=6.82$. Each of these graphs is composed of three graphs for $x=4$, $x=9$ and $x=17$, respectively. As shown in graphs, in the $m=1.0$ case, the ratio λ_x becomes larger with increasing θ in all range of Q_1/Q_3. However, in cases with larger m, the behavior of λ_x in the small Q_1/Q_3 range differs from that in the large Q_1/Q_3 range. For example, in the $m=3.18$ case, the ratio λ_x becomes larger with increasing θ in the $Q_1/Q_3 = 0.65 \sim 1.0$ range, while in the $Q_1/Q_3 = 0 \sim 0.65$ range, λ_x becomes smaller with increasing θ. Similarly, in the $m=6.82$ case, the border value of Q_1/Q_3 is about 0.4. At the small Q_1/Q_3 and the large m, the branch stream with high velocity jets through the main stream and impinges on the opposite wall to the branch opening, and then spreads along the pipe wall surface. Beside, the position of the impingement shifts toward the downstream with decreasing θ. On the other hand, in the $m=1.0$ case, flow separation and reattachment occur at the region just

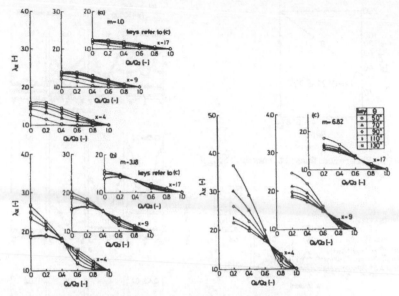

FIGURE 4. Normalized axial average Nusselt number, λ_x

downstream of the junction, and the branch stream does not impinge
strongly on the pipe wall except for the $Q_1/Q_3=0$ case. These complicated
flow patterns are conjectured to be the reason for these results.

Figure 5 plots the reference Nusselt number $\overline{Nu}_{ref}$ against the test
section Reynolds number Re. The values of $\overline{Nu}_{ref}$ closely agree with
the predictions of the Dittus-Boelter correlation[3] for circular tube,
which are drawn in the figure for the purpose of comparison.

Circumferential Variations of Local Nusselt Numbers

As mentioned above, at the small Q_1/Q_3, the branch stream flows
through the main stream and impinges on the opposite wall ,and generates
the strong circumferential secondary flow along the pipe wall surface. It
is considered that these flow patterns produce significant circumferential
variations of heat transfer coefficient.

Figure 6 shows representative results of the circumferential
variations of the local Nusselt numbers, which are normalized by those for
the $Q_1/Q_3=1.0$ case at every measurement position. In figure (a) for m=1.0
and in figure (b) for m=6.82. Both figures are composed of three graphs,
respectively, for $\theta=50°$, $\theta=90°$ and $\theta=130°$. In each graph, the ratio $Nu_x(\phi)$
/$Nu_{x,1.0}(\phi)$ is plotted as a function of ϕ; $\phi=0$ corresponds to the side of
the inlet port of the branch stream and $\phi=\pi$ (or $-\pi$) corresponds to the
side opposite to the inlet port.

In the m=1.0 case, the largest variations occur at the first
measurement station (x=1), and , by and large, the maximum is at $\phi=0$ and
the minimum at $\phi=\pi(-\pi)$. In most of cases, the circumferential variations
almost decay at the x=17 position, but in some case still remain there. On
the other hand, in the m=6.82 case, the most significant variations also
are at the x=1 position, but the maximum is at $\phi=\pi(-\pi)$ and the minimum at
$\phi=0$. The reason for these results is conjectured to be that in the m=1.0
case, the flow separation and reattachment cause substantial augmentation
of heat transfer coefficients, while in the m=6.82 case, the impingement
of the jet of the branch stream gives rise to enhance heat transfer

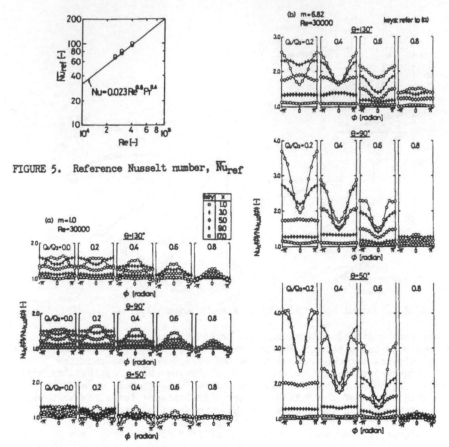

FIGURE 5. Reference Nusselt number, $\overline{Nu}_{ref}$

FIGURE 6. Representative results of circumferential distributions of $Nu_x(\phi)/Nu_{x,1.0}(\phi)$

coefficients and their circumferential variations.

CONCLUSIONS

The heat transfer characteristics for a turbulent airflow in a circular tube situated downstream of mixing wyes and tees were experimentally obtained, and the effects of the flow rate ratio Q_1/Q_3, the cross-sectional area ratio m and the branch angle θ on the local and the average heat transfer characteristics were investigated. The following conclusions were drawn.

The axial distributions of the circumferential average local Nusselt number $\overline{Nu}_x$ and the circumferential variations of the local Nusselt number $Nu_x(\phi)$ were strongly influenced by the cross-sectional area ratio m, the flow rate ratio Q_1/Q_3 and the branch angle θ.

In the case of m=1.0, the effect of the confluent flow on heat transfer coefficients became stronger with increase of θ for any floe rate ratio. On the other hand, in the cases of m=3.18 and 6.82, it became

202

stronger with decrease of θ for large flow rate ratio, while for small floe rate ratio the tendency was the same as the case of m=1.0.

The flow separation and reattachment and the impingement of the jet of the branch stream cause the circumferential variations of the heat transfer coefficient. The most significant variations occurs at the first measurement position. In the most case, the circumferential variations almost decay at the x=17 position, but in same case still remain there.

NOMENCLATURE

D   = pipe diameter, m
$h(\phi)$ = local heat transfer coefficient at angular position ϕ, $W/(m^2 K)$
l   = distance from the corner of mixing junction, m
m   = cross-sectional area ratio of the main pipe to the branch pipe, -
$\overline{Nu}$ = Circumferentially averaged Nusselt number, -
$Nu(\phi)$= local Nusselt number at angular position ϕ, -
Q   = flow rate, m^3/s
q   = heat flux, W/m^2
Re   = Reynolds number at the test section $(=\overline{V}_3 D_1/\nu)$, -
t   = temperature, °C, K
$\overline{V}$ = mean velocity, m/s
x   = dimensionless distance $(=1/D_1)$, -
x_s = dimensionless distance of the first measurement position, -
θ   = branch angle, °(degree)
κ_x = thermal conductivity of air, $W/(m K)$
λ_x = normalized axially and circumferentially averaged Nusselt number, defined by Eq.(2), -
ν = kinematic viscosity of air, m^2/s
ϕ = angular coordinate (see Figure 1), radian
<Subscripts>
1   = upstream in main stream
2   = branch stream
3   = downstream in main stream
b,x   = bulk condition at a distance x
ref   = reference station
w,x   = wall condition at a distance x
$\dot{x}$ = axial local condition at distance x
x,1.0= axial local condition at distance x when $Q_1/Q_3=1.0$

REFERENCES

1. JSME Data Book, Hydraulic Loss in Pipes and Ducts, pp.86-96, 1979

2. Kawashima, Y., Murai, K., Nakanishi, S. and Iuchi, S., Heat Transfer Characteristics in a Rectangular Mixing Junction, 2nd ASME-JSME Thermal Eng. Joint Conf., vol.2, pp.203-209, 1987

3. McAdams, W. H., Heat Transmission, 3n ed., McGraw-Hill, New York, p.219, 1984

4. Mills, A. F., Experimental Investigation of Turbulent Heat Transfer in a Entrance Region of a Circular Conduit, J. of Mech. Eng. Sci., vol.4, pp.63-77, 1962

5. Sparrow, E. M. and Kemink, R. G., The Effect of a Mixing Tee on Turbulent Heat Transfer in a Tube, Int. J. of Heat Mass Transfer, vol.22, pp.909-916, 1979

Friction Reduction and Heat Transfer Enhancement in Turbulent Pipe Flow of Non-Newtonian Liquid-Solid Mixtures

U. S. CHOI and K. V. LIU
Materials and Components Technology Division
Argonne National Laboratory
Argonne, Illinois 60439, USA

ABSTRACT

Argonne National Laboratory (ANL) has identified two concepts for developing advanced energy transmission fluids for thermal systems, in particular district heating and cooling systems. A test series was conducted at ANL to prove these concepts. This paper presents experimental results and discusses the degradation behavior of linear polymer additives and the flow and heat transfer characteristics of non-melting slurry flows. The test results furnished strong evidence that the use of friction-reducing additives and slurries can yield improved thermal-hydraulic performance of thermal systems.

1. INTRODUCTION

Many thermal systems are characterized by long runs of piping that convey a pumped energy transmission fluid between the source and sink heat exchangers of the system. Furthermore, these systems often operate with small temperature differences, so large volumes of fluid must be pumped and large heat exchangers must be used to satisfy load demands. In order to improve the performance of these thermal systems, Argonne National Laboratory is developing advanced energy transmission fluids based on the following two concepts [1-4]:

1. Utilization of very low concentrations (20-2000 wppm) of non-Newtonian additives to the appropriate carrier liquid to reduce frictional flow losses by 20-80%;

2. Utilization of a pumpable phase change slurry comprising particulates of a material with a high heat of fusion, conveyed by a liquid, to enhance both bulk convective energy transport and possibly heat transfer coefficients at heat exchanger surfaces.

The advanced energy transmission fluid thus comprises the appropriate friction-reducing additive, phase change particles, and carrier liquid. Some of the important technical issues related to implementation of the above concepts are (1) the degradation of non-Newtonian additives under continuous flow shear in a closed recirculating system, (2) the possibility of a significant increase in the pumping power required to move highly loaded slurries, (3) the feasibility of using friction-reducing additives with highly loaded phase change slurries, and (4) the influence of phase change particles on heat transfer in turbulent pipe flow.

Since these advanced fluids are non-Newtonian solid-liquid mixtures, little is known about the effects of the multiphase flow on frictional pressure drop, heat transfer, and longevity of friction-reducing additives. Therefore, pressure drop and heat transfer in turbulent pipe flows of advanced energy transmission fluids have been measured to generate experimental data to support the proposed concepts.

2. EXPERIMENTAL APPARATUS AND PROCEDURES

2.1 Friction-reducing Additive Degradation Test Facility

Many thermal systems are closed systems, and working-fluid recirculation is used to transport thermal energy. Some friction-reducing additives are known to be very effective in friction reduction but degrade fairly rapidly under recirculatory flow shear. To date, no systematic, carefully controlled experiments have been conducted on friction-reducing additives to characterize and compare tendencies for degradation. Therefore, it was essential to design and build a new test facility for this purpose.

The test apparatus that has been built is a pumped closed recirculatory loop system with a 6.4-m-long test section of 0.7747-cm-inside-diameter pipe in which both pressure drop and heat transfer can be measured as a function of time. Differential pressure was measured in the fully developed flow region by using two pressure taps separated by 4.27 m. A concentric-pipe heat transfer test section, simulating counterflow heat exchanger operation, was fabricated to measure the turbulent heat transfer. The heat transfer test section was preceded by a 2-m-long hydrodynamic development section. Thus, the hydrodynamically developed flow could be obtained at the inlet of the heat transfer test section. This apparatus will play a key role in quantifying the longevity of friction-reducing additives and identifying the most promising ones for closed-loop thermal systems.

Long-duration degradation testing was performed with two linear-polymer additives under the recirculatory flow shear conditions encountered in a typical thermal system. The additives were Polyox WSR-301, an ethylene oxide polymer manufactured by Union Carbide, and Separan AP-273, a polyacrylamide manufactured by Dow Chemical, with deionized water as the solvent.

2.2 Slurry Flow and Heat Transfer Test Facility

Proof-of-concept experiments on flow and heat transfer in highly loaded solid/liquid slurries and other non-Newtonian fluid flows have been conducted in the Slurry Flow and Heat Transfer Test Facility at Argonne. The major components of the test facility are a 3000-L (800-gal) supply tank, a 3000-L storage tank, a 380-L (100-gal) weighing tank, two progressing-cavity slurry pumps with pumping capacities of 0.063-3.786 L/s (1-60 gal/min), and two test sections. The test sections are straight tubes, approximately 6 m (20 ft) long. The test section used in the present work is made of stainless steel, with an inside diameter of 23.98 mm (0.944 in.). For heat transfer experiments, two DC electrical power supplies connected in series can supply up to 60 kW to this test section. Thus the slurry heat transfer tests were conducted under a constant-heat-flux boundary condition.

The instrumentation for the pressure drop and mass flow measurements consists of two pressure taps, a 68.9 kPa differential-pressure transducer, and a weighing tank. The pressure taps are placed 5.97 m (19.7 ft) apart in the tube wall of the stainless steel test section. An injection tap is placed in the tube wall upstream of the first pressure tap in order to ensure adequate mixing of the friction-reducing additive solution with test fluids in the test section. The differential-pressure transducer is used to measure the pressure drop between the two pressure taps. The instrumentation for the heat transfer measurements consists of 12 thermocouples and 6 voltage taps, all mounted on the outer tube wall. Thermocouples are also inserted in the supply tank to measure the slurry temperature in the tank. All test data, including pressure drop, mass flow, temperature, and voltage drop, are recorded by a Fluke Model 2285B data logger.

Experiments were performed to determine the effects of particle loading and particle size on frictional flow loss and heat transfer for a non-melting slurry. The particles used in this study were cross-linked form-stable high-density polyethylene (X-HDPE), 3.2 mm (1/8 in.) and 1.3 mm (1/20 in.) in diameter. The nominal particle volumetric loading, ϕ, was varied from 0 to 45%. In addition to tests with pure water and with a slurry of water and X-HDPE particles, pressure drop tests were conducted with water containing Polyox, and with an X-HDPE slurry (15% loading) containing Polyox.

3. RESULTS AND DISCUSSION

3.1 Degradation Tests on Friction-reducing Additives

A 76-kg (167.0-lb) quantity of 200-wppm Separan AP-273 solution in deionized water was circulated in the degradation test loop at a constant flow rate of 9.53 kg/min (21 lb/min) for a period of 730 hours at room temperature (25.0°C). The pressure drop and heat transfer data, expressed as percent friction and heat transfer reduction relative to pure water, are plotted as a function of time in Fig. 1.

At the start of the experiment, the fresh Separan solution exhibited significant friction and heat transfer reductions (over 60%) relative to water. The percent friction and heat transfer reductions decreased rapidly thereafter, and arrived at plateau values at about 240 hours of circulation. From 240 to 730 hours of shear, little additional change was observed in the friction and heat transfer coefficients. Even at 730 hours of shear, the pressure drop data represented about 25% friction reduction relative to pure water, as shown in Fig. 1. This is an important finding because a friction reduction of even 10% in thermal systems could result in large annual cost savings.

Initially, the percent heat transfer reduction (i.e., 78%) was larger than the percent friction reduction (about 65%). However, at 60 hours of shear, the percent friction reduction was about the same as the percent heat transfer reduction. After this point, the former was found

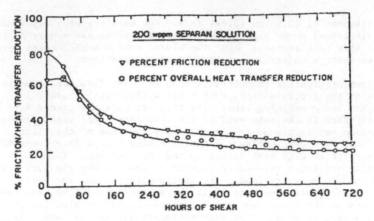

FIGURE 1. Percent friction and heat transfer reduction relative to pure
water as a function of circulation time for 200 wppm Separan solution at
25°C.

to be consistently larger than the latter by about 5%. The heat
transfer data at 730 hours of shear correspond to about 22% heat
transfer reduction relative to pure water. Although the heat transfer
reduction is not desirable, the advantage of friction reduction
outweighs the disadvantage of heat transfer reduction.

A 76-kg quantity of 200-wppm Polyox WSR-301 solution was prepared and
circulated at a constant flow rate of 9.53 kg/min in the same flow loop
at 25.0°C for 22 hours. Figure 2 shows the percent friction and heat
transfer reduction as a function of time. Although Polyox was as
effective as Separan when the solution was fresh, its friction-reducing
properties degraded very rapidly (compared with those of Separan
solution) as soon as circulation started; Newtonian values were reached
at about 18 hours of shear. The 200 wppm Separan solution took about
240 hours, or more than one order of magnitude longer, to reach
asymptotic values (Fig. 1). Clearly, Polyox solution was much more
susceptible to mechanical degradation than Separan solution at the same
concentration. This indicates that Polyox is not suitable at all for
closed-loop thermal system applications. As shown in Fig. 2, the
percent friction reduction for Polyox was consistently larger than the
percent heat transfer reduction, as was the case with Separan solution
after 60 hours of shear.

3.2 Slurry Pressure Drop and Heat Transfer Tests

The experimental data on slurry pressure drop as a function of particle
loading and size are shown in Fig. 3. The pressure drop associated with
the presence of 3.2 mm X-HDPE particles at low particle volumetric
loadings is not greatly increased over that for pure water at the same
mass flow rate. For loadings of less than 5%, in fact, a reduction in
pressure drop was observed. When the loading was greater than 30%, the
increase of pressure drop for the slurry relative to that of pure water
became great enough to affect pump performance.

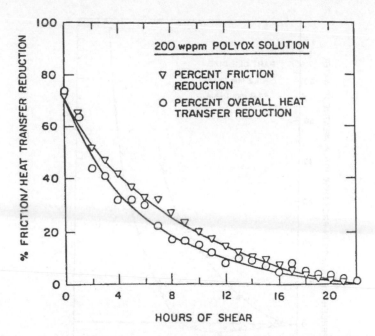

FIGURE 2. Percent friction and heat transfer reduction relative to pure water as a function of circulation time for 200 wppm Polyox solution at 25°C.

For slurry flows composed of 1.3-mm X-HDPE particles, a clear and consistent pressure drop reduction phenomenon was observed at low to medium volumetric loadings, below 30%. As in the case of the larger particles, the pressure drop increase became significant as the volumetric loadings exceeded 35-40%. The results from the current studies have demonstrated that (1) a pressure drop reduction phenomenon occurs at low volumetric loadings; (2) there exists a threshold volumetric loading (which is a function of particle size) above which the pressure drop reduction phenomenon disappears; (3) the flow regime and, hence, the pressure drop behavior appear to change from turbulent to quasi-laminar when the volumetric loading exceeds a threshold value; and (4) below the threshold loading value, the slurries can be pumped with little penalty in pressure drop relative to pure water.

For a turbulent slurry pipe flow with large particles, there is little consensus as to how the heat transfer is expressed. In the present study, the nonuniform electrical heating along the test section, the difference in thermophysical properties between phases, the degree of thermal equilibrium, the particle volumetric loading, and the particle size all affected the slurry heat transfer behavior. With limited experimental results obtained with one particle size (3.2-mm diameter), our first approach was to analyze the heat transfer data as one would for single-phase turbulent pipe flow. The related nondimensional parameters used were the Nusselt and Reynolds numbers.

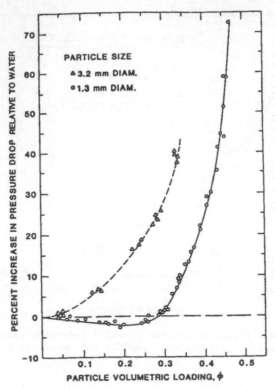

FIGURE 3. Percent pressure drop change as a function of X-HDPE particle loading and size.

With particle volumetric loadings of <30%, the relative slurry viscosity can be predicted fairly well by an empirical correlation [5] and expressed as a function of the particle volumetric loading alone. Therefore, a modified Reynolds number was defined on the basis of an effective slurry viscosity. The thermal conductivity used in the Nusselt number calculation was the fluid thermal conductivity, because no effective thermal conductivity data were available. The Nusselt number for slurry heat transfer, when plotted versus the modified slurry Reynolds number based on an effective viscosity, is larger than that for pure water and increases with increasing particle volumetric loading. Significant Nusselt number enhancement can be observed in Fig. 4. Over the entire Reynolds number range obtained with the 30% loading, the Nusselt number increases by a factor of 2.5 or more. In general, no satisfactory approach has been developed for correlating slurry heat transfer data of the type generated in this study. Much effort has been expended to utilize single-phase heat transfer parameters and methods. This approach is probably not adequate. The use of an effective slurry viscosity with conventional correlation parameters may have exaggerated the increase in heat transfer with particle volumetric loading.

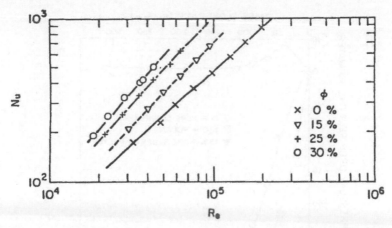

FIGURE 4. Variation of heat transfer Nusselt number (Nu) with effective pipe Reynolds number (Re) for slurries of X-HDPE and pure water at various slurry loadings (φ).

3.3 Combined-Concept Pressure Drop Tests

In order to demonstrate the feasibility of using friction-reducing additives with X-HDPE slurries, pressure drop tests were conducted with Polyox, both in water and in a 15% slurry of 3.2-mm-diameter X-HDPE particles. The solution was mixed to an initial concentration of 2000 wppm and injected into the pipe flow through the injection tap. The final concentrations of the additive varied from 30 to 200 wppm, depending on the flow rates of the test fluids. The maximum flow rate of 148 kg/min yielded the minimum additive concentration of 30 wppm. Although, as noted earlier, Polyox degrades too rapidly for long-term application, it was considered adequate for use in short-term experiments to test the proposed concepts.

Pressure drop data for a 15% X-HDPE slurry, water with Polyox, and a 15% X-HDPE slurry with Polyox were replotted in Fig. 5 to show the percent pressure drop change [i.e., the ratio between frictional pressure drop for the slurry flow and the frictional pressure drop for the corresponding single-phase (pure water) flow] as a function of mass flow rate. Some very interesting trends in pressure drop were observed for the slurry without Polyox. For example, at very low mass flows, the pressure drop was 7% greater than that for pure water. Since the X-HDPE particles are slighly less dense than water, this increase is believed to be due to stratified flow, with the particles concentrating toward the top of the pipe and causing nonuniform particle distributions across the pipe. Visual observation and video recordings of flow patterns confirmed partial stratification at low flow rates. In the stratified region, particles near the wall may cause increased friction, resulting in increased velocity and pressure drop. At high flow rates, the particles were uniformly distributed over the flow field owing to strong turbulence, and the pressure drop penalty relative to pure water became negligible. This pressure drop trend for a buoyant particle slurry is expected to be different from that of a neutrally buoyant particle slurry.

211

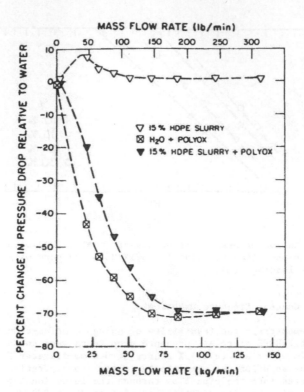

FIGURE 5. Percent pressure drop change relative to water vs. mass flow rate.

For flows with the friction-reducing additive, the mass flow rate of pure water or slurry was an important parameter in determining the effectiveness of the additive in reducing pressure drop. At low mass flow rates the friction-reducing additive was twice as effective for pure water as for slurry (see Fig. 5). However, at high mass flow rates, the percent pressure drop reduction due to the friction-reducing additive was almost identical for pure water and slurry. The maximum value of the pressure drop reduction was approximately 70%.

4. CONCLUSIONS

A series of pressure drop and heat transfer tests have been conducted to prove the concepts developed for advanced energy transmission fluids. The test results demonstrate the following:

1. Separan does not degrade completely but maintains a plateau value of about 25% friction reduction after the initial degradation period;

2. Polyox cannot be used in recirculatory systems because it degrades completely in a short period of time;

3. Friction reduction is always accompanied by heat transfer reduction;

4. Under turbulent pipe flow conditions, X-HDPE slurries at volumetric loadings below 30% show pressure drop increases of less than 25% relative to pure water (a conventional energy transmission fluid);

5. Highly loaded slurries exhibit increased heat transfer relative to pure water; and

6. A friction-reducing additive can be successfully combined with a phase change slurry for significant pressure drop reduction.

ACKNOWLEDGMENTS

Work supported by the U.S. Department of Energy, Conservation and Renewable Energy, Office of Buildings and Community Systems, under contract No. W-31-109-Eng-38. Appreciation is expressed to Mr. J. Kaminsky for creating supportive conditions for carrying out work of this kind. Special thanks to C. A. Bertino for typing of the manuscript and to E. M. Stefanski for her expert editorial assistance.

REFERENCES

1. Kasza, K. E., Choi, U. S., and Kaminski, J., "Optimal Energy Transmission Fluids for District Heating and Cooling Applications," Proc. 77th Int. District Heating and Cooling Association Annual Conference, Asheville, NC, Vol. 77, pp. 163-172, 1986.

2. Kasza, K. E., Choi, U. S., and Kaminski, J., "Reducing the High Costs of District Heating and Cooling," Specifying Engineer, Vol. 56, no. 4, pp. 38-42, October 1986.

3. Kasza, K. E., and Chen, M. M., "Development of Enhanced Heat Transfer/Transport/Storage Slurries for Thermal System Improvement," Argonne National Laboratory, ANL-82-50, June 1982.

4. Kasza, K. E., and Chen, M. M., "Improvement of the Performance of Solar Energy or Waste Heat Utilization Systems by Using Phase-Change Slurry as an Enhanced Heat-Transfer Storage Fluid," ASME Journal of Solar Engineering, Vol. 107, pp. 229-236, August 1985.

5. Thomas, D. G., "Transport Characteristics of Suspension: VIII. A Note on the Viscosity of Newtonian Suspensions of Uniform Spherical Particles," Journal of Colloid Science, Vol. 20, pp. 267-277, March 1965.

Heat Transfer Enhancement by Jet Array Impingement and Its Thermal Pattern Visualization

LI-GWO LI, JING JIANG, and JIAN-LONG LIN
Nanjing Aeronautical Institute
Nanjing, PRC

INTRODUCTION

As an effective method to enhance heat transfer, impingement by air jets with high velocity is widely used for the cooling of gas turbine airfoils and electronic equipments, as well as drying of paper and textiles. The jet air from the orifice plate, after impingement, is constrained to flow in a single direction along the channel between the cooled target surface and the jet orifice plate to its exit. Thus, in this configuration, the jets in the upstream impose a confined crossflow on the downstream of the jets. In some cases there is an initial crossflow passing the jet array, making the flow field much more complicated. Thus, due to complexity of problem, it is necessary to develop a method for measuring heat transfer distributions with high resolution.

A test model is provided with cholesteric liquid crystals, by virtue of their property of changing color with temperature for a continous measurement of temperature variations, and thus a clear visualization of thermal pattern can also be obtained. A heating film even coated with liquid crystals is attached tightly to the test surface, and the local heat transfer as well as their distributions can be measured from heat balance, so long as the electrical input power is

_____

Project aided by the National Commission of Natural Scientific Funds.

measured. The reference [1] gives the actual quantitative, continuous heat transfer coefficients on a turbine blade model with this method and shows that both the temperature pattern observed and the heat transfer coefficients found agree well with analytical heat transfer calculations.

In this paper, the impingement cooling for an array of 7×10 holes of 2.4mm in diameter in both inline and staggered arrangements with and without initial crossflow are investigated with nondimensional streamwise hole spacing Xn/d of 5, spanwise hole spacing Yn/d of 6.67 and the channel height Z/d ranged from 1 to 3. The relative cooling effectiveness of various jet impingement coolings can be indicated with the dark areas surrounded by the yellow isotherm of liquid crystals. Furthermore, heat transfer distribution is obtained from these measurements of dark areas, while the heat transfer coefficient represented by the yellow isotherm.

EXPERIMENTAL FACILITY AND PROCEDURE

Experimental Facility Experimental facility includes a jet orifice plate 1 of 2.4 mm thick, a U-shaped asbestos rubber spacer 2, an electrocaloric sheet of liquid crystals 3, and a transparent acrylic plastic plate 4 (Fig.1).All the jets after impingement exhaust from the opening of the U-spacer. Adjusting thickness of spacer will give some different non-dimensional distances of $Z/d=1$, 2, and 3. The jet orifice plate are made in two different arrangements, inline and staggered, with the same geometric parameters.

The total jet flow rates (m_j) are measured by a standard orifice plate flowmeter lacated at the flow metering section, upstream of the plenum.The initial crossflow air is supplied separately from the other plenum chamber. Electrocaloric liquid crystal sheet of 0.1 mm thick, acting as a target for impingement cooling, is attached to the transparent acrylic plastic plate 4, so that it is feasible to take color photograph of liquid crystal indication at a certain specified place.

The characteristics of the impingement heat transfer can be shown on the thermal pattern according to the color of the liquid crystals. The yellow color is used as a temperature standard and as a color standard for indicating heat transfer characteristics. Result of calibration shows that the temperature of the yellow color isotherm is about $42^{\circ}C$.

Experimental Procedure All tests are carried out with the same jet mass flow rate or with the same initial crossflow and jet flow rate for purpose of analytical comparison of heat transfer characteristics. After the flow rate is kept steady, the electrocaloric liquid crystal sheet is heated by DC current to a temperature at which the colors begin to appear. The power input is then tuned to fine adjustment until the desired value is reached. The phtographs of the thermal

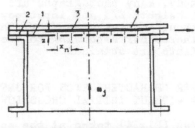

Fig.1   Test apparatus

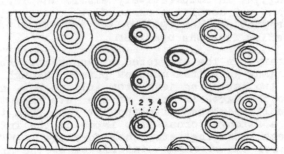

Fig.2 Streamwise distribution of average cooling area

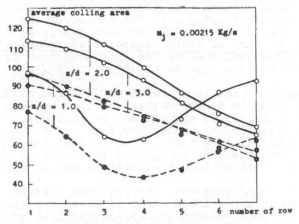

Fig.3   Isograms of heat transfer coefficient
distributions at different power inputs
$1-520W/(m^2.K), 2-415W/(m^2.K), 3-346W/(m^2.K), 4-260W/(m^2.K)$

patterns are taken from a specified point with a 35mm camera after the flow rate is steady for an interval of longer time, and then the voltage V, the current I are measured and recorded. Subsequently, other photographs are taken in the same way by changing Z/d.

In the present study, many photographs are also obtained for different channel heights Z/d and input powers under the same jet flow without initial crossflow. Thus, a series of yellow color isotherms corresponding to the local heat transfer coefficients are obtained.

HEAT TRANSFER CHARACTERISTICS FOR JET ARRAY IMPINGEMENT WITHOUT INITIAL CROSSFLOW

As all the pictures (Fig.4) taken at the specified point under the same power input (108w) with different hole arrangements and channel heights are enlarged in the same proportion, the cooling area surrounded by the yellow isotherm of the liquid crystal for each impinging hole is measured by means of AMSLER compensation planimeter and the algebraic-mean value along the spanwised rows is taken into consideration, the results as shown in Fig.2.

1. Influence of the Arrangement of Impingement Holes
    It is shown in Fig.2 that the average cooling area is larger for the inline arrangement of impinging holes than that for staggered one under the same condition for all values of Z/d (1,2, and 3). It means that for a given impinging flow rate the impingement cooling with inline arrangement is better than that with staggered arrangement.

In the inline arrangement, the spanwise distribution of the cross flow is non-uniform, most of the cross flows pass through the path between the streamwise rows of the holes with less flow resistance. The effect of the cross flow on any impinging jet is comparatively small. The impinging jets from the front row develop a cross flow to the jets from the rear row, protecting them against the influence of the strong crossflow from the upstream.

On the contrary, in the staggered arrangement, the air flow of the impinging jets from nth row passes through the space between the two impinging jets from (n+1)th row, and then is divided into two parts in front of the jets from the (n+2)th row, and finally combined with the jet flow from the (n+1)th row to form a stronger cross flow imposing the influence on the downstream impinging jets. Therefore the downstream impinging jets are strongly interfered by the cross flows in the staggered arrangement. Thus the cooling with inline arrangement is much better than that with staggered one.

2. Influence of the Distance to Diameter Ratio Z/d
    When other conditions keep the same, varying the channel height Z/d will change the velocity of cross flow and influence the the distribution of jet flow, resulting in the

218

change of the thermal patterns shown by the liquid crystal.

It is seen from Fig.2 that the cooling area is larger for
Z/d=2 than that for Z/d=1 and 3 regardless of the arrangement of impingement holes. For Z/d=2 and 3, the cooling area
is monotonously decreasing along the streamwise direction
with little difference between the both cases. The explanation of the decrease in cooling area is that the accumulation of the cross flow rate along the streamwise direction
imposes the interference on the jets.

At Z/d=1, the variation of cooling area versus the row
number was found to be quite different from that at Z/d=2
and 3: the cooling area at the upstream along the streamwise direction is decreased, first, to a minimum value, and
then, increased at the downstream regardless of the arrangement of impingement holes (Fig.2). There may exists a
certain critical ratio of the cross flow velocity to jet
velocity. When the ratio increases to a value greater than
the critical, the effect of the cross flow on the heat
transfer may become more significant than that of impinging
jet.

3. Distribution of Heat Transfer Coefficients
    The heat transfer coefficients represented by the yellow
isotherms at different power inputs may be determined by
means of the heat balance. The positions of yellow isotherms
at different power inputs are also equivalent to those of
heat transfer coefficient distributions. The heat transfer
coefficient can be obtained from heat balance as follows:

$$h = \frac{IV - Q_L}{A(T_y - T_r)} \qquad (1)$$

where T_y - temperature represented the yellow color of
liquid crystal.
    T_r - recovery temperature of gas, defined as:

$$T_r = T_j \left(1 + r \frac{K-1}{2} M_j^2\right) \qquad (2)$$

Because velocity of jet is not large, jet temperature is
approximately substituted for recovery temperature.
    A————effective area of electrocaloric liquid
            crystal sheet.
    Q_L————natural convective heat loss through trans-
            parent plastic plate to the ambient, deter-
            mined by heat transfer relations for vertical
            flat plate in air.

Figure 3 indicates the isograms of the heat transfer coefficient distributions at different power inputs for the
staggered array configuration at Z/d=1.0 (namely, the yellow
color isotherms of different power inputs). It will be seen
from Fig.3 that these figures truly show the features of
two-dimensional heat transfer for each jet, and the effect
of cross flow on impingement heat transfer may be directly
perceived.

219

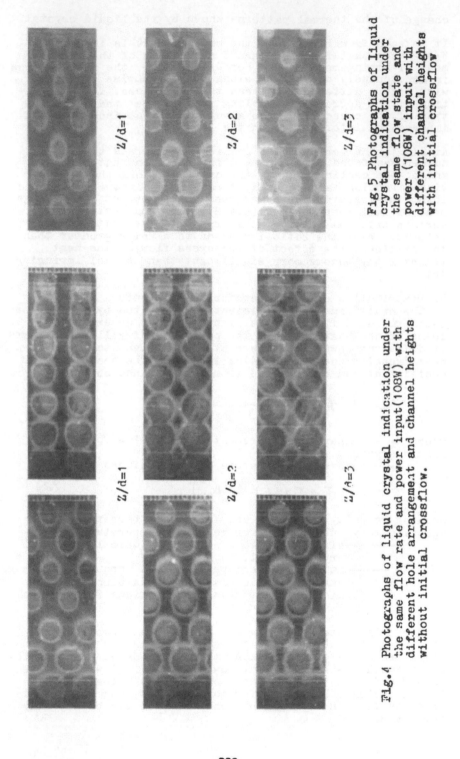

z/d=1

z/d=2

z/d=3

Fig.5 Photographs of liquid
crystal indication under
the same flow state and
power (108W) input with
different channel heights
with initial crossflow

z/d=1

z/d=2

z/d=3

Fig.4 Photographs of liquid crystal indication under
the same flow rate and power input(108W) with
different hole arrangement and channel heights
without initial crossflow.

Under the same power (108W) input, an initial crossflow
(m_c = 0.0003 Kg/s) imposes a strong interference on the
heat transfer of jet array impingement (m_j = 0.0017 Kg/s),
decreasing the relative cooling effectiveness, although both
are of the same temperature. Except the influence of the
arrangement of impingement holes, the initial crossflow also
significantly increases the influence of Z/d on impingement,
When Z/d is increased from 1 to 3, not only distributions
of cooling area along the streamwise direction are changed,
but also shapes of cooling areas are varied from elliptical
to tadpole-shaped as shown in Fig.5. Thus, adjusting Z/d
value may change the distribution of heat transfer by jet
array impingement.

CONCLUSIONS

1. The thermal patterns of liquid crystals can truly, and
   intuitively reflect both the two-dimensional heat
   transfer characteristics for each jet impingment cooling
   and crossflow effect on jet impingement. Thermal pattern
   shows that under influence of crossflow the uniform distribution of the jet does not mean uniform effectiveness
   of impingement cooling.

2. The comparison of these cooling areas indicates that the
   relative cooling effectiveness of inline arrangement is
   higher than that of the staggered under condition of the
   same flow rate and heat input and that the greatest cooling effectiveness occurs at Z/d=2.

3. For Z/d=2 and 3, the relative cooling effectiveness of
   arrays is monotonously decreasing along streamwise
   direction. For Z/d=1, however, the effectiveness is first
   decreased to a minimum, and than increased slowly.

4. The initial crossflow significatly increases the
   influence of Z/d on impingement. Thus, adjusting Z/d
   value may change the distribution of heat transfer by jet
   array impingement.

REFERENCES

1. Hippensteels, S.A., Russell, L.M., and Torress, F.J.
   "Local Heat Transfer Measurements on a Large Scale-
   Model Turbine Blade Airfoil Using a Composite of a Heater
   Element and Liquid Crystals", ASME Journal of Engineering
   for Gas Turbine and Power, Vol.107, PP953-960, 1985.

2. Florschuetz, L.M., Truman, C.R. and Metzger, D.E.
   "Streamwise Flow and Heat Transfer Distributions for Jet
   Array Impingement with Crossflow.", ASME Journal of Heat
   Transfer, Vol.103, PP337-342, 1981.

Local Characteristics of Heat Transfer from a Small Heater to an Impinging Round Jet of Liquid of Larger Pr Number

C. F. MA, Y. A. TIAN, H. SUN, and D. H. LEI
Beijing Polytechnic University
Beijing 100022, PRC

A. E. BERGLES
Rensselaer Polytechnic Institute
Troy, New York 12180, USA

ABSTRACT   A impingement heat transfer system was developed to investigate the local heat transfer characteristics of a impinging liquid jet on a small heater. Measurements with both submerged and free jets were made to determine the local heat transfer distributions. The effect of Pr number on impingement heat transfer was examined within the range of Pr=0.7~20. A set of semi-empirical formulae were recommended to estimate the local heat transfer distribution of impinging circular jets.

INTRODUCTION

As a effective active enhancement method, jet impingement heat transfer is a relatively new technological development that has attracted a growing interest of heat transfer engineers. Impinging gas jets have been commonly used in industrial cooling or heating where the high heat transfer rate is required and the space is restricted. Compared with the large number of technical reports on gas jet, there has been very little work done on the liquid jet impingement heat transfer. Recently, increased attention has been directed to the use of liquid jet. For example, water jet impingement has been employed in metallurgical industry [1] and jet impingement of fluorocarbons has been tested for microelectronic cooling [2-5].

In the case of liquids, the Pr number may be several orders of magnitude larger than that of gas. The influence of Pr number on jet impingement heat transfer has not been well understood at the present time. With oils of Pr number between 85 and 151, Metzger et al studied the effect of Pr number [6]. Kataoka and Mizushina measured the local heat transfer rate of large Pr number (2420-3300) liquid jet impinged on a horizontal plate [7]. To the best of author's knowledge, experimental data for working liquid of Pr number between 10 and 85 have not been available in any literature. In the present work kerosene (Pr≈20) is used as working fluid. Because of its flammability, kerosene is certainly not a good coolant for microelectronic equipment. However, the measured local heat transfer data of kerosene supplement the author's previous study with R-113 [2] and extend our knowledge about Pr-dependence in the correlation. Based on the present data, the correlations presented by Ma and Bergles in [2] were testified again and extended to the liquid of larger Pr number. Considering the incompletion of the information available on local heat transfer coefficients of impinging jet both in analytical and experimental aspects, the present study should provide further information for the heat transfer engineers concerning the use of liquid jets.

EXPERIMENTAL APPARATUS AND PROCEDURE

The experimental results reported in present paper were mainly obtained in Beijing Polytechnic University (BPU). Some data of R-113 or air in the paper were recorded in Iowa State University (ISU) previously. The apparatus set up in BPU is similar to that in ISU. The overall apparatus is shown in Fig. 1. The working fluid was circulated in a closed loop which had provision for metering, cooling and filtering. Fig. 2 shows the

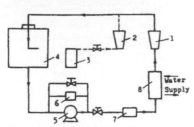

Fig. 1. Schematic layout of flow loop
1. Liquid rotameter   5. Pump
2. Gas rotameter   6. Spillage valve
3. High-pressed gas   7. Filter
   tank   8. Cooler
4. Test chamber

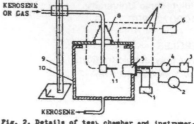

Fig. 2. Details of test chamber and instrumentation
1. MV meter   7. Thermocouple
2. AC voltage-steady power 8. Flexible plastics
3. Variac   9. Three dimensional
4. Amperometer   coordinate rack
5. Test section assembly 10. Chamber
6. Voltage meter   11. Jet tube

details of test chamber and instrumentation. The chamber was constructed of stainless steel. The jet tube of inside diameter of 0.987mm could be adjusted by means of a three dimensional coordinate rack with respect to the test section. The placement was accomplished to within 0.01mm. The details of test section is given in Fig. 3. The main part is a strip of 10μm thick constantan foil with heated section of 5mmx5mm. The active section of the foil was used as electrical heating element as well as heat transfer surface. The temperature of the inner surface of the heater was measured by a 40-gage iron-constantan thermocouple, which was electrically insulated from the heater yet in close thermal contact. The heat flux was calculated from a electrical power supplied to the test section. Alternating current of 220V was applied. The power determined by $P=I^2R$. The resistance R was measured accurately with direct current in preliminary experiments. The heat transfer coefficient was calculated by
$$h=q''/(T_w-T_j)$$

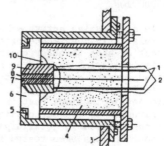

Fig. 3. Details of electrically heated test section
1. Power lead   6. Plexiglas
2. Voltage tap   7. Bakelite
3. Tank wall   8. 10μm constantan foil
4. Fiberglass   9. Copper block
5. O-ring   10. Epoxy

where heat flux $q''=P/A$, A is the area of the exposed foil surface, which was accurately measured with a microscope. T_w is the corrected wall temperature from the measured value, T_j is the jet temperature measured in the jet tube by a thermocouple. As the thickness of the constantan foil was extremely small, the conduction along the foil may be neglected so that the local heat transfer rate may be obtained experimentally.

RESULTS AND DISCUSSION

Heat Transfer at Stagnation Point.   For stagnation point heat transfer, the general form of the equation for the local Nusselt number has been well accepted [8]:

$$Nu = C_1 Pr^m Re_{d,u_o}^n \qquad (1)$$

The value of power n has been determined as 0.5 both from analytical and experimental results [2,8]. The value of m ranges from $\frac{1}{3}$ to 0.42 [8]. The power function of Pr with exponent 0.42 was found by comparison of the mass transfer measurements with the data on heat transfer to air and to water [9]. The constant C_1 depends on the turbulence level of the jet flow. The values of m and C_1 were determined experimentally as 0.4 and 1.29 respectively in the previous report by the authors [2]. The data obtained in ISU are presented in Fig. 4. It is obvious that all the data both for R-113 and air may be well correlated by equation (1) with C_1=1.29, m=0.4 and n=0.5 and are in agreement with the experimental results of Gardon and Cobonpue [10], Sparrow and Lovell [11], Schlünder and Gnielinski [9], Hrycak [15]. However, the correlation is somewhat higher than that presented by Martin [9] as shown in the Figure. It may be attributed to the higher turbulence level as long jet tubes used in the experiments both in ISU and BPU.

Gas Jet.

In order to make comparison of the experimental apparatus in BPU with that used in ISU previously, preliminary experiments were conducted with

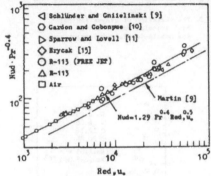

Fig. 4. Impingement heat transfer at stagnation point of circular jet

air and nitrogen gas. Excellent agreement was obtained between the data from the present study and those recorded in ISU as depicted in Fig. 5. By this fact the repeatability of the experimental data from the apparatus in BPU and ISU was well testified. All the gas jet data may be correlated with the R-113 data by the equation (1) using all the same constant.

Effect of Pr Number.

Heat transfer coefficients of kerosene at stagnation point are presented in Fig. 6 with the target within the potential core. It was found that the power on Pr number is 0.312 for kerosene (Pr ≈ 20) rather than 0.4 determined for

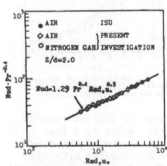

Fig. 5. Impingement heat transfer at stagnation point of circular gas jet

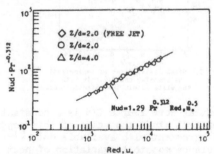

Fig. 6. Impingement heat transfer at stagnation point of kerosene circular jet

air, nitrogen gas (Pr ≈ 0.7) and R-113 (Pr ≈ 7). As shown in the Figure,
all the kerosene data may be well correlated by equation (1) with the same
values of C_1 and n and modified value of m compared with those for R-113
and gases. It is noted that the exponent on Pr number decreases with the
increasing of the Pr number itself. This general trend was also found by
Metzger in their investigation [6].

Variation of Heat Transfer Rates with Nozzle-to-plate Spacing.
     The local heat transfer rate at the stagnation point was measured as
a function of nozzle-to-plate spacing at different jet velocities. The
result is exhibited in Fig. 7. A better representation of the data can be
obtained when the Nusselt number ratio Nu/Nu_{max} is used in stead of Nu as
shown in Fig. 8. It is determined from Fig. 8 that the nondimensional

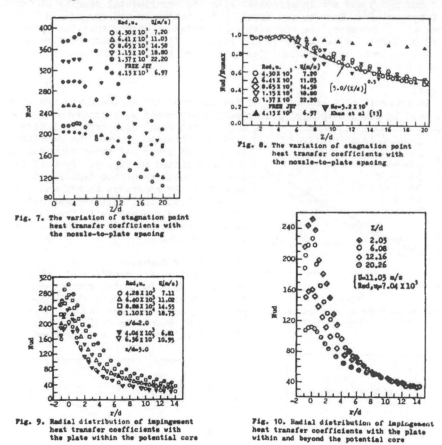

Fig. 7. The variation of stagnation point
heat transfer coefficients with
the nozzle-to-plate spacing

Fig. 8. The variation of stagnation point
heat transfer coefficients with
the nozzle-to-plate spacing

Fig. 9. Radial distribution of impingement
heat transfer coefficients with
the plate within the potential core

Fig. 10. Radial distribution of impingement
heat transfer coefficients with the plate
within and beyond the potential core

potential core length L/d is a constant regardless of the jet velocities.
In present study the value of this nondimensional length is 5 that is in
good agreement with available data [12]. Because of the high initial
turbulence monotonic variation of heat transfer coefficient with the
spacing was recorded as shown in the figure. This fact was also testified
by Ali khan et al [13]. As depicted in the figure, beyond the potential
core the heat transfer coefficient may be well expressed by the equation

$$Nu/Nu_{max}=[(L/d)/(Z/d)]^{0.5} \tag{2}$$

Radial Distribution of Heat Transfer Coefficients.   Fig. 9 shows the radial distribution of measured local heat transfer coefficients with the plate located within the potential core. The plots refer to 6 different Re numbers ranging from 4.04×10^3 to 1.10×10^4. With the plate beyond the potential core, the distribution of heat transfer coefficients were measured with different nozzle-to-plate spacings. As shown in Fig. 10, with increasing of the spacing, the stagnation point heat transfer rates decreased and the curves became planer.

Stagnation Zone.
From reference [14] the velocity distribution near the stagnation point of an impinging circular jet may be expressed by the equation:

$$\frac{U_{max}}{U_s} = \tanh(0.88r/d) \tag{3}$$

Utilizing equation (3) with the plate within the potential core, the local heat transfer coefficient distribution in the stagnation zone may be obtained:

$$Nu_d/Nu_{max}=(r/d)^{-0.5}\tanh^{0.5}(0.88r/d) \tag{4}$$

The data plotted in Fig. 9 may be represented in Fig. 11. It is found that all the data in stagnation zone $(r/d<2)$ are in very good agreement with the predicted curve of equation (4).

Wall Jet Zone.
Hrycak, using results from cold-jet hydrodynamic measurement [8], showed that a solution for the average heat transfer in the wall jet zone could be obtained through the Colburn analogy $(St\,Pr^{2/3}=C_f/2)$:

$$\overline{Nu}=1.95Pr^{1/3}Re_d^{0.7}(2r/d)^{-1.23} \tag{5}$$

The mean Nusselt number $\overline{Nu}$ is independent of the Z/d term for $Z/d<7$.

Considering the relationship between the local and mean heat transfer coefficients, the local heat transfer may be obtained from equation (5):

$$Nu_d=0.75Pr^{1/3}Re_d^{0.7}(2r/d)^{-1.23} \tag{6}$$

The exponent of Reynolds number in equation (5) has also been verified in present study with kerosene. The local heat transfer rates were measured at different locations. It is found that the local Nusselt number Nu_d is proportional to Re^n and the value of power n increases from 0.5 at $r/d=0$ to 0.723 at $r/d=14$ as shown in Fig. 12. For the whole wall jet zone $(r/d>2.0)$, value of 0.7 may be accepted as the

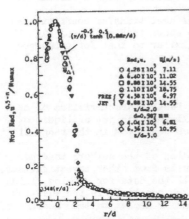

Fig. 11. Radial distribution of impingement heat transfer coefficients with the plate within the potential core

average power on the Reynolds number in agreement with the value reported by Hrycak. With regard to the power of (r/d) in equation, Ma and Bergles, using the result of Glauert analysis [16] and Colburn analogy, determined the exponent $k=-1.25$ which is very close to -1.23 in equation (5). As illustrated in Fig. 11, all the data for wall jet zone may be well correlated by:

$$Nu_d/Nu_{max}=0.348Red,u_o^{n-0.5}(r/d)^{-1.25} \qquad (7)$$

where Nu_{max} may be obtained by equation (1) and exponent n is presented in Fig. 12.

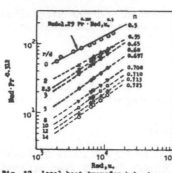

Fig. 12. Local heat transfer behaviour at different points on the heated surface

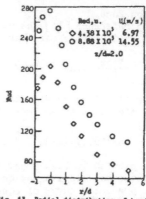

Fig. 13. Radial distribution of heat transfer coefficients of circular free jet

Free Jet.

Heat transfer rates at stagnation point with R-113 or kerosene free jet are presented in Fig. 4 and Fig. 6 respectively. They are in good agreement with those of submerged jet and may be expressed by the same formulae with same constants. But for free jet the heat transfer coefficients are weakly affected by the nozzle-to-plate spacing as illustrated in Fig. 7 and Fig. 8.

Measurements of radial distribution of heat transfer coefficients were also made with kerosene free jet. As shown in Fig. 13 and Fig. 11, the shape of distribution curves is very similar to that for submerged jet.

CONCLUSIONS

Measurements were made to investigate local characteristics of heat transfer from a small heater to a single axisymmetric jet of liquid of larger Pr number (≈ 20). The experiment was conducted in the range of $Re=4.04 \times 10^3 - 1.37 \times 10^4$.

Based on the present and previous studies of the authors together with some experimental and theoretical results from other investigations, Eq. (1) may be recommended to correlate all the data for working fluids of Pr number ranging from 0.7 to 20 with same constants in the equation with the exception of varying the exponent on Pr number to take account of the effect of that property.

The local heat transfer coefficients of kerosene in stagnation zone and wall jet zone may be well correlated by Eqs. (4) and (7) respectively. Another semi-empirical formulae Eq. (2) is recommended to predict the variation of stagnation point heat transfer with the nozzle-to-plate spacing.

ACKNOWLEDGEMENTS

This study was supported by the National Natural Science Foundation of China. The assistance of Mrs. S.F. Lee of the foundation is appreciated.

REFERENCES

[1] Kohrlng, F.C., "Waterwall Water-cooling Systems" Iron and Steel Engineer, June 1985, pp.30-36.

[2] Ma, C.F. and Bergles, A.E., "Convective Heat Transfer on A Small Vertical Heated Surface in An Impinging Circular Liquid Jet", 2th Beijing International Symposium on Heat Transfer, August 1988, Beijing.

[3] Ma, C.F. and Bergles, A.E., "Boiling Jet Impingement Cooling of Simulated Microelectronic Chips", in "Heat Transfer in Electronic Equipment", HTD-Vol. 28, ASME, New York, 1983, pp.5-12.

[4] Kiper, A.M., "Impinging Water Cooling of VLSI Circuits", International Communications in Heat and Mass Transfer, Vol. 11, pp.517-526, 1984.

[5] Jiji, L.M. and Dagan, Z., "Experimental Investigation of Single Phase Multi-Jet Impingement Cooling of An Array of Microelectronic Heat Sources", International Symposium on Cooling of Electronic Equipment, Hawaii, 1987.

[6] Metzger, D.E. et al, "Effects of Prandtl Number on Heat Transfer Characteristics of Impinging Liquid Jets", Heat Transfer — 1974, FC 1.5.

[7] Kataoka, K. and Mizushina, T., "Local Enhancement of the Rate of Heat-transfer in An Impinging Round Jet by Free-stream Turbulence", Heat Transfer — 1974, FC 8.3.

[8] Hrycak, P., "Heat Transfer from Impinging Jets — A Literature Review', AFWAL-TR-81-3054, Flight Dynamics Laboratory,Wright-Patterson AFB, Ohio, 1981.

[9] Martin, H., "Heat and Mass Transfer Between Impinging Gas Jets and Solid Surfaces", in Advances in Heat Transfer, Vol. 13, eds. J.P. Hartnett and T.F. Irvine, pp 1-60 Academic Press, New York, 1977.

[10] Gardon, R. and Cobonpue, J., "Heat Transfer Between A Flat Plate and Jets of Air Impinging on It", International Developments in Heat Transfer, pp.454-460, ASME, New York, 1962.

[11] Sparrow, E.M. and Lovell, B.J., "Heat Transfer Characteristics of An Obliquely Impinging Circular Jet", Journal of Heat Transfer, Vol. 102, 1980, pp.202-209.

[12] Gauntner, J.W. et al, "Survey of Literature on Flow Characteristics of A Single Turbulent Jet Impinging on A Flat Plate", NASA TN-D-5652, 1970.

[13] Ali Khan, M.M. et al, "Heat Transfer Augmentation in An Axisymmetric Impinging Jet", Heat Transfer — 1982, Vol. 3, pp.363-368.

[14] Chin, Y., "Flow of Jet Impinging on Flat Plate", Ph.D. Thesis, University of Tokyo, 1969.

[15] Hrycak, P., "Heat Transfer from Impinging Jets to a Flat Plate with Conical and Ring Protuberances", International Journal of Heat and Mass Transfer, Vol. 27, 1984, pp.2145-2154.

[16] Glauert, M.B., "The Wall Jet", Journal of Fluid Mechanics, Vol. 1, pp.625-643, 1956.

Heat Transfer from Tubes which Extend Radially Outward from a Rotating Shaft

LIEJUN QIU, LIQIU WANG, and YUNFENG SUN
Shandong Polytechnic University
Jinan, PRC

INTRODUCTION

In gas-liquid heat exchangers,heat transfer performance is controlled by the thermal resistance on the gas side.Two remedies are available to reduce this resistance:—1) by increasing heat transfer surface area through implementation of fins,and 2) by enhancing the heat transfer coefficient through either increasing the bulk flow or generating the secondary flow.In many applications,however,extended surfaces may enlarge the size and weight,and increasing flow may result in high operating pressure.Various means have been applied to produce secondary flow,for example,oscillation of the bulk flow,vibration or rotation of the surface,introduction of accoustic waves into the bulk flow,etc.In this paper,the rotating tubes (FIGURE 1) were used to generate the secondary flow,thus,heat transfer enhancement was achieved.

Up to this date,only one paper [1] reported the experimental results of heat transfer from rotating rods in fluid.The experiments were performed using the naphthalene sublimation technique rather than by direct heat transfer measurements.In this article,the heat transfer results between the rotating rods and the fluid were obtained via the analogy between the heat and mass transfer processes,and the per-rod average Nusselt number formula.It was also verified that the shaft diameter,length and the number of rods have very weak effect on the heat transfer between the rod and the fluid.But there is no mentioning about the local Nusselt number along the rod.Furthermore,the naphthalene sublimation technique only simulated the heat transfer for a fluid of Prandtl number 2.5 and the boundary condition of the uniform wall temperature.However,the rotating tubes may be generally used to enhance gas-side heat transfer rather than liquid-side (Prandtl number of a gas is usually much less than 2.5),and the boundary condition seems more close to uniform thermal flux.In order to bring the tubes into more practice,a study of the heat transfer characteristics of tubes which extend radially outward from a rotating shaft was conducted in quiet air with a Prandtl number 0.7.Direct heat transfer measurements were used rather than the naphthalene sublimation method.The experiments encompassed a number of rotating speeds,and thermal fluxes.The local Nusselt number and the per-tube average Nusselt number were evaluated from the experimental data.The Reynolds number reflects the effects of rotating speed and the dimension ratio of the problem on Nusselt number.An universal local Nusselt-Reynolds-Prandtl number correlation,independent of the dimension ratio of the problem,was obtained.The experiments were made for uniform heat flux boundary in this investigation.Therefore,the results would be more useful and practical.

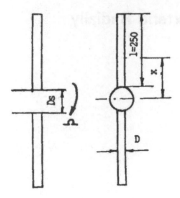

FIGURE 1. Rotating tubes.

EXPERIMENTAL APPARATUS AND MEASUREMENT SYSTEM

The experimental apparatus and measurement system are schematically shown in the FIGURE 2 The major dimensions are as follows:the rotating shaft diameter D=40 mm,its base-to-tip length l=250 mm,and the number of tubes N=2.The Prandtl number of surrounding quiet air was approximately 0.7.The rotating shaft was driven by a controlled electric motor which was operated at rotational speed between 0 and 400 rpm steplessly.From steady-state measurements of heating voltage,rotating speed,air temperature and the tube wall temperature at different radial positions,it was possible to determine the distribution of the local Nusselt number.

The heating current was provided by stationary meter(9)And the heat flux was varied by the voltage change using a variable transformer(8).A separate power slip-ring comprising two phosphor-bronze ring (5) with spring-loaded carbon brushes (5') enabled power to be transfered to the heated wire.The heat fluxes may be calculated from measurement of the voltage (10).In order to get high accuracy,the rotational speed was simultaneously recorded with two meters (12) and (13).The quiet air temperature was measured with five thermometers located at different positions in the laboratory room.

Eight pairs of thermocouples were mounted,equally spaced,along the effective length of the test section to enable that an accurate determination of the wall temperature distribution can be made.And two pairs of thermocouples were used in the same radial position x to moniter and insure the measurement accuracy.All these tube-mounted thermocouples could be monitored sequentially using a data logger (11),signals being taken from the rig via silver-silver graphite instrumetation slip-ring assemblies (7) mounted at one end of the rotating shaft.

The experiments were performed in a very large laboratory room (about 400 m^3) so that possible accumulations of heat were minimal.Several hours prior to the initiation of a data run,the windows of the laboratory were covered with sheets of polystyrene.Entrance to the laboratory was prohibited both prior to and during the run.The measures provided a very stable ambient that aided in the attainment of data.Consequantly,the data were both highly reproducible and virtually free of scatter.

232

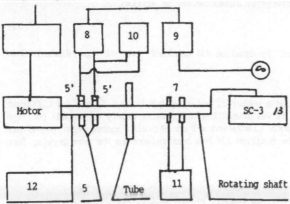

Change of speed

FIGURE 2. Experimental apparatus and measurement system.

Experimental procedure,the calibration of experimental apparatus and the data analysis can be found in reference [2].

DATA REDUCTION

To aid in the attainment of a simple correlation of the local heat transfer coefficient.hx,the functional dependence of hx on the other parameters may be written as:

$$hx = f(D,Ds,N,x,c,\Omega,\lambda,\rho,\mu)$$ (1)

where hx is the local heat transfer coefficient between the tube wall of radial position x and the fluid;D,Ds are tube and rotating shaft diameter respectively;x is the distance between the measurement point and the center line of the rotating shaft;N is the number of tubes;Ω is the angular velocity,and λ,ρ,μ,c are thermal conductivity,density,viscosity and specific heat of the fluid respectively.By employing dimensional analysis,there follows

$$hxD/ = f(Ds/D,N,x/D,\Omega D^{2}/\nu,Pr)$$ (2)

with D as characteristic dimension,the left side of the Eq.(2) can be identified as the local Nusselt number.

According to reference [1],the Nusselt number is independent of the number of tubes in the circumferential array N and has very weakly dependent on the Ds/D.Thus,the effects of N and Ds/D on the Nusselt number may be omitted,and the equation (2) becomes:

$$Nux = f (x/D,\Omega D^{2}/\nu,Pr)$$ (3)

In this investigation,Ds/D=2 and N=2 have been employed.

The group $\Omega D^{2}/\nu$ possesses the form of a Reynolds number,with D as characteristic dimension and ΩD having the units of velocity.In reality,the velocity of a tube varies

233

from ΩR_s ($R_s = D_s/2$) at its root to $\Omega(R_s +1)$ at its tip.The local Reynolds number at the radial position x,with D as characteristic dimension,may be written as:

$$Rex = \Omega xD/\nu = \Omega D^2/\nu \cdot x/D \tag{4}$$

Thus,the Reynolds number defined by equation (4) combined two of the dimensionless groups,namely,x/D and $\Omega D^2/\nu$.

From all experimental data including a number of rotating speeds,heat fluxes and different radial positions along the tube,it was found that Nux remains nearly unchanged for the same Rex.Thus,the effects of both $\Omega D^2/\nu$ and x/D on Nusselt number can be simply reflected by Rex.In consequence,the equation (3) has been reduced to the more popular form as follows:

$$Nux = f(Rex,Pr) \tag{5}$$

The experiments were performed with three thermal fluxes.Because the change of the Prandtl number of air under the test conditions is very small,the deviations of Nux are not noticed.As the result,the relationship between Nux and Pr has not been obtained in these experiments.However,the dependence of Prandtl number on the Nusselt number should still included for the gases whose Prandtl number is not too far from 0.7 according to reference [1].The detailed explanation will be given in the later section.

RESULTS AND DISCUSSION

The local Nusselt number encompassing part of the rotating speeds,heat fluxes and radial positions in this investigation are plotted in FIGURE 3 as a function of the Reynolds number defined by the equation (4).The symbols x , o , □ presented in the FIGURE correspond respectively to different heat flux experiments.It is seen that the deviation of Nux is very small due to the difference of heat flux.The reason that change in the Prandtl number of air caused by the difference of heat flux is very small.

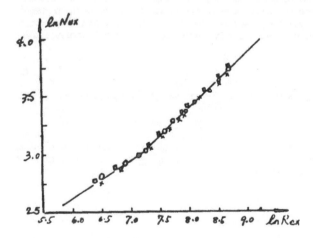

FIGURE 3. Local Nusselt number as a function of the Reynolds number.

From FIGURE 3,it is reasonable to consider separately

$$N_{ux} = CR_{ex}^{n} \tag{6}$$

for the Rex ranges below 1500 and above 1500.Strictly speaking,the change of slope at Rex =1500 is probably continuous rather than discrete,but for convenience in use,two simple curve-fitting representations have been chosen rather than one complicate higher-order fitting.

The numerical valuues of C and n for both ranges are determined by employing least-squares curve-fitting method.The results are given below:

$$N_{ux} = 2.012 \, Rex^{0.3204} \tag{7}$$

$$N_{ux} = 0.7905 \, Rex^{0.4522} \tag{8}$$

The comparisions with experimental data are shown in FIGURE 3.The maximum deviation of the data from these equations is only 2.5 percent.

Concerning the dependence of the Prandtl number on the results,guidance may be taken from the in-line tube bank and the single cylinder results in cross-flow.If the exponent of the Prandtl number 0.37 is taken,the equations (7) and (8) respectively become

$$N_{ux} = 2.296 \, Rex^{0.3204} Pr^{0.37} \quad ;Rex < 1500 \tag{9}$$

$$N_{ux} = 0.9020 \, Rex^{0.4522} Pr^{0.37} \quad ;Rex > 1500 \tag{10}$$

Equations (9) and (10) may be regarded as the most general form of the results of this investigation.

The per-tube average Nusselt number is defined as

$$Nu = \frac{1}{L} \int_{R_s}^{R_s + L} N_{ux} \, dx \tag{11}$$

On substitution of equation (9) or (10) together with $1/R_s = 12.5$ into the equation (11) and after integration we have

$$Nu = 2.296 \, [\Omega D (R_s + \partial 1)/\nu]^{0.3204} Pr^{0.37} \tag{12}$$

$$Nu = 0.9020 \, [\Omega D (R_s + \partial 1)/\nu]^{0.4522} Pr^{0.37} \tag{13}$$

where $0 < \partial < 1$,with
$\partial = 0.441$ [in equation (12)]
$\partial = 0.458$ [in equation (13)]

During the data reduction process,it is shown that $1/R_s$ has only a weak effect on ∂ (for example,the derivation of ∂ is less than 2.5 % for the change of $1/R_s$ from 5 to 12.5).In other words,the per-tube average Nusselt number is equal to the local Nusselt number located at $x = R_s + \partial 1$.

235

Following the idea in FIGURE 1 ,where the heat transfer surface consists mainly of a rotating tube system,a new rotating heat exchanger has been constructed.Using air and water as heat transfer fluids,the various performances of the heat exchanger,including heat transfer,air resistance,power required for rotating tube system, water flow rate, and water pressure difference between inlet and exit of the tube system, have been investigated experimently.Correlation equations and figures have been obtained [2].The results show that the heat exchanger not only performs at a high total heat transfer coefficient,but also has the advantage of automaticly water-absorbing function,and easy adjustment..The power required for the rotating tube system and the power consumed due to the increment of resistance are much less than the benifit obtained from heat transfer enhancement [2] [5]. And it has been confirmed that the correlations obtained from this investigation may be applicable to the practical engineering design .

ACKNOWLEDGMENT

Writers wish to express their appreciation of the valuable assistance given in the translation of the paper from Chinese into English by Prof Wang Zhenyi.

NOMENCLATURE

D = diameter of tube
Ds = diameter of rotating shaft
hx = local heat transfer coefficient
l = length of the tube
N = number of tubes
Nu = per-tube average Nusselt number
Nux = local Nusselt number
Pr = Prandtl number
Rex = local Reynolds number
Rs = radius of the shaft
μ = viscosity
ν = kinematic viscosity
ρ = density
Ω = angular velocity

REFERENCES

[1] Sparrow,E.M.,and Kadle,D.s.,Heat Transfer from Rods or Fins Which Extend Radially Outward from A Rotating Shaft,Journal of Heat Transfer ,vol.106,pp 290-296,1984.

[2] Wang,Liqiu,The Study on The Enhancement of Heat Transfer by Rotating Tubes,M.S. thesis,Department of Heat Power,Shandong Polytechnic University,1985.

[3] Zukauskas,A.A.,Heat Transfer from Tubes in Crossflow,Advances in Heat Transfer,vol.8,pp 93-160,1972.

[4] Whitaker,s.,Elementary Heat Transfer Analysis,Peryamon Press,Oxford,England,1976.

[5] Laboratory Report on High Efficient Rotating Heat Exchanger ,Heat Energy Laboratory,Shandong Polytechnic University,1987.

The Effect of Cold Wall Surface Conditions on Frost Deposit Processes under Forced Convection

HONGJI SHU, CHAO LIU, QINJIANG LIU, GUOJI XU, and HUIZHI LIU
Dalian Marine College
Dalian, PRC

1. INTRODUCTION

The frost arising from vapour in air on cold wall will increase the heat resistance of evaporator and block the air-flow.It is of realistic importance to prevent frost deposit on cold surface for both energy saving and increasing efficiency of refrigerating plant.

Having studied the process of frosting on hydrophobic film under natural convection(1), the authors have observed the depositing effect of hydrophobic film and vortex field behind cylindrical surface on frost processes under forced convection and the theory analysis is given. The authors have done experiments of frosting on hydrophobic film under different surrounding parameters.

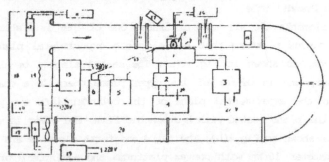

1. Semiconductor refrigerator 2. Heat exchanger 3. Refrigerator mains
4. Water vessel of constant temperature 5. Microcomputer 6. Reference

temperature of ice 7. Test element 8. Reading microscope 9. Side window opening 10. Upper window opening 11. Microammeter 12. Dry-bulb, wet-bulb thermometer 13. Cooler 14. Evaporator 15. Airtight air space 16. Heat supply of regulating temperature 17. Device of regulating humidity 18. Blower 19. Transformer of adjusting voltage 20. Wind channel 21., 22., 23.. Thermocouple 24. Instrument of controlling temperature 25. Instrument of wind velocity 26. Compensated micropressure instrument 27. Assistant light

Fig.1 - The sketch of experimental satup

2. EXPERIMENTAL APPARATUS

As shown in Fig.1, this experimental apparatus is made up of following apparatus and instruments: a closed wind channel, in which temperature, humidity and velocity of air can be adjusted, a semiconductor refrigerator of variable cold wall temperature, a system of measuring and showing air velocity, humidity, temperature and cryosurface temperature , a microcomputer control system, a microscope which is used to observe frosting process and thinness of frost and a photographic camera.

The wind channel which is made up of plexiglass is 5.2 meter long, its length of experimental section is 460 mm, its section area is 110x 80 mm. The inlet and exit of wind channel are connected to a space of 1500x900x410 mm and they constitute a closed channel. Air humidity can be adjusted by saturated solution of different salts and relative humidity controlled in range of 55-59%. The air velocity can be continuously adjusted in range of 0-10 m/s; the air temperature can be controlled in range of -258 -293 k, its error varies in range of ±0.2 k.

Air temperature and cryosurface temperature can both be measured with thermocouple, then the data can be controlled and shown by microcomputer. Wind velocity is measured by compensated micro-pressure instrument with Prandtl tube.

The test elements are all made of brass. One of them is a short cylinder with convex platform that has a smooth experimental plane of diameter 25 mm, as shown in Fig. 2(a). The second element is also a short cylinder, and in center of its upper plane there is a channel which divides the experimental plane of the cylinder into two parts. One part is the plane coated with hydrophobic film, the other is metallic plane, as shown in Fig.2(b). The third and forth elements are a cylinder of diameter 16 mm with convex platforms and semisphere of dia-

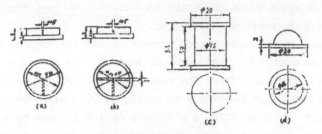

Fig. 2 - Test elements

meter 16 mm, respectively as shown in Fig. 2(c) and (d). The material
of hydrophobic film may be selected as silicone oil or silicone grease
with low volatility, highly hydrophobic property, high conductivity and
strong adhesion to metallic surface. The test element which is treated
ready is stuck on the cold platform by heatconductance silicone grease.
Since the small volume of test element and the high conductivity of
brass, the surface temperature of test element could be apporximately
treated as uniform.

3. EXPERIMENTAL RESULT AND THEORETIC ANALYSIS
(1) A series of experiments on hydrophobic film under conditions of
different air temperature, cryosurface temperature, air himudity and
wind velocity have been conducted, curves being shown as Fig. 3-6
respectively. It follows:
  (a) The higher the temperature of air, the lower the cryosurface
temperature, the higher the relative humidity and the higher the wind
velocity, then the faster the frosting layer growing, the earlier the
forming frost network.

The higher the sir temperature, the greater the driving force of
phase change. When the relative humidity is constant, the higher the
air temperature, the more water vapour content in air. The lower the
cryosurface temperature, the greater the driving force of vapour--solid
phase change. The higher relative humidity, the more water vapour in
air. The higher wind velocity, the bigger chance of getting vapour molecule in air by cryosurface. These factors can make the mass-transfer
reinforced totally.

  (b) The higher relative humidity and wind velocity, the earlier
frost ageing performing. See curves 2,3,4, in Fig. 6.

(2) Then, the experiments of comparing frosting on hydrophobic film

with on metallic plane under the same conditions have been conducted. The results are described as follows:

(a) The results under forced convection are the same as those under natural convection(2). Two kinds of surface pass a course of forming dew before frost is formed. The course of forming dew on metallic plane can be described as:dew pearls appearing on the plane--dew pearls growing joining in water film. The course of forming dew on hydrophobic film can be described as: dew pearls appearing--dew pearls growing--dew pearls growing continuously, but not forming water film. Then, the water film on metallic plane solidifies to ice layer. Later on, the frost dendrites generate on the ground of the ice layer and have a relatively uinform distribution. The dew pearls on hydrophobic film solidify as ice spheres. After, the frost dendrites generate on the surface of ice sphere and distribute non-uniformly. The forming frost network on hydrophobic film is earlier than that on metallic plane. See the curves 1, 2, in Fig. 6.

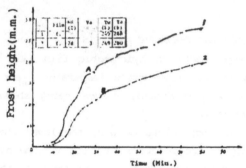

Fig. 3. The effect of air temperature on frost height

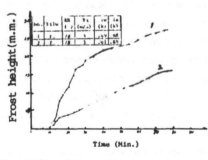

Fig.4 The effect of cryosurface temperature on frost height

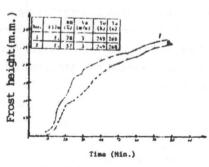

Fig.5 The effect of air humidity on frost height

240

The metallic surface has the polar. Thus, the metallic surface has absorbed more vapour molecules than those on the surface of hydrophobic film. The metallic surface for water is wet. The contact angle $\theta < 90°$, see Fig. 7(a). Therefore, the condensation on the cryosurface mentioned above can rapidly become the film condensation during the step of vapour--liquid phase change. This water layer solidifies to the ice layer during the step of liquid--solid phase change.

The side of the hydrophobic film toward air has not polar, see Fig.7 (b). The hydrophobic film has absorbed less vapour molecules. This surface for water is not wet. Its contact angle $\theta > 90°$, see Fig.7(c). Therefore, the condensation on the hydrophobic film always keeps dropwise condensation. Then the dew pearls solidify as the ice sphere during the step of liquid--solid phase change. Because the frost dendrites generate on the surface of ice spheres, the frost dendrites in this ocasion have "crossed property". This factor makes forming frost network earlier.

(b) The frost height on the hydrophobic film is always less than that on the metallic surface. See Fig. 6 and Fig. 8-11.

The relation between the surface curvature of liquid and the saturate pressure of vapour is

$$PT\,L_n\,\frac{P'}{P} = 2\gamma u/r$$

Where, T is temperature of the liquid, u is mol volume of the liquid, γ is boundary energy between liquid and vapour, r is radius of curvature of the liquid, P is saturate pressure of vapour when the liquid surface is plane, P' is saturate pressure of vapour when the liquid surface is curve. Therefore, at the same temperature the less the r, the larger the P'. Thus as the vapour partial pressure of the wet air on the cryosurface is the same, the supersaturation radio of the water layer on the metallic surface is greater than that of the water pearls on the hydrophobic film. Consequently the condensation process on the metallic surface proceeds more rapidly than on the hydrophobic film.

The surface curvature of the liquid has also influence on the temerature of solidification point. The shorter the radius of convex curvature, the lower the solidification point. Thus the water film on the metallic surface solidifies more raplidly than the water pearls on the hydrophobic film. The above-mentioned two factors are the main reasons of frost deposit lagging on the hydrophobic film.

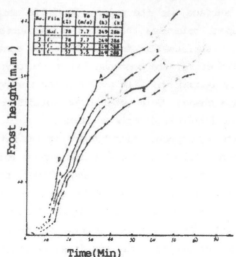

Time(Min)

Fig.6 The effect of wind velocity on frost height

(a) Water pearl on (b) Surface construction (c) Water pearl on
metal surface of phydrophobic film hydrophobic film

Fig.7 The water pearls on metallic and hydrophobic surface

(c) The greater the relative humidity of the ambient air, the more
evident the difference between the frost deposit on the metallic surface and that on the hydrophobic film. See the curves 1, 2 of Fig.6 and
the curves 1,2 of Fig.9.

The greater the relative humidity, the quicker the condensation process on the metallic surface, and the faster the solidification of
water film, and the faster the frosting on the basis of ice layer.

(3) Finally, the experiments have been proceeded on the surface of metallic cylinder and semi-sphere. First, it begins to dew on the area
of the surface of semi-sphere which is against the wind and the dew
drop becomes large gradually. The dew range expands to two side from
the vicinity of the forward stagnation point. Then, the dew drops begins to appear on the back stagnation point. Four minutes later, it begins to dew on the back side part of the semi-sphere and the dew drops
are tiny, see Fig.12(a). The frost situation is shown in Fig.12(b) in
ten minutes. Twenty minutes later, the frost situation of whole semi-
sphere is shown in Fig. 12(c). The frost height on the back side part
is far lower than that on the front half part and that on the vicinity

242

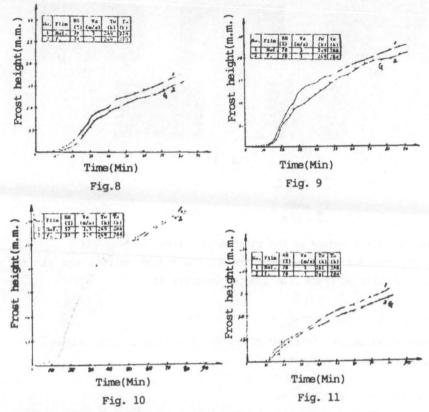

Fig.8

Fig. 9

Fig. 10

Fig. 11

of back stagnation point of semi-sphere. Here the frost is few and scattered. The same phenomenon can also be observed on the surface of cylinder as shown in Fig. 12(d), (e).

The reason of above phenomenon is that when the air flow atound the cylinder or semi-sphere, adverse pressure gradient is produced. The seperation of boundary layer happen on the back part of the cylinder or semi-sphere and the vortexes of large size are formed. Now, as an example, we discuss the flow around the cylinder (see Fig.13). In the experiment, the Reynolds' number of air flow is far greater than 40, so the Karman vortex street which is overlap arrangement on the back part of the cylinder is formed. It moves downward with the velocity which is far less than that of air flow, see Fig. 13.

In this paper, we will study the pressure distribution of the vortex region with the simplifying model of two-dimensional flow. The fluid inside the vortex rotates with angular velocity as the rigid body does. The velocity distribution is

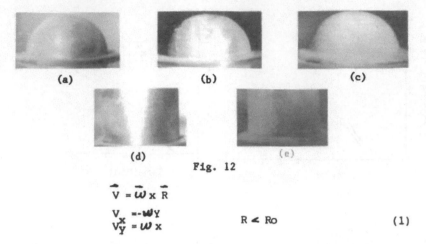

(a)　　　　　　　(b)　　　　　　　(c)

(d)　　　　　(e)

Fig. 12

$$\vec{V} = \vec{\omega} \times \vec{R}$$

$$\begin{aligned} V_x &= -\omega y \\ V_y &= \omega x \end{aligned} \qquad R < Ro \qquad\qquad (1)$$

Where, Ro is the radius of the rotational vortex, see Fig. 14.

Under the action of viscous force, the vortex carries the air to move in a circular trace. The induced velocity is

$$V_r = 0$$
$$V_\theta = \frac{\Gamma}{2\pi R} \qquad R > Ro$$

isthe circulation. In range of $R > Ro$, the flow is irrotational. According to Lagrangian equation:

$$P + \frac{1}{2}\rho V^2 = C$$

The out-boundary condition is

$$R = \infty \quad : \qquad P = Po \qquad V_\theta = \frac{\Gamma}{2\pi R} = 0$$

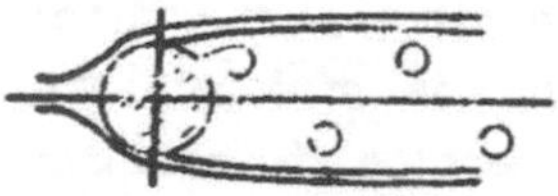

Fig. 13 The Karman vortex street

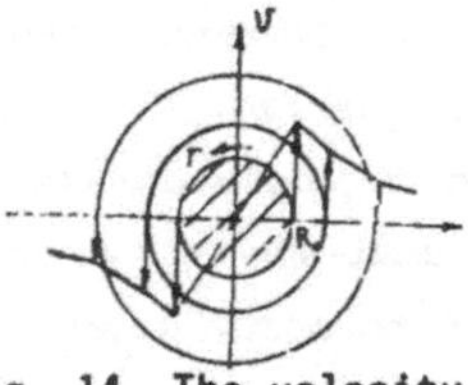

Fig. 14 The velocity distribution of the vortex

If the puripheral velocity of rotational vortex is Vo, its pressure P_y is

$$P_v = P_o - \frac{1}{2}\rho V_o^2 \qquad\qquad (2)$$

To determine the pressure distribution inside the vortex, use N-S equation

$$\frac{\partial \vec{V}}{\partial t} + (\vec{V}.\nabla)\,\vec{V} = \vec{F} - \frac{1}{\rho}\nabla P + \nu\nabla\vec{V} + \frac{1}{3}\nu\nabla(\nabla.V)$$

Because the velocity of wind is much lower than 1/3 that of sound in the experiment, the flow can be simplified as the flow of ideal incompressible fluid. Ignoring gravity, this equation can be simplified as

Euler equation

$$(\tilde{V} \cdot \nabla)\tilde{V} = -\frac{1}{\rho}\nabla P \tag{3}$$

Substituting(1) into (3), we can get the scalar notation

$$\omega^2 x = \frac{1}{\rho}\frac{\partial P}{\partial X}$$
$$\omega^2 Y = \frac{1}{\rho}\frac{\partial P}{\partial Y} \tag{4}$$

Miltiplying dx and dy through the above two equation, adding and integrating, we get:

$$P = 1/2.\rho V^2 + C \qquad R < R_0 \tag{5}$$

Using the out-boundary condition of rotational vortex: R = Ro, V = Vo, we get:

$$C = P_0 - \rho V_{O_2}^2 \qquad R < R_0 \tag{6}$$
$$P = P_0 + \tfrac{1}{2}\rho V^2 - \rho V_0^2$$

In the center of vortex, R = 0, V = 0, thus its pressure Pc is

$$P_c = P_0 - V_0^2 \tag{7}$$

Hence, the pressure in vortex center is the minimum. From the out boundary of vortex to its center, the pressure reduces gradually. A vacuum area exists around the center of vortex. The molecules of vapour and molecular groups near the cryosurface flow into the central area of the vortex. These vortexes are brought away by the main air flow and air humidity near the cryosurface decreases also. This factor plays a part in preventing forming dew and frost deposit. Simultaneously this factor make it easy for the part of water molecules having already formed dew and frost to vaporize or sublimate into the central area of vortex.

4. CONCLUSION

The study of the experiment and theoretic analysis in this paper is a trial, thus it has some limitations. But preliminary, we can prove preventing effect of the hydrophobic film and the vortex of large size formed by seperation of boundary layer on frost deposit processes. The surface conditions have obvious delay effect to growth rate of frost height. This gives a new way of preventing frost deposit and offers theoretic and experimental references to designing the shape and conditions of cryosurface.

5. REFERENCES

(1) Shu Hongji and Liu Huizhi. Analysis of the effect of cold surface condition on frost deposit in natural convection. The Journal of Dalian Marine College, Vol. 11, No.3, Aug. 1985.

(2) Shu Hongji, Liu Huizhi, Zhou Binru. The effect of cold wall surface conditions on frost deposit processes. Proceedings of XVIITH International Congress of Refrigeration. Vol. B.P. 286-291. 1987.

Two-Phase Flow Analysis of Natural Convection at a Heated Vertical Plate

C. BHASKER*
BHEL R&D Division
Vikasnagar
Hyderabad-500 593, India

MADHUSUDHAN BANGAD
Department of Mathematics
Osmania University
Hyderabad-500 007, India

ABSTRACT:

An analysis is presented to understand the effects of isothermal heated vertical plate which is initially cold and at rest under multiphase flow of particulate medium. The governing differential equations which describes Mass, Momentum and energy in two phases under the assumptions for small flow and thermal relaxation times are solved numerically through explicit finite difference method. Numerical results obtained for controlling parameters such as velocity, temperature and concentration are interpreted graphically and compared with that of clean fluid. Assessment of comparative study predicts that particle density in flowing fluid exhibits higher magnitude at leading edge during initial stages which can be attributed, particles may rebound/deposit on plate surface and becomes concern for erosion damage.

INTRODUCTION

Techniques for evaluating free convection effects caused by arranging heated vertical plates in combinations of rows and columns are of importance to electronics industry. Extensive review [1] for thermal process of buoyancy driven free convective fluid flows, numerical modells for heated vertical plate [2-3] and refinment of model [4] with viscous dissipation are the basis for present problem. It has been reported that presence of few micron sized spherical particles to air flowing turbulent motion in a pipe reduces the resistance coefficient, has drawn attention of several scientists to understand failure mechanisms [5]. To gain insight into complicated problems, author has reported preliminary results [6-7] for flow and thermal aspects in simple situations of rotating concentric and eccentric cylinders. The state-of-art [8-10] on two-phase flows for flat boundary layer theory aspects has given further insight to study the behaviour of isothermal vertical plate.

---

*Also Ph.D Research Scholar, Department of Mathematics, Osmania University, Hyderabad-500007 (India).

THEORY

The motion of gas containing solid spherical particles is critically dependent upon the interaction between fluid and particles. If the force accelerating the particle toward the fluid speed is given by Stoke's drag law, the process takes place within characteristic time, is termed as velocity equilibrium or relaxation time and can be represented mathematically as -

$$\tau_v = m/K \quad \text{Where} \quad K = 6\pi\mu a \quad (1)$$

This parameter will give some indication of gas-particle interaction process when compared with the time characterizing the flow. The corresponding velocity range is defined as -

$$\lambda_v = \tau_v u_o \quad (2)$$

Similar considerations hold for particle temperature T' which may be out of equilibrium with the temperature T of the fluid. When Stokes drag law is a reasonable approximation, the Nusselt number based on the particle radius is unity. Then the thermal relaxation time can be defined as

$$\tau_T = \rho c_p / Q \quad \text{where} \quad Q = 4\pi k a \quad (3)$$

and the corresponding thermal range is defined as

$$\lambda_T = u_o \tau_T \quad (4)$$

For the conditions

$$\tau_v > \tau \quad ; \quad \tau_v < \tau \quad ; \quad \tau_v \sim \tau$$
$$\tau_T > \tau \quad ; \quad \tau_T < \tau \quad ; \quad \tau_T \sim \tau \quad (5)$$

particle motion in fluid exhibits typical behaviour whose interpretation under $\tau_v > \tau$ and $\tau_T > \tau$ are difficult due to presence of non-linearities in the governing differential equations. However, when τ_v and τ_T become small, in limiting form, the particle slip velocity and gas-particle temperature becomes small.

As $\quad \tau_v \longrightarrow o \implies U' - U = V' - V = 0$

and $\quad \tau_T \longrightarrow o \implies T' - T = 0 \quad (6)$

But volumetric particle force, volumetric heat transfer rates between gas-particle phases remains finite.

FORMULATION

Free convective two phase flow in several mathematical regions of heated vertical plate, which is initially cold and at rest with boundary conditions are illustrated in Figure - 1.

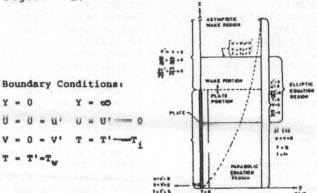

Boundary Conditions:

$Y = 0$ $\qquad$ $Y = \infty$

$U = U = U'$ $\quad$ $U = U' = 0$

$V = 0 = V'$ $\quad$ $T = T' \longrightarrow T_i$

$T = T' = T_w$

Fig. 1: Mathematical regions for natural convection flow

The governing differential equations in two phases describes Mass, Momentum and energy are -

$$U_X + V_Y = 0 \tag{7}$$

$$\rho\,[U_t + U\,U_X + V\,U_Y] = \rho g + \mu U_{YY} + Kn\,(U' - U) \tag{8}$$

$$\rho\,c_p\,[T_t + U\,T_X + V\,T_Y] = k\,T_{YY} + Qn(T' - T) \tag{9}$$

$$n_t + n\,U'_X + U'n_X + n\,V'_Y + V'n_Y = 0 \tag{10}$$

$$U'_t + U'U'_X + V'U'_Y = (U - U')/\tau_v \tag{11}$$

$$T'_t + U'T'_X + V'T'_Y = (T - T')/\tau_T \tag{12}$$

$$\rho = \rho_i\,[1 - \beta_i\,(T - T_i)] \tag{13}$$

$$Q'_q = \partial Q'/\partial q \;\; ; \;\; Q'_{qq} = \partial^2 Q'/\partial q^2 \tag{14}$$

where Q' can be any dependent variable such as U, V, T, U', V', T', n and q is function of (X, Y, t).

NOMENCLATURE

a   -   Radius of particle ;
c   -   Specific heat of fluid
 p -   at constant pressure;
g   -   Gravitational Constant;
Gr -   Grashof number;
k   -   Thermal conductivity;
K   -   Stokes drag law
L   -   Length of the plate;
m   -   Mass of fluid;
n   -   Number of Particles per
         unit volume;
Pr -   Prandtl   number;
Q   -   Heat exchange per unit
         volume;
t   -   Time;
T,T' - Fluid and particle
         Temperatures;
θ   -   Dimensional Temperature;
U,U' - Fluid and Particle
         velocities;
u   -   Characteristic velocity;
 o

V,V' - Fluid and particle
         velocities perpen-
         dicular to plate;
v   -   Dimensional
         velocity;
X,Y -   Coordinates along
         and perpendicular
         to plate;
x,y -   Dimensional
         coordinates;
τ_v, τ_T - Velocity and thermal
         relaxation times;
λ_v, λ_T - Velocity and thermal
         range;
β - Coefficient of vol.
         expansion of fluid;
μ - Viscosity of fluid;
ρ - Density of fluid;
τ - Characterisitc time;
i,w -   Subscripts, values
         of variables at Y=0
         and t=0;

ASSUMPTIONS

*   The work done by pressure is sufficiently small in comparision with the heat flow by conduction.

*   The fluid flow is laminar,unsteady,two-dimensional,free convective, incompressible and possess constant properties.

*   Temperatures in the system within a range approximately less than $1000°$ F that all radiation effects are neglected.

*   Flow and Thermal behaviour of heated vertical isothermal vertical plate has studied under $\tau_v < \tau$ and $\tau_T < \tau$.

GOVERNING EQUATIONS

The governing differential equations in non-dimensional quantities under assumptions and by equation (6) reduces to

$$(1+f) [u_{t'} + u u_x + v v_y] = Gr \theta + u_{yy} \qquad (15)$$

$$(1+f) [\theta_{t'} + u \theta_x + v \theta_y] = (Pr)^{-1} \theta_{yy} \qquad (16)$$

$$u_x + v_y = 0 \qquad (17)$$

$$f_{t'} + u f_x + f u_x + v f_y + f v_y = 0 \qquad (18)$$

with $u=v=0$: $\theta = 0$ at $x = 0$ and $y=\infty$
        $u=v=0$; $\theta = 1$ at $y = 0$ (19)
        $u=v=0$; $\theta = 0$; $f=f_o$; at $t' = 0$

and

$$t' = t \left[\left\{ g^2 \beta_i^2 (\Delta T)^2 \right\} / \vartheta \right]^{1/3} \; ; \; u = U/[\vartheta g \beta_i \Delta T]^{1/3}$$

$$v = V/ [\vartheta g \beta_i \Delta T]^{1/3} \; ; \; \theta = (T- T_i)/\Delta T; \; \Delta T = T_w - T_i \quad (20)$$

$$x = X \left[(g \beta_i \Delta T) / \vartheta^2 \right]^{1/3} \; ; \; y = Y \left[(g \beta_i \Delta T) / \vartheta^2 \right]^{1/3}$$

$$f = mn/\rho \; ; \; Pr = \mu c_p /k \; ; \; Gr = (g \beta \Delta T L^3)/\vartheta^2 ; \; \vartheta = \tfrac{\mu}{\rho}$$

The flow diagram Figure-2 describes solution for the coupled
equations (15-19) which has approximated by explicit finite
difference method in time domain.

RESULTS AND DISCUSSIONS

The space under investigation has been restricted to finite
dimensions $x=100$(max) and $y=25$(max) as corresponding to $y=\infty$
has been considered.   The components for grid size 100x100
at $t'=.5,1.5,..,80$ are computed and summarised $t'=10,20,..,80$.

1. The transient velocity profiles associated with particles
for different times are shown in Figure-3.   It is seen   at
$t'=10$, the profile path increases with the increase   in
distance and after obtaining peak, just few points after
leading edge, drops continuously.   The profile for clean
fluid shown in dotted line for velocity over y indicates
similar behaviour with deviation in peak values at higher
magnitudes.   It can also be observed from the distribution
pattern, the magnitude of particle loaded fluid velocity is
lower   than clean fluid velocity.   Profile trend   in
subsequent times shows same behaviour with a change in peak
values and falls to minimum with little descrepancies,which
can be attributed due to particle interaction.

2.   Figure-4 describes the behaviour of velocity component
perpendicular to the plate for fixed values of x.   In this
figure, continuous line indicate particulate laden fluid and
dotted line shows clean fluid.   The negative values of v at
all points in the mixture fluid reveals that v is directed
towards the plate.   As in the case of u, values of v
increase montonically to some extent and there after remains
unchanged. But, in the case of clean fluid, the velocity
component shows similar behaviour with higher values. On
comparision,   it   is observed that   particle   velocity
attains steady state faster than clean fluid in this case.

251

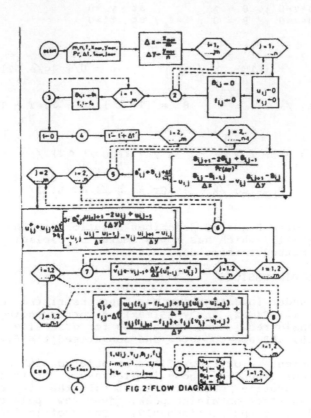

FIG 2 FLOW DIAGRAM

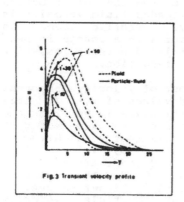

Fig. 3 Transient velocity profile

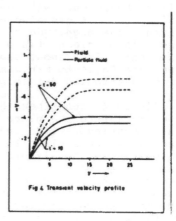

Fig 4 Transient velocity profile

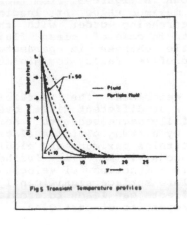

Fig 5 Transient Temperature profiles

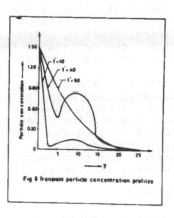

Fig 6 Transient particle concentration profiles

Comparision of steady state results for 10 X 10 grid with analytical or c numerical solutions of Clean fluid

sl. No	y .		0	2 5	5 0	7 5	10 0	12 5	15.0
1.	Numerical solution of fluid particles	Up	0	3.614	3.614	2 035	1 161	0.618	0.311
		Vp	0	-0.048	-0 106	-0 152	-0 183	-0 201	-0 217
		Tp	1	0.551	0 255	0 109	0 046	0.019	0.008
2	Numerical solution of fluid alone	U	0	4.470	5 027	4 067	2 862	1 855	1.123
		V	0	-0.050	-0.129	-0 209	-0 275	-0 323	-0.356
		T	1	0.704	0 450	0 269	0 155	0.087	0.047
3.	Analytical solution of clean fluid alone	U	0	4.650	5.290	4 150	2 780	1.650	0.980
		V	0	-0.051	-0.130	-0.210	-0 276	-0 325	-0.358
		T	1	0.705	0.451	0.270	0 156	0.088	0.049

TABLE - 7

253

3. Temperature distribution over y for fixed values of x at different t' s are plotted and shown in Figure-5. From these plots it is seen that θ attains maximum value at leading edge initially and varies in decreasing order with the increase in y. The profile path in case of clean fluid follows similar trend with little changes in parameter magnitudes. As t' increases, profile falls to minimum rapidly from the leading edge.

4. Plots in the Figure-6 describes the particle concentration for fixed values of x at different times over y. Particle concentration initially decreases upto some extent and vanishes as y changes by hitting plate surface. At later times, this profile attains maximum at middle of the plate and drops continuously. As time further increases, concentration which is a function of velocity, particles impact or deposit on plate surface significantly. However, the extent of impact effects can leads to erosion damage, depending upon material properties of particles.

5. Table-7 shows the data base generated in the present study, which are comparable with that of clean fluid results obtained through numerical and analytical solutions. Even though results for clean fluid are close in analytical and numerical techniques, the present numerical method reports significant reductions in the magnitudes of respective parameters due to presence of fine spherical sized particles distributed in flowing fluid.

REFERENCES

[1] Ede A.J.,Advances in Heat Transfer, Vol.4,pp 1-65,1965

[2] Hellums J.D and Churchill,SW, AIChE pp 690-692, 1962

[3] Carnahan B., Luther H.A., and Wilkies J.D Applied Numerical Methods, Willey, New York, 1969.

[4] Vajravelu K.,Numerical Heat Transfer,Vol.4,pp 71-83,1980

[5] Sproul W.T. , Nature, Vol.190, pp 463,1962

[6] Bhasker C., Proc. VIII Natl.HMT Conf. Andhra University, Visakhapatnam(India), pp 487-491,1985.

[7] Bhasker C., and Dr Bangad M, Proc.XV Natl. Conf. FMFP, REC/Srinagar (India) Vol.2 pp 671-675, 1987.

[8] Lee S.L , Advances in Applied Mechanics, Vol.22, pp 1-65, 1982.

[9] Soo, S.L., I&EC (Fundamentals) Vol.4, pp 426-433,1965

[10]Bootheryard, R.G., Flowing Gas-Solid Suspension, Chapman and Hall, London, 1971.

Enhancement of Heat Transfer Using Catalytic Surfaces

ADEL E. M. NASSER
Engineering and Physics Department
Arabian Gulf University
Bahrain

ABSTRACT :

The catalytic efficiency of a material for the recombination of certain atoms may affect the heat transfer through that material. The purpose of the present investigation is to present data about copper , nickel and silicon monoxide surfaces when subjected to hydrogen, oxygen and air at high temperatures. The experimental apparatus used for that purpose was a stainless steel shock tube of diameter of 127 mm. The model gauge is a flat plate made from pyrex and two adjacent nickel thin film gauges were mounted on top of it. The two gauges were then coated with SiO to act as the examined surface for one gauge and a base for coating a layer of Ni for the other gauge. The model was designed to measure the heat transfer rates from the two gauges in such a way that their measurements were under the same initial flow conditions. An independent method for calibrating the gauges has been used before and after the tests.

The gases used for this course of study were hydrogen diluted in argon, and Oxygen diluted in argon. The temperature ranges from 2500 K up to fully dissociation. Data for air is included in this paper.

The analysis of the data showed that hydrogen atom recombination on nickel enhances the heat transfer by about 12% than silican dioxide, while for oxygen, enhancement of about 15% is noticed than silicon dioxide. The data for air shows incerease in heat transfer of about 30% and 20% when subjected to copper and nickel surfaces with respect to silicon dioxide surface.

MAILING ADDRESS : Chairman of Engineering and Physics Department, Arabian Gulf University, P.O. Box : 26671, Manama, BAHRAIN.

INTRODUCTION :

In high temperature flows of dissociated gases, regimes exist in
which the heat transfer to a surface in contact with the gas may be
influenced by the extent of recombination at the surface and hence, by
the nature of the surface. As a contribution towards the study of hot
gas mixtures containing atomic gases an experimental study has been
made to measure the heat transfer from dissociated hydrogen/argon and
oxygen/argon mixtures to flat plate nickel and silicon dioxide surfaces in a shock tube.

The shock tube techniques permitted the study of heat transfer from
dissociated gases over a range of temperatures from 3000 K to 8000 k;
the argon being considered as an inert diluent. The experiments were
made in a 12.7 cm (5") diameter stainless steel shock tube. The flat
plate heat tranfer sensor was manufactured from pyrex on to which two
nominally identical thin film nickel resistance thermometers were
vacuum evaporated at the same stream wise distance from the leading
edge. A silicon monoxide layer was evapotated over both gauges, on
top of which a final nickel layer, acting as a catalyst to surface
recombination was deposited over one gauge only. The thicknesses were
chosen in order to minimise the difference in response time between
the two gauges. Valid comparison between the outputs of the two sensors were facilitated by independent calibrations and by measurements
in a known inert flow of shock heated argon. The experimental conditions were chosen to cover the range of shock Mach numbers necessary
to produce flows ranging from undissociated to full dissociated gases.
The inital channel pressure for most runs was 1 torr. Mixtures of 10%
H2 / 90% Ar and 10% O2 / 90% Ar mixtures were used .

From comparisons of the experimental measurements of the laminar heat
transfer from both H2/Ar and O2/Ar mixtures and pure argon to the different surfaces, it was possible to deduce that :

(i)   the recombination of hydrogen atoms at the nickel surface enhan-
        ces the heat transfer by about 12% more than silicon dioxide
        surface.

(ii)   the recombination of oxygen atoms at the nickel surface enhances
       the heat transfer by about 15% more than the silicon dioxide
       surface.

(iii) Data collected for air at copper and nickel surfaces show enhan-
      cements of about 30% and 20% more than the silicon dioxide sur-
      face.

EXPERIMENTAL APPARATUS :

A.   Shock Tube : The shock tube has been used during this course of
research. It is made from stainless steel. It has a 4.57 meter long
driver section and 12.75 meter long driven section, both of 127 mm
internal diameter and about 25 mm wall thickness.

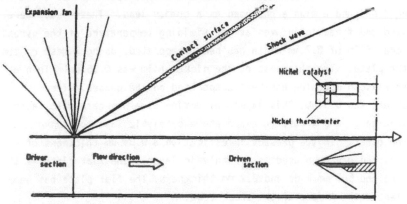

Fig (1) : The position of the model gauge inside the
           shock tube and its wave diagram.

     Diaphragms of various thicknesses made from aluminum are used.
The diaphragms are prescribed to ensure symmetric opening without
loss. The diaphragms are allowed to burst naturally using cold hydrogen as a driver gas at pressures upto 40 atmospheres. The channel or
low pressure section of the tube is normally evacuated to about 10
torr with a combined leak and degassing rate of about 0.9 x 10
torr/min.

257

The shock detection system consisted of thin film platinum resistance gauges fitted to the shock tube wall. These gauges were mounted along the entire driven section at 1.22 meter distance apart, giving maximum of 6 shock detection locations. In addition, there were further two gauges at distances 1.05 meter and 0.3 meter from the location of the model. Signals from these detectors were passed through pulse amplifiers and recorded as pulses on a raster display using a modified 535 Tektronix oscilliscope. The time makers were at 5 usec intervals and the time resolution of this recording system was approximately 1 usec. This system permits the determination of the shock speed history ahead of the model.

B.   The Model : The model used for the present investigation is a flat plate made from pyrex with three silver leads; two leads are along sides and a middle one used as a common lead. These leads were painted and fired in an oven at the yielding temperature of the pyrex. A nickel film of 0.7 mm width has been evaporated, using a mask on top of the plate. The thickness of the nickel film was 0.05 um. Then a layer silicon monoxide has been eraporated on the guages. The thickness was 0.5 um. This layer was acting as a non-catalytic layer for one gauge and a base to evaporate a catalytic layer on top of it on the other. In the present investigation a 0.05 um thickness of pure nickel has been used as a catalytic layer. The flat plate is 16 mm long and 5.7 mm wide and 2.2 mm thickness. The flat plate has been mounted in the middle of the shock tube.

Measurements of the heat transfer were obtained in the usual manner using T section analogue circuits. An independent method for calibrating the gauges has been used before and after the tests. Fig. 1 shows the model gauge inside the shock tube and the flow diagram after bursting the diophragm. Fig. 2 shows the T section analogue circuit while Fig. 3 shows the calibration of the catalytic and the non catalytic gauges calibration curves.

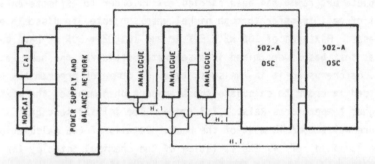

Fig (2) : The arrangement of the T section analogue
circuits.

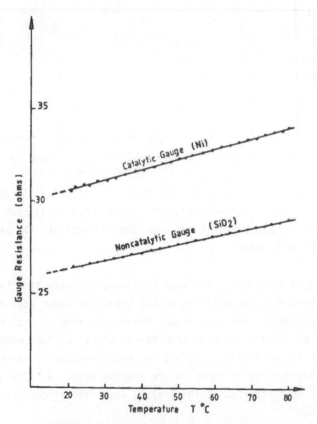

Fig (3) : Calibration curves for the catalytic (Ni)
and the noncatalytic (SiO_2) gauges.

RESULTS AND DISCUSSIONS :

A complete programme has been carried out in order to estimate enhancement of heat transfer through nickel when subjected to dissociated
the gases. Mixtures of 10% H2 + 90% Ar and 10% O2 + 90% Ar have been
used as test gases were mixed in a separate bottle for one hour, at
least, before using it in the shock tube. A computer program has been
developed in order to calculate the transport phenomena of the mixture
at higher temperatures Refs. 1,2,3 and 4. The initial test gas
pressure was 1 torr in most of the runs. Details of the calculation
may be found in ref. 4. The variation of the chemical enthalpy for
oxygen/Argon mixture with temperature is shown in Fig. (4).

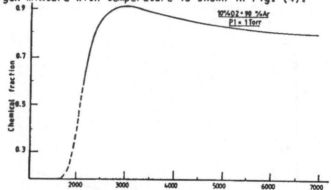

Fig (4) : Variation of chemical fraction due to
recombination of oxygen atoms with
temperatures.

The shock tube computer programme has been used to calculate the chemical equilibrium of reactant/dilutent system immediately down stream
of the incident shock. Ref. 6 has been used to estimate the chemical
enthalpy of the mixture.

The accuracy of determining the atom concentration with this technique
depends very much on the position of the catalytic layer with respect
to the leading edge. For this reason trials has been made to measure
the difference between the heat transfer on nickel surface which has
been made up to the leading edge and the heat transfer on nickel surface which has not been made up to the leading edge. A difference of
up to 2% has been observed at the present experimental conditions.

260

In this paper it is assumed that silicon dioxide surface may be considered as a non catalytic surface. With this assumption, it was possible to estimate the precentage enhacement in heat transfer for nickel surface when subjected to either hydrogen or oxygen atom recombination. Fig. 5 shows experimental data outputs of the enhancement of heat transfer percentages with the temperatures. At fully dissociated gases, 30% and 22% enhancement in heat transfer have been estimated for copper and nickel surfaces respectively when subjected to shock heated air Ref. 7. The Experimental conditions of the air data was different from the present course of study since a stagnation gauge has been used for the air measurements. For the present paper a flat plate gauge has been used for both hydrogen and oxygen measurements.

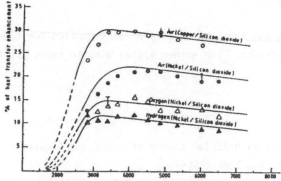

Fig (5) : Variation of the percentage of heat transfer enhancement of difference species on copper and nickel with respect to silicon dioxide with temperatures.

For dissociated gases, 12% and 15% enhancement in heat transfer through flat plate nickel surface have been recorded when subjected to dissociated hydrogen and oxygen respectively. It is worth mentioning here that the computed curves Fig. (5) are inclined down at higher temperatures. This is expected since the convective heat transfer is increasing with temperatures while the chemical enthalpy reached its maximum value at fully dissociation. The present data may add some information when dealing with combustion efficiency, combustion chamber design or pollution problems.

CONCLUSION :

Useful results have been obtained using the present techinque to estimate the enhancement heat transfer percentage due to recombination of certain atoms on a certain surface.

Since the technique depends on assuming that the silican dioxide is a non catalytic surface the technique may be used also with two similar models to give two simultaneous equations when solved give the absolute value of the catalytic efficiency of both materials.

ACKNOWLEDGEMENT :

The auther was has thank University of sorthamption for sponsering this research. The auther wishes also to thank Dr. R.A. East.

REFERENCES :

1 - Hirschfolder, et al.Moecular theory of gases and liquids " John Wiley and Sons, New York (1954).

2 - Dorrance. " Viscous hypersonic flow ",McGraw-Hill, New York (1962).

3 - A.E. NASSER and H. Sofrata " Viscosity, thermal conductivity and Diffusion of Hydrogen diluted in Argon at High Temperatmes " Warme-und stof fubertragung Vol. 4 Spring - Verlag 1980.

4 - A. Nasser " Heat Transfer measurements of shock heated oxygen/Argon mixture " Proceedings of the Fifth International conference for Mechanical Power Engineering, Ain Shams University, Egypt, 13-15 October 1984.

5 - " Heat transfer measurements with a nickel - silicon monoxide flat plate in dissociated hydrogen". AASU Report (1979).

6 - " Thermodynamic and transport properties for the hydrogen oxygen system NASA SP 3011 (1964).

7 - " Non Equilibrium stagnation point Heat Transfer Measurements to catalytic surfaces in shack Heated Air ". Proceedings of the Tenth International shoch tube symposium, Kyoto International Conference Hall, 1975.

An Experimental Investigation of Heat Transfer for Swirling Flow in a Vortex Tube

SUI LIN, CLYDE K. KWOK, GEORGIOS H. VATISTAS, and JIRUI CHEN
Department of Mechanical Engineering
Concordia University
Montreal, Quebec, Canada H3G 1M8

ZHEN LU
Department of Power Machinery Engineering
Shanghai Jiao-Tong University
Shanghai, PRC

ABSTRACT

An experimental investigation was made to study the heat transfer characteristics of a water-cooled vortex tube with air as the working medium. It was indicated that the inlet pressure and the cold mass ratio have significant influences on the heat transfer behaviour of the vortex tube. It was found that the heat transfer rate for the swirling flow in the vortex tube is 1.28 to 98.0 times higher than that for a flow without swirling in the turbulent entrance region in the same tube.

NOMENCLATURE

$C_{p\ell}$	specific heat of water at constant pressure
T	temperature
$\dot{m}_\ell$	mass flow rate of water
$\dot{m}_o$, $\dot{m}_h$, $\dot{m}_c$	mass flow rates of inlet air, hot air and cold air, respectively
h_ℓ, h_a, U	water-side, air-side and overall heat transfer coefficients, respectively
k_w, k_a	thermal conductivities of vortex tube wall and air, respectively
P_o, P_a	inlet air gauge pressure and atmospheric pressure, respectively
Q	steady-state heat transfer rate
y	mass ratio of the cold air to the inlet air
P	pressure ratio of the inlet air to atmospheric air
Pr	Prandtl number
Re	Reynolds number

Nu_a, Nu_r air-side Nusselt number and reference Nusselt number, respectively

Subscripts

c	cold air
h	hot air
0	inlet condition
ℓ	water
w	vortex tube wall
r	reference

INTRODUCTION

A vortex tube is a simple device operating as an energy separation machine without any moving components. It consists mainly of a simple tube. In this device compressed air is led tangentially with a high velocity into the vortex tube. While the air swirls rapidly around the inside of the vortex tube, an energy separation process takes place there. Right next to the entrance nozzle a cold air stream leaves the tube through a central orifice, while at the far end of the tube a hot air stream exhausts near the tube wall. This phenomenon of energy separation in a vortex tube was first reported by Ranque [1].

Since the flow pattern of the swirling flow in a vortex tube is significantly different from that of a flow without swirling in the same tube, it is expected that the convective heat transfered from the hot gas stream with rapid swirling to the wall of the vortex tube has to be significantly different from the convective heat transfered from the same hot gas stream without swirling. In the present paper, the heat transfer characteristics of a water-cooled vortex tube were studied with air as the working medium. Three inlet pressures (3, 4 and 5 bars gauge) were used in this experimental investigation. The cold mass ratio covered was in the range of 0.13 to 0.94, which is defined as :

$$y = \frac{\dot{m}_c}{\dot{m}_o} = \frac{\text{mass flow rate of the cold air}}{\text{mass flow rate of the inlet air}}$$

EXPERIMENTAL APPARATUS

A schematic layout of the apparatus used for the determination of the convective heat transfer coefficient for the swirling flow in the experimental vortex tube is shown in Fig. 1. Compressed air is led tangentially into vortex tube (1) through inlet valve (2) and filter (3). The entering air is expanded in the vortex tube and divided into a hot stream and a cold stream. The cold air leaves the central orifice right next to the entrance nozzle while the hot air discharges at the periphery at the far end of the tube. The flow rate of the hot air is controlled by valve (4). The flow rate of the cold air is controlled by valve (16). The flow rate of water is controlled by valve (17) and is measured by rotameter (18). The inlet and outlet temperatures of the water are measured by thermocouples (14) and (15), respectively. The mass flow rates of the hot air, the cold air and the inlet air are determined by measuring the

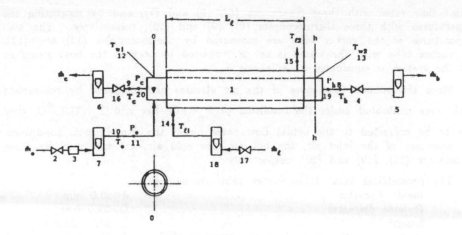

Figure 1. Schematic diagram of the
experimental apparatus

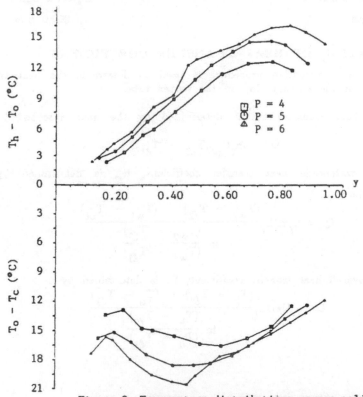

Figure 2. Temperature distributions versus cold
mass ratio with the pressure ratio as
a parameter

volume flow rates with three rotameters (5), (6) and (7), and by measuring the temperatures with three thermocouples (8), (9) and (10), respectively. The wall temperatures of the vortex tube are measured by thermocouples (12) and (13). The vortex tube with the water jacket is insulated to prevent the heat losses so that the system is considered as adiabatic.

Since the volume flow rates of the air streams are measured by rotameters which were calibrated under the condition of P_o=1.01 bar and T_o=21.1 $^\circ$C, they have to be corrected to the actual flow rates under the experimental conditions. The pressures of the inlet air, the hot air, the cold air, are measured by three manometers (11), (19) and (20), respectively.

The geometrical data of the vortex tube are as follows:

Inside diameter	d=20.0 mm
Outside diameter	D=22.0 mm
Length	L=18d=360.0 mm
Diameter of the inlet air nozzle (two nozzles)	d_o=2.5 mm
Diameter of the cold exit air orifice	d_c=6.5 mm

The following geometrical dimensions are used for the water jacket:

Inside diameter	D_ℓ=41.0 mm
Length	L_ℓ=260.0 mm

DETERMINATION OF HEAT TRANSFER COEFFICIENT

The following calculation procedure is used to determine the heat transfer coefficient, h_a, for the swirling flow in the vortex tube.

1. The heat transfer rate is determined by the heat absorbed by the water,

$$Q = \dot{m}_\ell C_{p\ell}(T_{\ell 2} - T_{\ell 1}) \tag{1}$$

2. The water-side heat transfer coefficient, h_ℓ, is determined by the following equation:

$$Q = h_\ell \pi DL_\ell \frac{(T_{w2} - T_{\ell 2}) - (T_{w1} - T_{\ell 1})}{\ln \dfrac{(T_{w2} - T_{\ell 2})}{(T_{w1} - T_{\ell 1})}} \tag{2}$$

3. The overall heat transfer coefficient, U, is determined by

$$Q = U\pi DL_\ell \frac{(T_h - T_{\ell 2}) - (T_o - T_{\ell 1})}{\ln \dfrac{(T_h - T_{\ell 2})}{(T_o - T_{\ell 1})}} \tag{3}$$

4. By making use of the definition of the overall heat transfer coefficient, we have

$$U = \cfrac{1}{\cfrac{1}{h_\ell} + \cfrac{D}{2k_w}\ln\cfrac{D}{d} + \cfrac{D}{h_a d}} \qquad (4)$$

therefore

$$h_a = \frac{D}{d}\cfrac{1}{\cfrac{1}{U} - \cfrac{1}{h_\ell} - \cfrac{D}{2k_w}\ln\cfrac{D}{d}} \qquad (5)$$

Taking into consideration of the change in the temperature of the swirling air along the vortex tube, the following dimensionless parameters are defined:

$$Nu_a = \cfrac{h_a d}{\frac{1}{2}(k_{a|o} + k_{a|h})}, \quad Re = \frac{1}{2}(Re|_o + Re|_h) \qquad (6)$$

$$Pr = \frac{1}{2}(Pr|_o + Pr|_h), \quad P = \frac{P_o + P_a}{P_a} \qquad (7)$$

Symbols "$|_o$" and "$|_h$" refer to the quantities which are evaluated at 0-0 and h-h sections, respectively, as indicated in Fig. 1.

The Nusselt number, the Reynolds number and the Prandtl number determined by Eqs. (6) and (7), respectively, are the mean values over the vortex tube.

RESULTS AND DISCUSSION

Experimental results indicate that the inlet pressure and the cold mass ratio have significant influences on the energy separation in the vortex tube. An increase of the inlet pressure, P_o, results in an increase of both the temperature differences $(T_h - T_o)$ and $(T_o - T_c)$ as shown in Fig. 2. The temperature differences $(T_h - T_o)$ and $(T_o - T_c)$ first increase with an increase of the cold mass ratio, y. They reach their respective maximum values and then decrease. With an increase of the inlet pressure, the maximum point of $(T_h - T_o)$ shifts to a larger value of the cold mass ratio and the maximum point of $(T_o - T_c)$ shifts to a smaller value of the cold mass ratio.

For a small cold mass ratio, because the mass flow rate of the cold air is small, the change in the temperature of the cold air stream is more sensitive to the change in the inlet pressure. In this case, the distributions of the temperature difference $(T_o - T_c)$ are well distinguished from one another. For a larger cold mass ratio, the situation is reverse: the change in the temperature of the cold air stream becomes insensitive to the change in the inlet pressure.

Eq. (3) indicates that the overall heat transfer coefficient is directly proportional to the heat transfer rate and inversely proportional to the overall log mean temperature difference (LMTD-O). Fig. 3 shows that the heat transfer rate increases with the inlet pressure as well as with the cold mass ratio. For a given value of the inlet pressure, the ratio of the heat transfer rate to the LMTD-O increases with an increase of the cold mass ratio as shown in Fig. 4. It can be seen that the variation of the overall heat transfer coeffcient for a fixed cold mass ratio is limited, except in the region of y>0.5, where the overall heat transfer coefficient for P=6 is lower than that for P=4 and 5. This is because that, in this region, for P=6, the ratio of the heat transfer rate to the LMTD-O for a fixed value of y is lower than that for P=4 and 5.

As indicated in Eq. (5), the air-side heat transfer coefficient increases with an increase of the overall heat transfer coefficient. Therefore, the air-side Nusselt number has a similar behaviour as the overall heat transfer coefficient, as shown in Fig. 5.

For the purpose of comparison, the following convective heat transfer relation for a flow in the turbulent entrance region in a tube [2] is used as a reference

$$Nu_r = 0.036 \ Re^{0.8} \ Pr^{\frac{1}{3}} \ (\frac{d}{L})^{0.055}$$ (8)

The value of L/d of the experimental vortex tube covered by the water jacket is

$$\frac{L_\ell}{d} = \frac{260}{20} = 13$$

Substituting this value into Eq. (8) yields

$$Nu_r = 0.031 \ Re^{0.8} \ Pr^{\frac{1}{3}}$$ (9)

The ratio of the experimental air-side Nusselt number to the reference Nusselt number as a function of the cold mass ratio for the inlet pressures investigated is given in Fig. 6. Calculating the representative values of the lower and upper limits of the experimental air-side Nusselt number, and substituting the corresponding values of the Prandtl number and the Reynolds number into Eq. (9) , we obtain

$$(\frac{Nu_a}{Nu_r})_{low} = 1.28$$ (10)

$$(\frac{Nu_a}{Nu_r})_{high} = 98.0$$ (11)

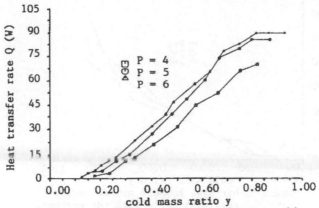

Figure 3. Heat transfer rate versus cold
mass ratio with the pressure ratio
as a parameter

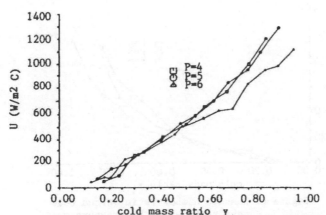

Figure 4. Overall heat transfer coefficient
versus cold mass ratio with the
pressure ratio as a parameter

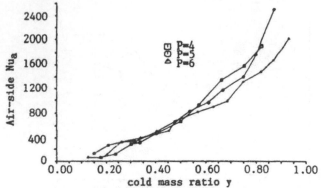

Figure 5. Air-side Nusselt number versus
cold mass ratio with the pressure
ratio as a parameter

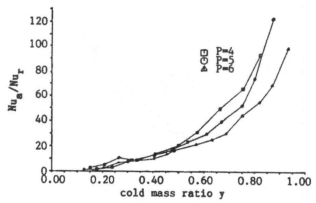

Figure 6. Nusselt number ratio versus cold
mass ratio with the pressure ratio
as a parameter

It is illustrated that the heat transfer rate for the swirling flow in the experimental vortex tube is 1.28 to 98.0 times higher than that for the flow without swirling in the turbulent entrance region in the same tube. The experimental results obtained indicate that the swirling flow in the vortex tube is an effective way for the enhancement of convective heat transfer.

REFERENCES

1. Ranque, G., "Expériences sur la Détente Giratoire avec Productions Simultanées d'un Échappement d'air Chaud et d'un Échappement d'air Froid", J. Phys. Radium 4, Ser. 7, 112-114, (1933).

2. Holman, J. P., "Heat Transfer", McGraw-Hill, New York, 6th Ed., (1986).

Heat Transfer to Highly Viscous Fluids in a Mechanically Agitated Vessel

WEIGUO MIN, BUQUAN XU, QIUAN SONG, and KAI WANG
Department of Chemical Engineering
Zhejiang University
Hangzhou, PRC

INTRODUCTION

Since high viscosity and the enormous heat released in polymer production, the heat transfer in agitated vessels is often akey problem. The research in this filed has been very active since 70's[1-14]. Alot of agitator systems have been studied. Some correlations have been proposed but have poor agreement with each other.

For increasing the heat transfer capacity, suitable heat transfer surface was tested except jacket, for example, the surface of impeller and draft tube ect.. A few apparatuses have been applied to commercial production, but more information is required. In the present paper the heat transfer has been studied for both the S-D and the S-D-P systems, including the correlations for industry design and disscussing the influencing elements.

1. EXPERIMENTAL APPARATUS AND METHODS

The experiments were carried out in the stainless steel agitated vessel of inside diameter 244 mm with double layers jacket. Fig.1 is the flow diagram. The details of agitator systems are show in Tab.1. The steam was fed into the jacket and cooling water was pumped into draft tube and pipe group to ensure that the experiments were carried out under steady state condition. The heat flux through the surface of the jacket is determined by measuring the guantity of condensate and the heat flux through the surface of the draft tube and the pipe group is determined by measuring the rate of flow and the rise of temperature of cooling water.

Eighteen copper-constantan thermocouples were installed in the vessel, and the six thermocouples of them turned round with impeller and voltage signal was brought out by the slip ring. The bulk temperature is the arithmetic mean of temperature measured. The wall temperature of the draft tube was measured by four thermocouples.

Newtonian fluid used is sugar solution. Carboxymethylcellulose(CMC) dissolved in water in various concentrations is used for pseudoplastic fluids, its rheological properties can be expressed by the power law model. The ranges of experimental variables rae shown in Tab.2.

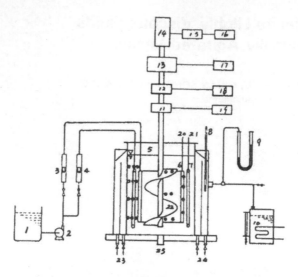

Fig. 1 Flow diagram of experiments

1.cooling water vessel 2.pump 3.4.flowmeter 5.agitated vessel 6.draft
tube 7.pipe group 8.mercury thermometer 9.U-pressure gauge 10.steam
boiler 11.reflection mark 12.torsion sensor 13.motor 14.slip ring
15.cold junction 16.voltmeter 17.speed controller 18.electric
resistance strain device 19.revolution counter 20.21.cooling water outlet
22.agitator 23.24.condensate 25.fluids outlet
 ⊖ copper-constantan thermocouples

Tab. 1 Geometry of Experimental
        Apparatus

System	S-D	S-D-P
d	135	108
p	135	108
l	216	222
h	16	25
D_{ti}	169	130
D_{to}	193	153
h_t	42	33
l_t	197	195
H	300	300
d_c		6
Number of pipes		24
Height of pipes		210
D_c		200

Tab. 2 Range of Variables

Variable	Range
K	1-100
n	0.3-0.8
μ	30-100
ρ	1410(sugar)
λ	1010-1050(CMC)
	0.38(sugar)
	0.60(CMC)
C_p	2500(sugar)
	4180(CMC)
t_s	99-101
N	20-240
t_b	40-60
t_i	17-24
t_o	25-30
CMC concentration(%)	2-6

2. RESULTS AND DISCUSSION

2.1 Calculation of Apparent Viscosity

The average shear rate in the agitated vessel is calculated by Metzner model

$$\dot{\gamma}_{av}=k_s N \tag{1}$$

where k_s is Metzner constant which can be obtained by measuring the power consumpation of Newtonian and non-Newtonian fluids in the laminar region. By means of experiments, the Metzner constants of both systems are as follows:

For S-D system: $k_s=9.7$

For S-D-P system: $k_s=11.5$

then, the apparent viscosity is given by

$$\mu_a=K(k_s N)^{n-1} \tag{2}$$

2.2 Heat Transfer Correlation

Heat transfer correlation may be obtained by using dimensional analysis and regressive analysis in the following form:

$$Nu=CRe^a Pr^b Vis^c \tag{3}$$

where the characteristic length in Nu is d for jacket side, $(D_{ti}+D_{to})/2$ for draft tube side and d for pipe group side, the characteristic temperature is t_b. In this paper, $b=1/3$ and $c=0.14$ are suitable, so the Eq.(3) may be written as

$$\frac{Nu}{Pr^{1/3} Vis^{0.14}}=CRe^a \tag{4}$$

The relation $Nu/(Pr^{1/3}Vis^{0.14})$ versus Re in log-log co-ordinates is plotted. Fig.2 and Fig.3 are the results of S-D system, those of S-D-P system are shown in Fig.4-6. It is seen that experimental data points around two straight lines when $Re<10$ and $Re>10$ respectively. Using the linear least-squares procedure, the correlations are obtained respectively (see Tab.3). It is seen that a is 1/3 when $Re<10$ and a is 1/2 when $Re>10$ approximately. Fixing $a=1/3$ when $Re<10$ and $a=1/2$ when $Re>10$, then the generalized correlations are obtained as follows:

For S-D-P system:
$0.2<Re<10$

$$Nu_j=1.56\ Re^{1/3}Pr^{1/3}Vis^{0.14} \qquad (S=6.8) \tag{5}$$

$$Nu_d=1.32\ Re^{1/3}Pr^{1/3}Vis^{0.14} \qquad (S=8.7) \tag{6}$$

$10<Re<300$

$$Nu_j=1.00\ Re^{1/2}Pr^{1/3}Vis^{0.14} \qquad (S=16.8) \tag{7}$$

$$Nu_d=1.01\ Re^{1/2}Pr^{1/3}Vis^{0.14} \qquad (S=11.1) \tag{8}$$

For S-D system:
$0.1<Re<10$

$$Nu_j=1.61\ Re^{1/3}Pr^{1/3}Vis^{0.14} \qquad (S=5.9) \tag{9}$$

$$Nu_d=1.43\ Re^{1/3}Pr^{1/3}Vis^{0.14} \qquad (S=9.1) \tag{10}$$

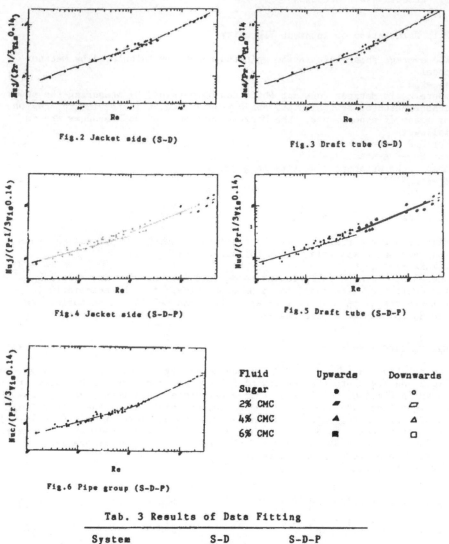

Fig.2 Jacket side (S-D)

Fig.3 Draft tube (S-D)

Fig.4 Jacket side (S-D-P)

Fig.5 Draft tube (S-D-P)

Fig.6 Pipe group (S-D-P)

Fluid	Upwards	Downwards
Sugar	●	○
2% CMC	◢	◿
4% CMC	▲	△
6% CMC	■	□

Tab. 3 Results of Data Fitting

System		S-D		S-D-P	
		Re<10	Re>10	Re<10	Re>10
Jacket	C	1.50	1.41	1.65	1.28
	a	0.35	0.40	0.30	0.39
	S	5.8	11.9	5.7	13.9
Draft	C	1.33	1.05	1.44	0.88
tube	a	0.32	0.49	0.31	0.57
	S	8.5	11.1	8.7	11.5
Pipe	C			0.12	0.071
proup	a			0.23	0.46
	S			10.3	3.8
Number of data		15	24	39	21

276

$Nuc=0.11\ Re^{1/3}Pr^{1/3}Vis^{0.14}$ (S=14.3)   (11)

$10<Re<500$

$Nuj=0.76\ Re^{1/2}Pr^{1/3}Vis^{0.14}$ (S=20.3)   (12)

$Nud=1.24\ Re^{1/2}Pr^{1/3}Vis^{0.14}$ (S=16.6)   (13)

$Nuc=0.059Re^{1/2}Pr^{1/3}Vis^{0.14}$ (S=6.3)   (14)

2.3 DISCUSSION

2.3.1 <u>Effect of rotational directions</u> The rotational directions refers to flowing upwards or downwards for the fluids inside the graft tube. Both the solid points (upwards) and the hollow points (downwards) fall in the same region (see Fig.2-6), which indicates that the rotational directions have no remarkable influence upon heat transfer. this implies that the flowing state of fluids is essentially the same regardless of flowing upwards or downwards at the same rotational speed.

2.3.2 <u>Comparision of α_j, α_d and α_c</u> The heat transfer intensity may be judged by α. At the same rotational speed α_d/α_j and α_c/α_j are calculated from Eq.(5) to (14) (see Tab.4). It is seen that both α_d and α_c are bigger than α_j. Particularly, α_c is about three times bigger than α_j. It is because the shear rate around the jacket wall is smaller than that of the others. The velocity profile in the annulus is parabolic. After installing the pipe group, the shear rate around it is bigger, therefore the heat transfer intensity of the surface of the pipe group is the highest of all surface.

Ishibashi studied the heat transfer in a agitated vessel with a centered screw, the correlations were proposed

$Nuj=1.0\ Re^{1/2}Pr^{1/3}Vis^{0.14}$ (50<Re<100)   (15)

$Nuj=0.65\ Re^{0.6}Pr^{1/3}Vis^{0.14}$ (100<Re<25000)   (16)

Comparision with Eq.(5) and (7), it is apparent that installing draft tube and pipe group has no obvious influence upon α_j, but the heat transfer area increases doubly. Therefore the heat transfer capacity is greatly increased.

2.3.3 <u>Effect of geometrical parameters</u> The influences of d, p, l, h and D_{ti} upon heat transfer are determined in this paper and the other works. It is found that d and D_{ti} have the greatest effects among all the parameters. The effects of p and h are relatively small in the range of the experiments. As Re increases, the effects of the geometrical parameters weaken.

Tab.4 Comparision of α_j, α_d and α_c

System		α_d/α_j	α_c/α_j
S-D	Re<10	1.14	
	Re>10	1.37	
S-D-P	Re<10	1.59	2.88
	Re>10	2.82	3.17

277

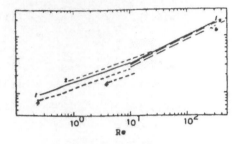

Line	Author	d/D
1	This work	0.553
2	Seichter	0.469
3	Seichter	0.323
4	Seichter	0.220
5	Mitsuishi (downwards flow)	0.525
6	Mitsuishi (upwards flow)	0.525

Fig. 7 Heat transfer correlation by
several authors (S-D)

2.3.4 Exponent a of Re in correlation It is seen from Eq.(5) to (14)
that a is 1/3 when Re 10 and a is 1/2 when Re 10, this has good agreement
with Mitsuishi's and Seichter's results. When Re is more high (Re 1000),
Mitsuishi reported that a is 2/3.
Actually, the slope of each point in the correlation curves is different
(see Fig.2-6), the influence of Re upon heat transfer weakens as Re
decreases. Coyle proposed that the heat transfer to highly viscous fluids
in a agitated vessel carried out by the conduction in the clearance
between the wall and the edge of agitator. It is deduced that exponent
a varies at low Reynolds number. There are the same conclusions for
helical ribbon and anchor agitators.

2.3.5 Comparision with literatures By present, only a few literatures
about the heat transfer of S-D system have been reported (see Fig.7). It
is seen that the author's results have good agreement with Mitsuishi's
and Seichter's.

CONCLUSION

1) The heat transfer film coefficient α_j, α_d and α_c were measured for
S-D system and S-D-P system. The heat transfer correlations are proposed
(Eq.(5) to (14)) which have good agreement with the literatures.
2) α_d and α_c are greater than α_j. It is confirmed that installing the
draft tube and pipe group as heat transfer surface is important measurement
to increase heat transfer capacity.
3) The rotational directions have no remarkable influence upon heat
transfer.
4) d and D_{ti} have the greatest influences upon heat transfer among all
the geometrical parameters.

NOMENCLATURE

a,b,c	=exponents in correlations	(-)
C	=coefficient in correlations	(-)
C_p	=specific heat	(W/kg K)
d	=screw diameter	(m)
d_c	=outside diameter of pipe	(m)
D	=vessel diameter	(m)
D_c	=diameter of pipe group	(m)
D_{ti}	=inside diameter of draft tube	(m)

D_{to}	=outside diameter of draft tube	(m)
h	=height of screw to bottom	(m)
h_t	=height of draft tube to bottom	(m)
H	=height of liquids	(m)
k_s	=Metzner constant	(-)
K	=consistency index	$(N\ S^n/m^2)$
1	=screw length	(m)
l_t	=length of draft tube	(m)
n	=flow behavior index	(-)
N	=rotational speed	(S^{-1})
p	=screw pitch	(m)
S	=average deviation between calculation and experiment	(-)
t_b	=bulk temperature	(K)
t_s	=steam temperature	(K)
$\dot{\gamma}_{av}$	=average shear rate in agitated vessel	(S^{-1})
ρ	=density	(kg/m^3)
α	=heat transfer film coefficient	$(W/m^2\ K)$
λ	=thermal conductivity	$(W/m^2\ K)$
μ	=viscosity	$(N\ S/m^2)$
μ_a	=apparent viscosity	$(N\ S/m^2)$

Subscripts

j	=jacket
d	=draft tube
c	=pipe group

Dimensionless Numbers

$Re=d^2N\rho/\mu$	Reynolds number
$Pr=Cp\lambda/\mu$	Prandle number
$Nu=\alpha L/\lambda$	Nusselt number
$Vis=\mu/\mu_w$	Viscosity ratio

LITERATURE CITED

1. Nagata, S., Mixing Principle and Application, Kodansha LTD Halsted Press (1975), Tokyo.

2. Prokopec, L., Int. Chem. Enging., 12(4), 712–719(1972).

3. Mitsuishi, N., et al, J. Chem. Enging. Japan, 6(5), 409–414(1973).

4. Mitsuishi, N., et al, J. Chem. Enging. Japan, 6(5), 415–420(1973).

5. Nagata, S., et al, J. Chem. Enging. Japan, 5(1), 83–85(1972).

6. Ishibashi, K., et al, J. Chem. Enging. Japan, 12(3). 230–235(1979).

7. Blasinski, H., et al, Int. Chem. Enging., 21(4), 679–683(1981).

8. Heinlein, H. E., et al, Int. Eng. Chem. Process Des. Develop., 11(4), 490–495(1972).

9. Kuriyama, M., et al, J. Chem. Enging. Japan, 14(4), 323–330(1981).

10. Kuriyama, M., et al, J. Chem. Enging. Japan, 16(6), 489–494(1983).

11. Chavan, V. V., A. I. Ch. E. J., 29(2), 177–186(1983).

12. Edwards, M. F., et al, Low Reynolds Number Flow Heat Exchangers, Edited by S. Kakac, Hemisphere(1983).

13. Ayazi Shamlou, P., et al, Chem. Enging. Sci., 41(8), 1957–1967 (1986).

14. Wang, H. Z., J. Chem. Ind. and Enging.(China), (4), 375–379(1984).

15. Wilson, E. E., et al, Trans. Am. Soc. Mech. Engrs., No1477, 47(1915).

16. Mizushina, T., et al, Kagaku Kogaku, 30, 819(1966).

17. Metzner, A. B., et al, A. I. Ch. E. J., 3, 3–10(1957).

18. Porter, J. E., Trans. Instn. Chem. Engrs., 49, 1–29(1971).

19. Coyle, C. K., Can. J. Chem. Enging., 48, 275–278(1970).

20. Min, W. G., Heat Transfer to Highly Viscous Fluids in a Agitated Vessel, M. Sc. Thesis, Zhejiang University, China(1988).

A Shear Rate Model Used in Heat Transfer Correlations of Agitated Vessels

KAI WANG and SHENGYAO YU
Department of Chemical Engineering
Zhejiang University
Hangzhou, PRC

INTRODUCTION

To calculate heat transfer coefficients for agitated non-Newtonian fluids, a suitable apparent viscosity is required. For pseudoplastic fluids obeying the power-law, a good estimate of shear rate is required for calculating the apparent viscosity. In this paper, the shear rate models established by previous workers are critically reviewed, and a new shear rate model, which is confirmed experimentally, is proposed. With this shear rate model, heat transfer correlations of non-Newtonian fluids are obtained for vessels equipped with various types of cooling tubes.

LITERATURE REVIEW

Gluz et al.(2) correlated heat transfer coefficients by calculation of μ_a with

$$\gamma = 4\pi N \tag{1}$$

Sandall et al.(4) defined Re* and Pr* by using the following equation:

$$\gamma = B(\frac{3n+1}{4n})^{n/(n-1)} N \tag{2}$$

where $B = 9.5 + 9/(1-(d/D)^2)$ for anchors.

Mizushina et al.(3) obtained heat transfer correlations by estimating shear rate at the wall surface:

$$\gamma = B\frac{\pi d\ N}{(D-d)/2} \tag{3}$$

For anchors $B=14$ was determined from experimental data. But it is not clear whether or not B would equal 14 for other agitated systems.

For agitated vessels without cooling tubes, Tang et al.(5) proposed:

$$\gamma = (P/\pi^2 D^2 NHK)^{1/n} \tag{4}$$

Equation (4) has been used successfully to correlate heat transfer coefficients in vessels without cooling tubes. But for vessels equipped with cooling tubes, further work would be needed. Bourne et al.(1) indicated that the proportionality between γ and N is valid only in the laminar regime. They further suggested:

$$\gamma \sim N^{(3-2q+qn)/(1+n)} d^{2(1-q)/(1+n)} \tag{5}$$

where q takes values 0 and 1 in the fully turbulent and laminar regimes, respectively.

From the above survey of shear rates, the following conclusions can be drawn: 1) The wall surface shear rate should be taken as the characteristic parameter for correlating heat transfer coefficients, because heat resistance resides mainly in the boundary layer; 2) in the expression $\gamma = f(d,D,n,N^b)$, the exponent b is expected to be greater than unity and is related to non-Newtonian behavior; 3) characteristic shear rate is affected by flow regime.

THEORETICAL ANALYSIS

It is assumed that the torque offered by a rotating impeller is proportional to the average shear stress τ_{aw} exerted on the wall surface of an agitated vessel:

$$M = k_1 V \tau_{aw} \tag{6}$$

where, k_1 is a proportionality coefficient.

For power-law fluids, $\tau_{aw} = K\gamma^n$. Thus, agitator power is

$$P = 2\pi k_1 \ VKN\gamma^n \tag{7}$$

Literature survey on agitator power for different regimes can be summarized in the following relations:

$$P/\rho d^5 N^3 = Kp_1/Re^f \tag{8}$$

where, the flow regime coefficient f is related to Re in the transitional regime and equals 0 and 1 in the fully turbulent and laminar regimes, respectively.

For non-Newtonian fluids, Eq.(8) is modified to:

$$P/\rho d^5 N^3 = K_p/Re_m^f \tag{9}$$

where, $Re_m = d^2 N^{2-n} \rho/K$ \qquad (10)

and, by assumption, $f = \exp(-mRe_m)$ \qquad (11)

Substitution of Eqs.(9), (10) and (11) into Eq.(7) gives

$$\gamma = (\frac{K_p d^5 \rho}{2\pi k_1 \ VK(d^2\rho/K)^f})^{1/n} \ N^{(2-f(2-n))/n} \tag{12}$$

282

Let

$$k = K_p \rho d^5 / 2\pi k_1 VK(d^2\rho/K)^f \tag{13}$$

Thus, a new shear rate model can be obtained:

$$\gamma = k^{1/n} N^{(2-f(2-n))/n} \tag{14}$$

$$f = \exp(-md^2 N^{2-n} \rho/K)$$

Experimental results demonstrated that model parameters m and k are slightly related to scale, flow regime, fluid viscosity and impeller type, and have been found through optimization to be 0.00705 and 0.4 respectively.

For power-law fluids, μ_a can thus be obtained as follows:

$$\mu_a = K \, 0.4^{(n-1)/n} N^{(2-f(2-n))(n-1)/n} \tag{15}$$

The above equation can then be used to define Re* and Pr* as follows:

$$Re^* = \frac{d^2 N\rho}{\mu_a} = \frac{d^2 N\rho}{K} \, 0.4^{(1-n)/n} N^{(2-f(2-n))(1-n)/n} \tag{16}$$

$$Pr^* = \frac{C_p \mu_a}{k} = \frac{C_p K}{k} \, 0.4^{(n-1)/n} N^{(2-f(2-n))(n-1)/n} \tag{17}$$

Then heat transfer coefficients for non-Newtonian fluids in agitated vessels can be described by the following correlation:

$$Nu = CRe^{*a} Pr^{*b} Vis^c \tag{18}$$

where C is related to the geometrical configuration of the agitator system, a,b,c and C are to be determined through regression of experimental data, and the viscosity ratio Vis for power-law fluids can be expressed:

$$Vis = K\gamma^{n-1}/K_w\gamma^{n-1} = K/K_w \tag{19}$$

It has been verified in experiments that the flow behavior index n is independent of temperature.

EXPERIMENTAL METHOD AND APPARATUS

Experiments were carried out in a jacketed stainless steel vessel with a flat bottom. The vessel diameter equals 0.23 m. Three types of cooling tube (inner helical coil, external helical coil and vertical tube) were installed inside the vessel. The vertical tubes acted as baffles as well as heat exchange surfaces. Figure 1 shows the configuration of cooling tubes. The geometrical parameters of impellers are shown in Table 1.

Heat transfer tests were conducted under steady-state conditions. The impeller rotations ranging from 60 to 1250 rpm were measured with a rotameter. The agitators were driven by a variable speed motor. Power consumption was determined by using a torque sensor and a strain gauge. Steam temperature, batch fluid temperature and cooling water temperature were measured with mercury-in-glass thermometers. Wall temperature was calculated by means of heat balance.

283

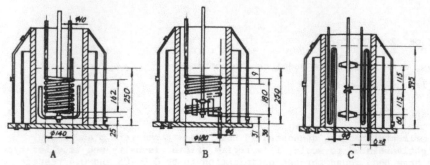

FIGURE 1. Schematic diagram of cooling tube configuration.
A: Inner helical coil with anchors; B: External helical coil with plate
paddles and a semi-elliptical impeller; C: Vertical tube with Pfaudlers,
turbines and MIG impellers.

TABLE 1. Geometrical configurations of impellers

The types of impellers	d	b	n_p	n_n	θ
Flat blade disc turbines	0.092	0.0184	3,6	2,3	0°
Pfaudler impellers	0.092	0.0184	3	2,3	15°
MIG impellers	0.089,0.1035,0.115	0.173d	2	3,5	30°
Plate paddles	0.075,0.085, 0.107	0.049	2	1	0°
The semi-elliptical impeller	0.142	0.142	2	1	45°
Anchors	0.181,0.194,0.2125	0.169	2	1	0°

TABLE 2. Heat transfer correlations for jacket,
$$Nu_j = CRe*^a Pr*^b Vis^c (\frac{d}{D})^e (\frac{\Sigma b \sin\theta}{H})^g$$

Impellers	C	a	b	c	e	g	Runs	Mean deviation	Re*
MIG impellers[1]	0.45	2/3	1/3	1/6	0.3	0.387	172	8.06%	200~18400
Turbines and Pfaudlers[1]	1.19	2/3	1/3	1/6		0.74	184	11.1%	25~6310
Plate paddles and semi-elliptical impeller[2]	0.557	2/3	1/3	1/6	0.4	0.2	175	8.46%	40~18600
Anchors[3]	0.723	2/3	1/3	1/6	0		110	10.45%	7~990

[1]Vertical tubes; [2]External helical coil; [3]Inner helical coil

Aqueous CMC (Carboxy Methyl Cellulose) solutions were used as pseudoplastic fluids. It has been shown that an aqueous solution with a concentration of CMC of less than 4% has almost the same C_p, k and ρ as that of
water (3). Hence, these properties of water were used here. Fluid rheological parameters were measured with a rotational cylinder viscometer. The
consistency index ranged from 0.02 to 12 Pa s^n, and flow behavior index
from 0.49 to 0.92. Heat flux was calculated by heat balance between

TABLE 3. Heat transfer correlations for cooling tube,

$$Nu_c = CRe*^a Pr*^b Vis^c \left(\frac{d}{D}\right)^e \left(\frac{\Sigma b \sin\theta}{H}\right)^g$$

Impellers	C	a	b	c	e	g	Runs	Mean deviation	Re*
MIG impellers[1]	0.048	2/3	1/3	1/6	0.477	0.347	184	6.82%	200~18400
Turbines and Pfaudlers[1]	0.077	2/3	1/3	1/6		0.528	183	9.9%	15~6310
Plate paddles & semi-elliptical impeller[2]	0.038	2/3	1/3	1/6	0	0	145	10.45%	40~18600
Anchors[3]	0.060	2/3	1/3	1/6	–0.6		96	9.82%	7~910

[1]Vertical tubes; [2]External helical coil; [3]Inner helical coil

TABLE 4. Comparison of present shear rate with those developed previously

No.[1]	1	2	3	4	5
Shear rate ⟍ Mean deviation(%)	Gluz $\gamma=4\pi N$	Mizushina $\gamma=14\frac{2\pi N}{D/d-1}$	Sandall $\gamma=11\left(\frac{3n+1}{4n}\right)^{\frac{n}{1-n}}N$	Tang $\gamma=\left(\frac{P}{\pi^2 D^2 NHK}\right)^{1/n}$	present work $\gamma=0.4^{1/n} \cdot N^{\frac{2-f(2-n)}{n}}$
Anchor	16.85	18.59	16.73	10.57	8.63
Multi-stage Pfaudler	11.93	14.29	11.90	10.83	9.00
Plate paddle	9.80	9.50	10.1	11.0	6.00

[1]Numbers refer to curves in Fig.6

cooling water and heating steam. Water flow rate was measured by a stopwatch and volume gauge. Preliminary experiments using syrup solution as Newtonian fluids were carried out in order to determine the total resistance external to the batch according to the modified Wilson graphical method.

RESULTS

The dimensionless group d/D, $\Sigma b \sin\theta/H$, D/H were introduced in order to take account of the influence of geometrical configurations on the film coefficients. where Σb is defined as the effective blade width:

$$\Sigma b = n_p \cdot n_n \cdot b \tag{20}$$

Then Eq.(18) can be expressed:

$$Nu = CRe*^a Pr*^b Vis^c \left(\frac{d}{D}\right)^e \left(\frac{\Sigma b \sin\theta}{H}\right)^g \tag{21}$$

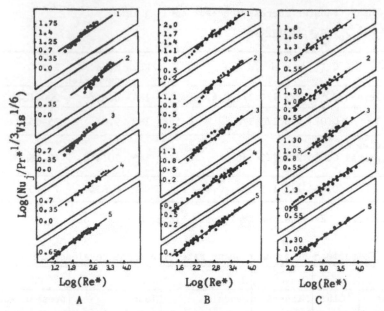

FIGURE 2. Comparison of present shear rate with those developes previously.
A: anchor; B: multi-stage Pfaudler; C: plate paddle. Arabic numberals
indicated is the same as those in Table 4.

where, b=1/3 has been derived from the boundary layer theory, c is taken
to be 1/6 in this paper (6). Values for a,e,g, and C can be obtained by
regression analysis of the data recorded. For various agitator systems,
the experimental results show that the exponents of Re* fluctuate around
the value 2/3. An exponent of 2/3 was thus taken for Re* as a fair value
in the final correlations. The results for jacket and cooling tubes were
presented in Table 2 and Table 3, respectively.

DISCUSSION

In order to compare the present shear rate model with those proposed previously, five shear rate models were used to correlate the film coefficients under the same agitator condition, as shown in Figure 2 and described in Table 4. As the results show, the present shear rate model (NO.5
in Figure 2) permits the best correlations for various agitator systems
with the smallest mean deviations.

CONCLUSIONS

1. A new shear rate model for determining apparent viscosity is proposed:

$$\gamma = k^{1/n} N^{(2-f(2-n))/n}$$
$$f = \exp(-md^2 N^{2-n} \rho / K)$$

This shear rate model can be used to correlate film coefficients in the

286

case of various types of impeller, different cooling tubes, baffled or unbaffled, for both the jacket and the cooling tubes.

2. For vertical tube, inner helical coil and external helical coil, heat transfer equations correlated by generalized Reynolds number Re* are respectively obtained. The equations can be used in the design of industrial agitated vessels.

NOTATION

b   width of a single blade, m

C_p specific heat, W/Kg k

D   vessel diameter, m

d   impeller diameter, m

d_c outside diameter of cooling tubes, m

f   flow regime coefficient, defined by Eq.(14)

H   liquid height, m

h_c film coefficient from fluids to cooling tubes, W/m^2 K

h_j film coefficient from fluids to jacket, W/m^2 K

K   consistency index at bulk temperature, Pa s^n

K_w consistency index at wall temperature, pa s^n

k   thermal conductivity, W/m K

m   model parameter

M   impeller torque, N m

N   impeller rotation, 1/s

n   flow behavior index

n_p the number of blades of an impeller

n_n the number of impellers on the axis

P   power consumption of the impeller, W

V   batch fluid volume, m^3

Greek symbols

γ representative shear rate, 1/s

μ_a apparent viscosity of non-Newtonian fluids, Pa s

ρ density, Kg/m^3

θ inclined angle of blade

Dimensionless group

Nu_j $h_j D/k$, Nusselt number from fluids to jacket

Nu $h_c d_c/K$, Nusselt number from fluids to cooling tube

Re* $d^2 N\rho/\mu_a$, generalized Reynolds number defined by Eq.(16)

Pr* $C_p \mu_a/k$, generalized Prandtl number defined by Eq.(17)

Vis   viscosity ratio, defined by Eq.(19)

REFERENCE

1. Bourne, J.R., M. Buerli and W. Regenass, Power and heat transfer to agitated suspensions: Use of heat flow calorimetry, Chem. Engng. Sci., vol.36, pp782-784, 1981.

2. Gluz, M. and I.S. Pavlushenko, Experimental studies on heat transfer to non-Newtonian fluids, Zh. Prikl. Khim., vol.11, pp2475-2483, 1966.

3. Mizushina, T., R. Ito, Y. Muradami and s. Tanaka, Experimental studies on film coefficients of non-Newtonian fluids in agitated vessels, Chem. Engng. Tokyo, vol.30,pp819-826, 1966.

4. Sandall, O.C. and K.G. Patel, Heat transfer to non-Newtonian pseudo-plastic fluids in agitated vessels, Ind. Engng. Chem. Proc. Des. Dev., vol.9, pp139-144, 1970.

5. Tang Furui, Gu Peiyun and Sun Jiangzhong, Studies on heat transfer to Newtonian and non-Newtonian fluids in agitated vessels, Chem. Ind. Engng. J. (China), No.4, pp389-394, 1983.

6. Yu Shengyao, Heat transfer to non-Newtonian fluids in agitated vessles, the Master Dissertation of Zhejiang University, China, pp29-35, 1986.

Heat Transfer to Pseudoplastic Fluids Using Dual Motion Impeller with Scraping Blades

KAI WANG, WENJUN ZHOU, BUQUIAN XU, and QIUAN SONG
Department of Chemical Engineering
Zhejiang University
Hangzhou, PRC

1. INTRODUCTION

In previous studies, it has indicated that the dual motion (rotation with simultaneous up-and-down shift) impeller with cross paddle is a very excellent mixer for high viscosity Newtonian and non-Newtonian fluids(1.2). At the same mixing quality, this impeller can save 70—90% energy comparing with double helical ribbon impeller. However, the heat transfer efficiency of this impeller is lower than the double helical ribbon impeller, and the heat transfer efficiency can be improved a lot by using the scraping blade instead of the cross paddle blade (CP).

In this work, the heat transfer coefficients for vessel wall, h_j of five kinds impellers with the fluid loaded scraping blades, which have various geometry and are made of Nylon and copper respectively, have been researched in non-Newtonian fluids. The installation method of these blades is floatable, that leads it to have some merits such as construction simplicity, durability, and being provided with operation flexibility.

2. APPARATUS AND EXPERIMENTAL METHOD

2.1. Apparatus

The experiments have been carried out in a 182 stainless steel vessel with jacket. The dual motion of the impeller is produced by means of a special transmission mechanism which consists of eccentric wheel, bevel gear box, connecting rod, universal joint and sliding bearing as shown in Fig.1. In Fig.2 the form of impeller with floatable scraping blade is shown. The geometric forms, dimensions, and materials of the give kinds of scraping blades are represented as Fig.2 and Tab.1.

TABLE 1. Dimensions of scraping blades (mm). $H \approx 60$, $\alpha = 30°$

Symbol of blades	Form in Fig.2	Material	L	w
SCu	(a)	copper	30	2.5
SHCu	(c)	copper	30	2.5
SWCu	(a)	copper	40	2.5
SNy	(a)	Nylon	30	4
SGNy	(b)	Nylon	40	4

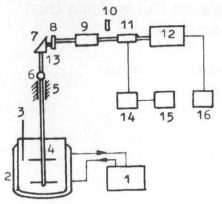

FIGURE 1. Schematic diagram of experimental apparatus
1—thermostat; 2—vessel; 3—thermocople; 4—impeller; 5—sliding
bearing; 6—universal joint; 7—bevel gear box; 8—eccentric wheel;
9—reduction gear box; 10—digital tachometer; 11—troque meter;
12—motor; 13—connecting rod; 14—strain gauge dynamometer; 15—recorder;
16—electronic governor

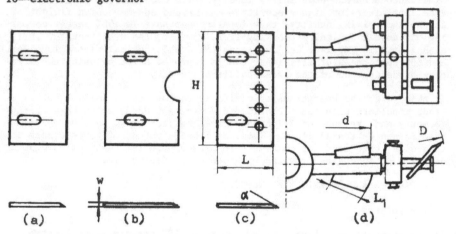

FIGURE 2. Form of the impeller with a floatable scraping blade.
Dimensions(mm): D=182; d=95; L_1=25

2.2 Physical properties of the fluids

The 2—4.5%(mass) aqueous CMC (Carboxy Methyl Cellulose) solutions were
used as pseudoplastic fluids. The rheological behavior of the fluids can
be represented by power-law model so that the apparent viscosity becomes

$$\eta_a = K(k_s \, N)^{n-1} \tag{1}$$

where, K, n, k_s are consistency index, flow behavior index and Metzner
constant, respectively. In this work, the range of K and n are respectively 3—38 $N.s^n/m^2$ and 0.4—0.7. The dependance of K on temperature can
be represented by

$$K = \exp(A + B \, t_b) \tag{2}$$

where, A and B are experimental constants. It was verified that CMC solution with concentration less than 4% can not appreciably change the basic properties of the water(3). Therefore heat conductivity and specific heat of the CMC solutions were proximately taken as those of water.

2.3. Heat transfer measurement

The heat transfer experiments have been carried out according to batch heat transfer method at atmospheric pressure. It was verified that because the dual motion impellers have very excellent mixing capability, the temperature distribution in bulk fluid is negligiable even if in laminar range. Therefore only one measuring point of bulk temperature was set. The area of heat transfer was calculated by means of measuring the depth of fluid. The volume of CMC solution used in each run was 5 l. The CMC solutions were heated by hot water which was forcedly circulated by thermostat, and the temperature at outlet of the thermostat was controlled in the range of $42\pm0.1°C$. In each run, the experimental period, $\Delta\theta$, is 30min. The beginning temperature, t_{b1}, and the finish temperature, t_{b2}, of the fluids in each experimental period are 27—29°C and 36–39°C, respectively. Because the discharge of thermostat relative to volume of the jacket is very large, the temperature difference between outlet and inlet of the thermostat is small, so that it is reasonable to suppose the mean temperature of the hot water in the jacket, t_j, as a constant during each experimental period proximally.

3. RESULTS AND DISCUSSION

3.1 Power consumption correlation and k_s

In laminar range, the power consumption can be correlated as

$$N_p \, Re^* = (P/\rho N^3 \, d^5)(D^2 \, N\rho/\eta_a) = K_p \tag{3}$$

For six dual motion impellers the values of K_p and k_s are summary represented in Tab.2.

TABLE 2. Values of K_p and k_s

No.	Blade	K_p	k_s	No.	Blade	K_p	k_s
1	CP	95	15.8	4	SWCu	147	17.3
2	SCu	140	18.0	5	SNy	137	15.8
3	SHCu	136	18.0	6	SGNy	125	15.0

3.2. Correlation of film coefficients of heat transfer

For high viscosity fluids, the agitation energy, Q_a, must be taken into account during making an energy blance.

$$Q = Q_j + Q_a \tag{4}$$

Where, Q and Q_j are the energy gained by the fluid in the vessel and the energy transfered into the fluid through the wall of jacket, respectively. The agitation energy can be calculated according to equation 5.

$$Q_a = 2 \pi N M \Delta\theta \tag{5}$$

291

Where, N is rate of shaft revolution measured by a digital tachometer; M is mean torque during a run, which was measured by a strain gauge dynamometer.

For the batch heat transfer, the following differential equation holds:

$$W\, C_p(dt_b/d\theta) = U\, A(t_j - t_b) + Q_a \tag{6}$$

Where, U and W are overall transfer coefficient and mass of the fluid in a batch. Rearranging and integrating the equation 6 from t_{b1} to t_{b2}, while the time passes from θ_1 to θ_2. One gets after rearrangement

$$U = \frac{W\, C_p}{A \Delta\theta} \ln \frac{t_j - t_{b1} + Q_a/U\, A}{t_j - t_{b2} + Q_a/U\, A} \tag{7}$$

In the integration it was assumened that there is insignificant or no variation in U with temperature, and there is no much variation in physical properties of the fluid in an experimental period. Using trial and error method with computer, the U can be obtained for various operation condition.

Then the heat transfer film coefficient for vessel wall, h_j, can be calculated from the equation

$$1/h_j = 1/U + \phi_j \tag{8}$$

Where, ϕ_j is the constant film thermal resistance, which can obtained experimentally according to modified Wilson plot(4) using Newtonian fluid.

In this work the ϕ_j is 0.00342 $m^2 J^{-1} K$.

The obtained h_j were correlated according to following two equations:

$$Nu = h_j\, D/k = C_1 (d^2 N \rho/\eta_a)^{a1} (c_p \eta_a/k)^{b1} (\eta_b/\eta_w)^j$$
$$= C_1 (Re^*)^{a1}\, Pr^{b1}\, Vis^j \tag{9}$$

$$Nu = C_2(\varepsilon D^4/\nu_a^3)\, Pr^{b2} \tag{10}$$

Where, dimensionless number $\varepsilon D^4/\nu_a^3$ is the Reynolds number which is represented by power consumption per unit mass. Using this type of Reynolds number can easy compare the heat transfer efficiency of impellers. In the equation 10, there is no Vis item, in this case the representarive temperature for calculation Physical properties of the fluids must take the arithmetic mean value of t_b and t_w.

For each impeller, 34--36 runs of experiment were done. Setting b1=1/3, j=0.14, regression analysis was performed according equation 9, it was found that the values of a1 of six impeller were in the range of 0.42--0.51. Then setting a1=0.5, b1=1/3, j=0.14, the regression was carried on once more, the values of C_1 were obtained with standard deviation less than 14% as shown in Tab.3.

TABLE 3. Values of C_1

Impeller	CP	SCu	SHCu	SWCu	SNy	SGNy
C_1	0.46	0.67	0.58	0.60	0.54	0.55

In order to compare with the heat transfer coefficients at same power consumption, the regression analysis was also performed according to equation 10 at setting b2=1/3. It was found that the range of a2 is 0.22--0.30. If setting a2=0.25, the values of C_2 were defined by regression with standard deviation less than 14% as shown in Tab.4.

TABLE 4. Values of C_2

Impeller	CP	SCu	SHCu	SWCu	SNy	SGNy
C_2	0.20	0.28	0.25	0.26	0.21	0.22

As seen in Tab.4, drilling some holes through the scraping blade like Fig.3(c) can lead the h_j to drop, it may be due to produce short pass of the fluid through the holes. Comparing C_2 of SNy and SGNy impellers, it is clear that there is no effect by making a gap on the scraping blade.

As for materials of the blades, it was found that Nylon blades could not keep close to the vessel wall. It is due to the floatable construction and the small disity difference between Nylon and CMC solution. It may be improved by combination Nylon with metal, in this case the merits of wearability and small friction coefficient of Nylon will full developed.

4. CONCLUSIONS

(1) At same power consumption, the heat transfer efficiency of the dual motion impeller with SCu blade is 40% higher than the dual motion impeller with cross paddle blade (CP) as shown in Tab.4.

(2) Two correlations of heat transfer film coefficients for vessel wall have been obtained as following:

$$Nu = C_1 \, (Re*)^{a1} \, Pr^{b1} \, Vis^{j}$$

$$Nu = C_2 \, (\varepsilon D^4/\nu_a^3)^{a2} \, Pr^{b2}$$

Where, the constants of C_1 and C_2 are shown in Tab.3 and Tab.4, and a1=0.5, a2=0.25, b1=b2=1/3, j=0.14.

(3) The k_s which is generally used for calculation power consumption can be also used for calculation heat transfer coefficients for pseudo-plastic fluids.

(4) Among the five kinds of scraping blades, the scraping blade made of cooper (SCu) shown in Fig.2(a) is superior to others.

NOMENCLATURE

C_p = specific heat   J/kg K
D = vessel diameter m
d = impeller diameter m

h_j = heat transfer film coefficient $W/m^2 K$

K = consistency index $N s^n/m^2$

K_p = constant ($K_p = N_p Re*$)

k = thermal conductivity $W/m K$

k_s = Metzner constant

N = rotating speed s^{-1}

n = flow behaviour index

P = power consumption W

Q = energy gained by fluid in vessel W

Q_j = energy transfered through vessel wall W

Q_a = agitation energy W

t_b = bulk temperature K

t_w = vessel wall temperature K

t_j = temperature of hot water in jacket K

θ = time s

η_a = apparent viscosity Pa.s

η_b = apparent viscosity at bulk temperature Pa.s

η_w = apparent viscosity at wall temperature Pa.s

ε = power consumption per unit mass W/kg

ν = apparent kinematic viscosity $N s m/kg$

REFERENCES

1. Murakami, Y., Hirose, T., and Ohshima, M., Mixing with an Up and Down Impeller, CEP, pp.78-82, May 1980.

2. Zhu Xiulin, Song Qiuan, Xu Buquan, and Wang Kai, Power Consumption and Mixing Characteristics of a Rotating Additional Up and Down Motion for Mixing Newtonian and Non-Newtonian Fluids, Synthetic Rubber Industry, vol.7, no.6, pp.421-425, 1984.

3. Ayazi Shamlou, P., Edwards, M.F., Heat Transfer to Viscous Newtonian and Non-Newtonian Fluids for Helical Ribbon Mixers, Chem. Eng. Sci., vol.41, no.8, pp.1957-1967, 1986.

4. Chapman, F. S., Holland, F. A., Heat Transfer correlations for Agitated Liquids in Process Vessels, Chem. Engng., pp.153-158, Jan.18, 1965.

CONDENSATION

Heat Transfer Enhancement with Phase Changes Research and Development in South China University of Technology

YING-KE TAN
Chemical Engineering Research Institute
South China University of Technology
Wushan, Guangzhou, PRC

ABSTRACT

Eight different kind of tubes and two novel tube inserts for heat transfer enhancement with phase changes which were studied and developed by SCUT in recent years are introduced. These tubes and tube inserts are specially designed for the improvement of "shell-and-tube" heat exchangers which are the main type of heat exchangers being widely used in different industries.

INTRODUCTION

In 1970's and 1980's the Chemical Engineering Research Institute of South China University of Technology has been doing some work on the field of heat transfer enhancement both for single-phase flow and for fluids with phase changes. This paper is only introducing the work of heat transfer enhancement with phase changes.

The main purpose of the work is for increasing the heat duty or reducing the size of "shell-and-tube" heat exchangers which are being widely used in different industries. Especially for chemical industry, it is necessary to treat with many different kinds of organic compounds. The heat transfer coefficients of organic compounds usually are very low, such as alkanes, olefines and Freons, the condensation heat transfer coefficients are only about 1/10 that of water vapour; and the boiling heat transfer coefficient is only 1/3 that of water. Therefore it is necessary to seek some methods to augment the heat transfer of boiling and condensation for organic compounds in order to reduce the main heat transfer resistance in a heat exchanger. The work which have been done by us may be classified as follows:

HEAT TRANSFER ENHANCEMENT OF VAPOUR CONDENSATION

The first part of our work is the heat transfer enhancement of vapour condensation which was done by the use of 4 kinds of special shaped tubes and two novel tube inserts.

1. Externally Fluted Tube (ZC tube)

The structure of the tube surface is shown in Fig. 1. It was invented by a

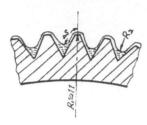

Fig. 1 Fluted Tube

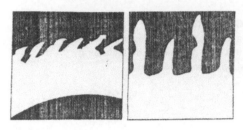

(a) Cross section　(b) Longitudinal

Fig. 2. Secional view of saw-
teeth-finned tube

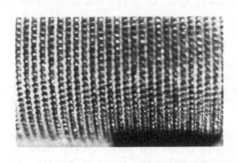

Fig. 3. Appearance of FL Tube

Fig. 4. A Sample of NL Tube

ASCLB

TSCLB

Fig. 5. Experimental tube inserts

Swiss named R. Gregoriq in 1954 [1], but was not used in our country until 1974. A portionof the tube cross figure shown in Fig. 1 is the first tube made by us in 1974. It is an aluminum tube with an outside diameter of 22 mm and a wall thickness of 2 mm. Along the periphery of the tube there are 48 triangular shaped flutes. The depth of the flutes is about 0.8–1.0 mm. When it is used in a vertical condenser for the condensation enhancement of isobutane, the coefficient of condensation heat transfer is (3500 – 4650 W/m²K) 4–6 times larger than with plain tubes. The temperature difference between the vapour and the tube wall could be lowered to 1.3–3.0°C.

The effect of flutes for condensation augmentation is due to the surface tension of the condensate. The condensate is driven by surface tension from the top portion of the flutes to the trough portion of the flutes. Condensate runs into the trough of the flutes and flows along the trough from the top to the bottom of the tube length by gravity force. The top surface of the flutes remain exposed to the vapour. The thickness of the liquid layer along the top of the flutes and the two sides is very thin so the heat transfer resistance is very low. In other words, the heat transfer coefficient of condensation is greatly increased by the fluted surface of the tube. Besides the surface tension effect, the surface area of heat transfer has been increased to about 70% by the fluted surface compared to the original tube surface.

This kind of tube is only suitable for vertical condensers. We designed and made an isobutane condenser of ZC tubes for a geothermal electric power station in Guangdong in 1980 [3].

2. Saw-teeth-finned Tube (JC tube)

It's cross and longitudinal sectional view is shown in Fig. 2. It is suitable for the enhancement of vapour condensation in horizontal "shell-and-tube" condensers. This kind of tube was firstly developed by Hitachi Co. of Japan [2].

The condensing film heat transfer coefficient could be raised 8–10 times over smooth tubes at the same operating conditions for refrigerants such as: R-11, R-113, R-12, etc.. The saw-teeth-finned tubes are much better than smooth tubes and common finned tubes, because:

(a) The tube has a larger heat transfer area and larger periphery than common finned tubes. In addition, saw-teeth shaped fins with pointed ends have better dropping ability for the condensate than common finned tubes.

(b) The surface of the saw-teeth is rather rough. This enhances ripple formation in the condensate as it flows over the surface, and thus augments the heat transfer from the condensate to the fin surface.

The temperature difference between the vapor and the wall for this tube could be lowered to 0.5°C. A correlation for calculating the condensing heat transfer coefficient has been found [4].

3. Radially-ridged-fin tubes (FL tube)

The FL tube was developed by SCUT just recently. The condensing film heat transfer coefficient of the FL tube with high ridges is 40% to 50% higher than JC tube. Because it has larger heat transfer area, making full use of

299

surface tension effect and promoting turbulent convective heat transfer in condensate films. The radially-ridged fins promote turbulence in condensate film by ridges on the fin surface, and 2-dimensional surface tension effect plays a dominant role for thinning the thickness of condensate film on the condensing surface. The latter gives a more important effect for heat transfer enhancement. The appearance of FL tube is shown in Fig. 3.

A dimensionless correlations for calculating the condensation heat transfer coefficient on the outside surface of FL tubes and the optimum parameters of FL tube have been proposed [5].

4. Internal Spiral Fluted Tube (NL tube)

This kind of tube has multiple spiral grooves and spiral fins in the internal tube wall. It can raise the tube side flow condensation heat transfer coefficient from 60% to 140% without increase of frictional loss compared with smooth tubes when R-113 was used as the working fluid. The mass velocity of R-113 was in the range of 100-450 (Kg/m² s) in a 3.35 meters long NL copper tube with a dimension of ϕ16x1.5. It was also found that the optimum structure for NL tube is with triangular fins and trapezoidal grooves, and the angle between the fins and the tube axis is around 18°[6].

Fig. 4 shows a sample of NL tube.

5. Novel Tube Inserts

As one of the passive augmentation heat transfer techniques, tube inserts have an advantage of conveniently improving the performance of conventional smooth tube condensers at a reasonable little extra cost. In general, twisted-tape inserts, which cause a swirl flow in condensate, demonstrated slight improvement in heat transfer coefficient but significant increase in frictional loss. The static mixer inserts, which can enhance mixing of the condensate and vapour, greatly increase heat transfer coefficients, but the pressure drop increase is also great.

How to greatly increase the heat transfer coefficient while at the same time minimize the rise of flow resistance is the purpose of this study.

Since the multi-interaction of interfacial shear, gravitation, surface tension and pressure gradient was exerted to condensate, flow regimes diverted from axisymmetrical annular flow. The effect of interfacial shear decreases while that of gravitational force increases along test tube as vapor qualities decrease from 1 to 0. The flow regimes are mist, annular, semi-annular, wavy and slug. In fact, annular, semi-annular and wavy are the major flow patterns for horizontal in-tube condensation.

The flow pattern with condensate being led to flow from the bottom to the top of the tube across tube core, increasing liquid entrainment may be one of the desirable flow patterns for in-tube condensation. Therefore, two novel tube inserts were developed by this idea, and were tested in a 3 meters long horizontal copper tube with a dimension of ϕ25x2.

(1) Arrow shape condensate leading band inserts (ASCLB)

300

As showed in Fig. 5, the band was of 6 mm width at the wide end and of 3 mm at the narrow end. The angle contained the band and axis of tube was 30° deg and the angle contained the connection section and axis of tube was 60° deg.

If the ASCLB inserts was installed with the wide end of the band at the top and the narrow end at the bottom of the tube (AN installing mode), the band exerted a large upwards force on vapor in upwards direction and created significant circumferential flow in condensate. In this case, the condensate was led upwards along the band, and the effect of prohibition against drainage of condensate was quite small.

(2) Trapezium shape condensate leading band (TSCLB) inserts

The narrow and wide end of the band were of 2 mm, 4 mm respectively. The angle contained the band and axis of tube was 30° deg. The connection section of TSCLB inserts was the remains after the bands were cut off from the tape of 6 mm wide. The connection section was pressed to the wall at the top of the tube by the elasticity of bands to reduce its prohibition against drainage of condensate. TSCLB inserts increased the mixing of vapor and condensate and led condensate to flow upwards along the bands, thus thinner condensate film both at the top and the bottom of the tube and significant droplets entrainment could be observed. The prohibition against drainage of condensate was quite small.

A systematic experimental study for flow regimes, heat transfer and pressure drops has been made for smooth tube and 2 novel tube inserts [7].

It could be concluded that the flow patterns with the condensate led to flow upwards across the core of the tube, increase liquid entrainment and reduce the resistance to the drainage of condensate are preferable to enhancing in-tube condensation. Under the same operating conditions, the condensing heat duty of the condenser with either TSCLB or ASCLB inserts could increase 35% to 70% over those of the smooth tube condenser of the same geometry [7].

HEAT TRANSFER ENHANCEMENT OF LIQUID BOILING

1. Mechanically made porous surfaced tubes (JK-1 tube)

This kind of tube was firstly developed by Hitachi Co. at 1978 [2]. It has a special surface structure which has tunnels circumferentially under the surface skin, with many openings to the outside. The liquid in the tunnels is heated rapidly and turns into vapour which leaves through the openings as bubbles. Some vapour nuclei are always remaining in the tunnels, and therefore, the boiling occurs continuously. The quantity of liquid which has become vapour as the bubbles escaping are compensated with liquid sucked into tunnels through adjacent openings. With the above described mechanism, the boiling occurs vigorously and gives a very high heat transfer efficiency. Fig. 6 shows the boiling heat transfer performance of our porous-surface tube compared with a plain tube. It shows that at the same temperature difference the heat flux of the porous-surface tube is about 4-12 times that of the plain tube for R-11 pool boiling.

The porous-surface tube (JK-1 tube) is also suitable for use in very low temperature difference between the tube wall and the boiling liquid. The temperature difference may be as low as 0.5°C. This property diminishes the

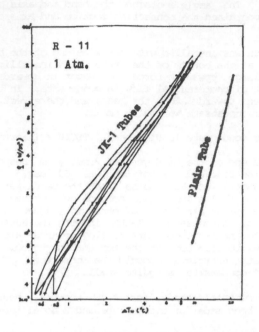

R - 11
1 Atm.

JK-1 Tubes

Plain Tube

Fig. 6. The Boiling Heat Transfer Performance
of JK-1 Tubes and Plain Tube

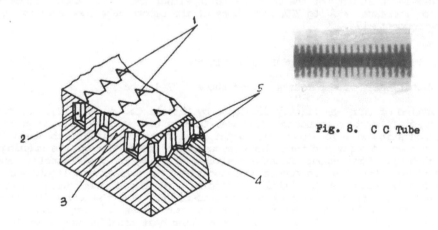

Fig. 8. C C Tube

1. small holes 2. helical tunnel 3. tube wall
4. groove   5. ribs

Fig. 7. Feature of JK-2 Tube

302

irreversible loss for energy recovery process.

Another merit of JK-1 tubes is that the critical heat flux could be raised about 60% compared with smooth tube for Freon refrigerants [8].

Recently a study of spray falling film boiling on a horizontally placed JK-1 tube was done [9]. The experimental results indicated that there was 11-79% increase of heat transfer coefficients of spray falling film boiling in comparison with pool boiling.

2. Finned Tunnel Porous Surface Tube (JK-2 tube)

It was recently developed by SCUT. This kind of tube is different from JK-1 tube by having some flutes on the wall of spiral tunnels under the tube surface. These flutes promote the heat transfer between the liquid film and the tunnel wall, and also may provide some nuclei sites for new bubble generation (Fig. 7).

An experimental comparing test was made between JK-1 and JK-2 tubes. The results showed that when the heat flux through the tubes was in the low and medium load range ($2-9 \times 10^4$ W/m^2) in pool boiling, the superheat for bubble generation may be lowered to 0.3-0.8°C for JK-2 tube which was only 1/4 to 1/2 that of JK-1 tube at the same condition [10].

3. T shape fin tube (TC tube)

This is a kind of Gewa-T tube which was firstly developed by Wieland-Werk company of West Germany. The boiling film heat transfer coefficient of TC tube is 2-10 times over smooth tubes proved by pool boiling experiments of R-113, and the critical heat flux could be raised 57% compared with smooth tubes. But it has a character of hysteresis, requiring higher superheat for starting the boiling. Therefore, for stable operation, it is better to operate at the maximum heat flux first, then adjust to the required heat flux [11].

We can make it in copper or steel tube material.

4. Internal Spiral Fluted Tube (NL tube)

The NL tube also could be used for enhancement of intube flow boiling. An experimental investigation was done on the characteristics of R-12 intube flow boiling in a NL tube. The result showed that the heat transfer coefficient was 1.6 to 2.2 times as smooth tube. The mass velocity of R-12 was in the range of 130 to 520 (Kg/m^{2}s) in a 3.05 m. long horizontal NL copper tube with a dimension of ϕ 16.5x2. The augmentation of flow boiling in NL tube is due to the increase of heat transfer area by multiple fins and the effect of secondary flow by spiral grooves [12].

DESUBLIMATING

The third kind of tubes is for increasing the rate of organic vapour desublimating which is called "Zigzag segmented fin tube" or CC tube (Fig. 8).

The Zigzag segmented fin tube is a new designed finned tube for enhancing both heat transfer and mass transfer of organic vapour desublimation. It has the advantages of compact fin structure, low contact heat resistance between fins and tube wall, and low manufacturing price. This kind of tube, after 4 years experiments, has been put into industrial use already. A switch condenser made of these tubes is being used in a phthalic anhydride plant. The phthalic anhydride vapour in a mixed gases is desublimated on the surface of the CC tubes. Its desublimating rate is over (200 g/m^* hr) which is double the rate of common high fin tubes [13].

REMARKS

Some functions of the above tubes may be combined together to get both tubeside and tube outside enhanced, such as we may combine the JC tube with NL tube to form a JC-NL tube for enhancing organic vapour condensing outside and liquid boiling inside the tube. The way of combination is up to your choice.

We have designed the manufacturing process and machine tools for the above 8 tubes which may be put in mass production with the lowest cost, and also the computer programs for finding optimal tube parameters for various fluids and different operating conditions.

REFERENCES

1. Gregorig, R., Zeit. Fur Ange. Math. und Phy, Vol. 5, 36, 1954.
2. Hitachi Cable Ltd., "High Flux Boiling and Condensation Heat Transfer Tube", HITACHI THERMOEXCEL, 1978.
3. Tan, Y. K. and Deng, S. J., "A Novel and High Efficiency Condenser Used in A Geothermal Electric Power Station", Proceedings of Condensed Papers of 8th Miami International Conference on Alternative Energy Sources, 1987.
4. Wang, S. P., et al., "Experiments of Saw Teeth Shape Finned Tubes Enhancing Condensation and its Dimensionless Correlation", J. of Engineering Thermodynamics, Vol. 4, No. 5, pp. 374-378, 1984.
5. Yin, Q. H., Wang, S. P., et al., "Investigation of Enhancing Condensation Heat Transfer With Radially Ridged Fin Tube", Journal of Chemical Engineering of Chinese Universities, Vol. 1, No. 1, pp. 377, 1985.
6. Zhou, X. Q., "Investigation of tubeside condensation heat Transfer Enhancement of R-113 in Internal Spiral Fluted Tubes", Master thesis of SCUT, 1987.
7. Liu, Z. Q., Tan, Y. K. and Cui, N. Y., "The Enhancement of R-113 Condensation Inside Horizontal Tubes With Novel Tube Inserts", Proceedings of International Symposium on Phase Change Heat Transfer, pp. 448-453, 1988.
8. Fang, Q. Y., "Investigation of the Pool Boiling Heat Transfer Mechanism on Mechanically-made Porous-Surfaced Tubes", Master thesis of SCUT, 1984.
9. Wang, G. Q., "A Study of Spray Falling Film Boiling on Mechanically Made Porous Surfaced Tubes", Proceedings of ISHTEEC, 1988.
10. Zhong, L. "Investigation on the Heat Transfer Characteristics of Finned Tunnel Porous-surfaced Tubes", Master thesis of SCUT, 1988.
11. Luo, G. Q., "Investigation of the Boiling Heat Transfer Characteristics of Gewa-T Tubes", Master thesis of SCUT, 1985.
12. Yang, W. N., Tan, Y. K., et al., "Investigation on Augmentation of R-12 Tubeside Flow Boiling Heat Transfer in the Internally Spiral-Fluted Tubes", Proceedings of ISHTEEC, 1988.
13. Yang, X. X., "Investigation of Desublimation Process of Phthalic Anhydride Vapour Mixture on Single Finned Tubes", Proceedings of ISHTEEC, 1988.

Free Convective Condensation
of Fluoroalcohol-Water Mixtures

KUNIO HIJIKATA, NOBUHIRO HIMENO, YAO-QUI ZHOU, and SHIGEYUKI GOTO
Department of Physical Engineering
Tokyo Institute of Technology
2-12-1 Ohokayama, Meguro-ku
Tokyo 152, Japan

1.INTRODUCTION

Rankine cycle using a binary mixture has attracted much attention because of its high efficiency of thermal energy conversion in middle or low temperature range. In this cycle a binary mixture of Freons is generally considered to be a suitable mixture for the working fluid. It is reported [1], however, that some of Freons, such as R11, R113 and R114, decompose at temperature above 200°C. In this temperature range, other organic fluids must be explored. Recently, fluoroalcohol-water mixtures are proposed for this end, because they are chemically stable and have superior thermo-physical properties in the temperature range from 200°C to 300°C.

However, when a non-azeotropic mixture is used as a working fluid, a volatile vapor accumulates near the condenser surface and creates a diffusion layer which reduces the heat transfer performance of the condenser. Even though an azeotropic mixture is used for the reducton of the diffusion resistance, the deterioration of the heat transfer might still occur.

There have been many researches on the condensation of a binary mixture of vapors. Sparrow [2] obtained a similar solution of the free convective condensation on a vertical plate. Fujii [3] also analyzed the same problem by using algebraic equations determined at the vapor-liquid interface. The forced convective condensation was numerically analyzed by Denny [4]. As for the experimental work, Goto [5] reported the experimental results of free convective condensation of R114-R12 and R114-R11 mixtures on a horizontal cylinder. The authors [6] made theoretical and experimental studies on the free convective condensation on a vertical plate using R113-R114 and R113-R11 mixtures.

The condensation heat transfer of fluoroalcohol-water mixtures has not been reported so far. In this report the free convective condensation of two fluoroalcohol-water mixtures, that is pentafluoropropanol ($C_3H_2F_5OH$)-water and trifluoroethanol(CF_3CH_2OH)-water mixtures, is investigated. The former mixture ($C_3H_2F_5OH-H_2O$) becomes azeotropic at the water concentration of 34.9 mol%, while the latter is non-azeotropic mixture. The experiments are performed using smooth and finned condensing surfaces, and the results are compared with the theory.

2.THEORETICAL ANALYSIS

In the condensation of a binary mixture of vapors on a vertical plate, a boundary-layer of the free convection, where the concentration

of the volatile component is high, is formed in the vapor near the condenser surface. When the molecular weight of the volatile component is larger than that of the other component, the boundary-layer develops upward, in opposition to the flow direction of the liquid film. For the case of pentafluoroalcohol-water mixture, which is an azeotropic mixture, the vapor-side boundary-layer develops upward, when the water concentration of the mixture is lower than the azeotropic point. In this case, a similar solution cannot be obtained, since the temperature of the vapor-liquid interface becomes varied along the flow direction. However, the mean heat transfer coefficient can be estimated well by assuming an imaginary constant temperature at the interface. The similar solution based on this assumption is called "pseud-similar solution" by the authors [6]. The same solution is applied in the present analysis. However, the Raoult's law is not used at the vapor-liquid interface, since the fluoroalcohol-water mixtures are far from an ideal solution. The assumptions used in the present analysis are as follows:
(1) Nusselt solution can be applied to the liquid film.
(2) Convective heat transfer by a temperature difference in the vapor phase and the velocity at the vapor-liquid interface are neglected.
(3) Vapors are saturated and stagnant far from the surface.
Based on these assumptions the fundamental equations are solved analytically using the integral method. As a result, the mass flux, $\dot{m}$, obtained from the vapor phase equations can be expressed as follows:

$$\dot{m} = \frac{8}{15}D^2\rho_i{}^*\frac{(W_{2i}-W_{2\infty})^5}{(W_{2i}-W_{2\ell i})^4}M^*\{1+\frac{15}{28}M^*(W_{2i}-W_{2\infty})\}$$

$$/[W_i(\frac{W_{2\infty}-W_{2\ell i}}{W_{2i}-W_{2\ell i}})\{Sc+\frac{\frac{20}{21}+\frac{25}{63}\cdot M^*(W_{2i}-W_{2\infty})}{1+\frac{15}{28}\cdot M^*(W_{2i}-W_{2\infty})}$$

$$\times W_i(\frac{W_{2\infty}-W_{2\ell i}}{W_{2i}-W_{2\ell i}})\}]\frac{1}{x} \tag{1}$$

where

$$M^* = (M_2-M_1)/\{M_1+(M_2-M_1)W_{2\infty}\} \tag{2}$$

On the other hand, the mass flux obtained from the liquid phase equations, Nusselt solution, is

$$\dot{m} = -\frac{\rho_\ell{}^2}{\mu_\ell}g\delta_\ell{}^2\frac{d\delta_\ell}{dx} = -\lambda_\ell\frac{T_i-T_w}{L\delta_\ell} \tag{3}$$

where, g is the gravitational acceleration, D is the mass diffusivity, ρ is the density of the vapor, W_2 is the mass fraction of the component 2, M is the molecular weight, Sc is the Schmidt number, x is the vertical distance from the upper edge of the condenser plate, and L is the latent heat of condensation. The subscripts i, 1, ∞ represent interface, liquid phase, bulk vapor far from the condenser surface, respectively.

In the calculation of the heat transfer coefficient, m is determined from Eq.(1), for given T_i. Then T_w is calculated from Eq.(3). Using these values, the mean heat transfer coefficient, h, is determined from the following equation.

$$h = \frac{4\dot{m}_{x=\ell}L}{3(T_\infty-T_w)} \tag{4}$$

where ℓ is the vertical length of the condenser plate and $\dot{m}_{x=\ell}$ is the mass flux at the lower edge of the plate.

3. EXPERIMENTAL APPARATUS AND METHOD

The experiments were carried out in an enclosed chamber of 400mmx 400mmx800mm. The experimental apparatus is illustrated in FIGURE 1. Since pentafluoropropanol(CH_2F_5OH) and trifluoroethanol(CF_3CH_2OH) used as test fluids are very corrosive, acrylic resin could not be used in the apparatus. Therefore, the chamber and the viewing window were made of stainless steel and glass, respectively. The test mixture was put into the chamber and heated by sheath heater located at the bottom of the chamber. The generated vapor was condensed on a vertical condensing plate attached to the side wall of the chamber. The condensing plate was made of copper and had 50mm square as a projected surface area. Its side surfaces were thermally insulated by teflon, and the back surface was

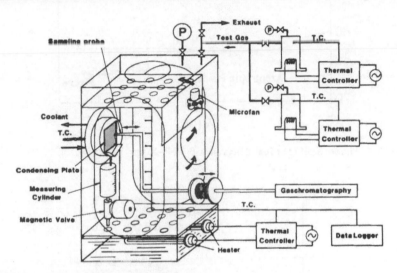

FIGURE 1. Experimental appratus.

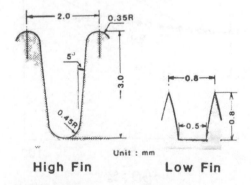

High Fin   **Low Fin**

FIGURE 2. Condensing surfaces.

307

cooled by water. The surface temperature was controlled by changing the temperature of the cooling water. The heat flux at the condensing surface was calculated from the direct measurement of the condensate flow rate.

The surface temperature of the condensing plate was measured by five thermocouples embedded inside the copper plate, approximately 3mm from the surface. The bulk temperature of the vapor was measured by the

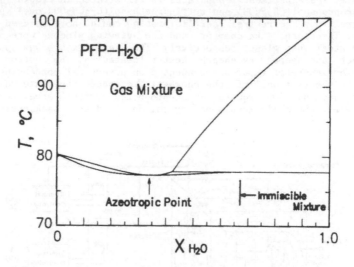

FIGURE 3. Phase equilibrium diagram of PFP-water mixture.

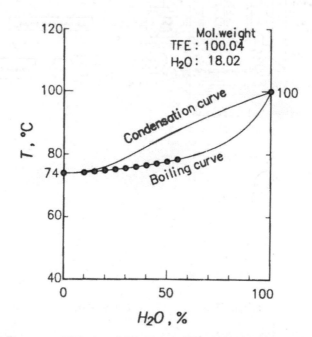

FIGURE 4. Phase equilibrinm diagram of TFE-water mixture.

thermocouples at 7 locations 50mm from the cooling surface. The mole fraction of the volatile vapor was analyzed by a gas chromatograph. In order to prevent the separation of mixed vapors due to the buoyancy force caused by the difference of the molecular weights of the components, the vapor was mixed by a micro-fan. The forced convection effect by the fan on the heat transfer was minimized by adjusting the power input to the fan.

The condensing surfaces used in the experiment were smooth and finned surfaces. The shapes and dimensions of the finned surfaces are illustrated in FIGURE 2. The "high fin" has a tall fin height to reduce the effect of the vapor diffusion layer. On the other hand, "low fin" was designed as a typical enhanced surface for pure vapor condensation. The ratios of the surface areas of "high fin" and "low fin" to their projected areas are 3.45 and 2.66, respectively. The performances of these finned surfaces for R113-R114 and R113-R11 mixtures was reported in reference [6].

Hereafter pentafluoropropanol and trifluoroethanol are abbreviated to "PFP" and "TFE", respectively. Although these fluids are considered to be the most promising fluids for the working fluids of Rankine cycle [1], the physical properties are not extensively known. Therefore, some of the properties had to be estimated using correlating formulae. The estimated phase equilibrium diagrams are shown in FIGURE 3 and FIGURE 4. Several correlating formulae, such as Margules, van Laar, Wilson [7], LEMF (local effective mole fraction) [8], and NRTL (non random two liquids equation) [9] formulae, were tried to correlate the phase equilibrium data. Among these formulae, Wilson and NRTL formulae were the best for the PFP-water and TFE-water mixtures, respectively. The PFP-water mixture becomes azeotropic at the water concentration of 34.9 mol%, while TFE-water mixture is non-azeotropic. The former phase equilibrium diagram in FIGURE 3 was calculated using the azeotropic data, while the latter in FIGURE 4 was calculated using the boiling point data reported by Halocarbon Products Corp. As shown in FIGURE 3, PFP-water mixture becomes immiscible at high water concentration. This was verified by the experiment. Even though a large amount of water was added to the mixture, the water concentration of the vapor remained constant at about 42 mol%, which agreed with the calculated value shown in FIGURE 3. The calculated diagram in FIGURE 4 also correlated the boiling data well.

4. EXPERIMENTAL RESULTS AND DISCUSSION

4.1 PFP-water Mixture

The experiment of PFP-water mixture was carried out for the mole fraction of water being 13%, 25%, 33%, 38%, and 42%. The heat transfer coefficients obtained for the smooth condensing surface are plotted against the bulk-to-wall temperature difference in FIGURE 5 and FIGURE 6. FIGURE 5 shows the results for the mole fraction of water which is lower than the azeotropic point, and FIGURE 6 shows the results for the higher. Nusselt solutions are shown by the lines for each water concentration. The heat transfer coefficient is reduced by the diffusion resistance in the small temperature difference region, but it approaches to the Nusselt solution with an increase of the temperature difference. This is a characteristic feature of the condensation of a binary mixture. However, the reduction in heat transfer is negligible except for the mole fraction of water 13%. Even in the case of 13% in water concentration, the reduction is not serious. One of the most important reason is related to the characteristic of the phase equilibrium. The temperature difference

309

between the boiling point and the condensation point, namely, the difference between the boiling curve and the condensation curve at the same concentration, causes the heat transfer reduction due to the diffusion resistance. As seen in FIGURE 3, the both curves are relatively close. Another reason, which is less important, is the large difference of the molecular weights between PFP (molecular weight: 150) and water (molecular weight: 18). Since the buoyancy force in the diffusion layer is proportional to it, the boundary-layer becomes thin and the larger difference of molecular weight brings about the reduction in the thermal resistance of the diffusion layer.

These results are confirmed by the theory. FIGURE 7 and FIGURE 8 show the comparison between the experiment and the theory for 13% and 42% in the water concentration, respectively. The broken line represents the

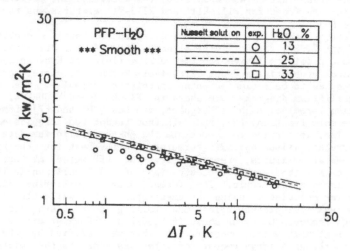

FIGURE 5. Heat transfer coefficient of PFP-water mixture on the smooth surface for the mole fraction of water lower than the azeotropic point.

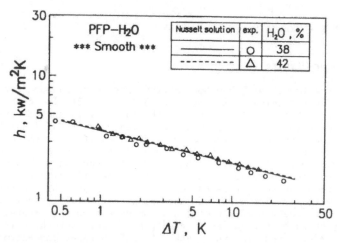

FIGURE 6. Heat transfer coefficient of PFP-water mixture on the smooth surface for the mole fraction of water higher than the azeotropic point.

Nusselt solution and the solid line represents the theoretical heat transfer coefficient calculated using Eqs.(1)-(4). The theory and the experiment coincide with each other within the experimental error. Thus, the reduction in the heat transfer is not serious even in the small bulk-to-wall temperature difference region. The theoretical calculation of the heat transfer coefficient was performed for the other concentrations of water. In these cases, the theoretical values of the heat transfer coefficient seem to be less than the experimental results, but the agreement was satisfactory when we consider the accuracy of the experiment. From these comparison between the theory and the experiment, it is revealed that the reduction in heat transfer due to the diffusion

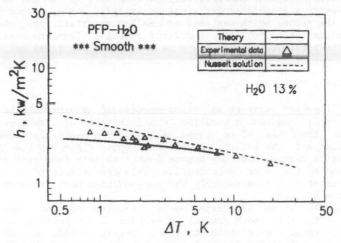

FIGURE 7. Comparison of the heat transfer coefficient of PFP-water mixture between the experiment and the theory for the mole fraction of water 13%.

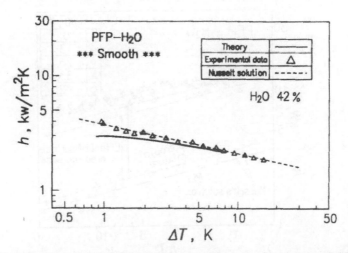

FIGURE 8. Comparison of the heat transfer coefficient of PFP-water mixture between the experiment and the theory for the mole fraction of water 42%.

resistance is not serious for PFP-water mixture, and the heat transfer coefficient increases with an increase of weight fraction of water because of the higher value of the latent heat and the conductivity of water.

The heat transfer reduction in the small temperature difference region is not conspicuous even for the finned condensing surface, as shown in FIGURE 9 where "high finned surface" was used. This means that the diffusion layer is thin enough for the "high finned surface" to enhance the heat transfer in the diffusion layer by the effect of the surface area increase of the fin.

From these results, it is concluded that PFP-water mixture has superior heat transfer characteristics. However, it should be noticed that PFP-water mixture corrodes aluminum, acrylic resin, other plastics, and also the rubber which was used as sealing material. Thus, though PFP-water mixture seems to be a excellent material from the heat transfer aspect, the corrosion problem might be serious in practical use.

4.2 TFE-water Mixture

The TFE-water mixture is a non-azeotropic mixture unlike the PFP-water mixture, and has a peculiar characteristic of the phase equilibrium, far from that of an ideal solution, as shown in FIGURE 4. The condensation and the boiling curves are quite close in the low water concentration region, and they become depart rapidly from each other with an increase of the water concentration. This characteristic of the phase equilibrium affects substantially the condensation heat transfer performance of the mixture.

FIGURE 10 shows the experimental results of the heat transfer coefficient for the smooth condensing surface. The experiment was carried out for the various water concentration, 10-11%, 15-18%, 23-28%. Nusselt solutions are also shown for each water concentration in the figure. The dependency of the heat transfer coefficient on the bulk-to-wall temperature difference is the same with PFP-water mixture. As an increase of the

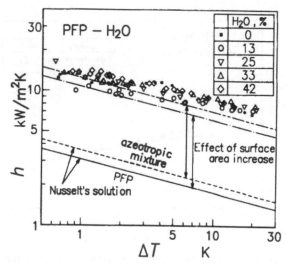

FIGURE 9. Heat transfer coefficient of PFP-water mixture on the "high finned surface".

temperature difference, the heat transfer coefficient approaches to the Nusselt solution. On the other hand, the reduction in the heat transfer coefficient becomes significant as an increase of the water concentration. As mentioned earlier, the temperature difference between the condensation point and the boiling point causes the diffusion resistance. It is found from FIGURE 4 that the temperature difference between both curves increases with the water concentration. The dependency of the heat transfer coefficient on the water concentration is related to the characteristic of the phase equilibrium.

The comparison of the heat transfer coefficient between the experimental results and the theoretical results obtained from Eqs.(1)-(4) are shown in FIGURE 11 and FIGURE 12 for different concentration of water. FIGURE 11 shows the result for the water concentration of 15-18%, and FIGURE 12 for 23-28%. The experimental results are a little higher than the theory in the small temperature difference range, but agreement is quite satisfactory for the both cases. Thus, the heat transfer coefficient of the free convective condensation of the binary vapor mixture is predicted well by the present theory, if the exact values of the physical properties, including the phase equilibrium diagram, are known.

In FIGURES 10-12, the experimental heat transfer coefficients have maximum values in the small temperature difference region, where the theory predicts a little smaller value than the experimental value. This feature is explained as follows. When we compare the latent heat of condensation between water and TFE, they are quite different; 23.088×10^2 kJ/kg for water; 3.713×10^2 kJ/kg for TFE. The latent heat of condensation of the mixture is, therefore, varied significantly by the water concentration. As seen from FIGURE 5, the water concentration largely varies with the interface temperature, T_i, especially in the low water concentration range. As a decrease of the wall temperature, namely, an increase of the bulk-to-wall temperature difference, T_i also decreases but it cannot become lower than the boiling point. When the bulk-to-wall temperature difference becomes larger than the temperature difference between the condensation and the boiling curves at the same concentration, T_i approaches practically to the boiling point, and T_i can be assumed as the boiling point temperature. Namely, T_i and, therefore, the latent heat of condensation varies according the change of the concen-

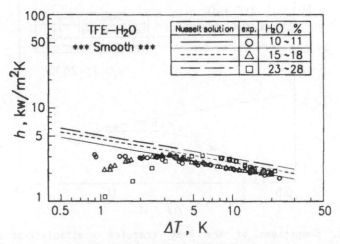

FIGURE 10. Heat transfer coefficient of TFE-water mixture on the smooth surface.

tration within this temperature region. Therefore, the change of the latent heat by the concentration makes the maximum in the heat transfer coefficient in the small temperature difference region.

The heat transfer coefficients was obtained by using "high fin" and "low fin" surfaces, and are plotted against the bulk-to-wall temperature difference in FIGURE 13 and FIGURE 14, respectively. In order to clarify the effect of the finned surface on the condensation heat transfer, the experimental data of the smooth surface and Nusselt solutions with and without considering the effect of the surface area increase of the fin are plotted by solid symbols and by lines. As an increase of the temperature difference, the heat transfer coefficient for each water

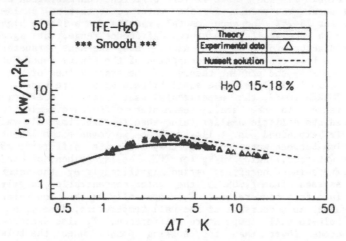

FIGURE 11. Comparison of the heat transfer coefficient of TFE-water mixture between the experiment and the theory for the mole fraction of water 15-18%.

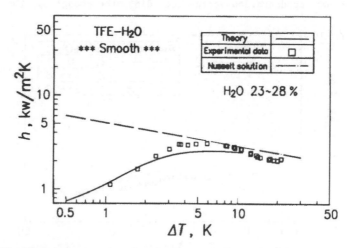

FIGURE 12. Comparison of the heat transfer coefficient of TFE-water mixture between the experiment and the theory for the mole fraction of water 23-28%.

314

concentration approaches to almost the same value. Since the thermal resistance of the liquid film is dominant in this temperature difference region, the over-all heat transfer coefficient is determined primarily by the liquid film. Therefore, the heat transfer coefficient is not much different in the large temperature difference region. The increase in the heat transfer coefficient from the Nusselt solution is considered to be caused by the surface tension effect. As seen from FIGURES 13 and 14, the "low fin", which has a superior performance in enhancing the heat transfer for a pure vapor, has very high heat transfer performance. On the other hand, in the small temperature difference region, the heat transfer coefficient decreases rapidly with decreasing the temperature difference. This reduction in heat transfer is caused by the diffusion

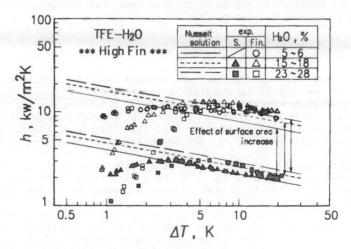

FIGURE 13. Heat transfer coefficient of TFE-water mixture on the "high finned surface".

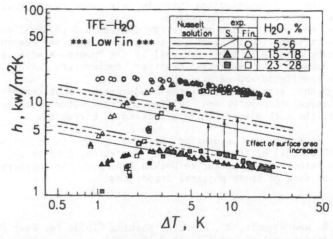

FIGURE 14. Heat transfer coefficient of TFE-water mixture on the "low finned surface".

315

layer. In both cases of the two kinds of finned surfaces, the heat transfer coefficients approach to those for the smooth surface. This means that the thermal resistance of the diffusion layer is dominant in the small temperature difference region. The fin of 3 mm in height hardly recovers the effect of the diffusion layer. However, this heat transfer reduction is practically negligible when the temperature difference is larger than 3-5 K. In a practical condenser, the working condition of the temperature difference is supposed to be in this range. The TFE-water mixture of 15% in water concentration, known as "FLUORINOL 85", has suitable thermodynamic properties for the working fluid of the Rankine cycle; its saturation curve in T-S diagram is vertical in vapor-side. From these facts TFE-water mixture is considered one of the best working fluid of the Rankine cycle for the utilization of the natural energy and waste heat. However, it should be noticed here that TFE-water mixture has the same problem of corrosion with PFP-water mixture.

5. CONCLUSIONS

In order to clarify the performance of the condensation heat transfer of the pentafluoroalcohol-water and the trifluoroalcohol-water mixtures, which are considered to be the most promising working fluids of the Rankine cycle for moderate temperature thermal energy utilization, the free convective condensation on the flat and finned plates is investigated experimentally and theoretically. The following conclusions are obtained:
1. The pentafluoroalcohol-water mixture becomes azeotropic at the water concentration of 34.9 mol%. The reduction in the condensation heat transfer due to the vapor-side diffusion resistance is negligible near the azeotropic point. Even at other mixing ratio, the heat transfer coefficient is not affected seriously by the diffusion layer, since the condensation curve and the boiling curve are relatively close.
2. The reduction in the condensation heat transfer of the trifluoro-alcohol-water mixture due to the vapor-side diffusion resistance is practically negligible when the temperature difference is larger than 3-5 K, if the water concentration of the mixture is less than 25%. This characteristic is closely related to the phase equilibrium diagram that the condensation curve is very close to the boiling curve in this water concentration region.
3. By using the proper phase equilibrium diagram and physical properties, the heat transfer coefficient of the free convective condensation of the binary mixture is predicted well by the "pseudo-similar solution", which assumes an imaginary constant temperature at the interface between the liquid and vapor.
4. Although the performances of the condensation heat transfer of the pentafluoroalcohol-water and trifluoroalcohol-water mixtures are excellent, the both mixtures have a problem of corrosion.

This research was supported by the grant in aid of Scientific Research (60040061) of the Ministry of Education, Science and Culture, Japan. The authors are grateful to Daikin Ind. Ltd. for supplying PFP and TFE, and the data of their physical properties.

REFERENCES

1. Enjou, N. and Noguchi, M., Suitable Working Fluids for Heat Pumps, *Refrigeration*, vol.59, no.676, pp.9-16, 1984.

2. Sparrow, E.M. and Marchall, E., Binary, Gravity-Flow Film Condensa-

tion, *Trans. ASME, J. of Heat Transfer*, vol.91, pp.205-211, 1969.

3. Fujii, T. and Kato, Y., Laminar Film Condensation of a Binary Miscible Vapour on a Flat Plate, *Trans. JSME*, vol.46, no.402, pp.306-312, 1980.

4. Denny, V.E. and Jusionis, V.J., Effects of Forced Flow and Variable Properties on Binary Film Condensation, *Int. J. Heat Mass Transfer*, vol.15, pp.2143-2153, 1972.

5. Goto, M. and Fujii, T., Film Condensation of Binary Refrigerant Vapours on a Horizontal Tube, Proc. 7th Int. Heat Trans. Conf., Munich, vol.5, pp.71-76, 1982.

6. Hijikata, K., Mori, Y., Himeno, N., Inagawa, M. and Takahasi, K., Free Convective Condensation of a Binary Mixture of Vapors, Proc. 8th Int. Heat Trans. Conf., San Francisco, vol.4, pp.1621-1626, 1986.

7. Wilson, G.M., A New Expression for the Excess Free Energy of Mixing, *J. Am. Chem. Soc.*, vol.86, pp.127-130, 1964.

8. Morisue, K., Noda, K. and Ishida, K., A Discussion of the Excess Gibbs Energy of Liquid Mixtures Based on the Local Composition, *J. Chem. Eng. Japan*, vol.5, pp.219-223, 1972.

9. Renon, H. and Prausnitz, J.M., Derivation of the Three-Parameter Wilson Equation for the Excess Gibbs Energy of Liquid Mixtures, *AIChE J.*, vol.15, pp.785, 1969.

Investigation on Performance of Freon Condenser with Both-Side Enhanced Tubes

SHIPING WANG, PEISEN LIN, and HUAILIANG WEI
Chemical Engineering Research Institute
South China University of Technology
Guangzhou, PRC

ABSTRACT

The condensations of Freon-113 on a both-side enhanced tube (B.E. tube) and a common finned tube with 25 F.P.I. were studied respectively. The overall heat transfer coefficient on the former is 50% to 60% greater than that on the latter. For further research, two condensers with the same construction but with different tubes——one with B.E. tubes but the other with sawtooth-finned tubes (S.T. tube) —— were made and tested in the same refrigeration system. The results showed that the condenser with B.E. tubes has overall heat transfer coefficient 22% to 35% higher than the condenser with S.T. tubes.

INTRODUCTION

In refrigerant condensers with water-cooling, the dominant resistance to heat transfer is on the refrigerant side. In order to improve the performance of these condensers, many researchers have developed different kinds of high efficient surfaces [1, 2]. Among these special surfaces, sawtooth-finned tube or "THERMOEXCEL-C" tube is one of the best.

For Freon-12 condensing on S.T. tube, the heat transfer coefficient on out-side surface (sawtooth-finned) is about 12,000-14,000 Kcal/$m^2 \cdot$ hr$^\circ$C while the coefficient on tubeside where the cooling water flows at the velocity of 1.5 - 2.5 m/s is about 6,000 - 9,000 kcal/$m^2 \cdot$ h$^\circ$C [3]. In that case, the dominant resistance transforms from refrigerant side to water side. To make full use of the advantage of the excellent performance of S.T. tubes, we must augment the heat transfer in tubeside. Increasing the water velocity can make higher heat transfer coefficient in tubeside, but the pressure loss will increase greatly as well. Therefore, a better way to solve this problem is to use the rough surface techniques into tubeside.

After having successfully used S.T. tubes into refrigerant condensers, we developed the B.E. tube and tested their performances. Although the both-side enhanced technique decreases the active heat transfer area on outside by about 20%, the tubeside rough surface will have higher heat transfer coefficient to compensate for the area loss on outside. Therefore, the overall heat transfer coefficient on the B.E. tube will be higher than that on S.T. tube. On the other hand, the rough surface in tubeside augments turbulence that reduces scaling and fouling on the wall of tubeside.

319

The apparatus is shown in FIGURE 1. The working fluid was Freon-113. The tested tube was set inside a transparent vessel for observation. Temperature of cooling water and condensate were controlled with two TWD - 702 type temperature auto-controllors respectively. The temperatures of condensate, cooling water and tube wall were measured with thermocouples. The tests were conducted after the noncondensing gas in the apparatus was strictly eliminated. The heat balance errors between data obtained from refrigerant side and those from water side were within ±3%.

The photograph of B.E. tube is shown in FIGURE 2. TABLE 1. shows the features of the tested tubes. Among them, Tube 1 and Tube 2 are B.E. tubes and Tube 3 and Tube 4 are finned tubes with 25 F.P.I.. Before testing the tubes were cleaned and inactivated.

TABLE 1.   Features of the tested tubes

Tube No.	Blank tube		Shaped tube				Outside area	Active
	D_o (mm)	D_i (mm)	d_o (mm)	d_i (mm)	h(mm)	p(mm)	m^2/m	length(mm)
1#(B.E.)	15.8	12.8	17.46	12.8	1.35	0.75	0.153	270
2#(B.E.)	18.8	15.8	20.60	15.8	1.45	0.75	0.195	278
3#(finned)	15.8	12.8	15.7	10.6	2.0	1.02	0.179	272
4#(finned)	18.8	15.8	18.8	13.5	2.1	1.02	0.224	278

For convenience, the overall heat transfer coefficient K and the condensation coefficient h_e on refrigerant side were calculated based on the dimension of the blank tubes and the heat transfer coefficient in tubeside (water side) h_i based on the exact dimension of the shaped tubes.

Based on the test results, we used Wilson's plot technique and obtained the correlations of heat transfer coefficient in tubeside, h_i, as follows,

For Tube 1, $h_i = 0.181 (k/d_i) Re^{0.658} Pr^{0.4}$ (1)

For Tube 2, $h_i = 0.194 (k/d_i) Re^{0.668} Pr^{0.4}$ (2)

For Tube 3, $h_i = 0.0247 (k/d_i) Re^{0.8} Pr^{0.4}$ (3)

For Tube 4, $h_i = 0.0259 (k/d_i) Re^{0.8} Pr^{0.4}$ (4)

The factors in eqn.(3) and eqn.(4) are 7.4% and 12.6% greater than the factor, 0.023, in the Dittus-Boelter equation respectively. This is caused by the shorter tested tubes. Consider the entrance effect of short tubes, we can get the correct factors, 1.07, for Tube 3 (L/d_i = 25.7) and 1.11 for Tube 4 (L/d_i = 20.6)[20] . Therefore, the results based on the eqn.(3) and eqn.(4) are almost the same as those on the Dittus-Boelter equation. Because of the effect of rough surface, turbulence is easier to be induced in B.E. tubes than in smooth tubes. For this reason, the power on the Reynolds number in eqn.(1) and eqn.(2) is less than the normal value, 0.8.

From FIGURE 3, it can be found that the heat transfer coefficient, K, of B.E. tube is 50% - 60% higher than that of 25 F.P.I. finned tube when both of these tubes were made with the same blank tubes. At lower water velocity, the slope of K-V curve is greater than that at higher velocity. Therefore, B.E. tubes are especially suitable for using into condensers with low water velocity.

FIGURE 4 showed the relationship between heat transfer coefficient in tube-

320

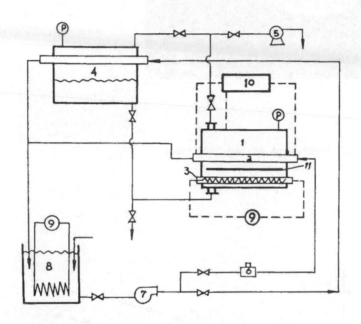

1. Test vessel 2. Test tube 3. Evaporator 4. Auxiliary condenser
5. Vacuum pump 6. Turbine flowmeter 7. Pump 8. Water tank
9. Temperature controller 10. Data collector
11. The plate separating liquid from gas

FIGURE 1. Experimental apparatus

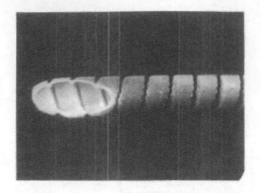

FIGURE 2. Photograph of the P.E. tube

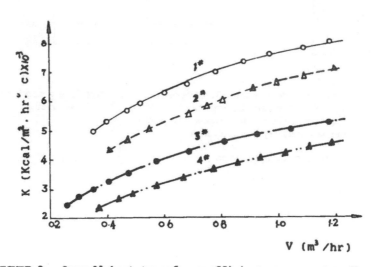

FIGURE 3. Overall heat transfer coefficient versus water flowrate

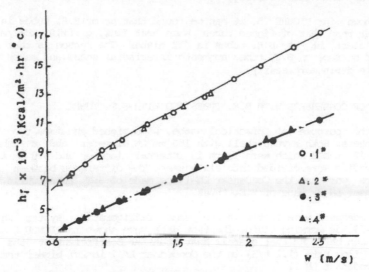

FIGURE 4. Heat transfer coefficient in tubeside

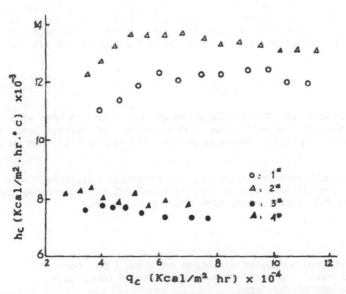

FIGURE 5. Heat transfer coefficient on refrigerant side

side, h_i, and water velocity, w.

As shown in FIGURE 5. it can be found that h_c of B.E. tubes is over 50% greater than that of finned tubes. When heat flux, $q_c > 6x10$ kcal/m²·h, the coefficient h_c of B.E. tubes is 65% higher. The reason is that sawtooth-finned surface on B.E. tubes augments interfacial shear and makes the condensate drain more easily.

TESTS OF CONDENSERS WITH B.E. TUBES AND WITH S.T. TUBES

For the purpose of practical usage, we designed and made two shell-tube condensers. Both have a shell with 180 mm in diameter and a tube bundle with 12 tubes which were 13 mm in internal diameter and 670 mm in length. The bundles were divided into six passes. The difference between two condensers was that the Condenser LN-1 was made of S.T. tubes while the other Condenser LN-2, made of B.E. tubes.

Two condensers were tested in the same refrigeration system which used Freon-12 as working fluid. The test data were shown in FIGURE 6 and FIGURE 7. It can be found that overall heat transfer coefficient at the designed water flow rate (1.2 t/h) in the Condenser LN-2 is 20% higher than that in the Condenser LN-1.

Based on analysis of the data, we can find that the overall heat transfer coefficient, K, in the Condenser LN-2 is 22% to 35% higher than in the Condenser LN-1 at the same water flow rate, while the former is 10% to 20% greater than the latter at the same tubeside pressure drop.

To make full use of the advantage of B.E. tubes, suitable water velocity should be chosen. Set the resistance to heat transfer on both sides be equal, we can figure out this optimal velocity. The coefficient, h_c, on condensation side for Freon-12 condensers with B.E. tubes is about 9,500-10,000 kcal/m²·h°·C in practical uses. At this condition, it can be calculated that the optimal water velocity is 1.6 m/s to 1.7 m/s as shown in FIGURE 8.

CONCLUSION

1. Compared with 25 F.P.I. finned tube, B.E. tube can more efficiently enhance heat transfer in refrigerant condensation. The overall heat transfer coefficient on B.E. tube can be over 50% higher than that on finned tube.

2. Because of both-side augmentation, the condenser compacted with B.E. tubes has the overall heat transfer coefficient 22% to 35% greater than the condenser with S.T. tubes at the same cooling water rate.

NOMENCLATURE

D_{io} —— the internal and outer diameters of blank tube, (mm)
d_{io} —— the internal and outer diameters of shaped tube, (mm)
h —— the height of fin, (mm)
h_c —— heat transfer coefficient on condensation side, (kcal/m²·h°C)
h_i ——heat transfer coefficient in tubeside, (kcal/m²·h°C)
K —— overall heat transfer coefficient, (kcal/m²·h°C)

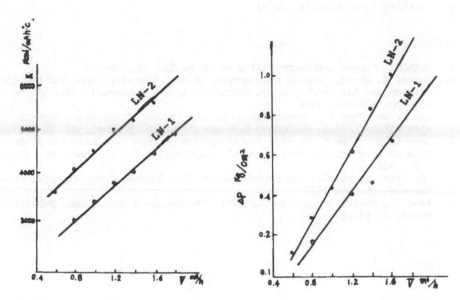

FIGURE 6. Overall heat transfer coefficient in condensers

FIGURE 7. Pressure drop in
condensers

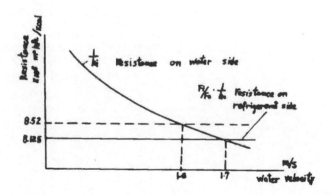

FIGURE 8. Resistance to heat transfer on the condenser LN-2

k   —thermal conductivity of water, (kcal/m .h.$^{\circ}$C)
ΔP—pressure drop, (Kg$_f$/cm^2)
p   — pitch of fing, (mm)
Pr — Prandtl number
q   — heat flus, (kcal/m^2.h)
Re —Reynolds number, V — water folwrate, (m^3/h)
w   —water flow velocity, (m/s)

REFERENCES

1. HITACHI, Copper and Copper-alloy Products Cat. No. EA-501.
2. Hozumi H. et al., Recent Development of Heat Transfer Tubes for Refrigeration and Air Conditioning, Refrigeration(in Japanese), vol. 59, no.
   683, pp. 854-866, 1984.
3. Lin P. and Wang S., Investigation of Refrigerant /condenser (Type CS-1)
   with Water Cooling, Guangdong Refrigeration (in Chinese), No. 2, 1983.
4. Deng S. J., Tan Y. et al., Investigation on Heat /transfer and Pressure
   Drop in Corrugated-tubes, Chemical /engineering (in Chinese), No. 6, pp.
   1-8, 1983.
5. Hijikata K. et al., The Proceeding of 20th Japanese Heat Transfer Conference. pp. 85, 1983.
6. Yang S., Heat Transfer (in Chinese), The People's Educational Publishing
   House, P. 54, 1985.

Dropwise Condensation on Teflon Coated Porous and Hobbed Surfaces

XIUFAN GUO, LING BAI, ZHENYE CAI, and JIFANG LIN
Institute of Chemical Engineering
Dalian University of Technology
Dalian, PRC

ABSTRACT

Dropwise condensation(DWC) of steam outside vertical tube was investigated. The condensing surfaces were coated with teflon on electrochemically etched(ECE) and mechanically hobbed (MH) substrates. The test tubes were made of copper, brass and stainless steel. All the teflon coated(TC) surfaces were successful in promoting good quality DWC. The DWC heat transfer coefficients are about 9, 4 and 1 times higher than that of filmwise condensation(FWC) respectively and the results have relation to the thermal conductivity of the substrate. The factors affecting DWC were analysed and the following dimensionless equation was obtained

$$Nu^* = A(1 - cos\beta)^{2.5}(Ks/Kl)^{0.22} Re^{*-1.12}$$

Where A equals 1×10^{-7} for TC-ECE tubes and 5.5×10^{-8} for TC-MH tubes.

1. INTRODUCTION

DWC shows the highest heat transfer coefficient of all the heat transfer phenomena have been known. Although it has been studied for more than 50 years and investigators have done a lot of research on it[1], this unique field of heat transfer has remained as a laboratory curiosity, applying it to industrial devices is still one of the most important tasks assigned to heat transfer specilists.

The necessary condition for a vapor to be condensed on a solid surface in the form of drop instead of film is that the surface must possesses rather low surface energy. As we know, some of the organic polymer possesses very low surface energy and can not be wetted by some liquids, such as water. The typical polymer is teflon. Some investigators[2,3,4] have studied the DWC on the TC surfaces. Recently Marto et al[5] have produced DWC on the surfaces with some other polymer coating on different substrates. But there exit two difficulties in applying polymer coating to sustain long-term DWC. The first is to make the film thin enough not to excessively retard heat transfer. The seccond is to form a film with a good adhesion to the substrate, few voids and sufficiently high mechanical

327

strength. The surfaces with teflon coating inside the ECE cavities was applied to nucleate pool boiling heat transfer, the enhancement was very evident(6). The present test was carried out on the surfaces with teflon coating on ECE and MH substrate. Most of the teflon was inside the cavities or grooves or pits, so the teflon coating have good adhesion to the substrate and possess small thermal resistance. Therefore the two difficulties mentioned above can be surmounted by using this kind of surfaces.

The objective of the present work is to use a new kind of surfaces to sustain good quality DWC and investigate the performance of these surfaces. Furthermore, the influence of substrate thermal conductivity, surface charateristic and contact angle on DWC heat transfer was to be analysed.

2.   EXPERIMENT AND PROCEDURE

2.1 Surface Preparing

The tubes of 400 mm length were processed to be either porous or grooved substrate by ECE and MH methods respectively first, then the test surfaces were formed by fusing the teflon powder to the roughened substrates. The thickness of teflon coating is about 15 μm for ECE surfaces and about 45 μm for MH surfaces. The characteristics of all test surfaces were shown in Table 1.

TABLE 1.   Surface characteristics

Surface No.	Tube dimension and material	Surface processing
1	$\varnothing$25X1.5, copper	ECE, TC
2	$\varnothing$25X1.5, copper	ECE, TC
3	$\varnothing$25X1.5, copper	ECE, TC
4	$\varnothing$25X1.5, copper	MH(groove), TC
5	$\varnothing$25X1.5, copper	MH(pit), TC
6	$\varnothing$24X1, brass	MH(groove), TC
7	$\varnothing$24X1, brass	MH(pit), TC
8	$\varnothing$25X1, stainless steel	ECE, TC
9	$\varnothing$25X1.5, copper	Abraded

2.2 Experimental Loop and Measurement

The system of experimental apparatus was shown in Figure 1. The steam was condensed on the external surface of the vertical tube at normal atomsphere and its average velocity is about 1.5 m/s. Cooling water was flowed upward through the inside of the tube. Its velocity was ranged from 0.66 to 6.6 m/s.

The steam-side condensation heat transfer coefficient was derived from the overall heat transfer coefficient. The resistance of teflon coating was not included in the wall resistance, but it was incorporated into the steam-side condensation resistance. Other details are similar to reference(7).

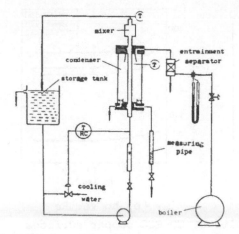

FIGURE 1.   Schematic of apparatus

3.   EXPERIMENTAL RESULTS AND DISCUSSION

3.1 Contact Angle

The contact angle between a liquid and a surface shows the non-wetting ability of this surface. The contact angles of water on different surfaces at room temperature were measured by Length Method. The results were presented in Table 2. Table 2 also gave the contact angle on teflon surface.

TABLE 2.   The contact angles of water on different surfaces

Surface No.	3	4	5	6	7	8	Teflon(8)
β	94.7°	97°	102.9°	101.9°	101.2°	87.6°	111°

3.2 Heat Transfer Results and Discussion

All the TC surfaces haved sustained excellent DWC. The photos of DWC on three surfaces of different substrate material were presented in Figure 2.

The curves of heat flux and the steam-side coefficient versus Re for three different TC-ECE copper surfaces were presented in Figure 3. The result of FWC on surface No. 9 under the same condition was also shown in Figure 3. The DWC heat flux is 1.5 to 4 times as high as that of FWC, while the steam-side coefficients of DWC are 6 to 15 times as large as that of FWC.

(a) Surface No.3   (b) Surface No.7   (c) Surface No.8
FIGURE 2.   Pictures of DWC

329

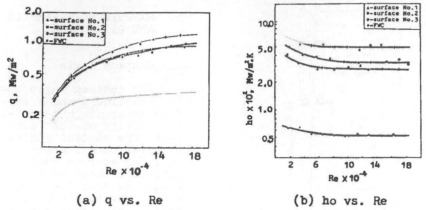

(a) q vs. Re   (b) ho vs. Re

FIGURE 3.   Heat transfer results of TC-ECE copper surfaces

The cavity densities of ECE copper substrates for surface No.1, 2 and 3 are different. Surface No.3 is the highest, surface No.2 is in the middle and surface No.1 is the lowest. From Figure 3, it can be concluded that the higher the substrate cavity density, the higher the heat transfer performance of the TC surface.

Figure 4 to 6 showed the results of different TC surfaces. For the TC surfaces on MH copper, MH brass and ECE stainless steel substrates, the DWC heat flux is 1.5 to 3.5, 1.4 to 3 and 1.5 times as large as that of FWC respectively; the steam-side coefficient of DWC is 5 to 13, 3 to 5 and 1 times higher than that of FWC respectively.

From Figure 3 to 6, it can be concluded that the higher

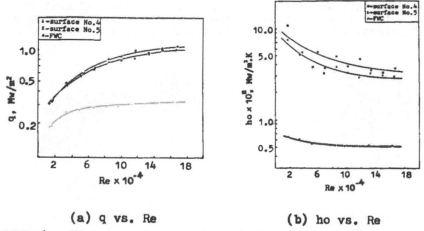

(a) q vs. Re   (b) ho vs. Re

FIGURE 4.   Heat transfer results of TC-MH copper surfaces

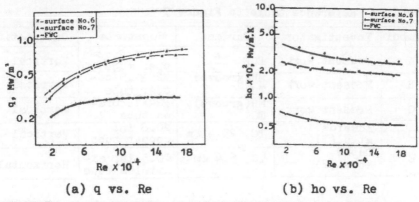

(a) q vs. Re   (b) ho vs. Re

FIGURE 5. Heat transfer results of TC-MH brass surfaces

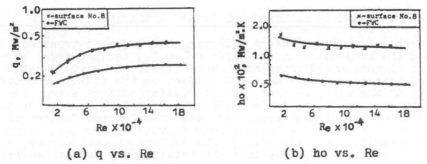

(a) q vs. Re   (b) ho vs. Re

FIGURE 6. Heat transfer result of TC-ECE stainless
steel surface

the thermal conductivity of the substrate, the larger the DWC heat transfer coefficient.

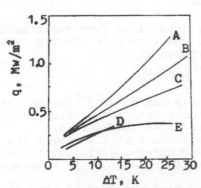

FIGURE 7. Comparison of different TC surfaces

The comparison of DWC heat transfer results on different TC surfaces was shown in Figure 7. At the same subcooling, the heat flux of the present work (TC copper and brass surfaces) is 1 to 2.5 times larger than those of the work by Depew et al(2) and Edwards et al(3). The symbols in Figure 7 were listed in Table 3. The main reason why their results were lower than the present one is that the coatings they used were evenly coated on the

TABLE 3. Reference table to Figure 7

Symbol	Investigator	Surface	Substrate	Orientation
A	Present work	ECE, TC	⌀25×1.5 copper tube	Vertical
B	Present work	MH(groove), TC	⌀25×1.5 copper tube	Vertical
C	Present work	MH(groove), TC	⌀24×1, brass tube	Vertical
D	Edwards et al(3)	TC, 25.4 μm	⌀0.8 in. copper tube	Vertical
E	Depew et al(2)	TC, 6.4 μm	⌀0.5 in. aluminium tube	Horizontal

smooth substrates, but coatings used in this study were coated on the porous or hobbed substrates.

4. HEAT TRANSFER ANALYSIS AND DATA CORRELATING

There are many factors affecting the DWC heat transfer. But when the condensing vapor, vapor velocity, cooling mode, surface dimension and orientation were fixed, the heat transfer coefficient of DWC is mainly dependent on the thermal conductivity of the wall, surface characteristics(surface energy and roughness) and heat flux(subcooling). Therefore, under the condition of the present work, the heat transfer coefficient of DWC is related to the thermal conductivity of the wall, surface characteristic and subcooling. The surface characteristic, surface energy and the roughness, is reflected by contact angle. So, the heat transfer coefficient of DWC can be expressed as follows

$$h = f(\sigma, T_v, \rho_1, \Delta H_v, K_1, K_s, \Delta T, \mu, \beta) \tag{1}$$

or in the dimensionless form

$$Nu^* = A(1 - \cos\beta)^a (K_s/K_1)^b Re^{*c} \tag{2}$$

Where A, a, b and c are constant determined from experimental data. According to equation (2), the following equqtion was obtained by correlating the experimental data:

$$Nu^* = A(1 - \cos\beta)^{2.5} (K_s/K_1)^{0.22} Re^{*-1.12} \tag{3}$$

Where A equals 1×10^{-7} for ECE tubes and 5.5×10^{-8} for MH tubes. The dispersities of the experimental data are within the limits of ±15% and ±22% respectively, as shown in Figure 8.

From equation (3), it is concluded that the larger the contact angle, the higher the heat transfer coefficient of DWC. The steam-side heat transfer coefficient of DWC is in proportion to $K_s^{0.22}$ and $\Delta T^{-0.12}$.

The endurance test for the TC surfaces used in this study was not carried out, but it can be predicted from the result by Marto et al(5) that these surfaces can sustain good DWC for a long period.

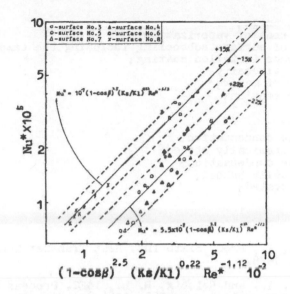

FIGURE 8. Comparison of experimental data with equation (3)

5. CONCLUSIONS

(1) The teflon coated surfaces on copper, brass and stain-less steel roughened substrates processed by either electroche-mically etched or mechanically hobbed methods sustained exce-llent dropwise condensation. The steam-side heat transfer coe-fficient was relative to the thermal conductivity of the subs-trate.

(2) The heat flux of dropwise condensation for teflon coa-ted surfaces of copper, brass and stainless steel are approxi-mately 1.5, 1 and 0.5 times larger than that of filmwise con-densation respectively, and the steam-side heat transfer coe-fficients are about 9, 4 and 1 times higher than that of film-wise condensation respectively.

(3) The factors influencing dropwise condensation heat tr-ansfer were analysed and the following dimensionless equation was obtained from experiments

$$Nu^* = A(1 - \cos\beta)^{2.5}(Ks/Kl)^{0.22} Re^{*-1.12}$$

NOMENCLATURE

ho, steam-side heat transfer coefficient including the resis-
     tance of teflon coating;
Kl, thermal conductivity of water;
Ks, thermal conductivity of substrate;
q,   heat flux;
Tv, vapor temperature;
Nu^*, δ Tv ho/($\rho_1 \Delta Hv$ Kl ΔT), Prandtl number;
Re^*, Kl ΔT/($\mu \Delta Hv$), Reynolds number;

333

β,   contact angle;
ΔH_v, latent heat of vaporization;
ΔT,   degree of surface subcooling including the temperature
      difference in teflon coating;
μ,   dynamic viscosity;
ρ_1,   density of water;
σ,   surface tension;

Abbreviations

DWC, dropwise condensation;
ECE, electrochemically etched;
FWC, filmwise condensation;
MH, mechanically hobbed;
TC,   teflon coated;

REFERENCES

1. Tanasawa, I., Proc. Sixth Int. Heat Transfer Conf., vol. 6, pp. 393, 1978.

2. Depew, C. A., and Reisbig, R. L., I&EC, Process Design and Development, vol. 3, pp. 365, 1964.

3. Edwards, J. A., and Doolittle, J. S., Int. J. Heat Mass Transfer, vol. 8, pp. 663, 1965.

4. Kosky, P. G., Int. J. Heat Mass Transfer, vol. 11, pp. 374, 1968.

5. Marto, P. J., Looney, D. J., Rose, J. W., and Wanniarachchi, A. S., Int. J. Heat Mass Transfer, vol. 29, pp. 1109, 1986.

6. Guo Xiufan, Cai Zhenye and Lin Jifang, To be presented to the 1988 Int. Conf. on Energy Conservation in Industry, Shenyang, Oct., 1988.

7. Zhang Dongchang, Lin Zaiqi and Lin Jifang, Proc. Eighth Int. Heat Transfer Conf., vol. 4, pp.1677, 1986.

8. Hinrichs, T., Hennecke, E., and Yasuda, H., Int. J. Heat Mass Transfer, vol. 24, pp. 1359, 1981.

Direct Contact Condensation of the Vapor of an Immiscible and Insoluble Substance on Falling Liquid Droplets

H. NAKAJIMA and I. TANASAWA
Institute of Industrial Science
University of Tokyo
Tokyo, Japan

INTRODUCTION

When direct contact condensation occurs between a low-temperature liquid drop and a vapor of different substance, and when the liquid phases of the both substances are immiscible, the binary drop has been observed to show some unique behavior, which is quite different from direct contact condensation of a single component system. Chang et al.[1] have reported that the condensed phase of refrigerant R113 does not spread uniformly over a falling water drop, but gather to form a cap-shaped protuberance. Such a phenomenon, possibly related to the surface or interfacial energies of the both substances, the circulatory flow inside the drop, deformation and oscillation of the drop and the thickness of the condensate film around the drop, must be effective in enhancing condensation heat transfer.

We have been carried out an experimental study under atmospheric pressure using falling water droplets as the dispersed phase and R113 vapor as the continuous phase. The liquid phases of these two substances are immiscible, and almost insoluble. In addition to this, the rates of condensation onto falling solid (glass and brass) spheres have been measured in order to estimate the effect of circulatory flow inside the liquid drop. The results have been compared with the result of numerical analysis by Jacobs[2] who has taken into consideration only the heat conduction resistances of both the condensate film of uniform thickness and the liquid drop and neglected deformation and oscillation of and circulation inside the drop.

METHOD OF EXPERIMENT

Shown in Fig.1 is the schematic diagram of the experimental setup.

The test chamber, 200mm(width)×200mm(depth)×700mm(height), is made of bakelite plates, with glass windows on the four vertical surfaces. R113 vapor generated in a boiler flows into the chamber from the bottom. The water droplets are formed and dropped into the chamber at a constant frequency of 0.1 per second from a stainless steel nozzle of 0.7mm I.D. and 1mm O.D. Supply of water from a container to the nozzle is regulated with a small value. The initial temperature of the water drop is controlled by circulating water from a constant-temperature bath around the nozzle. The temperature is measured with a copper-constantan thermocouple, the junction of which is placed upwardly just 0.5mm beneath the nozzle. This thermocouple junction is located in a small container which is for removing the water drops. When the mass of drop is to be measured, this small container with the thermocouple is moved aside. To prevent direct contact condensation to occur between R113 vapor and a water drop while it is still suspended on the tip of the nozzle, a cylindrical glass cover is placed around the nozzle and nitrogen gas is introduced to fill this space. The interface between nitrogen gas and R113 vapor has been visualized by means of the laser shadowgraph method and confirmed to be at a distance 3mm below from the bottom edge of the glass cover.

In the experiment using glass or brass spheres, a hollow cylinder with a shutter, with a copper-constantan thermocouple embedded on the side wall, is placed at the same position as the nozzle. The initial temperatures of the spheres are regulated just as in the case of water drops by means of the water circulation from the constant-temperature bath.

The masses of the initial water drops, formed at the tip of the nozzle, were almost constant, ranged between 18.3mg and 18.6mg (3.27mm–3.29mm in diameter) depending upon the initial temperatures. The diameters of glass and brass spheres were 3.22mm and 3.19mm, respectively.

To measure the amount of condensate accurately, a single falling drop was caught by a small catcher, which was made leak tight once a cover was closed, and weighed with an electronic balance. The catcher is

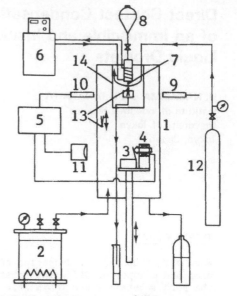

1 Test section	8 Water
2 R113 boiler	9 Light source
3 Droplet catcher	10 Photo sensor
4 Solenoid	11 Stroboscope
5 Time retarder unit	12 Nitrogen
6 Const. temp. bath	13 C–C Thermocouple
7 Nozzle	14 Cover

Fig.1 Experimental Setup

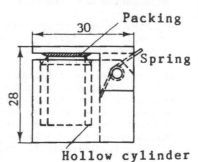

Fig.2 Droplet Catcher

shown in Fig.2. The detail of the measurement was as follows. The catcher was preheated in a warmer up to 50°C, which was a little hotter than the saturation temperature of R113 (i.e. 47.6 °C) at atmospheric pressure. The mass of R113 vapor filling the catcher was measured in advance. Then the timing of operation of a solenoid for closing quickly and tightly the cover on the top of the catcher was determined for different positions using a dummy catcher, while observing the falling drop by a Strobo flush light controlled by a time retarding unit. The timing was varied depending upon the distance of the catcher from the interface of nitrogen gas and R113 vapor. A light source and a photo sensor were used to detect a droplet at a position just below the nozzle, and the signal was then transmitted through a time retarder unit to the Strobo light source to illuminate the drop just entering the catcher. After this the dummy catcher was replaced by the real catcher, a falling single drop was caught, and the mass was measured by an electronic balance (Sartorius 1702) with the accuracy of 0.1mg. The time elasped between operation of solenoid and closing of the shutter cover was 7ms. A care was taken so that the droplet may not impinge on the bottom of the catcher during this period. The velocity of the falling droplet was measured by taking a doubly exposed picture of the drop by flashing twice the Strobo light at an interval of 5.25ms.

EXPERIMENTAL RESULTS AND DISCUSSION

Condensation of R113 Vapor on Water Droplets

Shown in Fig.3 is the result of experiment on direct contact condensation of R113 vapor on the falling water droplets. The ratio of the volume of condensate ΔV to the initial volume of the water droplet V_i is taken as the ordinate and the distance from the nitrogen gas–R113 vapor interface (or the time elasped for the drop to travel this distance) is taken as the abscissa. The initial subcooling of the water droplets are varied as 38°C, 29°C, 20°C and 11°C. Shown also on this figure by a broken and a solid lines are the results calculated from the model used by Jacobs[2]. The calculations are done for the two cases of $\Delta T=38°C$ and 11°C.

The measurements are restricted to rather narrow range of the distance from the gas-vapor interface because of the following reasons. The measurements at smaller distances were difficult owing to the structure of the experimental setup, while the measurements at larger distances were almost impossible since the paths of the falling droplets swerved greatly from the vertical making it difficult to catch the droplet.

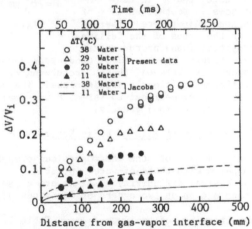

Fig.3 Condensation Rates of R113 Vapor on Water Droplets

The measured amounts of condensate in Fig.3 are higher than those calculated. Also the rates of increase of the amount of condensate, for the cases of $\Delta T=38°C$ and 11°C, are found to exceed the calculated values up to the falling distance of 200mm, indicating that the heat-transfer coefficients are quite large. One of the possible reasons may be that the mixing inside the drop is enhanced due to the circulatory flow and deformation when the water drop enters into the R113 vapor. On the other hand, the rate of increase of the amount of condensate becomes smaller at distances exceeding 200mm due possibly to accumulation of condensate on the surface of water drop, retarding the circulatory flow in the drop. A possible cause that the amounts of condensate at smaller distances are lower than the calculated values in the case of ΔT=11 °C may be that the noncondensable gas, i.e. nitrogen, is carried along with the droplets. In the case of ΔT=38°C, the rate of increase of the condensate is expected to approach the calculated value, but it has not been confirmed by experiment.

Shown in Fig.4 are the photographs of falling droplets taken at different traveling distances, denoted H,from the gas-vapor interface. When a water droplet fell through air (upper left photograph), its shape remained spherical, though a small elongation-shortening oscillation in the vertical direction was observed. On the other hand, when condensation took place, the profiles of the binary droplets varied at different positions and for different degrees of subcooling, as shown by the remaining pictures. It is very interesting to note that, in many instances, protuberances of various shapes were observed near the tops of the falling droplets. These protuberances

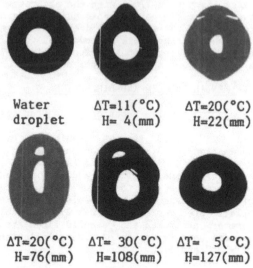

Water droplet   ΔT=11(°C)
H= 4(mm)   ΔT=20(°C)
H=22(mm)

ΔT=20(°C)
H=76(mm)   ΔT= 30(°C)
H=108(mm)   ΔT= 5(°C)
H=127(mm)

Fig.4   Condensation on Falling Water Droplets

were thought to be R113 condensate accumulated together, because the boundaries between these protuberances and the other parts of the droplets were clearly distinguished. The shapes of such binary drops were almost axisymmetrical so far as H was smaller than 70 - 80mm, but after that they deformed greatly.

Condensation of R113 on Solid Spheres

Some values of the thermophysical properties of glass and brass are compared with those of water in Table 1. The thermal conductivity and the thermal diffusivity of glass are close to those of water. Those of brass are about two orders of magnitude higher.

The results of measurements on condensation of R113 vapor on the falling glass spheres are shown in Fig.5 for the cases of ΔT = 38°C and 20°C. The results already shown in Fig.3 for water drops are plotted for comparison.

Table 1. Comparison of Properties

		Water	Glass	Brass
ρ	kg/m^3	998	2590	8560
c	kJ/kgK	4.18	0.75	0.39
λ	W/mK	0.60	0.74	99
a	m^2/s	1.44×10^{-7}	3.81×10^{-7}	2.97×10^{-5}

However, the distance from the gas-vapor interface is modified using a result of measurement of the falling velocity.

The results for glass spheres agree fairly well with the calculated values based on the Jacobs model. The reason why the measurements show a little higher values at large distances may be that the thickness of the condensate film becomes nonuniform. At any rate, the rates of increase of the amount of condensate are lower than those for water drops, suggesting importance of circulatory flow inside the drops. Deflection of the falling paths from the vertical was not observed in either case of glass or brass spheres.

The results of condensation on brass spheres are shown in Fig.6. The degrees of subcooling are again 38°C and 20°C. The rates of increase of the amount of condensate are a little higher than the calculated values in this case. Such a tendency is remarkable in the case of $\Delta T=38°C$, and the amount of condensate differs considerably from the calculation. The reason for this can be guessed as follows by observing the photographs shown in Fig.7. The brass sphere, possibly with a thin film of R113 condensate covering the surface uniformly, keeps its shape spherical so long as the traveling distance remains small ($H=44mm$). On the other hand, the upper half of the sphere is seen to be covered by R113 condensate of a considerable thickness where the distance is large ($H=248mm$). The rate of condensation in this case may be governed by condensation onto the lower half

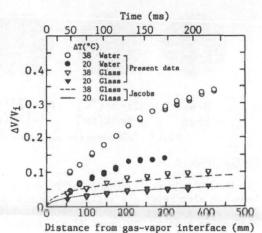

Fig.5 Condensation Rates of R113 Vapor on Glass Spheres

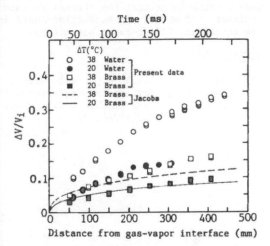

Fig.6 Condensation Rates of R113 Vapor on Brass Spheres

part of the sphere where the thickness of the condensate film is much smaller. Since the thermal conductivity of brass is quite high, the average heat-transfer rate over the entire brass sphere becomes much higher than the one calculated by the Jacobs model.

It is estimated that the distribution of thicker film of condensate on the upper half of the brass sphere is related to where on the sphere separation of flow does occur, for the Reynolds number at $H=248mm$ is about 5000. It should be added that the amount of condensate on the brass sphere was much smaller than that on water droplet in spite of high thermal conductivity of brass.

339

Brass   ΔT=38(°C)   ΔT= 38(°C)   ΔT= 38(°C)
bead   H=44(mm)   H=126(mm)   H=248(mm)

Fig.7 Condensation on Brass Spheres

CONCLUSION

(1) When R113 vapor is condensed directly on falling water drops, higher heat-transfer rates are available during the initial stage. Later heat-transfer rate approaches the one calculated on the bases of the heat conduction model. The cause of the higher heat-transfer rate may be the circulatory flow inside and the oscillatory deformation of the drops.

(2) When R113 vapor is condensed on falling solid spheres, the rate of condensation is almost the same as is calculated by the conduction model, which means that the thermal resistance of the condensate film is dominant. However, the conduction model fails when the distribution of the condensate film becomes nonuniform.

REFERENCES

(1) Chang, C.-S., Tanasawa, I. and Nishio, S.: Proc. 1st ASME/JSME Joint Thermal Engineering Conference, Vol.3(1983), 305.
(2) Jacobs, H.R. and Cook, D.S.: Proc. 6th Int. Heat Transfer Conference, Vol.2(1978), 389.

Heat Transfer Enhancement by Direct Contact Condensation

MARIO DE SALVE, BRUNO PANELLA, GIUSEPPE SCORTA, and DINO RAVIOLO
Dipartimento di Energetica
Politecnico di Torino
C.so Duca degli Abruzzi 24
10129 Torino, Italy

ABSTRACT

In order to shed light on the direct condensation mechanims an experimental research on the fragmentation and the heat transfer of a liquid jet, flowing downward from a nozzle in an air or saturated steam ambient, has been performed. Experimental results obtained by a stroboscopic light technique of the jet breakup length are presented and best fit correlations are proposed and compared with other authors correlations. Another technique based on the laser light intensity analysis has been utilized for the breakup length in air at ambient pressure. Data of direct condensation heat transfer from saturated steam to the subcooled liquid jet are also presented in terms of Stanton number and condensation ratio . A comparison between the direct and indirect condensation heat transfer data obtained for the same thermalhydraulic conditions has been done too.

INTRODUCTION

Direct contact condensation is of great interest for the water reactor nuclear industry and for other applications like the direct contact heat exchangers (that are more compact and with less severe corrosion problems than closed ones),the thermal degasifiers,the sea-water desalting plants, the OTEC plants. The investigation of the heat transfer mechanisms in such conditions requires the knowledge of the jet hydrodynamics. In many cases the jet is characterized by an unbroken section followed by a drop flow one, so the heat and mass transfer depends on the jet breakup length, which in turn is affected by the bulk turbulence, the superficial instability, the edge effects at the nozzle exit, the interface shear-stress between the steam and the liquid. Present authors previously investigated the direct contact condensation heat transfer and the liquid jet fragmentation [1],[2]. In order to better understand the physical mechanisms further tests have been performed to investigate the breakup length of a vertical liquid jet flowing downward either in air at pressures from 1 to 3 bars or in steam ambient at pressures of 2 and 3 bars, at flow rate ranging from 0.003 to 0.02 kg/sec, at inlet temperature ranging from 20 to 90°C. The heat transfer has been investigated for different jet lengths at flow rate ranging from 0.008 to 0.02 kg/s at pressure of 2 and 3 bars and at 50°C inlet temperature. The nozzle (I.D. 0.002 m and length 0.02 m) is cylindrical with a conical shape at the upper edge. In the same facility indirect condensation heat transfer has been investigated across a stainless steel tube. The results are compared with the direct contact heat transfer data at the same thermalhydraulic conditions at the nozzle exit.

EXPERIMENTAL FACILITY, TEST PROCEDURE AND RESULTS

Two similar test rigs have been utilized : the first one is very simple and concerns the tests of the jet flowing in air at ambient pressure, the second one (fig.1) concerns the tests at higher pressure either with air or with saturated steam.The apparatus shown in fig.1 consists of the test chamber A, the liquid collector CL, the pump P , two heat exchangers SC1 and SC2, the liquid jet supply line LL, the steam generator GV, the steam supply line LV and the nozzle N.The inner diameter and heigth of the test chamber are 0.2 m and 0.6 m respectively; three pyrex glass windows allow the visualization of the jet. The liquid collector CL is placed on a bar that can continously travel along the jet axis. The degassed steam flows into the chamber from the bottom; the desired pressure is maintained constant by an automatic change of the electric power in the steam generator. In the experiment with pressurized air the gas flows through the valve V5. The instrumentation consists basically of thermocouples T to measure the liquid and steam temperatures, of a flow meter W to measure the liquid flow rate and of a pressure transducer PP to measure the pressure in the vessel. The jet fragmentation has been estimated by utilizing two techniques : the first one is based on the photographic observation of the jet that was lighted by a 2 μs strobe light; as regards the second one (only for air at ambient pressure) the jet is crossed by a laser beam at different distance from the nozzle exit and the light intensity is measured by a photodetector behind the liquid jet. The device to study the indirect heat transfer consists of a 0.002 m I.D., 0.004 m O.D. and 0.45 m long stainless steel tube that is attached to the nozzle; the liquid flows downward from the nozzle along the tube and is collected in the LC collector at the tube outlet. As regards the breakup of the jet flowing in the air at ambient pressure the laser beam is located perpendicular to the jet axis and lined up with the photodetector with the same light intensity for all runs. The photodetector signal, proportional to the laser light intensity, is digitally sampled and filtered by a second order Butterword digital filter to cut the higher frequency components. The signal is then processed by a Smith digital trigger to obtain an "ON- OFF" type output where the "ON" state corresponds to the presence of the liquid along the laser beam. The triggered signal is characterized by the time intervals where the light intensity is strongly attenuated by the liquid (level 1) and the time intervals where the laser beam is not attenuated (level 0).The ratio R between the signal durations at level 1 and the total duration has been adopted as jet fragmentation index. R is a function of the distance x from the nozzle exit and the flow rate : R (fig.2) decreases as the jet length increases and the flow rate decreases. At lower flow rate R presents a

rather sharp decrease , followed by a nearly constant value, at a distance
from the nozzle that can be identified as the jet breakup length in agreement with the pictures . As regards the fragmentation of the liquid jet
flowing in pressurized air or steam the breakup length has been estimated
by the pictures taken with a strobe light technique.The experimental
results are reported in ref. [2] and in spite of their spread out, owing
to the statistical mechanism of the fragmentation, the breakup length presents some clear trends :it increases strongly as the flow rate increases,
it is slightly higher at higher pressure, while the effect of the temperature is not clear. From the comparison between air and steam data at the
same pressure it is shown that the condensation process tends to stabilize
the jet. Fig. 3 reports the breakup length L_B as a function of the flow
rate W at two pressure p values for air and steam. The steady state direct
contact condensation heat transfer has been investigated from the measurements of the jet temperature axial distribution at the distance from the
nozzle exit ranging from 0.15 to 0.45 m. The effect on the heat transfer
of the pressure ,flow rate and inlet temperature has been evaluated .
Typical results are presented in fig. 4 as Stanton number versus flow
rate. The Stanton number is defined by :

$$St = -0.25 \ (d/x) \ \ln((T_s - T_j)/(T_s - T_{lo})) \qquad (1)$$

where d is the nozzle diameter, T_j the local jet temperature, T_s the saturation temperature, T_{lo} the exit nozzle temperature.

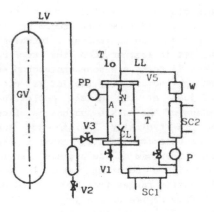

FIGURE 1.Experimental facility   FIGURE 2.Jet fragmentation index R
                                        vs. jet length at different flow rate.

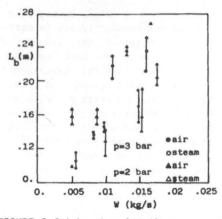

FIGURE 3.Jet breakup length vs.
flow rate

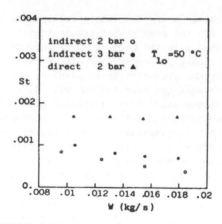

FIGURE 4.Stanton number vs. flow rate
for direct and indirect heat transfer

The indirect condensation heat transfer has been estimated from the measurements of the liquid temperature just at the outlet of the tube 0.45 m
from the nozzle exit. The comparison of the indirect condensation with the
direct one, at the same jet length and thermalhydraulic conditions at the
inlet, shows a much less effective heat transfer.

DISCUSSION.

Breakup length of a cylindrical liquid jet.

The jet fragmentation occurs at a certain distance from the nozzle exit
(breakup length) that depends mainly on the liquid flow rate according to
refs. [2] and [3].The flow regime of a free jet can be laminar , turbulent
or atomized. According to ref. [4] the jet is laminar when the surface
looks rather smooth whilst it is turbulent if there are some protuberances
on the surface that disturb the surroundings too; finally the jet is atomized when it looks like a droplet cloud just at the nozzle outlet.The
flow regime of the jet depends on the shape of the nozzle and on the flow
rate. At lower flow rate the flow regime is typically laminar and the
breakup length increases linearly with the velocity [3]. The transition to
the turbulence is characterized by a sharp decrease of the breakup length
that increases with the flow rate but not with a constant rate. In the
present tests the Reynolds number in the nozzle ranges from 3500 to 26000,
so it can be reasonably assumed that the flow regime of the jet is turbulent, as the more common correlations [3],[5],[6],[7] give critical
Reynolds number values from 2500 to 6000. The fragmentation is mainly due
to the disturbances that arise at the nozzle exit and amplify with the
distance from the nozzle because of the drag force depending on the relative velocity between the liquid and the gas. A particular instability
mode, that has been obtained in most of the present tests, is characteri-

zed by a helicoidal shape of the jet, in agreement with ref. [4]; the
fragmentation is due just to the amplification of the helicoidal mode
induced by the aerodynamic drag at the interface, while the friction
between the phases does not seem to be significant. The effect on the jet
stability of the gas density is not clear [7]. The pictures [4] show that
the jet breakup length increases if the gas flows cocurrently with the
liquid .Present data agree with other authors findings [8] as regards the
increase of the jet breakup length in presence of a mass transfer, namely
the condensation of the steam on the liquid. There are several correlations of the jet breakup length (see the bibliography of ref. [3]). In
fig. 5 the present data with steam obtained by the strobe light technique,
are compared with some correlations ; the Iciek experimental correlation
[9] seems to work better also because the nozzle and the thermalhydraulic
conditions are similar to the present ones. Present experimental breakup
length L_B at inlet temperature between 30 and 90 °C have been correlated
with the liquid Weber number We (ranging from 70 to 2200) :

$$L_B/d = 25.4 \; We^{0.21} \quad (p=2 \; bar) \; , \qquad L_B/d = 20.4 \; We^{0.24} \quad (p=3 \; bar) \tag{2}$$

with a square mean deviation of 16.5% and 13.3% respectively.

Direct contact heat transfer

In the direct contact heat transfer both the mass and the heat transfer
occur. Among the theoretical models that give the axial bulk temperature
of a cylindrical jet Lyczkowski [10] assumes either a constant heat transfer coefficient or an uniform liquid velocity along the jet axis; Celata
et al. [11] give the solution of the conservation differential equations
for laminar flow, then they take in account the turbulence contribution by
assuming an equivalent liquid thermal conductivity k .Another approach is
to assume the linear condensation rate C_R proportional to the jet local
subcooling $(T_s - T_j)$ through a constant "a". The integration of the energy
conservation equation, taking into account the liquid mass conservation
equation gives:

$$(\frac{c_p}{r})^2 \left[\ln(1 + \frac{r}{c_p(T_s-T_j)}) + \frac{T_s - T_j}{(T_s - T_j + r/c_p)} \right] = C + \frac{a \, x}{W_{lo}(T_s - T_{lo} + r/c_p)} \tag{3}$$

where c_p is the specific heat, r the latent heat of vaporization, W_{lo} the
flow rate at the nozzle exit. The integration constant is found by assuming $T_j = T_{lo}$ at x=0. In order to compare prediction with data, correlations
of the heat transfer coefficient h (or Stanton number) or of the condensation rate or of the turbulent conductivity are needed. As regards the present data the fig.6 shows the logarithm of $\vartheta = (T_s - T_j)/(T_s - T_{lo})$ as a function of the distance from the nozzle. For the jet length investigated ran-
ge (x > 0.15 m) the slope is rather constant and the following expression
can be assumed:

$$\ln \vartheta = - C_1 + C_2 x \qquad ; \qquad \vartheta = A \, \exp(-C_2 \, x) \tag{4}$$

345

where the value of A=exp(-C_1) is between 0.44 and 0.58 while C_2 is about 2.39 m . So at x=∞ St_∞ = 0.0012. The "a" coefficient of the present model decreases as the distance from the nozzle increases and it tends to an asymptotic value, but it depends also on the flow rate and the pressure. Fig. 7 presents a typical comparison between the jet temperature distribuition along the axis predicted by the three models: the model of ref. [11] seems to work better; the "a" coefficient needs to be expressed as a function of the pressure and the flow rate.

Jet axial temperature distribution and breakup length

In order to relate the jet heat transfer to the flow regime regions the axial temperature distribution can be expressed as a series of exponential terms,where the coefficients have to be correlated to the fluiddynamics of the jet. If the jet, as for the present tests, is characterized by a continuous region and a drop region a good approximation seems to be the sum of two terms, the second of which becomes negligible at a distance from the nozzle higher than the breakup length:

$$\vartheta = A \exp\left(- \frac{4}{d} St_\infty x\right) + (1 - A) \exp\left(- \frac{4\beta}{d} St_\infty x \right) \tag{5}$$

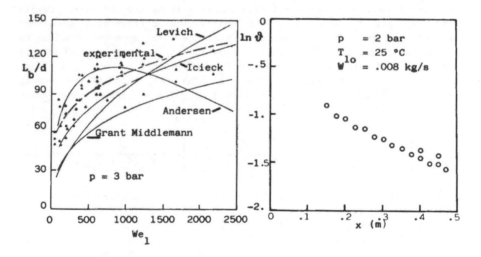

FIGURE 5. Jet breakup length vs. Weber number: present data and correlations.

FIGURE 6. Jet temperature vs. jet length.

346

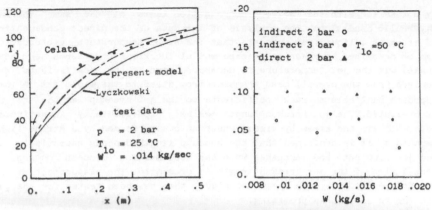

FIGURE 7.Jet temperature vs. length; FIGURE 8.Condensation ratio for present data and prediction;h=12kW/m², direct and indirect heat transfer. a=4.0*10⁻⁵ Kg/sm°C; K=4 W/m °C.

The constant A can be found from eq. (5) at x= L_B where L_B can be expressed as a function of the Weber number (eq.2) and $\vartheta_B = 1 - \eta_B$ can be expressed as a function of Kutateladze, Reynolds, Prandtl and Weber numbers [2], (η_B is the thermal efficiency at x=L_B). The coefficient β must be sufficiently high to get negligible the second term of eq. (5) at x= L_B. On the basis of the accuracy of the temperature measurement ($\pm 1°C$) a value of β = 6/(4 $St_\infty L_B$/d) can be assumed.

Comparison between direct and indirect condensation heat transfer

The index of the condensation process effectiveness $\varepsilon = (W_{steam} / W_{lo})$, that has been introduced by the present authors [1] for the direct contact heat transfer, can be used also for the indirect condensation and the comparison between the two heat transfer modes can be made by comparing the condensation ratio ε . Fig. 8 shows such comparison for the same thermalhydraulic conditions at the nozzle exit, the same length and the same steam pressure. The higher efficiency for the direct condensation process looks manifest owing to the thermal resistance of condensing film on the outer wall and to the steel internal thermal resistance for the indirect condensation heat transfer.In such a case the condensing steam flow rate decreases as the liquid flow rate inside the pipe increases, probably because the thickness of the condensed film is higher at higher flow rate, so the heat transfer from the steam to the outer wall is lower and the overall heat transfer decreases. This finding is confirmed by the Stanton number values presented in fig.4 where the heat transfer improvement for the direct contact condensation can be appreciated.

CONCLUDING REMARKS

The direct contact condensation heat transfer enhancement has been shown by comparison with the indirect condensation tests at the same therma-lhydraulic conditions. The analysis of the data of the direct condensation of steam on a liquid jet suggests that the axial temperature distribution can be described by a series of exponential functions (two terms seem adequate) but the jet temperature is underpredicted if the coefficients are derived from the overall heat transfer from steam to the whole jet. A new approach that relates such coefficients to the jet breakup length and St is presented. The jet breakup length has been experimentally investigated both with air and steam by visual observations performed by a strobe light technique :it is confirmed that the breakup length depends essentially on the jet flow rate and increases when the condensation process occurs. A correlation of the jet breakup length is presented. The measurement of the jet fragmentation by the laser technique, that has been tested for the jet flowing in air, looks promising.

REFERENCES

1. De Salve,M., Panella,B. and Scorta,G., Heat and Mass Transfer in Direct Condensation of Steam on a Subcooled Turbulent Water Jet, Eight Int. Heat Transfer Conf., San Francisco, vol.IV, pp.1653-1658,1986.
2. De Salve,M., Panella,B., and Scorta,G., Fluid-Dynamics and Heat Transfer in Direct Condensation of Steam on a Liquid Jet, 4th Int. Symposium on Multiphase Transport & Particulate Phenomena , Miami Beach, Session A6, December 1986.
3. Grant, R.P., Middlemann,S., Newtonian Jet Stability, A.I.Ch.E.J., vol. 12,p.669,1966.
4. Hoyt, J.W., Taylor,J.J., Waves on Water Jets, J. Fluid Mech. ,vol.83, part 1,pp.119-127,1977.
5. Van De Sande,E.,Smith J.M., Jet Breakup and Air Entrainment by Low Velocity Turbulent Water Jets, Ch. Eng. Sci.,vol. 31,p.219,1976.
6. Celata,G.P., Cumo,M., Farello,G.E. and Focardi,G., Esperienze di Condensazione per Contatto Diretto su getti di Liquido, ENEA- Dipartimento Reattori Termici,O.I., NEBS 7,April 1987.
7. Phinney,R.E., Breakup of a Turbulent Jet in a Low Pressure Atmosphere, A.I.Ch.E.J.,vol. 21,p.996,1975.
8. Coyle, R.W., Niwa, J.C.,Liquid-Liquid Jet Breakup under Conditions of Relative Motion, Mass Transfer and Solute Adsorbtion, Ch. Eng. Scie. vol.36, p.19,1981.
9. Iciek, J., The Hydrodynamics of a Free, Liquid Jet and Their Influence on Direct Contact Heat Transfer-III, Int. J. Multiphase Flow,vol.9,pp.167-179,1983.
10. Lyczkowski,R.W, A First Order Model for Non Equilibrium ECC Pressure Suppression, Multi-Phase Flow and heat Transfer III, Part B:Applications,ed.,Veziroglu and Bergles,pp.243-258, Elsevier Science Publishers,Amsterdam,1984.
11. Celata,G.P,Cumo,M.,Farello,G.E.,and Focardi,G.,A Comprehensive Analysis of Direct Contact Condensation of Saturated Steam on Subcooled Liquid Jets,forthcoming in Nuclear Eng. and Des.,1988.

Augmentation of Condensation Heat Transfer from Non-Condensable Gases with the Kenics Static Mixer

JIANZHOU YANG, XIAOJIAN MA, and WEIFAN FANG
Chemical Engineering Department
Zhenzhou Institute of Technology
Zhenzhou, PRC

ABSTRACT

The compative experiments of heat transfer and pressure drop of air, pure saturated steam, and their mixture were made in vertical tubes with and without the kenics static mixing elements. The influences of the twist ratio of the mixing elements and the contents of non-condensable gases on condensation were investigated. The results clearly indicated that the kenics static mixer could not only enhance the rates of heat transfer of gases and condensation with and without non-condensable gases, but also stabilize the process of condensation from non-condensable gases for all the ratios of the mixing elements. According to the investigation, it was suggested that heat transfer would be augmented through reduced flow resistance, by appropriately increasing the twist ratio of the mixing element.

1. INTRODUCTION

There are many heat transfer processes involving condensation from non-condensable gases in the petrochemical industry and some other fields. Investigation and practice have indicated that the capacity of the condenser would diminish prominently even if the content of a non-condensable gas is about 0.5 percent (volume ratio) and, moreover, the rates of the condensation would vary radically with the content. For example, under the condition of flow vapour velocities, the condensation heat transfer coefficient can be reduced by over 40 percent if the content of a non-condensable gas is only 1.07 percent, and reduced by about a factor of 10 if the content is 6.21 percent[1]. Thus, enhancing the rate of condensation heat transfer from non-condensable gases is of significance. It has been demonstrated that the Kenics static mixer is effective on augmenting single phase flow, condensation, and boiling heat transfer[2-8]. This paper reports the result of augmenting condensation heat transfer from non-condensable gases with the Kenics static mixers. The result is satisfactory and can be also used for the reference of other condensations of multi-component mixtures.

2. EXPERIMENTAL FACILITY AND THE MIXER

2.1 Experimental Facility

Fig. 1 shows a schematic diagram of the experimental facility used. It includes two identical double-pipe heat exchangers mounted vertically. Each inner

349

tube of them has an I.D of 22 mm and a length of 1200 mm. For gas heat transfer, air was heated by steam in one of the double pipes and then was cooled by cooling water in another double pipe. For condensation heat transfer, air and steam were mixed in the mixing chamber and, then cooled and condensated by cooling water in a double pipe. The mass flow rate of condensing fluid was calculated by the graduated cylinder. All temperatures were measured by the copper-constantan thermocouples. The pressure drops were measured by the U-tube and microbarometer.

2.2 Static Mixer

The Kenics static mixer is constructed of a number of short elements of right-helixes and left-helixes. In general, each element is twisted through 180 degrees and has a twist ratio of 1.5. These short helixed elements are alternated and welded together, oriented such that their leading edges are at 90 degrees to the trailing edge of the one ahead. Then, the element assembly is inserted axially insided a tube, which has a bore diameter nominally equal to the width of the elements. Fig. 2 shows an isometric view of the assential feature of the Kenics static mixer. Because of these geometric features, the flow stream divides at the leading edge of each element and flow through the semi-circular channel created by the element's shape. At each succeeding elements, the two streams are further divided and the directions of the fluid rotation are changed again. This sequence of flow division and rotation results in thorough and efficient radial mixing and tends to minimize the radial gradients in temperature, velocity and composition.
The two twist ratios of the element in this experiment were used. The length of the element was either 33 mm or 44 mm respectively corresponding to twist ratios of 1.5 or 2. The codes of the mixers of the twist ratios of 1.5 and 2 were A and B respectively and the code of the smooth tube was S.

3. RESULTS AND DISCUSSION

Nine sets of average heat transfer coefficient and three sets of friction factor were taken for the present investigation. The average heat transfer coefficient h inside the tubes is given by

$$h = Q / (F\Delta t) \tag{1}$$

In Eq. (1), the heat transfer rate Q was calculated from the total enthalpy change rate of the fluid inside the tube, the heat transfer area F is the inside surface area of the inner tube, and the mean temperature difference Δt was the logarithmic-mean temperature difference. The friction factor f of the single phase flow is given by,

$$f = \Delta p / (L/d \cdot \rho \cdot u \cdot u/2) \tag{2}$$

3.1 Experiments on Frictional Resistance and Heat Thansfer of Air

Between Re=6000—20000, the measured resistance data for all experimental tubes deviated generally less than ±5 percent from the corresponding values estimated by the following equations,

Mixer A $\qquad f = 6.789\ Re^{-0.15}$ $\hfill(3)$

Mixer B $\qquad f = 5.366\ Re^{-0.19}$ $\hfill(4)$

Smooth tube S $\qquad f = 0.3164\ Re^{-0.25}$ $\hfill(5)$

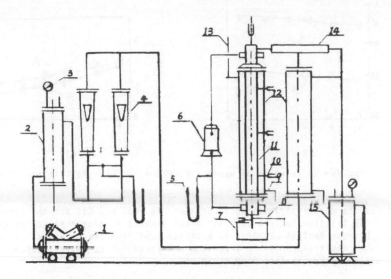

FIGURE 1. Experemenal facility; 1. air compressor; 2. pressure stabilizer;
3. pressure tap; 4. flow meter; 5. U-tube; 6. microbarometer;
7. Graduated cylinder; 8. mixing vapour outlet; 9. cooling water inlet;
10. thermocouple;11. experimental section; 12. double-pipe heat exchangers;
13. cooling water outlet; 14. mixing chamber; 15. boiler.

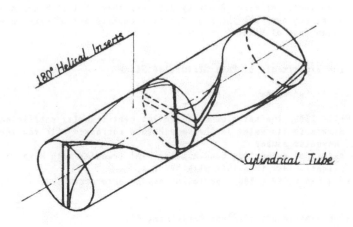

FIGURE 2. Essential features of the kenics static mixing system.

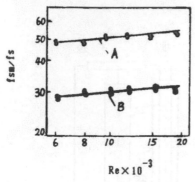 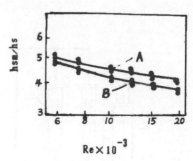

FIGURE 3. fsm/fs—Re relation of gases.   FIGURE 4. hsm/hs—Re relation of gases.

Eq. (5) is the empirical equation of Blasius. Eq. (3) and (4) are the correlations of the friction facters of Mixer A and Mixer B on single phase convective flow[8]. Between Re=6000—20000, the measured heat thansfer data for all experimental tubes deviated generally less than ±6 percent from the corresponding values estimated by the following equations,

Mixer A	$Nu = 0.18 \ Re^{0.729} \ Pr^{0.333} \ (\mu/\mu w)^{0.14}$	(6)
Mixer B	$Nu = 0.28 \ Re^{0.665} \ Pr^{0.333} \ (\mu/\mu w)^{0.14}$	(7)
Smooth tube S	$Nu = 0.021 \ Re^{0.8} \ Pr^{0.333} \ (\mu/\mu w)^{0.14}$	(8)

Eq. (8) is generally recommended for the correlation of gas heat transfer in tubes. Eq. (6) and (7) are the correlations of heat   transfer of Mixer A and Mixer B on single phase convestive flow[8].
Fig. 3 and Fig. 4 illustrate respectively the (fsm/fs) -Re and (hsm/hs) -Re relations. The result is found from Fig. 3 and Fig. 4; At the same Reynolds number, the friction factor of Mixer B was by far less than that of Mixer A in whole experimental regions. But the heat transfer coefficient of Mixer B was not much less than that of Mixer A.

3.2 Condensation Experiments of Pure Staturated Steam

The results are listed below;

   (1). At Rel < 2300, the ratio of condensation heat transfer coefficient with the mixers to its value in the smooth tube increased with the increase of Reynolds number
   (2). The effect of augmenting condensation heat transfer with Mixer B was just slightly less than that with Mixer A.
   (3). Between Rel = 550 - 980, the ratios hsm/hs were about 1.45 - 1.95.

3.3 Condensation Experiments of Steam Containing Air

Because there is a layer of gas film on the liquid film interface at the wall, which impedes the diffusion of vapour molocules to the wall and the sensible heat transfer to the wall, the rate of condensation from non-condensable gases is lower and affected intensively by the contents of non-condensable gases. This effect can be expressed in the (h/hc) -X relation, where X is mole fraction of non-condensable gases at the inlet and hc is average condensation coefficient without any non-condensable gas. According to the present investigation, the (h/hc) -X relations can be expressed as,

Mixer A $\qquad$ h/hc = 1.85 Exp (-4.1 $X^{1.4}$) $\qquad$ (9)

Mixer B $\qquad$ h/hc = 1.75 Exp (-4.3 $X^{1.4}$) $\qquad$ (10)

Smooth tube S $\qquad$ h/hc = 0.1 $X^{-0.5}$ $\qquad$ (11)

Eq. (11) is the empirical equation of smooth tubes. The coefficient 1.85 and 1.75 in Eq. (9) and (10) were the ratios h/hc of Mixer A and Mixer B at X=0 and Rel=920. The hc is given by the following well-known equation,

$$hc = 1.88 \ \lambda_L [(\rho_L^2 \ g)/(\mu_L \ Rel)]^{1/3} \qquad (12)$$

The measured data for all experimental tubes deviated generally less than ±15 percent on the corresponding values estimated by Eq. (9)—(11) under the conditions,
   (1) At the inlet, the mixing vapour temperature was 90°C—103°C with superheat 0°C—8°C; at the outlet, the mixing vapour temperature was 75°C—100°C with atmospheric pressure.
   (2) At the inlet, the mole fraction content of air was 0—0.4; at the outlet, the mass velocity of the condensed liquid was 4.4—15 kg/(m s) and Rel was 250—920.
Fig.5 illustrates the (h/hc) -X relation. The result in Fig. 5 shows that the influence of the content of non-condensable gases on condensation heat transfer in the mixers was far less than that in the smooth tube. For example, at X = 0.03 and X = 0.1, the ratio for the smooth tube was decreased respectively by about a factor of 2 and a factor of 3 at least. But the ratio h/hc for the mixers was only reduced respectively about 1/20 and 1/6. This character of the mixers was of significance for some condensable processes.
Fig. 6 illustrates the (hsm/hs) -X relation. The following observation can be made from the results in Fig. 6 ,
   (1) The heat transfer coefficient of Mixer B was only slightly less than that of Mixer A in all experimental regions.
   (2) The ratios of the heat transfer coefficient with the mixer to its value in the smooth tube were greater than 1.75 and increased with the content of non-condensable gases, reached a maxium value in the neighborhood of X = 0.17.
On single phase convective flow, the pressure drop of Mixer B was about 55 percent that of Mixer A from the preceding experiment in air resistance. On the flow with changes of phase, the measured data showed that the pressure drop of Mixer B was also considerablly lower than that of Mixer A. This indicates that if the twist ratio of the mixing elements is increased properly, the pressure drop of the mixer can be greatly reduced, but the influence on condensation of the mixer with or without non-condensable gases is slight.
The condensation from a non-condensable-gas is a non-isothermal complicated process governed by the rates of diffusion and heat transfer, and often called cooling-condensing process. There are many design methods for cooling-condensing process. For exmple, Colburn—Hougen method, Cribb-Nelson method, Mizushina method, Gilmour method, etc. The main differences are how to treat mass transfer and sensible heat transfer, and how to divide them into sections. However, the cooling-condensing coefficient can be analysed by the following equation,

$$1 / h = 1 / hc + 1/hg \ (1+Y) = 1/hc + 1/hg' \qquad (13)$$

where, Y = the degree of influence of mass transfer process to gas heat transfer process,
      hg'= the effective or equivalent gas film coefficient.

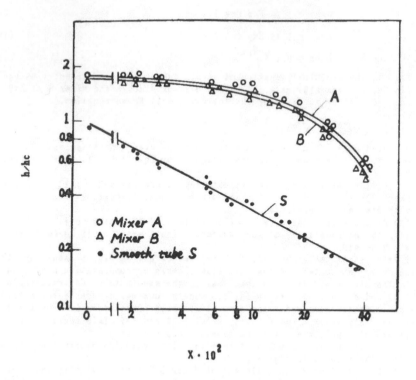

FIGURE 5. h/hc—X relation of the cooling-condensing process

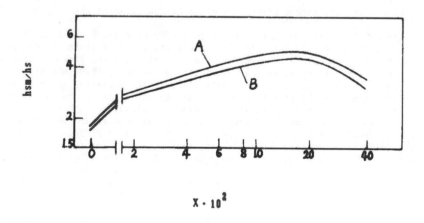

FIGURE 6. hsm/hs—X relation of the cooling-condensing process

From Eq. (13), we can see that the cooling-condensing coefficient is determined by the condensation film coefficient, gas film coefficient and mass transfer. Because the mixer can greatly augment the condensation heat transfer and gas heat transfer at the same time, the effect on augmenting cooling-condencing process with the mixer was very prominent.

4. CONCLUSION

(1) The Kenics static mixer would considerably augment the gas heat transfer, condensation heat transfer, and cooling-condensing heat transfer. Between $Reg=6000-20000$, the ratio hsm/hs for gas heat transfer was about $3.4-5$; between $Rel=550-980$, the ratio hsm/hs for condensation was about $1.45-1.95$, between $X=0.02-0.4$, the ratio hsm/hs for cooling-condensing heat transfer was about $2.5-5$.
(2) The influence of the content of non-condensable gases on condensation heat transfer in the mixer was far less than that in the smooth tube. At $X=0.03$, the ratio h/hc for the mixer was only reduced by about 5 percent, but that for the smooth tube was reduced by about 50 percent.
(3) The decrease of friction factor was more than 40 percent but that of heat transfer coefficient was less than 20 percent under all operational conditions of the present experiments while the twist ratio of the mixing element was increased from 1.5 to 2.

NOMENCLATURE

d	inside diameter	Subscripts	
k	thermal conductivity		
L	tube length	c	condensation
Nu	Nusselt number	g	gas
Pr	Prandtl number	l	liquid
u	mean velocity	s	smooth tube
Δp	pressure drop	sm	Kenics static mixer
ρ	density	w	wall
μ	dynamic viscosity		

REFERENCES

1. Fan Zhixin, Engineering Heat Thansfer, pp. 201, Chemical Industry Press, Beijing, 1982.

2. Palt, M. H. et al., Chemic-Ingenieur-Technik, vol. 51, no.5, pp.347, 1979.

3. Azer, N. Z. et al., Proc. Heat Transfer Fluid Mech. Inst., pp. 512, 1976.

4. Lin S. T. et al., Proc. Heat Transfer Fluid Mech. Inst., Stanford California, pp. 117, 1978.

5. Marner, W. J. et al., Proceedings of 6th Int. Heat Transfer Conference, Toronto Canada, pp. 583, 1978.

6. Azer, N. Z. et al., IEC. PDD., vol. 26, no. 11, pp. 246, 1980.

7. Huang Goughao et al., Joural of Chemical Industry and Engineering (China), no. 1, pp. 23, 1983.

8. Yang Jianzhou et al., Joural of Chemical Industry and Engineering (China), no. 3, pp. 366, 1988.

Investigation of Condensation Heat Transfer Enhancement Mechanisms of Particularly-Shaped Fin Tube with Numerial Method

SHIPING WANG, QINGHUA YIN, YINGKE TAN, and SONGJIU DENG
Chemical Engineering Research Institute
South China University of Technology
Guangzhou, PRC

ABSTRACT

A numerical method for the calculation of local condensation heat transfer coefficient of particularly-shaped fin tube is presented. The result indicates that surface tension effect and form drag induced turbulence in the condensate film are the two main functions causing heat transfer enhancement. The calculated results can be satisfactorily explained by the experimental data and observed phenomena.

INTRODUCTION

Particularly-shaped fin tube is the joint name for saw-teeth tube (S. T. tube or C. tube) and radially ridged fin tube (R. tube). This kind of tube is different from low-finned tube, the fins of which are smooth; the fin fringe of S.T. tube is saw-teeth in shape, and the fin sides of R. tube are grooved and ridged.

S. T. tube and R. tube were studied experimentally with the actuating media of R11 and R113 [1], [3], [4]. The results showed that the single-tube condensation heat transfer coefficients (α_c) of S.T. tube were 8 to 12 times as high as those of smooth tube, and 1.5 to 2.2 times as high as those of low-finned tube, its optimum structure parameters were 0.6 to 0.7 (mm) for fin pitch (p) and 1.0 to 1.2 (mm) for fin height (h); α_c of the optimal R. tube were 10% to 30% higher than those of the optimal S.T. tube, its optimum structure parameters were 0.8 to 0.85 (mm) for p and 1.3 to 1.4 (mm) for h. Their mechanisms of heat transfer enhancement have been proposed [3].

The dimensionless number correlations were established to evaluate out-side single horizontal S.T. tubes and R. tube. The correlation of S. T. tube has been successfully applied to design condensers of refrigerators and air-conditioners with freon refrigerants for practical usage.

The former investigation had been carried out experimentally to measure the average heat transfer coefficient of a whole tube. To make a deepgoing inquiry into the condensation heat transfer enhancement mechanisms of particularly-shaped fin tube, it is necessary to find out the change of local condensation heat transfer coefficients (α_{cl}) along the circumference outside the tube. A numerical method has been applied to achieve this purpose.

* Project supported by the national nature science fund of China.

1. Experiment Scheme and Measurement of Wall Temperature

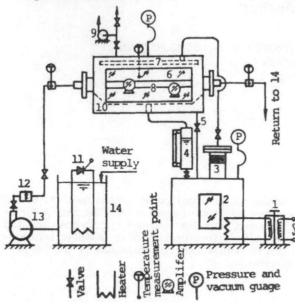

Water supply

FIGURE 1. Experiment Scheme

1. Transformer 2. Evaporator 3. Form remover
4. Condensate measurement cylinder 5. Drain
for the condensate formed on the condenser wall
6. Condenser and glass window 7. Vapor
distribution tube 8. Test tube
9. Vacuum pump 10. Collector for the
condensate flowing down from the test tube
11. Temperature controller 12. Flowmeter
13. Water pump 14. Water tank

Figure 1 shows the experimental scheme. There are two closed cycles for the actuating medium and cooling water respectively. The condensate flowing down from the test tube is separated from that condensed on the condenser wall by a collector, then flows into a measuring cylinder, which is used to measure the flow rate of the condensate. It flows back into the evaporator and evaporates again.

Two glass windows are installed on the sides of the condenser and two optic amplifiers are fixed in front of the upper and lower parts of the test tube for observing condensation phenomena.

The flowrate of cooling water is measured by a flowmeter, the inlet and the outlet temperature of it are measured by the thermocouples. The absolute deviation of measured temperature is smaller than 0.1 (°C), and the relative deviation of the flowrate is less than 1%.

For the measurement of local temperature on each end of the tube, 8 holes are peripherally uniformly distributed on the circumference of radius r_2, as shown in Fig. 2, and thermocouples are buried into these holes.

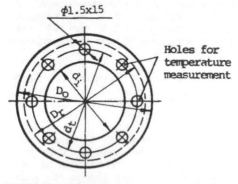

$\phi 1.5 \times 15$

Holes for temperature measurement

FIGURE 2. Distribution of the temperature measurement holes

2. Some Experimental Phenomena

In the experiments, three patterns of condensate drainage were observed in the different heat flux regions, which were named dropwise, linewise and drop-line joint drainage for low, high and middle heat flux regions respectively.

It was also discovered that α_c could be raised 16% to 30% by cutting a slot on the bottom of an R. tube and embedding a copper flake bound with a little cotton yarn into the slot.

THE NUMERICAL METHOD

Based on the local temperature measured, the field of temperature in the tube wall can be obtained according to the following step.

1. Plotting the Temperature Curve in the Middle of the Tube

The mean of the two local temperature values on two ends of the tube is taken for the wall temperature at the corresponding location in the middle of the tube. The temperature distribution curve on the circumference where the measurement holes are located can be obtained by curve fitting based on the mean temperature. In consideration of the temperature distribution symmetry, the multinomial without odd power terms is adopted for the curve fitting, shown as follows:

$$T(\theta) = Z_0 + Z_1(\theta/\pi)^2 + Z_2(\theta/\pi)^4 + Z_3(\theta/\pi)^6 + Z_4(\theta/\pi)^8 + Z_5(\theta/\pi)^{10} \qquad (1)$$

$$Z_5 = -(Z_1 + 2Z_2 + 3Z_3 + 4Z_4)/5 \qquad (2)$$

where, Z_1 to Z_5 are fitting constants.

Fig. 3 shows the temperature distribution curves fitted and the temperature values measured (shown as the symbols "□" in the figure) of two test tubes at various heat fluxes (q_m).

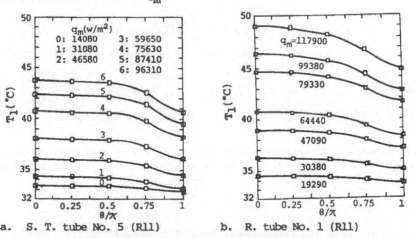

a. S. T. tube No. 5 (R11)   b. R. tube No. 1 (R11)

FIGURE 3. Wall temperature distribution curves of two test tubes

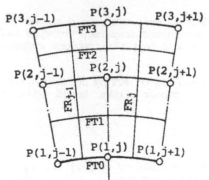

FIGURE 4. Sketch map of knots and network square

P(i,j)—Knots, i—Radial No., i=1,2,3
j—Circumferential No., j=1 to 21
FT—Camber of circumferential network square
FR—Plane of radial network square

2. Establishing and Solving the Difference Equation Sets

The knots and network square for establishing the difference equation sets are shown in Fig. 4. The semicircumference is divided into 20 equi-parts. The knots $P(1,j)$ are located on the inner wall surface of the tube, the radius of which is r_1 ($r_1 = d_i/2$), $P(2,j)$ at the circumference with radius r ($r_2 = d_t/2$), and $P(3,j)$ on the fin base with radius r_3 ($r_3 = D_r/2$). From heat balance of a control volume, the difference equation sets can be established as follows.

$$T_{1,j} = \frac{A_1^2(T_{1,j+1} + T_{1,j-1}) + A_2 r_1 (\frac{\pi}{20})^2 T_{2,j} + 2r_1^2 A_1 (\frac{\pi}{20})^2 \frac{\alpha_1}{\lambda_{Cu}} T_w}{2A_1^2 + \eta A_2 (\frac{\pi}{20})^2 + 2r_1^2 A_1 (\frac{\pi}{20}) \frac{2\alpha_1}{\lambda_{Cu}}} \quad (3)$$

$$T_{3,j} = \frac{(\frac{A_4}{A_3} + \frac{800A_3}{\pi^2 r_2} + \frac{A_2}{A_1})T_{2,j} - \frac{400(r_3 - r_1)}{\pi^2 r_2}(T_{2,j+1} + T_{2,j-1}) - \frac{A_2}{A_1}T_{1,j}}{(A_4/A_3)} \quad (4)$$

$$\alpha_{3,j} = \frac{\lambda_{Cu}[\frac{10A_3}{r_3}(2T_{3,j} - T_{3,j+1} - T_{3,j-1}) + \frac{A_4}{40A_3}(T_{3,j} - T_{2,j})]}{(\pi/20)r_3(T_c - T_{3,j})} \quad (5)$$

where, $A_1 = r_2 - r_1$, $A = r_2 + r_1$, $A_3 = r_3 - r_2$, $A_4 = r_3 + r_2$

$\alpha_1 = Q/2\pi r_1 L(T_{1m} - T_w)$, $T_{1m} = T_{2m} - Q(r_2 - r_1)/\lambda_{Cu}L(r_2 + r_1)$.

The equation set (3) is tridiagonal and can be solved by iteration method or running method. Then, $T_{3,j}$ and $\alpha_{3,j}$ can be obtained by solving the equations (4) and (5).

ANALYSIS

The study is based on the experimental data of five particularly-shaped fin tubes chosen from [4]. Tab. 1 lists their structure parameters. The results acquired with the above numerical method are shown in Fig. 5 and Fig. 6.

TABLE 1. Structure parameters of test tubes

Tube No.	P (mm)	D (mm)	D (mm)	h (mm)	s (mm)	R (mm)	b (mm)	c R11	R113	Remark
1	1.00	23.72	20.78	1.47	0.4	0.10	0.35	0.731	0.74	
2	1.00	23.72	20.78	1.47	0.4	0.10	0.35			A slot on bottom
3	1.50	23.33	21.13	1.10	0.85	0.15	0.28	0.806		
4	0.90	23.54	20.89	1.32	0.54	0.10		0.721		S.T. tube
5	1.00	23.22	21.00	1.11	0.62	0.10		0.739		S.T. tube

Remarks: 1. Spaces between two ridges of R. tubes are all 0.85 (mm).

2. $\beta_c = \arccos[\dfrac{2\sigma}{\rho_f \cdot g(s + b/2)R} - 1)]$.

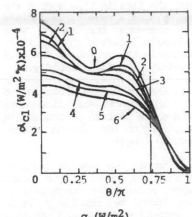

$q_m(W/m^2)$

0: 16020   3: 74760   6: 135140
1: 30380   4: 79330
2: 47090   5: 99380

a. R. tube No. 1 (R11)

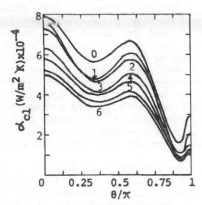

$q_m(W/m^2)$

0: 19290   3: 64440   6: 117900
1: 38150   4: 90980
2: 55740   5: 117100

b. R. tube No. 2 (R11)

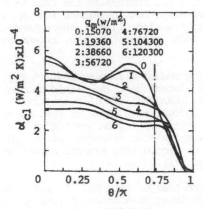

$q_m(W/m^2)$
0:15070   4:76720
1:19360   5:104300
2:38660   6:120300
3:56720

c. R. tube No. 1 (R113)

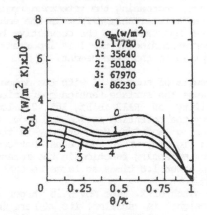

$q_m(W/m^2)$
0: 17780
1: 35640
2: 50180
3: 67970
4: 86230

d. R. tube No. 3 (R11)

FIGURE 5. Curves of   versus θ/π at different heat fluxes of R. tubes

From Fig. 5a, it can be seen that in the low and middle heat flux regions (q_m<5×10^4 W/m^2), curves of α_{cl} versus (θ/π) of tube No. 1 present the form of double wave crests. In these regions, the force f (f=ρ_f.g.sinθ) driving the condensate at fin base to flow along the circumference with radius r (base circumference) varies along the circumference. In the region of θ/π <=0.25, both f and the passage for condensate drainage between two fins are very small, the condensate at fin base thickens rapidly along the base cir-cumference and the area influenced by 2-dimensional surface tension effect also decreases rapidly, therefore, α_{cl} falls drastically. In the region of θ/π=0.25 to 0.4, the driving force increases, the thickness of the conden-sate film at fin base increases less rapidly; but the flow velocity and turbulence intensity in the condensate film becomes larger, thus, α_{cl} changes gently. In the region of θ/π=0.4 to 0.6, the driving force and tur-bulence intensity increase to its maximum value; the flow velocity becomes larger, the boundary of retention region fluctuates down and up (caused by unsteady condensate dropping and draining), the fins in this sphere are scoured periodically by the condensate flowing by, as a result, α_{cl} rises progressively and the curves of α_{cl} versus θ/π presents the second wave crest. For the drastic increase in the condensate flowrate and capillary effect of narrow fin gap, the condensate retains in the region of θ/π >0.6 and covers nearly all the fin surface, consequently, α_{cl} falls rapidly.

For the high heat flux region (q_m>5×10^4 W/m^2), the condensate flowrate and velocity increase with q_m, the ridges disturb the condensate flow more intensely, therefore, α_{cl} decreases gently. Since the pattern of condensate drainage becomes linewise and the fluctuation of retention region disap-pears, the curves don't show the 2nd wave crest no longer.

The structure parameters of tube No. 2 in Fig. 5b are the same as those of tube No. 1 but a slot is cut on the bottom of it and a copper flake bound with a little cotton yarn embedded into the slot. Because the condensate in the original retention region is attracted away (down) by capillary force of liquid on the cotton yarn, the area of the retention region and the film on sides of fins decrease; therefore, in the region of θ/π>0.5 are larger than R. tube No. 1. In high heat flux region, the curves also presents the second wave crest. Due to the additional heat conduction of the copper flake, α_{cl} rise at the location of θ/π=1 again. In low heat flux region, 2-dimensional surface tension effect plays the main role in the enhance-ment, thus, decreasing the retention region, i.e., increasing the thin film region has a remarkable effect in the enhancement. But the higher the q_m, the more important role the convective heat transfer plays in the enhance-ment, correspondingly, the less important role decreasing the retention region plays in the enhancement.

The results of tube No. 1 with the actuating medium R113 are shown in Fig. 5c. Because the surface tension coefficient, latent heat and liquid thermal conductivity of R113 is 5%, 18%, 21% lower than that of R11 respectively, α_{cl} above the retention region are about 25% lower than those in Fig. 5a at the same q and θ/π. However, α_{cl} in the retention region are higher than those in Fig. 5a, which indicates that convective heat transfer plays the dominant enhancing function in the retention region (for Prandtl number of R113 are about 1.8 times as large as that of R11).

The fin pitch of R. tube No. 3 is larger than that of tube No. 1 and its ridge height is smaller, its α_{cl} are much lower than those of tube No. 1 and the values of its 2nd wave crest lower than those of tube No. 1. This indicates that the R. tube with low ridge height not only makes less full use of 2-dimensional surface tension effect, but also produces less remark-

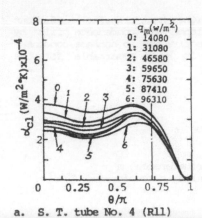

a. S. T. tube No. 4 (R11)

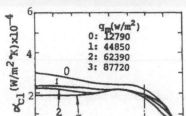

b. S. T. tube No. 5 (R11)

FIGURE 6. Curves of versus q of S. T. tubes

ably convective heat transfer. Since its fin pitch is large, its α change gently along circumference.

The results of S.T. tube No. 4 are illustrated in Fig. 6a. Having no ridges on its fin sides, it has less area of thin film than R. tube. Therefore, its α_{c1} near the top is much smaller than R. tube No. 1. At any heat flux, the curve of it has the 2nd wave crest. In high heat flux region, α_{c1} values at the 2nd wave crest are larger than those at the top of the tube. This shows that saw-teeth on the fins interrupt the condensate boundary layer continuously, which produces more intense convective heat transfer than the ridges.

Fig. 6b illustrates the results of S. T. tube No. 5 with larger fin pitch, smaller fin height and larger fin tip curve radius than tube No .4. As surface tension effect greatly weakens, its α_{c1} above the retention region are 30% lower than tube No. 4. But its mean condensation heat transfer coefficients in the retention region are nearly equal to tube No. 4, which shows that in the retention region, the heat transfer rate is dominated by the turbulence intensity in condensate film and less related to the heat transfer area.

CONCLUSION

1. Numerical method is a good mean for the detailed study of mechanisms of enhancing condensation heat transfer of horizontal tubes with non-uniform heat flux.

2. Extending heat transfer area, making full use of surface tension effect and promoting turbulent convective heat transfer in condensate film are the three mechanisms for particularly-shaped fin tube to enhance condensation heat transfer. In the region of $\theta/\pi <0.25$, the former two mechanisms play the dominant role in the enhancement. From $\theta/\pi>0.25$ to the retention region, all the three have enhancement function. In the retention region, the last is dominant.

3. The higher the q_m, the more important role the last mechanism plays in the enhancement. Furthermore, the larger the Prandtl number of actuating medium, the better the effect of promoting convective heat transfer.

4. In general, 2-dimensional surface tension effect plays the dominant role in the enhancement of R. tube, and for S. T. tube, the latter two mechanisms play a commensurate in the enhancement.

363

5. Turbulent condensate film has a much higher heat transfer coefficient than the laminar film. Promoting turbulence in condensate film by roughness on the fin surface is an effective way to enhance condensation heat transfer, the effect of which is more remarkable for the actuating media of large Prandtl number.

NOMENCLATURE

b : Height of ridges on fin sides of R. tube, m
D_o : Diameter of the fin tip circumference, m
D_r : Diameter of the fin base circumference, m
d_t : Diameter of temperature measurement circumference, m
f : Force driving unit condensate to flow along circumference, N/m^3
g : Gravity acceleration, m^2/s
h : Height of fin, m
L : Effective length of test tube, m
p : Pitch between two fins, m
Q : Heat transfer quantity in unit time, W
q_m : Mean heat flux of condensation, W/m^2
R_d : Radius of fin tip, m
s : Space between two fins, m
$T_{1,j}$: Wall temperature at knot P(i,j), °K
T_{1m}: Mean temperature of inner wall, °K
T_{2m}: Mean temperature on the circumference of temperature measurement holes, K
T_c : Condensation temperature, °K
T_w : Cooling water temperature, °K
α_1 : Heat transfer coefficient of in-tube cooling water, W/m^2 °K
α_{cl}: Local heat transfer coefficient of condensation based on the area of fin base*, W/m^2 °K
$\alpha_{3,j}$: Local heat transfer coefficient of condensation based on the area of fin base, W/m^2 °K
α_c : Average heat transfer coefficient of condensation based on the area of fin base, W/m^2 °K
β_c : Coefficient of capillary effect
θ : Angle measured from the top of tube, rad
λ_{Cu}: Thermal conductivity of copper, W/m °K
ρ_f : Density of condensate, kg/m
σ : Surface tension coefficient of condensate, N/m
*The area of fin base is $\pi D_r L$.

REFERENCE

1. Wang Shiping, Deng Songjiu et. al., Material of the National Chemical Engineering Conference of China, C-2-3, 1981.
2. HITACHI, Copper and Copper-alloy Products Cat., No. EA-501.
3. Wang Shiping et al., Experiments of Saw-teeth-shape Finned Tubes Enhancing Condensation and its Dimensionless Correlation, Journal of Engineering Thermo-Physics, vol. 4, No. 5, pp. 374-378, 1984. (Chinese)
4. Yin Qinghua et al., Investigation of Condensation Heat Transfer enhancement with Radially ridged Fin Tube, Journal of Chemical Engineering of Chinese Universities, vol. 1, No. 1, pp. 32-40, 1986.
5. K. K. Yan, et. al., Journal of Heat Transfer, No. 107, pp. 377, 1985.

An Investigation on Heat Transfer to Air Water Mixture in Improved Hi Static Mixer

XIAO JIAN MA, JIANZHOU YANG, and WEIFAN FANG
Department of Chemical Engineering
Zheng Zhou Institute of Technology
Zheng Zhou, PRC

ABSTRACT

The drag coefficient and heat transfer coefficient in single-phase flow and two-phase flow for air and water have been measured and it is examined that the intermediate chamber of improved Hi Static Mixer element has an effect on the heat transfer. The following correlations have been obtained ϕ_L-X, Nu_{GL}-Re_{GL}, and h_{GL}/ϕ_L-X.

The mechanism for enhancing heat transfer in improved Hi Static Mixer element is investigated through the test of flow visulization, and the relationship of h_{GL}, h_L-V_G, V_L has been discussed in this paper.

INTRODUCTION

In many industrial systems, the majority of the Static Mixer are used as mass transfer equipments and reactors. Phenomena like heat transfer are often seen in the procedure of mass transfer and reaction. For example, gas-liquid reactants need to be premixed and heated before they reach reactor, or the heat of reaction need to be removed in the reaction. In this case, using the Static Mixer will achieve two things at one stroke: it enhances not only mass transfer but also heat transfer. Recently, many reports have been published on the Static Mixer used for heat transfer in single-phase flow[1], boiling[2] and condensation[3]. Some investigations have been carried out using different media on different types of the Static Mixers and various correlations of drag coefficient and heat transfer coefficient have been obtained[4,5]. But so far, few reports have been seen on heat transfer for two-phase flow in which a lot of non-condensable gas is contained. The improved Hi Static Mixer element developed by ourselves is an element which has a strong ablity for heat transfer augmentation and less pressure drop[6]. The effects are are satiafactory if the element is used for absorption and gas-liquid reaction[7]. In order to establish the prediction of heat transfer for gas-liquid two-phase flow in which a lot of non-condensable gas is contained in the element, in this paper, the investigations have been conducted by using air-water mixture on fluid-drag and heat transfer, and relation between drag performance and heat transfer performance has been found.

EXPERIMENTAL SECTION

Experimental flowsheet is shown in Fig.1. It consists of providing water, providing gas and heat transfer test section. The temperature for water in

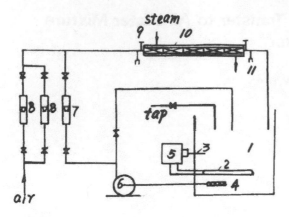

steam

9   10   11

1. tank
2. electric heater
3. temperature sensor
4. filler
5. temperature controller
6. pump
7. flowmeter
8. flowmeter
9. pressure tap
10. heat transfer test section
11. thermal couple

tap

air

FIGURE 1 Experimental flow chart

tank 1 keep constant through temperature sensor 3, electric heater 2, temperature controller 5 and appropriate amount tap water added into the tank
continually. In the precedence of test section 10, there is 1 m long gas-
liquid mixing section and its internal diameter is 20 mm with 12 the Kenics
Static Mixer elements installed inside. Before entering the test section,
gas-liquid mixture is adequately mixed so that temperature of both phases is
the same. The test section consists of several improved Hi Static Mixer elements. The construction of the element is shown in Fig.2. Based on Masanori's
investigation[6] the element diameter hardly affects the performance of its
transfer when nominal diameter of the element is larger than 20 mm and the
size effect can be ignored during scale-up. Hence, nominal diameter of the
element used in the test is 20 mm. The test section is 500 mm long and is
horizontally set. Combinations of the elements are length. The elements are
made up of brass. Four thermal couples mounted within the wall are uniformly
distributed along the test section to measure the wall temperature. Water
flow rates vary from 0.03 kg/s to 0.69 kg/s and air flow rates from 0 to
0.005 kg/s. Heating is done by steam, and condensate is collected to make
the heat balance. The error of the heat balance is within 5% in the test.

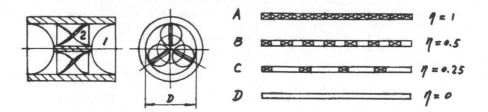

A $\eta = 1$
B $\eta = 0.5$
C $\eta = 0.25$
D $\eta = 0$

1. intermediate chamber
2. spiral vane
FIGURE 2 Construction of the element

FIGURE 3 Combinations of the
             elements

RESULTS AND DISCUSSION

Fluid drag

The drag coefficient for single-phase flow in the Static Mixer can be described by the following equation

$$f = aRe^{-n}$$

where a and n are regressing coefficient.
Based on drag data which air or water flows alone in the elements, the following correlation is obtained:

$$f = 13.2\,Re^{-0.149} \qquad (3000 < Re < 70000,\ \eta = 1.0\ ) \tag{1}$$

The correlation of data is shown in Fig.4.
At the same Re, f value given by equation (1) is 110 times higher than that by Blasius equation, but 80% lower than that by the equation used for the Hi Static Mixer element[9]

$$f = 131\,Re^{-0.21} \qquad (\ Re > 1000\ ) \tag{2}$$

For the drag in two-phase flow, following two variates are difined that, according to general method dealing with the drag for two-phase flow.

$$X^2 = (\ dp/dz\ )_L / (\ dp/dz\ )_G \tag{3}$$

$$\phi_L^2 = (\ dp/dz\ )_{GL} / (\ dp/dz\ )_L \tag{4}$$

When both phases are in turbulent flow if alone, the equation (5) and (6) can be obtained by the test data.

$$X^2 = (\rho_L/\rho_G)^{0.851} (\mu_L/\mu_G)^{0.149} (\ v_L / v_G\ )^{1.85} \atop (0.4 < X < 30,\ \eta = 1) \tag{5}$$

$$\phi_L^2 = 2.69 + 31.2/X + 86.8 / X^2 \tag{6}$$

The data plotted on log-log chart are shown in Fig.5. The relative error between data and equation (6) is within 11% . The pressure drop for two-phase flow in improved Hi Static Mixer element can be calculated by using equation (1), (5) and (6).

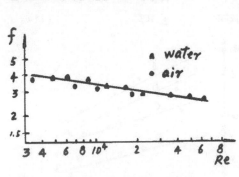

FIGURE 4 Drag coefficient for A test section in single phase flow

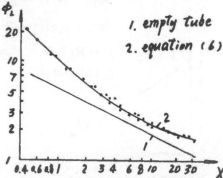

FIGURE 5 Pressure drop correlation for gas-liquid two-phase flow in the element

Performance of heat transfer and affecting factors

It is fairly effective that improved Hi Static Mixer is used to enhance heat transfer for single phase flow, which is clearly shown in Fig.6. Equation (7) is obtained from the data.

$$j_{HL} = 3.25 \, Re_L^{-0.575} \tag{7}$$

where $j_{HL} = Nu_L / (\, Re_L \, Pr_L^{0.333})$

Under the same conditions, heat transfer coefficient in improved Hi Static Mixer element is 5~6 times higher than that in empty tube.

Two-phase heat transfer coefficient is much more than single phase heat transfer coefficient in the empty. But heat transfer coefficient for two-phase flow does not increase very much more than that for single phase flow in improved Hi Static Mixer element. The reason is probably that turbulence in the element is very violent while single phase flow is taking place, therefore, the increase of heat transfer coefficient provided by adding of gas is limited. This can be seen in Fig.7, where A line is a correlation which is obtained by A test section in Fig.3 and D line by D test section (empty tube). Obviously, enhancing ability for two-phase flow heat transfer is lower than that for single phase in improved Hi Static Mixer element and the former is 50% lower than the latter.

The intermediate chamber between the adjacent elements is a place in which fluids are congregated and dispersed. It is a important construction for improved Hi Static Mixer element. The drag and heat transfer is greatly affected by the forms and sizes of the intermediate chamber . In order to examine the effects, B and C test section are arranged in this paper. Their inter-mediate chamber are lengthened to different extents. The results given by the test are shown in Fig.7. It can be seen that heat transfer coefficient decreases with the increase of the length of intermediate chamber. Following correlations are obtained from the test data.

test section	correlation	relative error	
A	$j_{HGL} = 1.75 \, Re_{GL}^{-0.485}$	$\pm 7\%$	(8)
B	$j_{HGL} = 0.789 \, Re_{GL}^{-0.429}$	$\pm 7\%$	(9)
C	$j_{HGL} = 0.171 \, Re_{GL}^{-0.301}$	$\pm 6.5\%$	(10)
D	$j_{HGL} = 0.0463 \, Re_{GL}^{-0.222}$	$\pm 10\%$	(11)

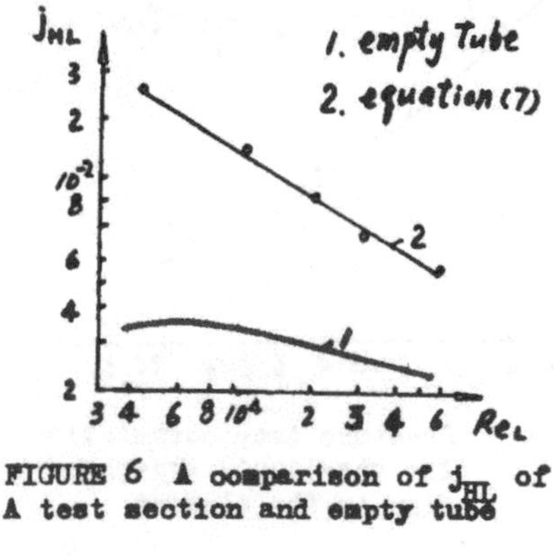

FIGURE 6 A comparison of j_{HL} of A test section and empty tube

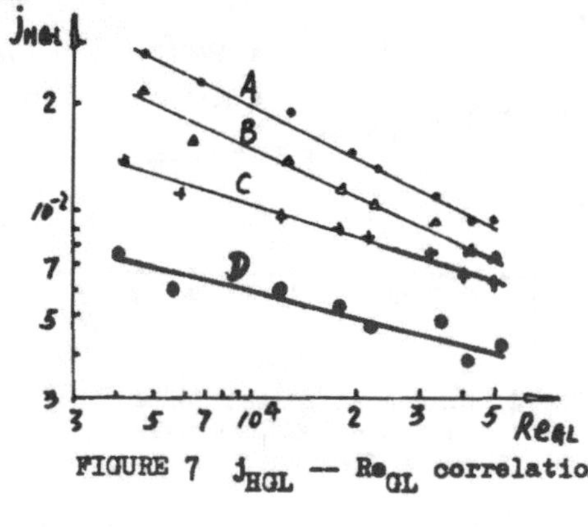

FIGURE 7 j_{HGL} — Re_{GL} correlation

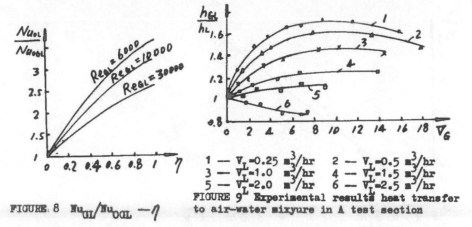

FIGURE 8 $Nu_{GL}/Nu_{OGL} - \eta$

1 — V_L=0.25 m³/hr   2 — V_L=0.5 m³/hr
3 — V_L=1.0 m³/hr   4 — V_L=1.5 m³/hr
5 — V_L=2.0 m³/hr   6 — V_L=2.5 m³/hr

FIGURE 9 Experimental results heat transfer to air-water mixyure in A test section

The physical properties used in the Nusselt and Prandtl numbers are those of the liquid, for the heat transfer is controlled mainly by the fluid adjacent to the wall. The following correlation can be obtained by regressing from equation (8) to (11), so that calculation and analysis are carried on conveniently.

$$Nu_{GL} = (1 + 42.9 \, Re_{GL}^{-0.319} \eta^{0.71}) \, Nu_{OGL} \tag{12}$$

where $Nu_{OGL} = 0.0463 \, Re_{GL}^{0.778} \, Pr_L^{0.333}$ \tag{13}

Equation (12) is plotted in Fig.8. From Fig.8, we can see that, when $\eta < 0.5$, with the increase of η, Nu_{GL} / Nu_{OGL} increases faster; when $\eta > 0.5$, Nu_{GL} / Nu_{OGL} increases slower.

Superficial velocities of liquid and gas have a significant effect on two-phase heat transfer. Fig.9 shows the relationship between h_{GL} / h_L and the superficial velocities of liquid and gas in A test section. From Fig.9, it is found that, when the liquid flow rate is given, with the increase of gas flow rate, h_{GL} / h_L increases gradually until it reaches a maximum, then h_{GL} / h_L decreases gradually. When liquid flow rate is lower, the phenomenon is even more obvious in particular. This is probably because the addition of a small mount gas can enhance the heat transfer, but, when gas flow rate reaches a certain value, the overall heat flow rate reduces because the heat transfer surface occupied by gas. Thus, the heat transfer coefficient decreases with the increase of gas flow rate. When gas flow rate is given, the ratio h_{GL} / h_L will decrease with the increase of liquid flow rate, and when the liquid flow rate is 2.5 m³/hr, the h_{GL}/h_L rate is less than 1. The reason causing the phenomenon is that, the turbulence is intensified with the increase of liquid flow rate, h_L rised rapidly, and increase of heat flow rate caused by the addition of gas is less than the decrease of heat flow rate caused by heat transfer surface taken. Finally, the ratio h_{GL} / h_L is less than one.

The relationship between drag performance and heat transfer performance

From a heat-momentum analogy, it is expected that the two-phase heat-transfer coefficient would be related to the two-phase friction parameter. In this paper, the data are plotted in Fig.10 in the form h_{GL} /ϕ_1 against X. The test results are shown in Fig.10, from which some relationship can be found between them. Equation (14) has been obtained by regressing the test

369

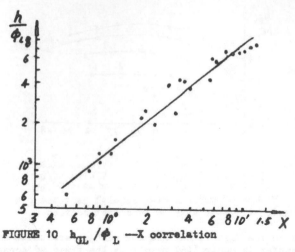

FIGURE 10 h_{CL}/ϕ_L —X correlation

data. The relative error between the data and equation (14) is within ±10%.

$$h_{CL}/\phi_L = 1270 \ X^{0.791} \qquad (\ 0.4 < X < 15 \ \text{and} \ \eta = 1 \ ) \qquad (14)$$

MECHANISM FOR ENHANCING HEAT TRANSFER

From the test of flow visualization, there only appear buble flow, churn flow and mist flow in the flow patterns because of good mixing and dispersing in the element. When the flow pattern in empty tube is stratified flow, the flow pattern in the element is churn flow and gas-liquid mixture is uniform in cross-section of the element. When the flow pattern in empty tube is the bubble flow, the flow pattern in the element is the bubble flow, too, but the bubble diameter in the element is much smaller than that in the empty tube and the bubble is uniform in the cross-section of the element. When the flow pattern in the empty tube is annular mist flow, the flow pattern in the element is mist flow. Under various flow pattern, the turbulence in the element is more violent than that in the empty tube. Liquid adjacent to the wall is rapidly renewed. Because the channel in the element is not continuous, boundary-layer can not be completely developed. Those factors above have effectively enhanced heat transfer process.

CONCLUSIONS

In the two-phase heat-transfer, there is a very obvious effect on enhancing heat transfer by using the improved Hi Static Mixer element, and the heat transfer coefficient in the element is three times greater than that in the empty tube. Though the increase of the pressure drop is obvious, the element still has pratical value in industrial applications because not only heat transfer but mass transfer is enhanced at the same time.

The economical performance in application can be increased if the intermediate chamber of the element is appropriately lengthened.

The pressure drop and heat transfer coefficient for the two-phase flow in the element can be calculated by using correlations given in this paper.

The drag performance and heat transfer performance in the element are related to each other in two-phase flow, and this relation is described by

370

equation (14).

NOMENCLATURE

D = nominal diameter of the element, m
f = drag coefficient, $f = 2\Delta PD / (L\ u^2)$
h = heat transfer coefficient, $w / (m^2 k)$
j_H= j_H—factor for heat transfer, $j_H = Nu / (Re\ Pr^{0.333})$
L = length of the test section, m
u = velocity, m / s
V = volumetric flow, m^3 / hr
X = variate defined by equation (3), dimensionless
Re= Reynolds number, dimensionless
Nu= Nusselt number, dimensionless
Pr= Prandtl number, dimensionless
ΔP= pressure drop, N / m^2
dp / dz = pressure gradient along the test section, N / m^3

GREEK SYMBOLS

ρ = density, kg / m^3
μ = viscosity, Ns / m^2
η = ratio of length, dimensionless
ϕ = variate defined by equation (4), dimensionless

SUBSCRIPTS

L = liquid phase
G = gas phase
GL= gas-liquid mixture
O = empty tube

REFERENCES

[1] Sumumu, J. H.,Kenics Company Technical Report 1970, No. 1002.

[2] Azer, N. Z., et al, Ind. Eng. Chem. process Des. Dev. vol.19, 246, 1980.

[3] Azer, N. Z., et al, Proc. Heat Transfer Fluid Mech. Inst. 512, 1976.

[4] Genetti, W. E., Chem. Eng. Commun. vol. 14, 47, 1982.

[5] Hang, G. H., et al., J. Chem. Ind. Eng. (China) no. 1, 23, 1983.

[6] Ma, X. J., Fang, W. F., J. Zhen Zhou Institute of Technology (China) no.1, 69, 1986.

[7] Yang, J. Z. et al., Proc. 1987 annual meeting chem. Eng. Soc. of China, Dalian, 695, 1987.

[8] Masanori, A., Hiromu, T., Chem. Plant (Japan) no.2, 65, 1980.

[9] Wang, D. C., Static Mixer, Textile Industry Press, Beijing, 1985.

Experimental Analysis of Heat Transfer in Spray Columns

ZONG XIN MENG, LUNG XUAI CHU, and GIN XIN GUO
Guangzhou Institute of Energy Conversion
Guangzhou, PRC

INTRODUCTION

Researchs on heat transfer in spray columns have been carried out widely. Ferrarini [1] proposed a heat transfer model for surface heat transfer coefficient of drops. Since there are several theoretical assumptions in the model but with no experimental support, there has not yet been satisfactory result in its practical application. Letan [2,3] suggested a theoretical model of wake of drops. If an accurate solution is to be got, one have to make experiments for a particular system in order to get the constants involved in the model. Moresco etc.[4] measured the temperature field for large single drop (9mm) and derived the heat transfer coefficient expressions for the inside and outside of the drop. However, these formulas do not agree with our data measured in a dense packing column with smaller drops. Leo etc.[5], Rosenthal [6] and Woodward [7] obtained experimental formulas of volume heat transmission coefficient that can only predict individual systems.

In actual direct contact heat exchangers, heat transfer data are normally reported in terms of the volumetric heat transmission coefficient, defined as

$$U_v = \frac{Q}{\text{Vol LMTD}} = k \, A_s \tag{1}$$

For spherical drops (drops of diameter less than 4.5 mm) the total transfer area per unit column can be given by

$$A_s = \frac{6H}{d_p} \tag{2}$$

The interfacial heat transmission coefficient k consists of an internal and an external heat transfer coefficient:

$$k = \frac{a_o \, a_i}{a_o + a_i} \tag{3}$$

According to the defination of Nusselt number, the internal heat transfer coefficient of drops is:

373

$$a_i = \frac{Nu_i \; K_d}{d_p} \qquad (4)$$

and external heat transfer coefficient of drops is:

$$a_0 = \frac{Nu_0 \; K_c}{d_p} \qquad (5)$$

Combining equations (3), (4) and (5) provides:

$$k = \frac{Nu_i \; K_d}{d_p \left(1 + \dfrac{Nu_i}{Nu_0} \dfrac{K_d}{K_c}\right)} \qquad (6)$$

A dimensionless expression of the interfacial heat transmission coefficient is derived by transforming equation (6):

$$M = \frac{k \; d_p}{K_d} = \frac{Nu_i}{\left(1 + \dfrac{Nu_i}{Nu_0} \dfrac{K_d}{K_c}\right)} \qquad (7)$$

Since the continuous phase in the column is a forced flow, the external Nusselt number is a function of the Reynolds number and the Prandtl number.

$$Nu_0 = f(Re_c , Pr_c) \qquad (8)$$

Because the circulation of the dispersed phase is caused by the interfacial shearing stress and the gravity difference caused by temperature difference is less significant, the flow behavior inside the drops can also be considered as a forced flow.

$$Nu_i = f(Re_d , Pr_d) \qquad (9)$$

Kd/Kc in equation (7) is not a independent argument. Therefore M is only the function of Reynolds number and Prandtl number.

$$M = \frac{k \; d_p}{K_d} = f(Re_d, Pr_d, Re_c, Pr_c) \qquad (10)$$

Water is the unique continuous phase used in the work. For simplification, the Reynolds number and Prandtl number were assumed to have a same exponential. Equation (10) can be rewriten as follows:

$$M = \frac{k \; d_p}{{}_i K_d} = f(Re_d, Pr_d, Pe_c) \qquad (11)$$

here

$$Re_d = \frac{v_s\, d_p}{\nu_d} \tag{12}$$

$$Pr_d = \frac{c_d\, r_d\, \nu_d}{K} \tag{13}$$

$$Pe_c = \frac{v_s\, \dot{c}_c\, r_c\, d_p}{K_c} \tag{14}$$

In a countercurrent flow the slip velocity is related to the superficial velocity of the two phase as [8]:

$$v_s = \frac{v_d}{H} + \frac{v_c}{1 - H} \tag{15}$$

This equation is helpless for a spray column design. However, it can be used to predict the slip velocity after measuring the flow rates of two phases and holdup ratio in an exist column.

Richardson [9] suggested that the slip velocity of rigid spheres in a dense packing column is a function of a terminal velocity of single rigid sphere and a holdup ratio. A semiempirical expression derived from their experimental results for a liquid - solid sedimentation system:

$$v_s = v_t\, (1 - H)^n \tag{16}$$

where $n = 3.65$ for $Re_t < 0.2$

$n = 4.35\, Re_t^{-0.03} - 1$ for $0.2 < Re_t < 1$

$n = 4.45\, Re_t^{-0.01} - 1$ for $1 < Re_t < 500$

$n = 1.39$ for $500 \leqslant Re_t$

The terminal velocity v_t of the particle is required in the flow characteristics of the system and is obtained as follows. The balance between the drag and gravity forces yields:

$$f\, \frac{\pi}{4}\, d_p^2\, \frac{\rho_c\, v_t^2}{2} = \frac{\pi}{6}\, d_p^3\, (\rho_d - \rho_c)\, g \tag{17}$$

However the particle terminal Reynolds number Re_t is defined as:

$$Re_t = \frac{v_t\, d_p}{\nu_c} \tag{18}$$

Combining equations (17) and (18) provides:

$$f \, Re_t^2 = \frac{4}{3} \, d_p^3 \, \frac{\rho_c \, (\rho_d - \rho_c)}{\mu_c^2} \tag{19}$$

In general the relation between Reynolds number and drag coefficient for a smooth rigid sphere is [10]:

Creeping flow $Re_t < 0.1$ $f = \dfrac{24}{Re_t}$

Intermediate $2 < Re_t < 500$ $f = \dfrac{18}{Re_t^{0.6}}$

Newton's Law $500 < Re_t < 2 \times 10^5$ $f = 0.44$

Turbulent $2 \times 10^5 < Re_t$ $f = 0.2$

Combining equations (18) and (19) the terminal velocity Vt can be solved.

EXPERIMENTAL APPARATUS

Two spray columns tested at the Guangzhou Institute of Energy Conversion was shown in figure 1. The first one had a 300 mm diameter and a 1380 mm height. There were 1353 holes with 1.5 mm Dia on a nozzle.

The second column had a 148 mm inner diameter and a 2600 mm height. Its nozzle had 742, 1.5 mm holes.

A scheme of the liquid-liquid spray column test loop is shown in figure 2. Hot water from an automatic control electric water heater 4 was pumped by a piston pump 5 through a turbine type flowmeter 6'into the bottom of the column. The water rose up along the column and exited at the top then returned to the water heater.

Freon 11 or 113 was used as dispersed fluid and held in a tank 9. It was pumped from the tank by another piston pump 7 through a turbine type flowmeter 8 into the nozzle at the top of the column. The freon received heat from the hot water during it fell down along the column and was discharged from the bottom. Then it was cooled in a cooler 1 before it was returned to the tank. The columns worked at a saturated pressure corresponding to its temperature.

A DPF 100-D differential meter with 0.25 degree accuracy was used as holdup ratio meassuring instrument.

There were three pairs of sightglass on the column for photo or monitoring. A 4 mm diameter steel pellet was mounted inside the column and served as reference size when measure the drop diameters on the pictures. The maximun variation of

this measurement method is 10% according to Pierce etc.[11].

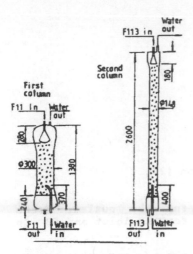

FIGURE 1. Testing Columns

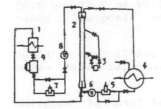

FIGURE 2. Scheme of Test Loop

All temperature measurements for the exit and entrance of
both phase and temperature field in the water column used
copper - constantan thermocouples. The temperature field was
used to determine the effective heat transfer length and the
mean water temperature.

EXPERIMENTAL RESULTS

49 runs were carried out in two spray columns. After a
equilibrium condition was established the temperatures,
flow rates as well as holdup ratio were recorded. The
records were taken three times for each run and the data
used in this report were their mean values. The time
interval between two records were 15 minutes.

33 set of data were chosen for the data correlation. Among
them 13 set of data derived from first column using F11 as
the dispersed phase. The other 9 set and 11 set of data came
from the second column using F11 and F113 respectively.

In our experiment water temperature range from 62 to 72 °C

377

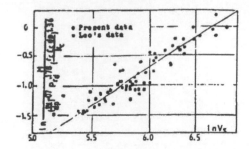

FIGURE 3. The Expression of Equation (22)

for the inlet and 52 to 66 °C for the outlet. Freon temperature range from 28 to 35 °C for the inlet and 53 to 72 °C for the outlet.

The drop diameters varied from 2.2 to 3.5 and holdup ratioes ranged from 0.018 to 0.28 for different runs.

For all runs the flow rate ranged from 0.72 to 2.67 and 0.32 to 2.8 m3/h for the continuous and dispersed phase respectively.

The higher volumetric heat transmission coefficient of 50000 to 53000 Kcal/m3h °C appeared in the runs with holdup ratio 0.2 to o.24.

Data correlation yielded an experimental form of equation (11) using least square method:

$$M = \frac{k\, d_p}{K_d} = 0.00022 Re_d^{-0.07}\, Pr_d^{0.178}\, Pe_c^{1.36} \tag{20}$$

Replace equations (12), (13) and (14) into equation (20) yields:

$$\frac{M}{(\frac{\nu_d}{d_p})^{0.07}\, Pr_d^{0.178}\, (\frac{r_c\, C_c\, d_p}{K_c})^{1.36}} = 0.00022\, V_s^{1.29} \tag{21}$$

Thus

$$Ln\, \frac{M}{(\frac{\nu_d}{d_p})^{0.07}\, Pr_d^{0.178}\, (\frac{r_c C_c d_p}{K_c})^{1.36}} = -8.422 + 1.29 Ln(V_s) \tag{22}$$

This experssion is presented as figure 3.

Present experimental data and Leo's data [5] are also presented in figure 3.

378

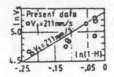

FIGURE 4. Slip Velocity
vs. Holdup
Ratio

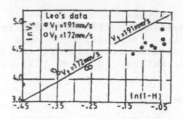

FIGURE 5. Slip Velocity vs.
Holdup Ratio

The standard deviation between equation (20) and the experimental data is 13.6% and the maximun deviation is 26%. A standard deviation between Leo's data and equation (20) is 18% and the maximun deviation is 62%. Leo etc.used benzene as dispersed phase and water as continuous phase in their experiment. The relative difference of kinematrical viscosity, conductivity, specific heat and density between Leo's benzene and our freon is 130%, 100%, 98% and 80% respectively. The drop diameters of present work were in a range of 2.2 to 3.5 mm and Leo's were 6.7 to 8.9 mm.

The computed results of Vs from equation (15) were not agree with those from equation (16) using present data. A new correlation yielded after correlating the data.

$$Vs = Vt(1-H)^{3.4} \qquad \text{for } Re_t \geqslant 500 \qquad (23)$$

When equation (23) was correlated, the terminal velocities of a single drop came from equations (18) and (19) and slip velocities of drops in the dense packing column Vs came from equation (15). The drops in the experiments were considered as rigid sphere.

Equation (23) can be rewriten as follows:

$$\ln Vs = \ln Vt + 3.4 \ln (1-H) \qquad (24)$$

This equation is drawn in figure 4 and 5. A part of present data and Leo's data are also presented in the figures. The standard deviation between equation (23) and the experimental data is 17% for present work and 28% for Leo's work respectively.

CONCLUSIONS

The results obtained in this investigation can be summarized as follows:

1. Equation (20) inferred from the present work representes the actual heat transfer characteristics of the liquid spray columns with water as a continuous phase and may be used for

379

sizing purposes.
2.   Correlation (16) proposed by Richardson for predicting the slip velocity of the solid particles in the sedimentation beds is not valid for liquid-liquid spray columns. Rather, equation (23) may be used for the purposes.

NOTETION

a	Heat transfer coefficient, Kcal/m s C
A_s	Total transfer area per unit column volume, 1/m
C	Specific heat, Kcal/kg
d_p	Drop diameter, m
f	Drag coefficient
H	Holdup ratio, volumetric fraction of dispersed phase in a column
k	Interfacial heat transmission coefficient, Kcal/m s C
K	Heat conductivity, Kcal/m s C
LMTD	Log mean temperature difference, C
M	$= k\, d_p / K_d$ Dimensionless expression of interfacial heat transmission coefficient
n	Constant in equation (16)
Nu	Nusselt number
Pr	Prandtl number
Pe	Peclet number
Q	Heat transfered, Kcal/s
r	Gravity, kg/m^3
Re	Reynolds number
Re_t	Reynolds number at terminal velocity
U_v	Volumetric heat transmission coefficient, $Kcal/m^3 s\ {}^{\circ}C$
V_t	Terminal velocity of a single drop, m/s
V_s	Slip velocity, m/s
ρ	Density, $kg\ m^2/m^4$
μ	Dynamic viscosity, kg/m s
ν	Kinematrical viscosity, m^2/s

SUBSCRIPTS

i	Internal
c	Continuous phase
d	Dispersed phase
o	External

REFERENCES

1. Ferrarini R., VID-Forschungaheft 551, 1972.
2. Letan R. and Kehat E., AIChE J. v14,n3, 1968, pp 398 - 405.
3. Letan R., Transaction of the ASME Vol.130, 1981, pp 586.
4. Moresco, L.L. and Marschall E., J. of Heat Transfer, v102,1980, pp 591 - 601.
5. Leo Garwin and Buford D. Smith., Chemical Engineering Progress, Vol.49, No.11, 1953.
6. Treybal, R.E., "Liquid Extraction" (1st Edition) McGraw Hill, New York, 1951, pp 328.
7. Teynham Woodward, Chemical Engineering Progress, Vol.57, No.1, January 1961.

8. L.Lapidus., and J.C.Elgin, AIChE J. Vol.3, No.1, 1957, pp 63.
9. J.F.Richardson and W.N.Zaki, TRANS. INSTN CHEM. ENGRS, Vol.32, 1954, pp 35 - 53.
10. R.Letan, "Design of a Particulate Direct Contact Heat Exchanger: Uniform, Countercurrent Flow" Presented at the ASME-AIChE Heat Transfer Conference.St.Louis.Mo., Aug. 9-11, 1976.
11. D.R.Pierce, O.E.Dwyer and J.J. Martin, AIChE J. Vol.5, No.2, June 1959.

Steam Condensation in Ordinary Plate Heat Exchangers

ZHONG-ZHENG WANG, ZHEN-NAN ZHAO, and QING-CHUN LU
Thermophysics Engineering Department
Tianjin University
30072, Tianjin, PRC

ABSTRACT

The paper identifies the remarkable energy saving of ordinary type of plate heat exchangers by means of testing study when the exchangers are used in water—stram condensation. It analyzes the relation of steam condensation heat transfer coefficient with steam velocity, pressure and condensation temperature difference under the condition of high velocith flow in plate heat exchangers. A correlation formula for solving condensation heat transier coefficient has been obtained. The effects of concurrent and countercurrent flow on pressure drop and steam condensation coefficient are also analyzed.

INTRODUCTION

Plate heat exchangers arising in recent several decades have obviously exhibited advantages in use of liquid—liquid heat transfer and shown tendency in place of shell—tude heat exchangers in several fields. inasmuch as energy saving problem has been severe and plate heat exchangers have the characteristics of high efficiency, the plate heat exchanges originally used in liquid—liquid heat transfer have also shown marked energy saving effects when their application is extended to condensation heat transfer. But until now special published papers of research on condensation in plate heat exchangers are a few (1, 2, 3, 4) and reports on testing results are not included in detail. The reason for that basically is the complexcity of fluid flow and phase change in plate heat exchangers so that testing measurement, theortical analysis and calculation are rather difficult. The authors did water—steam condensation testing at various steam pressure and flow rate with a BR—05 type of plate heat exchanger oreginally used in liquid—liquid heat transfer (called "ordinary type", here). The paper reports experimental equipment, calculation of total heat transfer and steam conednsation heat transfer coefficient, correction of pressure drop. It analyzes the relation of steam condensation coefficient with steam velocity, pressure and condensation temperature difference. The effect of concurrent and countercurrent flow on pressure drop and steam condensation coefficient is also analyzed. Finally, the remarkable energy saving of ordinary plate heat exchangers and the problem which is needed to solve are pointed out. Meanwhile, the prospects of appliation are predicted.

EXPERIMENTAL EQUIPMENT AND OPERATING CONDITION

The test system of water—steam condnsation which consisist of electrical boiler, BR—05 plate heat exchanger and back condenser is shown in Fig. 1

1. Boiler
2. Valve
3. Thermometer
4. Meter for steam quality
5. Manometer
6. U—tube of manometer
7. Meter for liquid column
8. Plate heat exchanger
9. Back condenser
10. Measure container
11. Container
12. Pump
13. Container for counterflush

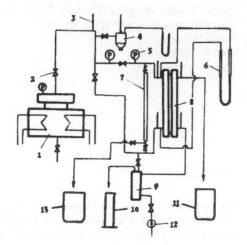

Fig. 1 schematic diagrem of the test system.

There are three channels in the plate heat exchanger, which includes the steam channel in the middle and the colling water on two sides. The plate pattern is horizontally herringbone with 90 ° (see Fig. 2). The heat transfer projected area is 0.045m² for one plate. The distance between two plates is 2.5mm. The cross section area of a channel is 0.0003625m² The plate material is stainless steel. To elimiante the influence of noncondensable gas on steam condensation heat transfer, a tube which is used for counterflush is mounted for the purpose of air drain.

The operating conditions of water—steam condensation testing are:

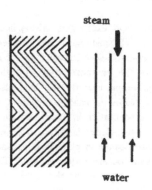

Fig. 2 plate pattern and channels.

1. keep the steam inlet pressure of the exchanger at 0.13, 0.16, 0.2 MPa respectively, change flow rate of the cooling water and make the condensate outlet temperature of the exchanger possibly reach to saturated temperature at outlet pressure for water—steam countercurrent flow.

2. It is the same as 1), but keep the steam inlet pressure of the exchanger at 0.2Mpa for water—steam concurrent flow only.

DATA ARRANGEMENT

Based on the basic equation of heat transfer, total heat transfer coefficient is calculated as:

$$U = \frac{Q}{F \cdot \Delta t_m} \qquad (1)$$

The Q is arithmetic average value of the heat released by steam and the heat received by cooling water and relative error $\Delta Q = ((Q_s - Q_c)/Q) < 5\%$ is required. F is heat transfer projected area. Δt_m is LMTD calculated by inlet and outlet saturated temperature of steam and inlet and outlet temperature of cooling water.

Steam condensation heat transter coefficient h_s will be obtained from relation of thermal resistance:

$$h_s = \left(\frac{1}{U} - \frac{1}{h_c} - \frac{\delta_p}{K_p} \right)^{-1} \qquad (2)$$

Convective heat transfer coefficient h_c is calculated by using convective heat transfer correlation formala obtained by us in Ref. (10). Determination of heat conductivity K_p of the plate is that evaluation of average wall temperature $\bar{t}_w$ at the thickness in the middle of the plate at first:

$$\bar{t}_w = \frac{Q}{F} \left(\frac{\delta_p}{2K_p} + \frac{1}{h_c} \right) + t_c \qquad (3)$$

and then k_p value would be found from table of thermal physics data (8) t in equation (3) is average temperature of water inlet and outlet.

Calculation of pressure drop Δp is dependent upon $\Delta p'$ obtained from reading of a mercury manometer and the corection of influence caused by ΔL_w, the difference between column of condensate on ineterface of mercury in two capillary, and ΔL_p, the difference betwween steam inlet and outlet height, that is

$$\Delta p = \Delta p' - (\Delta L_w + \Delta L_p)/13.6 \qquad (4)$$

RESULTS AND DISCUSSION

Following view—points have been obtained after the test data of water—steam condensation measure from the 78 runs of testing were processed and analyzed.

1. The grid type of corrugated channel and high velocity flow of steam induce intense two phase turbulent flow in plate heat exchangers. Consequently the steam condensation heat transfer coefficient increase greatly. The testing results show the condenstion heat transfer coefficient in the plate exchanger is 2 to 3 times as higher as that one of filmwise condensation on vertical surface in extensive space. The highest value of coefficient reach to 4.5×10^4 w/m^2 °C (correspondent to steam inlet velocity of 32.7m/s and mean condensation temperature difference of 6°C) in the testing, which is 4 times the film—wise condensation coefficient on vertical surface in extensive space under the same condition.

2. As to calculation of steam condensation coefficient in plate heat exchangers, a proper theoretical or empirical formula has not been found because of complexity of this problem. L. D. Boyko [3] proposed a calculation formula of condensation coefficient averaged along plate length :

$$Nu = C \cdot Re_{2.0}^{0.73} \cdot Pr_2^{0.43} \frac{1}{2} \left[\sqrt{1 + x_1 \left(\frac{\rho_\ell}{\rho_v} - 1 \right)} + \sqrt{1 + x_2 \left(\frac{\rho_\ell}{\rho_v} - 1 \right)} \right] \tag{5}$$

It can be seen by analysing formula (5) that steam condensation heat transfer coefficient would increase with enhancement of steam velocity. However, this conclusion obtained from analysis of testing is correct only under certain conditions.

3. The testing is proceeded under various conditions. By analysis and comparison of all the test data. an important law has been revealed: not only does steam condensation coefficient h_s depend on steam velocity in channel, but also it is related to condensation temperature difference $\Delta t =$ ($t_s - t_w$). A relation between them is existent. While all the results from different operating conditions are shown on two—logarithm coordinate as Fig. 3, a linear relstion of Nu_s with Re_L of condensate and ($Cp\Delta t/H$), called supercooling parameter, can be seen. Re_L can be replaced by Re_{sl} of steam at inlet whereas relation is the same since all the testing is limited to full condensation of

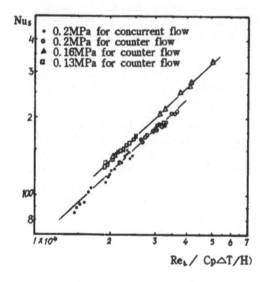

Fig. 3 $Nu_s = f(Re_L, Cp\Delta T/H)$

saturated steam. Obviously, variation of condensation temperature difference $\triangle t$ should be very small for the purpose which is condensation heat transfer coefficient h_s increases with enhancement of steam velocity.

4. Condensation heat transfer coefficient is influenced by steam pressure to a certain extent. In Fig. 3 the upper Line is correspondent to an averaged condensation pressure of 0.12–0.14MPa, and the lower one to pressure of 0.172–0.198MPa which shows Nu_s or h_s decreases with a raise of condensation pressure. The conclusion is the same if formula (5) or following E. P. Ananiev's formula

$$h = h_f \sqrt{\frac{P_f}{P_m}} \qquad (6)$$

of local heat transfer coefficient [1] is analyzed. The tendency of pressure effect has appeared in the testing although the testing has not been operated in wide range because of the limitation of equipment condition.

5. It can be seen from Fig. 3 that arrangement of flow direction will not influence the law shown by Fig. 3 so the same correlation formula can be used. in consideration of the difference in pressure drop through the channel, concueernt arrangement is favourable for most cases of operation.

6. Heat transfer effect intensified is present on both sides of liquid and steam when plate heat exchangers are used in steam condensation. Usually, steam condensation transfer coefficient is above 1×10^4 w/m². °C and the thermal resistance on steam side is very small and correspondent to the resistance of plate material. Since the maximum velocity of colling water is around 0.42 m/s and correspondent heat transfer coeffieient on colling water side is 1.56×10^4 w/m² °C. Meanwhile, h_s is 2.35×10^5 w/m² °C(steam velocity of 37.1m/s) and total heat transfer coefficient reaches to 7076w/m² °C. which is much greater than in shell–tube heat exchanger. In point–View of heat transfer dominative thermal

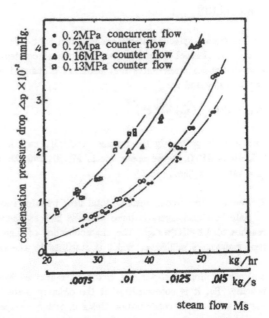

Fig. 4 condensation pressure drop changes with steam flow Ms. and inlet pressure P_{si} for full–condensation.

resistance is on water side under the condition of high velocity in plate heat exchanger. Therefore, if water velocity is raised, heat transfer would be intensified further.

7. Steam condensation pressure drop $\triangle$p has relation to steam flow rate, steam pressure and flow direction. Steam pressure increasses notably with increase of steam flow rate and pressure drop differs as inlet pressure is different (Fig. 4). If flow rate is kept to the same, the higher the steam inlet pressure the smaller the pressure drop beeause the steam density increases. In addition, steam pressure drop of countercurrent flow is larger than that of concunent flow at the same steam inlet pressure and flow rate. For instance, compared with concurrent flow the averaged pressure drop of countercurrent flow increases 17%. The reason for this is that it can be seen from fluid temperature distribution along plate length (Fig. 5) that temperature difference of two fluids at steam inlet is small for cunter—

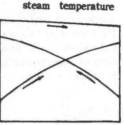

steam temperature

Plate length

Fig. 5 Temperature differece comparison between counter flow and concurrent flow

current flow so that most of steamcondenses on lower part of plate. For concurrent flow, a great part of steam condenses reversedly on upper part, hence averaged steam velocity along plate length is lower than that for contercurrent flow and pressure drop decreases.

CONCLUSION

1. The factors affecting steam condensation heat transfer in plate heat exchangers are steam velocity in channel, temperature difference of condensation and steam pressure etc. When saturated steam condensates fully under an unchanged pressure, a general correlation formula can be expressed as:

$$Nu_s = C\, Re_L{}^n\, (C_p \Delta T/H)^{-n} \qquad (7)$$

for the palte pattern in the testing, n=0.961; c=0.1, at averaged condensation pressure of 0.12—0.14MPa(i. e, the upper line in Fig. 3), C=0.09, at averaged pressure of 0.172—0.198 (i. e. the lower line).

2. Based on the testing results, total heat transfer coefficient for water—steam condensation of plate heat exchangers is three times as high as shell—tube heat exchangers. The heat flux reaches to $3.8 \times 10^5 w/m^2$. The characteristics of high efficiency and compactness for plate heat exchangers still have, which is testifiled by the testing.

3. Pressure drop on steam side in the plate heat exchanger is big. especially at rather higher velocity. But it is noteworthy that the ordinary plate heat exchangers are very suitable for use in some steam condensation fields in which no special requirement for pressure drop is needed because their heat transfer effects increase evidently.

4. When the plate heat exchangers are used, if a given duty is statisfied, operating conditions with concurrent flow are selected rather than with countercurrent flow since pressure drop

with countercurrent flow is larger, which causes the reduction of effective heat transfer temperature difference.

5. It gives a direction that plate heat exchangers are extended to use in wide liquid—vapor (gas) heat transfer fields. But, special plate pattern should be designed, and ports and channel construction improved (5).

6. The problem pertaining to theory and calculation of plate heat exchangers with phase change is still unsolved. Research related to the characteristics of two phase flow and heat transfer is needed.

ACKNOWLEDGMENT

Thanks for Mr. Li Lu Jun, Wu Jin Siang and Chi Chang He's help during construction of experimental equipment and testing.

REFERENCES

1. Cooper. A., Condensation of Steam in Plate Heat Exchangers, AICHE Symp. Ser. 138, Vol. 70.

2. Tovazhyanskiy, L. L. et al., Coeffcients of Heat Transfer for Condensation of Low—pressure Steam in Plate Condensers with Slotlike Channels in a Grid Pattern, Heat Transfer Soviet Reash, Vol. 9, No. 2, March—April, 1977.

3. L. D. Boyko et al. Heat Transfer and Hydraulic Resistance Duning Condensation of Steam in a Herizontal Tube and in a Bundle of Tubes. Int. J. Heat Mass Transfer Vol. 10 (1967).

4. Kumar. H., Condensation Duties in Plate Heat Exchangers, Symposium on Condenser: Theory and Practice, 1983/3/22—23, Univ. of Manchessters.

5. Uehara. H., et al. Heat Transfer Coefficients for a Vertical Fluted—plate Condenser with R113 and 114, Inter. Joural of Referigeration, vol. 8, No. 1, 1985.

6. Collier. J. G., Convective Boiling and Condensation, Mcgraw—Hill, Inc., 1972.

7. Wang Zhong—Zheng, Zhao Zhen—Nan., Heat Transfer at Low Re in a Grid—type of Corrugated Channels, 1988 (to be published).

8. Welty. J. H., Engineering Heat Transfer.

9. Westwater. J. W., Compact Heat Exchangers with Phase Change, 8th Inter. Heat Transfer Conference, San Francisco, 1986.

10. Wang Zhong Zheng, Zhao Zhen Nan et al., An Experimental Study on Plate Heat Exchangers for Building Heating with Geothermal Water, Proceeding of the 7th Geo-thermal Workshop, New Zealand, 1985.

BOILING AND EVAPORATION

Enhanced Heat Transfer Tubes for the Evaporator
of Refrigeration Machines

W. NAKAYAMA, K. TAKAHASHI, and H. KUWAHARA
Mechanical Engineering Research Laboratory
Hitachi, Ltd.
502 Kandatsu, Tsuchiura, Ibaraki, Japan

INTRODUCTION

The purpose of the present study is to develop high performance heat transfer tubes for the flooded-type evaporator of
refrigeration machines. The authors have already reported in
the previous papers [1,2] that nucleate boiling heat transfer
is significantly enhanced on the surface named Thermoexcel-E.
Thermoexcel-E is composed of parallel tunnels runnig under-
neath the surface and pores which serve as communication ports
between the tunnels and the exterior liquid pool. Since then,
attempts have been made to obtain further enhancement of heat
transfer in different operative environments by introducing
modifications of the surface structure. A companion paper
reported earlier [3] reports the improvement of the performance produced on the modified version called Thermoexcel-E2 in
low system pressure environment and small wall superheats.

With the achievement of significant heat transfer enhancement
on the shell side, it has become necessary to direct attention
to the heat transfer resistance on the tube side. Among many
schemes of enhancement of heat transfer for single-phase heat
transfer inside the tube, the one by spiral ribs offers a
great advantage in the economy of manufacturing tubes, particularly when the so-called roll formation is employed. In the
roll formation the spinning rollers are applied on the outer
surface of the tube to produce spiral ribs on the internal
surface of the tube. The method ensures a very fast speed of
manufacturing. As for the heat transfer performance of ribs
produced by this method, there are several reports in the
literature [4 - 6].
In the meantime the authors have investigated the performance
of three-dimensional ribs, in an expectation to obtain an
equally high heat transfer enhancement with less penalty of
pressure drop [7]. The manufacturing of three-dimensional
ribs was done by the work of internal mandrels inserted inside
the tube, which has proved not attractive from the commercial
viewpoint. In the present study, the roll formation is employed to form discrete protuberances on the internal surface
of the tube, with the rolls equipped with the teeth which are
to be pressed against the external surface of the tube. The
roll formation method, by its working principle, causes a loss
of the Thermoexcel structure on the external surface of the
tube on the spots where the teeth of the rollers are applied.

Consequently, one has to take a balance between the enhancement on the inside and the reduction in the enhancement on the outside of the tube. This paper reports the results of the study on the optimization of the placement pitch of internal protuberances.

TEST TUBES AND METHOD OF EXPERIMENT

Internal Surface Structure

Fig.1 shows the two types of ribbed tube tested in this study, and the schemes of providing ribs or protuberances on the internal surface of the tube. Table 1 shows the dimensions of the surface structure of the type Fig.1(b), where the term 'rib' is synonymous with 'protuberance', and β is the angle formed between the neighboring teeth on the forming disk. The circumferential pitch of the protuberances, z in Fig.1(b), is correlated with the angle β . The axial pitch, p, is varied by changing the angle of the disk against the tube axis. The inside diameter is the maximum diameter, namely, the distance between the spots located off the protuberances. Table 2 shows the dimensions of the two-dimensional ribs of Fig.1(a). The material of the tubes is copper. Fig.2 shows the pictures of some of the tubes.

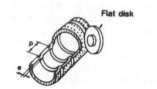
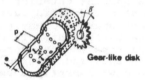

(a) 2-D ribs   (b) 3-D ribs (protuberances)

Fig.1 Schemes of manufacturing ribbed tubes

Table 1   Dimensions of 3-D ribs

	Rib height e (mm)	Axial pitch p (mm)	Disk angle β°	Inside diameter Di (mm)
C-1	0.44	7	10	14.4
C-2	0.44	7	15	14.4
C-3	0.45	7	20	14.4
C-4	0.46	7	15	15.8
C-5	0.52	7	15	15.8
C-6	0.57	7	15	15.8
C-7	0.50	5	15	15.8
C-8	0.54	10	15	15.8

Table 2 Dimensions of 2-D ribs

	Rib height e (mm)	Axial pitch p (mm)	Inside diameter Di (mm)
D-1	0.5	7	14.4
D-2	0.5	14	14.4

Tube No.

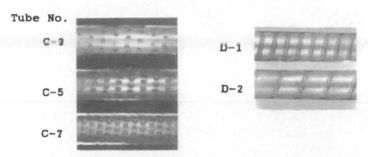

C-3   D-1

C-5   D-2

C-7

Fig.2 Photographs of internal surface structure

External Surface Structure

Fig.3 shows the sketch of the structure of Thermoexcel-E2, where the pore is partially blocked by the protuberance called 'small rib' in the figure.   The partial blockage is intended to enhance the stability of vapor packets in the subsurface cavities, and indeed, its effectiveness in heat transfer enhancement in low system pressure environments and at small ΔT has been proved [3].

Fig.4 shows the photograph of a sample of the test tubes. The outside diameter of the tube is 18 mm.   The experiments were conducted using R-11 as the working fluid, and setting the system pressure at 0.04 MPa, which corresponds to 0° C of the saturation pressure of R-11.

The Method of Experiment

In order to measure the effectiveness of internal ribs in the tube-side heat transfer, the ribbed tubes without Thermoexcel structure on the outer surface were used.   Constantan wires having electrical insulation coating were wound around the tube to be used as the heater.   Copper-constantan thermocouples were used to measure wall temperatures, the inlet and outlet temperatures of the water.   The estimated accuracies of the temperature and heat flux measurements are ± 0.1 °C and ± 20 W/m^2 , respectively.

395

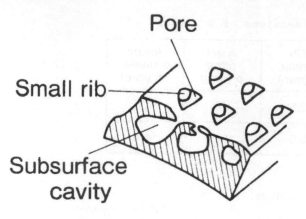

Fig.3 Microporous structure on the outer surface of tube

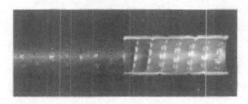

Fig.4 Photograph of the tube having simultaneous enhancement
of heat transfer on the shell- and tube-sides

The pressure drop data were obtained by providing the pressure
taps at 8 locations on each cross section of the inlet and
outlet of the test tube.   The pressure was measured by means
of the U tubes.

RESULTS AND DISCUSSION

Performance of Internal Ribs

In reporting the experimental data the heat transfer coefficient, α , is defined on the basis of the surface area of the
tube having a straight internal surface with the diameter D_i .
Namely, the increase of real surface area produced by the ribs
is not taken into account.   The data of heat transfer and
pressure drop will be presented in terms of the Nusselt number,
the friction factor, and the Reynolds number.
Their definitions are given as follows.

$$Nu \equiv \frac{\alpha D_i}{k} \tag{1}$$

$$f \equiv \frac{\Delta P\, D_i}{L(\rho v^2/2)} \tag{2}$$

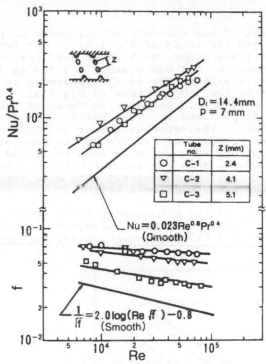

Fig.5 Nusselt number and friction factor in 3-D rib tubes;
the effect of the placement pitch of protuberances along
the path of rolling disk

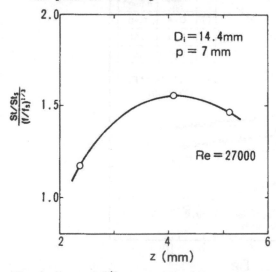

Fig.6 The $St/f^{1/3}$ ratio and the pitch z

$$Re \equiv \frac{v \, D_i}{\nu} \qquad\qquad (3)$$

Fig.5 shows the effect of the circumferential pitch of protuberances, z, in the 3-D rib tubes of Fig.1(b).   The tube
no. C-2, where z = 4.1 mm, shows the highest heat transfer
performance.   The tube no. C-1 has a small z (2.1 mm); its
friction coefficient is high at the level comparable to the
friction coefficient in the 2-D rib tube of Fig.1(a).
Fig.6 shows a cross section of the data, in terms of the
parameter St/ St_s /$(f/f_s)^{1/3}$ and z.   The suffix s is used for
the smooth tube values.   The Reynolds number is 2700, corresponding to the water velocity of 2.2 m/s in the tube of 14.4
mm diameter.   The optimum value of z is located around 4 mm.
This optimum z is produced when the angular pitch of the teeth
β is set at 15° .   All the subsequent data were obtained
by the tubes processed with this teeth angle.

Fig.7 shows the effects of the rib height, e, on heat transfer
and pressure drop.   The Nusselt number increases with the increase in e from 0.46 to 0.52 mm, however, a further increase

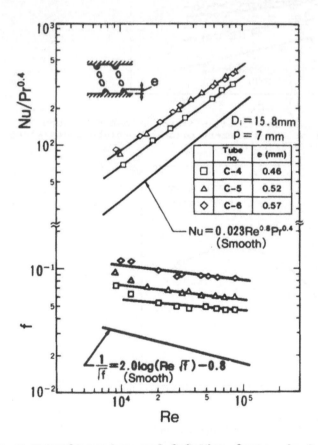

Fig.7 Nusselt number and friction factor in 3-D rib tubes;
the effect of rib height

398

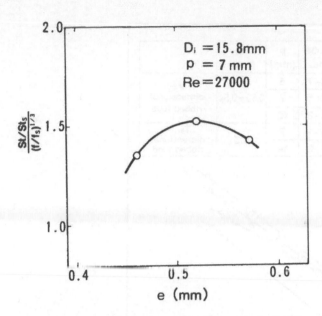

Fig.8 The St/f$^{1/3}$ ratio and the rib height e

of e to 0.57 mm invites the rise of friction factor alone.
Fig.8 shows a cross section of the data in terms of the St/f$^{1/3}$
ratio and the rib height e.   It is found that the optimum
value of e is around 0.5 mm.

The variable parameter of the data plotted in Fig.9 is the
axial pitch of the ribs.   The data of the 2-D rib tubes are
also shown.   The 3-D ribs reduce the friction factor, while
producing a comparable effectiveness in heat transfer enhancement to that obtained with the 2-D ribs.   For instance, the
comparison of the data of the tube no.C-5 and those of the
tube no. D-1 reveals that almost equal Nusselt numbers are
obtained with both tubes, however, the friction factor in the
2-D tube is higher than that in the 3-D tube by about 70 %.
Fig.10 shows the St/f$^{1/3}$ vs. the axial pitch p.   The figure
shows the comparison of heat transfer performance on the basis
of equal pumping power for the tubes C-5 and D-1.   The heat
transfer performance of the 3-D rib tube (C-5) is higher than
that of the 2-D rib tube by 18 %.

Nucleate Boiling Heat Transfer On The Outer Surface

Experiments were conducted using the tubes having partial loss
of Thermoexcel structure due to the roll formation of internal
ribs.   The boiling fluid was R-11, and the system pressure is
set at 0.04 MPa.   Fig.11 shows the scene of boiling on the
tube C-5.   Visual inspection of the actual boiling surface
reveals that the density of bubble formation sites is reduced
in the area where Thermoexcel structure has been crushed under
the teeth of the roller.   Fig.12 shows the boiling curves
obtained with the tubes having different proportions of the

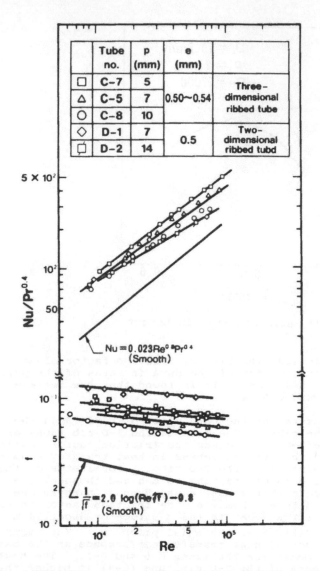

Fig.9 Nusselt number and friction factor in ribbed tubes;
the effect of axial pitch of ribs

crushed area.   As the axial pitch of ribs is decreased, the
crushed area increases, hence, the heat transfer coefficient
on the outer surface of the tube decreases.   Compared with
the data of the tube having no crushed area (p = ∞) on the
basis of equal ΔT, the data of p = 14 mm and 7 mm show the
decrease of heat flux by 17 % and 30 %, respectively.   This
reduction of heat flux corresponds roughly to the proportion
of lost area to the total surface area, implying that the loss
of effective boiling area is a cause of the heat transfer
deterioration.

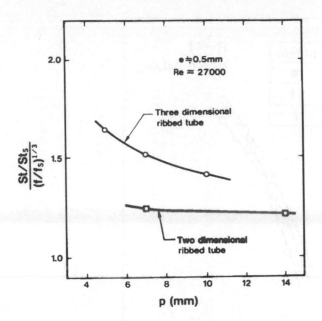

Fig.10 The $St/f^{1/3}$ ratio and the axial rib pitch p

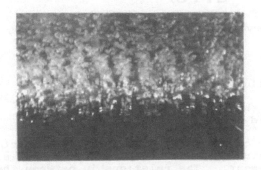

Fig.11 Boiling of refrigerant R-11 on the tube having enhanced
surfaces; p = 7 mm, q = 2.1 x 10 W/m²)

Optimization Of Axial Pitch Of Ribs

The design analysis is performed to show the effectiveness of
the 3-D rib tubes, where the determination of the axial pitch
of ribs is a focal issue. The reference heat exchanger is
composed of Thermoexcel-E2 tubes with no internal ribs. The
specifications on heat load and pumping power of water are
assumed given; also, the following parameters are fixed in the
design analysis.

the outer diameter of the tube   18 mm
the inner diameter of the tube   15.8 mm

401

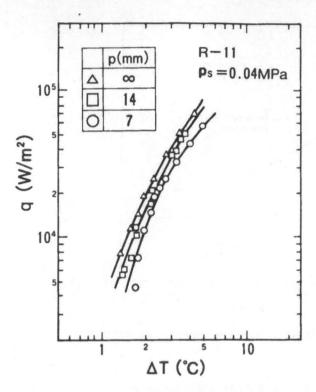

Fig.12 Boiling curves obtained with the tubes having internal ribs

the reference value of water velocity   2.5 m/s
                                          (Re = 3 x 10^4)
the pressure in the shell   0.04 MPa
the working fluid   R-11

The fouling factor on the water side is not taken into account here to best illustrate the counterplay of the tube-side and shell-side enhancement. The relationship between the heat transfer coefficient on the shell-side and the axial pitch of ribs is determined on the basis of experimental data, or from the ratio of crushed area to total area where the data are not available. The tube-side heat transfer coefficient is based on the experimental data shown in Fig.9.

Fig.13 shows the ratio of heat transfer area to that of the reference heat exchanger, A/As. The horizontal axis shows the axial pitch of ribs. The curves are the results of design analysis for the cases of heat flux q = 2.0 x 10^4 and 3.0 x 10^4 W/m². The maximum saving of heat transfer area to be gained by the use of the 3-D rib tubes is 18 % at q = 3.0 x 10^4 W/m², and the optimum pitch exists near 7 mm. This optimum value of the rib pitch depends little on the heat flux. In the case of the 2-D rib tubes the optimum rib pitch seems to be located near 14 mm, where the area saving is about 14 % at q = 3.0 x 10^4 W/m².

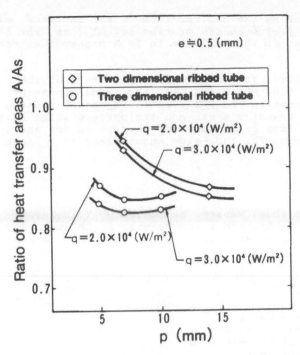

Fig.13 Effect of the axial rib pitch on the reduction of heat
transfer area of the R-11 evaporator

CONCLUSIONS

Enhancement of heat transfer is achieved on both the shell-
side and tube-side of the evaporator of refrigeration machine.
The shell-side enhancement is produced by the modified version
of Thermoexcel-E which has proved its effectiveness in promotion of nucleate boiling heat transfer particularly in low
pressure environment and at small ΔT.   Three dimensional
protuberances are provided on the internal surface of the tube
to enhance single-phase heat transfer to water.   The optimization of the surface structure on each of the shell- and
tube-side has been achieved by conducting laboratory experiments.   Besides, the method developed to manufacture enhanced
tubes at low cost introduced a problem of striking a balance
between the loss of external porous structure and the gain in
the tube-side heat transfer.   The following conclusions are
obtained.

(1) The optimum values of the height of the protuberance and
the placement pitch of protuberances in the circumferential
direction are 0.5 mm and 4 mm, respectively.   This conclusion
holds for the Reynolds number in a range $8 \times 10^3 - 8 \times 10^4$.

(2) The pressure drop in the the 3-D rib (protuberance) tube
is less than that in the 2-D rib tube which is produced by the
similar roll forming method.   The comparison of heat transfer
performance on the basis of equal pumping power illustrates

the advantage gained by the 3-D rib tube. For instance, with
the axial pitch of protuberances or ribs set at 7 mm, the heat
transfer coefficient in the 3-D tube is 18 % higher than that
in the 2-D tube.

(3) The design analysis for the evaporator of R-11 at the
system pressure of 0.04 MPa shows that the axial placement
pitch of protuberances in the 3-D tube is important for the
saving of the heat transfer area, and its optimum value exists
at 7 mm, where the loss of Thermoexcel surface on the shell-
side is compensated by the tube-side enhancement in a right
proportion.

Acknowledgement

The authors wish to thank Messrs. S. Sugimoto, Y. Nakayama,
Tsuchiura Works, Hitachi, Ltd., and Messrs. K. Ohizumi, Y.
Shinohara, H. Furuto, Hitachi Cable Ltd., for their encourage-
ment of this study.

Nomenclature

A = heat transfer area m
D_i = internal diameter of tube m
e = rib (protuberance) height mm
f = friction coefficient
k = thermal conductivity W/(m K)
L = length of tube m
Nu = Nusselt number
p = axial pitch of protuberances or ribs mm
p_s = system pressure MPa
Pr = Prandtle number
Δp = pressure drop Pa
Re = Reynolds number
St = Stanton number
ΔT = wall superheat °C
v = average velocity of water m/s
q = heat flux W/m²
z = placement pitch of protuberances along the spiralling
     path of forming roll mm
$\propto$ = heat transfer coefficient W/(m²K)
β = angular pitch of the forming teeth deg
ν = kinematic viscosity m²/s
ρ = density kg/m³
Suffix
s = smooth tube

References

1. Nakayama, W. Daikoku, T., Kuwahara, H., and Nakajima, T.,
   Dynamic Model of Enhanced Boiling Heat Transfer on Porous
   Surfaces, Part I: Experimental Investigation, ASME J. Heat
   Transfer, Vol.102, No.3 (1980) pp.445-450.
2. Nakayama, W. Daikoku, T., Kuwahara, H., and Nakajima, T.,
   Dynamic Model of Enhanced Boiling Heat Transfer on Porous
   Surfaces, Part II: Analytical Modeling, ASME J. Heat

Transfer, Vol.102, No.3 (1980) pp.451-456.

3. Nakayama, W., Kuwahara, H., Nakajima, T., and Yoshida, Y.,
   Evaporator Tubes with Enhanced Boiling Surfaces: The
   Effect of Partial Blockage of Pores with Protuberances,
   International Symposium on Phase Change Heat Transfer,
   May 20-23, 1988, Chongqing, China.

4. Withers, J. G., Tube-Side Heat Transfer and Pressure Drop
   for Tubes Having Helical Internal Ridging with Turbulent/
   Transitional Flow of Single-Phase Fluid. Part I. Single
   Helix Ridging, Heat Transfer Engineering, Vol.2, No.1
   (1980) pp.48-58.

5. Li, H. M., Ye, K. S., Tan, Y. K.,and Deng, S. J., Investigation on Tube-Side Flow Visualization, Friction and Heat
   Transfer Characteristics of Helical Ridging Tubes, HEAT
   TRANSFER-1982, Vol.3, pp.75-80, 1982, Hemisphere Publishing Corporation.

6. Nakayama, W., Takahashi, K., and Daikoku, T., Spiral
   Ribbing to Enhance Single-Phase Heat Transfer Inside Tubes
   Proc. ASME/JSME Thermal Engineering Joint Conference,
   Vol.1 (1983) pp.365-372.

7. Takahashi, K., Nakayama, W., and Kuwahara, H., Enhancement
   of Forced Convective Heat Transfer in Tubes Having Spiral
   Ribs of Three-Dimensional Structure, Trans. JSME, Series B
   Vol.51, No.461 (1985) pp.350-355.

Two-Phase Flow Boiling Instabilities and Oscillation Thresholds in a Vertical Single Channel with Heat Transfer Enhancement

A. MENTES*, S. KAKAÇ, and T. N. VEZIROĞLU
Clean Energy Research Institute
University of Miami
Coral Gables, Florida 33124, USA

H. Y. ZHANG
Beijing Institute of Technology
Beijing, PRC

X. J. CHEN and Z. H. LIN
Engineering Thermophysics Research Institute
Xian Jiaotong University
Xian, PRC

INTRODUCTION

The phenomenon of thermally induced two-phase flow instability is of interest for design and operation of many industrial systems and equipment, such as steam generators, thermosiphon reboilers, refrigeration plants, and other chemical process units.

Because of the economical reasons and space constraints, these is a tendency to reduce the size of the heat exchanging systems by increasing the heat transfer coefficient. This is being achieved by using two-phase flows, and enhanced surfaces, such as fin-like prodrudings, grooves in the heat transfer surfaces, and also roughened and specially treated surfaces.

Oscillations of flow rate and system pressure are undesirable as they can produce mechanical vibrations, problems of system control. Under certain circumstances, large flow oscillations can lead to burn-out and failure due to solely by the increase in wall temperature. But a more likely cause of failure would be thermal fatigue resulting from a continual cycling of the wall temperature.

The effect of various parameters, such as inlet and exit restrictions, inlet subcooling, heat flux on two-phase flow instabilities in a single and parallel channel upflow system with smooth tubes have been studied experimentally and theoretically [1-7].

In recent years, there has been an increased emphasis on techniques to enhance the two-phase flow heat transfer, an survey of the subject is presented by Bergles and Joshi [9,10]. The resulting increase in the use of heat transfer augmentation and the engineering importance of the subject caused the present investigation to be taken up in order to study the effect of different heater surface configurations on two-phase flow instabilities.

EXPERIMENTAL INVESTIGATION

An open-loop, forced convection boiling, upflow system, operating between two constant pressures has been designed and built to generate the main types of two-phase flow oscillations. Tests were carried out using Freon-11 as the test fluid. The set up allows the determination of the effect of various parameters, such as heat input, mass flow rate, heater surface characteristics and inlet temperature. A schematic diagram of the

---

*The Kent Corp., Birmingham, Alabama.

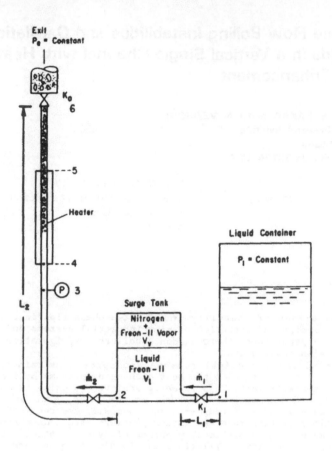

Figure 1. Schematic Diagram of the System for Mathematical Modeling.

experimental set up, is shown Figure 1. More detailed information on the set up and the instrumentation can be found in Ref.[1].

Six different augmented heater surfaces have been prepared and their description is summarized in Table 1. The experiments have been performed with each tube at a constant inlet temperature and constant heat flux for various values of mass flow rate. This had been repeated for six different values of heat input and for various values of inlet temperatures. Each experiment was composed of a sufficient number of tests to cover a wide range of boiling regimes, from single-phase liquid up to very high qualities; mass flux, heat input and inlet temperature are varied from 35 kg/m²s, 0 W/m² and -9°C to 680 kg/m²s, 3x10⁴ W/m² and 38°C respectively.

Each heater tube had of 7.493 mm inside and 9.525 mm outside diameter and was 605 mm long; these were Nichrome tubes. Tubes are classified according to their effective diameter which is defined as

$$de = \sqrt{\left(\frac{4V}{\pi L}\right)} \tag{1}$$

where V is the net inside volume and L is the length of the heater tube.

For a given heater tube, several sets of experiments with various heat inputs and/or inlet temperatures were conducted. Each set was composed of a sufficient number of tests to cover the available flow range. Stability boundaries were located for each case. Oscillations were identified by cyclic variations in pressures, flow rate and temperatures and, also by observing the transparent section of the set up, the pressure gauge pointers and the recorder. In defining the boundaries short-life transients were ignored, only the sustained oscillations were considered. Detailed information on the experimental procedure and experiments is given in reference [11].

Table 1. Description of the Heater Tubes.

Tube	Description of the tube	de mm
A	Bare	7.493
B	Threaded, 7.938mm - 16 threads per 25.4 mm	7.619
C	With internal spring of 0.794 mm wire diameter and 19.05 mm pitch	7.446
D	With internal spring of 0.432 mm wire diameter and 3.175 mm pitch	7.401
E	With internal spring of 1.191 mm wire diameter and 6.350 mm pitch	7.192
F	Coated with Union Carbide Linde High Flux Coating	7.073

EXPERIMENTS

Steady-State Characteristics

Pressure drop from the surge tank exit to the system exit is referred to as the system pressure drop. Experimental and theoretical data for the heater tube A are presented on pressure drop versus mass flow rate coordinates in Fig.2. The results show characteristic parabolic behavior with no heat input and a tilted and inverted "S" shape for different values of heat inputs.

Oscillations

The first oscillations were observed just after boiling started. In general these were sinusoidal, but changed shape with the operation point along the system pressure-drop vs. mass flow rate curve. Oscillations were observed with every heater tube in succession. However, there were differences in magnitude, amplitude and threshold of oscillations. Occasionally the number of superimposed oscillations were more for certain tubes.

Thresholds for pressure-drop type oscillations and for the density-wave type oscillations superimposed on pressure-drop type oscillations were located during the experiments and are indicated on the pressure-drop mass-flow rate coordinates (Fig. 2). The stability boundary maps are obtained by connecting these threshold points. Boundaries for the pure pressure-drop type oscillations for different heater tubes are collected in Fig. 3. An

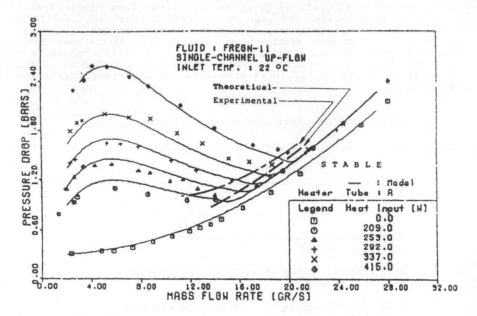

Figure 2. Comparison of Theoretical and Experimental Steady-State Characteristics and Oscillation Boundaries for the System with Heater Tube A at Different Heat Inputs.

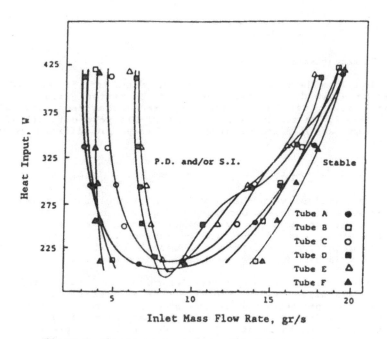

Figure 3. Boundaries for Pressure-Drop Type Oscillations.

410

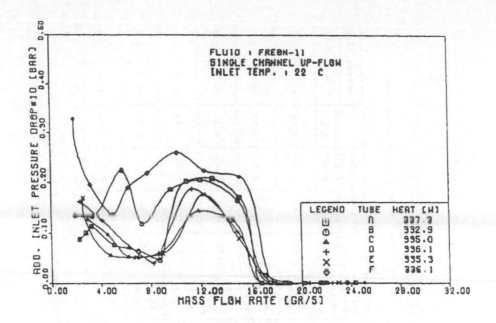

Figure 4. Inlet Throttling Necessary to Stabilize the System.

examination of the figure shows that the curves for different heater tubes exhibit a similar pattern, with internally springed tubes having narrower unstable regions. The same order of the tubes was obtained for the superimposed oscillation boundaries. A similar order is observed in Fig. 4, which plots the inlet throttling necessary to stabilize the system versus the mass flow rate for a heat input of 335 W.

The inlet throttling necessary to stabilize the system during oscillations is found by subtracting the inlet pressure drop (which would be observed if the inlet throttling valve were fully open) from the inlet pressure drop. This is found after stabilizing the system by closing the inlet throttling valve. The former fictitious pressure drop was calculated using Eq. (2), which is obtained by the least squares ploynomial fitting of the inlet frictional drop data for zero heat input experiments:

$$\Delta p_{if} = 9.450 \times 10^3 + 64.031 \, \dot{m} + 14,390 \, \dot{m}^2 \qquad (2)$$

where $\dot{m}$ is the mass flow rate in gr/s and Δp_{if} is the fictitious pressure drop in N/m^2.

Table 2 presents the values of the system parameters seen during oscillations, and was prepared to compare the system with different heater tubes. Mass flow rate gives the average value obtained after stabilizing the system, and exit quality is the calculated value at the orifice inlet. The temperature values are given in two rows. Top row shows the steady-state values, while the bottom row gives the overall variation of heater wall temperature at the given location.

411

Liquid Container Pressure : 6.53 Bar

Pc.Vc : 193.1 N.m (in the surge tank)

TUBE	Heat Input W (Kw)	Mass Flow (Gr/s)	Exit x %	Temperatures Steady State & Amplitude (C)							Pressures (Bar) Surge System Amplitude				Periods (Sec)	
				Inlet	2	3	2	3	4	Exit	Surge Tank	Exit	P.D.	D.W.	P.D.	D.W.
A	425.0	8.32	5.1	22.1 / N.A.	66.7 / N.A.	N.A. / N.A.	70.6 / N.A.	73.3 / N.A.	72.2 / 27.7	61.7 10.0 / 2.5	3.41	1.26	6.89	7.65	55.0	1.5
B	422.3	9.45	6.3	22.7 / N.A.	62.3 / 10.0	63.9 / 7.0	67.0 / 7.0	65.7 / 7.5	66.1 / 15.0	58.5 7.5 / 2.5	3.41	1.08	3.63	4.98	40.0	2.0
C	411.8	10.45	3.7	22.7 / N.A.	61.2 / N.A.	64.3 / N.A.	66.1 / N.A.	68.1 / N.A.	66.1 / 8.8	57.6 / 8.8	3.32	1.05	4.69	4.13	48.0	2.0
D	413.3	9.90	4.8	22.5 / N.A.	59.2 / 5.0	62.8 / N.A.	64.3 / N.A.	69.4 / N.A.	69.2 / 7.5	57.1 / 8.8	3.34	1.12	4.70	3.42	54.0	2.0
E	416.8	8.72	8.2	22.5 / 2.5	56.2 / 17.6	63.4 / 17.8	65.9 / 12.5	66.5 / 12.5	67.4 / 8.8	58.0 / 8.8	3.38	1.10	8.16	2.08	88.0	4.0
F	411.9	8.20	9.7	22.2 / N.A.	62.5 / N.A.	62.3 / N.A.	64.5 / N.A.	62.3 / N.A.	63.6 / N.A.	58.7 / 7.5	3.41	1.02	4.41	3.31	36.0	4.0

Table 2. Sustained Oscillation Data for Different Tubes During Superimposed Oscillations.

412

Experimental Results

The conclusions drawn from the tables and figures, generated from the experimental data can be summarized as follows:

1. instabilities are not affected by slight changes of heat transfer. Heat transfer coefficient were in the same range for the same type of heater surfaces and the oscillation boundaries and the additional pressure drop are found to be approximately the same for the heater tube surfaces of C,D and E

2. The system becomes more unstable as nucleate boiling increases. Tube F was coated with Linde High Flux Nucleate Boiling coating, which increased the heat transfer coefficient more than tenfold, and it was found to be the most unstable tube . However, the amplitude and periods of the oscillations were smaller for this tube.

3. The amplitude of the oscillations increases as the axial temperature gradient over the heater increases.

4. For the same type of heater surfaces, system stability increases with decreasing equivalent diameter.

5. Period of oscillations depends on heater surfaces. For heater tube E oscillation periods were the longest. In some cases , periods in tube E were four times the periods of oscillations for tube F.

MATHEMATICAL MODELING

One-dimensional homogeneous equilibrium flow model is used to describe the flow in the system under consideration [11].

Conservation equations are rewritten by setting the time dependent terms to zero. An implicit backward finite difference scheme and an iterative approach have been used for the solution. The system under study is shown in Fig. 1.

Part of the system between the points 2 and 6 in the Fig. 1 is sectionalized in z-direction parallel to the fluid flow by dividing it into n segments. Solution starts with assumption of an inlet pressure for a given mass flow rate, heat input and inlet temperature combination, and marches downstream. If the resulting exit pressure matches the boundary condition, the computations are over for the given mass flow rate and a smaller mass flow rate is taken and computations are repeated for the new mass flow rate. If the values do not match, the assumed inlet pressure is modified and computations are repeated until a match is obtained.

Computations are carried out at various mass flow rates and heat inputs for constant inlet temperature. The results are presented as continuous lines in Fig. 2.

For all heat inputs the results of the model follows the experimental data points closely. Maximum deviation occurs around the start of boiling and has a value of 10% for tube A. The average deviation is within ±3% for the tube A.

The experimental and the theoretical values match exactly for the single-phase liquid flow region of heat input experiments as well as for no heat input case. The match somewhat breaks down around the minimum of each heat input curve, which corresponds to the start of boiling and then becomes good again. The misfit around the start of boiling is as expected, since the homogeneous equilibrium model is used, and around the minimum subcooled boiling takes place and thermodynamic non-equilibrium between the phases occurs.

The instability thresholds are obtained by linearizing the governing equations with small perturbation analysises for the tube A and the results are presented in the Table 3. Calculated thresholds are also marked in Fig. 2 [11].

413

Table 3 indicates that the pressure-drop type oscillation thresholds can be located with linearized analysis and steady-state data. The prediction is within 1% for 415 W but error increases at lower heat inputs. For every case theoretical values are conservative.

Table 3. Pressure-Drop Type Oscillation Thresholds for the Heater Tube A

Heat Input Threshold [gr/s]	209 W	253 W	292 W	337 W	415 W
Experimental Value	9.0	13.5	16.0	17.8	19.4
Theory with Experimental Data	14.0	15.0	17.0	19.0	20.0
Theory with Theoretical Data	14.6	16.3	17.4	18.5	19.6

CONCLUSIONS

Results of experimental and theoretical study can be summarized as:
1. Instabilities are not affected by small changes in heat transfer.
2. Nucleate boiling increases system instability.
3. Amplitude of the oscillations increases as the axial temperature gradient over the heater increases.
4. For the same type heater surfaces system stability increases with decreasing equivalent diameter.
5. Period of the oscillations depend on the heater surface.
6. Homogeneous equilibrium flow model can be used for the analysis and a steady-state behavior can be found accurately.
7. Linearized analysis and steady-state data can be used to determine the oscillations thresholds.

ACKNOWLEDGEMENTS

The authors gratefully acknowledge the financial support of the National Science Foundation.

NOMENCLATURE

d   diameter, m
L   length, m
$\dot{m}$ mass flow rate, kg/s
V   volume, m^3

Greek Letters

Δp pressure drop, N/m^2

Subscripts

e   equivalent, exit
if inlet fictional

REFERENCES

1. Veziroglu, T.N., and Kakac, S., "Two-Phase Flow Instabilities", Final
    Report, N.S.F. Project CME 79-20018, Clean Energy Research Institute,
    Coral Gables, Florida, 1983.
2. Kakac, S., and Veziroglu, T. N., "A Review of Two-Phase Flow
    Instabilities". Kakac, S., and Ishii, M. (eds.): in Advances in Two-
    Phase Flow and Heat Transfer. Fundamentals and Applications, Martinus
    Nijhoff, The Hague, The Netherlands, Vol 2, pp. 577-669, 1983.
3. Stenning, A., and Veziroglu, T.N., "Flow Oscillation Modes in Forced
    Convection Boiling", Proceedings of the 1965 Heat Transfer and Fluid
    Mechanics Institute, Stanford University Press, Calfironia, pp.301-316,
    1965.
4. Saha, p., Ishii, M., and Zuber, N., "An Experimental Investigation of
    the Thermally Induced Flow Oscillations in Two-phase Systems", Journal
    of Heat Transfer, Trans. ASME, Vol. 98, No. 4, pp. 616-622, 1976.
5. Aritomi, M., Aoki, A., and Inoue, A., "Instabilities in Parallel
    Channel of Forced-Convection Boiling Upflow System, (III) System with
    Different Flow Conditions Between Channels", J. Nuclear. Sci. Tech.,
    Vol. 16, pp. 343-355, 1979.
6. Veziroglu, T.N., and Kakac, S., "Two-Phase Flow Instabilities and
    effect of Inlet Subcooling", Final Report, N.S.F. Project ENG 75-16618,
    Clean Energy Research Institute, Coral Gables, Florida, 1980,.
7. Cumo, M., Plazzi, G., and Rinaldi, L., "An Experimental Study on Two-
    Phase Flow Instability in Parallel Channels with Different Heat Flux
    Profile", CNEN-RT/ING (81)1, 1981.
8. Mentes, A., Yildirim, O.T., Gurgenci, H., Kakac, S., and Veziroglu,
    T.N. "Effect of Heat Transfer Augmentation on Two-Phase Flow
    Instabilities in a Vertical Boiling Channel", Warme und Stoffuber-
    tragung, Vol. 17, pp. 161-169, 1983.
9. Bergles, A.E., "Principles of Heat Transfer for Augmentation-II, Two-
    Phase Heat Transfer", Kakac, S., Bergles, A.E. and Mayinger, F. (eds.),
    in Heat Exchangers Thermal Hydraulics Fundamentals and Design,
    Hemisphere, New York, 1981.
10. Bergles, A.E. and Joshi, S.D., "Augmentation Techniques for Low
    Reynolds Number in Tube Flow", Kakac, S. and Shah, R.K. (eds.) in Low
    Reynolds Number Flow Heat Exchangers, Hemisphere, New York, 1983.
11. Mentes, A., Two-Phase Flow Instabilities: Pressure-Drop Type
    Oscillation Thresholds, Ph.D. Thesis, University of Miami, 1985.
12. Dogan, T., Kakac, S., and Veziroglu, T.N., "Analysis of Forced-
    Convection Boiling Flow Instabilities in a Single-Channel Upflow
    System", Int. J. Heat & Fluid Flow, pp. 145-156, 1983.
13. Lin, Z.H., T.N. Veziroglu, S. Kakac, H. Gurgenci, A. Mentes, "Heat
    Transfer in Oscillating Two-Phase Flows and Effect of Heater Surface
    Conditions", Proc. of the 7th Int. Heat Transfer Conf., Munchen,
    F.R.G., Vol. 5, pp. 331-336, Hemisphere, New York, 1982.
14. Stenning, A.H., Veziroglu, T.N., and Callahan, G.M., "Pressure-Drop
    Oscillations in Forced Convection Flow with Boiling", Proc. of
    Symposium on Two-Phase Flow Dynamics, Eindhoven, the Netherlands, 1967.

Enhancement of Nucleate Boiling Heat Transfer with Additives

TIANQING LIU, ZHENYE CAI, and JIFANG LIN
Institute of Chemical Engineering
Dalian University of Technology
Dalian, PRC

ABSTRACT

Base on the tests of more than ten kinds additives, several effective additives were obtained. The boiling heat transfer behavior can be improved greatly when the additives were put in water with trace amount. The mechanism of the enhancement of nucleate boiling with additives has been investigated in this paper. The results show that one of the important reasons for the enhancement of nucleate boiling is that the nucleation sites can be increased.

1. INTRODUCTION

Enhancing nucleate boiling with additives has been investigated for many years (1-4). Nevertheless there few studies on the mechanism of this subject. At present the experiments only indicate that trace additives of the right kind and the right amount can considerably increase the heat transfer (5). So both experimental and mechanismic studies are necessary.

In this paper, ten kinds additives were chosen and tested. In order to analyse the mechanism of the enhancement of boiling heat transfer with additives, the surface tension and the bubble departure frequencies for pure water and its solution containing additives were measured.

2. EXPERIMENTAL APPARATUS

The experimental apparatus is shown in Fig. 1. The boiling experiments were carried out under normal atmospheric pressure.

The main and auxiliary heaters were supplied with A. C. electrical source. Copper-constantan thermocouples were inserted into the copper block along the axis at three different locations. The surface temperature can be computed from the measured temperatures.

The bubble departure frequency was measured by means of an optical fiber probe.

3. EXPERIMENTAL RESULTS

The experimental results are plotted in Fig. 2 to 6, which show that:

(1) With trace amount of additive BA-1, BA-2 or BA-3, the heat flux of nucleate boiling can be increased by 2 to 7 times.

(2) There exist optimum concentrations for additive BA-1, BA-2 or BA-3 respectively, which means that right amount additives can enhance nucleate boiling best.

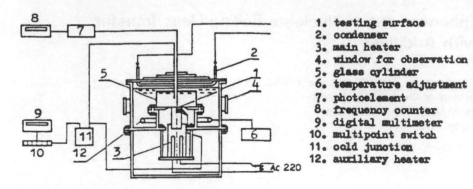

1. testing surface
2. condenser
3. main heater
4. window for observation
5. glass cylinder
6. temperature adjustment
7. photoelement
8. frequency counter
9. digital multimeter
10. multipoint switch
11. cold junction
12. auxiliary heater

FIGURE 1. Experimental apparatus

(3) Besides above three additives, other six additives tested do not
have effect on boiling heat transfer which indicates that only right kind
additive can increase heat transfer.

(4) The heat flux of nucleate boiling on the surface with artificial
nucleation sites can be further increased by 1 to 2 times with 50 to 100
ppm BA-4 and BA-2.

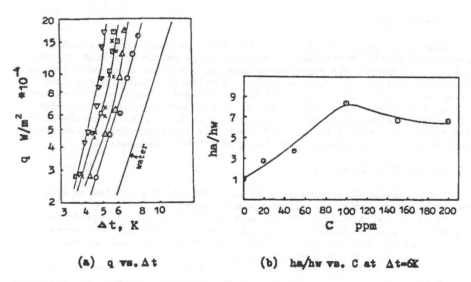

(a) q vs. Δt   (b) ha/hw vs. C at Δt=6K

FIGURE 2. The effect of additive BA-1 on boiling heat transfer of water
concentrations: O   20 ppm   Δ   50 ppm   ▽ 100 ppm
                   □ 150 ppm   × 200 ppm

418

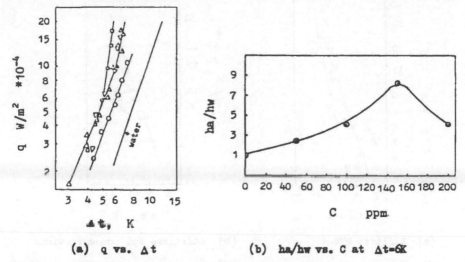

(a) q vs. Δt   (b) ha/hw vs. C at Δt=6K

FIGURE 3. The effect of additive BA-2 on the boiling heat transfer
concentrations: ○ 50 ppm   △ 100 ppm
                         □ 150 ppm   ▽ 200 ppm

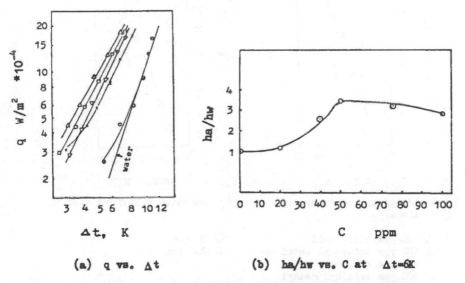

(a) q vs. Δt   (b) ha/hw vs. C at Δt=6K

FIGURE 4. The effect of additive BA-3 on the boiling heat transfer
concentrations: ○ 20 ppm   × 40 ppm   △ 50 ppm
                         □ 75 ppm   ▽ 100 ppm

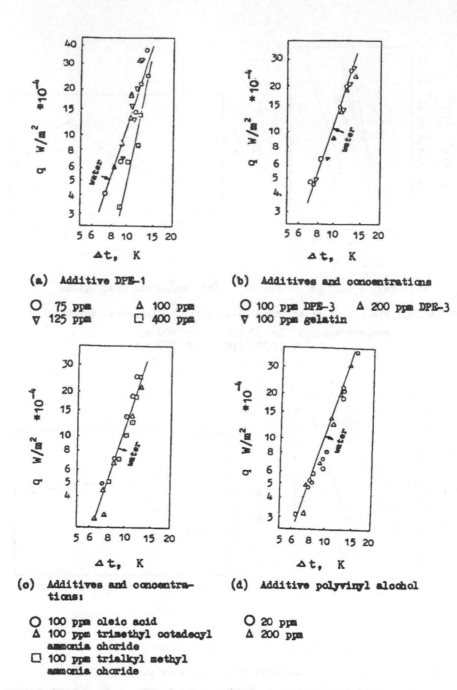

(a) Additive DPE-1

O 75 ppm   △ 100 ppm
▽ 125 ppm   □ 400 ppm

(b) Additives and concentrations

O 100 ppm DPE-3   △ 200 ppm DPE-3
▽ 100 ppm gelatin

(c) Additives and concentrations:

O 100 ppm oleic acid
△ 100 ppm trimethyl octadecyl
   ammonia choride
□ 100 ppm trialkyl methyl
   ammonia choride

(d) Additive polyvinyl alcohol

O 20 ppm
△ 200 ppm

FIGURE 5. The test results of the additives with which heat transfer
was not enhanced

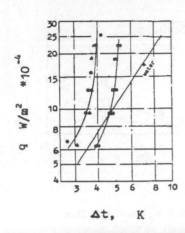 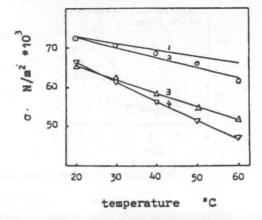

FIGURE 6. The effect of additives
on heat transfer on an
enhancing surface (nucleation sites $N=32$ cm^{-2})

□ 50 ppm BA-4 ○ 50 ppm BA-2
△ 100 ppm BA-2

FIGURE 7. The influence of additives
on the surface tension
1. pure water;
2. 50 ppm BA-3;
3. 100 ppm BA-1;
4. 150 ppm BA-2.

4. MECHANISM ANALYSIS

The reason for the enhancement of nucleate boiling with additives was
resulted in the decrease of surface tension by some researchers (1,3).
However, the actual effect of surface tension on boiling heat transfer is
not convincing (2). The surface tension of some aqueous solutions tested
in this paper had been measured, as shown in Fig. 7. Although additive BA-
1 can not drop surface tension as remarkable as BA-2, it indeed enhances
heat transfer more obviously than BA-2. In addition, it has been shown by
Yu Min Yang (4) that the decrease in surface tension by sodiium lauryl sulfate and sodium lauryl benzene sulfonate is almost same, but the ability
of enhancing heat transfer with these two surfactants is very different.
So the effect of additives on nucleate boiling is determined by not only
the surface tension reduction but also other factors such as the kinds and
the structures of additives.

In order to study the influence of additives on bubble behavior and
boiling heat transfer, the bubble departure frequencies of pure water and
the aqueous solution containing 100 ppm BA-2 were measured using optical
fiber probe. The corresponding bubble departure diameters were calculated
using the model (6). From Fig. 8, it can be seen that the bubble departure
frequency increases, and the bubble departure diameter becomes smaller because of the existing of additive BA-2.

Moreover, with the data of bubble departure frequency and diameter
mentioned above the latent heat flux q_1 and the sensible heat flux q_s which
stems from the partial departure boundary layer when bubbles depart from
a nucleation site can be calculated in the formulas (6):

$$q_1 = \frac{1}{6} \pi \cdot D_d^3 \cdot \rho_v \cdot h_{fg} \cdot f \tag{1}$$

421

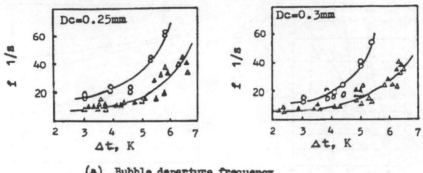

(a) Bubble departure frequency

△ pure water   ○ 100 ppm BA-2 aqueous solution

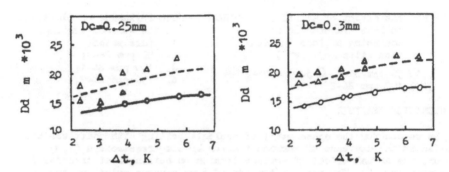

(b) Bubble departure diameter

△ measured results for pure water
---- calculating results for pure water
-○- calculating results for 100 ppm BA-2 solution

FIGURE 8. The comparisons of the bubble departure frequency and diameter for water and its solution

$$q_s = \frac{1}{4} \pi \cdot (D_d^2 - D_o^2) \cdot \delta \cdot \rho_1 \cdot C_1 \cdot \frac{\Delta T_w}{2} \cdot f \qquad (2)$$

The calculating results of q_1 and q_s for pure water and its solution with 100 ppm BA-2 are shown in Fig. 9. It can be seen that if the nucleation sites did not change, the heat flux with additive would not increase although the bubble dparture frequency raises. The increase of nucleation sites with additives is clearly revealed by the photos of the nucleate boiling surface, as shown in Fig. 10. This represents that one important reason for the enhancement of boiling heat transfer with additives is the increase of nucleation sites.

As for the reason why the number of nuclei can be increased by trace amount additives, there are probably three answers: (1) the minium nuclei radia will be reduced since the surface tesion is lowered, this will produce more active sites; (2) the properties of the interface between solid and liquid may be changed due to the adsorption of additives on the solid

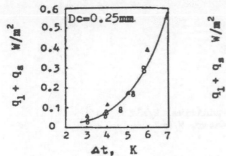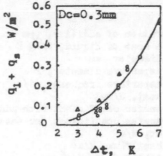

FIGURE 9. The comparison of the heat fluxes for water and its solution

   △ water   ○ 100 ppm BA-2 solution

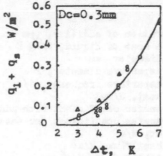

(a)   (b)   (c)   (d)

$q=47157 \ W/m^2$

FIGURE 10. The comparison of nucleation sites

  (a) pure water   (b) 100 ppm BA-1 solution
  (c) 150 ppm BA-2   (d) 50 ppm BA-3

surface; (3) large molecule additives in the boundary layer near heating surface may play the role to become nucleation sites.

5. CONCLUSIONS

(1) Several additives such as BA-1 have been selected, and the heat flux of nucleate boiling can be increased by 2 to 7 times with trace amount additives.

(2) The bubble departure frequencies of pure water and its solution containing 100 ppm BA-2 have been measured. The results show that bubble departure frequency increases because of the exist of the additive which can reduce the surface tension, but heat flux would not increase if the nucleation sites did not change.

(3) one important reason for the enhancement of nucleate boiling heat transfer with additives is the nucleation sites to be increased.

NOMENCLATURE

C concentration of additive, ppm
C_l specific heat of liquid, J/kg K
D_c cavity diameter, mm
D_d bubble departure diameter, m
f bubble departure frequency, 1/s
h_{fg} latent heat, J/kg
h_a heat transfer coefficient with additive, $W/m^2 K$
h_w heat transfer coefficient for water, $W/m^2 K$
q heat flux, W/m^2
q_l latent heat flux, W/m^2
q_s sensible heat flux, W/m^2
Δt temperature difference, K
δ thermo-boundary layer thickness, m
ρ_l density of liquid, kg/m^3
ρ_v density of vapor, kg/m^3
σ surface tension, N/m

REFERENCES

1. Saltanov, G. A., Kukushkin, A. N., 8th. Int. Heat Transfer Conf., Vol.5, pp.2245, 1986.
2. Lowery, A. I., et al, Ind. Eng. Chem., Vol.49, No. 9, pp.1445, 1957.
3. Morgan, A. I., Ind. Eng. Chem., Vol.41, pp.2767, 1949.
4. Yu Min Yang and Jer Ru Maa, Trans. of the ASME J. of Heat Transfer, Vol.105, No.1, pp.190, 1983.
5. Stephan, K., 7th. Int. Heat Transfer Conf., Vol.1, pp.21, 1982.
6. Chen Jiabin, Cai Zhenye and Lin Jifang, J. of Chem. Ind. and Eng. (China), Vol.2, No.2, pp.190, 1987.

A Study of Spray Falling Film Boiling on Horizontal Mechanically Made Porous Surface Tubes

GUOQING WANG, YINGKE TAN, SHIPING WANG, and NAIYING CUI
Chemical Engineering Research Institute
South China University of Technology
Guangzhou, PRC

ABSTRACT

Totally 16 different tubes were tested including a smooth tube. R-113 was used as the working fluid. The results indicated that there was 11%-80% increase of spray boiling heat transfer coefficients for the enhanced tubes in comparison with those of their pool boiling; and there was 2.7-6.2 times increase in comparison with those of the spray boiling for the smooth tube. Based on the experiments and mechanism analysis, a correlation was proposed which could predict spray falling film boiling heat transfer coefficients for horizontal mechanically made porous surface tubes within the relative deviation of ±8%.

NOMENCLATURE

C — constant
D — tube outside diameter, m
d — diameter of pore, m

d — bubble departur

    diameter, m

f — bubble generation

    frequency, 1/s

h — heat transfer

    coefficient, $w/(m^2k)$
k — heat conductivity of
    liquid, $w/(m\ k)$
N — pore density, 1/dm
Na — active pore density,
    $1/dm^2$
Nu — Nusselt number

$$h(\frac{\mu_l^2}{g\ \rho_l^2\ k^3})^{\frac{1}{3}}$$

Pr — Prandtl number, $\dfrac{C_p\mu_l}{k}$

q — heat flux, w/m^2
Rel — liquid film Reynolds
    number, $4\Gamma/\mu_l$

Rev — vaper Reynold numble, $\dfrac{q\ D}{\lambda\ \mu_l}$

ΔT — wall superheat, k

ρ_l — liquid density, kg/m^3

ρ_v — vapor density, kg/m^3

μ_l — liquid viscosity, Ns/m^2

μ_v — vapor viscosity, Ns/m^2
σ — surface tension, N/m
λ — liquid latent heat of
    evaporation, J/kg
Γ — mass flow rate per unit
    length, kg/ms

Subscripts
a — active
b — bubble
l — liquid
v — vapor

425

INTRODUCTION

Horizontal spray falling film evaporators are used in desalination, refrigeration, chemical engineering processes and so on. In order to reduce the size of equipment and increase the heat duty, it is necessary to improve the horizontal spray evaporators by using enhanced surfaces. Several enhancement techniques have been studied by some investigators. But, as for the mechanically made porous surface, it was studied scarcely. In this paper, at 0.1 MPa system pressure, R-113 was used as the working fluid, the heat transfer performance of 15 different mechanically made porous surface tubes (JK-1 tubes) for horizontal spray falling film boiling was tested, in order to explore the performance and mechanism of spray boiling, and to obtain a correlation which can be used in application.

APPARATUS

The test facility is shown in Fig.1. Saturated R-113 liquid located at the bottom of the test chamber(3) was transported to the top of the chamber through the flowmeter(4) by the pump(1). The liquid got to the top was a bit subcooled. Then, after heated to saturate temperature by the heater(11), it sprayed out from the liquid feeder(7). we took the first row tube as the test tube for boiling test. The liquid dropped into the liquid collector(15) came back to the bottom of the chamber through the measuring cylinder(16). The vapor entered the condenser(20). The liquid obtained by condensation came back to the bottom of the chamber through the measuring cylinder(14).

The porous layer structure of JK-1 tubes is shown in Fig.2. Pore pitch and tunnel pitch was given as manufacturing. Then, density N could be obtained by calculation. Pores are triangular in shape. Pore diameter d_* is represented by the equivalent diameter of the triangular area which can be obtained by measuring the width x and height y.

An alternating current power supplied to the electrical heater in the test tubes was adjusted by a variac(18) and monitored by a wattmeter(19). The vapor pressure was measured with a precise barometer(13).

TEST RESULTS AND DISCUSSIONS

1.   First of all, a reliability test was carried out with a polished smooth tube. The test proceeded with no nucleate boiling. Tts data are shown in Fig.3. Experimental data of Fujita[1], correlations suggested by Lorenz-Ying[2], Owens[3] and Chun-Seban[4], and Nusselt's theoretical formula are also plotted in Fig.3. According to Fig.3, when liquid film Reynolds number $Rel \leqslant 500$, the present data are between Fujita's data and Owens's prediction; when $Rel > 500$, the present data are slighly over both Fujita's and Owens's. The data are only about 5% higher than Fujita's. Because there were some visible scars and cavities on the smooth tube surface, its boiling heat transfer performance had been increased slightly. Therefore, the present test facility and the instrumentation were reliable.

2.   Data for the best JK-1 tube No.4, the worst JK-1 tube No.2 and a smooth tube are shown in Fig.4. According to Fig.4, the heat transfer coefficients for JK-1 tubes are about 2.7-6.2 times higher than those for the smooth tube. Kuwahara's[5] experimental results for a JK-1 tube

are also plotted in Fig.4 with dash and dot curve. From this comparision, the spray boiling heat transfer performance of JK-1 tubes developed in our institute have reached the level of Japanese similar enhanced tubes. Fang's[6] experimental results of the pool boiling for the same JK-1 tubes and a smooth tube are also plotted in Fig.4 with broken curves. At heat flux q from 7.72×10^{3} to 58.19×10^{3} w/m², the heat transfer coefficients of JK-1 tubes for the spray boiling are 11%-80% higher than those for the pool boiling; while at q larger than 58.19×10^{3} w/m², the heat transfer coefficients for the spray boiling will be lower than those for the pool boiling. The higher heat transfer coefficients of JK-1 tubes for the pool boiling, the higher their heat transfer coefficients for the spray boiling; but the increasement tends to decrease. This is in agreement with Kuwahara's experimental results.

3.   Heat transfer coefficients of JK-1 tubes for the spray boiling are much higher than those of the smooth tube by two major reasons. Firstly, in a large extent of heat flux, the vapor joined together in the tunnel of JK-1 tubes. There was only a layer of thin liquid film remaining on the wall surface of the tunnel. The thermal resistance of the thin liquid film was very small, it only requires a small temperature difference. Thus, quick evaporation was caused. Secondly, the cavities of the porous layer could intercept and retain a small amount of vapor to form new nucleation and keep the boiling continuously.

4.   At higher heat flux, the tunnel was in dried-up mode. The dried-up of the tunnel made the wall temperature risen. On the other hand, the external surface of JK-1 tubes was moistened with the liquid film, the rising of the wall temperatures were restricted and the wall temperatures were stabilized at a higher value. The liquid static pressure in the spray boiling was very low in comparison with that in pool boiling, the liquid of the outer surface of JK-1 tubes could not enter the tunnel in time by overcoming the rush force of the vapor and the friction resistance of the pore mouths. As a result of this, the tunnel has turned into the dried-up mode even the heat flux of the spray boiling was much lower than that of the pool boiling. This is the reason that at higher heat flux the heat transfer coefficients for the spray boiling was lower than those for the pool boiling.

5.   At lower heat flux, the heat transfer coefficients of JK-1 tubes for spray boiling are remarkably higher than those for the pool boiling also for two major reasons. Firstly, when the sprayed liquid flowed over the test tubes, the collision between the sprayed liquid and the top of the test tubes, and the roughness of the surface of the tubes caused the pulsation of the liquid film; therefore, the liquid film has entered the turbulent flow even liquid film Reynolds number was very low. The liquid, however, around test tubes in the pool boiling is in natural convection. Secondly, because the liquid film of the outer surface of the test tubes in the spray boiling was very thin, its liquid static pressure needed to be overcome for the generation of vapor bubbles was much smaller than that in pool boiling. Both of the reasons improved the heat transfer performance of the spray boiling.

6.   The test results of JK-1 tubes No.2, No.4, No.7 and No.9 are shown in Fig.5. The effects of the porous layer structure parameters on heat transfer performance are found be remarkable. JK-1 tubes of middle pore diameter and bigger pore density (e.g. tube No.4) have better heat transfer performance in whole heat flux extent, while JK-1 tubes of bigger pore diameter and smaller pore density (e.g. tube No.2) have

427

worse heat transfer performance in whole heat flux extent. JK-1 tubes of smaller pore diameter (e.g. tube No.9) have better performance at lower heat flux extent and worse performance at higher heat flux extent, while JK-1 tubes of bigger pore diameter (e.g. tube No.7) have worse performance at lower heat flux extent and better performance at higher heat flux extent. In conclusion, the bigger pore density of JK-1 tubes, the better of their heat transfer performance, and the optimum pore diameters increase with the increase of the heat fluxes. because the vaporization nuclei increase with the increase of the density, the boiling heat transfer performance is improved. The rate of formation, growth and separation of the bubbles depends on the extra pressure caused by the surface tension and the friction resistance of the pore months. At lower heat flux, the extra pressure is important, so the smaller pore diameter is favourable to the boiling heat transfer; while at higher heat flux, the friction resistance of the pore months is important, so the bigger pore diameter is favourable to the heat transfer. Therefore, the optimum pore diameters mainly depend on the heat flux.

7. The effects of the heat flux q and spray density Γ on the heat transfer coefficients are shown in Fig.6. The effects are obvious according to Fig.6, so this results aren't in agreement with Kuwahara's[5] results that is the heat transfer coefficients didn't depend on the heat flux and the spray density. Actually, the increase of the heat flux made the wall temperatures risen, thus the superheat of the liquid film increased and the viscosity of the liquid film decreased. Therefore, the heat transfer coefficients could be increased. On the other hand, the increase of the heat flux made the generation frequency of the bubbles increased, thus the turbulence of the liquid film was strengthened. So, the heat transfer coefficients were also increased. Although the increase of the spray density made the liquid film thickened and this was unfavourable to the heat transfer, but at the same time this increase also made the liquid film disturbed and this was favourable to heat transfer. It is obvious that the favourable factor is greater than the unfavourable factor, so the heat transfer coefficients were improved.

CORRELATION

Because of the complexity of the boiling mechanism, up to now it is impossible to solving mass, energy and momentum equations with the analytical solutions in association with the boundary conditions. Generally, the correlations are established with the dimensional analysis in association with the experimental data. J. C. Chen[7] established a precise correlation by saturate boiling heat transfer plus forced convection heat transfer. The present paper also adopts this idea.

The total quantity of the heat transfer can be divided into three parts:
1. Convective vaporization of the liquid film surface
2. Liquid film vaporization in the tunnels
3. Flash of the superheated liquid at the bottom of the test tubes

According to the similarity between evaporation and condensation, the relationship about the laminar evaporation over the horizontal tubes can be assumed as follow:

$$h(\frac{\mu_i^2}{g\,\rho_i^2\,k^3})^{\frac{1}{3}} = C_1(\frac{4\Gamma}{\mu_i})^m \tag{1}$$

The liquid film becomes turbulent for disturbance caused by the bubbles and the roughness of the porous surface. The Prandtl number Pr may be used to revise the equation(1) which becomes:

$$h(\frac{\mu_l^2}{g\,\rho_l^3\,k^3})^{\frac{1}{3}} = C_1(\frac{4\Gamma}{\mu_l})^m Pr^t \tag{2}$$

Owens[3] obtained the Pr exponent of t=0.5 based on the experimental data of ammonia and water. Sernas[8] obtained the Pr exponent of t=0.66 based on those of water. The average 0.58 is taken as the present Pr exponent for a lack of the experimental data.

$$h(\frac{\mu_l^2}{g\,\rho_l^2\,k^3})^{\frac{1}{3}} = C_1(\frac{4\Gamma}{\mu_l})^m Pr^{0.58} \tag{3}$$

The first part of heat transfer is obtained by rearrangement of the equation(3).

$$q_1 = C_1(\frac{4\Gamma}{\mu_l})^m Pr^{0.58}(\frac{\mu_l^2}{g\,\rho_l^2\,k^3})^{-\frac{1}{3}}\Delta T \tag{4}$$

The sconed part of the heat transfer can be described as follow:

$$q_2 = \frac{\pi}{6}d_b^3\rho_v\lambda fNa \tag{5}$$

According to Chesters's[9] analysis, one obtains:

$$d_b = C_b[\frac{\alpha d_0}{g(\rho_l - \rho_v)}]^{\frac{1}{3}} \tag{6}$$

Assume the relationship between the bubbles departure diameter d_b and the bubbles generation frequency f is the same as Jakob's[7] relationship about smooth surface.

$$fd_b = 0.59[\frac{\alpha g(\rho_l - \rho_v)}{\rho_l^2}]^{\frac{1}{4}} \tag{7}$$

Na, the number of active pores, majorly depends on the heat flux. This can be expressed by the vapor Reynolds number Rev. Furthermore, when heat flux has a certain value, the bigger the pore density, the more the active pores. So, one can assume:

$$Na = CaN(\frac{q\,D}{\lambda\,\mu_v})^n \tag{8}$$

Equation(6),(7) and (8) are introduced in equation(5). Then, the second part of the heat transfer is obtained:

$$q_2 = C_2\rho_v\lambda N[\frac{\alpha d_0}{g(\rho_l-\rho_v)}]^{\frac{1}{3}}[\frac{\alpha g(\rho_l-\rho_v)}{\rho_l^2}]^{\frac{1}{4}}(\frac{q\,D}{\lambda\,\mu_v})^n \tag{9}$$

Because the liquid is in turbulent state and the laminar layer is very thin, and the superheat liquid majorly located in the laminar layer. So, the superheat liquid is very little, thus the third part of the heat transfer is negligible.

Let the first part of the heat transfer multiplied by its second part, One obtains:

$$h(\frac{\mu_l^2}{g\rho_l^2k^3})^{\frac{1}{3}} = C_3\rho_v\lambda N[\frac{\alpha d_0}{g(\rho_l-\rho_v)}]^{\frac{2}{3}}[\frac{\alpha g(\rho_l-\rho_v)}{\rho_l^2}]^{\frac{1}{4}}Pr^{0.58}(\frac{4\Gamma}{\mu_l})^m(\frac{qD}{\lambda\mu_v})^n \tag{10}$$

Equation(10) is regressed in least square method with 234 experimental data. One obtains:

$$h(\frac{\mu_i^2}{g\rho_i^2 k^3})^{\frac{1}{3}}=2.43\times10^{-8}\hat{R}\lambda N[\frac{\sigma d_o}{g(\rho_i-\rho_v)}]^{\frac{1}{3}}[\frac{\sigma g(\rho_i-\rho_v)}{\rho_i^2}]^{\frac{1}{4}}Pr\ (\frac{4\Gamma}{\mu_i})^{a_u}(\frac{q_D}{})^{a_{31}} \quad (11)$$

Equation(11) is expressed in dimensionless numbers as follow:

$$Nu = 2.43\times10^{-8}\rho_v\lambda N[\frac{\sigma d_o}{g(\rho_i-\rho_v)}]^{\frac{1}{3}}[\frac{\sigma g(\rho_i-\rho_v)}{\rho_i^2}]^{\frac{1}{4}}Pr^{a_{uu}}Rel^{a_{u}}Rev^{a_{31}} \quad (12)$$

This correlation synthesizes the experimental data satisfactorily. The largest relative deviation is -10%. Except for 6 points, this correlation can predict the spray falling film boiling heat transfer coefficents for the horizontal mchanically made porous surface tubes within the relative deviation of ±8%.

CONCLUSION

1.   The spray falling film boiling heat transfer performance of JK-1 tubes developed in our institute have reached the level of Japanese similar enhanced tubes.

2.   The heat transfer coefficients for spray boiling of JK-1 tubes are much larger than those of the smooth tube. At the lower heat flux, the heat transfer coefficients for the spray boiling of JK-1 tubes are remarkably higher than those for the pool boiling of JK-1 tubes. Therefore, JK-1 tubes can be used to improve the heat transfer performance of the horizontal spray falling film evaporators.

3.   The correlation established in this paper has a higher precision, so it can be recommended for the device calculation.

REFERENCES

1.   Fujita, H., Takahama, H. and Asano, H., "Heat and Mass Transfer in Countercurrent Flow of Air and Water Film (Effects of Cylinders Installed in the Air Flow)," Transactions of the Japan Society of Mechanical Engineers, vol. 48, pp. 518-525, 1982.

2.   Lorens, J. J. and Yung, D., "A Note on Combined Boiling and Evaporation of Liquid Films on Horizontal tubes," J. Heat Transfer, vol. 101, pp. 178-180, 1979.

3.   Qwens, W. L., "Correlation of Thin Film Evaporation Heat Transfer Coefficients for Horizontal Tubes," Proc. 5th. OTEC Conf., vol. 4, pp. 71-89, 1978.

4.   Chun, R. R. and Seban, R. A., "Heat Transfer to Evaporating Liquid Films," J. Heat Transfer, vol. 93, pp. 391-396, 1971.

5.   Kuwahara, H., Yasukawa, A., Nakayama, W. and Yanagida, T., "Enhanced Heat Transfer Tubes for the Evaporator of Refrigerating Machines (3nd Report: Evaporative Heat Transfer from Horizontal Tubes in Thin Film Flow)," 22nd National Heat Transfer Symposium of Japan, Tokyo, pp. 49-51, 1985.

6.   Fang, Qiyuan and Deng, Songjiu, "A Study of Boiling Heat Transfer Mechanism for Mechanically Made Porous Surface Tubes," Master Disse-

rtation, South China Universty of Technology, 1984.

7. Collier, J. G., "Convection Boiling and Condensation," Hemisphere, New York, 1982.

8. Sernas, V., "Heat Transfer Correlation for Subcooled Water Films on Horizontal Tubes," J. Heat Transfer, vol. 101, pp. 176-198, 1979.

9. Chesters, S. K., "Bubble Sizes in Nucleate Pool Boiling," BOILING PHENOMENA, edited by S.J. Vun Stralen and R. Cole, Hemishere, New York, vol. 2, pp. 879-899, 1979.

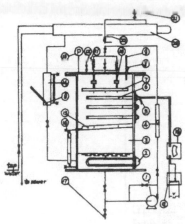

1. Pump
2. Auxiliary heat
3. Test chamber
4. Flowmeter
5. Thermocouple
6. Taxt
7. Liquid feeder
8. Thermocouple
9. Mercury thermometer
10. Filter
11. Heater
12. Inlet
13. Barometer
14. Measuring cylinder
15. Liquid collector
16. Measuring cylinder
17. Outlet
18. Variac
19. Wattmeter
20. Condenser
21. Exhaust pipe
22. Superheater

Fig.1 Test facility

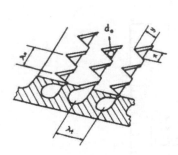

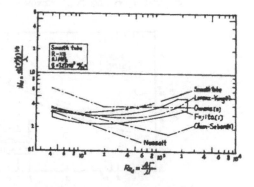

Fig.2 Porous layer
structure of JK-1 tubes

Fig.3 Evaporation heat transfer
performance of smooth tubes

431

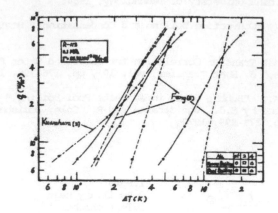

Fig.4 Boiling curves of JK-1 tubes

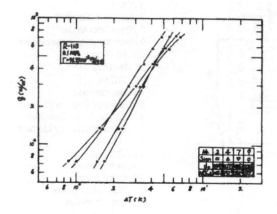

Fig.5 Effects of the porous layer structure parameters
d_o and N on the heat transfer performance

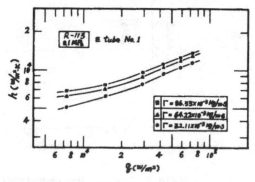

Fig.6 Effects of the operation parameter q and Γ
on the heat transfer coefficients

Heat Transfer for Evaporation in Vertical Tube Turbulent Falling Film Systems

QI ZHAO, ZAIQI LIN, and JIFANG LIN
Research Institute of Chemical Engineering
Dalian Institute of Technology
Dalian, PRC

ABSTRACT

A model of flow and heat transfer for evaporation in vertical tube turbulent falling liquid film systems is presented. An experimental study of heat transfer characteristics for a long tube vertical falling film evaporator was made under general conditions used in industry. Comparison of predicted and experimental data shows that they are in satisfactory agreement.

1. INTRODUCTION

The long tube vertical (LTV) falling film evaporators are extensively used in many industries because these evaporators have higher heat transfer coefficients at lower temperature differences and have smaller hold-up. Since Nusselt gave the specification for the hydrodynamics of a liquid film flowing down a wall in laminar motion in 1916, there have appeared almost innumerable researches on the theoretical and experimental aspects of the falling film problem. However, many aspects of the problems are still not sufficiently well understood.

Dukler(1) analyzed heat transfer characteristics in vertical falling film systems and gave a heat transfer model of flowing liquid film. Theoretical values of heat transfer coefficients were about 30% higher than experimental values. This is because he overestimated the effect of turbulent transport near the free liquid/vapor interface. Levich (2) supposed that the turbulent structure of two-phase film flow near the liquid/gas interface can not be indentical with the sturcture of one-phase pipe flow. From theoretical consideration he concluded that the surface tension damps the turbulent transport near the free liquid/gas interface. Based on the basic idea, several models for heat transfer in a falling liquid film were built up. (3,4,5) Seban (6,7) indicated some defects in the models and concluded that a reliable eddy diffusivity equation was very important for the establishment of a heat transfer model.

The purpose of this paper is to study heat transfer characteristics of the LTV evaporator both experimental   and theoretically and to give a model of flow and heat   transfer for evaporation in vertical tube turbulent falling   film systems.

2. EXPERIMENTAL EQUIPMENT

To give a reference for industrial design and to   check the models, heat transfer experiments of falling film evaporation of water inside a vertical copper tube were   made under conditions used in industry.

The experimental equipment is shown schematically in Fig. 1. The diameter of evaporating tube is 28/22 mm. and the length of the tube is 4 meter.

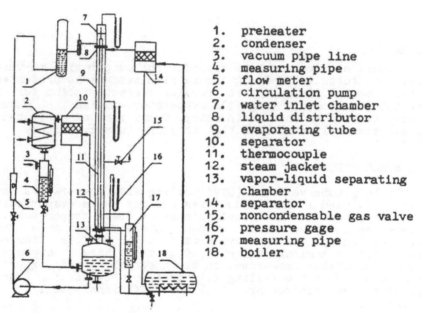

1. preheater
2. condenser
3. vacuum pipe line
4. measuring pipe
5. flow meter
6. circulation pump
7. water inlet chamber
8. liquid distributor
9. evaporating tube
10. separator
11. thermocouple
12. steam jacket
13. vapor-liquid separating chamber
14. separator
15. noncondensable gas valve
16. pressure gage
17. measuring pipe
18. boiler

FIGURE 1. Test unit

The Reynolds number of liquid film is varied from 2000 to 10000 and heat flux is varied from 10 to 80 kw/m^2.Temperature of evaporation is from 70 to 100 °C.

3. ANALYSIS OF TURBULENT FILM FLOW AND HEAT TRANSFER

3.1 The Basic Equation

The flow is assumed to be of constant properties, of constant flow rate, and of uniform thickness, momentum   equation and energy equation are:

$$\frac{\partial}{\partial y}(v + \varepsilon_M)\frac{\partial u}{\partial y} + \left(1 - \frac{\rho_v}{\rho_L}\right)g = 0 \qquad (1)$$

434

$$\frac{\partial}{\partial y}(\alpha + \varepsilon_H)\frac{\partial T}{\partial y} = 0 \qquad (2)$$

The key to the solution of these equations is a suitable expression for the turbulent transport terms, ε_M and ε_H .

3.2 Turbulent Model

For falling film flow the film is divided into two regions. Near the wall van.Driest model is used and near the liquid /gas surface the model of eddy diffusivity given by author is used.

van.Driest model. For one-phase pipe flow, van.Driest (8) gave a model of mixing length:

$$l = Ky\left[1 - \exp\left(-\frac{y^+}{26}\right)\right] \qquad (3)$$

By use of equation 1 and 3, the modified van.Driest model of diffusivity which is suitable for the region of near wall is obtained:

$$1 + \frac{\varepsilon_M}{\nu} = \frac{1}{2}\left(1 + \sqrt{1 + 0.64 y^{+^2}\left(1 - \eta^2 \frac{y^+}{\delta^+}\right)\left(1 - \exp\left(-\frac{y^+}{26}\right)\right)^2}\right) \qquad (4)$$

The eddy diffusivity model near liquid/gas surface. Due to the action of surface tension, the eddy diffusivity is assumed to decrease continuously from a high value in the bulk liquid to zero at the liquid/gas interface.(2) On the basis mass transfer coefficients equation and eddy diffusivity equation are:

$$\frac{1}{k_o} = \int_o^\infty \frac{1}{D + ay'^2}dy' \qquad (5)$$

$$\varepsilon_D = ay'^2 \qquad (6)$$

The key to the establishment of eddy diffusivity model is a reliable mass transfer coefficients expression. Recently Koziol (9) measured mass transfer coefficient for gas absorption into a turbulent falling water film and correlated their results for mass transfer coefficient :

$$sh_o = 8.923 \times 10^{-4} Re_L^{0.71}Sc^{0.5} \qquad (7)$$

From above equations, the eddy diffusivity equation is obtained:

$$\varepsilon_D = 8.3 \times 10^{-4} Re_L^{1.42}(\delta - y)^2 \qquad (8)$$

The equation is based only on results for water at 11°C, so that it can not be used directly. Using method of Mills and Chung (3) to generalize the equation to temperature other than 11°C and to liquid other than water, the dimensionless form of eddy diffusivity is obtained:

$$1 + \frac{\varepsilon_D}{\nu} = 1 + 4.91 \times 10^{-3} Ka^{\frac{1}{3}} Re_L^{1.42} \frac{(\delta^+ - y^+)}{\delta^{+\frac{1}{3}}} \qquad (9)$$

Considering the effect of vapor shear stress (4) , we get:

$$1 + \frac{\varepsilon_D}{\nu} = 1 + 4.91 \times 10^{-3} Ka^{\frac{1}{3}} Re_L^{1.42}\left[\tau_i^* + \delta^*\left(1 - \frac{y^+}{\delta^+}\right)\right]\frac{(\delta^+ - y^+)}{\delta^{+\frac{1}{3}}} \qquad (10)$$

435

Interfacial shear stress. Significant interfacial shear exists in LTV falling film evaporator. In order to calculate interfacial shear a method for determing two-phase pressure drop is required. To date, the only method of evaluation is by empirical means. Using equation of two-phase pressure drop correlated by Dukler [1] , the dimensionless interfacial shear stress equation is deduced:

$$\tau_i^* = A(Re_{L,i} - Re_{L,s})^{1.4} Re_{L,s}^{0.4}$$

$$A = \frac{0.25 \mu_L^{1.172} \mu_G^{0.10}}{g^{\frac{2}{3}} d^2 \rho_L^{0.503} \rho_G^{0.70}}$$

(11)

$$Re_{L,i} - Re_{L,s} = \frac{Re_{g,s} \mu_g}{\mu_L}$$

(12)

Turbulent Prandtl number and Schmidt number. According to recent boundary layer practice, it is assumed that turbulent Prandtl number Pr_t or $\frac{\varepsilon_M}{\varepsilon_H} = 0.9$ and turbulent Schmidt number Sct or $\frac{\varepsilon_M}{\varepsilon_D} = 1$.

By foregoing research we eventually get the turbulent model which is suitable to the whole liquid film:

$$\varepsilon^* = \frac{1}{2}\left(1 + \sqrt{1 + 0.64 y^{*2}\left(1 - \eta^2 \frac{y^*}{\delta^*}\right)\left(1 - \exp\left(-\frac{y^*}{26}\right)\right)^2}\right)$$

(13)

$$0 < y^* < y_i^*$$

$$\varepsilon^* = 1 + 4.91 \times 10^{-3} Ka^{\frac{1}{3}} Re_L^{1.42}\left[\tau_i^* + \delta^*\left(1 - \frac{y^*}{\delta^*}\right)\right]\frac{(\delta^* - y^*)}{\delta^{*\frac{1}{3}}}$$

(14)

$$\varepsilon^* = 1 + \frac{\varepsilon_M}{\nu} \qquad\qquad y_i^* < y^* < \delta^*$$

where y_i^* is intersection point of equation 13 and equation 14.

3.3 Numerical Solutions

The equations evolved in the model are too complex for analytical solution. Numerical solution is, therefore, accomplished with a computer.

Velocity distributions. The intergration of equation 1 gives the velocity distribution:

$$u^* = \int_0^{y^+} \frac{1 - \eta^2 \frac{y^*}{\delta^+}}{\varepsilon^*} dy^*$$

(15)

Local Reynolds number is defined as:

$$Re_{L,s} = 4\int_0^{\delta^+} u^+ dy^+$$

(16)

From the equations the velocity distribution for any film Reynolds number and interfacial shear stress can be obtained. Fig. 2,3 and 4 show some typical velocity distributions. As the value of τ_i^* or Reynolds number increases, the velocity at

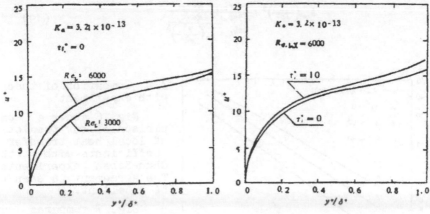

FIGURE 2. The effect of $Re_{L,x}$ on u^+

FIGURE 3. The effect of τ_i^* on u^+

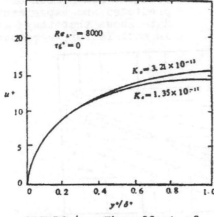

FIGURE 4. The effect of Ka on u^+

any value of y^+/δ^+ increases, as shown in Fig. 2 and Fig. 3. Fig. 4 shows that Kapitza number has a little effect upon velocity distribution near region of wall.

Film thickness. Using the equations $\delta^+ = \frac{\delta u^*}{\nu}$ and $\eta^* = \frac{\rho g \delta}{\tau_w}$

dimensionless film thickness equation is obtained:

$$\delta^* = \eta \delta^{+\frac{3}{2}} \qquad (17)$$

Fig. 5 shows that increased interfacial shear causes decreased film thickness.

Local heat transfer coefficient. The equation 2 is intergrated dimensionless local heat transfer coefficient equation is obtained:

$$h_x \left(\frac{\nu^2}{gk^3} \right)^{\frac{1}{3}} = \frac{Pr\delta^{+\frac{3}{2}}/\eta}{\int_0^{\delta^+} \dfrac{dy^+}{\dfrac{1}{Pr} + \dfrac{1}{Pr_t}(\varepsilon^+ - 1)}} \qquad (18)$$

Average heat transfer coefficient. In order to calculate the average heat transfer coefficient, it is necessary to integrate along the length of the tube. The applicable relation is shown as following equation:

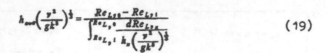

$$h_{\infty\sigma}\left(\frac{\nu^2}{gk^2}\right)^{\frac{1}{3}}=\frac{Re_{L,s}-Re_{L,s}}{\int_{Re_{L,s}}^{Re_{L,s}}\frac{dRe_{L,s}}{h_{x}\left(\frac{\nu^2}{gk^2}\right)^{\frac{1}{3}}}}$$

(19)

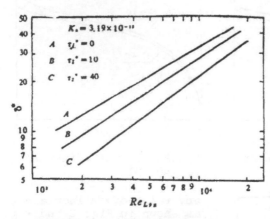

FIGURE 5. Film thickness

3.4 Comparion of theory with experiment

Fig. 6 shows a comparison of our prediction of local heat transfer coefficients with the Chun-Seban experiments. The agreement is seen to be excellent.

Fig. 7 compares our prediction of average heat transfer coefficient with our own experimental results. Comparison of predicted and experimental data shows that they are in satisfactory agreement.

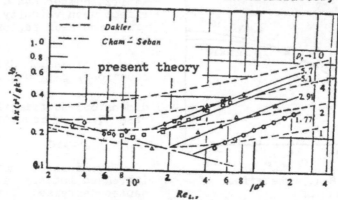

FIGURE 6. Comparison of theory with the experiments of Chun and Seban for local heat transfer coefficients.

4. CONCLUSION
4.1 A model of flow and heat transfer for evaporation in vertical turbulent falling film systems is developed, which enables to evaluate velocity distributions, film thickness local heat transfer coefficients and average heat transfer coefficients etc.

4.2 The eddy diffusivity model for falling film flow is developed.

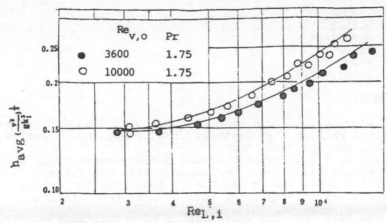

FIGURE 7. Comparison of theory with our own
experimental results for average heat
transfer coefficients.

NOMENCLATURE

a	constant in equation 5, m^{2-n}/s
D	molecular diffusion coefficient, m^2/s
d	tube diameter, m
g	gravity acceleration, m/s^2
h	heat transfer coefficient, w/m^2 K
k	thermal conductivity, w/m K
k_c	liquid phase mass transfer coefficient, m/s
l	mixing length , m
u	velocity in the film in x-direction, m/s
u^*	friction velocity, m/s
u^+	dimensionless velocity, (u/u^*)
y	distance normal to the wall, m
y^+	dimensionless distance, (yu^*/v)
α	thermal diffusivity, m^2/s
δ	film thickness, m
δ^*	dimensionless film thickness
δ^+	dimensionless distance
ε_M	eddy diffusivity for momentum, m^2/s
ε_H	eddy diffusivity for heat, m^2/s
ε_D	eddy diffusivity for mass, m^2/s
η	defined in equation 17.
μ	dynamic viscosity, kg/m s
ρ	density, kg/m^3
τ	shear stress, N/m^2
τ^*	dimensionless shear stress,
Ka	Kapitza number, defined as $(\frac{\mu^4 g}{\rho \sigma^3})$
Pr	Prandtl number
Re_L	film Reynolds number
Re_v	vapor Reynolds number
Sh_z	Sherwood number, defined as $\frac{k_c}{D}(\frac{v^2}{g})^{\frac{1}{3}}$

Subscripts

i	inlet; interface.

439

l	liquid.
o	outlet.
t	turbulent
v	vapor
w	wall
x	local value

REFERENCES

1. Dukler, A.E., Fluid Mechanics and Heat Transfer in Vertical Falling Film Systems, Chem. Engng. Symp. Series, Vol. 56, pp. 1-10, 1960
2. Levich, V.G., Physicochemical Hydrodynamics,Chap. XII, Prentice-Hall Inc., Englewood Cliffs New Jersey, 1962.
3. Mills, A.F., and Chung, D.K., Heat Transfer Across Turbulent Falling Film. Int. J. Heat Mass Transfer,Vol. 16, pp. 694-696, 1973.
4. Blangeffi, F., and Schlunder, E.U., Local Heat Transfer Coefficients on Condensatiom in a Vertical Tube, 6th Int. Heat Transfer Conference, Toronto, General Papers, Vol.2, pp. 437-442, 1976.
5. Hubbard, G.L., Mills, A.F., and Chung, D.K., Heat Transfer Across a Turbulent Falling Film With Cocurrent Vapor Flow, J. Heat Transfer, Vol. 98, No. 2, pp. 319-320 1976.
6. Seban, R.A., and Faghri, A., Evaporation and Heating With Turbulent Falling Liquid Films, J. Heat Transfer Vol. 98, No. 2, pp.315-318, 1976.
7. Seban, R.A., Transport to Falling Films, The Proceeding of Sixth Int. Heat Transfer Conference, Vol. 6, pp.417-428, 1978.
8. van Driest, E., Joural Aero. Science, Vol. 23, pp. 1007-1011, 1956.
9. Koziol, K., Broniarz, L., and Nowicka, T., Int. Chem. Eng. , Vol. 20, No. 1, p.136, 1980.

Heat Transfer to Falling Films on Electrochemically Etched Porous Surfaces

ZHENMIN LI, ZHENYE CAI, and JIFANG LIN
Chemical Engineering Institute
Dalian University of Technology
Dalian, PRC

INTRODUCTION

Falling film evaporator is a very common evaporation equipment in chemical industry, it is very important to improve its heat transfer efficient for the saving of energy. In recent years, our Research Labratory has made out a new type of heat transfer element, the electrochemically etched porous(ECEP) surfaces. The object of this paper is to investigate falling film boiling mechanism on the ECEP Surface, to establish a model and to verify the model by experimental results.

Experimental Apparatus & Results

The apparatus is shown in Fig.1. The testing tubes were of $\Phi 25 \times 1mm$ stainless steel with smooth surface and ECEP surfaces. The thermally developed section of the testing tube was of 280mm length and the testing sections were of 280mm and 700mm length respectively. The top of the testing tube was provided with an annulus distributor which ensures uniform film flow downwards outside the tube.

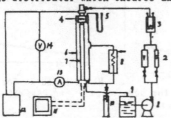

1. pump 2. flowmeter 3. preheater 4. annulus distributor
5. pressure gauge 6. glass tube 7. test tube 8. condenser
9. liquid vessel 10. volume meter 11. potential gauge
12. D.C. generator 13. current meter 14. voltage meter

Fig.1 Schematic of apparatus

TABLE 2. λ values of surfaces

Tube No.	2	3	4	5	6
λ value	5.98	5.46	2.02	6.53	2.00

441

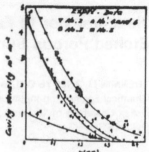

TABLE 1. Surface charact. of tubes and testing liquids

tube NO	surfacs charact	length mm	cravity density N.cm⁻²	Testing Liquide		
				water	10%wt suger aq	5%wt glycerin aq
1	smooth	280	/	✓	/	/
2	porous	280	357	✓	✓	✓
3	porous	280	337	✓	✓	✓
4	porous	700	140	✓	✓	✓
5	porous	280	330	✓	✓	✓
6	porous	280	140	✓	✓	✓

Fig. 2 The distribution of cavity diameter on porous surfaces

Electrical heating with D.C. source for the testing tubes was employed. Six Copper-constntan thermocouples were employed for measuring the internal wall temperature. The external wall temperature can be calculated from the known internal wall temperature. The characteristics of the testing tubes and the testing liquid are shown in Table 1.

The distribution of the cavity density on ECEP surfaces are shown in Fig.2. Using deionized water as testing liquid, the experimental results on ECEP surfaces and smooth surface are shown in Fig.3. The h° of the porous surfaces for water is at the range of 0.31 to 0.62, it is about 1.7 times higher than that of smooth surface, which is shown in Fig.3.

Fig.4 shows the experimental results of sugar aq.(10%wt) on both smooth and ECEP surfaces. The h° of ECEP surfaces is at the range of 0.24 to 0.45, it is about 1.4 times higher than that obtained by Trommelen[2] on smooth surface. The h° of ECEP surfaces for glycerin aq.(5%wt) is at the range of 0.19 to 0.35, it is about 1.1 times higher than that obtained by wilke[3] on smooth surface, as shown in Fig.5.

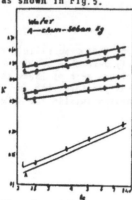

Fig.3 h×and Re for water

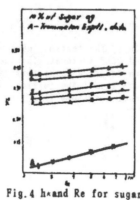

Fig.4 h×and Re for sugar aq.(10%wt)

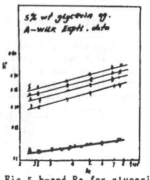

Fig.5 h×and Re for glycerin aq.(5%wt)

Theoretical analysis In this paper, a physical model has been developed to predict the falling films heat transfer on the porous surface.

(1). Basic equation of falling film boiling on porous surface For a evaporating film flowing downwards a heated porous surface, the momentum and energy equations can be expressed as follows:

Momentum eq.: $(1-\alpha)\frac{\partial}{\partial y}(V_L+\mathcal{E}_m)\frac{\partial u}{\partial y}+(1-\frac{\rho_g}{\rho_L})=0$ (1)

B.C.:

$$u\Big|_{y=0}=0 \qquad \frac{\partial u}{\partial y}\Big|_{y=\delta}=0$$

Energy eq.:

$$(1-\alpha)\frac{\partial}{\partial y}(V_H+\mathcal{E}_H)\frac{\partial T}{\partial y} \quad 0$$ (2)

B.C.:

$$\frac{\partial T}{\partial y}\Big|_{y=0}=-\frac{1}{k}(q_w-q_{fg}) \qquad T\Big|_{y=\delta}=T_{sat} \qquad q_{fg}=f\,h_{fg}\,\rho_g\,N_0\,V_b$$ (3)

In order to solve the above equations, the eddy diffusivities $\mathcal{E}_m$ and $\mathcal{E}_e$ should be determined first. From the analogy of momentum, mass and energy transport, the universally used relations of $\mathcal{E}_m$ and $\mathcal{E}_e$ should be determined first. From the analogy of momentum, mass and energy transport, the universally used relations of $\mathcal{E}_m$ and $\mathcal{E}_e$ are [4]:

$$\mathcal{E}_H/\mathcal{E}_m=0.9\left[1-exp(-\frac{y^+}{26})\right]/\left[1-exp(-\frac{y^+}{B^+})\right]$$ (4)

$$B^+=P_r^{-1/2}\sum_{j=1}^{5}C_j(log_{10}P_r)^{j-1}$$ (5)

For turbulent falling film, $\mathcal{E}_e$ equals $\mathcal{E}_m$. So in the condition of knowing $\mathcal{E}_m$ and $\mathcal{E}_e$, $\mathcal{E}_e$ can be determined.

(2). The determination of $\mathcal{E}_e$ for falling film boiling on porous surfaces During the nucleate boiling on porous surfaces, the motion of bubbles and the turbulence caused by liquid itself have contribution to the flow turbulent heat transfer. On the basis of Sato's study [5] on bubbly flow, here, it is assumed that in falling film the turbulence caused by the motion of bubbles is indepedent of that caused by liquid phase turbulence, and by combining these two comptentsthe total flow turbulence can be evaluated.

For the falling films on free surface, it may be reasonable to take account of the damping effect of the solid wall on the turbulent motion and the damping effect near the interfacial region caused by surface tension. on the turbulent heat and mass transfer, the falling film is considered as two parts and $\mathcal{E}_e$ is determined respectively.
(a)near wall region For liquid, from Van Drist mixing length theory [6] neglecting the interfacial shear stress, one yeilds:

$$1+\mathcal{E}_{mL}/V_L=0.5\left\{1+\sqrt{1+0.64y^{+2}(1-\frac{y^+}{\delta^+})\left[1-exp(-\frac{y^+}{26})\right]^2}\right\}$$ (6)

for bubble, with the modifying to Seto's eq., the following eq. can be obtained:

$$\mathcal{E}_{mg}/V_L=1.2\,\alpha D_0/2\; u/V_L\left[1-exp(-\frac{y^+}{26})\right]$$ (7)

(b) interfacial region For liquid, from Nowicka's falling film mass transfer coefficient eq. [7], the surface evaporating $\mathcal{E}_e$ can be expressed as fellows [12]:

$$1+\mathcal{E}_{eL}/V_L=1+4.91\times10^{-3}Ka^{1/3}Re^{1.42}(\delta^+-y^+)^2/\delta^{+2/3}$$ (8)

for bubble, obtained as [12],

$$\varepsilon_{mg}/\upsilon_L = 0.6 \alpha \upsilon D_B \left[1 - exp\left(-\frac{Y_C^+}{26}\right) \right]/\upsilon_L \tag{9}$$

Y^+. is the intersection point of two regions. in order to solve eq. (6,7,8 & 9), one should determine Ne, f, D_* and $\varepsilon_{..}$.

(3) **Determination of some parameters** On the porous surfaces, the cavity density is of the order of 10^*m^{-*}. The distribution of cavity diameters can be seen in Fig.2. From Fig.6, it is found that on the porous surface, some of the cavities, especially the smaller ones, cannot become active sites. So Ne should be determined first during the calculation.

Webb[8] proposed that cavity having liquid contact angle θ close to 90°, if $\beta > \pi/2-\theta$, will be active and those having $\beta \geqslant \pi$ cannot retain vapor nucleus and will be unactive. Fig.7 shows the cross section of an active site, Fig.8 shows the experimental results of which were measured from many enlarged pictures of ECEP surfaces cross section. Assuming the distribution of β angle is a normal one, the distribution function is,

$$\phi(\beta) = \begin{cases} \frac{1}{\sqrt{2\pi}S} \, exp\left[-\frac{(\beta-\bar\beta)^2}{2S^2}\right] & \beta \geqslant 0 \\ 0 & \beta < 0 \end{cases} \tag{10}$$

based on the experimental results with Maxium likelihood,yeilds, $\bar\beta = 0.4415$, s=0.4262. After with x^2-test, it is known that the distribution of β is normal one, the results can be shown as,

$$\phi(\beta) = \int_{\frac{\pi}{2}-\theta}^{\pi} \frac{1}{0.4262\sqrt{2\pi}} \, exp\left[-\frac{(\beta-0.4415)^2}{0.3633} \right] d\beta \tag{11}$$

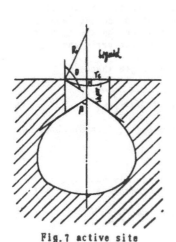

Fig.7 active site

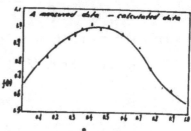

Fig.6 Enlarged picture of the porous surface cross section

Fig.8 Distribution of β angle

From Fig. 2, the cavity radius has a distribution of exponent,

$$\phi(r) = \begin{cases} \lambda e^{-\lambda r} & r > 0 \\ 0 & r \leq 0 \end{cases} \tag{12}$$

Using Maximum likelihood, the λ values of every porous surface can be obtained, the calculated results were listed in table 2. Under the equilibrium condition, the resultant forces acted on the vapro-liquid interface must equal to zero, obtained as [12],

$$r_{cmax} \left[\frac{12 \sigma \sin\theta}{\Delta \rho g (\sec\theta - tg\theta)[3 + (\sec\theta - tg\theta)]} \right]^{1/2} \tag{13}$$

if $r_c > r_{cmax}$, the equilibrium condition of the vapor-liquid interface would be broken, the liquid will fill the cavity. So the cavity with $r_c > r_{cmax}$ cannot become active. From the present measurement, the smallest cavity radius is 0.035, so

$$\phi(r_c) = \int_{0.035}^{r_{cmax}} \lambda e^{-\lambda r} \, dr \tag{14}$$

The density of active cavity, Ne, is,

$$Ne = N\phi(r_c)\phi(\beta) \tag{15}$$

It is know from eq. (15) that different surfaces have different Ne. From the previous paper [10], the bubble departure diameter during the falling film on ECEP surfaces has follow relation,

$$1.75 \times 10^{-4} D_s^3 R_e^{0.8} \rho / \mu^2 (\pi - \theta + \cos\theta\sin\theta) - 0.17 \rho g D_s^3 (2 + 3\cos\theta + \cos^3\theta) - 6(\cos\beta_1 - \cos\beta_2)\sin\theta = 0 \tag{16}$$

where β_1 and β_2 are the bubble advanced and retreated contact angles respectively. Using experimental results of β_1 and β_2, D_s can be solved by the above eq..

For a bubble growing in faling films, Sernas [11] proposed,

$$\frac{dR}{dt} = \frac{k}{\delta_0 \rho_1 h_{fg}} \left\{ \frac{\theta_w}{2} + \frac{2P\theta_{sup}}{\pi R} \left[\sum_{k=0}^{\infty} (\frac{Rk}{\lambda^1 k}) \sum_{k=0}^{\infty} (\frac{Rk}{\lambda^1 k}) \int_0^{\pi/2} \exp(- \frac{2R\cos\phi}{\rho}) \cos\phi \, d\phi \right] \right\} \tag{17}$$

By comparing bubble growing time t_g and waiting time t_w, sernas found that $t_g \gg t_w$, that is to say, t_w can be neglected. So the bubble depature frequency f can be obtained. After knowning, Ne, D_s and f, α, we can obtain the numerical solution of momentum and energy eq..

Fig. 9 shows the relation between ε^*_m and y^*/δ_v, it is known ε^*_m increases with α. Fig. 10 shows the relation between U^* and y^*/δ_v, in present experiment, α has a little influence on U^*.

Fig. 11 shows the relation between T^* and y^*/δ_v. The predicted results obtained by proposed model are compared with experimental data in Fig. 12 to 14. Fair agreement between tem shows that the model is satisfactory.

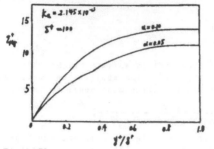

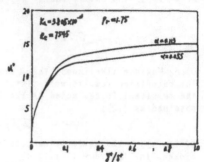

Fig.9 The relation of $\epsilon^\circ_{..}$ and y°/δ.

Fig.10 The relation of U° and y°/δ.

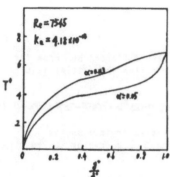

Fig.11 The relation of T° and y°/δ.

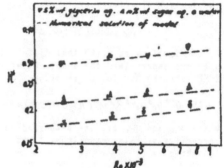

Fig.12 comparison of exptl. and predicted results

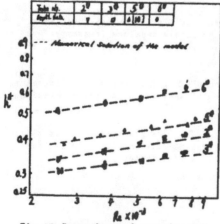

Fig.13 Comparison of experimental and predicted results

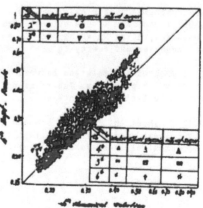

Fig.14 Comparison of experimental and predicted results

Conclusion

(1). Falling film boiling experiments were carried out on ECEP surfaces. The experimental results show that the heat transfer coefficient of ECEP surfaces for several testing liquids is about 1.5 times higher than that of smooth surface. (2). A model has been developed and a correlation of falling film boiling heat transfer on porous surfaces has been obtained. Agreement between the predicted values and teh experimental data has proved to be quite satisactory.

NOMENCLATURE

h^*, dimensionless heat transfer coefficient;
h_{r_*}, latent heat of vaporization;
M, mass fraction of vapor;
N, cavity density;
T^*, dimension less temperature;
V_*, volume of bubble;
W, mass velocity;
a, volume fraction of vapor;

β, cavity top angle;
δ, dimensionless distance;
ε_*, eddy mass diffusivity;
ε_*, eddy thermal diffusity;
ε_*, eddy viscosity;
σ, surface tension;
V_*, thermal diffusity;
V_*, kinematic viscosity;
subscripts, 1, liquid; g, vapor;

References

(1), Chun, K. R. & Seban, R. A., J. Heat Transfer, Vol. 93, pp 394--96, 1971
(2), Trommelen, D. I., AICHE. Srmp., Vol. 64, No. 62. pp103--9.
(3), Wilke, J. M. ect., Inter. Develop. Ht. Trans., pp212-13, 1963
(4), Kabib, L. & Na. I., J. Ht. Trans. ASME., Vol. 960, 1974
(5), Sato, Int. J. Mulitiphase Flow, Vol. F., 1981
(6), Van Driest., J. Aero. Sci., Vol. 23, pp1007-11, 1956
(7), Koziol, K., ect., Int. Chem. Eng., Vol. 20, No. 1, pp136-42, 1980
(8), Webb, R. L., Ht. Trans. Eng., Vol. 4, No. 3-4, 1983
(9), Nishikawa, ect., Ht. Trans. In Eng. Problem, pp111-18, 1982
(10), Chen Jiabin & Cai Zhenye ect., The 2th reactor thermal
     Engineering flowing conference, 1983, 4. Xiamen
(11), Sernas, U., ect., Int. J. Ht. Mass Trans., Vol. 28, No. 7. 1985
(12), Li Zhenmin, Dalian University of Technology Postgraduate Thesis, 1987.7

Bubble Departure Frequency during Boiling from Porous Surfaces

MING QI and YIHAN CAI
Tianjin Geothermal Research and Training Center
Tianjin University
Tianjin, PRC

D. H. FREESTON
Geothermal Institute
University of Aukland
New Zealand

ABSTRACT

An experiment is described in which the bubble departure frequency on a horizontal porous surface heated from below is measured as a function of the applied heat flux. The working fluid is R113.

Results are presented for six porous surfaces and a plain surface for a range of heat fluxes from 0.2-4 W/CM².

It is concluded that the frequency verses wall superheat curve can be divided into four stages. At high heat fluxes a horizontal flow of boiling fluid is identified which results in an enhanced heat transfer over that of a plain surface. Finally a simple formula is presented which is considered sufficiently accurate for engineering purposes to calculate bubble departure frequency as a function of wall superheat.

INTRODUCTION

The early patents of Milton(1968), Kakizaki et al (1975) and Fujie et al (1977) introduced the ideas of using porous surfaces to enhance the heat transfer during nucleate boiling[1,2,3]. There has been a number of experiments performed on various types of porous surfaces[4], also the use of such surfaces has been reported in many fields, for example, the work of Czikk et al (1970), O'Neill et al (1971) on sintered surfaces and Nobukatsu et al (1977) on manufactured surfaces[5,6,7].

Despite these applications, the heat transfer mechanism and the reasons for the improved performance over a standard smooth tube are not well understood. There are many variables involved in the boiling processes and the control of these variables, experimentally, has proved difficult, so that a data base on which to build a detailed theoretical model has not been possible.

In recent years the work of Nakayama et al (1980,1982) has been prominent in providing data and theoretical analysis to aid in understanding the processes[8,9]. The theoretical work is based on some simplyfying assumptions which have introduced a lack of generality of the final result.
One of the major problems in developing a general theory has been the lack of quality data, particularly with respect to the bubble departure

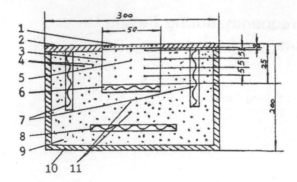

FIGURE 1. The structure of a test block (not to scale)
(1) porous surface (2) lid (3) nickle block (4)(5)(11) thermalcouples
(6) main heater (7)(8) auxilliary heaters (9) insulation (glass wool)
(10) container

frequency. To provide this data the authors set up an experiment to
measure frequencies as a function of heat flux. This paper describes
these experiments, presents the data and discusses the results with a
description of the heat transfer processes involced.

EXPERIMENTAL APPARATUS

Figure 1 shows the structure of the test block. The porous surface (1) is
facing up and horizontal. It is made of two parts: the copper test
surface and the nickle base. The design is similar to that used in
reference[10], however the means of joining the cupper to the nickle
surface differed, a vacuum dispersion technique was used. This is
considered superior to the silver soldering technique used in [10], since
the copper and nickle surfaces are in direct contact with each other and
do not have a layer of metal between them. However, a copper nickle
connection layer formed by molecular movement of not more than 0.005mm
exists which in the context of thermal resistance is considered to be
negligible.

Also included in the test block are the main heater(6) and auxiliary
heaters (7)(8). The auxiliary heaters maintain temperature in the
insulating material (9) to control heat losses from the test area. Buried
thermocouples, (4)(11), indicate the dirrection of heat transfer, which
can be corrected by use of these auxiliary heaters. The apparatus is
encased in a container(10).

Power input to the test section is measured on a calibrated power meter.
Thermocouples (5) inserted into the nickle block, as illustrated on
figure 1 give the test surface temperature by extroplation of the
temperature distribution to the surface. The saturated liquid temperature
above the test surface was also measured by a thermocouple.

The pressure is controled to atmospheric by manipulating the cooling
water through the condenser tubes. Figure 2 is the bubble departure
frequency measurement system which consists of five parts: (1) fibre
optics, (2) laser source, (3) light-eletricity convertor, (4)
preamplifier and (5) data acquisition system(DAC including ADC+APPLE
II+). The test fluid is R113.

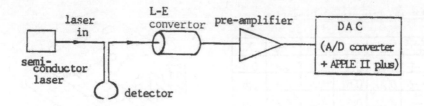

FIGURE 2. Frequency measurment system

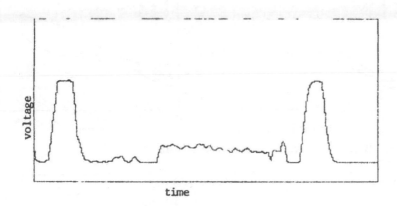

FIGURE 3. Voltage pulse of bubble

The light emitted by the laser is transmitted through the fibre optics system to the boiling zone. A density change caused by a bubble being released from the surface is detected producing a voltage pulse which is recorded by the DAC.

The detector's dimentions are such that all bubbles released from one pore, over a given time, are recorded as voltage pulses by the DAC. Also because of its size there is little interruption to the bubble movement.

Figure 3 shows a typical voltage pulse recorded by the computer.

EXPERIMENTAL RESULTS AND DISCUSSION

Six test surfaces were examined. Table 1 and figure 4 give the structural dimentions of the test surfaces. These were tested over a range of heat fluxes from 0.2 W/CM^2 up to 4 W/CM^2,however above 4 W/CM^2 difficulty was experienced in recording pulse frequency due to the output from the detector indicating a large turbulent mixing zone at the test surface.

TABLE 1. Sturctural Dimention

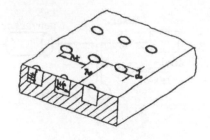

(mm)

surface No.	d_o	H_t	W_t	λ_o	λ_t
1—1	0.20	0.9	0.4	0.8	0.6
1—2	0.20	0.6	0.4	0.8	0.6
1—3	0.20	0.3	0.4	0.8	0.6
2—1	0.15	0.9	0.4	0.8	0.6
2—2	0.15	0.6	0.4	0.8	0.6
2—3	0.15	0.3	0.4	0.8	0.6

FIGURE 4. Geometry of porous surface

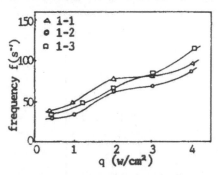

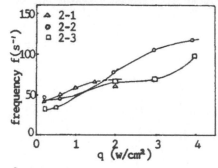

FIGURE 5. Bubble departure frequency v's heat flux q

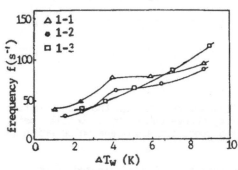

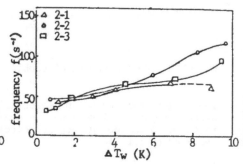

FIGURE 6. Bubble departure frequency v's wall superheat Tw

The test technique was to vary the heat flux and observe changes in wall
surperheat while taking measurements of bubble departure frequency.
Figure 5 and 6 show both plots of deperture frequency against heat flux q
and against wall superheat ▲Tw for the range of test surfaces(Table 1).

Figure 7 shows diagramatically the trend of these experimental results in
which the wall superheat defined as ▲Tw(surface temp.-saturated liquid
temp.) is plotted against bubble departure frequency f. The bubble
departure frequency f increases as wall superheat ▲Tw is increased and
four stages of boiling are identified. For a paticular test surface, the
geometry is fixed and for a given system pressure the wall superheat
required for bubble departure will also be fixed.

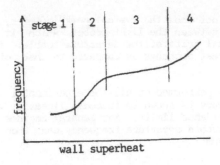

FIGURE 7. The trend of frequency

Initially the wall superheat has a low value and no bubble is generated at the surface. As the superheat is increased the bubble departure frequency is rose and a number of pores became active. These active pores carry the increased heat flux away from the surface, which does not allow the individual pore bubble departure frequency to increase. As described in reference [10], about 10% of pores are active at this stage,stage 1.

When increasing the wall superheat still further, the frequency rises at an increaseing rate which is labled stage 2 on figure 7. It is suggested that the vapor in the channels prefers to find its way to the surface through the activated pores rather than those not active, that is an increased heat flux, during this stage, results in an increased rate of bubble departure frequency. This stage occurs over a relativelly small range of superheat.

With increases of heat flux and consequently wall superheat, the resistance to flow through the active pores becomes greater than the resistance to flow through the non-active pores, these pores then became active. So the rate of increase of bubble frequemvy with ΔTw becomes less. This is stage 3 and is termed the "post-activating" stage. At the end of this stage more pores have been activated and to maintain the mass balance of liquid entry to the channels with vapor release, the limit to the number of active pores has been reached. This is stage 4. Increasing heat flux results in a increased rate of bubble departure frequency up to a heat flux of 4 W/CM^2, after which as mentioned earlier, it was not possible to record departure frequency correctly.

Figure 8 shows the boiling curves, heat flux q v's wall superheat ΔTw, for the six test surfaces and a plain surface. All porous surfaces show improved heat transfer over that of the plain surface up to about 6-7 W/CM^2, after which all data trends to collapse onto a single curve. There is a slope transition between heat fluxes of 1-3 W/CM^2. The same phenomena was identitied in reference [8], however the authore were not able to offer an explanation at that time. A comparison of figures 5 and 6 with figure 7 shows this region to correspond to stage three, the post-activating stage. the heat flux required for this stage is a function of the geometry of the heat transfer surface as demonstrated in figure 8.

Figure 9 shows pulse traces recorded at high fluxes(>4 W/CM^2). Because of disturbances caused by the speed of departure and the number of bubbles there would appear to be some horizontal movement of bubbles at the bubble release point. A-B on figure 9 shows a typical trace for vertical

453

bubble departure, while C-D does not show the typical peak. however, the trace reflects an interference between the light probes which it is suggested is caused by a horizontal drift of the departure bubble. This results in an improvment of the heat exchange as compared to that of a plain surface.

Finally a regression analysis is performed on all the experimental data and results in figure 10. Frequency is snown to increase linearly with wall superheat within the 99% confidence limits. For general engineering design purpose with R113, the bubble departure frequency can then be expressed as an equation of the form:

$F=C \cdot \Delta Tw$

for the test geometry the equation is $F=130 \cdot \Delta Tw$ for ΔTw between 2--12K.

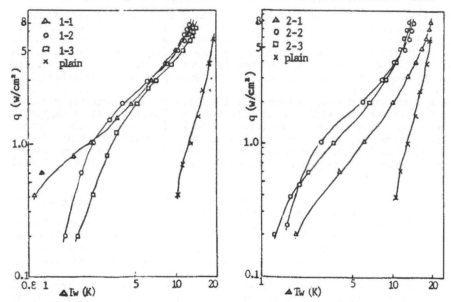

FIGURE 8. Boiling curves

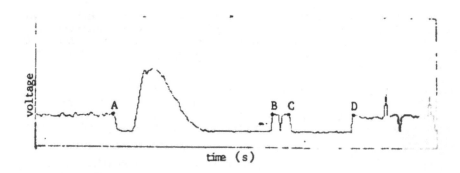

FIGURE 9. Voltage pulse of bubbles recorded at high heat flux(>4 W/CM^2)

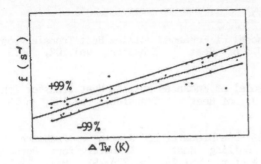

FIGURE 10.   Frequency regression result

CONCLUDING REMARKS

An nucleate boiling experimental study of the bubble departure frequency from heated porous surface has led to the following conclusions:

1. Bubble departure frequency increases with applied heat flux. Four stages can be identified by the rate of the increase of frequency with heat flux. A "post-activating" region was defined as occuring when an increase in the number of activepores reduces the slope of the frequency-heat flux curve.

2. A horizontal flow at high heat flux (>4 W/CM^2) was identified which showed an increased heat transfer over that of the plain surface.

3. For engineering use the bubble frequency, based on these experiments, can be calculated from a simple linear equation: $F = C \cdot \Delta Tw$. More experimental measurments are necessary to determine wether this formula is general enough to be applied for all manufactured porous surfaces.

REFERENCE

1. R.M.Milton: U.S. Patent, No.3384154, May, 1968.

2. Kakizaki et al: U.S. Patent, No.3906604,Sept.,1975.

3. Fujie et al: U.S. Patent, No.4060125, Nov., 1977.

4. R.L.Webb: Heat Transfer Engineering,1982.

5. A.M.Czikk et al: Performance of Advanced Heat Transfer   Tubes   in Refrigerant-Flooded Liquid Coolers,   ASHRAE, vol. 76, part I, pp.96-109, 1970.

6. P.S.O'Neill et al: Heat Exchanger for NGL, Chemical   Engineering Progress, vol.67, no.7, pp.80-82, July, 1971.

7. Arai Nobukatsu et al: Heat Transfer Tubes Enhancing   Boiling and Condensation in Heat Exchangers of   Refrigerating Machine, ASHRAE,

Trans., vol.83, part 2, 1977.

8. M.Nakayama et al: Dynamic Model of Enhanced Boiling Heat Transfer on Porous Surface--Part I, J. of Heat Transfer, vol.102, no.3, pp.445-450, 1980.

9. M.Nakayama et al: Dynamic Model of Enhanced Boiling Heat Transfer on Porous Surface--Part II, J. of Heat Transfer, vol.102, no.3, pp.450-456, 1980.

10. M.Nakayama et al: Effects of Pore Diameter and System Pressure on Saturated Pool Nucleate Boiling Heat Transfer form Porous Surface, J. cf Heat Transfer, Vol.14, pp.286-291, May, 1982.

Improvement of Boiling Heat Transfer Performance for Vertical U-Shaped Tubes

TING-KUAN CHEN and XIU-ZHUANG JIA
Xi'an Jiaotong University
Xi'an, 710049, PRC

ABSTRACT

In order to improve the boiling heat transfer performance in vertical U-shaped tubes, a modified U-shaped tube with inclined upward bottom was tested in our high pressure electrically heated water loop under the conditions of p=7.5-14 MPa, G=500-1350 kg/m^2.s and q=90-330 kW/m^2. This paper describes the experimental results on the boiling heat transfer deterioration in the bend of modified U-shaped tube with R_1/d =3, R_2/d=6 and the inclined angle of the bottom straight tube of 18 degree. According to the tests the mass velocity at which the wall temperature rises is much decreased than that in ordinary U-shaped tubes. So the modified U-shaped tube may be recommended for waste heat boilers in synthetic ammonia plants for ensuring the safe operation of the boiler.

INTRODUCTION

Waste heat boilers with vertical U-shaped tubes are used in larger synthetic ammonia fertilizer plants with daily output of a thousand tons to recover heat from high temperature technological gas and generate high pressure saturated steam of 180 t/h at 10.6 MPa. This is significant for increasing the economic efficiency of the plants. However, in operation of the boiler the accident of tube failure in the bend due to overheating occurs frequently, seriously affecting the safety of production. In order to find out the cause of the tube failure, the heat transfer performance in the bend of vertical U-shaped tubes have been studied experimentally by authors [1,2] and it has been obtained that in the bend of vertical U-shaped tubes at lower mass velocities and low steam qualities, under the action of both centrifugal and gravitational forces, the steam-water stratification occurs, leading to elevated wall temperature in the inner side of the bend. The region of heat transfer deterioration is from the middle of the downward section of the bend to 15° of the upward section and the wall temperature rises to a maximum at 15° upstream the bottom of the tube. Fig.1 shows the maximum temperature difference Δt between the inner side and outer side of the bend at p=10.6 MPa for various mass velocities and heat fluxes. In waste

The project was supported by the NSF of China.

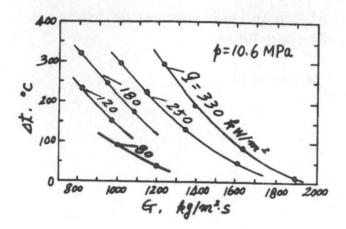

Fig.1. The maximum temperature difference between the
inner side and outer side of the bend at various mass velocities and heat fluxes

heat boiler design of synthetic ammonia plant, the mass velocity in tube was taken about 1200-1300 kg/m².s and the maximum heat flux is about 300 kW/m², so it may be seen that the
wall temperature rise is quite great, thus the accident of
tube failure in the bend of U-shaped tube occurs frequently.

According to author's study on two-phase flow pattern
in vertical U-shaped tubes[2], at lower mass velocities and
low steam qualities, the two-phase flow is in stratified flow
region. Because of steam-water stratification, the inner side
of the tube is in contact with steam generated and can't be
cooled by water, resulting in heat transfer deterioration and
wall temperature rise. It also may be seen from bubble moving
characteristics that in the downward section of vertical U-
shaped tube bend, since the bubble velocity is slower than
the liquid under the action of buoyance which direction is
opposite to the flow direction, thus the bubble easily accumulates in the inner side of the tube, leading to wall temperature rise. When the bubble moves to the upward section of
the tube bend, the buoyance direction acting to the bubble is
same with the flow direction, so the bubble moving velocity
is faster than the liquid, the bubble no longer accumulates
on the wall of the tube, thus the wall temperature rise disappears soon. In vertical upward or downward flow, when the
bubble generated on the wall grows to certain size, it will
depart from the wall surface to the flow, so there is a better
cooling condition on the wall. In horizontal or curved tube,
the bubble flows along the upper side or inner side of the
tube under the action of gravitational and centrifugal forces,
so the heat transfer deterioration easily occurs on the upper
side or inner side of the tube. It may be known from the above
analysis that in various direction flow tubes, the flow condition of the slight inclined downward tube is the worst, so
the heat transfer deterioration occurs easily. Thus in vertical U-shaped tubes, the heat transfer in the downward section

near the bottom of the tube is the worst, so the wall temperature rises most seriously.

From above analysis, in order to improve the heat transfer performance in vertical U-shaped tubes, the downward section of the tube in which the fluid flows downward slowly should be eliminated, so a modified tube type with inclined upward bottom is presented instead of the ordinary U-shaped tube to improve the heat transfer performance.

EXPERIMENTAL APPARATUS AND PROCEDURE

The schematic diagram of the high pressure test loop is shown in Fig.2. Feed water is pumped by the high pressure plunger pump through the regenerative heat exchanger and main preheater to the test tube. The steam-water mixture from the test tube flows through the regenerator, cooler, and then returns to the tank. The main preheater and test tube are directly heated by alternating-current power supplies with maximum heating capacities of 500 kW. The test tube is made of stainless steel tube of $\phi 25 \times 2$ mm and the central distance between both straight tubes is 1400 mm.

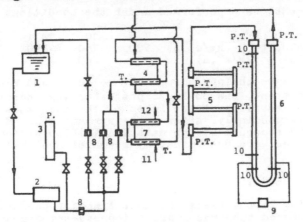

Fig.2. Schematic diagram of the high pressure test loop

1-water tank 2-plunger pump 3-pressurizer 4-regenerative heat exchanger 5-main preheater 6-test tube 7-cooler 8-orifice flowmeters 9-differential pressure transmitter 10-electrodes 11-inlet of cooling water 12-outlet of cooling water T-thermocouples P-pressure gauges

The test section structure of modified U-shaped tube and the arrangement of thermocouples are shown in Fig.3. The heating length is of 2.38 m and that is same with the original U-shaped tube. 36 chromel-silicon thermocouples with 0.2 mm diameter are arranged at 12 sections of the test section to measure the outside wall temperature profile along the tube. Three thermocouples on each section are arranged at the top, bottom and side of the tube.

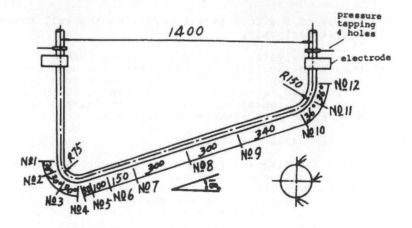

Fig.3. The modified U-shaped tube test section
and the thermocouple arrangement

Experiments were performed under the conditions as
follows:

Pressure P (MPa)	7.5	10	14	
Mass velocity G (kg/m^2.s)	500	750	1050	1350
Heat flux q (kW/m^2.s)	90	260	290	330

In the tests the pressure, mass velocity and heat flux
were kept constant, the power to preheater was increased in
steps, so the enthalpy at the inlet of the test tube was increased correspondingly. After each increasing all data were
collected and recorded by the data acquisition system. The
steam qualities were determined by means of heat balance method.

EXPERIMENTAL RESULTS

According to the tests the heat transfer performance in
modified U-shaped tube has a great improvement comparing with
the ordinary U-shaped tube. The mass velocity at witch the
heat transfer deterioration occurs at inlet bend of the tube
in low steam quality region is much decreased than that in
ordinary U-shaped tube. Fig.4. shows the wall temperature profile when the heat transfer deterioration occurs at various
mass velocities under the conditions of p=10 MPa and q=330
kW/m^2. It may be seen from the Fig.4 that at low mass velocities the heat transfer deterioration occurs at inlet bend of
the tube and eliminates gradually after the inlet bend. With
increasing the mass velocities, the wall temperature rise decreases. When G≥1050 kg/m^2.s, the heat transfer deterioration
doesn't occur. Fig.5 shows the maximum temperature difference
between the inner side and outer side of the bend at various
mass velocities and heat fluxes and gives the data of U-shaped
tube simultaneously for comparison. It maybe seen that the
heat transfer performance in modified U-shaped tube is much
better than that in ordinary U-shaped tube. If the synthetic
ammonia waste heat boiler uses the modified U-shaped tubes,

then the problem of heat transfer deterioration won't occur under the ordinary design mass velocity condition of 1200-1300 kg/m^2.s, and it will be quite safe in operation.

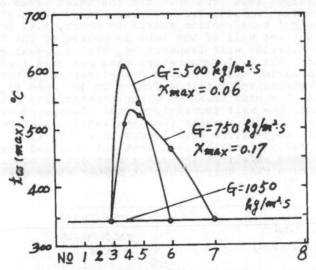

Fig.4. The wall temperature profile at various mass velocities

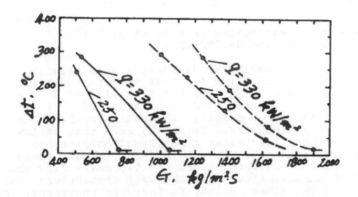

Fig.5. The maximum temperature difference between the inner side and outer side of the bend at various mass velocities and heat fluxes
solid line -- for modified U-shaped tube
dotted line -- for ordinary U-shaped tube

The heat transfer performance in modified U-shaped tube may be divided to three parts, i.e. inlet bend, inclined straight tube section and outlet bend. In the test parameter range, the heat transfer deterioration at the outlet bend doesn't occur. Fig.6 shows the heat transfer deterioration

461

performance at p=10 MPa, q=330 kW/m^2 and G=1050 kg/m^2.s. It may be seen from the Fig.6 that at low steam qualities, the heat transfer deterioration doesn't occur at both inlet bend and inclined straight tube section. When the inlet steam qualities x_{in} increases to 0.17, the heat transfer deterioration at inclined straight tube section starts to occur, i.e. the liquid film on the top wall of the tube is broken by the flow, thus leading to elevated wall temperature. With increasing the steam quality, the wall temperature rise has some increase, but the region occuring heat transfer deterioration is basically constant between the sections from 8 to 10. When x_{in} increases to 0.4, the heat transfer deterioration also occurs at the inlet bend. The wall temperature rise disappears after the inlet bend and it rises again from the section 8, forming the especial wall temperature profile for this tube type. From the test results it may be seen that the modified U-shaped tube can be used safely at the waste heat boiler conditions, the wall temperature rise only occurs at higher steam qualities.

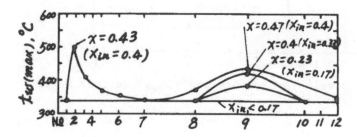

Fig.6. The wall temperature profile at p=10 MPa, q=330 kW/m^2 and G=1050 kg/m^2.s

Fig.7 shows the results at decreasing mass velocity to 750 kg/m^2.s. At this condition the heat transfer deterioration starts to occur at the inlet bend steam quality of 0.08 and the wall temperature reaches a maximum value 520°C at x=0.17. With increasing the steam quality, the wall temperature rise decreases and then disappears. It may be seen that at low mass velocities the steam-water stratification type heat transfer deterioration occurs at inlet bend, as the steam quality increases to certain value, the two-phase flow changes to liquid film annular flow, thus the wall temperature rise disappears when the steam quality furthermore increases, the liquid film on the top wall of the inclined straight tube section will be broken, leading to elevated wall temperature. With increasing the steam quality, the wall temperature rise has some increase first, and then has some decrease due to increasing the steam velocity.

With decreasing the heat flux, the heat transfer performance will be better. It is shown in Fig.8 that when the heat flux decreases to 260 kW/m^2, the heat transfer deterioration doesn't occur at inlet bend. As x_{in} increases to 0.27, the heat transfer deterioration starts to occur at inclined straight tube section. With increasing the steam quality, the wall temperature rise increases until x_{in} to 0.42, after then the wall temperature decreases due to increasing the steam velovity. Comparing with the condition of q=330 kW/m^2, the steam quality at which the heat transfer deterioration starts to occur increases, and the wall temperature rise decreases.

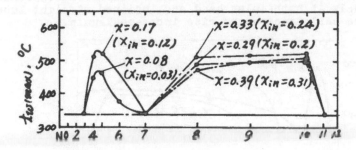

Fig.7. The wall temperature profile at p=10 MPa, q=330 kW/m^2 and G=750 kg/m^2.s

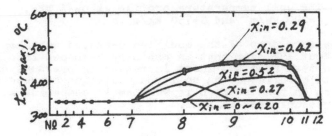

Fig.8. The wall temperature profile at p=10 MPa, q=260 kW/m^2 and G=750 kg/m^2.s

Fig.9 and Fig.10 show the test results at different pressures. It may be seen that with increasing the pressure, the heat transfer performance at inlet bend tends to be better at low steam qualities, this is similar with the test results of ordinary U-shaped tube. That is in curved tubes, the steam-water stratification type heat transfer deterioration will be

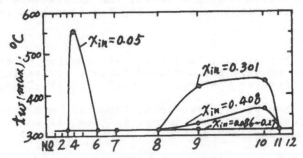

Fig.9. The wall temperature profile at p=7.5 MPa, q=290 kW/m^2 and G=750 kg/m^2.s

better with increasing the pressures. But in inclined straight tube section, the heat transfer deterioration is enlarged with increasing the pressure. For example as p=14 MPa, the heat transfer deterioration starts to occur at x_{in}=0.03 and with increasing the steam qualities, the wall temperature rise decreases. As x_{in}=0.5, the heat transfer deterioration occurs simultaneously at both inlet bend and inclined straight tube section, the wall temperature rise increases again.

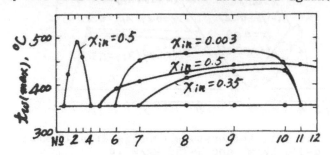

Fig.10. The wall temperature profile at p=14 MPa, q=290 kW/m² and G=750 kg/m².s

The pressure drop of the modified U-shaped tube bend is increased about 19% over that in ordinary U-shaped tube bend based on the test results. These increases in pressure drop has only little influence on boiler circulation because the pressure drop in bend section is very small comparing with the total pressure drop in coil.

CONCLUSION

It may be obtained from the test results that using the modified U-shaped tubes with inclined upward bottom instead of the ordinary U-shaped tubes may much improve the heat transfer performance in synthetic ammonia waste heat boiler. Under the original design conditions of p=10.6 MPa, q_{max}=300 kW/m², G=1200-1300 kg/m².s and x<0.1, the heat transfer deterioration may be avoided to occur. If the thermal insulating measure is used at the inlet bend of the tube to decrease the heat flux, then the operation of the modified U-shaped tube will be more safe. Using modified U-shaped tubes, the manufacture technology is also more simple due to using the same bend radius.

REFERENCES

1. Ting-kuan Chen et al., "Boiling heat transfer deterioration in the bend of vertical U-shaped tubes", Proceedings of 8th Int. Heat Transfer Conference, Vol.5, pp.2179-2184, 1986.

2. Ting-kuan Chen et al.,"Two-phase flow and heat transfer in vertical U-shaped tubes" Journal of Chemical Industry and Engineering (English Edition of China) Vol.1. No.2 pp.1-34, 1986.

Boiling Heat Transfer in Liquid Saturated Porous Bed

M. H. SHI and L. L. JANG
Department of Power Engineering
Nanjing Institute of Technology
Nanjing, PRC

ABSTRACT

An experimental and theoritical investigation has been conducted to determine boiling heat transfer performance in a liquid saturated porous bed heated from below. The influence of
porous materials on nucleate boiling process was analysed.
According to the model of resistance to vapor flow the limiting heat flux for the region of boiling heat transfer enhencement in porous bed was determined. The effects of particle
material on boiling heat transfer using glass beads, silicon
sand and steel balls as the test media were investigated.

INTRODUCTION

Boiling heat transfer in liquid saturated porous bed is an
object of intensive study in recent years. It is encountered
in many important applications, such as high-flux boiling
heat transfer surface, heat pipe, geothermal energy system
and post-accident heat transfer performence in liquid cooled
nuclear reactors.

A number of researches have been conducted to determine the
boiling heat transfer characteristics in liquid saturated
porous layer. Tsychiya et al(1) investigated boiling heat
transfer performance in different layers, consisting of particles such as steel balls, bronze balls, sand and glass beads.
Ferrel(2) studied the mechanism of boiling heat transfer from
wick covered surface. Corwell(3) made a visual observation of
boiling in porous media by use of a polyurethane foam layer
of 6 mm thick. Torrance(4) investigated boiling processes in
a vertical cylinder heated from below and cooled from above.
They found that a two-phase zone formed near the heating surface when the local temperature reached saturation. Sugawara
et al(5) determined the effects of particle size, particle
properties on boiling heat transfer. Fukusako et al(6) investigated transition and film boiling heat transfer in a liqid
saturated porous bed.

All these previous works are instructive, but uncertainties
exist about many aspects of this boiling process, especially

in bubble dynamics and heat transfer characteristics. In this investigation, therefore, the nucleate boiling heat transfer mechanism in porous bed is analysed, the special attention is focused on the determination of the limiting heat flux for the region of boiling enhencement and the effects of diameter of the particle, height of porous layer and particle properties on boiling heat transfer characteristics in porous bed.

THE LIMITING HEAT FLUX FOR THE REGION OF BOILING ENHENCEMENT

The boiling heat transfer characteristics in porous bed is significantly different from that in an unrestrained continuum fluid, namely in pool boiling without particles. The major effects of porous material on pool boiling heat transfer mechanism are as follows:

1. Additional active nucleation sites are available due to existance of the porous materials being contacted with the heating surface, the boiling heat transfer from the surface is enhenced. The porous medium will shift the boiling curve to the left side in the q-ΔT_s diagram. A boiling curve of water in glass beads bed is given in Fig.1. It shows that the enhencement of boiling heat transfer in lower heat flux range is obvious.
2. After a bubble nucleus is formed at the cavity it will grow up within the gaps between particles, the heat is transferred to the bubble not only from the heating surface but from the particles. The bubble growth rate would be increased if the porous material has better thermal conductive character. That is often the case, therefore the boiling heat transfer is enhenced.
3. The particles being contacted with the heating surface have the fin-like effect and cause the increase of the heating surface area.
4. The departure of the bubble from the heating surface and movement in porous bed will be restricted and become difficult. The frictional resistance to vapor flow through the small passages causes the vapor bubble to collect in some area at the heating surface and the vapor zone will be formed, the boiling heat transfer rate will decrease.
5. The liquid flow passing through the small passages and moving toward the heating surface is also resisted. The liquid-vapor exchange would be affected and thanthe heat transfer rate will decrease.

The general effect of porous materials on boiling heat transfer depends on all these factors mentioned above. At lower heat flux range the number of nucleation sites, the bubble frequency and growth rate are smaller, the flow resistance of bubbles passing through the porous bed is not significant, the first three factors mentioned above govern the boiling process and cause the heat transfer to enhence. By increasing the heat flux the hydrodynamic resistance for vapor bubbles flow passing through the porous bed will increase until the effect of resistance becomes a governing factor at high heat flux range and boiling heat transfer rate will reduce. Thus the boiling curve in porous bed will intersect the boiling curve without particles at a certain heat flux. We call the heat flux corresponding to this intersecting point the limiting heat flux for the region of boiling enhencement as indicated in Fig. 1.

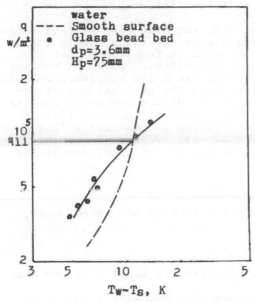

Fig. 1 Boiling curve in porous bed

To estimate this limiting heat flux, it is necessary to consider the movement of the vapor in porous bed. A visualization of heat transfer in porous bed shows that the bubbles departing from the heating surface are first collected at some areas which are in close to the heating surface and formed the local vapor zones. When the vapor pressure in the vapor zone is large enough to overcome the frictional resistance to vapor flow through the small passages, the vapor zone will discharge periodically the vapor bubble passing through the porous layer to free surface of the liquid.

The flow through the porous layer is a very complex phenomenon. The basic characteristic is that the porous material confine the fluid flow in some definite passages and cause a drag force acting on the fluid. Some physical models were proposed in the literatures to determine the flow characteristics in porous bed. A resistance model suggested by Rumer(10) will be used to analyse the vapor bubble flow in porous bed here.

According to this model the local flow resistance in porous bed can be a analogy with the resistance caused by fluid passing through a sphere, thus the local flow resistance in porous bed can be expressed as

$$D = \lambda \mu v d_p w \qquad (1)$$

where λ is the coefficient of resistance depending on the interaction between the sphere and other particles.

It is obvious that λ depends on the situation around the particle, thus λ is a function of the structure of porous bed and for a single sphere in infinite fluid $\lambda = 3\pi$.

The vapor velocity increases when the heat flux becomes larger, consequently the flow resistance for vapor bubble moving upward in porous bed will increase. It is assumed that when this flow resistance approaches a critical value the boiling heat transfer in porous bed would reach the reduction region. Thus the heat flux corresponding to the critical resistance is equal to the limiting heat flux for the region of boiling enhancement. It is expected that the limiting heat flux in small particle bed is smaller because it has narrow flow passages and the local resistance of vapor flow will increase sharply as the heat flux increases. The critical resistance can be expressed as

$$D_{cr} = \lambda \mu_v d_p w_{cr} \qquad (2)$$

If the heat is transferred in the form of the latent heat, the energy equation is

$$q = n A s w \rho_v h_{fg} \qquad (3)$$

where n --- number of small passages for vapor flow per unit surface;
  A --- equivalent flow area of a passage;
  s --- coefficient considering the fin-like effect of the particles being contact with surface.
solving the average vapor velocity w from equation(3) and inserting it into equation(2), we get

$$D_{cr} = \lambda \mu_v d_p q / (\rho_v h_{fg} n A s) \qquad (4)$$

Thus the limiting heat flux can be written as

$$q_{li} = D_{cr} \rho_v h_{fg} n A s / (\lambda \mu_v d_p) \qquad (5)$$

Assuming that $D_{cr} n A / \lambda$ is a function of the porous bed structure and can be expressed by

$$D_{cr} n A / \lambda = C_1 \varepsilon^{n1} d_p^{n2} \qquad (6)$$

The coefficient s mainly depends on the structure and thermal conductivity of the porous media, it can be expressed by

$$s = C_2 \varepsilon^{n3} d_p^{n4} F(K_p / K_1) \qquad (7)$$

Where $F(K_p / K_1)$ is a function of thermal conductivity K_p and K_1. Inserting Equ.(6) and Equ.(7) into Equ.(5), finally we get

$$q_{li} = C h_{fg} d_p^n \varepsilon^m F(K_p / K_1) \nu^{-1} \qquad (8)$$

where C is a constant depending on the combination of liquid
and surface, it can be gotten by experimental data. The exponents n and m, as well as the expression of $F(K_p/K_l)$ can also
be obtained from the experiments. We shall do this in next.

EXPERIMENTAL WORK

Experimental apparatus and procedure

The experimental apparatus is shown in Fig.2. It consists of
an inner Pyrex tube 90 mm in diameter and 150 mm in height.
A guard heater around the inner tube is used to keep the test
liquid in saturation temperature. The inner tube is filled
with a liquid saturated porous materials. 10 mm liquid level
is held above the porous bed. To prevent fluidization of the
porous bed,a thick punched brass plate is used on porous bed
surface. The test section is consists of a cylindrical copper
block, a film heater put on bottom and three 0.2 mm Chromel-
alumel thermocouples positioned at three levels of centerline
in the copper block. The heat transfer surface is 60 mm in
diameter. The surface temperature is estimated by extrapolating
the temperature distributions obtained by the thermocouples.
Four particles are used in the test. They are listed in Table
1. Distilled water is used as test liquid. The bed height is
changed over a range of 30-100 mm. Starting with pool boiling
without particles the experiments are carried out for different particles and bed parameters at steady saturation states.

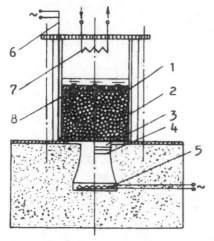

1. porous bed
2. glass tube
3. test section
4. thermocouples
5. film heater
6. auxiliary heater
7. condenser
8. brass plate

Fig.2 Experimental apparatus

Table 1 Test particles and their physical properties

Materials	d_p, mm	ρ_p, kg/m^3	K_p, w/m.k	ε
sand	1.0	1700	0.3	0.38
glass beads	1.1	2430	0.74	0.38
glass beads	3.6	2430	0.74	0.39
steel balls	3.0	7840	47.5	0.39

Results and discussion

1. Comparison with boiling on porous surface

As mentioned above, boiling heat transfer in porous bed can
be enhenced at lower heat flux region because of additional
active nucleation sites and larger bubble growth rate. But
compare with the porous surface, the experimental results show
that the degree of boiling enhencement in porous bed is much
smaller than on the porous surface indicated in Fig.3. This
means that the boiling in porous bed with a given bed height
would be influenced by the resistance of porous bed to vapor
flow.

2. Effect of particle diameter on boiling heat transfer

The Fig.4 shows that the boiling heat transfer rate for small
particles is somewhat greater than that for larger particles
at lower heat flux, but with increase of heat flux the situation will be changed, the boiling heat transfer rate for large
particles is much larger than that for the small particles
because the effect of the resistance to vapor flow. It is also
shown that the limiting heat flux decreases with decreasing
the particle diameter. For the nonuniform particle bed,such
as mixed sand bed, the heat transfer rate may be increased
because of the fluidization of small particles.

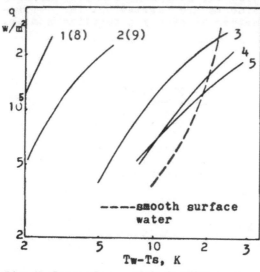

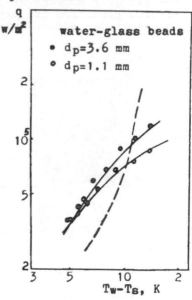

Fig.3 Comparison with boiling on
the porous surface
1. sintering surface 2. porous
surface 3. steel ball bed
4.glass beads bed 5.sand bed

Fig.4 Effect of parti-
cle diameter

3. Effect of bed height on boiling heat transfer(Fig.5)

For large particles, no distinguishable effect of bed height
is found on heat transfer. This means that the boiling heat
transfer rate is mainly determined in the comparatively thin

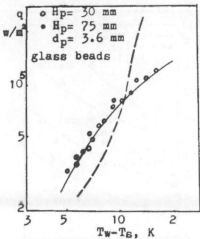
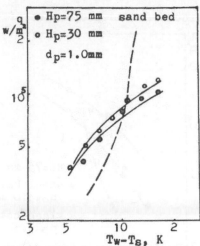

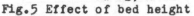

Fig.5 Effect of bed height   Fig.6 Effect of fluidization

layer of particles near the heating surface because the resistance effect plays a important role only in this layer. But
for small particles the situation would be changed because of
fluidization of the particles. With increase of bed height,
small particle's fluidization decreases, hence the heat transfer rate becomes smaller as shown in Fig.6.

4. Effect of particle material on boiling heat transfer (Fig.7)

It appears that for the same particle diameter the higher heat
transfer rate belongs to the material which has larger thermal
conductivity. It can be expounded by the fin-like effect as
well as the top heating effect to the bubbles.

5. The limiting heat flux for the region of boiling enhencement

According to the experimental data of this work and literatures
(5,6,11,12), a semi-empirical equation based on Equ.8 for the
limiting heat flux can be expressed as follows:

$$q_{11} = C \, h_{fg} d_p^{0.4} \nu_v^{-1} \varepsilon \, (1-e^{-0.7(K_p/K_1)}) \tag{9}$$

where C is a constant depending upon the physical properties
of the liquid and the base surface. For water-copper base,
$C = 2.85 \times 10^{-5}$, for F-11-copper base, $C = 1.36 \times 10^{-5}$. Fig.8
shows that the agreement about the experimental data and the
calculated by Equ.(9) is satisfactory.

CONCLUSIONS

1. There exists a limiting heat flux for boiling enhencement
in porous bed. It can be calculated by Equ.(9).

2. The boiling characteristics in porous bed is much different from pool boiling on the porous surface, it depends not
only on liquid-surface combination but on the structure of

porous bed and properties of porous material.

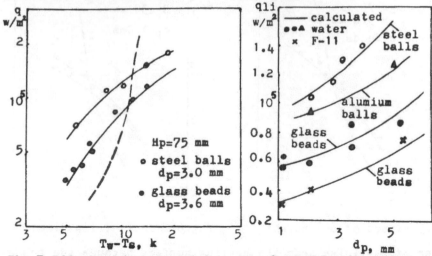

Fig.7 Effect of bed material   Fig.8 Limiting heat flux for
                                     boiling enhencement

NOMENCLATURE

d_p particle diameter, mm, m; q heat flux, w/m^2
K thermal conductivity, w/mk; ε porosity
μ dynamic viscosity, pa·s; ρ density, kg/m^3
H_p height of porous bed, mm; ν kinematic viscosity, m^2/s
w average vapor velocity in porous media, m/s

REFERENCES

1. Tsuchiya,M, Shimizu,s., and Takeyama,T.
   6th National heat Transfer Symposium of Japan,1969,pp25-28.
2. Ferrell,J.K., and Alleavitch,J. Chem.Eng.Sym., Vol66, 1970.
3. Cornwell,K., Nair,B.G.,and Patter,T.D.
   Int.J. Heat Mass Transfer, Vol.19, 1976, pp.236-238.
4. Torrance,K.E., 1983 ASME-JSME Thermal Engineering Joint
   Conference, 1983, Homolulu, Hawaii, USA.
5. Sugawara,A.,and Takahashi,I., 18th National Heat Transfer
   Symposium of Japan, 1981, pp. 352-354.
6. Fukusako,S., Komoriya,T.,and Seki,N., J. Heat Transfer,
   Feb., Vol.108, 1986, pp117-124.
7. Fukusako,S., Seki,N.,and Komoriya,T., 1983 ASME-JSME
   Thermal Engineering Joint Conference, Vol.2, 1983, Hawaii.
8. Zhuan L.X., 1985 National Heat Transfer Conference, China.
9. Chen J.B., et al, 1984 National Heat Transfer Conference,
   China, 1984.
10.Rumer,R.R., Flow through porous media, Academic Press,
   New York, 1969.
11.Fukusako,S,N.Eguchi and N.seki, 2nd ASME-JSME Thermal
   Engineering Joint Conference, Vol.2, 1987, Hawaii.
12.Jang,L.L., M.S. Thesis, Nanjing Institute of Technology,
   China, 1988.

A Theoretical Model for the Prediction of Void Fraction in Two-Phase Channel Flow

LIE-JIN GUO and XUE-JUN CHEN
Engineering Thermophysics Research Institute
Xi'an Jiaotong University
Xi'an, PRC

Abstract

A theoretical model for the prediction of void fraction in gas-liquid or vapour-liquid two-phase flow is proposed. It is an extension of a two-region variable-density single fluid model, which can be applied to both adiabatic and diabatic cases in variety of channels including straight tubes and curved tubes such as helical coils and bends. The affects of buoyancy and centrifugal forces on the steady-state relative phase concentration and velocity distributions are considered. In comparison with the model, the experimental data and other correlations for the calculations of void fraction are collected, and some experiments are conducted to measure the local and average void fraction of air-water and steam-water two-phase flow in horizontal helical coils, the average void fraction correlation for horizontal helical coils are empirically obtained.

1. Introduction

Void fraction is one of the important parameters, which have to be used to determine the pressure drop and the instability conditions of two-phase flow. Hence the prediction of steady-state relative phase velocities and void fraction become one of the important problems. Much work was done in straight tube, great improvements were achieved specially in vertical straight tubes. (1,2). Some experimental measurements in vertical helical coils were also done (3), but only a little work was done in horizontal helical coils (4). Up to now, little work has been found to provide the prediction correlation of the void fractions and the relative phase velocities in the curved tubes, especially in the horizontal helical coils.

The present paper proposes a two-region variable-density single fluid theoretical model for cocurrent two-phase flow channel specially for curved tubes, which, although containing more adjustable parameters than the earlier model for straight tubes, can be checked directly by radial void fraction and velocity profiles measurements, and can be applied to a wide range including straight and curved tubes. The effects of the buoyancy and centrifugal forces on the void fraction profile and the steady-state velocity are taken into account. In comparison with the model, some experimental data in straight tube are chosen, and an experiment is conducted to measure the average and local void fraction in air-water and steam-water two-phase flow in horizontal helical coils by high-speed closing valves and high-speed photography.

2. The Theoretical Model

There are two effects which must be taken into account in analysis of the void fraction problems. One is the effect of the local relative velocity between the phases, which causes the local slip between the phases; and the other is the effect of the nonuniform flow and concentration distributions across the duct, which lead to the entire slip, the two kinds of slip all affect the average void fraction across the duct. In the previous literatures, these two effects were considered separately; Zuber and Findlay (5) defined the average value of a scale or of a vector quantity F over the cross-sectional area A of the duct

$$< F> = \frac{1}{A}\int_A F \ dA \tag{1}$$

and its weighted mean value

$$\bar{F} = \frac{\frac{1}{A}\int_A \alpha F \ dA}{\frac{1}{A}\int_A F \ dA} \tag{2}$$

then provided a general expression for prediction the average void fraction

$$<\alpha> = \frac{<\beta>}{C_0 + \frac{<\alpha U_g>}{<\alpha><U_m>}} \tag{3}$$

where: C_0 =empirical distribution parameter used to revise the one-
  dimensional uniform phase model, determined by experiments;

  α =local void fraction;

  β =flowing volumetric concentration of gas;

  U_g =gas local velocity;

474

$\langle U_m \rangle$=average volumetric flux density of the mixture.

It is a good general expression, but can't be directly used to calculate the average void fraction because of the lack of experimental data about the flow and concentration distributions especially in the curved tubes.

Here, a general method is provided to predict the void fraction, analyze and interpret experimental data in various channels specially in curved tubes, which consider the effects of buoyancy and centrifugal forces. One can note, first of all, that there are very few void fraction distribution measurements, and nearly all of these are in the bubbly flow regime in straight tubes. These data indicate, however, that in fully developed adiabatic bubbly flow, the radial void fraction distribution can be expressed as a power law. As $\langle \beta \rangle$ increases, the flow pattern becomes slug, annular and annular-dispersed. The experimental data show, that in the bubbly flow, the local relative velocity contribute only a small amount to the total slip, while in separated flow regimes (including stratified, annular and annular-dispersed flow), there is little or no local phasic relative velocity, only the entire slip. Hence, it is reasonable to omit the contribution of local slip to the calculation of the total slip between phases, and the power law can be extended to these flow patterns, except the stratified or wave flow regimes in horizontal tubes. While the intermittent flow is rather transtory, and can be frequently treated by interpolation between bubbly and annular flow. Hence, we propose an extension of the two-region model, shown in Fig. 2. in which the outer annular water layer contains bubbles, while the inner gas core contains liquid droplets. This model can express all the flow patterns except the intermittent and stratified flows. We assume that a power law void fraction distribution still holds for the outer region, but that the void fraction in the inner region is a constant or follow another law; the inner core region increases monotonically with increasing volumetric quality   .

In the set of eccentric coordinate (r, θ) as shown in Fig. 1

O: the centre of cross-sectional area of duct;

O_1, O_2: respectively, the position where the void fraction and flow velocity is maximum;

P: any point in cross-sectional area of duct.

let, $\overrightarrow{OO_1} = \hat{R}_1$, $\overrightarrow{OO_2} = \hat{R}_2$, $\overrightarrow{OP} = \hat{r}$, $\overrightarrow{O_1P} = \hat{r}_1$, $\overrightarrow{O_1A} = \hat{r}_{1w}$, $\overrightarrow{O_2P} = \hat{r}_2$,

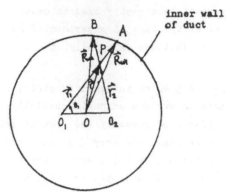

Fig. 1 The set of eccentric coordinate (r, θ)

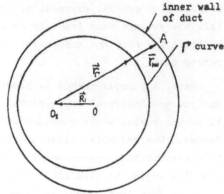

Fig. 2 The two-region model

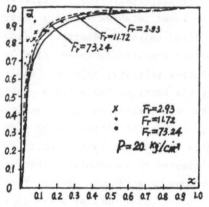

Fig. 3 Comparison with Experimental Data from Ref.(6)

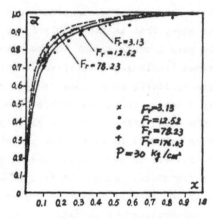

Fig. 4 Comparison with Experimental Data from Ref.(6)

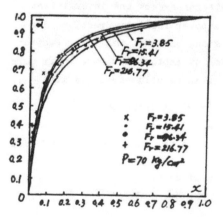

Fig. 5 Comparison with Experimental Data from Ref. (6)

$$\overrightarrow{O_2B} = \overrightarrow{r}_{2w} \ , \quad \overrightarrow{OA} = \overrightarrow{R}_{w1} \ , \quad \overrightarrow{OB} = \overrightarrow{R}_{w2} \ ;$$

the two-region model as shown in Fig. 2: the curve Γ is the line of demarcation of the outer and inner regions. Assume that Γ curve follows:

$$\left| \overrightarrow{r}_1 \right| \Big/ \left| \overrightarrow{r}_{1w} \right| = r_c^* \tag{4}$$

Under certain flow conditions, $r_c^* = $ constant. Because the inner region increase monotorically with increasing $<\beta>$, we assume

$$r_c^* = <\beta>^p \tag{5}$$

where p is a non-negative constant.

For $r_c^* = 0$, the bubbly flow; $r_c^* = 1$, the disperesed flow; $0 < r_c'' < 1$, the annular, annular dispersed flow.

$$\text{let,} \quad r_1^* = \left| \overrightarrow{r}_1 \right| \Big/ \left| \overrightarrow{r}_{1w} \right| \tag{6}$$

$$r_2^* = \left| \overrightarrow{r}_2 \right| \Big/ \left| \overrightarrow{r}_{2w} \right| \tag{7}$$

Consistent with the above discussion, the radial void fraction is assumed as follow:

for adiabatic flow, $\alpha = 0$ at inner wall of duct, then

$$\begin{aligned}
\alpha/\alpha_m &= ((1-r_1^*)/(1-r_c^*))^{\frac{1}{n}} = (s_1/q)^{\frac{1}{n}} & 0 \leqslant s_1 \leqslant q \\
\alpha/\alpha_0 &= 1 & q < s_1 \leqslant 1
\end{aligned} \tag{8}$$

For diabatic flow, $\alpha = \alpha_w \neq 0$ at inner wall of duct, instead of $\alpha_w > 0$, then using following distribution form:

$$\frac{(\alpha - \alpha_w)}{(\alpha_m - \alpha_w)} = \begin{cases} (s_1/q)^{\frac{1}{n}} & 0 \leqslant s_1 \leqslant q \\ (\alpha_m - \alpha_w)/(\alpha_0 - \alpha_w) & q < s_1 \leqslant 1 \end{cases} \tag{9}$$

the velocity distribution in above two cases are assumed as follow:

$$(u/u_0) = s_2^{\frac{1}{m}} \qquad\qquad 0 \leqslant s_2 \leqslant 1 \tag{10}$$

$$\text{where:} \quad s_1 = y_1/r_{1w} = (r_{1w} - r_1)/r_{1w} = 1 - r_1^* \tag{11}$$

$$s_2 = y_2/r_{2w} = (r_{2w} - r_2)/r_{2w} = 1 - r_2^* \tag{12}$$

$$q = 1 - r_c^* \tag{13}$$

α, α_0 is the void fraction at P and O_1 respectively, and α_m is the void fraction value of the assumed distribution $(s_1/q)^{\frac{1}{n}}$ at O_1, not certainly equal to α_0;

u, u_0 is the velocity at P and O_2, respectively;

n, m are positive constant.

According to the definition of average value by Zuber and Findlay, we can get for adiabatic case:

$$<\alpha> = \frac{1}{A}\int_A \alpha\, dA = 2\int_0^1 \alpha\cdot(1-s_1)ds_1 = 2\alpha_0 \left(\left(\frac{\alpha_m}{\alpha_0}\right)q\left(\frac{n}{1+n}+\frac{nq}{1+2n}\right) + \frac{1}{2}(1-q)^2\right) \tag{14}$$

let,

$$(\alpha_m/\alpha_0) = t \tag{15}$$

Hence, the local void fraction distribution across the duct for adiabatic case is following:

$$\alpha = \begin{cases} \alpha_m(s_1/q)^{\frac{1}{n}} = (s_1/q)^{\frac{1}{n}} \cdot \dfrac{<\alpha>t}{2tq\left(\frac{n}{1+n}+\frac{nq}{1+2n}\right)+(1-q)^2} & 0 \leq s_1 \leq q \\[4mm] \dfrac{<\alpha>}{2tq\left(\frac{n}{1+n}+\frac{nq}{1+2n}\right)+(1-q)^2} & q < s_1 \leq 1 \end{cases} \tag{16}$$

$$<u_m> = \frac{1}{A}\int_A u\, dA = \frac{2}{A}\int_0^\pi r_{2w}^2\, d\theta_2 \int_0^1 u r_2^* dr_2^* = \frac{2u_0}{(\frac{1}{m}+1)(\frac{1}{m}+2)} \tag{17}$$

$$u = \frac{1}{2}\left(\frac{1}{m}+1\right)\left(\frac{1}{m}+2\right)<u_m> s_2^{\frac{1}{m}} \tag{18}$$

$$<u_g> = \frac{1}{A}\int_A u_g\, dA = \frac{1}{A}\int_A \alpha U dA$$

$$= \frac{2}{A}\int_0^\pi <\alpha> \frac{\frac{1}{2}(\frac{1}{m}+1)(\frac{1}{m}+2)<u_m>}{2tq(\frac{n}{1+n}+\frac{nq}{1+2n})+(1-q)^2} r_{1w}^2\, d\theta_1 \int_0^{r_c^*} s_2^{\frac{m}{m}} r_1^* dr_1^*$$

$$+ \frac{2}{A}\int_0^\pi <\alpha> \frac{\frac{1}{2}(\frac{1}{m}+1)(\frac{1}{m}+2)<u_m>}{2tq(\frac{n}{1+n}+\frac{nq}{1+2n})+(1-q)^2} r_{1w}^2\, d\theta_1 \int_{r_c}^1 (s_1/q)^{\frac{1}{n}}(s_2)^{\frac{1}{m}}r_1^* dr_1^* \tag{19}$$

$$<u_g>/<u_m> = \frac{1}{A} \cdot \frac{\frac{1}{2}t(\frac{1}{m}+1)(\frac{1}{m}+2)}{2tq(\frac{n}{1+n}+\frac{nq}{1+2n})+(1-q)^2}\left(\int_0^\pi r_{1w}^2 d\theta_1 \int_0^{r_c^*} s_2^{\frac{1}{m}}r_1^* dr_1^* \right.$$

$$\left. + t\int_0^\pi r_{1w}^2 d\theta_1 \int_{r_c}^1 (s_1/q)^{\frac{1}{n}}\cdot s_2^{\frac{1}{m}}r_1^* dr_1^* \right) \tag{20}$$

$$<\beta> = \frac{Q_g}{Q_1 + Q_g} = \frac{<u_g>A}{u_m A} = \frac{<u_g>}{<u_m>}$$

478

$$\langle \alpha \rangle = \cfrac{\langle \beta \rangle}{\cfrac{1}{A} \cfrac{(\frac{1}{m}+1)(\frac{1}{m}+2)}{2tq(\frac{n}{1+n}+\frac{nq}{1+2n})+(1-q)^2}(\int_0^\pi r_{1w}^2 d\theta_1 \int_0^{r_c^*} s_2^{\frac{1}{m}} r_1^* dr_1^*}$$

$$+ t\int_\theta^\pi r_{1w}^2 d\theta_1 \int_{r_c^*}^1 (s_1/q)^{\frac{1}{n}} s_2^{\frac{1}{m}} r_1^* dr_1^*)} \tag{21}$$

It is difficult to directly make this integration in the denominator of above equation. By using the power series to spread out the integrating function, we can obtain following form:

$$\langle \alpha \rangle = \cfrac{\langle \beta \rangle}{C_0 - \Delta} \tag{22}$$

where:

$$C_0 = (\frac{1}{m}+1)(\frac{1}{m}+2)\cfrac{r_c^{*2}+(1-r_c^{*2})t/q^{\frac{1}{n}}}{2tq(\frac{n}{1+n}+\frac{nq}{1+2q})+(1-q)^2} \tag{23}$$

$$\Delta = \frac{(\frac{1}{m}+1)(\frac{1}{m}+2)}{2tq(\frac{n}{1+n}+\frac{nq}{1+2n})+(1-q)^2} \frac{1}{A}(-\frac{2}{3n}(1-r_c^{*3})t/q^{\frac{1}{n}}$$

$$+ \int_0^{\frac{1}{m}} r_{1w}^2 d\theta_1(\int_{r_c^*}^1 t/q^{\frac{1}{n}} r_2^* r_1^* dr_1^* + \int_0^{r_c^*} r_2^* r_1^* dr_1^*)) \tag{24}$$

The final two terms can be integrated by numerical or calculate by empirical method.

Generally speaking, the eccentric degrees of the maximum void fraction and velocity positions $(2R_1/D_{in})$, $(2R_2/D_{in})$ depend on the buoyance, centrifugal and inertia forces; the exponents m and n are function of the Reynold number Re, Froude number Fr, and Weber number We, when the geometric position and flow conditions are fixed. In equilibrium vapour-liquid flow, when the pressure is given, the physical and thermodynamic properties of the flow are fixed, therefore, the exponent m and n should be the function of dimensionless pressure and one other dimensionless number, these functions can only be determined empirically. Here, we assume that $(2R_1/D_{in})$ depends on $Fr'=$(buoyance/inertia force)$=(G^2/\rho_1 \cdot D_{in})/((\rho_1 - \rho_g) \cdot g \cdot \cos\gamma)$; $(2R_2/D_{in})$ depends on De=(centrifugal force/viscous force)$=Re(D_{in}/Dc)^2=(GD_{in}/\mu)(D_{in}/Dc)^2$; m depend on the Froude number $Fr=G^2/(\rho_1 \cdot gD)(1+x(\rho_1/\rho_g-1)$; and n depend on the pressure only. Because $\langle\alpha\rangle/\langle\beta\rangle=1$, when the pressure goes to the critical pressure P_{cr}. Therefore, the following functional forms are chosen:

$$\left(\frac{2R1}{D_{in}}\right) = aF_r'^b$$

$$\left(\frac{2R2}{D_{in}}\right) = cD_e^d \tag{25}$$

$$m = eF_r^f$$

$$n = k\left(\frac{P}{P_{cr}-P}\right)^h$$

where γ is the angle between the fluid flow direction and vertical line; a, b, c, d, e, f, k, h, are the empirical coefficients.

3. Correlations and Comparisons

In the vertical straight tube, take $t=1$, $(2R_1/D_{in})$, $(2R_2/D_{in})$ equal to zero, equation (22) can be simplified as follow:

$$\frac{<\alpha>}{<\beta>} = \frac{1+1\left(\frac{1}{1+2n}q-\frac{2}{1+n}\right)}{1+q\left(\frac{1}{m}+1\right)\left(\frac{1+m}{m+2mn+n}q-\frac{1+2m}{m+mn+n}\right)} \tag{26}$$

it were derived by Lixun Chen (2); For diabatic $\frac{\alpha-\alpha_w}{\alpha_s-\alpha_w}$ was used to substitute the $\frac{\alpha}{\alpha_m}$, then, the following equation is obtained:

$$\frac{<\alpha>}{<\beta>} = \frac{1+q\left(\frac{1}{1+2n}q-\frac{2}{1+n}\right)\left(1-\frac{\alpha_w}{\alpha_m}\right)}{1+q\left(\frac{1}{m}+1\right)\left(\frac{1+m}{m+2mn+n}q-\frac{1+2m}{m+mn+n}\right)\left(1-\frac{\alpha_w}{\alpha_m}\right)} \tag{27}$$

Upon comparison with experimental data, take

$$\frac{\alpha_w}{\alpha_m} = \frac{C_1}{<\beta>} \tag{28}$$

where: C_1 is an empirical constant, equal to 0.04 for straight tubes. From the experimental data in steam-water flow, $e=1$, $f=1/3$, $k=1.56$, and $h=1$. Equation (26), (27) were compared with some experimental data (6,7,8), and other correlations (9), the result shows a good agreement as shown in Fig. 3,4,5,6,7,8,9,10.

For comparison with the theoretical model in curved tubes, an experimental is conducted to measure the average void fraction of air-water two-phase flow in horizontal helical coils by high-speed closing value, the experimental data show the average void fraction can be calculated by using following equation:

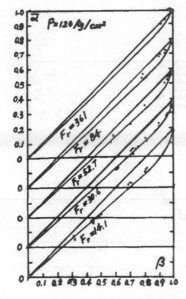

Fig. 6 Comparison with Experimental Data from Ref.(7)

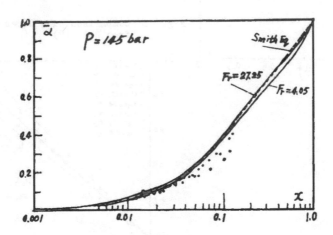

Fig. 8 Comparison with Other Correlation from Ref. (9)

Fig. 7 Comparison with Experimental Data from Ref.(8)

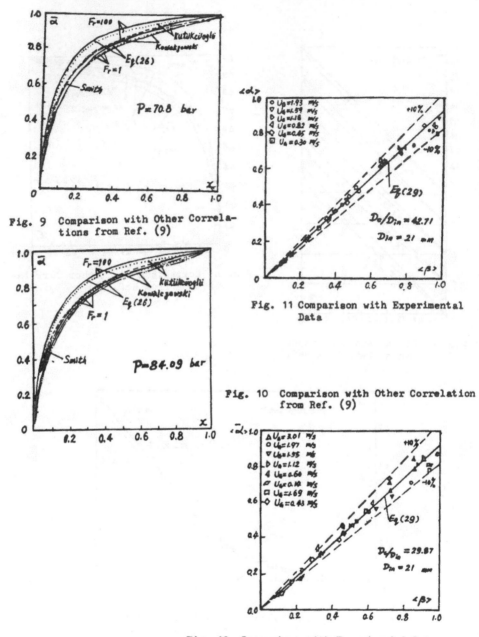

Fig. 9 Comparison with Other Correlations from Ref. (9)

Fig. 11 Comparison with Experimental Data

Fig. 10 Comparison with Other Correlation from Ref. (9)

Fig. 12 Comparison with Experimental Data

$$\frac{\langle\alpha\rangle}{\langle\beta\rangle} = 0.930 \tag{29}$$

the experimental data and equation are showed in Fig. 11,12.

Further experiments conducted to measure average and local void fraction in high pressure steam-water two-phase flow by high-speed photographic method, and local void fraction distributions in air - water two-phase flow by electric conductance needle method in horizontal helical coils are doing. Some numerical calculation will be done to determine all of the empirical coefficients in equation (25) and to obtain a simpler form of equation (24).

4. Conclusions:

A theoretical model which is an extension of the two-region variable density single fluid model is proposed, from it, the general expressions of local void fraction radial distribution, relative velocity radial distribution and the average void fraction are derived. For vertical straight tubes case, the empirical correlation of average void fraction is obtained and compared with the other author's data and correlations. For horizontal helical coils, experiments are done, and correlation is empirically obtained, further work need to do.

References:

1. Bergles, A.E.et al. "Two-phase Flow and Heat Transfer in the Power Process Industries", McGYaw-Hill, Book Co. 1981.

2. Chen Lixun. "An Extension Variable-Density for Steam-water Flow" Proceeding of the 1st International Multiphase Flow and Heat Transfer Symposium", Xi'an Jiaotong University,1984.

3. Zhou Fangde. "Void Fraction in Helical Coils Two-Phase Flow"Ph.D Thesis, Xi'an Jiaotong University, Xi'an,China, 1985.

4. Osamu Watanabe et al. "Flow and Heat Transfer of a Gas and Liquid Two-phase Flow in Helical Coils (For the Cases When the Coil Axis is horizontally placed), Proceeding of JSME(Ed.B) Vol. 52,No. 476, pp. 1857-1864.

5. Zuber N. and Findlay J.A. "Average Columetrical Concentration in Two-Phase Flow System" J. Heat Transfer, Trans. ASME Series C,

Vol. 87, pp. 455-468, 1965.

6. Miropolskii, Z.L. et al "Study of the Phase Composition of the Vapor-Water Mixture in a Pipe" Undergoing Heat Temperature.

7. Kosterin, S.I. et al "Relative Velocity of a Steam-Water Flow in Vertical Unheat Pipes", Teploerergetika, Vol.8.

8. Smith, S.L. "Void Fractions in Two-Phase Flow: A Correlation Based upon An Equal Velocity Heat Model" Proc. Instn. Mech. Engrs. Vol.184,Pt. 1,667, 1969-70.

9. Friedel, L. "Momenton Exchange and Pressure Drop in Two-Phase Flow" Two-Phase Flow and Heat Transfer Vol.1, (Proc. of NATO Advance Study Institute, August 16-27, 1976, Istanbul Turkey, Edited by Kakac, S Maying, F) 239.

10. Guo Liejin Zhou Fangde Chen Xuejun "Investigation on Flow Pattern Transition of Gas-Liquid Two-Phase Flow in the Horizontal Helical Coils" Proc. 4th Miami Inter National Symposium on Multi-Phase Transport & Particle Phenomena ,1986.

11. Chen Xuejun Zhou Fangde Guo Liejin "The Flow Patterns and Their Transition Characteristics of Two-Phase Downflow in Vertical Helical Coils" J. Engineering Thermophysics, China, VOL.8, No. 1,pp. 55-59, 1987.

An Experimental Study on Heat Transfer Enhancement of Freon Dry Evaporator

LIXIAN ZHUANG, YINGSHENG LU, and ZHIQIAN RUAN
Chemical Engineering Research Institute
South China University of Technology
Guangzhou, PRC

YULING XIE and HANZHAO LIU
Wu Han Refrigerator Manufactory
Wuhan, PRC

ABSTRACT

This paper is an experimental report on heat transfer enhancement of freon dry evaporator using the integral internal spiral fin tube. The experimental medium is freon-12. In this experiment, a variable speed compressor was adopted to test the variation of overall heat transfer coefficient and pressure drop with the variation of mass flux of refrigerant in tube side at constant water velocity in sheel side and with the variation of water volocity at constant mass flux of refrigerant. The experimental results show that the integral internal spiral fin tube is very effective for heat transfer enhancement of freon dry evaporator. It can increase the overall heat transfer coefficient of evaporator by about 60% than that of a smooth tube at the same conditions, or reduce the heat exchange surface by about 37%. The experimental results of bundles have provided a basis for the design of dry evaporator of unitary air-conditioners.

INTRODUCTION

There are two kinds of evaporators being used in unitary air conditioner. The first one is flood evaporator in which the refrigerant boils in sheel side and another one is dry evaporator, where the refrigerant flows and boils in tube side.

The flood evaporator have some advantages such as higher heat transfer coefficient, smaller exchange area required since the mechanically fabricated porous surface tube was adopted(1). Unfortunately, the flood evaporaior have several own defects such as larger filling volume of refrigerant, difficult to return oil to compressor and dangerous ice breakdown of tube if a wrong operation was made(2). These defects are not existent for dry evaporator. Usually, the flood evaporator was adopted for larger unitary air conditioner and the dry evaporator was used for middle and small one.

However, the controlling resistance is in tube side for dry evaporator. It has a low heat transfer coefficient in tube side, so that a larger heat transfer area is requisited. Many people had paid an attention to heat transfer enhancement of dry evaporator, for example, in 1970's, both the aluminium fin inserted tube and integrate straight fin tube (thick in wall, higher in fin and less in starts) had been developed at home and abroad(2),

---

Projects supported by national natural science foundation of China.

(3) These techniques have not been extensively applied in industries because of higher pressure drop, heavyer weight per unit lengh of tube and consepuent more expensive cost, although they can get some improvement for heat transfer. Xian Jiatong University(2) had adopted the small diameter tubes and varied the evaporator structure to enhance heat transfer, but they can not still change the state of controlling resistance in tube side. James. G et al(4), Hwang Shaolie et al(5) and Hitachi cable Ltd.(3) had developed helical corrugated tube in flow boiling. Corrugated tube is low in cost but its heat transfer performance is not sufficient. At the same time, its pressure drop is still relative high.

Recent years, the enhanced heat transfer techniques have a great advancement since the integral internal spiral fin tube (brief IISF) was developed. This kinds of tubes have a lower fin height, concentrated fin starts (define the fin pitch) and certain inclination (see figure 1).

Massaki Ito et al (6), Hedeyuki Kimura et al (7), Kichitashon et al (8), Hitachi(3) and Yang Weinian(9) reported the test results of single tube. The results show that integral internal spiral fin tube have some special features such as higher heat transfer coefficient, much lower in pressure drop.

According to the reports, the pressure drop is almost the same as that of a smooth tube.

The experimental tests of tube bundle of this kinds of tubes seem not to have been seen in reports. So, to do some experimental work are significant before industrial application.

EXPERIMENTAL SET-UP AND PROCEDURE

Experimental set-up was designed and fixed according to two purposes of present test (see fig.2). One of them is to test the variation of overall heat transfer coefficient and pressure drop with mass flux of refrigerant in tube side at the constant water velocity in shell side. Another purpose is to test the variation of overall heat transfer coefficient with water velosity in shell side at the constant mass flux of refrigerant. So, a variable speed motor was adopted in experimental set-up. The mass flux of refrigerant can be changed in a wide range (120 - 280 Kg/㎡ s) by means of varying revolution of compressor.

There are four evaporators (I,II,III,IV) being fixed side by side in expermen-tal set-up. The kinds of evaporators and tube parameters are listed in table 1. The refrigerant and water can be supplied to indivitual evaporator by means of controlling valves. All four evaporators were designed according to the thermal load of 29,075 W. For comparison, the performances of smooth tube and aluminium fin inserted tube was also measured.

The freon-12 (R12) was adopted as experimental medium.

The experimental operation was carried out as follows. High pressure R12 vapour coming from compressor 3 was condensed in condenser 4 after de-oil. The liquid refrigerant was filtered, dried and its temperature was measured, then it passed through the visual rotameter 10. The exchange area of condenser is so large that it can condense all the vapour of R12 in-to subcooled liquid. So, the refrigerant flowing through the rotameter is whole liquid and there by reading of rotameter will be more stable. The glass 15 is used to see if it is a whole liquid. The liquid refrigerant

486

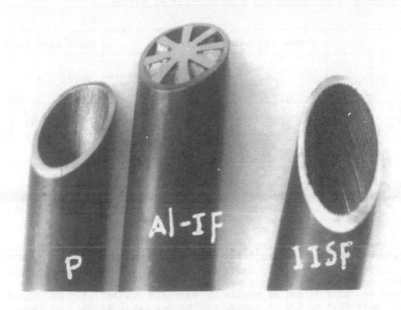

FIGURE 1. Configuration of tested tubes

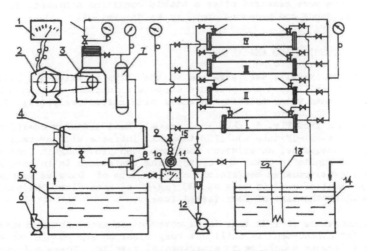

FIGURE 2. Schematic diagram of experimental set-up

1. Conditioner of speed   2. Variable speed motor   3. Compressor
4. Condenser 5. Cooling water tank 6.12. Pump 7. Separator of oil
8. filter   9. expanse valve   10. Visual rotameter   11. Rotameter
13. electric heater   14. Refrigerated water tank   15. glass.
Ⅰ. Evaporator of IISF (ϕ 12x1)   Ⅱ. Evaporator of smooth (ϕ 12x1)
Ⅲ. Evaporator of AL-IF (ϕ 16x1).   Ⅳ. Evaporator of IISF (ϕ 16x1.5)

TABLE 1. Kinds of evaporators and tube parameters tested

No.	tube side (mm)							shell side (mm)	
	$d \times \delta$	L	P	n	fin parameter			$D \times \delta'$	s
					N	h	β		
I	12×1	2057		50	50	0.2-0.25	18°		140
II	12×1	3270		50	smooth tube			159×6	
III	16×1	2940	3	25	8 starts AL-IF tube				205
IV	16×1.5	2914		25	60	0.337*	30°		

* practical value measured using projective method.
$d \times \delta$, $D \times \delta'$—diameter and thickness L—lengh p—pass n—number of tube
h—height β—inclination s—distance between baffles N—number of start

coming fron rotameter 10 flowed in to expanse valve 9 and expanded to form
two phases of vapour and liquid. The liquid was evaporated fully in
evaporator since it was heated by water coming from tank 14. The evaporated
vapour return to compressor 3, So a new cycle begins again. The cooled
water returns to tank 14 and is heated by electric heater 13. For first
purpose, the mass flux of R12 is change in proper order at the constant
condensation temperature(313°k), inlet evaporate temperature (278°k-define
by pressure), inlet water temperature (285°k) and water flow rate (5 m³/h)
When owtlet evaporate pressure and water temperature and parameters
mentioned above were measured after a stable condition achieved. So, the
overall performance and pressure drop can be determined.

EXPERIMENTAL RESULTS AND ANALYSIS

The experimental results were arranged to the forms in which the overall
heat transfer coefficient and pressure drop as a function of mean mass flux
in tube side, mean water velocity is shell side respectively (Fig. 3,4,5,)

It can be found from fig. 3 that the overall heat transfer coefficient
K of plain tube, AL-IF tube and IISF tube all increase with average mass
flux G, in consequance, an existent fact was been proven again from the
standpoint of bundle in which the controlling resistance is in tube side.
The overall heat transfer coefficient of $IISF_2$ tube of ϕ12×1mm is 60%
higher than that of a plain tube at 200 $(kg/m^2 s)$ mass flux. The performance
of tube No.4 is the best of four tested tubes.

Why the performance of IISF tube is improvement so much? The mechanism of
heat transfer enhancement is still not very clear so far.people can only
make some inference according to experimental results. These inferences
contain : (1) increasing internal heat exchange area after the tube was
fabricated; (2) for low flux region of stratified flow, exist of capillarity
of tiny grooves can allow liquid to rise along the wall and consequently
the wall was wetted; (3) for high flux region of a annular flow, the spiral
flow and separation flow of boundary layer induced from spiral fin and
groovs can reduce thickness of boundary layer or separate it; (4) the sharp
corner of fin can perform a role of vapour nucleus. It seems that inference
(1) and (3) are more suitable for present test. This rate of 60% increase
of heat transfer coefficient means that the heat transfer area can be cut

488

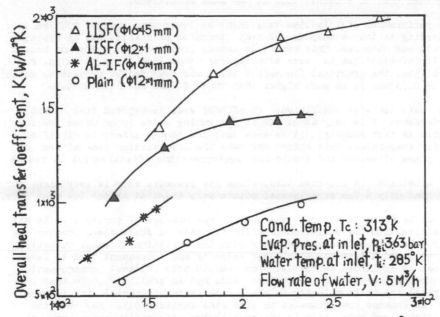

FIGURE 3. Overall heat transfer Coefficent K versus average mass flux G

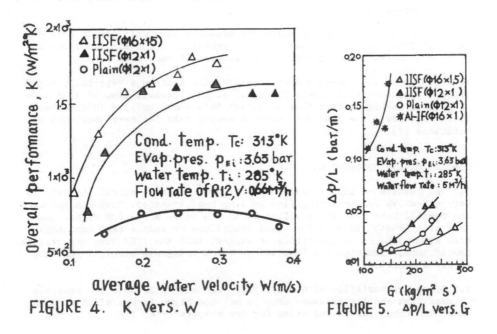

FIGURE 4. K vers. W

FIGURE 5. Δp/L vers. G

37% than that of a smooth tube at the same conditions.

The performance of ϕ16x1.5mm tube ought be lower than that of ϕ12x1mm tube according to the standpoint of heat transfer principle, but an opposite result was obtained. This result is induce from that the shape and height of fin of ϕ16x1.5mm is more advantageous than that of a ϕ12x1mm. For ϕ16x1.5mm, the practical fin height value was measured by projective method to be 0.337mm. It is much higher than that of fins in ϕ12x1mm tube.

The heat transfer coefficient α_i of tube side decomposed from overall performance K is only about 74% of designning value (when water fouling regard as zero roughly). It is show that the bundle effect is significant in dry evaporator. This effect may make the distridution non-uniform of two phase of vapour and liquid and consequent the effectiveness is reduce.

For aluminium fin inserted tube, since its pressure drop is very high, So, that only a few experimental points were measured at the low heat flux.

It can be found from fig.4, the overall preformance of smooth thbe is not sensitive with the variation of water velocity in shell side, However the IISF tube is sensible considerably with the variation of water velocity. The value of K increase with water velocity and consequent tend to level. It can be explained as follow. For smooth tube without enhancement, controlling resistance is in tube side not in shell side, So that to increase the water velocity in shell side is insubstantial. The internal fin can enhance heat transfer to tube side considerably. May be the resistance of tube side is in equilibrium with shell side even the controlling resistance changes to shell side. There by, increase water velocity in shell side can improve the overall performance for IISF tube. For example, when the water velocity was increased from 0.16 (m/s) to 0.2 (m/s), the overall performance can be improved by a percentage of 17%. For IISF tube, to increase properly the water velocity by means of increase the number of baffle after balanced between heat transfer and pressure drop is considerable too.

Fig. 5 show the relations in which the pressure drop unit length (it can be found from fig. 2 the pressure drop, practically contain the local lost of inlet, outlet and tube ends) as a function of mean mass flux.

As fig.5 show, the pressure drop of AL$_3$IF tube achieve a "specially high" degree. When mass flux equals 200 Kg/m^2 s, the pressure drop of IISF tube of ϕ12x1mm is higher than that of smooth tube and ϕ16x1.5mm tube respectively. However, the pressure drop of smooth tube is lower than that of reference (7).

CONCLUSION AND SUGGESTION

1. The results of bundle test show, integral internal spiral fin tube is very effective for enhancing flow boiling heat transfer. They can increase the overall heat transfer coefficient to the value about 60% higher than that of a smooth tube at the same conditions or reduce heat exchange area about 37%. It is reasonable to suggest that the IISF tube can replaces smooth tube, aluminium fin inserted tube and integrate streight fin tube in dry evaporator.

2. Since the variation of overall performance of IISF tube is sensible to mass flux and its pressure drop is not too high, An operation at high mass flux condition is suggested for dry evaporator of IISF tube. The

490

recommendable mass flux is about 180-200 $(kg/m^2 s)$ for the tube of ϕ12x1mm. It can be much higher for the tube of ϕ16x1.5mm.

3. For IISF tube, it is advisable to increase properly the water velocity by means of increasing the number of baffle after balanced between heat transfer and pressure drop.

4. From the fact of ϕ16x1.5mm tube showed superior performance to ϕ12x1mm, tube, it is important to ensure structure parameter of tube in fabricated process.

5. It is necessary to test the bundle effect of much larger unit because the experimental results have shown the existent of bundle effect obviously.

ACKNOWLEDGMENT

The authors wish to acknowledge the support of national nature science fund of china.

REFERENCE

1. Hitachi cable Ltd. High Flux Boiling and Condensation Heat Transfer Tube......HITACHI THERMOEXCEL. 1978.

2. Shanghai Air Conditioner Factory at al, Heat Transfer Enhancement of Freon Dry Evaporator, The Journal of Xian Jiaotong University, No.2, 1975. (in Chinese)

3. HITACHI, Hitachi High-performance Heat Transfer Tubes, 1985.

4. James G. Withers et al, Heat Transfer Characteristics of Helical-Corrugated Tubes for intube Boiling of Refrigerant R12, AICHE Symposium series, Heat Transfer-Research and Design, Vol.70, NO.138, 1973.

5. Huang Shao Lie et al, Investigation on Friction and Heat Transfer Characteristics of the Horizontal-Helical-Ridging Tubes for Tube Side Boiling of R113, Proceedings of Condensed Paper of "China-U.S. Seminar on Two Phase Flow and Heat Transfer", Xian, China, May. 1984.

6. Masaaki Ito et al, Boiling Heat Transfer and Pressure Drop in Internal Spiral-Grooved Tubes, Bulletin of the JSME, Vol.22, No.171, 1979.

7. Hideyuki Kimura et al, Evaporating Heat Transfer in Horizontal Internal Spiral-Grooved Tubes in the Region of Low Flow Rates, Bulletin of the JSME, Vol.24, No.195, 1980.

8. Kichitashon et at, Investigation on Boiling Heat Transfer Enhancement of Refrigerant in Horizontal Tube, Proceedings of 20th national Heat Transfer Conference of Japan, pp154-156, 1983. (in Japanese)

9. Yang Weinian, The Flow Boiling Heat Transfer of Freon-12 in Tube Side of Multiple Starts Screw Tubes, The Paper of Graduate Student of Master's Degree, South China University of Technology. June. 1986. (in chinese).

SUBLIMATION
AND DESUBLIMATION

Investigation of Desublimation Process of Phthalic Anhydride Vapor Mixture on Single Finned Tube

XIAO-XI YANG, GUO-XING YE, JIAN-DONG CAI, YIN-KE TANG,
and SONG-JIU DENG
Chemical Engineering Research Institute
South China University of Technology
Guangzhou, PRC

ABSTRACT

The desublimation process of phthalic anhydride vapor mixture
containing large proportion of non-condensable gas is
investigated. This is an unsteady process that both heat
and mass transfer coefficient is changed with time . In this
paper we have explored the mechanism of desublimation
process, developed some suitable types of finned tubes. The
experimental results showed that the vapor mixture velocity,
the pitch of finned tube and the shape of fin influence
directly the process of desublimation.

INTRODUCTION

Gaseous mixture containing phthalic anhydride (PA) at about

150 °C with low consistency at about 40 g/nm^3 (0.07 % mole)
comes into the Switch condenser. Here PA desublimates on the
outside surface of finned tubes, it is changed directly from
gasous to solid state due to its low consistency. After some
hours it is melted to liquid state by heating with hot oil.
In the switch condenser large proportion of non-condensable
gas is forced to flow across the outside surface of the
finned tubes, in the gas phase outside the solid layer heat
transfer process is complicated with simultaneous heat and
mass transfer. Five different types of finned tubes were
tested for desublimating process. Zig-zag segmented finned
tube was developed specially for desublimating process. The
mechanism of desublimation and the way to enhance heat and
mass transfer on the outside surface of the finned tube was
studied. The experiment results verified that the heat
transfer coefficient of zig-zag segmented finned tube with
selected fin pitch is over 50 % higher than that of circular
finned tube formly used in the factories, and the mass
transfer coefficient has an increase over 90 %.

ANALYSIS

The gaseous PA containing large proportion of non-condensable
gas desublimated on the outside surface of finned tube. Its
heat transfer resistence is increased with the desublimating
time. The hot PA vapor (as component 1) is diffusing at

steady state through a stagnant film of non-condensable gas (as component 2) to a cold gas solid interface at Z=0 where PA desublimated as shown in FIG 1. The differential equation of continuty for multi-component gas phase in steady flow condition without chemical reaction may be written as :

$$N_{1z} - x_1 \cdot N_{1z} = - C_x D_{12} \frac{dx_1}{dz} \qquad (1)$$

After integrating :

$$N_{1z} = \frac{C \times D_{12}}{X_{2m} \times \delta} \times \Delta X_1 \qquad (2)$$

where

$$X_{2m} = \frac{(X_{2o} - X_{2\delta})}{Ln \frac{X_{2o}}{X_{2\delta}}} \qquad (3)$$

If $X_1 \ll 1$, $X_{2m} \approx 1$

$$N_{1z} = \frac{C \times D_{12}}{\delta} \times \Delta X_1 \qquad (4)$$

$$\delta = \frac{C \times D_{12}}{N_{1z}} \times \Delta X_1 \qquad (5)$$

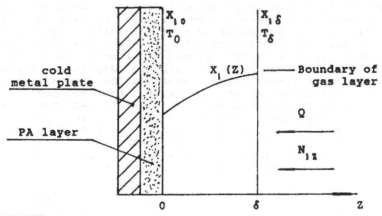

FIGURE 1   PA vapor diffuse through a stagnant film to cold metal plate

With the data obtained in this experiment:

$C = 41.94 \times 10^{-6}$ (gmole/cm^3) $D_{12} = 0.15822$ (cm^2/s)

$N_{12} = 8.61 \times 10^{-8}$ (gmole/cm^2.s)   at $\Delta X_f = 7.58 \times 10^{-5}$

It is calculated to be : $\delta = 0.584$ (cm)
The Blasius equation can be used to calculate the thickness of the laminar boundary layer on a plate.

$$\frac{\delta_f}{X} = \frac{4.94}{\sqrt{Re_x}} \tag{6}$$

When circular fin is used to be heat transfer surface, equivalent distance L should be applied as X in equation (6). Some information introduce that outer diameter of the fin is used as the equivalent distance L. It seems that it is more appropriate to apply the equivalent distance Le instead of L, where the equivalent distance is calculated by:

$$L_e = \frac{S_1}{h} \tag{7}$$

$$S_1 = \frac{1}{2} \; (\ell r_2 - a r_1) \tag{8}$$

where : ℓ -- arch length (mm)
      r_1 -- the fin inner radius (mm)
      r_2 -- the fin outer radius (mm)
      S_1 -- arch surface area (mm)
      h -- fin height (mm)
In this work: $r_1 = 16$ (mm) . $r_2 = 45$ (mm)

$$\alpha = 2 \times arccos \; \frac{16}{45} = 2.4146$$

$$a = 2 \pi \sqrt{45^2 - 16^2} \; = 84.12 \, (mm)$$

$$\ell = r_2 \times \alpha = 108.66 \, (mm)$$

$$h = r_2 - r_1 = 29 \, (mm)$$

$$S_1 = 1771.806 \, (mm^2)$$

$$L_e = \frac{S_1}{h} = 61 \, (mm) = 0.061 \, (m)$$

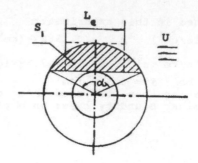

FIGURE 2 schematic diagram of the fin

When U = 0.35(m/s)

$$Re_L = \frac{L_e \times U}{\nu} = 782.5 < Re_c \qquad (9)$$

It is demonstrated that laminar boundary layer covered the whole distance of fin. The maximum thickness of the boundary is :

$$\delta_{max} = \frac{4.91 \times L_e}{\sqrt{Re_L}} = 0.0109(m) = 10.9(mm) \qquad (10)$$

In order to compare with the thickness of the mass transfer boundary layer. The mean thickness of the laminar boundary layer can be calculated by:

$$\delta_f = \frac{4.91 \ X}{\sqrt{\frac{X \ U}{\nu}}} = 0.042859 \ \sqrt{X} \qquad (11)$$

$$\delta_m = \frac{\int_0^L \delta_f \, dx}{L_e} = 6.8(mm) = 0.68(cm) \qquad (12)$$

It is demonstrated that the calculated mean thickness of boundary layer is close to the mean thickness of mass transfer boundary layer , as shown in equation (5). The heat transfer coefficient of the boundary layer can be calculated by following equation :

$$Nu_L = 0.664 \times Pr^{.33} \times Re_L^{0.5} \qquad (13)$$

Substituting the data of this experiment into equation (13)

$$Nu_L = .664 \times 0.7^{.33} \times 782.5^{0.5}$$

$$h_0' = Nu_L \times \frac{\lambda_2}{L_e} = 8.5 \times 10^{-4} \quad (W/cm^2 \cdot °K)$$

The expreiment heat transfer coefficient is

$$h_0 = 9.63 \times 10^{-4} \quad (W/cm^2 \cdot °K)$$

Comparing the experimental results with the calculated value using equation (13) , it can verified that the definition of equivalant distance L_e used in this paper is reasonable.

If the fin pitch H is smaller than 2δmax given by equation (10), the laminar boundary layers on the adjecient fins will merge together. The transport of PA in the laminar flow gas-vapor mixture layer is mainly by the means of molecular diffusion. It caused the long needle shape PA crystal and forming nets and bridges between the fins. When the pitch is bigger than $2 \times \delta m$, the phenomena of forming nets and bridges is not observed in the experiments. Thus it desublimates granularly and forming solid PA crystal layer with higher density on the outside surface of finned tube, Zig-zag segmented fin has a discontinue surface with shorter Le (shown in FIG 3). It can reduce the mean thickness of the laminar boundary layer and enhance the heat and mass transfer. The mean heat transfer coefficient increases as shown in FIGURE 4.

EXPERIMENTAL RESULTS

The analysis given above is only based on steady heat and mass transfer. But the desublimation process is an unsteady process that both heat and mass transfer coefficient is changed with time. When PA vapor desublimated on the outside surface of finned tube. It formed a layer of PA crystal on the surface of fins and the thickness of the layer

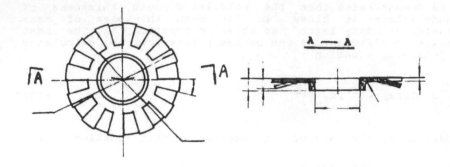

FIGURE 3   Zig-zag Segmented Fin

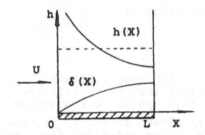 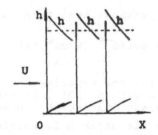

FIGURE 4   A. Continue fin   B. Zig-zag Segmented fin

increase with time. The experimental results for different finned tubes can be seen in FIGURE 5.

For Zig-zag segmented finned tube :

$$h\text{\textbullet} = (11.8975+5.66079 \times t+2.0582 \times t^2 +0.37721 \times t^3) \times 10^{-10} \quad (W/cm^2 \cdot {}^\circ K)$$

$$\bar{G} = 0.6576 \times t^{-0.1966} \qquad (kg /m^2 hr)$$

CONCLUSION

1. Select suitable pitch of fin according to the velocity and concentration of PA vapor mixture is very important to avoid

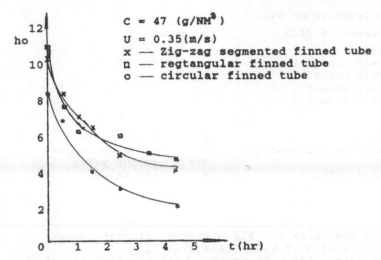

FIGURE 5   heat transfer coefficient with times

forming nets and bridges on the fins.
2. Select suitable higher Renolds number that can reduce   the
thickness of laminar boundary layer and increase   local   heat
and mass transfer coefficient.
3. Developing new type of fins to enhance the   local
turbulence on fin surface and make laminal boundary layer
seperation is good means to enhance the heat and mass
transfer.

NOMENCLATURE

C	overall concentration	$(gmole/cm^3)$
D_{12}	diffusion coefficient	(cm^2/s)
G	desublimating weight	$(g/m^2.hr)$
h_0	heat transfer coefficient	$(W/cm^2.{}^{\circ}K)$
h	fin height	(mm)
H	fin pitch	(mm)
L	fin height	(mm)
M	molecular weight	
N	mass flux	$(gmole/cm^2.s)$
t	desublimating time	(hr)
T	absolute temperature	$({}^{\circ}K)$
T_0	temperature on wall	$({}^{\circ}K)$
T_1	PA vapor mixture temperature	$({}^{\circ}K)$
U	velocity of PA mixture	(m/s)
x	molecular fraction	
x_{1S}	mole fraction in vapor mixture	

x_{10} mole fraction on wall

δ thickness of film   (mm)

ρ density (kg/m^2)
λ thermal conductivity $(W/cm.^{\circ}K)$
Pr   Prantle number
Nu   Nusselt number
Re   Renolds number

subscripts:

1   PA vapor   2   air
1z   PA vapor in Z dimension   e   equivalent value
f   fluid   max maximun value
c   critical value   m   mean value

REFERENCE

1.Sublimation From Disk to Air Streams Flowing Normal to There Surface. J of H..T.ASME. 1958.PP61-71
2.Experiments on the Transfer Characteristics of a corrugation Fin and Tube Heat Exchanger Configuration.J.of H.T.ASME, 1976.No.1 Feb.p26
3.I.E White & C.J.Cremers,Prediction of Growth Parameters of Frost Deposits in Forced Convection
4.Dr.HERMANN SCHLICHTING, Boundary-Layer Theory
5.T.Senshu,T.Hatada and K.Ishibare, Surface Heat Transfer Coefficient of Fins Utilized in Air-Cooled Heat Exchangers. Refrigeration, Vol 5,No 4, 1979
6.Akira Arai, Takao Seushu. Yoshihisa Hosoe, Development of High Efficiency Air conditions. Hitachi Review Vol.30 No 1,1981.
8.Harry Kassat, Luft Kuhlt aggressive Medien .
9.Song-jiu Deng, Qi-en Li,Fundmentals of Transport Process. SOuth China University Press. 1988.(In Chinese)

Condensation Process of Dilute Phthalic Anhydride Vapor Mixture on a Vertical Surface

HUA SHENG ZHOU*, GUO XING YE, XIAO XI YANG, JIAN DONG CAI,
and SONG JIU DENG
Chemical Engineering Research Institute
South China University of Technology
Guangzhou, PRC

ABSTRACT

The process for condensation of phthalic anhydride vapor
mixture containing large proportion of noncondensable gas is
investigated.A correlation for the presentation of the heat
transfer coefficint for this kind of simultaneous heat and
mass transfer process is provided.The final computed and
experimental measured results showed that the main heat
tranfer resistance is concentrated in the gas phase.

INTRODUCTION

The heat transfer performance of the partial condenser for
the phthalic anhydride vapor mixture in the factories
porducing phthalic anhydride from naphthalene in our country
is much lower than the designed value.For the development of
the phthalic anhydride industry in our country, it is
necessary to improve the design of the condenser of phthalic
anhydride vapor mixture in the factories.For this purpose
better understaning of the mechanism of this kind of
condensation process is required.

ANALYSIS

A Heat Transfer Film Coefficient Of Gaseous Phase With
Simultaneous Heat And Mass Transfer.

The condensation model of phthalic anhydride vapor containing
large proportion of non-condensable gas is shown in figure 1.
The assumptions and simplification employed in our analysis
are as follows:
1.The total heat transfer rate for the process are equal to
the sensible heat transfer rate of gas-vapor mixture plus the
rate of latent heat transfer from vapor condensation.
2.The heat transfer of bulk of gas-vapor mixture to the
interface obeys the rules of single phase fluid heat transfer
to a solid surface.
3.If the condensate film on the cold vertical surface is in
laminar flow, the temperature distribution in the condensate
film is linear to film thickness.
4.The condensables in the gas-vapor mixture passes through
the imaginal stagnant gas film only by diffusion.
* Present address,Wuhan Institute of Chemical Technology

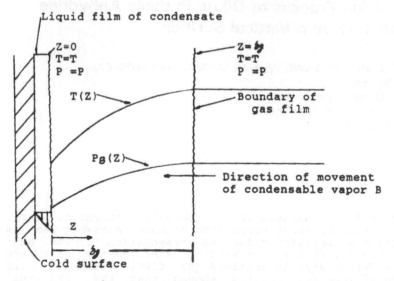

Figure.1. Condensationofhot vapor B on a cold surface in the presence of a noncondensable gas A

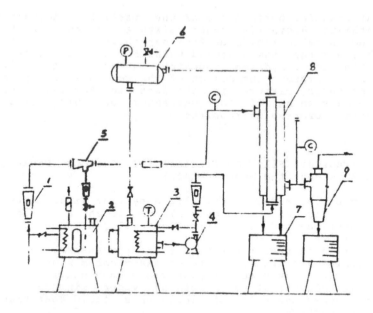

1. flow meter . 2.the heating vessel for melting phthalic anhydride. 3 water heating vessel. 4.pump. 5.jet. 6.steam-water separator. 7.volum cylinder. 8.partial condenser. 9.separator
Figure 2. Experimental procedure

Under these assumptions, the differential equations of continuity and conservation of momentum and energy for mulicomponent gas phase in steady flow condition without chemical reaction may be written as:

$$\nabla * (\rho_i u + j_i) = 0 \tag{1}$$

$$\nabla * (\rho u u + \pi) = \sum \rho_i g_i \tag{2}$$

$$\nabla * e = \sum (n_i g_i) \tag{3}$$

Where e represents the energy flux

$$e = \rho(E + \frac{1}{2} u^2)u + q + (\pi \cdot u) \tag{4}$$

When $(\pi \cdot u)$ and $(\frac{1}{2}\rho u^2)u$ are negligible, equation (4) can be simplified as

$$e = -k\nabla T + \sum n_i H_i \tag{5}$$

Bird (9) used these equations to solve the problem that a hot condensable vapor B is diffusing at steady state through a stagnant film of non-condensable gas A to a cold gas-liquid interface at z = 0 where B condenses and obtained the temperature distribution in the gas film :

$$\frac{T - T_i}{T_g - T_i} = \frac{1 - exp[(\frac{N_B \times C_{PB}}{K})Z]}{1 - exp[\frac{N_B \times C_{PB}}{K})\delta_g]} \tag{6}$$

The heat flux at the cold inter-surface is

$$q_{sg} = -k(\frac{dT}{dZ})_{z=0} = -\frac{N_B \times C_{PB} \times (T_g - T_i)}{1 - exp[\frac{N_B C_{PB}}{K}]\delta_g} \tag{7}$$

From equuation (1), we have

$$N_B = \frac{C \times D_{AB}}{P_{AM} \times \delta_g} \Delta P_B = -k_a \Delta P_B \tag{8}$$

Or

$$\frac{N_B C_{PB} \delta_g}{K} = -\frac{CD_{AB} C_{PB}}{K \times P_{AM}} \Delta P_B = \frac{-\Delta P_B}{P_{AM} \times Le} (\frac{C_{PB}}{C_{PM}}) \tag{9}$$

Where $Le = \frac{K}{CD_{AB} C_{PM}}$, $\Delta P_B = P_{Bg} - P_{Bi}$

505

Substitute equation (9) to equation (7),we can get

$$q_{sg} = \frac{(K/g)\ (T_g - T_\ell)}{1 - \frac{1}{2!}\left(\frac{P}{L_e\ P_{AM}}\ \frac{C'_{PB}}{C'_{PM}}\right) + \frac{1}{3!}\left(\frac{P_B}{L_e\ P_{AM}}\ \frac{C'_{PB}}{C'_{PM}}\right)^2 \cdots}$$ (10)

Assuming the forced convection heat trasfer film coefficient of the gas phase with simultaneously heat and mass trasfer to be h ,by definition

$$h_{sg} = \frac{q_{sg}}{T_g - T_\ell} = \frac{K/\delta_g}{1 - \frac{1}{2!}\left(\frac{P_B}{L_e\ P_{AM}}\ \frac{C_{PB}}{C'_{PM}}\right) + \frac{1}{3!}\left(\frac{P_B}{L_e\ P_{AM}}\ \frac{C_{PB}}{C'_{PM}}\right)^2}$$ (11)

That means heat transfer resistance of the gas stream is equivalent to the resistance of a stagnant gas film of thickness δ_g with simultaneous heat and mass transfer.The term k/δ_g can be written as h_g which is the forced convection heat transfer film coefficient of single phase with same operation Re and Pr number as the gas-vapor mixture. The value of h can be calculated by conventional correlation such as Dittus-Boetters equation with known Re and Pr number. The total heat flux pass through the condensate film is :

$$dq_t = dq_{sg} + dq_c$$ (12)

Gas mixture sensible heat flux pass through the gaseous film is:

$$dq_{sg} = - (W_A \times C_{PA} + W_B \times C_{PB}) \times dT_g - W_A \times C_{PB} \times dH \times (T_g - T_\ell)$$ (13)

Vapor latent heat flux is

$$dq_c = - W_A \times i \times dH$$ (14)

B. The Relation Between Operating Temperature of Gas-vapor Mixture and Partial Pressure of Condensing Vapor.

With the vapor continuosly condensed, the concentration of vapor in the bulk of mixture of gas-vapor goes down. So the mixture of gas-vapor must releases sensible heat to set up new system equilibrium. If the concentration of non-condensable gas in the mixture is large, the sensible heat becomes the main part of total heat flux. Taking account of equilibrium of heat and mass transfer, we can write approximately:

$$h_g \times (T_g - T_\ell) \times dA = (W_A \times C_{PA} + W_B \times C_{PB}) \times dT_g$$ (15)

$$K_q \times M_B \times (P_{BS} - P_{B\ell}) \times dA = W_A \times dH$$ (16)

From equation (8): $K_q = \dfrac{D_{AB} \times h_g \times P}{R \times T \times K \times P_{AM}}$ (17)

Humidity defined as $\quad H = \dfrac{W_{\mathit{g}}}{W_A} = \dfrac{P_{\mathit{g}} \times M_{\mathit{g}}}{(P-P_{\mathit{g}}) \times M_A}$ (18)

Making the first derivative of H with respect to P we have

$$\frac{dH}{dP_{\mathit{g}}} = \frac{P \times M_{\mathit{g}}}{(P-P_{\mathit{g}})^2 \times M_A}$$ (19)

Soloving the equations (15), (16), (17),and (19) simultaneously, we can get the differential equation for the relation between operating gas-vapor mixture temperature and partial pressure of condensing vapor

$$\frac{dT_{\mathit{g}}}{dP_{\mathit{g}}} = \left(\frac{P_{AM}}{P-P_{\mathit{g}}}\right)\left(\frac{T_{\mathit{g}} - T_{\ell}}{P_{\mathit{g}} - P_{\mathit{g}\ell}}\right).L_{\mathit{e}}$$ (20)

Taking account of force balance in the condensate film,relation between the thickness of liquid film and condensating load can be expressed by.

$$\Gamma = \frac{\rho_{\ell}(\rho_{\ell}-\rho_{\mathit{g}})g\delta^3}{3u_{\ell}} + \frac{\rho_{\ell}\tau_{\mathit{g}}\delta_{\ell}^2}{2u_{\ell}}$$ (21)

C,Calculation Procedure

Using these equations,we have done numerical calculation by iteration.
The whole testing length was divided into many small sections and the temperature distribution in gas and liquid film along both the axial and also the radial distance,the amount of the heat and mass transfered and mean heat transfer coefficient of gas and liquid film were calculated by a computer.The calculated results were compared with the exprimental results

EXPERIMENTAL RESULTS AND DISCUSSIONS

For the purpose of checking the theory presented in this paper,a set of experimental apparatus was designed which is shown in figure 2.The partial condenser was a concentric tube,the inner tube is a copper tube with 22 mm outside diameter,the outer tube is a steel tube with 38 mm inside diameter.
The effective test section was 2000 in length and divided into five small test sections.
Phthalic anhydride vapor containing large proportion of air was used as working medium,boiling-water as cooling agent.
The condensate was collected and measured by volume cylinder and balance.The temperature was monitored by the system

consisting of copper-constantan thermocouples and multipath
driect current digit voltmeter.
Following results was obtained by our experiments and
theoretical calculation.

A,The Effect of Velocity of Vapor-gas Mixture on The Heat and
Mass Flux:
Experiments have been done under four kinds of gas-vapor
mixture velocity.The rate of total heat and mass transfer at
each velocity were measured.The experimental results is shown
in table 1 with the comparison to calculated results.
The results in table 1 indicate that the rate of heat
transfer and mass transfer increase repectively as the
gas-vapor mixture flow rate increases and approximately
proportionl to 0.8 power of the rate of gas-vapor mixture.
The reasons for this are that while the gas-vapor mixture
flow rate increases the bulk of mixture brushes the interface
and the stagnant gas film thickness becomes thiner

Table 1.Heat and mass transfer at different gas vapor mixture
velocities

Velocity of gas-vapor mixture m/s	Calculated Results				Experimental Results				Derivation	
	q_{sg} kcal/h	q_c kcal/h	q_t kcal/h	m_B g/h	q_{sg} kcal/h	q_c kcal/h	q_t kcal/h	m_B g/h	Heat q_t	mass m_B
9.91	75.22	36.06	111.28	344.70	83.11	42.3	125.42	404.5	11.3	14.8
12.12	93.11	42.30	135.41	404.40	104.94	49.11	154.05	496.6	12.1	13.9
13.32	102.98	45.47	148.45	434.38	115.56	52.49	168.05	501.8	11.7	13.4
14.10	110.04	47.16	158.00	450.76	120.95	55.14	176.09	527.20	10.3	14.5

Table 2,$(T_g - T_\ell)/(T_\ell - T_w)$ at the exit of different sections

Velocity of gas-vapor mixture m/s Iters	9.91	12.12	13.32	14.10
first section	62.48	50.74	49.75	46.71
second section	47.48	40.74	37.81	36.79
third section	38.37	35.27	31.52	31.00
fourth section	32.03	29,55	26.57	25.26
fifth section	24.04	22.32	26.28	20.27

B,Effect of Flow Rate on Distribution of Heat Transfer
Resistance and Total Heat Transfer Coefficiont.
Using these equations we can calculated out the ratio of
temperature drop in the stagnant gas film to that in
condensate film.Results in table 2 show that heat transfer
resistans in the stagnant gas film is as 20-60 times greater
as that in the condensate film approximatelly and the
magnitute of the ratio decreases appreciably with the
increase in flow rate of the gas-vapor mixture.

508

Table 3,heat transfer coefficient

Velocity of gas-vapor mixture — heat transfer coefficient kcal/m.h.c	9.91	12.12	13.32	14.10
mean liquid film heat transfer coefficient h_L	2908.84	2902.14	2911.98	2940.3
mean gas film heat transfer coefficient h_g	41.31	50.16	54.94	58.86
mean total heat transfer coefficient calculated h_c	61.96	73.62	79.20	84.36
mean tolat heat transfer coefficient by experiment, h_c'	75.46	90.80	96.41	99.32
deviation %	17.89	18.9	17.85	15.26

On the basis of assumptions above,the toatl heat transfer coefficient of both the gas and liquid film on condensating side is:

$$\frac{1}{h_c} = \frac{1}{h_L} + \frac{dq_{sg}/dq_t}{h_{sg}} \tag{22}$$

Under the experimental conditions $h_L \gg h_{sg}$ and $(dq_{sg}/dq_t) > 0.5$ equation (22) may be approximatly written as

$$\frac{1}{h_c} = \frac{dq_t}{dq_{sg}} \times h_{sg} \tag{23}$$

The results of total heat transfer coefficieat h_c calculated by the theory and obtained from the experimental dates given in table 3.
The outcomes in table 3 show that the condensating process of vapor containing large proportion noncondensable gas includes both process of vapor condensation and gas mixture cooling.The former so prevails over the latter that the total heat transfer coefficient of condensating side approches to the heat transfer film coefficient of single gas phase.

CONCLUSIONS:

Based on the theoretical and experimental results,We can come to following conclusions:
1.The controlling resistance of heat transfer of phthalic anhydride vapor condensation in the presence of large proportion of noncondensable gas lay in the stagnant gas film.
2.The rate of heat and mass transfer is increased with the increase of gas-vapor mixture velocity.

3.The calculating methods presented in this paper is simple
and feasible.
The predicted results are lower than the experimental results
by deviation less than 20%.It may be resulted by the heat
losses from the outer tube of the partial condenser in the
experimental apparatus.

NOMENCLATURE:

A   Area of heat transler   Le Lewis number
C   molar concentration   M   mass flux
C_p Heat   capacity   Me Molar flux of component B
C_p Molar heat capacity   P   Total pressure
D   Diffusivity   p. Partial pressure
De Equivalent diameter   Re Reynolde number
h   Heat transfer confficient   T   Temperature
i   Latent heat of condensation   W   Mass flow rate
K_q Mass transfer film coefficient Y,Z Coordinates
K   Heat conductivity

GREEK:

Γ Mass flow rate of condensate π pressare tensor, = +p
δ falling film thickness ρ Fluid density
μ viscosity τ Shear stress tensor

SUBSCRIPT:

A   Noncondensable substance   1   liquid phase
B   vapor   M   mean value
C   Condensating   sg sensible heat of gas
                                          mixture
g   gas phase   T   total
i   interface   W   wall

REFERENCES

1.Silver L., Trans.Inst.of Chem.Eng.,25,30,1974
2.Bell,K.L.and M.A.Chaly,AICHE Symp.Ser.,69(131),72,1972
3.Minkowycz,W.J.and E.M.Sparrow,Int.J.of heat and mass
transfer,9,1125,1966
4.Sparrow, E. M. and Minkowycz, W. J. Int. J. of heat mass
transfer,10,1829,19667.
5.Webb, R. L. and A.B. Wanniayachchi, ASHRAE Trans., 86, part
A,142,1980.
6.Fujii,T.etal.,Trans.JSME(B),50 (455),1716,1984.
7.Huans,H.D.and F.Z.Wang,Chemical Eng.(in Chinese),2,1,1986.
8.Wang,C.Y.,C.J.Tu and K.X.Shien,Chinese Nationl Heat
Transfer Conference,1986.
9.Bird,R.B.etal.,Transport Phenomena,Wiley and Sone,1960.

Heat Transfer Enhancement under Frosting Conditions through Sublimation by High Voltage Electrostatic Field

FANJIONG MENG, HONGBIN MA, and YANLING PAN
Department of Marine Engineering
Dalian Maritime University
Dalian, PRC

ABSTRACT

The paper presents a combined experimental and theoretical study on heat and mass transfer of forst formation through sublimation of water vapor under high voltage electrostatic fields. A horizontal test plate precooled to subfreezing temperatures was exposed to the ambient air and the effects of electric fields were studied through the application of a high voltage between the grounded test plate and coroua wire. It was discovered that high voltage electrostatic fields have strong effect on heat and mass transfer under frosting conditions. The phenomena of needle frost and thin frost layers on cold plate surface were discussed. A simple theoretical model was developed for predicting theeffects of electric fields upon the heat and mass transfer processes associated with frost formation.

INTRODUCTION

It is quite complicated, that the influence of electric field on heat and mass transfer processes between humid air and solid surface under frosting conditions. First, frost formation is essentially a transient process. As the frost grows not only the heat transfer rate and the forst-air interface temperature vary continuosly with time, but also the frost properties change unpredictably from one instant to another. Second, the frost layer is a heaterogeneous porous medium whose structure strongly depends on the conditions under which it is formed. Furthermore, in the absence of some fundamental knowledge, such as the microscopic study of the physical process of frosting. Difficulties are encoutered in predicting the possible effects of electric fields on the formation of frost. In 1970, document [1] studied the phenomena of frost formation on vertical cold surface under electrostatic effect provided by single wire electrode-plate electrostatic field, presenting that the non-uniform electrostatic has strong effect to frost formation, and could enhance heat and mass transfer in frost formation through sublimation under electrostatic effect. The paper studied the frost formation through sublimation on a horizontal surface and heat and mass transfer in electrostatic field, utilizing a experimental system of frost formation in high voltage electrostatic field formed by single wire electrode-plate and many wire electrodes-plate respectively.

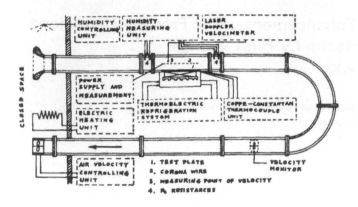

FIGURE 1. Schematic diagram of test loop

EXPERIMENTAL APPARATUS AND EXPERIMENTAL RESULTS

The system of frost formation under forced convection used in experimental
study of heat transfer enhancement under electrostatic fields, showed in
Fig.1, it was made up by a polymethyl-methacryate wind tunnel and a closed
space with controllable air temperature, air velocity, humidity, with
measure and refrigeration system [2] . The electrostatic fields and
thermoelectric refrigerator were placed in test section. The electrical
field during the tests were provided by a high voltage power source with a
capacity of 0-40kv and up to 20 milliampere. By means of observation and
photography with microscope and comparison of experimental research on the
cold surface of frost formation through sublimation of water vapor under
electrical and non-electrical field. It was discovered that high voltage
electrostatic field have strong effect on frost structure, properties and
growth rate of frost deposit processes and heat and mass transfer. It was
presented that experimental results of heat and mass transfer coefficients
of frost formation, frost density and weight under electric field, which
were formed by using single wire electrode-plate and many wire electrodes-
plate, were compared to non-electric field.

In order to measure the coefficients of heat and mass transfer in the frost
formation situation under electrostatic effect, a experimental cold surface
was designed, showed in Fig.2.

Utilizing the above experimental equipment under following conditions,
temperature of the cold surface T_w = -25 - 0°C, temperature of humid air
T_a= 16 - 10°C, velocity of air flow V_∞ = 0.3 - 8 m/s, relative humidity
RH = 60 - 80%, electrostatic voltage V = 0 - 40 kv, the heat and mass
transfer of frost formation on the horizontal cold surface was studied.
In this paper the heat transfer coefficient and mass transfer coefficient
are defined as following:

$$h^* \equiv \frac{q}{T_\infty - T_w} \qquad (1)$$

$$h^*_M \equiv \frac{m_t}{\rho_{v\infty} - \rho_{vw}} \qquad (2)$$

512

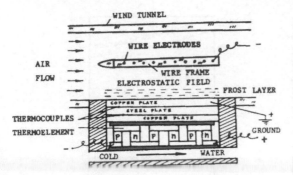

FIGURE 2. Test section

The formula of h* can be written as following as

$$\frac{1}{h^*} = \frac{1}{\sigma h} + \frac{y_s}{k_f} \tag{3}$$

The   is seen to be the ratio of h* $(T_\infty - T_w)$, the sum of convective heat and mass transfer to h $(t_\infty - t_s)$, the convective heat transfer, showed in Fig.3, where

$$\sigma = \frac{h^*(T_\infty - T_w)}{h(T_\infty - T_s)} = \frac{h^* (T_\infty - T_w)}{h^* (T_\infty - T_w) - Lh^*_M (\rho_{v\infty} - \rho_{vw})} \tag{4}$$

the coefficient of convective heat transfer of frost layer can be expressed as

$$h = k_f \frac{h^* - h^* (\rho_{vw} - \rho_{vw}) L/ (T_\infty - T_w)}{K_f - y_s h^*_M} \tag{5}$$

definition (3) shows that the measured parameter h* reflects not only the heat transfer of frost surface but also the effect of frost to heat transfer.

In the electrostatic fields formed by many wire electrodes-plate the needle frost will growth on the cold surface. It is very loose and could not

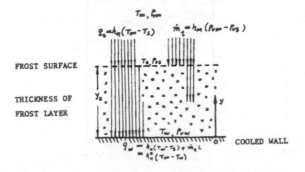

FIGURE 3. Heat and mass transfer of frost deposition

513

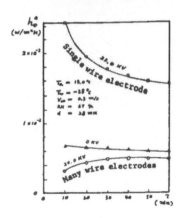
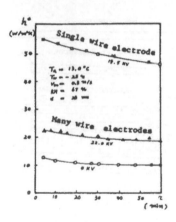

FIGURE 4. Mass transfer coefficient FIGURE 5. Heat transfer coefficient

cover the whole surface within the first three hours of frost formation, similiarly as "dropwise condensation". During the growth of frost, its needle is very unstable as if "electric hit", is affecting it. It may disappear suddenly, but in very short duration, new needle will grow on the primary place accompanied with several branches. These phenomena continue repeatedly and the frost needle disappared does not drop on the cold surface, but attracted to the electrostatic net under the electrical force, the forst on the cold surface, then is loss than that of nonelectrostatic effect, the h_M^*, the mass transfer coefficient decreases to 40%, showed in Fig.4, the h*, the heat transfer coefficient increases by 1.8 times in comparision to that of non-electrostatic effect, showed in Fig.5, The heat transfer coefficient is in proportion with the electrostatic voltage provided, the higher the coefficient, showed in Fig.6. In the electrostatic fields formed by single wire electrode-plate, while the voltage cxceeds certain value, blue spark will appear on the electrode corona wires distinctly, directing to the cold surface the electric wind

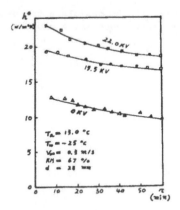
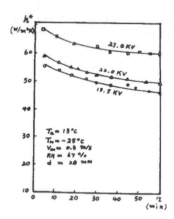

FIGURE 6. Heat transfer coefficient FIGURE 7. Heat transfer coefficient

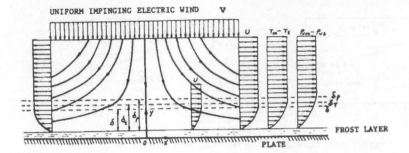

FIGURE 8. Two-dimensional stagnation flow under impingement of uniform electric wind

velocity becomes higher. The frost layer on the cold surface is closely heapped, its surface quite tidy. Under the influence of electric wind, heat and mass transfer coefficients increase to 5.8 times and 4.0 times respectively in comparision to that of non-electrostatic effect, showed in Fig. 4, 5, 7.

ANALYSIS TO HEAT AND MASS TRANSFER

In the electrostatic field, while the voltage exceeds certain value, the electric wind will appear on the electrostatic net so as to enhance the heat and mass transfer in forst formation. Theretically, of the heat and mass transfer coefficient in frosting under the effect of electric wind. Since the electric wind is caused by the electric force in the narrow space from the electric corona wires to the receiving plate electric corona. In the case of a fluid injecting to a plate, the boundary layers, the two-dimensional stagnation Hiemenz flow (3) , form on the frost layer surface of the plate surface.

Above frost layer the conbined momentum, energy and mass transport in a two-dimensional incompressible steady laminar boundary layer of a binary mixture, which is humid air for this case, are , showed in Fig.8, as given by Chuang (1)

Continuity: $\dfrac{\partial u}{\partial x} + \dfrac{\partial v}{\partial y} = 0$ (6)

Momentum: $u\dfrac{\partial u}{\partial x} + v\dfrac{\partial u}{\partial y} = U_\infty\dfrac{\partial U_\infty}{\partial x} + \nu\dfrac{\partial^2 u}{\partial y^2}$ (7)

Energy: $u\dfrac{\partial T}{\partial x} + v\dfrac{\partial T}{\partial y} = \alpha\dfrac{\partial^2 T}{\partial y^2}$ (8)

Diffusion: $u\dfrac{\partial \rho_v}{\partial x} + v\dfrac{\partial \rho_v}{\partial y} = D\dfrac{\partial^2 \rho_v}{\partial y^2}$ (9)

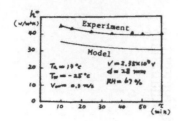 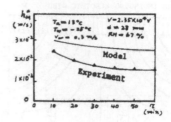

FIGURE 9. Comparision between experimental and numerical

The boundary conditions appropriate to the flow over a frost layer surface are

At $y = y_s$: $u = 0$, $v = V_s$, $T = T_s$, $\rho_v = \rho_{vs}$ (10)

At $y = \infty$: $u = U_\infty$, $v = 0$, $T = T_\infty$, $\rho_v = \rho_{v\infty}$ (11)

in the above model assume that: the surface of frost layer is smooth and has not friction.

In frost layer the conbined mass and energy transport in a one-dimensional form of the channel porous medium model of frost layer, are as given by reference [4]

Mass transport: $\dfrac{d}{dy} (nD\tau^* \dfrac{d\rho_v}{dy}) - \dfrac{d\rho_f}{dt} = 0$ (12)

Energy transport: $\dfrac{d}{dy} (K_f \dfrac{dT}{dy}) + L \dfrac{d}{dy}(nD\tau^* \dfrac{d\rho_v}{dy}) = 0$ (13)

The initial and boundary conditions appropriate to the frost layer on a frosting surface are

At $t = o$: $T = T_w$, $\rho_f = \rho_{fo}$, $y = y_0$ (14)

At $y = 0$: $T = T_w$, $K_f (dT/dy) = h (T_\infty - T_s) + \dot{m}_t L$, $(d\rho_v/dy)_o = o$. (15)

At $y = y_s$: $T = T_s$, $K_f(dT/dy)_s = h (T_\infty - T_s) + L\rho_f (dy_s/dt)$ (16)

$\{ nD\tau^* (d\rho_v/dy)\}_s = \dot{m}_t - \rho_f(dy_s/dt)$ (17)

where tortuosity of frost layer

$\tau^* = n \exp[-\dfrac{2}{3} (T_w/T - 1)]$ (18)

the amount of water vapor being transported to the layer:

$\dot{m}_t = h_M (\rho_{v\infty} - \rho_{vs}) = \rho_f(dy_s/ dt) + (d\rho_f/dt) y_s$ (19)

We have set up the equations of the heat and mass transfer during frost
formation through sublimation under electrostatic effect. Solving
equations (6) - (19), we can present the velocity distribution of boundary
layer and density of water vapor, temperature distribution in different
point, at different time. Further, the heat and mass·tranfer coefficient
can be obtained. In this paper the coordinate conversion was made to
above mathematical model appropriately. Utilizing Terrill numerical method
[5] the numerical solutions were obtained, showed in Fig.9.

CONCLUSIONS

In the electrostatic field formed by many wire electrodes - plate, needle
frost grows on the horizontal cold surface, its weight reducing by 40%
while the heat transfer coefficient increasing to 1.8 times in comparision
to that of nonelectrostatic effect.

In the electrostatic field formed by single wire electrode-plate, close
frost layer grows on the horizontal surface. The heat transfer coefficient
increase by 3.0 - 5.8 times, the mass transfer coefficients increass by
2.5 - 4.0 times in comparision to that of non-electrostatic effect.

A non-uniform field characterized by corona discharge can enhance considerably the frost formation.

NOMENCLATURE

D   mass diffusivity
h   convective heat transfer coefficient
h* reduced convective heat transfer coefficient
hM convective mass transfer coefficient
h$_M^*$ reduced convective mass transfer coefficient
L   latent heat of sublimation
$\dot{m}_v$ amount of mass transfer of water vapour
n   porosity of frost layer
q   heat flow density
RH humidity
T   temperature
U_∞ free stream velocity in x direction
u   velocity in x direction
v   voltage or velocity in y direction
y   d⊥rection normal to surface

GREEK LETTERS

δ thickness of velocity boundary layer
δ_T thickness of thermal boundary layer
δ_ρ thickness of concentration boundary layer
ν kinematic viscosity
ρ density

SUBSCRIPTS

f   frost
s   frost surface
v   water vapour

w   test plate surface
∞   ambient conditions

REFERENCES

1. Chung, T.H., and Velkoff, H.R., Heat and Mass Transfer to a Subfreezing Surface in a Non-Uniform Electric Field, AD- 680185.

2. Meng, F.J., Gao, W.G. and Pan, Y.L., Growth Rate of Frost Formation through Sublimation — a Porous Medium Physical Model of Frost Layers, Heat Transfer Science and Technology, Hemisphere Pub. Corp., Washington, pp. 584 - 593, Feb. 1987.

3. Schlichting, H., Boundary - Layer Theory, McGraw - Hill Book Comp., pp.95 - 101, 1979.

4. Meng, F.J., The conceptual Model of Porous Media and Porosity and Tortuosity of Frost Layer, JETP (in Chinese) Vol.8, No.1, pp.64-68, 1987

5. Ma, H.B., Experimental Study of Heat and Mass Transfer of Frost Formation through Sublimation on Horizontal Surface under Electrostatic Field and Its Numerical Analysis, Master's Paper (in Chinese), Dalian Maritime University, Feb. 1987.

The Influence of Frosting on the Design of Finned-Tube Evaporators

KEITH CORNWELL and JAMAL AL-SAHAF
Energy Technology Unit
Heriot-Watt University
Edinburgh, Scotland, UK

ABSTRACT

Many refrigeration finned-tube heat exchangers are designed for optimum operation under dry conditions, but used under frosted conditions. In this work the effect of frosting on the heat transfer is examined and suggestions are made regarding more thermally efficient design. The frosting of a single cylinder is examined theoretically and experimentally and the analysis is then extended to the finned-tube heat exchanger. Indications are that closer fin spacing could be tolerated and fewer tubes, but with larger diameters, could be beneficial under frosting conditions.

1   INTRODUCTION

Conventional finned-tube heat exchangers are commonly used for extracting heat from ambient air in cold stores, refrigerated display units and heat pumps where frost builds up on the surface. Under this condition the frost layer is periodically melted away by defrosting. The energy losses associated with the thermal resistance of the frost layer and the heat required to melt the frost during defrosting typically amount to 15 to 20% of the energy throughput. In spite of this, the thermal and mechanical design is generally based on the dry, unfrosted exchanger. The only concession to the frosting is the wide spacing of the fins, leading to very large heat exchangers.

The work described here is the first phase of a programme aimed at optimising the exchanger to give maximum heat transfer including the frosting and defrosting effects. The frosting of a cylinder is experimentally studied and theoretically analysed to identify the sensitivities of the important parameters. The theoretical analysis is then extended to the finned-tube heat exchanger and conclusions are drawn on the effects of fin spacing, fin thickness and tube diameter on the thermal performance.

Several researchers have studied the fundamentals of ice growth and theproperties of frost on heat transfer surfaces. Some of the more recent (>1970) include Biguria and Wenzel (1970), Brian et al (1970), White and Cremers (1974), Jones and Parker (1975), Hayashi et al (1977), Varma and Charan (1978), Abdel-Wahed et al (1984), Barrow (1985), O'Neal and Tree (1985) and Sami and Doung (1988). Experimental studies of growth on a cylinder include Chung and Algren (1958), Schneider (1978) and Aoki et al (1979) and on a complete finned-tube exchanger, Stoecker (1957), Hasoda and Uzuhashi (1967), Gatchilov and Ivanova (1979)and Sami and Doung (1988). The longitudinal development of frost along a plate is covered by Chuang (1976) and flow reduction due to frost blockage by Stoecker (1957) Gates et al (1967), Huffman and Sepsy (1967) and Hosada and Uzuhashi (1967). While much of this previous work is useful in understanding the mechanism, no development into design optimisation based on modelling of the frosting is reported.

2   THERMAL CONDUCTIVITY AND THICKNESS OF THE FROST LAYER

Several of the papers mentioned in Section 1 make reference to thermal conductivity and Harraghy and Barber (1986) review both the experimental and theoretical results. Over the range of time (τ = 5 - 300 minutes) and ice density (ρ = 100 - 300kg/m^3) relevant in this work the experimental results for thermal conductivity k_i are well represented by

$$k_i = 0.87 \times 10^{-3}\rho. \text{ W/mK} \tag{1}$$

The tolerance on this equation is necessarily about 20% due to variations of frost structure.

The frost density is a function of the infiltration of moisture into the porous layer and thus a strong function of time. It has been less widely studied, although Hayashi et al (1977) and Aoki et al (1979) give experimental results. For a metal surface in the range −5 to −10°C and an air velocity of around 3m/s (used as a datum in our work) the density is represented by

$$\rho = 100 + 120\tau^{2/3} \text{ kg/m}^3 \tag{2}$$

where τ is in <u>hours</u>. Combination of these equations yields

$$k_i = 0.087 + 0.104 \ \tau^{2/3} \text{ W/mK} \tag{3}$$

which compares favourably with the only results available at a similar velocity (but much lower surface temperature) in Yonko and Sepsy (1967).

The growth of the frost layer with time is widely reported in the references of Section 1. Growth is dependent on humidity and temperatures but fairly insensitive to air velocity. For a humidity range of around 60-90% and the surface temperature range of about −5 to −10°C used in this work the general form of the dependence (Schneider, 1977) is

$$s = C\tau^{1/2}\Delta T_{iw}^{1/2} \quad \text{m} \tag{4}$$

where ΔT_{iw} is the temperature difference across the frost layer between the surface of the ice (i) and the wall (w). Literature values lead to a value of C in the range 0.3 to 0.7 $\times 10^{-3}$ (s in m, τ in hours). Our results in Section 3 yield a value of $C = 0.37 \times 10^{-3}$ and this value is used in the calculations.

It is interesting to note that, in the analysis of the finned-tube heat exchanger, k_f and s are involved as the ratio (s/k_f) or approximately this ratio when radial terms are used. This ratio is given by equations (3) and (4) as

$$\frac{s}{k_f} = [0.235\tau^{-0.5} + 0.281\tau^{0.167}]^{-1} \Delta T_{iw}^{1/2} \tag{5}$$

The bracket [ ] only varies from 0.58 to 0.48 over the period from $\tau = 0.5$ to 2 hours, thus indicating the insensitivity of this term in the time period of interest in exchangers. (Defrosting usually occurs every 1 to 2 hours) in heat exchangers).

3   FROSTING ON A SINGLE CYLINDER

3.1   Experimental Analysis

The rig shown in Fig.1 consisted of a closed-cycle air flow duct in which the velocity, temperature and humidity were controlled. The duct cross-section measured 350 x 350mm and the straight length of duct upstream of the test cylinder 1.95m. Two refrigeration systems were used, one for air temperature control and the other for the cylinder. The test section had Perspex sides to enable viewing and photography and housed the measurement probes. The air velocity was measured using a pitot-tube and precision micromanometer, the humidity by an electronic sensor and the temperature by 0.25mm diameter Chromel-Alumel thermocouple wires and a Hewlett-Packard 3421A data transfer unit.

The test cylinder was a commercial copper rod of 25.4mm diameter, positioned horizontally across the duct with a heat sink cooled by refrigerant at each end as shown in Fig.2. The flow of refrigerant was tapped from the main coil and manually adjusted for each test to give the desired surface temperature. The rod temperature distribution was measured at seven positions and 1D differential analysis was used to determine the heat flow. The heat transfer varied slightly along the rod and a point 6mm from the side of the duct was selected for photographically measuring the frost thickness; this being near enough to the heat sink to be virtually at the heat sink temperature and far enough away from the wall to be unaffected by the boundary layer. While the heat flow from the surrounding air at this point could in principle be found from the end temperature gradient, this involved dependence on a very small ΔT. For this reason the mean heat transfer Q_{rod} along the rod was used, thus slightly underestimating the local value. Since, as will be seen, the heat transfer coefficient to the frost is not accurately known and relative effects are of primary importance, this is allowable. 1D analysis along the rod , with ends at T_1, and centre at

T_2 in warmer air at T_a is found to yield:

$$\frac{T_a - T_2}{T_a - T_1} = \frac{1}{\cosh(m\ell/2)} \tag{6}$$

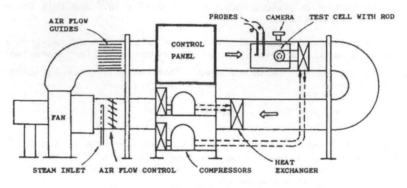

Fig.1. The Experimental Rig

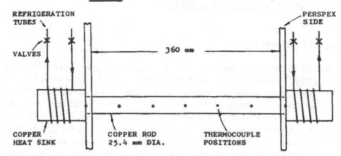

Fig.2. Cooled Copper Rod Arrangement

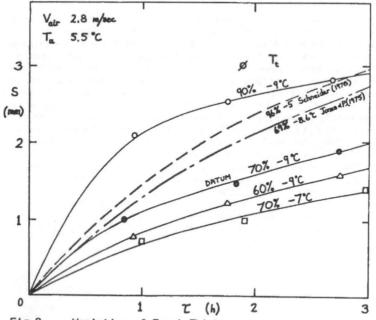

Fig.3. Variation of Frost Thickness on the Rod with Time

$$\frac{Q_{rod}}{\ell} = 4(T_a - T_2)\frac{k_r A_r}{\ell^2}\left[\frac{m\ell}{2}\right] \sinh\left[\frac{m\ell}{2}\right] \qquad W/m \qquad (7)$$

Here k_r and A_r are the rod conductivity and cross-sectional area, ℓ is the length of the rod and the product $(m\ell/2)$ is determined from (6) and then used in (7).

Tests were conducted over a range of conditions and around DATUM values of velocity $V = 2.8 m/s$, relative humidity $\phi = 70\%$, air temperature $T_a = 5.5°C$ and surface temperature $T_t = -9°C$. Sample results are shown in Fig.3 together with those from Jones and Parker (1975) for similar conditions, but on a flat plate and Schneider (1978) for rather different conditions but on a tube. The strong dependence of s on ϕ (at high values of ϕ) and on T_t is very evident. As mentioned earlier the dependence on T_a and V (not shown) was found to be rather less. The experimental variation of Q_{rod} with time for the DATUM conditions is shown in Fig.4 and discussed in Section 3.2.

3.2 Analysis of Heat Transfer

Heat transfer to the surface of the rod from the air is by diffusion and phase change, convection and conduction as shown in Fig.5. Here A is the effective surface area of the frost at radius (R+s), ΔC is the concentration difference between the moisture in the air and the saturated liquid at the interface, T_f, and L is the latent heat from vapour to ice of 2834.3 kJ/kg. If the mass/heat transfer analogy is applied with Lewis No, Le≈1, then the mass transfer coefficient becomes

$$h_D = \frac{h}{\rho_g c_{pg} (Le)^{2/3}} = 0.9815 \, h \qquad (8)$$

Conventional heat transfer analysis based on Fig.5 and substitution for h_D and L then yields, for the heat flow rate at the surface,

$$\frac{Q_{sur}}{\ell} = \left[\frac{1}{2\pi(R+s)h\left[1+2782(\Delta C/\Delta T)_{si}\right]} + \frac{\ell n[(R+s)/R]}{2\pi k_i}\right]^{-1} (T_a - T_t) \qquad (9)$$

Substitution of h as obtained from the conventional cross-flow relationship (for air at the DATUM conditions given earlier)

$$Nu = 0.24 \, Re^{0.6},$$

s from the photographically measured mean frost thickness and k_i from equation (3), then gives the variation with time shown in Fig.4.

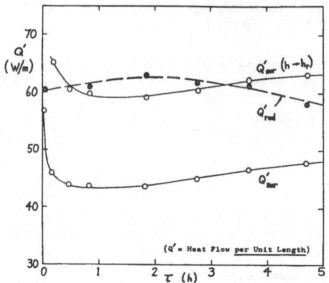

Fig.4. Comparison of Surface Heat Transfer found by Conduction along the Rod with that Calculated for the Surface from Equation(9) for DATUM conditions.

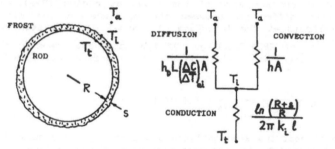

Fig.5. Heat Flow Path through the Frosted Surface

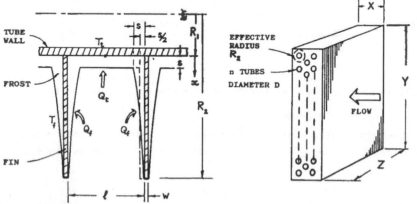

Fig.6. Geometrical Parameters used in Modelling of a Finned-Tube Evaporator.

For the rod with no frosting ($\tau = 0$) there is a discrepancy of about 4W/m between the heat transfers per unit length from conduction and convection. This is due to experimental error and analysis simplifications. As the frost layer develops there is a larger difference which may be explained as follows. The initial decrease in Q_{sur} is due to the predominant effect of the frost effective conductivity as the first crystals are formed. When the layer becomes more established the conductivity rises due to the filling of the interstitial spaces and the partial melting and refreezing mechanism noted by many researchers. At the same time the roughness of the surface increases considerably, thus increasing h by typically a factor of 1.5 to 2x, see Chen and Rohsenow (1964) and Hosoda and Usahashi (1967). If the present data are replotted with h replaced by effective value h_r where $h_r \approx 2h$ the new curve falls in the region of the conduction results. More detailed study of this area is required, but the general corroboration is sufficient to allow extension of this approach to the plate-fin evaporator.

4   ANALYSIS OF FROSTING IN A PLATE-FIN EVAPORATOR

4.1 The Model

An exact analysis of frosting in a complex heat exchanger is not possible at the present stage, but the following model is sufficient for determining the effects of geometrical parameters on overall heat transfer. The model is based on a simple finned-tube evaporator as shown in Fig.6. There are n tubes in a staggered arrangement and the effective fin radius R_2 associated with the tubes is given by

$$n\pi R_2^2 = XY \tag{11}$$

From observation frosting takes the form shown in Fig.5 (which is not to scale as $R_2 \gg \ell$). It is assumed for the normal arrangement where X contains only 2-4 rows of tubes that all the tubes are equally frosted. In practice the upstream tubes will have rather more frosting as shown by Chuang (1976). Again from observation the frost thickness varies from a maximum at the tube to virtually zero at the extremity of the fin (except for the leading edge where it is thicker). In the model a mean frost thickness on the fin of s/2 is used.

As time progresses the frost thickness on the fins increases until the flow passage is blocked. The flow variation depends on the fan characteristics as pointed out by Barrow (1985) and others. If we preclude from our analysis the latter stages just before blockage (by which time defrosting will have occurred) it is reasonable to assume that the mass flow rate of air is approximately constant. In this case the air velocity increases with dimension ($\ell - s$) and thus with time. A mean convective heat transfer relationship, rearranged from Blundell (1978), for fin spacings of about 2 to 6mm is

$$Nu = 0.1 \, Re^{0.65} Pr^{0.33} \tag{12}$$

525

based on a mean hydraulic radius of

$$r_h = \frac{4Y(\ell-s)}{2(\ell-s) + 2Y} \approx 2(\ell-s) \qquad m \tag{13}$$

Substitution of air properties at a nominal temperature of $0^\circ C$ yields

$$h = 2.51 \, V^{0.65} \, (\ell-s)^{-0.35} \, W/m^2 k \tag{14}$$

where V is the mean air velocity through the exchanger, given by

$$V = \frac{\dot{m}_a}{\rho_a YZ[\{P_t - 2(R + s)\}/P_t]}$$

where P_t is the transverse tube Pitch (Zukauskas and Ulinskas (1983)).

Further information is required on the temperature distribution in order to solve the heat transfer equations and this is provided by considering the frost surface temperature. Under conditions of interest here, where the air temperature is above $0^\circ C$, the surface tends towards $0^\circ C$ once the layer is established. Melting at the frost surface was noticed in our experiments and Biguria and Wenzel (1970) show a similar trend with time at much lower wall temperatures. In our model the frost surface temperature is taken as $0^\circ C$ and the fin tip temperature (at R_2), where the frost just fails to form, is also $0^\circ C$.

4.2 Heat Flow to the Fins and Tubes

The general equation for heat flow in a circular rectangular fin as shown in Fig.6 is

$$\frac{d^2 \Delta T_{af}}{dr^2} + \frac{1}{r} \frac{d \Delta T_{af}}{dr} + \frac{2U_{af}}{k_f w} \Delta T_{af} = 0 \tag{15}$$

Here suffix a indicates the surrounding air and f the fin, and

$$U_{af} = \left[\frac{1}{h_r \, [1 + 2782(\Delta C/\Delta T)_{ai}]} + \left(\frac{s}{k_f} \right) \right]^{-1} \tag{16}$$

The term (s/k_f) is a function of time and an analytical solution is very cumbersome. The following approximation to a non-radial fin is adequate for our needs and leads to an error of only a few %. The equation for a similar plain fin of length R_1 is

$$\frac{d^2 \Delta T_{af}}{dx^2} = m^2 \Delta T_{af}$$

where $$m = \sqrt{\frac{2U_{af}}{k_f w}}$$

This equation is adapted (see for example, Becker (1986)) by replacing m by m', where in our case

$$m' = m\sqrt{\frac{1}{2}\left(1 + \frac{R_2}{R_1}\right)} \tag{17}$$

The solution is then

$$\Delta T_{af} = M e^{-m'x} + N e^{+m'x}$$

$$Q_f = -k_f A \left(\frac{d\Delta T_{af}}{dx}\right)_{x=0} = k_f A m'(M-N)$$

With the boundary conditions:

$$\text{at } x = 0 \quad \Delta T_{af} = \Delta T_{at}$$
$$\text{at } x = R_2 - R_1 \quad \Delta T_{af} = T_a - 0°C = T_a (°C)$$

we find

$$Q_f = \frac{2\pi R_1 k_f m' w}{\sinh[m'(R_2 - R_1)]} \left[\cosh[m'(R_2 - R_1)] - \frac{T_a}{T_a - T_t} \right] (T_a - T_t) \tag{18}$$

Heat flow from the air to the tube between the fins is given by

$$Q_t = U_{at}(\ell - s)(T_a - T_t) \tag{19}$$

where

$$U_{at} = \left[\frac{1}{2\pi(R_1 + s)h_r[1 + 2782(\Delta C/\Delta T)_{ai}]} + \frac{\ln[(R + s)/R]}{2\pi k_i} \right]^{-1} \tag{20}$$

527

The total heat flow to the whole evaporator is then given by reference to Fig.6 as:

$$Q_E = n\left[\frac{Z}{\ell} + 1\right]\left[Q_f + Q_t\right] \tag{21}$$

Evaluation of equation (21) involves use of the assumptions of Section 4.2 and equations (14) to (20). The mass transfer potential term, $\Delta C/\Delta T$ requires evaluation from property data or charts. For the datum conditions of (ø = 70%) used here it is found that

$$\left(\frac{\Delta C}{\Delta T}\right) \approx 0.27 \times 10^{-3} \quad {}^{\circ}C^{-1} \tag{22}$$

This value is not very sensitive to ø. The frost surface roughness is allowed for by using $h_r = 2h$ where h is the normal value given by equation (10) for U_{at} and equation (14) for U_{af}.

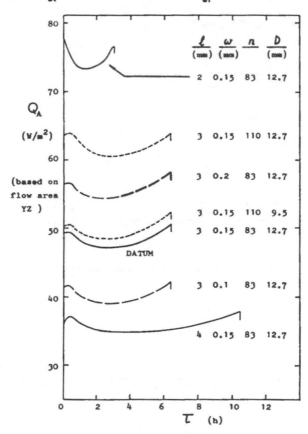

Fig.7.   Effect of Geometrical Parameters on the Variation of Heat Transfer with Time in a Frosted Finned-Tube Evaporator

5   RESULTS OF ANALYSIS

Some of the results of the analysis are presented here by showing the effects of design changes on a DATUM design.   The DATUM conditions are as follows:

Geometry   Y = Z = 1m   X = 0.1m
            n = 83 (3 columns; 28, 27, 28)
            D = 12.7mm l = 3mm (fin spacing)
            w = 0.15mm (Aluminium fins)

Conditions
            T_a = 5.5°C, T_f = 0°C
            T_t = -9°C   ø = 70%
            $\dot{m}_a$ = 3.58kg/s   (Velocity through YZ, 2.8m/s)

Fig.7 shows the total heat transfer Q_A <u>per unit frontal area</u> (ie Q_E/YZ) at this datum, together with the effect of geometrical variations from this datum.   These variations include changes of fin spacing, fin thickness, number of columns of tubes and tube diameter.

The shape of the curves is due to a balance between the frost surface roughness (from an initially smooth tube), the insulating effect of the frost layer and air velocity.   The gradual rise in Q_A after some hours is due to the influence of increasing velocity as the free flow area in the exchanger decreases.   In practice, after long periods the mass flow rate would drop for most air blowers and Q_A would drop towards zero.

The cut-off in the model is taken as the point when the air gap between the frost layers drops to 0.2mm.   The curves bear out the observation that the heat exchange rate varies very little during the normal defrost period of about 1½ to 2 hours.   With careful design it would appear that this period could be extended.

Fin spacing appears to be the single most important geometrical factor and closer spacings than normally used could be tolerated if the defrost time is not extended.   Fin thickness has less effect, unless a considerably thicker fin is used, but tube packing and diameter are important.   A closer packing from 3 to 4 columns of tubes (for same X) increases Q_A by over 25% while a reduction in diameter from 12.7 to 9.5mm (1/2 to 3/8 inch) reduces Q_A by a similar amount.   Approximately the same heat transfer may be obtained by 110 tubes of 3/8 inch diameter or 83 tubes of 1/2 inch diameter.   Since the latter case involves less manufacture time and less material, and the refrigerant side flow area (not covered here) is greater, it is much to be preferred.

6   CONCLUSIONS

1 Comparison of experimental results and theoretical analysis of frosting on a single rod indicates the importance of frost surface roughness on the heat balance.

2　Theoretical analysis of frosting in a finned-tube heat exchanger under normal ambient air flow conditions is feasible using a conceptual model.

3　Results of analysis indicate the importance of fin spacing, tube packing and tube diameter on heat transfer.

4　Heat exchanger arrangements with closer-spaced fins and fewer tubes, but of larger diameter, than conventionally used could be beneficial and would repay experimental investigation under frosting conditions.

7　REFERENCES

Abdel-Wahed, R.M., M.A.Hifni and S.A.Sherif (1984), Heat and Mass Transfer from a Laminar Humid Air Stream to a Plate at Subfreezing Temperature, Int.J. of Ref. 7, 49-55

Aoki, K., M.Hattori and Y.Hayashi (1979), Frost Formation on a Cylinder Surface under Forced Convection, XVth Int. Congress of Refrig., B1/55, 389-394

Barrow, H. (1985), A Note on Frosting of Heat Pump Evaporator Surfaces, Heat Recovery Systems, 5, 195-201

Becker, M. (1986), Heat Transfer - A Modern Approach, Plenum, New York

Biguria, G. and L.A.Wenzel (1970), Measurement and Correlation of Water Frost Thermal Conductivity and Density, I & EC Fundam., 9, 129-138

Blundell, C.J. (1978), Heat Exchanger Design for a Domestic Dehumidifier, The Electricity Council Research Centre, ECRC/M116, 28

Brian,P.L.T., R.C.Reid and Y.T.Shah (1970), Frost Deposition on Cold Surfaces, Ind.Eng.Chem.Fundam., 9, 375-380

Chen, M.M. and W.Rohsenow (1964), Heat, Mass and Momentum Transfer Inside Frosted Tubes - Experiment and Theory, J.Heat Transfer, 86, 334-340

Chung, P.M. and A.B.Algren (1958), Frost Formation and Heat Transfer on a Cylinder Surface in Humid Air Cross Flow, Parts I and II, Heating , Piping and Airconditioning, 30, 171-177, 115-122

Chuang, M.C.(1976), The Frost Formation on Parallel Plates at Very Low Temperature in a Humid Stream, ASME, paper 76-WA/HT-60, 1-5

Gatchilov, T.S. and I.S.Ivanova (1979), Characteristics of Extended Surface Air Coolers During Operation Under Frosting Conditions, Int.J.Refrig., 2, 233-236

Gates, R.R., C.F. Sepsy and G.D.Huffman (1967), Heat Transfer and Pressure Loss in Extended Surface Heat Exchangers Operating under Frosting Conditions, ASHRAE Trans., 73, 2:I.3.1

Harraghy, P.G. and J.M.Barber (1986), Frost Formation, The Institute of Refrigeration, Proc.Inst.R. 1986-87, 1-13

Hayashi, Y., A.Aoki, S.Adachi and K.Hari (1977), Study of Frost Properties Correlating with Frost Formation Types, J.Heat Transfer, Trans.ASME, 99, 239-245

Hosada, T. and H.Uzahashi (1967), Effect of Frost on Heat Transfer Coefficient, Hitachi Review, 16, 254-259

Huffman, G.D. and C.F Sepsy (1967), Heat Transfer and Pressure Loss in Extended Surface Heat Exchangers Operating under Frosting Conditions, ASHRAE Trans. 73, 2:I.2.1

Jones, B.W. and J.D.Parker (1975), Frost Formation with Varying Environmental Parameters, J.Heat Transfer, Trans.ASME, 97, 255-259

O'Neal, D.L. and D.R.Tree (1985), A Review of Frost Formation in Simple Geometries, ASHRAE Trans. 91, 267-281

Sami, S.M. and T.Duong (1988), Numerical Prediction of Frost Formation on Cooled Heat Exchangers, Int.Comm. Heat Mass Transfer, 15, 81-94

Schneider, H.W. (1978), Equation of Growth Rate of Frost Forming on Cooled Surfaces, Int.J.Heat and Mass Transfer, 21, 1019-1024

Stoecker, W.F. (1957), How Frost Formation on Coils Affects Refrigeration Systems, Refrig.Eng., 65, 42-46

Varma, H.K. and V.Charan (1978), Simultaneous Heat and Mass Transfer to a Flat Plate in Humid Air Stream under Frosting Conditions, Letters in Heat and Mass Transfer, 5, 297-305

White, J.E. and C.J.Cremers (1974), Prediction of Growth Parameters of Frost Deposits in Forced Convection, AIAA/ASME Thermophysics and Heat Transfer Conference.

Zukauskas, A. and R.Ulinskas (1983), Heat Exchanger Design Handbook, 2, 2.2.4-2, Hemisphere Publishing Corporation, U.S.A.

HEAT EXCHANGERS

The Optimization Analysis Calculation for High Performance Heat Exchanger

DAVID SONG
China National Offshore Oil Co.
China Offshore Oil Development and Engineering Co.
Beijing, PRC

WEIZAO GU
The Institute of Engineering Thermophysics
The Chinese Academy of Sciences
P.O. Box 2706
Beijing 100080, PRC

ABSTRACT

The optimization criteria P/Q and P/Q f of the offset-strip-fin heat exchanger performance were proposed and one computer program used to calculate the best core sizes was developed. Some best core sizes and optimum values of P/Q, heat transfer coefficients, fin efficiencies were calculated as an example.

INTRODUCTION

One of the main ways for raising the cyclic efficiency of gas turbine plant is using high performance recuperator--a kind of heat exchanger--to recover exhaust energy for heating the air from the compressor, thereby decreasing the qantity of oil consumed in the combustor. Offset-strip-fin recuperator, i.e. OSF recuperator, is a kind of high performance gas-to-gas heat exchanger which is being widely studied and applied. The recuperator core sizes, such as the fin length x, fin thickness δ, fin height h and transverse spacing w between two neighbouring fins are important factors which influence the OSF recuperator performance (1-4). How to select the best core sizes being aimed at different requirement is the main research purpose of this paper.

OPTIMAL DESIGH PROBLOM FOR CORE SIZES AND CALCULATING METHOD

Due to the similarity of the momentum exchange and the heat ex-

535

change of fluid, in OSF recuperator, the enhancement of heat
transfer certainly accompanies with the increase in pressure
drop.   It leads up to a criterion problem about rating the
heat exchanger performance.   Some researchers have proposed a
variety of the rating criteria (5-7).   These criteria were for
the different applications having different objects.   The aim
of this paper is to obtain the more heat transfer rate Q with
the less fluid pumping power P, Through adjusting the sizes
of offset-strip-fins.   Therefore, P/Q was used as a rating
criterion to recuperator operating performance and an objective
function of the optimization problem here.   In order that the
fin efficiency η_f was not to be too low, $P/Q\eta_f$ was also used
as an objective function for the optimization problem.

If there are n rows of offset-strip-fins between two plates
of the recuperator and m fins are contained in each row of fins,
the pumping power and heat transfer rate by the fluid flow
can be expressed as follows, respectively,

$$P = (\frac{\dot{m}}{\rho}) \Delta p = \frac{2G^3 mnxw \, (1 + \frac{\eta_f}{a})}{\rho^2} \qquad (1)$$

$$Q = 4wxnmH(t_w - t_f)(1 + \frac{\eta_f}{a}) \qquad (2)$$

where a=w/h.
Substituting the Colburn heat transfer factor

$$j = \frac{H}{C_p G} P_r^{\frac{2}{3}}$$

into expression (2), it will be varied to the form

$$Q = 4wxmn(t_w - t_f) j G C_p P_r^{-\frac{2}{3}} (1 + \frac{\eta_f}{a}) \qquad (3)$$

From Eqs.(1) and (3), the correlation of P/Q can be reduced to

$$\frac{P}{Q} = \frac{G^2(1 + \alpha) P_r^{\frac{2}{3}}}{2C_p(t_w - t_f)(\alpha + \eta_f) \rho} \cdot \frac{f}{j} \qquad (4)$$

The fin efficiency η_f can be expressed as

$$\eta_f = \frac{\frac{\alpha}{w}\sqrt{\frac{2k\delta}{H}}\left[1-\exp(-\sqrt{\frac{2H}{k\delta}}\frac{w}{\alpha})\right]\left[1-\exp(\sqrt{\frac{2H}{k\delta}}\frac{w}{\alpha})\right]}{\exp(-\sqrt{\frac{2H}{k\delta}}\frac{w}{\alpha})-\exp(\sqrt{\frac{2H}{k\delta}}\frac{w}{\alpha})} \qquad (5)$$

The friction factor f and Colburn heat transfer factor j in
Eq. (4) can be determined by A.R.Wieting's empirical correlation formulae (3). The fin length x must be less than the
critical value at which the Nusselt number varing with distance
from entrance of the passage becomes a stable value. For the
air laminar flow, $2x/D_h \leq$ o.o1RePr (8), and in the turbulent
flow regime, $x/D_h \leq 1.8$ (8). In addition, for the different
flow regimes in the heat exchanger, the Reynolds number varing
range should be limited. All of these are the inqualtity constraint conditions of the optimization problem. The hydraulic
diameter D_h is defined as 2w/(1+a); the Reynolds number is
defined as GD_h/μ.

To sum up, the optimal design problem of the OSF recuperator
core sizes may be mathematically expressed as follows

For the turbulent flow: $\dfrac{P}{Q} = \dfrac{2.67G^{2.17}P_r^{0.667}(1+a)^{0.816}w^{0.184}\delta^{0.445}}{\rho^2\mu^{0.47}C_p(t_w-t_f)(\alpha+\eta_f)X^{0.459}}$

Objective function

Design variables $\vec{X} = (x,w,\alpha)^T$

Design parameters $G, k, \mu, \rho, P_r, (t_w-t_f), C_p, \delta$

Constraint conditions $g_1 = 2000 - \dfrac{2Gw}{\mu(1+\alpha)} \leq 0$

$g_2 = \dfrac{x(1+\alpha)}{2w} - 1.8 \leq 0$

$g_3 = -x < 0$

$g_4 = -w < 0$

$g_5 = -\alpha < 0$

For the laminar flow:

Objective function
$$\frac{P}{Q} = \frac{8.19\, G^{1.324} P_r^{0.667} W^{0.006} \alpha^{0.072} (1+\alpha)^{0.959}}{\rho^2 \mu^{0.176} C_p (t_w - t_f)(\alpha + \eta_f) X^{0.212}}$$

Design parameters $G, R, \mu, \rho, P_r, (t_w - t_f), C_p, \delta$

Constraint conditions
$$g_1 = \frac{2GW}{\mu(1+\alpha)} - 1000 \leq 0$$
$$g_2 = \frac{\mu(1+\alpha)^2 X}{2 P_r G W^2} - 0.01 \leq 0$$
$$g_3 = -X < 0$$
$$g_4 = -W < 0$$
$$g_5 = -\alpha < 0$$

This multidimensional constrained problem was solved by Interior Penalty Function Method and the Steepest Decent Method (a kind of Gradient Based Search Method) (9) was used to search the optimum solution of this problem. A computer program for optimization calculation has been developed.

COMPUTATIONAL EXAMPLE AND RESULTS

The numerical optization of the core sizes of the recuperator in one gas turbine plant was oerformed. The compute was carried out by HP-1000 electronic computer, it took about 3 minutes CPU time to perform for a time complete compute. The operating parameters and the computational results of the example were tabulated in Tables 1 and 2. In order to providing a reference for designer to select mass flow velocity G through the recuperator, a plot of the minimal value of P/Q and optimum value of the heat transfer rate per unit of plate area Q/F_B against mass flow velocity G was expressed in Fig.2. The results as shown in Fig. 2 were got under the objective function $P/Q\eta_f$. The variations of values P/Q, H, η_f, following fin parameters x, w and α were shown in Figs. 3,4,5, respectively.

TABLE !. The optimum results for OSF recuperator core sizes in the exhaust side, at p=1.02ata, t_f=506c, t_w-t_f=-22.8c, Pr=0.7, δ =0.3mm

Under the objective function P/Qη_f

$\frac{G}{kg/m^2s}$	P/Q $_f$	x mm	w mm	a	P/Q	η_f	H W/m^{2}k
5	0.0074	25.4	13.9	0.95	0.0059	0.81	59.24
7.5	0.0190	16.9	8.6	0.82	0.0158	0.83	92.2
10.0	0.0398	12.7	5.8	0.63	0.0330	0.83	126.
15.0	0.1111	9.0	4.5	0.63	0.0867	0.84	187.

Under the objective function P/Q

G kg/m^{2}s	P/Q $_f$	x mm	w mm	a	P/Q	η_f	H W/m^{2}k
5		33.4	18.1	0.95	0.0057	0.72	52.3
7.5		21.2	10.8	0.82	0.0153	0.73	83.1
10.0		19.6	9.0	0.63	0.0312	0.74	103.
15.0		15.8	7.2	0.63	0.0788	0.74	147.

TABLE 2. The optimum results for OSF recuperator core sizes in the air side, at p=9.0ata, t_f=460c, t_w-t_f=22.8C, Pr=0.7, δ =0.3mm

Under the objective function P/Qη_f

$\frac{G}{kg/m^2s}$	P/Q $_f$	x mm	w mm	a	P/Q	η_f	H W/m^{2}k
40	0.0086	4.7	2.1	0.627	0.0077	0.90	478
60	0.0217	3.9	1.8	0.626	0.0195	0.90	666

Under the objective function P/Q

G kg/m^{2}s	P/Q $_f$	x mm	w mm	a	P/Q	η_f	H W/m^{2}k
40		11.2	5.1	0.627	0.0069	0.72	320
60		8.2	3.8	0.627	0.0178	0.72	474

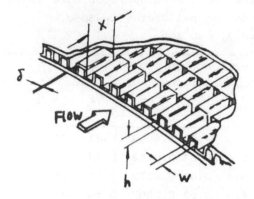

FIGURE 1. Typical offset-strip-fin heat exchanger.

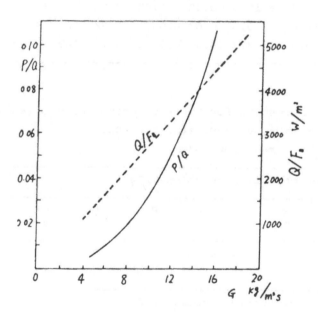

FIGURE 2. The variation of the minimal P/Q value and the optimum Q/F_B value with the mass flow velocity G.

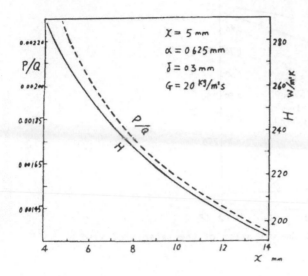

FIGURE 3.   The influence of the variation of fin length x.

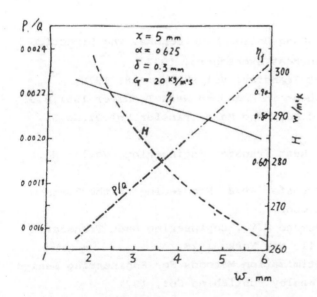

FIGURE 4.   The influence of the variation of the transverse spacing w
between two neighbouring fins.

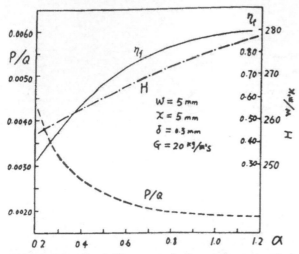

FIGURE 5. The influence of the variation of the ratio a of w to h.

REFERENCES

1. E.M. Sparrow et al, Trans. ASME, J. Heat Transfer, Vol.99, No.1, 1977

2. China Shanghhai Si-Fang Boiler Menufacture, The Direction About The Plate-Fin Heat Exchangers, 1983

3. A.R.Wieting, J.Heat Transfer, Vol.97, pp.488, 1975

4. S.V.Patankar, Advances in Enhanced Heat Transfer-1981,p.51

5. Adrian Bejan, Int.J.Heat and Mass Transfer,Vol.21,No.5, 1978

6. B.M.Johnson et al, Heat Transfer Engineering, Vol.4, No.1, 1983

7. R.K. Shah , Heat Transfer 1978, Proceeding of the Sixth Int. Heat Transfer Conf.

8. Karlekar B.V., Desmond R.M., Engineering Heat Transfer, 1st ed., McGraw-Hill, New York, 1978

9. Rlchard L. Fox, Optimization Methods For Engineering Design, 1st ed., Addisson-Wesley Publishing Co., 1971

Optimization of Phase-Transitional Heat Exchangers for Maximum Effective Heat Flux

KEGUANG WANG, DEMIN SHANG, ZHENPING NING, and ZHAOQIN ZHU
Harbin Institute of Technology
Harbin, PRC

ABSTRACT

The purpose of energy conservation and the skill of heat transfer enhancement are realized with a new kind of heat energy recovery devices, the phase--transitional heat exchangers, which have high rate of heat transfer and reliable operation but less corrosion and fouling.

Heat transfer enhancement can increase the heat energy recovered, Q, but the consumed power, W, increases more quickly than Q. We suggest the effective heat flux $q = (Q - ZmW)/A$ be a new criterion to evaluate the performance of heat exchangers used in waste heat recovery.

Optimization of the phase--transitional heat exchangers is carried out to obtain the maximum effective heat flux and the optimum parameters.

1. INTRODUCTION

There are a large number of industrial boilers and furnaces in our country. Most of them are in small or middle size and low thermal efficiency. The fuel they consumed each year accounts for 140 million tons of standard coal. Many devices have been created to recover the residual heat energy of these thermal equipments. Since 1983 we have applied ourselves to developing the phase--transitional exchangers,

FIGURE 1.   Phase--transitional heat exchanger

Fig.1. Many years experiments and operation results have shown the advantages of this kind of devices over other kinds, [1].

The principle of the phase--transitional heat exchangers is sketched in Fig.2. The flue gas flows through the spirally corrugated tubes in the evaporator,giving heat to the boiling medium water outside the corrugated tubes. The evaporated vapor moves up to the condenser,where it condenses and gives out heat to the water inside the copper tubes. Thus,the heat is finally recovered in the form of hot water. The medium water is maintained at such an optimum temperature [2] that the production cost is the lowest. The optimum temperature is usually between 85°c and 105° c. Consequently,the wall temperature of the corrugated tube is about 50° c higher than that of the conventional heat exchangers. This advantage enables the phase--transitional heat exchangers to recover more heat from the flue gas without the problems of sulphur corrosion and soft ash fouling.

Spirally corrugated tubes have been proved to be an effective measure of heat transfer enhancement for the phase--transitional heat exchangers. Their advantages are (1) easier to be fabricated, (2) less material and production cost, (3) limited fouling and wearing in operation,and (4) high rate of heat transfer enhancement than other heat transfer elements. However,as pointed by Bergles and other researchers [3] the enhancement of heat transfer brings flow resistance as well.

Many criteria have been proposed to evaluate the performance of heat exchangers. For the best economy of the operation,we suggest the effective heat flux q be a new criterion.

$$q = (Q - ZmW)/A \tag{1}$$

Where,balance coefficient Zm equals to the ratio of unit electricity to unit heat.

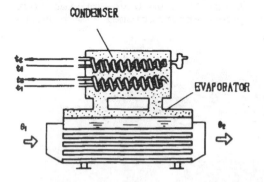

FIGURE 2. Structure of phase--transitional heat exchanger

2. HEAT RECOVERED

From the equation of heat transfer, the total heat recovered by a phase--transitional heat exchanger can be expressed as

$$Q = AK \, \Delta tm \tag{2}$$

The heat transfer coefficient of the corrugated tubes, k, is usually expressed as

$$K = a_i \{1 + a_i [(d_o - d)/(2\lambda_w) + d/(a_o d_o) + r]\}^{-1} \tag{3}$$

Since the sum of the thermal resistances of the tube wall, the boiling convection and the fouling is much less than that of the convection of flue gas with the inner wall of the corrugated tube, it is true that

$$a_i [(d_o - d)/(2\lambda_w) + d/(a_o d_o) + r)] \ll 1$$

Therefore we may get a temporary constant

$$C = \{1 + a_i [(d_o - d)/(2\lambda_w) + d/(a_o d_o) + r]\}^{-1} \tag{4}$$

For the corrugated tube in the phase--transitional heat exchangers, C only varies between 0.88-0.97.

Reference [4] selected the correlation of convection heat transfer inside corrugated tubes as

$$Nu = 165 Pr^{0.4} (e/d)^{1/3} (p/d)^{-1/2} (Re/10^4)^{0.8 - 3.5 e/d} \tag{5}$$

Substituting equations (4) and (5) into (3), we got

$$K = 165C \, Pr^{0.4} (\lambda/d)(e/d)^{1/3} (p/d)^{-1/2} (Re/10^4)^{0.8 - 3.5 e/d} \tag{6}$$

Then the heat recovered, expressed in equation (1), can be written as a function of the logarithm mean temperature difference, the flow and structural parameters

$$Q = 165A \, \Delta tmC \, Pr^{0.4} (\lambda/d)(e/d)^{1/3} (p/d)^{-1/2} (Re/10^4)^{0.8 - 3.5 e/d} \tag{7}$$

The refered area A here is the heat transfer area of the evaporator, namely the total inside area of the corrugated tubes, $A = n\pi dl$.

3. POWER CONSUMED

The power consumed consists of two parts,

$$W = W_1 - W_2 \tag{8}$$

W_1 is the power consumed by the fan pulling flue gas through the evaporator. W_2 is the power consumed by the pump circulating the water through the condenser.

From the general equation $W_1 = G_1 \, \Delta p_1$, we can get

$$W_1 = (n\pi d^2/4)u_1 f_1 (1/d)(\rho_1 u_1^2/2) = A(\rho_1 f_1/8)(\gamma/d)^3 Re^3 \qquad (9)$$

Similarly, it may be written that

$$W_2 = A_2(\rho_2 f_2/8)u_2^3$$

However, because the specific structure and the selected design parameters, A_2 and U_2 can not be varied independently. They must obey the governing relations

$$A_2 = K_1 A_1 \qquad \text{and} \qquad U_2 = K_2 U_1$$

Where k_1 and k_2 are ratio factors given from our former work [2], thus

$$W_2 = K_1 K_2^3 A(\rho_2 f_2/8)(\gamma/d)^3 Re^3 \qquad (10)$$

By substituting equations (9) and (10) into equation (8), the total power consumption can be given.

$$W = (A/8)(\gamma/d)^3 (\rho_1 f_1 + \rho_2 f_2 K_1 K_2^3) Re^3$$

The frictional factors f_1 and f_2 were chosen as [4]

$$f_1 = 11.3(e/d)(p/d)^{-0.7} Re^{-0.2}$$

$$f_2 = 0.316 Re^{-0.25} \qquad (11)$$

Finally, the total power consumption is derived out as

$$W = [1.41\rho_1 (e/d)(p/d)^{-0.7} + 0.04\rho_2 K_1 K_2^3 Re^{-0.05}]A(\gamma/d)^3 Re^{2.8} \qquad (12)$$

4. OPTIMUM PARAMETERS

Substituting equations (7) and (12) into (1) yeilds the detailed objective function for the optimization, the effective heat flux expression

$$q = 165C\,\Delta tmPr^{0.4}(\lambda/d)(e/d)^{1/3}(p/d)^{-1/2}(Re/10^4)^{0.8-3.5e/d} - [1.41\rho_1 (e/d)(p/d)^{-0.7} + 0.04\rho_2 K_1 K_2^3 Re^{-0.05}](\gamma/d)^3 Re^{2.8} Zm \qquad (13)$$

Since the first term in equation (13), the heat recovered, increases more slowly than the second term, the power con-

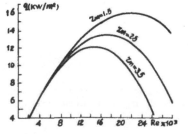

FIGURE 3. Calculated curve of q-Re

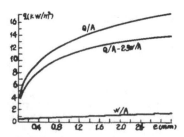

FIGURE 4. Calculated curve of q-e

sumed,when the 'Re' or the 'e' increases,the effective heat flux is a set of convex curves on the q--Re and q--e charts, Fig. 3 and 4. The climaxes of these curves correspond to the optimum values for the specific conditions.

Generally speaking,q is a multi-function of many variables, e,d,p, Δtm,Re and so on. However,in the consideration of optimum operation values,only the Reynolds number,Re,has significance. By setting the partial derivatives of the effective heat flux to the Reynolds number to zero we can find the expression of the optimum Reynolds number

$$Re^* = 10^4 \{[165*10^{-10}(0.8 - 3.5e/d)(e/d)^{1/3}(p/d)^{-1/2}(\lambda/d)C\,\Delta tm\;Pr^{0.4}]/\;[(62.6\rho_1(e/d)(p/d)^{-0.7} + 1.81\rho_2 K_1 K_2{}^3 Re^{-0.05})\;(\gamma/d)^3 Zm]\}^{1/(2+3.5e/d)} \quad (14)$$

Although there is a $Re^{-0.05}$ in the right side of equation (14), it is basically an explicit expression of the Re^*. Since the index of the $Re^{-0.05}$ is much smaller than unity,$Re^{-0.05}$ is very close to unity whatever the Re is. Therefore the $Re^{-0.05}$ may be taken as 1 at first,and the calculated Re^* is more than 97 percent close to its true value.If more accurate result is needed,one iteration would be enough.

Fig.5 gives out the change patterns of optimum Reynolds number.

5. DISCUSSION AND CONCLUSION

As a waste heat recovery device,the phase--transitional heat exchanger is aimed to achieve best economical benefit. The optimization for maximum effective heat flux help it to do so well.

We have noticed other criteria for the performance of heat exchangers,but the effective heat flux $q = (Q - ZmW)/A$ looks more suitable for the situation of waste heat recovery. The coefficient Zm varies considerably in different regions. It is basically a ratio of unit electricity to unit heat. In

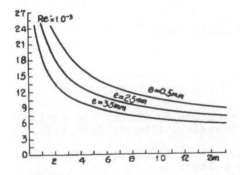

FIGURE 5. Calculated curve of Re-Zm

547

some part along the Yangtze River where electricity from hydropower is cheaper while fuel transported from mines long away is expensive,the Zm is small. In other regions the value of Zm may be larger than 10.

Much research work has been done on spirally corrugated tubes and many good correlations have been given by other researchers. For simplicity,we only used the simple ones in this paper. If more strict result is needed,it would be better to use the more recent correlations,such as that of R. Sethumadhavan and M. Raja Rao [5] for the corrugeted tubes.

The optimization method and results here may be served as reference for similar works and we also appreciate comment and discussion from our collegues.

NOMENCLATURES

A--reference heat transfer area
C--temporary constant
d--inside diameter
d_e--outside diameter
e--roughness height
f_1,f_2--friction factors
G_1,G_2--volume flux
K--heat transfer coefficient
K_1,K_2--ratio factors
l--tube length
n--number of tubes
Nu--Nusselt number
p--pitch
Pr--Prandtl number
q--effective heat flux
Q--heat recovered
r--heat resistance of fouling
Re--Reynolds number
Re^*--optimum Re
Δtm--logarithm mean temperature difference
u_1,u_2--flow velocity
W--power consumed
Zm--balance coefficient
ρ_1,ρ_2--density
α,α_e--convection heat transfer coefficient
λ,λ_e--conductivity
γ--dynamic viscousity

REFERENCES

1. Wang Keguang,Zhang Jizhou,Shang Demin, Theory and Application of Vacuum Type Heat Transfer, Proc.of JHPA,1985.

2. Wang Keguang, Ning Zhenping, Optimization of PT Heat Exchangers, Saving Energy. No.1,1987.

3. Bergles, A.F.,Blumenkrantz,A.R.,and Taborek.J., Perfor-

mance Evaluation Criteria for Enhanced Heat Transfer Surfaces, Proc. of 5th IHTC, Vol 2,1974.

4. Wang Keguang,Shang Demin,Guo Feng,Yao Yuhua, Research and Application of Spirally Corrugated Tubes, Scientific Research Report of HIT, No. 179,1983.

5. R. Sethumadhavan,M.Raja Rao, Turbulent Flow Friction and Heat Transfer Characteristics of Sinlge and Multistage Spirally Enhanced Tubes, Journal of Heat Transfer ,Vol. 108,1986.

Experimental and Numerical Studies on Heat Transfer Augmentation of Radiation Exchanger with Inserted Cylinder

HUA-QIANG LI, BO-HUI BIAN, WEI-LIN HU, and ZENG-YUAN GUO
Department of Engineering Mechanics
Tsinghua University
Beijing 100084, PRC

1. INTRODUCTION

This paper presents experimental and numerical investigation on a new method for augmenting heat transfer. Unlike the ordinary method of using various types of extended surfaces, heat transfer augmentation has been obtained by inserting a cylinder in convective annular ducts. Simple structure and small pressure drop are its main advantages. Thermal roundabout flow phenomenon has been found in our experiment system and checked by numerical evaluation. This leads to the existence of an optimum diameter of inserted cylinder, with which the heat transfer augmentation reaches its maximum.

2. CONCEPTS AND DEFINITIONS OF EFFICIENT OF HEAT TRANSFER AUGMENTATION

In order to illustrate quantitatively the effect of augmentation of radiation heat exchanger, the coefficient of heat transfer augmentation, Em, was defined in literature[1].

$$Em = Q_{total}/Q_2 \tag{2.1}$$

where Q_{total} means the total heating power to the fluid, Q_2 represents the heating power from wall 2 to the fluid by convective.

Because the absorptivity of the working gas, air, is very small, the radiation heat transfer from wall surface to the fluid can be regarded as zero. therefore, we have

$$Q_{total} = \begin{cases} Q_2 + Q_4 & \text{without inserted cylinder} \\ Q_2 + Q_3 + Q_4 & \text{with inserted cylinder} \end{cases} \tag{2.2}$$

where Q means the amount of heat transfer by convective, subscripts 2, 3, 4 represent inner wall surface, the inserted cylinder surface and the external surface, seen in Fig. 1(b), respectively.

The efficiency of heat transfer augmentation is defined as follow.

$$E_f = \frac{Em_a - Em_b}{Em_b} \times 100\% \tag{2.3}$$

where subscripts a and b represent the cases with and without inserted cylinder respectively.

551

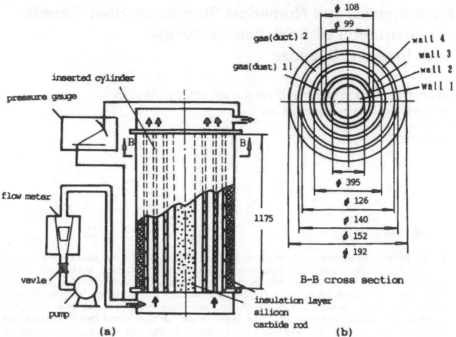

Fig 1. Experiment system

For the convenience of comparing the experimental results with numerical results, the external wall must be adiabatic. Nevertheless, there is heat loss from external wall to ambient through insulation layer in our experiments. The efficiency of heat transfer augmentation should be modified. Under the assumption that the heat loss is only through heat conduction of the insulation layer. The following equation can be derived.

$$h_1 \ E_f \ (TW_2(x)-Tg_1(x))\Delta A_2 + \frac{k \ \Delta T}{\delta}\Delta A_4 = const \qquad (2.4)$$

where k is the thermal conductivity of insulation material, δ is the thickness of the insulation layer.

We have ΔT_r present the adiabatic temperature difference, and Em $h_1\Delta T_r$ show the total heating power, we have

$$\Delta T_r = (Tw_2(x)-Tg_1(x)) + \frac{k \ \Delta T \ d_{out}}{\delta \ Em \ h_1} \qquad (2.5)$$

where d_{out} is the equivalent diameter of insulation layer, Average adiabatic temperature difference is:

$$\Delta T_r = \frac{\int \Delta T_r dA_2}{A_2} \qquad (2.6)$$

We use ΔT_r to estimate the efficiency E_f. The less ΔT_r, the higher the efficiency. So that, from (2.3), we have the E_f expression in adiabatic

552

case.

$$E_f = \frac{(\Delta T_r)_b - (\Delta T_r)_a}{(\Delta T_r)_b} \times 100\% \tag{2.7}$$

In numerical studies, it is assumed that the profile of silicon carbide rod temperature and the flow rate are kept the same whether or not with inserted cylinder. From (2.1) and (2.3),in this case, the E_f is presented as follow:

$$E_f = \frac{(Tg_{out}-Tg_{in})_a - (Tg_{out}-Tg_{in})_b}{(Tg_{out}-Tg_{in})_b} \tag{2.8}$$

where the Tg_{out} is the outlet temperature of heated gas, and Tg_{in} is the inlet temperature of heated gas.

3. EXPERIMENTAL STUDY

Experimental system(shown in Fig.1(a)) consists of three parts: 1). power system, including air pump and electric heating system. Thermal radiation source is silicon carbide rod. 2). Measurement system. 3). Experiment section. There are two cases in experiments (1). Without inserted cylinder; (2). With inserted cylinder. In both cases, inlet parameters

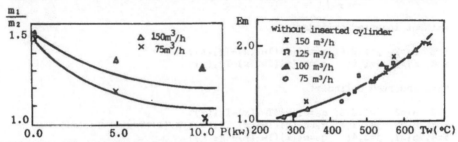

Fig. 2. Ratio of mass flow rates of inner duct to outer varies with heating power (in experiment)

Fig. 3. Em varies with temperature wall 2

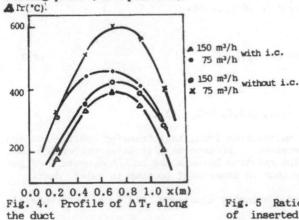

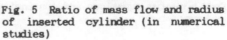

Fig. 4. Profile of ΔT_r along the duct
(i.c.:inserted cylinder)

Fig. 5 Ratio of mass flow and radius of inserted cylinder (in numerical studies)

553

(flow rate, temperature, pressure etc.) and heating power were kept the same. The values of heating power for each case are 5.5 kw, 10.5 kw, the flow rates (25°C) are 150 m³/h, 125 m³/h, 100 m³/h, 75 m³/h respectively. In order to calculate E_f, ΔT_r, we use the assumption[1][2] that the temperature distribution along the duct of fluid is linear.

4. NUMERICAL ANALYSIS

Considering the interaction between heat transfer and mass transfer including thermal roundabout flow caused by the difference of heating power per unit mass in the two sides of the inserted cylinder, the effects of diameter and emissivity of inserted cylinder, heating power and gas flow rate on heat transfer augmentation were studied. Since this kind of problem is very complicated, we have to make the hypotheses as follow: (1). The temperature and the velocity of gas flow only vary along the axis. (2). The surface of the walls and silicon carbide rod are grey body, gas is non-participant medium. (3). The heat conduction is negligible. (4). External wall is adiabatic. The heat transfer equation is simplified and discreted subject to the boundary and continuous, we can obtain nonlinear algebra coupled equations as follows.

For the wall surface section:

$$n_i \epsilon_i \sigma Tw_i{}^4(x) - \epsilon_i \sum_j \int B_j(x_j)K(x,x_j)dA_j + q_c = 0 \qquad (4.1)$$

without inserted cylinder:

i=2, n_1=2, j=1,2,4; q_c=$h_1(x)(Tw_2(x)-Tg_1(x))$
i=4, n_1=1, j=2,4; q_c=$h_1(x)(Tw_4(x)-Tg_1(x))$

with inserted cylinder:

i=2, n_1=2, j=1,2,3; q_c=$h_1(x)(Tw_2(x)-Tg_1(x))$
i=3, n_1=2, j=2,3,4; q_c=$h_1(x)(Tw_3(x)-Tg_1(x))+h_2(x)(Tw_3(x)-Tg_2(x))$
i=4, n_1=1, j=3,4; q_c=$h_2(x)(Tw_4(x)-Tg_2(x))$

where, $K(x,x_j)$ is exchange coefficient, and

$$B_j(x) = \epsilon_j \sigma Tw_j{}^4(x) + (1-\epsilon_j) \int B_k(x_k)K(x,x_k)dA_k \qquad (4.2)$$

For the gas section:

$$m_i Cp_i(x) \Delta Tg_i(x) = \sum_j h_i(x)(Tw_j(x)-Tg_i(x)) \qquad (4.3)$$

without inserted cylinder: i=1, j=2,4;

with inserted cylinder: i=1, j=2,3; i=2, j=3,4.

Physical parameters for calculating local heat transfer coefficient are given by local film temperature. Because mass transfer can effect heat transfer, we must know the relation between the bulk temperature of gas and the ratio of mass flow rate in inner duct to that in outer duct.

From momentum equations and continuity equations, and from reference[4], we have:

554

$$(P_{1,in} - P_{1,out}) = V_{1,out}^2 \rho_{1,out} - V_{1,in}^2 \rho_{1,in} + \frac{F_1}{A_1} \qquad (4.4)$$

$$m_1 = A_1 \rho_{1,in} V_{1,in} = A_1 \rho_{1,out} V_{1,out} \qquad (4.5)$$

$$m = m_1 + m_2 \qquad (4.6)$$

$$\frac{F_1}{A_1} = \left(\frac{m_1}{A_1}\right)^2 \frac{\Delta l}{2 d_{a1}} \sum_{j=1}^{N}\left(\frac{f_j}{\rho_j}\right) \qquad (4.7)$$

where, $i=1,2$ and $(P_{1,in}-P_{1,out})$ is caused by the change of density. i.e. these pressure drops are caused by thermal drag[7].

Both at inlet and outlet, the pressures in two ducts are equal. And from (4.4) to (4.7), we have:

$$\frac{m_1}{m_2} = \frac{A_1}{A_2} \sqrt{ \frac{ \frac{\Delta l}{2 d_{a2}} \sum_{j=1}^{N}\left(\frac{f_{2j}}{\rho_{2j}}\right) + \left(\frac{1}{\rho_{2,out}} - \frac{1}{\rho_{2,in}}\right) }{ \frac{\Delta l}{2 d_{a1}} \sum_{j=1}^{N}\left(\frac{f_j}{\rho_{1j}}\right) + \left(\frac{1}{\rho_{1,out}} - \frac{1}{\rho_{1,in}}\right) } } \qquad (4.8)$$

where $\Delta l = L/N$.

Because $P_{1,in}$, $P_{2,in}$, $P_{1,out}$, $P_{2,out}$ are near atmosphere pressure, $\rho_{1,in}$, $\rho_{1,out}$, $\rho_{2,in}$, $\rho_{2,out}$ are considered to be the function of the gas temperature only. The formula of convective heat transfer coefficient h and friction coefficient f are quoted from literature[6].

Monta-Carlo method is used to define thermal radiation exchange coefficient $K(r_k,r_j)$. If physical properties and geometric shape are

TABLE 1. Considering drag, efficiency has relation to flow rate, radius of inserted cylinder and emissivity

E_f flow rate	radius emissivity	0.056m	0.061m	0.064m	0.068m
	0.64	-0.9%	11.6%	18.5%	18.0%
150 m³/h	0.54	-2.7%	13.2%	17.0%	17.1%
	0.45	-4.7%	9.8%	16.1%	16.1%
	0.35	-7.4%	7.4%	14.0%	14.6%
	0.64	3.5%	18.5%	24.0%	22.4%
125 m³/h	0.54	1.7%	16.6%	21.9%	21.5%
	0.45	-0.5%	14.5%	21.0%	20.4%
	0.35	-3.6%	11.5%	18.4%	18.5%
	0.64	10.0%	24.8%	32.8%	28.8%
100 m³/h	0.54	8.0%	22.4%	31.0%	27.7%
	0.45	5.6%	19.8%	29.0%	26.5%
	0.35	2.2%	16.0%	25.1%	24.2%
	0.64	20.0%	39.2%	45.9%	36.7%
75 m³/h	0.54	17.9%	36.4%	43.0%	35.6%
	0.45	15.2%	32.9%	41.0%	35.0%
	0.35	11.3%	28.3%	34.8%	32.5%

known, subjecting to preceding hypotheses, we can get exchange coefficient. In this paper, the statistic basic number is 5000, which is used to calculate exchange coefficient one arbitrary element to the others.

5. RESULTS AND ANALYSIS

The experimental and numerical results show that the heat transfer of radiation heat exchanger can be augmented by inserting a cylinder into the duct. When the heating power is 10.5 kw and the flow rate is 150 m³/h, the experimental and numerical value of E_f are 10% and 16% respectively. As the flow rate decreases to 75m³/h, the experimental result of E_f is 26% and the numerical one is 41%. The difference of E_f between experimental and numerical results is caused by the assumption of linear temperature distribution of the silicon carbide rod in numerical study. There are two reasons for heat transfer augmentation. First, the area of convective heat transfer is enlarged by inserting a cylinder. Second, the velocity gradient at wall surface is increased. When the flow rate is 75m³/h, the velocity gradient at wall surface is increased by two times because the flow is laminar. If the flow is turbulent, e.g. the flow rate is 150m³/h, the increase of velocity gradient at wall surface is much smaller than laminar flow. It means that inserting a cylinder into turbulent flow will lead to weaker heat transfer augmentation compared with laminar flow.

In our experiments, the phenomenon of thermal roundabout flow was found. The difference of heating power per unit mass between two ducts makes the bulk temperatures different at the same height. That results in different variation of flow drag in the two sides of the inserted cylinder. Nevertheless, the inlet pressure and outlet pressure in different ducts are equal to each other. Therefore, the mass flow will be redistributed between two ducts because the heating power per unit mass in inner duct is larger than that in outer duct. The experimental and numerical results of thermal roundabout flow are shown in Fig. 5 and Fig. 10. In our experiment, the ratio of the mass flow in inner duct to that in outer duct is 1.5 if without heating. When the heating power is 10.5 kw, the ratio reduces to 1.2. It means that about 10% of the mass flow in inner duct flows into the outer duct. The numerical studies show the same result. The numerical studies also show that there must be an optimum diameter of inserted cylinder whether considering the effects of drag or not. The dimensionless optimum diameter is about 1.0. The dimensionless diameter equals d/d_{av} ($d_{av}=0.5(d_2+d_4)$). In addition, numerical studies also show that the greater the emissivity, the more obvious the effect of heat transfer augmentation.

6. CONCLUSIONS

1. Both experimental and numerical results show that the heat transfer of radiation heat exchanger can be augmented by inserting a cylinder into the duct.

2. The efficiency of augmentation depends on the emissivity of the inserted cylinder, the inner wall temperature and Re number. High inner wall temperature large emissivity and low Re number correspond to high efficiency of augmentation.

3. There exists an optimum diameter of the inserted cylinder for heat

transfer augmentation. The thermal roundabout flow phenomenon will influence the value of the optimum diameter of the inserted cylinder.

NOMENCLATURE

English

A	area
C_p	specific heat
d	diameter
E_f	efficiency of heat transfer augmentation
E_m	coefficient of heat transfer augmentation
$K(r_k, r_j)$	exchange coefficient
F	friction
h	convective heat transfer coefficient
k	thermal conductivity
L	length of experiment section
m	mass flow rate of fluid
N	the number of the elements in numerical studies
P	pressure of fluid
Q	amount of heating power
r	radius
T	temperature
V	velocity of fluid

Greek

ε	emissivity
ρ	fluid density
σ	Stefan-Boltzmann constant

Subscript

1	Silicon carbide rod(for wall),duct between walls 2 and 3(for gas)
2	inner wall(for wall), duct between walls 3 and 4(for gas)
3	inserted cylinder
4	external wall
h	hydraulic
g	fluid
w	surface
av	average
in	inlet
out	outlet

REFERENCE

(1). Y. Yamada, Y. Mori, *J. Heat Transfer,* Nov., 1984, Vol. 106/735
(2). Y. Yamada, Y. Mori, *ASME J. Trans.* 1975, Vol. 97, No. 3, pp. 400-405
(3). S. Mori, M. Kataya, A. Tanimoto, *Heat Trans. Engg.* 1980, Vol.2,No.1
(4). E. G. Keshock, R. Siaged, *J. Heat Trans.* 1964, pp.341-350
(5). Bo-hui Bian, Hua-qiang Li, Numerical Study on thermal radiation heat exchanger, J. Tsinghua University, rept. TH87009, No. 203, 1987
(6). *VDI-WÄRMEATLAS,* Verlag des Vereins Deutscher Ingenieure, Dusseldorf 1977
(7). Z. Y. Guo, Thermal Drag and Thermal Roundabout Flow in Convective Problem, *Proc. 8th Int. Heat Transfer Conf.* San Francisco, CA U.S.A., Vol. 1, pp59-68, 1986

Solutions to Temperature Field of Porous Heat Exchangers

X. A. LIN*, F. Z. GUO, and HUILING WANG
Department of Dynamic Engineering
Huazhong University of Science and Technology
Wuhan, Hubei, PRC

ABSTRACT

A new combined orthogonal collocation-perturbation method is used to deal with the cylindrical porous heat exchanger, which involves incoming medium flowing through the porous medium in axial direction and heat brought in only by the outer circumference or conversely. The concisely analytical correlations of fluid and solid temperatures relying on parameters of geometrical structure, flowing state and the essential physical properties, are obtained. The optimum design of such heat exchanger is discussed. And it follows that, for obtaining the excellent performance of porous heat exchanger, we'd better devise the remarkable anisotropic porous materials or structure, i.e. very high radial and extremely low axial thermal conductivities, and make it run at small radius of porous heat exchanger, and low mass flow rate etc.

INTRODUCTION

Porous heat exchangers are of great interest in the design of heat exchangering units in low temperature researches, thermal storage systems and chemical reactors. This articale presents solutions of the problem of incoming medium flowing through the cylindrical porous heat exchanger in axial direction. The results are quite a simple form for finding the stablized temperatures of the porous medium and flowing fluid as a function axial and radial coordinates.

The general problems were solved in different casses by Pust[1] and Maiorov et al.[10,11] with the approach of separation of the variables, and by Montakhab[2] with Riemann method[3,4]. And also done by Dixon et al.[6], Finlayson[7] with the orthogonal collocation techniques. Siegwarth and Radebough[9] directly used the numerical integral, differentiation and many times iterations to solve the problem with variable physical properties.

With the separation of variables apprough, it'll result in finding the sum of an infinite series functions including complex coefficients[1,10, 11]. Due to their slow convergences and complexity, the computation needs much time. Furthermore, it makes correlation between parameters more obscure and greatly lessens the significance of theoretical analysis. Also will it need skills to deal with the boundary conditions in some cases, for instance, in Pust's paper[11], correlation of the solid dimensionless temperature via dimensionless axial coordinate, as Fig.3[1] shows, isn't subject to its boundary condition at the inlet section, which might arise

* The author is now with Industrial Corp.,
Ningbo Economic-Technical Development Zone, Ningbo, P. R. China

from the inappropriate treatment of boundary conditions at inlet section.

Riemann method can only be applied to the general type of Riemann problem. Its results are also expressed in an infinite series polynomial terms. And the computation isn't simple enough.

Numerical method needs too much time for great number of iteration procedures under certain accuracy. And the results obtained give us less guidance in theoretical prediction.

Comparatively, however, the orthogonal collocation method as an efficient semianalysis tool is simpler and more flexible to handle either the linear or nonlinear differential equation and boundary conditions. Its accuracy is comparable to that of Galerkin method. Collocation method was first applied to differential equations by Frazer, Jones and Skan, and developed by Willadsen and Stewart[5] with orthogonal collocation techniques, and have been applied to packed bed reactor analysis by Finlayson [7]. Ferguson [8] dealt with application prospects of this method in chemical engineering. Dixon and Cresswell[6] used the method together with perturbation skills to packed bed problem and got some exciting results.

Here, the combined orthogonal collocation-perturbation method to porous heat exchanger is presented and quite simpler results are derived, which siggests its vitality as an analysis method in studying some heat transfer problems.

MATHEMATICAL DESCRIPTION

As Fig.1 shows, fluid flows through the porous cylindrical board in axial direction with the inlet temperature T_{fo} which can be higher or lower than the board circumference temperature T_{so}

If we assume:
a) Physical properties are constants.
b) The porous substance is regarded as a continuous media.
c) The heat is brought in solely by the outer circumference of porous board and removed by fluid or conversely.
d) The fluid thermal conductivities is negligible comparing with the solid thermal conductivities.
e) For convenience, we may as well consider $T_{so} > T_{fo}$. In the opposite case, the same relations will remain valid.

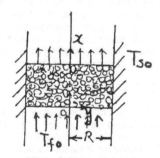

Fig. 1 Porous heat exchanger considered

The following differential equations may be set up based on local heat balances:

Solid Phase:
$$\frac{\partial^2 t}{\partial y^2} + \frac{1}{y}\frac{\partial t}{\partial y} + \gamma\frac{\partial^2 t}{\partial x^2} = A\frac{\partial T}{\partial x} \qquad (1)$$

Fluid Phase:
$$\frac{\partial T}{\partial x} = B\,(t-T) \qquad (2)$$

With the boundary conditions:

$$\frac{\partial T}{\partial y} = \frac{\partial t}{\partial y} = 0, \quad y = 0 \qquad (3)$$

$$T = t = 1, \qquad y = 1 \qquad (4)$$

$$T = 0, \qquad x = 0 \qquad (5)$$

560

$$T = t = 1, \quad X \longrightarrow +\infty \tag{6}$$

Where (3) represents the geometrical symmetry, (4) consider both the solid and fluid temperature at circumference as T_{so}, (5) reflects inlet temperature of fluid. We can image, if the length of heat exchanger spreads to infinite, the outlet temperature at the very section will be T_{so}, (6) suggests it.

SOLUTION

Now, let's find the analytical solution of the problem as described above. First, the radial derivatives in this system are discretized by one-point collocation and polynomial interpolation techniques. Here, the collocation point is chosen by the zero point of orthogonal polynomial $P1(y^2)$ described by Villadsen et al. [6]. Denote this by $Y1(= 1/\sqrt{3})$, and further $Y2=1$. Then we can write $T(x,y_i)=T_i$, and $t(x, y_i)=t_i$, i=1, 2.

Considering conditions (3), (4), we can suppose

$$t = a_1 + a_2 y + a_3 y^2 \tag{7}$$

Substituting conditions (3),(4) and $t|_{y=y_1}=t_1$ into (7), we obtain

$$t = \frac{1}{2}(3t_1 - 1) + \frac{3}{2}(1-t_1)y^2 \tag{8}$$

Similarly $\quad T = \frac{1}{2}(3T_1 - 1) + - (1- T_1)y^2 \tag{9}$

From (8), we obtain,

$$(\frac{\partial^2 t}{\partial y^2} + \frac{1}{y}\frac{\partial t}{\partial y})_{y=y_1} = 6-6t_1$$

Hence equation (1) turns to be

$$\gamma \frac{d^2 t_1}{dx^2} - 6t_1 - A\frac{dT_1}{dx} = -6 \tag{10}$$

at y1 point
Similarly, equation (2) may be rewritten as follows

$$\frac{1}{B} \cdot \frac{dT_1}{dx} + T_1 - t_1 = 0$$

Let $\epsilon = 1/B$, then

$$\epsilon \frac{dT_1}{dx} + T_1 - t_1 = 0 \tag{11}$$

Due to the small value of ϵ (approximately 1×10^{-3} quantative scale), we can take ϵ as perturbation parameter and expand T into:

$$T_1(x,\epsilon) = \sum_{m\geq 0}^{m} \epsilon^m \psi_m(x) \tag{12}$$

Substituting (12) into (11) and equating coefficients of power of , we get

$$\psi_0 = t_1 , \psi_m = -\psi_{(m-1)}' (-1)^m t_1^{(m)}$$

If we truncate (12) after K+1 terms and substitute it into (10), we get an K+1 order differential equation. Hence, it's practical to consider K=1, and getting:

$$T_1 = t_1 - \epsilon t_1' \tag{13}$$

Substituting (13) into (10), we gain:

561

$$(\gamma + A\epsilon)\frac{d^2 t_1}{dx^2} - A\frac{d\,t_1}{dx} - 6t_1 = -6 \qquad (14a)$$

Equation(14a) is a second order ordinary differentia; equation. It has two eigenroots P_1, P_2 for eignequation

$$(\gamma + A\epsilon)P^2 - AP - 6 = 0 \qquad (14b)$$

$$P_1 = [A - \sqrt{A^2 + 24(\gamma + A\epsilon)}]/[2(\gamma + A\epsilon)] \qquad (\;0)$$

$$P_2 = [A + \sqrt{A^2 + 24(\gamma + A\epsilon)}]/[2(\gamma + A\epsilon)] \qquad (\;0)$$

Hence, the solution of equation (14a) takes the form

$$t_1 = ae^{P_1 x} + be^{P_2 x} + 1 \qquad (15)$$

Where a,b are constants to be determined by boundary conditions. Substituting (15) into (13), yields

$$T_1 = a(1 - \epsilon P_1)e^{P_1 x} + b(1 - \epsilon P_2)e^{P_2 x} + 1 \qquad (16)$$

Considering (15) with condition (6), we have b=0
Substituting (5) into (16), yields

$$a = -1/(1 - \epsilon P_1)$$

So far, we've determined the constants a,b, hence

$$t_1 = 1 - \frac{1}{1 - \epsilon P_1}\, e^{P_1 x} \qquad (17)$$

$$T_1 = 1 - e^{P_1 x} \qquad (18)$$

Finally, substituting (17), (18) into (8) and (9), we have

$$T = 1 - \frac{3}{2}(1 - y^2)e^{P_1 x} \qquad (19)$$

$$t = 1 - \frac{3}{2(1 - \epsilon P_1)}(1 - y^2)e^{P_1 x} \qquad (20)$$

and

$$T_{av} = \frac{1}{\pi}\int_o^1 [1 - \frac{3}{2}e^{P_1 x}(1 - y^2)]y\,dy \cdot \int_o^{2\pi} d\theta = 1 - \frac{3}{4}e^{P_1 x_o} \qquad (21)$$

Discussion of Results

Equations (17)-(21) are final results we want. They are plotted in Fig. 2-4 in different situations. From these formula or figures, the thermal efficiency of porous heat exchanger and the total heat transfered to fluid or conversely can be derived.

Fig. 2 shows the solid temperature at collocation point varies with axial coordinate at different γ values. Curves 1 and 7 indicate the limit case when values are $+\infty$ and 0, it's quite clear that the temperature of t_1 increases with decreasing γ value. And in the limit case $\gamma = 0$, the t value is the closest to temperature $T_{s\theta}$. Temperatures of solid and fluid are almost the same in this case (B=2000), therefore the thermal efficiency is greatly increased by use of remarkable anisotropic porous materials or structure of low axial and high radial thermal conductivities. It matches well in comparison with Pust[1] expect the short inlet passage.

From Fig.2, we might also notice that line 4 is a critical line (correspondingly $\gamma = 0.2$). Above that (denoting I area), t_1 is quite sensitive to γ. A small changes of γ will cause greater changes of t_1, especially at smaller γ area. On the contrary, below line 4 (denoting II area), changes of γ will cause smaller changes of t_1. As a whole, t_1 is more sensitive to γ with decreasing γ value. Hence, it's practicle to manage to decrease γ value for gainning excellent performance of heat exchanger.

Fig.3 presents the dependences of fluid temperature at the collocation point y_1 on axial coordinate in different A value situations. Curve 8 is the limit situation, where A=0. It's apparent that the temperature T_1

increases with decreasing A value. Similar to Fig.2 curve 4 (correspondingly A=1) is a critical line. Above curve 4(denoting I area) T_l is more sensitive to A value than II area(below curve 4). And within I area, the same changes of A cause approximately the same changes of T_l. So for the good performance of heat exchanger, we should design the heat exchanger to work exactly in I area. This may be done by decreasing fluid flux, reducing the radius of heat exchanger, and enlarging the radial thermal conductivity of solid phase. The conclusion is also quite in agreement with Pust [1] deduced.

Fig.4 indicates the relation of dimensionless temperature t via dimensionless distance referred to the center of the heat exchanger, on the upper plane(drawing with dotted lines), and lower plane of the porous heat exchanger (with solid lines, 6 curves almost merging together) in different A value situations. The pair of curves corresponding to the same value of A are interconnected by the same number. The dependence of t_l at the upper section on dimensionless parameter A is quite similar to T_l via A in Fig.3. We may also find that at the short inlet period, the result has great error, e.g. at the inlet section (x=0), $t|_{y=0,x=0} = -1/2$. Similarly, two points orthogonal collocation at this spot will cause an error of 1/3. This phenomenan is in agreement with that Dixon at al.[6] discussed. Great A value would take longer inlet period length e.g. corresponding to line 1 (A=10) the inlet length is longer than 0.32 (dimensionless length).

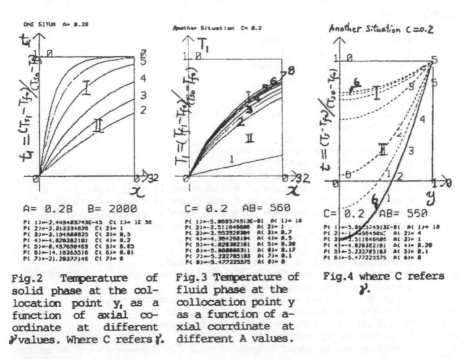

Fig.2 Temperature of solid phase at the collocation point y_l as a function of axial coordinate at different y values. Where C refers y.

Fig.3 Temperature of fluid phase at the collocation point y as a function of axial coordinate at different A values.

Fig.4 where C refers y.

CONCLUSION

An analysis of the problem, heat transferred radially through cylindrical board circumference, through porous media removed by fluid flowing axially, has been made in this article. The expressions(19), (20) are presented in a nondimenssional way that stablized temperature fields of both

solid and fluid are described as a function of axial and radial coordinates
. The results are quite simple and greatly reduce the computation time
involved in designing the porous heat exchangers. The method may be naturally extended to the situation in which radial and axial thermal conductivities of fluid phase are required considering, porous heat exchanger
working in superflowing He II temperature areas is the case. But the case
that physical properties vary greatly with temperature needs further investigation.

The results are expressed in three dimensionless parameters A,B, and
γ. The effect of anisotropy, axial and radial thermal conductivities of
heat exchanger, on solid temperature is shown in Fig.2. The effect of A
value, i.e. the value of fluid flow flux, radius and radial thermal conductivity of porous heat exchanger, on temperatures is considered and
graphically presented in Fig. 3, 4, and inlet period error is pointed out
and put it across in Fig. 4.

From the discussion, it follows that, the ideal geometrical structure
for high efficient porous heat exchangers is proved to be axially discrete and radially dense one, such as multilayer metal net etc., to get
high radial and low axial thermal conductivities; for gaining the excellent performance of porous heat exchangers, it's also necessary to design
a smaller radius of the heat exchanger, and make it run at smaller fluid
flow flux.

NOMENCLATURE

W	specific mass throughflow
C_p	specific heat of fluid
R	radius of cylindrical board
r	radial coordinate
z	axial coordinate
T_s	solid temperature
T_{so}	solid temperature at the board circumference
T_f	fluid temperature
T_{fo}	fluid inlet temperature
P_1, P_2	two eigenroots of equation(14b)
A	dimensionless parameter, $WC_p R/\lambda_r$
B	dimensionless parameter, $\alpha\beta R/WC_p$
t	dimensionless temperature of solid phase, $(T_s-T_{fd})/(T_s T_{fo})$
T	dimensionless temperature of fluid phase, $(T_f-T_{fo})/(T_{so}-T_{fo})$
t_1	dimensionless temperature of solid phase at collocation point
T_1	dimensionless temperature of fluid phase at collocation point
X	normalized axial coordinate, Z/R
Y	normalized radial coordinate, r/R
xo	normalized length of porous heat exchanger
α	convective heat transfer coefficient between solid and fluid phase
β	specific area of heat transfer
γ	ratio of the axial and radial thermal conductivity, $\lambda z/\lambda_r$
λ_z	axial thermal conductivity of solid phase
λ_r	radial thermal conductivity of solid phase
ϵ	perturbation parameter, 1/B
T_{av}	dimensionless averaged temperature of fluid phase at outlet sectio

References

1. Ladislav Pust, Investigation of Temperature Fluid in Heat Exchanger
   of Porous Cylindrical Board.
   J. Heat Mass Transfer 20, 1255-1277 (1977)

2. Ali Montakhab, Covective Heat Transfer in Porous Media.
J. Heat Transfer 101,507 (1979)

3. H. M. Lieberstein, Theory of Partial Differential Equation.(1983)

4. P.R. Garabedian, Partial Differential Equations.
John Wiley and Sons (1964).

5. J.V. Villadsen and W.E. Stewart, Solution of Boundary-Value Problems
by Orthogonal Collocation.
Chemical Engineering Science 22,1483-1501 (1967)

6. A.G. Dixon, and David L. Cresswell, Theoretical Prediction of Effective Heat Transfer Parameters in Packed Beds.
AICHE J. 25,663-676(1979).

7. Bruce A. Finlayson, Packed bed reactor analysis by orthogonal collocation. Che. Eng. Sci. 26,1081-1091 (1971).

8. H.B. Ferguson, Orthogonal collocation as a method of analysis in
chemical engineering, Ph.D. Thesis, Univ. of Washington, Seattle
(1971).

9. J. D. Siegwarth and R. Radebaugh, Analysis of Heat Exchangers for
Dilution Refrigerators
Rev. Sci. Instrum. 42,111-19(1971)

10.V.A. Maiorov, V.M. Polaev, L.L. Vasiliev, A.I. Kiselev, Augmentation
of convective heat transfer in conduits with high temperature porous
matrix. Part 1. Heat Transfer in case of local heat balance in matrix.
J. Eng. Phys. (in Russian) 47,13-24(1984).

11. V.A. Maiorov, V.M. Polyaev, L. L. Vasiliev,
Augmentation of convective heat transfer in conduits with high
temperature porous matrix. Part 2. Heat transfer in case of high heat
flux in matrix.
ibid. 47,199-205 (1984).

Experimental Verification about a Quick and Effective Method for the Design Calculation of a Phase-Change Heat Exchanger

BO YI CHEN and XIAN ZHI DU
The Naval Academy of Engineering
Wuhan, PRC

ABSTRACT

For a heat exchanger with change in phase, this paper employs a quick and effective method for its design calculation, i.e. Method in Sections. With the use of the method, the theoretical charts of temperature distribution and thermal efficiency of the heat exchanger may be afforded. In order to verify accuracy of computational result an experimental apparatus with the phase-change heat exchanger is established. The experimental results agree well with those of the theoretical computation.

INTRODUCTION

A phase-change heat exchanger is one of key components of dehumidifying apparatus. In certain cases, water vapor of air-water vapor mixture has great effect on people's life and production. For example, moist air makes people feel uncomfortable, food rot and machines rust easily. Especially, in air conditioning apparatus of an airplane, it is imperative that air-water vapor mixture coming into the apparatus should be dehumidified effectively, which prevents the refrigerating accessories from icing up, provides electronic instrument cabin with dry air and makes the electronic elements be in suitable environment.

SUMMARIZATION OF THE METHOD IN SECTIONS

Based on dew point of hot fluid (i.e. air-water vapor mixture), heat-transfer area of the phase-change heat exchanger is separated into a dry region where the hot fluid temperature is higher than the dew point, and a wet region where the hot fluid temperature is lower than the dew point, then the dry and wet regions may be also divided into a certain number of small areas with equal temperature differences respectively. Let thermal efficiency of the phase-change heat exchanger be defined as follows:

$$\varepsilon = \frac{\text{actual temperature drop of hot fluid}}{\text{maximum possible temperature drop of hot fluid}}$$

$$= (t_{h,in} - t_{h,ou})/(t_{h,in} - t_{c,in}) \tag{1}$$

By means of the Method in Sections, separation of heat-transfer area of the heat exchanger is shown in Fig. 1. For a multipass heat exchanger, it is suppositionally straightened into a single-pass heat exchanger.

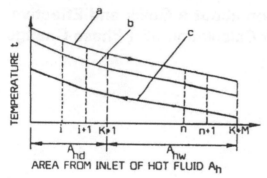

FIGURE 1. Scheme for the Method in Sections: a——Temperature of hot fluid, b——Temperature of finned wall, c——Temperature of cold fluid

The heat-transfer rate of the small wet area number n is equal to the sum of sensible and latent heat-transfer rates,i.e.

$$Q_{hw(n,n+1)} = m_h C_{ph} [t_{h(n)} - t_{h(n+1)}] + m_h r [d_{h(n)} - d_{h(n+1)}] \qquad (2)$$

where r is specific latent heat of vaporization corresponding to the dew point.

Because the heat-exchanger walls are finned, the heat-transfer rate from the hot fluid to the finned wall is given by

$$Q_{hw(n)} = A_{hw(n)} \alpha_{hw} \eta_{hwt} [t_{h(n)} - t_{wa(n)} + r (d_{h(n)} - d_{wa(n)})/C_{ph}] \qquad (3)$$

where α_{hw} is given by Ref. [1] as follows:

$$\alpha_{hw} = \alpha_{hd} (0.164 \ln Re_h - 0.02) \qquad (4)$$

where Re_h is Reynolds number of hot side of the heat exchanger.

On the other hand,the heat-transfer rate from the finned wall to cold fluid (i.e. dry air) of the heat exchanger is given by

$$Q_{cw(n)} = A_{cw(n)} \alpha_c \eta_c (t_{wa(n)} - t_{c(n)}) \qquad (5)$$

From Eqs. (3) and (5) we may find the finned wall temperature distribution $t_{wa(n)}$, through combination with Eqs. (2) and (3) we may obtain the heat-transfer area $A_{hw(n)}$ of the small wet area number n. Similarly, we may also obtain the heat-transfer area $A_{hd(i)}$ of the small dry area number i, thus the total heat-transfer area A_{ht} measured from inlet end of the hot fluid is given by

$$A_{ht} = \sum_{i=1}^{K} A_{hd(i)} + \sum_{n=K+1}^{K+M} A_{hw(n)} \qquad (6)$$

The main computations of the Method in Sections are simply described by the technological process, as shown in Fig. 2, thereby the theoretical chart of the thermal efficiency of the phase-change heat exchanger is drawn in Fig. 3. It shows that the results obtained by the present Method in Sections agree with those of the present experiment measurements to a close approximation.

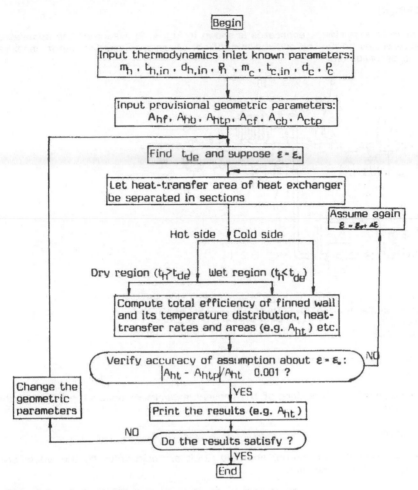

FIGURE 2. Technological process for Method in Sections

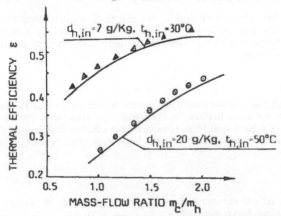

FIGURE 3. Thermal efficiency of phase-change heat exchanger

EXPERIMENT

The present experiment apparatus is shown in Fig. 4. By means of the apparatus, air stream may be heated and be humidified, whereas air-water vapor mixture may be dehumidified.

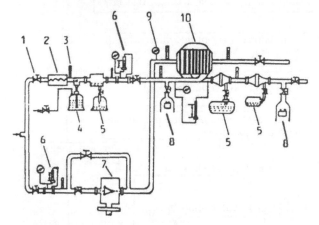

1—VALVE, 2—ELECTRIC HEATER, 3—THERMOCOUPLES
4—HUMIDIFIER, 5—WATER SEPARATOR, 6—FLOWMETER
7—TURBINE, 8—HSI-1 TYPE PSYCHROMETER, 9—PRE-
SSURE GAUGE, 10—HEAT EXCHANGER WITH FINNED WALL

FIGURE 4. Experiment apparatus scheme

Main components or functions of the present experiment apparatus are described as follows:

Flow Control

The mass-flow rates of the hot and cold fluids are controlled by the valves and are measured by flowmeters.

Temperature Control

The inlet temperature of the hot fluid is raised by an electric heater. The inlet temperature of the cold fluid is dropped down to 0°C or even lower by a turbine. All the temperatures are measured by combination with thermocouples and a numerical voltmeter with a printer.

Humidity Control

The heated air stream is humidified by a humidifier which makes water spray fog-wise upon the air stream from a nozzle 0.5 mm in diameter. In this way, the air stream with high pressure (P_h - 4 bar) and high temperature causes the sprayed water to be vaporized into water vapor so that the inlet specific humidity of the hot fluid is given by

$$d_{h,in} = d_0 + w_g/(\tau m_h) \tag{7}$$

where d_0 is specific humidity of air stream before humidification, w_g is weight of the vaporized water, τ is time for humidifying.

Measurements of dry and wet bulb temperatures (T_{DB} and T_{WB}) are made by a model HSI-1 telemetering and ventilating psychrometer. From Carrier's equation (Ref. [3]), water-vapor pressure P_v is given by

$$P_v = P_{gs} - (P_h - P_{gs})(T_{DB} - T_{WB})/(2800 - T_{WB}) \tag{8}$$

in which P_{gs} is water-vapor saturated pressure corresponding to wet bulb temperature T_{WB}, units of all the pressures and temperatures are [lb/in²] and [°F] respectively. In this case specific humidity of air-water vapor mixture is determined by

$$d_h = 622 \, P_v/(P_h - P_v) \tag{9}$$

Heat Exchanger

This is a cross-flow heat exchanger with finned walls and unmixed fluids, it is assumed that its thermal efficiency is calculated by

$$\varepsilon = \beta \varepsilon_{cou} \tag{10}$$

where ε_{cou} is a computational value of the thermal efficiency for counterflow condition, β is a correction factor which is the same as that of the logarithmic mean temperature difference.

Water separator

When the hot fluid (i.e. air-water vapor mixture) and the cold fluid (i.e. dry air) flow in the heat exchanger, water vapor in the mixture is mostly condensed into water which is removed by water separators due to gravity action on water.

The present experimental results show the fact that the thermal efficiency drops down with increasing inlet specific humidity $d_{h,in}$ of the hot fluid for a certain flow ratio of the cold and hot fluids. For example, when $d_{h,in}$ increases from 7 g/Kg to 20 g/Kg, ε cuts down by about 15 percent, as shown in Fig. 3.

CONCLUSIONS

1. The present experiment apparatus with a phase-change heat exchanger may be used to produce air-water vapor mixtures with various specific humidities, it is possessed of a dehumidifying function. By means of the apparatus, the thermal efficiency of the heat exchanger may be really measured.
2. The theoretical and experimental results show that thermal efficiency of a phase-change heat exchanger tends to drop down obviously with increasing inlet specific humidity of the hot fluid.
3. The present experiment measurements may verify the fact that the Method in Sections is a quick and effective method for design calculation of a heat exchanger with change in phase.

NOMENCLATURE

A	heat-transfer area
C	specific heat of fluid
d	specific humidity
m	mass-flow rate
P	pressure of fluid
Q	heat-transfer rate
t	temperature

Greek Symbols

α convective heat-transfer coefficient
ε thermal efficiency of a phase-change heat exchanger
η total efficiency of finned wall

Subscripts

b   fin base
c   cold side of heat exchanger
d   dry region of heat exchanger
de   dew point of air-water vapor mixture
f   fin
h   hot side of heat exchanger
in   inlet end
ou   outlet end
p   provisional
t   total
w   wet region of heat exchanger
wa   finned wall of heat exchanger

REFERENCES

1. Guillory, J. L. and F. C. Mcquiston, ASHRAE Transactions, vol. 29 June 1973.

2. ASHRAE Transactions, vol. 82, part 2, 1976.

3. Holman, J. P., Heat Transfer, 4d ed., PP. 481-485, McGraw-Hill, New York, 1976.

The Effect of Matrix Longitudinal Heat Conduction on the Gases and Matrix Temperature Distributions in the Rotary Heat Exchangers

TEODOR SKIEPKO
Technical University of Bialystok
Wiejska 45A, 15-351 Bialystok, Poland

ABSTRACT

In the paper two variants of models according to transport phenomena of energy in rotary heat exchangers are considered: one disregarding and the other including longitudinal heat conduction in the matrix. Each model is described by required the system of energy-balance equations which are solved by means of analytical methods. On the basis of these solutions the effect of matrix longitudinal heat conduction on the gases and matrix temperature distributions is shown.

NOMENCLATURE

A_k coefficient of the series,

c specific heat,

c_p gas specific heat at constant pressure,

d distance between temperature fields,

h matrix height,

r_1 real root of a characteristic equation,

r_2, r_3 real and imaginary part of complex conjugate root of a characteristic equation, respectively.

S total section of the metal matrix for longitudinal heat conduction [13],

t matrix temperature,

T gas temperature,

v velocity of gas in matrix,

W heat capacity rate,

Y matrix heat transfer area per unit volume,

GREEK SYMBOLS

α heat transfer coefficient,

β coefficient,

δ variable of integration,

λ thermal conductivity,

μ variable of integration,

μ^2 root of a transcendental equation,

ε porosity, ρ density,

η exchanger heat transfer Ψ zone angle,
     effectiveness,
                                          ω rotational speed,

ζ, Φ coordinates, one along the matrix in the direction of
     the gas flow and the other in the direction of
     rotation of the matrix, respectively.

Subscripts   Superscripts

j   =1(2) heating(cooling)   '   at the inlet,
    zone,

k   serial number,

m   matrix,

m,j matrix in the jth zone.

DIMENSIONLESS QUANTITIES

$NTU = \alpha Y h/(\varepsilon \rho c_p v)$ number of transfer units for gas,

$NTU_m = \alpha Y \Psi/[(1-\varepsilon) \rho_m c_m \omega]$ number of transfer units for
                                         metal matrix,

NTU_o overall number of transfer units,

$Pe = [\lambda_m \psi/(\rho_m c_m \omega h^2)]^{-1}$ Peclet number of metal matrix,

$z = \zeta/h$ longitudinal coordinate,

$\theta = (t-T_2')/(T_1'-T_2')$ matrix temperature,

$\theta = (T-T_2')/(T_1'-T_2')$ gas temperature

$\varphi = \Phi/\Psi$ coordinate in direction of
                                         rotation,

$\lambda^* = S \lambda_m /(W_{min} h)$ conduction parameter [13].

1.   INTRODUCTION

     Applications of the rotary heat exchangers as thermal
regenerators for the steam boilers,gas turbine instalations or
ventilation and air conditioning systems are generally
known.This is due to large and inexpensive heat transfer area
per unit volume in the rotary regenerators.Thus the
counterflow rotary heat exchangers combine compactness with
high performance.

     Transport phenomena of energy in the rotary heat
exchangers have been modelled by systems of partial
differential equations formulated with various simplifying
assumptions.These models supplemented by appropriate boundary
conditions,serve as mathematical models of heat transport
phenomena occuring in the rotary heat exchangers.For steady

574

state operation the models can be classified into two basic
categories:
(i) taking into considerations both convection and exchange
     terms[1-12],
(ii) considering not only the terms mentioned above but also
     the term of matrix longitudinal heat conduction [13-18].

The thermal conduction effect on the temperature
distributions in rotary heat exchangers was evaluated
in[14,18].MONDT in [14] found that it causes a reduction of
the temperature drops.Whereas [18] presents the results of
investigations on the effect of matrix longitudinal heat
conduction on the gases and matrix temperature distributions
in rotary heat exchangers at $NTU_1 = NTU_2$ ranging from 2. to 10.
for $NTU_{m,1} = NTU_{m,2} = .2$ and 1., respectively.

In the paper attention is directed to showing of the
effect of the matrix longitudinal heat conduction on the gases
and matrix temperature distributions.This is done by comparing
of the temperature fields calculated from the solutions of
conservation equations.

2.   THE SOLUTIONS

2.1. Matrix Thermal Conduction Effect Excluded.

In this case ,after introducing the dimensionless
coordinates and parameters defined in the Nomenclature and
using the assumptions usually formulated [19],the system of
energy conservation equation describing the temperature
distributions in a rotary heat exchanger may be written in a
nondimensional form

$$\partial\theta_j / \partial\varphi = NTU_{m,j} \cdot (-\theta_j + \theta_j)$$
$$\partial\theta_j / \partial z = NTU_j \cdot (\theta_j - \theta_j)$$
$$, \quad j = 1, 2 \tag{1}$$

where the coordinates and some denotations are shown in
Fig.1. Simultaneosly the above equations are supplemented by
following boundary conditions

$$\theta_1(\varphi, z=0) = 1 \tag{2}, \quad \theta_2(\varphi, z=0) = 0 \tag{3}$$

$$\theta_1(\varphi=0, z) = \theta_2(\varphi=1, 1-z) \tag{4}, \quad \theta_1(\varphi=1, z) = \theta_2(\varphi=0, 1-z) \tag{5}$$

The solution to the problem formulated above is
constructed in [18] being based on the general solution of
the partial differential equation system (1) given by LACH and
PIECZKA in [20] for crossflow recuperator and expressed by
means of special functions denoted in [20] as Bs and Bes.It is
used to find the temperature distributions for each the
crossflow gas-matrix heating and cooling zone when matrix
thermal conduction effect is excluded.

After substituting (2) and (3) into the solutions [20]
can show that the temperature distributions in the former zone
are described by the following expressions:

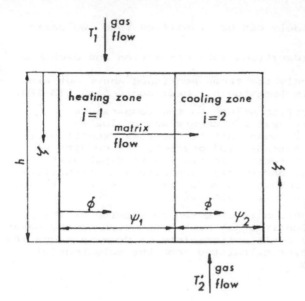

FIGURE 1.
The coordinates and some denotations related to the rotary heat exchanger.

matrix temperature

$$\theta_1(\varphi,z) = \theta_1(0,z) \cdot e^{-NTU_{m,1}\varphi} + \int_0^z \theta_1(0,\mu) \cdot e^{-NTU_{m,1}\varphi - NTU_1(z-\mu)} \cdot$$

$$\cdot Bes_1(NTU_{m,1}NTU_1\varphi, \ z-\mu \ ) \ d\mu \ + \tag{6}$$

$$+ \ e^{-NTU_{m,1}\varphi - NTU_1 z} \cdot Bs_1(NTU_{m,1}\varphi, \ NTU_1 z \ )$$

gas temperature

$$\theta_1(\varphi,z) = e^{-NTU_{m,1}\varphi - NTU_1 z} \cdot Bs_0(NTU_{m,1}\varphi, \ NTU_1 z \ ) \ +$$

$$+ \ NTU_1 \int_0^z \theta_1(0,\delta) \cdot e^{-NTU_{m,1}\varphi - NTU_1(z-\delta)} \cdot \tag{7}$$

$$\cdot Bes_0[NTU_{m,1}NTU_1(z-\delta), \varphi] \ d\delta$$

In the matrix cooling zone they are given as follows:

matrix temperature

$$\theta_2(\varphi,z) = \theta_2(0,z) \cdot e^{-NTU_{m,2}\varphi} + \int_0^z \theta_2(0,\mu) \cdot e^{-NTU_{m,2}\varphi - NTU_2(z-\mu)} \cdot$$

$$\cdot Bes_1(NTU_{m,2}NTU_2\varphi, \; z-\mu \;) \; d\mu \tag{8}$$

gas temperature

$$\theta_2(\varphi,z) = NTU_2 \int_0^z \theta_2(0,\delta) \cdot e^{-NTU_{m,2}\varphi - NTU_2(z-\delta)} \cdot$$

$$\cdot Bes_0[NTU_{m,2}NTU_2(z-\delta),\varphi] \; d\delta \tag{9}$$

The unknown functions $\theta_1(0,z)$ and $\theta_2(0,z)$ must be found basing on the conditions (4) and (5). By substituting the expressions (6) and (8) into those of (4) and (5) one arrives at a system of Volterra integral equations [18].

2.2 Matrix Thermal Conduction Taken Into Consideration.

Several simplifying assumptions usually formulated for this case have been used. These are:

(i)   the temperature variation in a radial direction is not considered,
(ii)   the thermal properties of both fluids and matrix also surface heat transfer coefficients are regarded as temperature independent,
(iii) the mass flow rate of the fluid in each zone is constant,
(iv)   the transport of energy with fluids as a result of carryover in direction of rotation is ignored.

From the assumptions given above the governing equations (for coordinate system see Fig.1) for the problem may be written in a nondimensional (see Nomenclature) form as follows

$$\partial\theta_j/\partial\varphi = NTU_{m,j} \cdot (-\theta_j + \theta_j) + Pe_j^{-1} \; \partial^2\theta_j/\partial z^2$$

$$\partial\theta_j/\partial z = NTU_j \cdot (\; \theta_j - \theta_j \;) \quad , \; j = 1, \; 2 \tag{10}$$

The corresponding dimensionless boundary conditions are

$$\theta_1(\varphi,z=0) = 1 \tag{11}, \quad \theta_2(\varphi,z=0) = 0 \tag{12}$$

$$\theta_1(\varphi=0,z) = \theta_2(\varphi=1,1-z) \tag{13}, \quad \theta_1(\varphi=1,z) = \theta_2(\varphi=0,1-z) \tag{14}$$

$$\partial\theta_j[\varphi, (z=0 \; and \; z=1)]/\partial z = 0 \; , \; j = 1, \; 2 \tag{15}$$

The solution of the above problem is given in [17,18]. It has been found there that in the matrix heating zone the temperature distributions are described by following series:

matrix temperature

$$\theta_1(\varphi,z)=1-\sum_{k=0}^{\infty} A_k \ e^{-\mu_k^2 \varphi + r_{2,k} z} \cdot \left[e^{(r_{1,k}-r_{2,k})z} \cdot \left(1+\frac{r_{1,k}}{NTU_1}\right) + \right.$$

$$+ \left(\frac{\beta_k r_{3,k}-r_{2,k}}{NTU_1} - 1\right) \cdot \cos(r_{3,k}z) + \tag{16}$$

$$\left. + \left[\beta_k + \frac{\beta_k r_{2,k}+r_{3,k}}{NTU_1}\right] \cdot \sin(r_{3,k}z)\right]$$

gas temperature

$$\theta_1(\varphi,z)=1-\sum_{k=0}^{\infty} A_k \ e^{-\mu_k^2 \varphi + r_{2,k} z} \cdot \left[e^{(r_{1,k}-r_{2,k})z} + \right.$$

$$\tag{17}$$

$$\left. - \cos(r_{3,k}z) + \beta_k \cdot \sin(r_{3,k}z)\right]$$

Similar expressions are given in [17,18] to the temperature distributions in the matrix cooling zone. The unknown coefficients A_k are determined on the basis of the boundary conditions (13) and (14) by utilizing a collocation method.

3. THE EFFECT OF MATRIX HEAT CONDUCTION

As an example this effect is illustrated in Fig.2 where are shown the temperature distributions in a rotary heat exchanger correspondingly to various of the models above presented. The results show the trends of the temperature

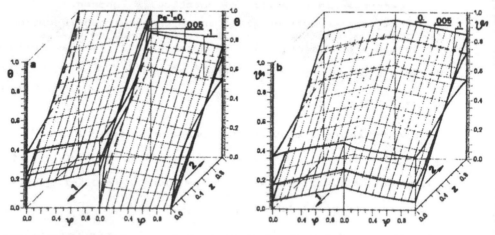

FIGURE 2. Effect of matrix longitudinal heat conduction on the tempearature distributions in rotary heat exchanger at $NTU_{m,1}=NTU_{m,2}=1$ and $NTU_1=NTU_2=8$; a) gas temperature distributions, b) matrix temperature distributions.

changes as a function of the coordinates both when heat
conduction is neglected (correspondingly to $Pe^{-1}=0.$) or taken
into account (correspondingly to $Pe^{-1}>0.$). It is seen from the
figures that the effect of longitudinal heat conduction makes

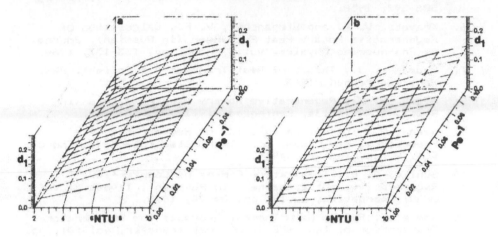

FIGURE 3. Effect of NTU and Pe values on the distance d_1
between the gas temperature fields in the heating zone at
$NTU_1=NTU_2=NTU$ and $Pe_1=Pe_2=Pe$; a) $NTU_{m,1}=NTU_{m,2}=.2$,
b) $NTU_{m,1}=NTU_{m,2}=1$.

the gas and matrix temperature at the outlet of the heating
zone higher and the cooling zone lower as compared to the
temperatures when the heat conduction is left out from the
account. The effect of matrix longitudinal heat conduction has
been evaluated numerical by calculating of a distance between
temperature fields. To this end the formulas presented in [18]
are used. Numerical results are shown in Fig.3 at $NTU_m=.2$ and
$NTU_m=1$. It is seen from Fig.3 that the distance d_1
between temperature fields of gas increases with increase in
Pe^{-1} and NTU values at NTU_m=constant.

Hence and the results presented in [18] we may conclude
that the effect of longitudinal matrix heat conduction is
essential at small NTU_m values. Moreover the effect becomes
greater as Pe^{-1} and NTU_m values increase.

REFERENCES

1. Nusselt, W., Die Theorie des Winderhitzer, Zeitschrift des
    Vereines deutscher Ingenieure, vol.71, pp. 85-91, 1927.

2. Iliffe, C. E., Thermal Analysis of the Contra-Flow
    Regenerative Heat Exchanger, Proceedings of The
    Institution of Mechanical Engineers, vol.159, pp. 363-371,
    1948.

3. Saunders, O. A. and Smoleniec, S., Heat Regenerators,
    Proceedings 7th International Congress for Applied
    Mechanics, vol.3, pp. 91-105, 1948.

4. Coppage, J. E. and London, A. L., The Periodic-Flow

Regenerator: A Summary of Design Theory, Transaction of the ASME, vol.75, pp. 779-787, 1953.

5. Lambertson, T. J., Performance Factors of a Periodic-Flow Heat Exchanger, Transaction of the ASME, vol.80, pp. 586-592, 1958.

6. Kravetz, V. F. and Stepanchuk, V. F., Calculation of Regenerative Rotary Heat Exchanger (in Russian), Journal of Engineering Physics, vol.3, no.3, pp. 133-137, 1960.

7. Madejski, J., Theory of Heat Transfer (in Polish), PWN, Warszawa-Poznan, 1963.

8. Migay, V. K., Regenerative Rotary Air Preheaters (in Russian), Energija, Leningrad, 1971.

9. Neaga, C., Thermal Calculation of Regenerative Air Preheaters, Rev. Roum. Sci. Tech.-Electrotech. at Energ., vol.19, pp. 193-206, 1974.

10. Schilo, A. F., Temperature Fields of Counterflow Regenerative Heat Exchanger (in Russian), Izvestija VUZ USSR: Energetika, no.10, pp. 89-94, 1972.

11. Romie, F. E., Periodic Thermal Storage: The Regenerator, Transaction of the ASME J. of Heat Transfer, vol.101, pp. 726-731, 1979.

12. Hill, A. and Willmott, A. J., A Robust Method for Regenerative Heat Exchanger Calculations, Int. J. Heat Mass Transfer, vol.30, no.2, pp. 241-247, 1987.

13. Bahnke, G. D. and Howard, C. P., The Effect of Longitudinal Heat Conduction on Periodic-Flow Heat Exchanger Performance, Transaction of the ASME J. of Engineering for Power, vol.86, pp. 105-120, 1964.

14. Mondt, J. R., Vehicular Gas Turbine Periodic-Flow Heat Exchanger: Solid and Fluid Temperature Distributions, Transaction of the ASME J. of Engineering for Power, vol.86, pp. 121-126, 1964.

15. Brodowicz, K., Theory of Heat and Mass Exchangers (in Polish), PWN, Warszawa, 1982.

16. Chung-Hsiung Li, A Numerical Finite Difference Method for Performance Evaluation of a Periodic-Flow Heat Exchanger, Transaction of the ASME J. of Heat Transfer, vol.105, pp. 611-617, 1983.

17. Skiepko, T., Solution of the Heat Transport Equation for the Rotational Regenerator (in Polish), Archiwum Termodynamiki, vol.8, no.1-2, pp. 35-53, 1987.

18. Skiepko, T., The Effect of Matrix Longitudinal Heat Conduction on the Temperature Fields in the Rotary Heat Exchanger, Int. J. Heat and Mass Transfer (in press).

19. Shah, R. K., Thermal Design Theory for Regenerators, in Heat Exchangers: Thermal-Hydraulic Fundamentals and Design, ed. S. Kakac, A.E. Bergles, F. Mayinger, pp. 721-763, Hemisphere, Washington, 1981.

20. Lach, J. and Pieczka W., On the General Solution to a Certain Class of Heat and/or Mass Transfer Problems, Int. J. Heat Mass Transfer, vol.28, no.10, pp. 1976-1981, 1985.

Stacked Perforated Plates Gas-Gas Heat Exchanger

GUOXING YE, JIANDONG CAI, SHAOXIAN HUANG, XIAOXI YANG, and GUIQIANG LI
Chemical Engineering Research Institute
South China University of Technology
Guangzhou, PRC

ABSTRACT

This paper briefly describes the experimental results of the stacked perforated plates Gas-Gas Heat Exchanger. The heat exchanger consists of
a block formed by the stacking of perforated metal plates whose apertures
create the fluid cross--flow passages. The passages have two stacking
modes: staggered/contract-expanded and staggered/lined up. The first mode
has been tested with a small heat exchanger in laboratory. The heat transfer coefficients up to 70 kcal/m^2hr °c were obtained at the air velocities of 10 m/s. The compactness(with specific area up to 100 m^2/m^3) and
modular structure of the heat exchanger have shown the installation flexibility and possible investment saving. Based on the experimental data,
we have established the correlations of heat transfer coefficientes and
pressure drops related to air flowrates for design purpose.

INTRODUTION

Various industries such as metallurgy, chemistry, petrochemical, glass,
ceramic paper, textile and food industries all need gas-gas heat exchanger
for recovery heat energy from gas and gas-heating and cooling. Gas-gas
heat exchangers generally are bigger and heavier than those for other
fluids, because gases have a lower heat transfer coefficient than other
fluids. Therefore, their uses are restricted by higher installation cost.
Sometimes, there are a series of difficulties resulted from narrow space
to set up a heat exchanger [1,2,3].
The problem of high efficiency and compactness, as has been said, has
been solved by stacked perforated plates Gas-Gas Heat Exchanger, which

not only has a higher heat transfer coefficient and specific area but also can fit the operating conditions and space limit by adjusting the sizes of the perforated metal plates.

GEOMETRICAL FEATURES OF THE HEAT EXCHANGER TESTED

The heat exchnager tested consists of a block of superpositioned perforated metal plates whose apppertures create the fluid cross-flow passages by sealing one row out of two row apertures on the end plates. Because of the differences of the shape and arrangement of the apertures, the cross-flow passages have two constructions. In the first one(defined as "staggered/contract-expanded" mode) each perforation in each plate communicates with two perforations of the following plate(Fig.1). In the second mode("staggered/lined up" mode) rows of staggered perforations and rows whose perforations are simply lined up are arranged alternately (Fig.2).

The heat exchanger is readily constructed by stacking and tightening of

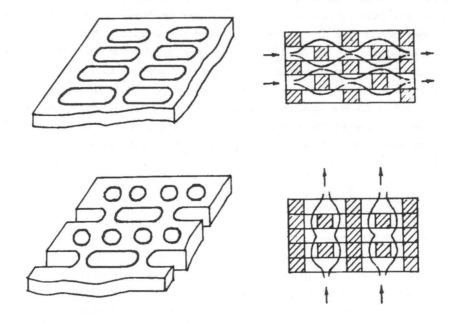

Figure 1. Staggered/contract-expanded mode(Left: perforated metal plates; Right: vertical section of passages)

the plates. The heat exchanger investigated has plates with elliptic perforations stacked according to the "staggered/contract-expanded" mode
(Fig.3)
Main specifications are as follows:
1. Size of the heat exchanger: 155x155x150(mm^3);
2. The heat transfer area: o.36 m^2;
3. The staggered duct mean cross section: 4.09x10^{-3} (m^2);
4. The contract-expanded duct mean cross section: 4.54x10^{-3}(m^2).

EXPERIMENTAL DEVICES

A schematic diagram of the device used for this test is given in Fig.4.
Before the air, which comes from the roots blower(1), being introduced
into the heat exchanger(2), flow rates are measured by two flow-meters(3)
respectively in local conditions. One stream(cool air) flows in a direction from bottom to top perpendicular to the plane of the plates, and
the other stream after being heated in electric preheater(4) is fed parallel
to these planes. The experimental device also includes the U-tube water
manometers(5) for pressure drop measurements and the copper-constantan
thermocouples(6) for measuring the inlet and outlet temperatures of the
two streams. A thermal insulation reduces external heat loss around the
heat exchanger.
The main purpose of this experiment is to determine the heat transfer
coefficients and pressure drops for the heat exchanger with staggered/
contract-expanded mode and establish their correlations related to the
cold and hot gas velocities.

EXPERIMENTAL RESULTS

Heat Transfer Coefficient
12 set of experimental data of the heat exchanger with staggered/contract-
expanded stacking mode, as an example, are given in Table 1. From this
Table, it can be seen that the measured heat energy released from the
hot air are greater than those received by the cold air. As no significant leaks were detected between the two compartments of the exchanger,
this thermal balance defect (< 6%) could only be explained by thermal
losses from the heat exchanger itself to the ambient air. The corrected
heat transfer coefficient values in the Table were calculated by assuming

583

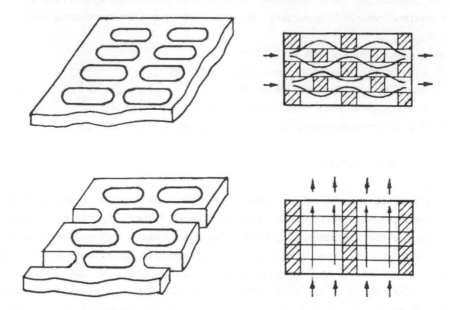

Figure 2. Staggered/lined up mode

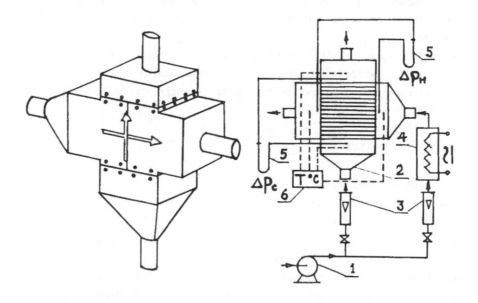

Figure 3. Appearance of SPP heat exchange

Figure 4. Schematic diagram of test system

that these losses are respectively direct proportional to the differences
between the temperatures of the cold and hot air and the ambient air
temperature.

TABLE 1. SPP heat exchanger experimental data

Wh	Wc	Ti	To	ti	to	Qh	X	Q	K
9.41	15.23	182.4	121.5	51.2	83.7	2049.4	5.2	1974.7	68.9
9.39	13.53	184.9	125.4	50.7	87.2	1999.1	2.9	1958.1	66.9
8.30	13.22	88.5	73.6	56.7	64.9	439.6	4.0	428.9	62.8
18.47	7.46	110.9	99.0	54.4	79.9	777.8	3.9	759.6	58.7
9.49	8.55	183.8	134.1	47.6	96.6	1686.9	2.2	1661.6	56.3
11.22	7.45	103.9	88.8	54.2	74.2	599.6	2.6	590.0	53.7
9.52	7.72	108.5	87.3	46.5	69.8	714.3	1.1	708.9	52.4
9.49	6.86	185.2	140.1	47.2	103.0	1532.8	1.6	1516.2	51.1
5.61	8.27	97.0	76.1	54.9	67.0	416.0	5.5	401.9	47.5
7.40	6.64	95.2	78.9	51.9	68.0	428.0	1.9	422.9	46.1
6.50	5.80	102.3	82.5	51.9	70.9	457.5	5.2	443.0	42.7
5.60	5.00	95.8	77.6	49.8	67.7	360.9	2.8	354.5	38.0

By means of the multiple regression of these experimental data, we established the relationship between the heat transfer coefficients and mass
velocities of cold and hot air, which can be expressed as:

$$K = \cfrac{1}{\cfrac{4.7833 \times 10^{-2}}{Wc^{0.65}} + \cfrac{3.1412 \times 10^{-2}}{Wh^{0.65}} - 9.9165 \times 10^{-4}} \qquad (1)$$

Where

K: The heat transfer coefficient (kcal/m^2hr °c);

Wc: The cold air mass velocity(kg/m^2s);

Wh: The hot air mass velocity(kg/m^2s).

The deviations are within ±4%, comparing the heat transfer coefficients
predicted by Eq.(1) with all experimental data. Eq.(1) only provided a
simple empirical formula for engineering design. It covered up many factors including the Re number defined from an equivalent diameter, Pr number; mode of the passages; main structure parameters of perforated plates,
for example the shape and porosity of perforations, which affect the heat
transfer process. For convenience, we compared Eq.(1) with the data provided by literature [4] in Fig. 5 and listed porosity [r] and equivalent
diameter [de] of perforated plates in Table 2.

First of all, the film heat transfer coefficient is influeced by the formed mode of passages and perforation shape. When the first and second terms in the denominator of Eq.(1) are respectively considered as both heat transfer resistances of the contract-expanded and staggered passages have

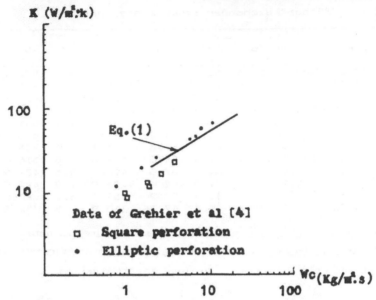

Figure 5. Comparison of Eq.(1) with data from literature [4]

Table 2. Porosity [σ] and Equivalent Diameter [de]

	perforation shape	σ	de
This paper	Elliptic	0.36	0.16
Literature [4]	Elliptic	0.34	0.12

greater heat transfer resistances than the staggered ones. In other words, the staggered passages have higher film heat transfer coefficients. Such facts proved that they are better for heat transfer enhancement. By analogy, the contract-expanded passages should have higher film heat transfer coefficients than the lined up. Therefore, the heat transfer coefficients of our devices are higher than those of staggered/lined up mode with square perforations in literature [4](see Fig. 5). Of course,

586

the different shape of perforations is another reason. Elliptic perforations are superior to squares have also been confirmed by [4].

However, the porosity of perforated plates is a more important structure parameter than the factors mentioned above. From Table 2, it can be seen that the equivalent diameter is 33% greater than that of [4] because of the increase of the porosity σ about 6%. Now, we only analyze the influence of equivalent diameter on film heat transfer coefficient. The functional equation may be expressed as:

$$NU= f(\text{ Re, Pr, passage mode, perforation shape, } \sigma) \tag{2}$$

and which can be expressed by the general empirical relationship, as for air:

$$NU= C \text{ } Re^m \tag{3}$$

From Eq.(3) the following equation can be derived:

$$h= C' \text{ } U^m/de^{(1-m)} \tag{4}$$

Eq. (4) shows that when the flow velocity is constant, the film heat transfer coefficient is inversely proportional to the equivalent diameter. Therefore, the heat exchanger we used with staggered/contract-expanded, in reasoning, should have higher heat transfer coefficients than that of lined up mode in [4], but actually it was lower. For the same reason, the results of square perforations tested in literature [4] also were lower than these in this test.

On the other hand, after increasing σ, the specific area increase and the flow resistances will descend at a certain extent. Obviously, making a thorough study is necessary for how to select a optimum size and density of perforations.

Pressure Drop

The linear relation deduced from the experimental data between the pressure drop measured and the kinetic energies of each fluid under isothermal flow showed that the prevailing influence was dynamic losses. The experimental relation related to the air velocity can be expressed as:

$$\Delta P = C \frac{\rho U^2}{2} \tag{5}$$

The values of constant C are obtained for both duct as follows:

staggered duct: C = 53.1;

contract-expanded duct: C = 4.0.

If a pressure drop at the inlet and outlet of the air ducts is 1.5 times of the kinetic energy, the general pressure drop and velocity relation for ary air ducts length L can be shown as follows:

$$\Delta P = (C'x L + 1.5) \times \frac{\rho U^2}{2}$$
(6)

where

staggered passages: $\qquad$ C' = 333;

contract-expanded passages: C = 11.9

It can be seen that the pressure drop of staggered passage is 13 times as great as the contract-expanded ones. Both passage sections and flow lines were plotted in Fig. 1., which shows the reason of producing such a big difference in ΔP.

CONCLUSION

1. This study showed that the heat transfer coefficient of heat exchanger with staggered/contract-expanded mode up to 70 ($kcal/m^2$ hr °c) were obtained at the air velocities of about 10 m/s, with corresponding maximum pressure drops lower than 0.003 kgf/cm^2 (for contract-expanded) and 0.035 $kgf.cm^2$ (for staggered passages).

2. Generally, maximum pressure drop allowed is about 0.1 kgf/cm^2 for a heat exchanger recovering waste heat from exhaust fume. Therefore, it is possible to scale up a heat exchanger according to the experimental results of the modeling study by means of adjusting the relative sizes of two passages length. We should arrange the exhaust fume to pass the contract-expanded or lined up passages, the latter is easier to clean.

3. The stacked perforated plates heat exchanger as compared with a tubular heat exchanger has obvious advantages, for example, the larger specific area; the higher heat transfer coefficient and the compactness; and is a great promising gas-gas heat exchanger for industrial application.

NOMENCLATURE

de equivalent diameter (m)

h   partial heat transfer coefficient ($kcal/m^2$hr °c)

k   heat transfer coefficient (kcal/m^{2}hr °c)

L   passage length (m)

ΔP   pressure drop (kgf/m^2)

Q   heat transfer rate (corrected) (kcal/hr)

T   hot air temperature(°c)

U   air velocity in the mean passages cross section (m/s)

w   mass velocity (kg/m^2)

x   thermal balance defect (%)

ρ air mass density (kg/m^3)

σ porosity of perforated metal plate (%)

subscripts

h   hot air

c   cold air

i   inlet

o   outlet

REFERENCES

1. Afgan, N.H. and Schlunder, E.U., "Heat Exchanger: Design and Theory Sourcebook", Mc Graw-Hill, New York, 1974.

2. Shak, R.K. and Pearson, J.T., "Advances in Heat Exchanger Design", HTD-vol.66, ASME, 1986.

3. Rabas, T.J. Myers, G.A. and Eckels, P.W., "Comparison of the Thermal Performance of Serrated High-Finned Tubes Used in Heat Recovery System", HTD-vol.59, ASME, pp.33-40, 1986.

4. Grehier, A., Raimbault, C., Rojey, A., Chligue, B. and Dreuilhe, J., "Compact Gas-Gas Heat Exchanger", Proceedings of the contractors' meetings of CEO, 1982.

Recovery of Latent Heat from Flue Gas by Cold Air in Condensing Heat Exchangers

CHAOYANG WANG, CHUANJING TU, and LING YAN
Department of Thermoscience and Engineering
Zhejiang University
Hangzhou, PRC

ABSTRACT
It is proposed in this report that condensing heat exchangers with enhancement surfaces are used to recover latent heat in flue gases and to heat cold air. Augmenting heat transfer on both sides of heat transfer tubes is investigated, and the obtained heat transfer results are incorporated in the numerical simulation for predicting thermal performance of condensing heat exchangers. Smooth as well as enhancement tubes are computationally tested, and the results indicate that it is a practical program to recover considerable amounts of latent heat contained in flue gas for preheating of air by using enhancement surfaces.

1. INTRODUCTION

There exist many exhaust gases with high content of steams released from chemical engineering, power engineering and drying industry. Fired boilers, particularly those fired with natural gas and coal-water slurry, produce a significant amount of waste heat in the exhaust gas stream in the form of latent heat of vaporization of water. Obviously, recovering a part of this kind of latent heat is of great significance for energy conservation and has remarkable economic benefits.

The experimental and theoretical studies have been done on the recovery of latent heat from flue gases by condensing heat exchangers (CHE) for sereral years in our laboratory[2]. The condensing heat exchanger is a tube bank exchanger in which flue gas containing steams crosses tube bank and heat released by flue gas is transferred to the low-temperature fluids (water or air) flowing in tubes. By making the temperature of tube walls lower than the dew point of flue gas, the partial condensation of steams in flue gas occurs, and as a result latent heat is recovered by coolants. From heat transfer point of view, the partial condensation enhances conective transfer of sensible heat in flue gas. If the coolant is water, and therefore the thermal resistance in gas phase is dominant over the total resistances, then the coexistence of convection and condensation on outwall of tube will substantially improve the performance of the exchangers, with thermal efficiency being increased by as 100 percents as much [1].

If the coolant is air, then the thermal resistance inside tubes would be the same or even less order of magnitude compared with that outside tubes(in flue gas). In such a case, the augmentation technique has to be applied to convection process of cold air inside tubes so as to correspond to the enhancement of heat transfer outside tubes(in flue gas) due to the partial condensation of steams on tube walls. In other words, convective heat transfer inside tubes must be enhanced when condensing heat exchangers are used to heat air, or the thermal performance of the exchangers is poor.

591

To this end, the present paper designs two condensing heat exchangers to recover latent heat from flue gas by cold air, one consists of internally finned tubes, and the other of internally rough tubes. In these exchangers, heat transfer is augmented not only by the mechanism of combined forced convection and condensation outside tubes, but also by rough or extended surfaces inside tubes. Such an augmentation technique on both sides of heat transfer areas will greatly increase the thermal performances of the exchangers, as compared with convectional heat exchangers, and make it practical to preheat cold air demanded in combustion processes.

2. HEAT TRANSFER ANALYSIS

2.1 Heat and Mass Transfer outside Tubes

The typical configuration of a condensing heat exchanger is schematically shown in Fig.1. Flue gas containing small amounts of steams loses heat during flow in banks, and cold air flowing inside tubes gains heat. The exchanger is designed so that wall temperature on a portion of its surfaces is lowered below the dew point of flue gas, and in this way the partial condensation of steams occurs on the walls, and subsequently enhances convective transfer of sensible heat.

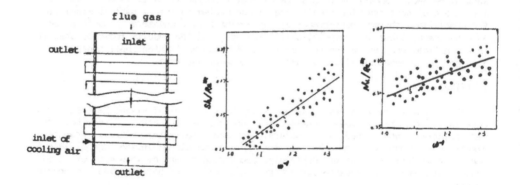

FIGURE 1. Condensing Heat   Fig. 2. Experimental data for heat and mass
           Exchanger   transfer outside

Heat and mass transfer data are necessary for the prediction of thermal performance of condensing heat exchangers, as well as for evaluations of heat transfer augmentation outside tubes. But few studies[2,3] have been conducted on convective transfer of sensible heat as well as mass transfer of steams during the condensation process on walls of tubes. In the range of parameters falling into the practical applications of CHE, Taniguchi et.al [3] conducted the pioneer work on this problem by experiments, but their experiments corresponded to a simple configuration, namely single cylinder, and can not be applied to practical configurations, tube banks. Therefore, the comprehensive experiments for heat and mass transfer in tube banks have recently been conducted in our laboratory. Experimental conditions are the following: mass fraction of steams in flue gas is between 5 and 25 percents; inlet temperature of flue gas varies from 110℃ to 140℃; inlet temperature of coolant ranges from 20℃ to 27℃; Reynolds numbers of flue gas and coolant are 10°–10° and 1600–10°, respectively. The choice of these experimental parameters is according to the consideration that the experiments in laboratory can simulate practical operation conditions of CHE as much as possible. Because of the limitation of the paper length, a detailed introduction to experimental apparatus and procedures is not presented here, and can be referred to in[2].

The experimental data suggest the following correlations for heat and mass transfer coefficients α_s and β:

$$Nu = \varepsilon f_h (C\, Re^m) \tag{1}$$
$$Sh = 0.91\, \varepsilon f_m (C\, Re^m) \tag{2}$$

where C and m are the same constants as those in the correlation for single-phase heat transfer in tube banks, and have been well documented in [4]. ε is a correction factor related with row number of tube bank.

The experimental data for heat and mass transfer are shown in Fig.2, and the factor f_h and f_m can be correlated within $\pm 30\%$ by the following equations [2]:

$$f_h = 0.4 + 0.6/\omega \tag{3}$$

and

$$f_m = 0.412 + 0.588/\omega \tag{4}$$

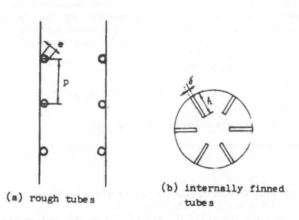

(a) rough tubes

(b) internally finned tubes

Fig. 3. Configurations of enhancement elements

where ω is the ratio of noncondensable gas concentration in the bulk of flue gas to that at the gas-liquid interface.

Having these equations for heat and mass transfer coefficients outside tubes, the later sections can evaluate the augmentation effect resulting from partial condensation of steams in flue gas, and calculate quantities of latent heat recovered by CHE.

2.2 Heat Transfer Augmentation inside Tubes

Because heat transfer coefficients of air are low, large flowrates of air are required in order to recover considerable amounts of latent heat; on the other hand, air can be heated only to low temperature by using large flowrates in CHE. As a result, the quality of recovered heat is poor even though its quantity is abundant. This kind of large-flowrate and low-temperature air is not much useful for serving the purpose of combustion. The augmentation technique for heat transfer inside tubes is absolutely needed if the coolant is air.

Various forms of surface roughness and extended surfaces (fins) have been studied for enhancement of gas and liquid flow in tubes [6]. Here two kinds of enhancement condensing heat exchangers are designed as shown in Fig .3. one consisting of repeated-rib rough tubes, and the other of longitudinally finned tubes. Based on the guidelines for selecting preferred enhancement dimension proposed by Webb and Eckert [7], and Carnavos [8], respectively, we designed two sets of enhancement dimensions for rough and finned tubes, respectively, as shown in Table 1.

Table 1 Enhancement Sizes of Rough and Finned Tubes

No.	inner diameter of tube mm	sizes of enhancement elements		
design I (rough tubes)	55	wire diameter	wire spacing	
		0.25 mm	2.5 mm	
design II (finned tubes)	55	fin height	fin thickness	No.
		10 mm	2 mm	32

Heat transfer coefficient data are necessary for the thermal performance analysis of condensing heat exchangers with enhancement surfaces. In our analysis of the next section, the following three heat transfer correlations, which are anailable in the literature [6-8], are used:

$$Nu_c = 0.023 \ Re^{0.8} Pr^{0.4} \tag{5}$$

for smooth tubes.

$$Nu_c = 0.023 \ Re^{0.8} Pr^{0.4} (A_e/A_r)^{0.1} (A_e/A)^{0.5} \tag{6}$$

for internally finned tubes and

$$Nu_c = \frac{Re \ Pr \ f/2}{1 + \sqrt{f/2}[\tilde{g}(e^*)Pr^{0.44} - B(e^*)]} \tag{7}$$

for rough tubes, where the functions of $\tilde{g}(e^*)$ and $B(e^*)$ were defined in [7], and the term e^* is the roughness Reynolds number, which may be written as

$$e^* = \frac{e}{D} \ Re\sqrt{f/2} \tag{8}$$

The gas flow friction factor in rough tubes is given by the correlation [7]

$$2/f = B(e^*) - 2.5\ln(2e/D) - 3.75 \tag{9}$$

Simultaneously solving eqs.(8), (9) and (7), one may obtain the heat transfer coefficient inside the rough tubes with the specified dimension of rough elements.

2.3 Thermal Performance of CHE

This section will analyze heat load of the exchanger, shown in Fig.1, using row-by-row difference method, in which the exchanger is divided into N sections of heat transfer surfaces along the flow direction of flue gas, and each section involves a row of same tubes.

Apart from the known flowrates of the mixture and coolant, three independent variables, i.e. T_{gj}, T_{cj} and W_{vj}, are unknown in each section. As the heat load and mass transfer rate of the condensing heat exchanger depend on the distribution of these physical variables in the exchanger, first it is necessary to set up three governing equations to describe the trends of these variables along the flow direction.

Noting that the condensing heat exchanger consists of two portions on which heat transfer has different mechanisms, namely wet and dry portions. Condensation process occurs only on the wet portion. The parameter ω, the ratio of noncondensable gas fraction in main stream to that near the wall, can characterize the transition from dry portion to wet one of heat transfer surfaces. The transition point is where ω is equal to 1. In other words, a section is wet if $\omega < 1$, and is dry if $\omega > 1$. Thus, in order to build up the unified governing equations that are suitable for any section of heat transfer surfaces, whether wet or dry, we define the parameter ω in the governing equations as follows:

$$\omega = 1.0 \qquad \text{if } \omega > 1 \qquad (10)$$

The following three equations are the difference form of the equations on each discreticized section which are all derived by heat and mass balances.

Equation of mixture temperature variation, $T_{e,j}$

$$-G_e C_{p_e}(T_{e,j+1} - T_{e,j}) = \alpha_e(\overline{T}_{e,j} - T_{o,j})A_j$$
$$j=1,2,\ldots,n \qquad (11)$$

Conservation law of vapor mass

$$-G_e(W_{v,j+1} - W_{v,j}) = \beta(1 - \overline{W}_j/W_{o,j})A_j$$
$$j=1,2,\ldots,n \qquad (12)$$

where W_j is air concentration in the main stream, thus $\overline{W}_j=1-W_{o,j}$. $W_{o,j}$ is air concentration at gas-liquid interface, and is determined solely by the interface temperature $T_{o,j}$.

Equation of coolant temperature variation, $T_{c,j}$

$$-G_c C_{p_c}(T_{c,j+1} - T_{c,j}) = \alpha_c(T_{w,j} - \overline{T}_{c,j})A_j$$
$$j=1,2,\ldots,n \qquad (13)$$

In order to make accurate computations, the arithematic mean difference for T_e, W_v and T_c is adopted in all three difference equations, that is

$$\overline{T}_{e,j} = (T_{e,j+1} + T_{e,j})/2 \qquad (14)$$
$$\overline{W}_{v,j} = (W_{v,j+1} + W_{v,j})/2 \qquad (15)$$
$$\overline{T}_{c,j} = (T_{c,j+1} + T_{c,j})/2 \qquad (16)$$

The unknown interface temperature $T_{o,j}$ and wall temperature $T_{w,j}$ appearing in equation (11-13) can be determined by the heat balance relations in thermal network, which are expressed as follows:

$$\alpha_e(T_{e,j}-T_{o,j})+\beta(1-W_j/W_{o,j})h_{f_e} = \alpha_1(T_{o,j}-T_{w,j})$$
$$= \alpha_c(T_{w,j}-T_{c,j}) \qquad (17)$$

where α_1 is heat transfer coefficient of condensate film, which can be calculated by the following Fujii correlation [5]:

$$Nu_1 = 0.91(g \cdot d\rho_1 \mu_1 h_{f_e}^*/u^* q^*)^{0.1} Re_{2\phi}^{1/3} \qquad (18)$$

where $Re_{2\phi}$ is two-phase Reynolds number, and defined as ud/ν_1.

The governing equations (11-13), subject to two-point boundary conditions, namely the known inlet temperatures of coolant and mixture, and the auxiliary equations were solved numerically by explicit finite difference method and iteration methods. The numerical simulation for thermal performance of condensing heat exchangers is necessary because of the highly nonlinear nature of the governing equations. In the governing equations, the term associated with condensation heat transfer is not linear function of temperature difference between stream and wall, also outlet temperature of coolant required to start the solution is missing. Thus two levels of iteration processes emerge in the numerical simulation. Fisrt level involves estimating outlet temperature of coolant as encountered in most counterflow exchanger analyses. This was dealt with by shooting method [2]. The second level of iteration is associated with the determination of interface temperature, and is coped with by under-relaxation method [2]. The computer program has been recognized as an easy-operation, great suitability, and efficient code. All computations were conducted on a DPS8 computer.

After the distributions of the three variables in CHE were predicted, the total heat load and condensate mass flowrate can be simply obtained by

$$Q = G_e C_{p_e}(T_{e,1} - T_{e,n+1}) + G_e(W_{v,1} - W_{v,n+1})h_{f_e} \qquad (19)$$
$$G_d = G_w(W_{v,1} - W_{v,n+1}) \qquad (20)$$

3.RESULTS AND DISCUSSION

In order to make the results and discussion presented in this paper valuable to engineering applications, we stimulate the thermal performance of a condensing heat exchanger designed for a 10 T/H boiler fire with coal-water slurry in Yongrong mining company. Table 2 lists the design parameters of the exchanger. Three kinds of heat transfer tubes, namely smooth, rough and internally finned tubes, are tested for comparison. such a comparison would provide the answer to the feasibility of the technical program that air is used as a coolant to recover latent heat contained in flue gas in condensing heat exchangers.

The simulation results of air temperature, heat load of the exchanger and condensate rate are depicted in Fig.4-6. As expected, with flowrate of air increasing, condensate rate (subsequently latent heat) and heat load will increase because of the improvement of heat transfer conditions inside tubes by large

Table 2 The design parameters of the CHE for a 10 T/H boiler

Flowrate: flue gas 4.0 kg/s	Inlet temp. : flue gas 210℃
	air 20℃
Steam content in flue gas : 0.109	
Number of tube rows: 25	Number of tube lines: 32
Tube diameter: outer 60 mm	Tube spacing: longitudinal 1.8d
inner 55 mm	transverse 1.8d
Arrangement: in-line	
Cross section area : 3.456 m²	

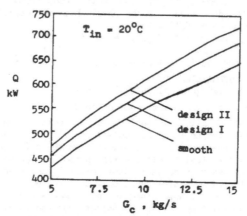

Fig.6 Comparison of heat load for three kinds of tubes

Reynolds number gas flow and because of the low temperature level of coolant when G_c is large. But exit temperature of air decreases though heat gained by air is increasing. Take an example, for smooth tubes when G_c=5kg/s, the exit temperature can be larger than 100℃. G_c, however, is as small as 13g/s, and the exchanger operating at this point recovers nearly no latent heat in flue gas. Obviously, our original objective to recover latent heat from flue gas with high content of steams failed by this design. If the flowrate of air is raised as high as 15kg/s, indeed, G_a can be increased up to 75kg/s, and the efficiency of the exchanger increases by 42% because considerable latent heat contained in flue gas is recovered in this case. Unfortunately, the exit temperature of air would be lower than 63℃. The air of large-flowrate and low-temperature is useless as inlet air of boilers.

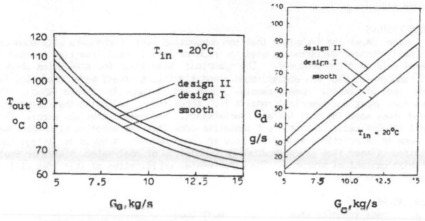

Fig.4 Comparison of exit temperature for three kinds of tubes

Fig. 5. Comparison of condensate rate for three kinds of tubes

More important than the aforementioned general trends in Figs.4-6 are the effects of enhancement by applying rough and internally finned tubes. From these figures we see that thermal performances of design I and II are better than the smooth tubes with design II having the highest efficiency. Quantitatively, for a given flowrate of cold air, in the case of design II, the exit temperature of air can increase by 8.5%; the condensate rate is raised by at least 35%; heat load of the exchanger is increased by 10%. In the case of design I, these figues are 5.4%, 17% and 6%, respectively. It is thus concluded that the substantial increase in condensate rates(namely latent heat recoverd) can be led to by using internally finned or rough tubes as heat transfer surfaces in condensing heat exchangers when cold air is the coolant. Much more enhancement of CHE's thermal efficiency is arrived at by these enhanced tubes if the comparison is based on the criterion of the same exit temperature of air. For instance.

Requirement: air is needed to be preheated to 80°C before it enters combustion
                        chambers
By means of Fig.4,
                G_c = 8.56 kg/s   for smooth tubes
                G_c = 9.44 kg/s   for rough tubes
                G_c = 10.31 kg/s   for finned tubes
The values of G read in Fig.5 which correspond the above flowrates, respectively, are
        increased by
                G_d = 35.5 g/s   for smooth tubes
                G_d = 52.0 g/s   for rough tubes   46.5%
                G_d = 68.0 g/s   for finned tubes   91.6%
The figures in the last column indicated that using enhancement tubes is an effective way to recover considerable amounts of latent heat contained in flue gas

by cold air.

4. CONCLUSIONS

In the paper the technique that the condensing heat exchangers with enhancement heat transfer surfaces are used to recover latent heat from flus gas and to preheat cold air was proposed. The numerical simulation for predicting thermal performances of CHE was accomplished, and simulating tests were conducted for three kinds of tubes, namely smooth, rough and internally finned tubes. The comparison among the numerical results for three tubes indecated that internally finned tubes have the best thermal performance, and smooth ones the worst. This report theoretically supports our tentative idea that the adoption of enhancement surfaces is a practical method to recover considerable amounts of latent heat contained in some flue gases produced by combustion of coal-water slurry or fuels with high content of hydrogen. The experimental verification of this idea will be pursued in the future.

NOMENCLATURE

A	heat transfer area	Q	heat load
A_e	flow area	Re	Reynolds number
$B(e^{\circ})$	the function	Sh	Shewood number
C	constant	T	temperature
C_p	capacity heat	W	noncondensable gas concentration
d	diameter of tubes	Greek symbols	
f	friction factor	α	heat transfer coefficient
G	mass flowrate	β	mass transfer coefficient
$h_{f.g}$	latent heat of steams	μ	viscosity
L	length of tubes	ν	kinetic viscosity
Nu	Nusselt number	ρ	density
Pr	Prandtl number	ε	factor
q	heat flux	ω	the ratio of noncondensable gas concentration

REFERENCES

[1] Wang Chaoyang and Tu Chuanjing, Recovery of latent heat from flue or exhaust gases by condensing heat exchangers, 8th Miami Conf. on Alternative Energy Sources, Dec. 1987, Miami Beach, U.S.A.

[2] Wang Chaoyang, The theoretical and experimental study on heat transfer system to recover latent heat of steams from flue or exhaust gases, M.S. Thesis, Zhejiang University, 1987

[3] Taniguchi, et.al, 6th Miami Conf. on Alternative Energy Sources, 12-14 Dec., Miami Beach, U.S.A., 1983

[4] Zukauskas, A.A., Heat transfer in tube banks, in Adv. in Heat Transfer, 1968

[5] Fujii, T. et.al, 18th Nat. Heat Transfer Conf., San Diego, ASME/AICHE, pp. 35-43, 1979

[6] Webb, R.L., Special surface geometries for heat transfer augmentation, in Dev. in Heat Exchanger Tech.—I.D. Chisholmed., Applied Science Publishers, 1980, Chap.7

[7] Webb, R.L., and Eckert, E.R.G., Heat transfer and friction in tubes with repeated-rib roughness, Int. J. Heat Mass Transfer, V.14, 1971, pp.601-617

[8] Carnavos, T.C., Heat transfer performance of internally finned tubes in turbulent flow, Adv. in Enhanced Heat Transfer, ASME, 1979 pp.61-67

FOULING AND ANTICORROSION

An Experimental Study of CaCO$_3$ Scaling

R. ISHIGURO, H. SAKASHITA, and K. SUGIYAMA
Hokkaido University
Sapporo, Japan

INTRODUCTION

Control of fouling is very important in maintaining good performance of
heat exchangers. Among the various ways in which fouling occurs, scaling, the deposition of soluble matter on heat transfer wall surfaces, is
one of the major causes. Many studies regarding scaling have been
carried out. They can be classified into two groups, based on their
approach to the problem. One group involves studies made in response to
practical needs; experiments that derive empirical relationships are
performed over several months or several years in real industrial situations. This approach is suitable for obtaining data relating to design
of heat exchangers that will operate under similar condition. The second
approach of laboratory study, where experiments can be carried out under
well-controlled conditions, allows for the simplification of complex
problems. This approach is preferable for the study of the essential
mechanisms of scaling.

With these goals in mind, various studies have been carried out. For
example, Hasson et al. studied the scaling of CaCO$_3$[1] and CaSO$_4$[2] from
a turbulent flow of aqueous solution through an annular passage with a
uniformly heated inner rod. Watkinson et al.[3,4] also studied CaCO$_3$
scaling problems with smooth tubes and enhanced heat transfer tubes under
conditions of uniform wall temperature. Ritter[5] attempted to find
correlating parameters affecting scaling behaviors with a series of
experiments using CaSO$_4$ and Li$_2$SO$_4$ as deposited matter.

In spite of these varied studies, fundamental mechanisms of scaling are
not completely understood. Furthermore, the greater efforts of past
studies have been devoted to the scaling process that follows primary
deposition of matter responsible for the commencement of scaling on a
clean surface. It is well known that scaling begins with nucleus formation on a wall surface, followed by growth and coalescence of these
nuclei. Greater knowledge of nucleus formation and its growth process is
essential if scaling is to be understood and its performance predicted.
However, no report that quantitatively measures nucleus formation and the
growth process is known, with the exception of a study of Chandler[6],
who measured the nucleation rate of Na$_2$HPO$_4$·7H$_2$O and Na$_2$SO$_4$·10H$_2$O on
cooling wall surface.

The purpose of the study described here is to observe the early stages of
the scaling phenomenon to obtain essential information that will clarify

the scaling mechanism. The deposition of CaCO₃ from a carbonate solution
is taken up in the present study. CaCO₃ has been widely used in past
experiments because it is representative of scaling in real industrial
situations. However, the selection of this matter also presents some
methodological problems. CaCO₃ must be present in a supersaturated
condition to allow for scaling within a realistic period for laboratory
testing, and the instability of the solution makes it difficult to
maintain constant conditions throughout the experiment. In order to
overcome this inconsistency, a new experimental procedure was developed
for this study. This study also employed microscopic observation of
crystal nucleus formation and growth as CaCO₃ was deposited onto the heat
transfer wall surface. Results of this observation permitted the evaluation of the growth and nucleation rate of CaCO₃ crystals.

EXPERIMENTAL METHOD

This study utilized a new method developed to achieve a specified degree
of supersaturation and maintain that state over a long period of time.
This method takes advantage of the dependency of CaCO₃ solubility on the
concentration of CO_2 dissolved in solution. The experimental system is
shown in Fig. 1. The solution fed from the bottom tank to the bubble
column comes into contact with N_2 gas blown through the filter which
strips the solution of CO_2 to increase the degree of supersaturation in
the column. The resultant highly supersaturated solution flows through a
1 μm mesh filter that eliminates any foreign matter, then flows into the
test section where CaCO₃ is deposited on a heat transfer surface. After
the solution passes through the test section, it absorbs CO_2 gas in the
bottom tank, which lowers the degree of supersaturation. The solution
was prepared in the following manner: deionized water with a conductivity of less than 0.1 μs/cm was filtered with a 0.45 μm mesh filter,
then calcium chloride ($CaCl_2$) and sodium bicarbonate ($NaHCO_3$) were

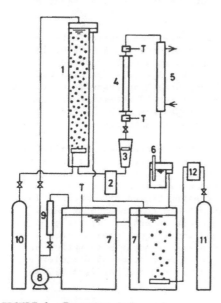

1. Bubble column
2. Filter
3. Flow meter
4. Test section
5. Cooler
6. pH meter
7. Bottom tank
8. Pump
9. Auxiliary heater
10. N_2 gas cylinder
11. CO_2 gas cylinder
12. Pressure gauge
T. Thermocouple

FIGURE 1   Experimental system

dissolved in the water. The concentration of calcium was 5×10^{-3} mol/ℓ in all experimental runs.

A cross-sectional view of the test section for $CaCO_3$ deposition is shown in Fig. 2. The flow channel had a rectangular cross section of 16 mm in width and 5 mm in height. To allow for optical measurements, one side of the channel was used for a heating surface while the opposite side was made of acrylic resin. The heating block was made of copper to increase the accuracy of temperature measurement. However, exposure of the copper surface to the solution causes corrosion and the dissolved copper ions inhibit measurement of calcium concentration. To prevent such a problem, the top of the copper block that served as a heating surface was gold plated. This block was placed in a bakelite housing to reduce heat loss. Thermocouples for measuring the surface temperature were placed in 1 mm dia. holes drilled from the bottom of the copper block to 1 mm below its surface. They were located 50 mm apart along the flow direction. The heating surface had a length of 500 mm and an entrance section 400 mm long is set upstream. Deposition behavior was observed by a microscope with a maximum magnification of 128. A camera was mounted on the microscope to photograph the heating surface during $CaCO_3$ deposition.

RESULTS

All experimental runs were performed with the solution at a temperature of 25°C and a Reynolds number of 7200 at the inlet of test section. The pH of the solution that flowed into the test section was adjusted to 7.60. The pH of 7.60 corresponds to 8.5 degrees of supersaturation, calculated according to the following equation.

$$S = \frac{[Ca^{2+}][CO_3^{2-}]}{K_s} f_D^2 \qquad (1)$$

Fig. 3 shows the change in thermal resistance over time measured at a location 350 mm downstream from the starting point of the heating section. The thermal resistance was calculated by measuring the wall to bulk temperature difference and dividing it by wall heat flux estimated

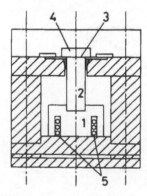

FIGURE 2   Test section

1, 2   Copper block
   3   Heating surface (gold plated)
   4   Flow channel
   5   Heater

from the total heat input to the copper block. A preliminary examination made by solving the heat conduction problem in the copper block verified that the heat transfer wall satisfied uniform heat flux conditions and that heat loss through the bakelite housing was negligible compared with the total heat input. Thus, the data on this figure show the change in thermal resistance caused by $CaCO_3$ deposition, as long as heat transfer characteristics of the wall remain unchanged.

However, there are some ambiguities involved in determining whether the initial temperature corresponds to the deposit-free surface because 1 to 1.5 hrs. were required to bring the test section to a steady state. For this reason, the arithmetic average of several measurements of wall temperature taken after about 2hrs. from heat supply was regarded to be the initial wall temperature, Tw_{in}, for each run. Within first 2hours of the heat supply, thermal resistance increasing due to $CaCO_3$ deposition was negligible because little number of fine crystals existed on the heating wall. At low Tw_{in}, thermal resistance showed moderate increases that continued over a long period and showed a small tendency to accelerate at about 15 hrs. At high Tw_{in}, an induction period of 4 to 5 hrs. was clearly observed before a linear increase in thermal resistance.

Fig. 4 and 5 show photomicrographs of the heat transfer surface which were taken at the same location as the measurement made for Fig. 3. It can be seen that fine crystals were formed on the clean surface and that they grew to cover the surface. From these figures it is also apparent that not only one type, but several types of crystals existed simultaneously. Within the range of the experimental conditions used here, the following three types of crystals were observable: 1) A needle-shaped crystal (A-type crystal). 2) A crystal with a semi-spherical or cubic shape that grew in that same shape (C-type crystal). 3) A thin crystal that grows along the heated wall surface (B-type crystal). Although no crystallographic analyses were carried out, it was assumed that A-type crystal is aragonite and C-type is calcite. B-type crystal could not be identified, and might have resulted from unusual growth of A or C-type crystals. At low Tw_{in}, nearly all crystals present on the wall surface were C-type as seen in Fig. 4. Fig. 5 shows those present at high Tw_{in}, under which conditions a large number of A and B-type crystals as well as C-type were observed. Observation of the wall surface at 9 hrs. shows that a major portion of it is covered with A-type crystals, since their rate of growth is considerably greater than the other crystal types.

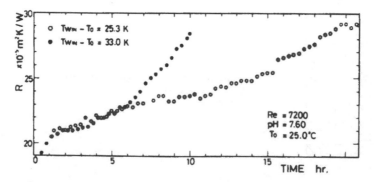

FIGURE 3   Change in thermal resistance over time

To quantitatively demonstrate the growth rate of each type of crystal, the change of crystal size over time was recorded with photography at a magnification of 64. For C-type crystals, the length of one side for cube-shaped crystals, or its diameter for spherical crystals was used as a measure of crystal size. The maximum length was measured for A-type crystals which grow in a particular direction. Since the dimensions of B-type crystals were difficult to determine, this crystal type was excluded from the discussion of growth rate. Fig. 6 shows the change of crystal size over time for C-type crystal at low Tw_{in}. This figure includes data from a total of fourteen crystals. To obtain the average growth rate from this figure, an attempt was made to correct for the deviation in nucleation time for each crystal in the following manner: At first, a interpolation line was drawn from the data for crystals No. 1 to No. 6, which were considered to nucleate on the wall surface at the same time, since they appear to lie on a single curve. Next, data points for crystals except crystals No. 1 to No. 6 were shifted horizontally in order to fit them satisfactorily on the interpolation line. The result of this correction is shown in Fig. 7. Since a fair fit was obtained for all the data, it seems that a good approximation of the average growth rate for C-type crystals can be given by this curve. The growth rate of C-type crystals tends to decrease slightly over time. The result obtained in the same manner for A-type crystals at high Tw_{in} is shown in Fig. 8. A-type crystals seemed to at a constant rate, since the data lie on a straight line.

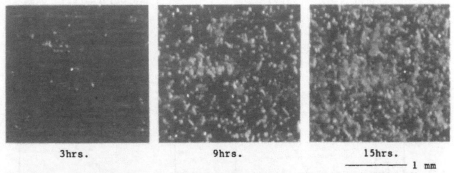

3hrs.   9hrs.   15hrs.
                                                  ——————— 1 mm
FIGURE 4. Photomicrographs of $CaCO_3$ deposition (Tw_{in} - To = 25.3 K)

3hrs.   6hrs.   9hrs.
                                                  ——————— 1 mm
FIGURE 5. Photomicrographs of $CaCO_3$ deposition (Tw_{in} - To = 33.0 K)

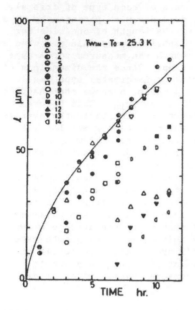

FIGURE 6 Change in crystal size over time (C-type)

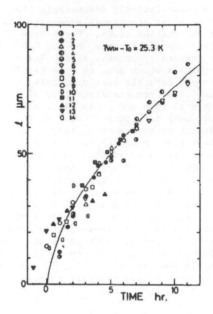

FIGURE 7 Corrected data-change in crystal size over time (C-type)

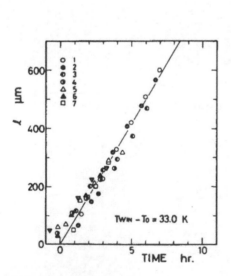

FIGURE 8 Corrected data-change in crystal size over time (A-type)

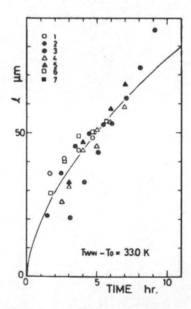

FIGURE 9 Corrected data-change in C-type crystal size over time at high Tw_{in}

Comparing Fig. 7 and Fig. 8, it was apparent that the growth rate of
A-type crystal is about ten times greater than that of C-type. Fig. 9
shows the growth behavior of C-type crystals observed with A-type crystals at high Tw_{in}. The growth curve almost coincides with that of Fig. 7
at low Tw_{in}. It is confirmed that the difference in growth rate between
A and C-type crystals was not due to wall temperature difference, but is
probably due to inherent characteristics of the crystals.

To determine the nucleation rate, fixed rectangular areas of 1.5 mm x
1 mm (real size) were selected from the photomicrographs that were taken
at intervals of approximately thirty minutes. The number of crystals
present in each area was carefully counted. The lower limit of detected
crystal size is about 5 μm. The results obtained were plotted in Fig. 10
in the form of cumulative number density of crystals against time. At
low Tw_{in}, the number of crystals, most of which are C-type, increases in
a linear fashion. Assuming the crystals have the same growth rate up
until the time they can be detected on a photograph after nucleation, a
nucleation rate of 0.36 nuclei/cm^{2}sec is obtained from the slope of the
interpolation line in Fig. 10. At high Tw_{in}, the cumulative number
density shows slightly higher values than that at low Tw_{in}. However,
there are still problems with these results because the number was
calculated without distinguishing between A and C-type crystals, and the
fast growth of A-type crystals obscures the visual count of fine crystals. The results at high Tw_{in} should, therefore, be checked by more
intensive measurements.

The effect of Reynolds number on scaling behavior, which is an important
and unsolved question, could not be examined in this study. However, in
this experiment at a Reynolds number of 7200, once the crystals were
deposited on the wall they remained at the same location and never
re-entered the flow.

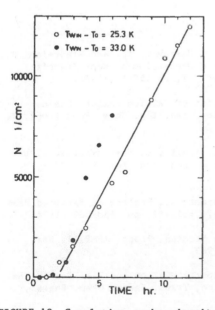

FIGURE 10 Cumulative number density of crystals

CONCLUSION

As a result of the present experimental study, the following conclusions
were obtained:
(1) The experimental method utilizing the dependency of $CaCO_3$ solubility
    on the concentration of CO_2 made it possible to achieve a specified
    degree of supersaturation and maintain that state over a long
    period.
(2) Within the range of the present experimental conditions, three types
    of crystals (A, B, and C-type) existed on the heat transfer surface.
(3) The crystals were almost C-type under the condition of low Tw_{in}. At
    high Tw_{in}, A and B-type crystals as well as C-type were observed,
    and the growth rate of A-type crystal was about ten times greater
    than that of C-type.
(4) From the photomicrographs of the heating surface during $CaCO_3$
    deposition, a nucleation rate of 0.36 nuclei/cm^2sec was obtained for
    C-type crystals at low Tw_{in}.

NOMENCLATURE

f_D = activity coefficient for divalent ions
Ks = solubility product for $CaCO_3$, mol^2/ℓ^2
ℓ = crystal size, μm
N = cumulative number density of crystals, cm^{-2}
R = thermal resistance, m^2K/W
Re = Reynolds number
S = degree of supersaturation
To = solution temperature at the inlet of test section, °C
Tw_{in} = initial wall temperature, °C
$[Ca^{2+}]$ = molar concentration of Ca^{2+} ions, mol/ℓ
$[CO_3^{2-}]$ = molar concentration of CO_3^{2-} ions, mol/ℓ

REFERENCES

1. Hasson, D., Avriel, M., Resnick, W., Rozenman, T., and Windreich, S.,
    Mechanism of Calcium Carbonate Scale Deposition on Heat-Transfer
    Surfaces, Ind. Eng. Chem. Fund., vol. 7, pp. 59-65, 1968.

2. Hasson, D., and Zahavi, J., Mechanism of Calcium Sulfate Scale
    Deposition on Heat-Transfer Surfaces, Ind. Eng. Chem. Fund., vol. 9,
    pp. 1-10, 1970.

3. Watkinson, A. P., and Martinez, O., Scaling of Heat Exchanger Tubes
    by Calcium Carbonate, Trans. ASME, J. Heat Transfer, vol. 97,
    pp. 504-508, 1975.

4. Watkinson, A. P., Louis, L., and Brent, R., Scaling of Enhanced Heat
    Exchanger Tubes, Can. J. Chem. Eng., vol. 52, pp. 558-562, 1974.

5. Ritter, R. B., Crystalline Fouling Studies, Trans. ASME, J. Heat
    Transfer, vol. 105, pp. 374-378, 1983.

6. Chandler, J. L., The Effects of Supersaturation and Flow Conditions
    on the Initiation of Scale Formation, Trans. Instn. Chem. Engrs.,
    vol. 42, T24, 1964.

Impact of Fouling in Design of Heat Exchangers

S. KAKAÇ, A. K. AGRAWAL, and H. Y. ZHANG
Department of Mechanical Engineering
University of Miami
Coral Gables, Florida 33124, USA

1. INTRODUCTION

A heat exchanger must affect the desired change in thermal conditions of the process streams within allowable pressure drops, and it must continue to do so for a specified time period. Fouling is an undesirable product of operation causing a reduction in the rate of heat transfer and increase in pressure drop. Most heat exchangers foul and require either onsite or offsite cleaning using chemical or mechanical means. At the design stage, an allowance for fouling is made to assure desired operation of the heat exchanger. In this paper, the basic equations in the heat exchanger design are given. The effects of fouling on the rate of heat transfer and pressure drop are shown through examples.

2. BASIC EQUATIONS IN DESIGN

Heat exchanger is a device in which heat is transferred from a hot fluid to a cold fluid. In most applications, the transfer of heat takes place through a separating wall which takes on a variety of geometries. The present discussion is limited to these heat exchangers, often called recuperators. Thermal analysis of heat exchangers involves two activities namely rating and sizing. Rating is the evaluation of the performance of a specified heat exchanger, while sizing is the determination of geometric parameters of a heat exchanger for given process conditions. Thermal analysis of a heat exchanger is governed by the conservation of energy in that the heat release by the hot fluid stream equals the heat gain by the cold fluid. The heat transfer rate, Q is given as:

$$Q = (\dot{m} \, c_p)_h (T_{h1} - T_{h2}) \qquad \text{; for the hot fluid stream} \qquad (1)$$

and

$$Q = (\dot{m} \, c_p)_c (T_{c2} - T_{c1}) \qquad \text{; for the cold fluid stream} \qquad (2)$$

The heat transfer rate is related to the heat exchanger and flow parameters as:

$$Q = U \, A \, \Delta T_m \qquad (3)$$

where U is the overall heat transfer coefficient based on the total surface area, A. Since the temperature difference along the heat transfer surface is not constant, a mean temperature difference, ΔT_m is used. Equations (1), (2) and (3) are the basic equations for thermal analysis of a heat exchanger under steady state conditions. For a rating problem, the heat transfer rate is obtained while for a sizing problem it is known.

609

TABLE 1. Order of magnitude of heat transfer coefficient, h

Fluid	h, W/m^2K
Gases (Natural convection)	5-25
Flowing Gases	10-250
Flowing Liquids (non-metal)	100-10,000
Flowing Liquid metals	5,000-250,000
Boiling Liquids	1,000-250,000
Condensing Vapors	1,000-25,000

In Eq. (3), the mean temperature difference depends upon the inlet and the outlet temperatures of both fluids, and flow arrangement. The overall heat transfer coefficient depends upon the heat transfer mechanisms on both sides of the separating wall and heat conduction through the wall. For an unfinned tubular heat exchanger, the overall heat transfer coefficient based on the outside surface area of the wall is given as:

$$U_o = \frac{1}{A_o/A_ih_i + + A_o\ln(r_o/r_i)/2\pi kL + 1/h_o}$$ (4)

where h_i and h_o represent the heat transfer coefficients on the inside and the outside of the tube, respectively.

The inside and outside heat transfer coefficients, h_i, h_o are usually obtained from a suitable heat transfer correlation based on flow and geometric parameters. Order of magnitude and range of heat transfer coefficient for various flow conditions are given in Table 1. As seen, h of flowing gases is considerably smaller than that of liquids. Higher heat transfer is obtained in two-phase flow (boiling or condensation) as compared to single phase flow. Least heat transfer is obtained in natural convection.

Pressure drop in the heat exchanger for each fluid is calculated by:

$$\Delta P = 4f \frac{L}{D_h} \rho \frac{u_m^2}{2}$$ (5)

where f is Fanning friction factor. Values of friction factor for fully developed flow can be obtained from the experimental data given by Moody [1]. Correlations based on geometric and flow parameters are also given for friction factor [2]. For turbulent flows, the friction factor for smooth and rough tubes can be predicted by, respectively [3].

$$f = 0.0014 + 0.125 \, Re^{-0.32}$$ (6)

$$f = 0.0035 + 0.264 \, Re^{-0.42}$$ (7)

3. EFFECTS OF FOULING

Fouling of heat transfer equipment can be defined as the deposition of unwanted material on the heat transfer surface. The material is transported to the surface by the fluid transferring heat. In practice, most heat exchangers foul. Various types of fouling may be present. Each type of fouling has different characteristics and results from different mechanism. Due to vast diversity of process conditions, most fouling situations are virtually unique.

Successive events which commonly occur in fouling situations are initiation, transport, attachment, removal and aging. Futher details are given by Epstein [4].

Fouling deposits reduce the effectiveness of a heat exchanger by reducing the heat transfer and by impacting the pressure drop, generally unfavourably but sometimes favourably. Effects of fouling on the design of heat exchangers, including both thermal and hydraulic considerations, are treated in this section.

3.1 Heat Transfer

Additional thermal resistance to heat transfer due to fouling can be related to the fouling thermal conductivity (k_f) and thickness (t_f) as:

$$R_f = \frac{t_f}{k_f}$$ (8)

The overall heat transfer coefficient under fouled conditions for an unfinned tubular heat exchanger, U_f can be obtained by adding inside and outside fouling resistances (R_{fi} and R_{fo}) in Eq. (4) :

$$U_f = \frac{1}{r_o/r_i h_i + r_o/r_i R_{fi} + r_o \ln(r_o/r_i)/k + R_{fo} + 1/h_o}$$ (9)

U_f can be related to clean surface overall heat transfer coefficient, U_c (given by Eq. 4) as :

$$\frac{1}{U_f} = \frac{1}{U_c} + R_{ft}$$ (10)

where R_{ft} is the total fouling resistance. The heat transfer rate under fouled conditions, Q_f can be expressed as:

$$Q_f = U_f A_f \Delta T_{mf}$$ (11)

where the subscript, f refers to fouled conditions. Process conditions usually set the heat duty and fluid temperatures at specified values. Therefore using $Q_f = Q$ and $\Delta T_{mf} = \Delta T_m$, it can be shown from Eqs. (4), (10) and (11) that :

$$\frac{A_f}{A_c} = 1 + U_c R_{ft}$$ (12)

Equation (12) represents the increase in area due to fouling. Percent increase in heat transfer surface area due to fouling as a function of clean surface overall heat transfer coefficient is shown in Figure 1. It can be seen that the percentage of added surface is small if U_c is low, regardless of high fouling resistance, R_{ft}. However, for high U_c, even a small value of R_{ft} results in large increase in heat transfer surface area. An increase of 100% in heat transfer surface area due to fouling is not uncommon.

Average total fouling resistances specified in the design of some 750 shell-and-tube heat exchangers are given in Table 2 [5]. It is noticeable that the used values of fouling resistances are smaller for vapor flows as compared to liquids flows. Average total fouling resistance for liquid on both tube and

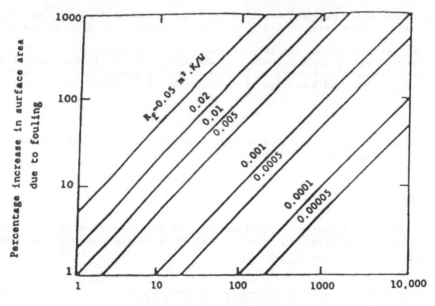

Figure 1. Effect of fouling on surface area.

shell sides of the heat exchanger is twice than that of vapor on both sides. Fouling resistances for other fluid type combinations lie in between the above two types. Using typical values of heat transfer coefficients from Table 1, and average total fouling resistance from Table 2, the percentage increase in heat transfer surface area for shell-and-tube heat exchangers is summarized in Table 3. The wall resistance has been neglected in these calculations. The increase in surface area due to fouling is high in two-phase flow applications because of their high heat transfer coefficient. In liquid flows, the increases is due to moderate U_c and high R_{ft}. In gas flows, percent area added due to fouling is small because of low U_c and relatively small R_{ft}.

TABLE 2. Specified Total Fouling Resistance, $R_{ft} \times 10^4$, m^2K/W

Tube Side	Shell Side		
	Liquid	Two-Phase	Vapor
Liquid	7.927	6.518	5.989
Two-Phase	6.694	5.108	5.108
Vapor	5.108	4.756	3.875

TABLE 3. Added Surface Area for Typical Fluid Combinations.

Tube Side		Shell Side		R_{ft}x 10^4,	U_c,	U_f,	Added Surface Area
Fluid	h_i, W/m^{2}K	Fluid	h_i, W/m^{2}K	m^{2}K/W	W/m^{2}K	W/m^{2}K	Percent
Liquid	100	Liquid	100	7.927	50.0	48.1	4.0
	100		1,000	7.927	90.1	84.1	7.1
	1,000		1,000	7.927	500.0	358.1	39.6
	100	Boiling	1,000	6.518	90.1	85.1	5.9
	100		5,000	6.518	98.0	92.1	6.4
	1,000		1,000	6.518	500.0	377.1	32.6
	1,000		5,000	6.518	833.3	540.0	54.3
	100	Vapor	10	5.989	9.1	9.1	.5
	100		100	5.989	50.0	48.5	3.0
	1,000		10	5.989	9.9	9.8	.6
	1,000		100	5.989	90.1	85.5	5.4
Boiling	1,000	Liquid	100	6.694	90.1	85.0	6.0
	1,000		1,000	6.694	500.0	374.6	33.5
	5,000		100	6.694	98.0	92.0	6.6
	5,000		1,000	6.694	833.3	535.0	55.8
	1,000	Boiling	1,000	5.108	500.0	398.3	25.5
	5,000		1,000	5.108	833.3	584.5	42.6
	5,000		5,000	5.108	2500.0	1098.0	171.7
	1,000	Vapor	10	5.108	9.9	9.9	.5
	1,000		100	5.108	90.1	86.1	4.6
	5,000		10	5.108	10.0	9.9	.5
	5,000		100	5.108	98.0	93.3	5.0
Vapor	10	Liquid	100	5.108	9.1	9.1	.5
	10		1,000	5.108	9.9	9.9	.5
	100		100	5.108	50.0	48.8	2.6
	100		1,000	5.108	90.1	86.1	4.6

Values for Vapor - Boiling and Boiling - Vapor combinations are same.

	10	Vapor	10	3.875	5.0	5.0	.2
	10		100	3.875	9.1	9.1	.4
	100		100	3.875	50.0	49.0	1.9

3.2 Pressure drop

In tubular heat exchangers, the fouling deposits roughen the surface, decrease the inside diameter, and increase the outside diameter of the tubes. Pressure drop effects on shell side are difficult to quantify because of complex flow passage. However, inside the tube, the fouling layer decreases the inside diameter and roughens the surface resulting in:

TABLE 4. Fouling Deposit Effect on Pressure Drop Example.

	Pressure Drop, Pa	% Change
Clean condition	1740	--
Fouled condition with area reduction only	2545	46
Fouled condition with area reduction and rough surface	2995	72

(a) Pressure drop increase due to the roughened surface.
(b) Pressure drop increase due to the reduced flow area.
(c) Velocity increase due to the reduced flow area.

There is a relationship between fouling layer thickness and thermal conductivity for a specified value of fouling resistance as given by Eq. (8). The fouling layer thickness can be determined by considering the fouling layer to be a cylindrical thermal resistance [6] i.e.

$$R_f/A_c = \ln(d_c/d_f)/(2\pi \, k_f L) \tag{13}$$

Eq. (13) rearranged to express the fouled diameter as a function of the fouling resistance is

$$d_f = d_c \, e^{(-2k_f R_f/d_c)} \tag{14}$$

For most liquid-side deposits, the thermal conductivity varies between 0.03 and 10 W/mK. Fouled diameter, and velocity under fouling conditions can be obtained for given fouling resistance, thermal conductivity and mass flow rate. From this, the pressure drop can be obtained from Eq. (5).

For a 25.4 mm OD, 16 BWG tubes using cooling tower water on the tube side (velocity = 1.83 m/s) and tube side fouling resistance of 0.000528 m^2K/W, the pressure drop contributions due to the increase in the velocity and the roughness of the deposit are summarized in Table 4.

4.0 CONCLUSIONS

Fouling in heat exchangers is the undesirable accumulation of insulating material on heat transfer surfaces. In general, when fouling occurs, the heat transfer decreases and the pressure drop increases. Fouling plays an important role in the design of heat exchangers. The effect of fouling on the increase in heat transfer surface area is more pronounced if the clean surface heat transfer coefficient is high. Although the clean surface heat transfer coefficients can be predicted with good accuracy, there is a greater uncertainty in the prediction of fouling resistances.

NOMENCLATURE

A   total heat transfer surface area on one side of a heat exchanger, m^2
c_p specific heat of fluid at constant pressure, J/kgK
D_h Hydraulic diameter, m

d	tube diameter, m
f	Fanning friction factor, dimensionless
h	heat transfer coefficient, W/m^2K
k	thermal conductivity, W/mK
L	length of the heat transfer surface, m
$\dot{m}$	fluid mass flow rate, kg/s
ΔP	fluid static pressure drop on one side of a heat exchanger, Pa
Q	heat transfer rate in the heat exchanger, W
R_f	fouling factor, m^2K/W
r	radius of the tube, m
Re	Reynolds number, $\rho u_m D_h/\mu$, dimensionless
T	fluid temperature, K
T_h	hot fluid temperature, K
T_c	cold fluid temperature, K
ΔT_m	true mean temperature difference, K
t	thickness of the wall, m
U	overall heat transfer coefficient, W/m^2K
u	fluid velocity, m/s
μ	fluid dynamic viscosity coefficient, Pa.s
ρ	fluid density, kg/m^3

SUBSCRIPTS

c	cold fluid side, or clean
f	fouling
h	hot fluid side
i	inner
m	mean
o	outer
t	total
w	wall
1	inlet
2	outlet

REFERENCES

1. Moody, L.F., Friction Factor for Pipe Flow, Trans. ASME, Vol. 66, pp. 671-684, 1944.

2. Kakac, S., Shah, R.K. and Aung, W., eds., Handbook of Single Phase Convective Heat Transfer, Chapters. 3, 4 and 18, John Wiley, NY, 1987.

3. Kern, D.Q., Process Heat Transfer, McGraw-Hill, New York, 1950.

4. Epstein, N., Fundamentals of Heat Transfer Surface Fouling: With Special Emphasis on Laminar Flow, in Low Reynolds Number Flow Heat Exchangers, eds. S. Kakac, R.K. Shah and A.E. Bergles, Hemisphere, New York, 1981.

5. Chenoweth, J.M., Fouling Problems in Heat Exchangers in Heat Transfer in High Technology and Power Engineering, Eds. Yang and Mori, Hemisphere, 1987.

6. Marner, W.J. and Suitor, J.W., Fouling with Convective Heat Transfer, in Handbook of Single-Phase Convective Heat Transfer, Eds. S. Kakac, R.K. Shah and W. Aung, Chapter 21, John Wiley, New York, 1987.

Heat Transfer and Drag Characteristics of a Cleaner-Augmentor in Tube Side of Surface Condensor

SHAN-RANG YANG and GUO-DONG ZAI
Northeast China Institute of Electric Power Engineering
Jilian City, PRC

ABSTRACT

The cleaner-augmentor is a spiral turbulator with 1-3 reinforcing wires, which is intended for enhancing heat transfer and cleaning the fouling in tube-side of surface condensors. This paper presents the results of the experimental investigation on the heat transfer and drag characteristics of the cleaner--augmentor and proposes a model for predicting heat transfer coefficient. A tentative heat transfer enhancement mechanism of the cleaner-augmentor is discussed.

INTRODUCTION

Surface condensors play an important role in steam power plant. An existing condensor incorporating enhanced heat transfer can reduce the turbine back pressure, and raise the thermodynamic efficiency of the plant. With fuel costs rising, even very small increase in efficiency can lead to significant reductions in plant operating cost.

Generally speaking, the major resistance to heat flow in conventional surface condensors is usually on the tube side, where single phase turbulent forced convection occurs. And, the accumulation of unwanted material on the internal surface of tubes further increase the resistance to heat flow. In such cases, tube side enhancement and fouling clean up may be expected to be of primary benefit in increasing condensor performance.

Concerning the heat transfer enhancement in surface condensors, Webb[1] and Marto and Nunn [2] provided the excellent review of it. In general, the enhancement method maybe divided into tube-side enhancement (on cooling water side) and shell-side enhancement (on steam side) technique . It is likely that the most promising technique in the tube-side enhancement is helically corrugated tube [3-7]. Another enhancement heat transfer technique is the displaced enhancement devices, Klaczak pointed out that the spiral turbulator (Fig.1a) used inside tube is more effective in enhancing heat transfer than helical turbulators (Fig.1b) do [8].And furthermore, it is very attractive for cleaning surface that all the turns of the spiral turbulator can nestle closely to the internal surface of the tubes. It seems that the most early investigation on spiral turbulator was made by Nagaoka and Watanabe [9] in 1936. Since then , some important progress has been procured [8.10.11].But in comparison with the study on helically corrugated tube, the study of spiral turbulator is still inadequate to meet the demand of application. The heat

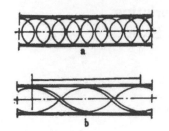

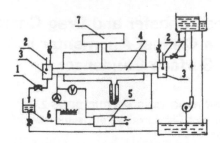

FIGURE 1. Schematic diagram of turbulator
a-spiral turbulator
b-helical turbulator

FIGUER 2. Schematic diagram of test rig
1-adjusting valve 2-mercury thermometer
3-header 4-heat exchanger 5-stabilizer
6-volumetric cylinder 7-multimeter

transfer coefficient, especially the friction factor are short of experiment data and analytical model, and so far, on application of spiral turbulator to condensor has been reported in literatures.

This paper intends to present the results of experimental investigation on influence of the cleaner-augmentor parameters on the heat transfer for turbulent flows and a model for predicting heat transfer coefficient as well as a tentative mechanism of heat transfer enhancement.

EXPERIMENTAL PROCEDURES

Figure 2 shows a schematic diagram of the experimental facility. The water was forced to flow through the tubes 4 and heated by an electrothermal strip wrapped on the external surface of the tubes. The water volume flow rate was measured by volumetric cylinder 6.To measure the tube wall temperature, three platinum resistance thermometers were bound at equal distance on the external surface of each tube. All water temperature were measured with mercurial thermometers of $0.1°C$ graduation. The hydraulic resistance measurements were made with differential manometers.

The cleaner-augmentor used for the experiments were made of red copper wire coverd with coat of synthetic resin. The reinforcing wires of same diameter was welded on the inside of cleaner-augmentor coil (Fig.3). The dimensions of the turbulators investigated in the present experiments are given in table 1.

TABLE 1.

| e / d | 0.0404 | 0.0452 | 0.0678 | 0.0848 |
| s / d | 0.7830 | 1.0435 | 1.5650 | 2.0870 |

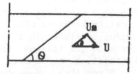

FIGURE 3. Key diagram of cleaner-augmentor

FIGURE 4. Spiral angle

The cleaner-augmentor is about the same length as the tube and its diameter is little less than the inside diameter of tube. It was mounted in tube and the outer surface of cleaner- augmentor nestle against the internal surface of tube. When the cleaner-augmentor is driven to move forth and back periodically, it will wipe the fouling off the internal surface of tube. Provided that the travel of cleaner-augmentor is greater than the coil pitch, the fouling on the whole internal surface of tube can be cleared off. The fouling cleaning effectiveness of cleaner-augmentor can be evaluated through the measurement of thermal resistance or drag [12].

To make the experiments more effectively, we fixed up the experiments with orthogonalizing method. The total number of runs were 28 with the Reynolds number varying from 3800 to 4.2×10^4.

RESULTS AND DISCUSSIONS

Based on the experimental data, we correlated the Nusselt number and Fanning friction factor with the Reynolds number and geometry parameters of cleaner--augmentor. The heat transfer correlation has the form,

$$Nu = 1.8357 \ Re^{0.887} \ Pr^{0.4} \ (s/d)^{-0.1888} \ (e/d)^{0.1888} \ \pm 10 \ \% \tag{1}$$

and the friction factor data were correlated within 20 % by equation

$$f = 62.094 \ Re^{-0.889} \ (s/d)^{-0.011} (e/d)^{0.444} \tag{2}$$

Equations (1) and (2) hold true in the following range, $3800 < Re < 42000$, $5.5 < Pr < 7.9$. Studying these results and the investigation which have been performed previousely [4-8, 10, 11], the following several points should be pointed out.

(1) The Effects of Cleaner-augmentor on Heat Transfer and Friction Factor,

In order to demonstrate the effects of cleaner-augmentor on heat transfer and friction factor, we contrast the heat transfer coefficient and friction factor data of the tube with cleaner-augmentor to the plain tube data. We selected the Dittus-Boelter correlation [13,14],

$$Nu = 0.023 \ Re^{0.8} \ Pr^{0.4} \tag{3}$$

and Drew-Koo-McAdam equation [15],

$$f = 0.0014 + 0.125 \ Re^{-0.32} \tag{4}$$

to be plain tube correlations. The comparison results are shown in Fig. 5, 6. The results show that the cleaner-augmentor can increase the heat transfer by around 1.5--3.0 times and cause a clear increment of friction factors. It is of interest to note that the effects of cleaner-augmentor on both of heat transfer and friction factors are larger than that of helically corrugated tube. By means of the orthogonalizing test analysis, considering the heat transfer and drag characteristics as well as fouling cleaning effectiveness comprehensively, we select a set of optimal geometry parameters of cleaner--augmentor from 9 sets of parameters [20]. They are s/d =1.565 and e/d =0.0452.

(2) The Effects of Reinforcing Wire on Heat Transfer and Hydraulic Resistance,

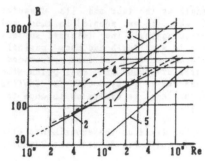

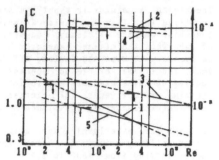

Fig. 5 Comparison of heat transfer characteristics

1.
$$\frac{Nu}{(s/d)^{-\bullet\cdot\bullet\bullet}(e/d)^{\bullet\cdot\bullet\bullet\bullet}Pr^{\bullet\cdot\bullet}}=1.836Re^{\bullet\cdot\bullet\bullet\bullet}\quad\text{(This paper)}$$

2.
$$\frac{Nu}{(s/d)^{-\bullet\cdot\bullet\bullet}(e/d)^{\bullet\cdot\bullet\bullet}Pr^{\bullet\cdot\bullet\bullet}}=1.04Re^{\bullet\cdot\bullet\bullet}\quad[3]$$

3.
$$\frac{Nu}{(s/d)^{-\bullet\cdot\bullet\bullet\bullet}(e/d)^{\bullet\cdot\bullet\bullet\bullet}Pr^{\bullet\cdot\bullet}}=1.138Re^{\bullet\cdot\bullet\bullet\bullet}\quad[15]$$

4.
$$\frac{Nu}{(s/d)^{-\bullet\cdot\bullet}(e/d)^{\bullet\cdot\bullet}Pr^{\bullet\cdot\bullet}}=165(Re/10^{\bullet})^{(\bullet\cdot\bullet-\bullet\cdot\bullet\bullet/\bullet\bullet)}\quad[14]$$

5.
$$\frac{Nu}{Pr^{\bullet\cdot\bullet}}=0.023Re^{\bullet\cdot\bullet}\quad[10]$$

Fig. 6 Comparison of drag characteristics

1.
$$\frac{f}{(s/d)^{-\bullet\cdot\bullet\bullet\bullet}(e/d)^{\bullet\cdot\bullet\bullet\bullet}}=62.09Re^{-\bullet\cdot\bullet\bullet}\quad\text{(This paper)}$$

2.
$$\frac{f}{(s/d)^{-\bullet\cdot\bullet}(e/d)^{\bullet\cdot\bullet\bullet}}=23.1Re^{-\bullet\cdot\bullet\bullet\bullet}\quad[15]$$

3.
$$\frac{f}{(s/d)^{-\bullet\cdot\bullet}(e/d)^{\bullet\cdot\bullet}}=1.3(Re(5\times10^{\bullet})^{-\bullet})^{-\bullet\cdot\bullet}\quad[14]$$

4.
$$\frac{f}{(s/d)^{-\bullet\cdot\bullet\bullet}(e/d)^{\bullet\cdot\bullet\bullet}}=20.6Re^{-\bullet\cdot\bullet\bullet\bullet}\quad[15]$$

5.
$$f=0.0014+0.125Re^{-\bullet\cdot\bullet\bullet}\quad[10]$$

To find out the effects of reinforcing wire, we made two sets of contrast experiments. Its results are shown in table 2. From the table we can draw a conclusion that the reinforcing wire enhance the heat transfer slightly and increase the friction factors obviously. And these effects become more distinct when the cleaner-augmentor parameters s/d and e/d turn large. This phenomena confirm that the swirl flow, as Newson and Hodgson proposed [16], does play a role in enhancing heat transfer, because the reinforcing wire orientation coincides with the direction of tube axis, so it nearly does not affect the axial flow. Then it only obstructs the swirl stream from flowing and stir up eddy flow, this is the reason why the reinforcing wire have influence on the heat transfer and hydraulic resistance.

Moreover, the phenomena that the larger the parameters s/d, e/d, the intensive the swirl flow are quite similar to the helically corrugated tube.

PREDICTION OF HEAT TRANSFER

Based on the previous investigation [17], we can consider the heat transfer in tube with cleaner-augmentor to be superposition of the following three effects:
(a) The heat transfer attributed to turbulent flow in the spiral channel formed by spiral wire, q_s. For the convenience of statement, we call the turbulent flow in the spiral channel a swirl flow. It should be noted that the swirl flow

620

TABLE 2.

		s/d = 1.565 e/d = 0.0452			s/d = 2.078 e/d = 0.0678		
Re	1	5062	9958	17126	5028	9670	16650
	2	5088	10095	17521	5181	10423	18130
Pr	1	6.85	7.17	7.63	7.05	7.51	7.87
	2	6.69	6.95	7.22	6.61	6.83	7.02
Nu	1	156	160	174	172	181	188
	2	153	158	166	158	163	169
f	1	0.127	0.0976	0.0868	0.134	0.102	0.092
	2	0.117	0.0842	0.076	0.124	0.093	0.081

1—The   cleaner-augmentor   2—The   spiral turbulator

TABLE 3.

No.	sets of parameter	Nu_m	Nu_1	Nu_2	Nu_3	error %
1	$Re_1 s_1 e_1$	315	329.0	18.7	347.7	10.4
2	$Re_1 s_1 e_2$	309	362	22	384	24.3
3	$Re_1 s_1 e_3$	321	417	21.2	438	36.0
4	$Re_3 s_1 e_3$	82.7	55.3	5.0	60.3	-27
5	$Re_3 s_1 e_1$	98	64.1	5.2	69.3	-29.3
6	$Re_2 s_2 e_2$	212	199.1	13.2	212.3	0.14
7	$Re_2 s_1 e_1$	222	230.9	13.0	244	9.9
8	$Re_2 s_1 e_3$	200	176.7	11.6	188.3	-5.9
9	$Re_4 s_2 e_2$	152	102.2	7.7	110	27.7
10	$Re_2 s_2 e_1$	239	264.8	17.2	282	18.0

is different from the rotating flow generated by twisted tapes [18]. The swirl flow is confined within the near wall area, it does not spread to the main core flow in tube with cleaner-augmentor.
(b) The heat transfer caused by the centrifugal convection effect of swirl flow, q_a.
(c) The heat transfer due to conduction of cleaner-augmentor spiral wire attached to the internal surface of tubes, q_o.
In surface condensor, for protecting inside surface of tube against corrosion, usually smear the tube inside surface with coat of ferrous sulphate($FeSO_4$). To maintain the protective film, we coated the spiral wire of cleaner-augmentor with coat of synthetic resin. The thermal conductivity of the synthetic resin coat is generally negligible. So we can ignore the third part of heat transfer. Then the total heat transfer, assuming no cross coupling, is given by,

$$q=q_1+q_2 \tag{5}$$

Let us define the total heat transfer coefficient h as,

$$h = \frac{q}{(Tw-Tb)\pi \cdot DL} = \frac{q_1}{(Tw-Tb)\pi \cdot DL} + \frac{q_2}{(Tw-Tb)\pi \cdot DL} = h_1 + h_2 \tag{6}$$

or multiplying the above equality by d/k_b yields the following equation,

$$Nu = Nu_1 + Nu_2 \tag{6a}$$

For predicting the first part of the heat transfer, we consider the whole flow to be along the spiral channel, then as proposed by Newson [16], this part of heat transfer is given by,

$$Nu_1 = [1 + \pi^2 / (s/d)^2]^{0.4} Nup \tag{7}$$

where, Nup - the Nusselt number for plain tube, it can be determined according to equation (3)

$$Nup = 0.023 Re^{0.8}Pr^{0.4} = 0.023 (Um \cdot d/\nu)^{0.8} Pr^{0.4} \tag{8}$$

Here, Um -the mean velocity along the spiral channel; it can be evaluated by,

$$Um = \frac{4G \cdot \cos\Theta}{\rho \cdot \pi \cdot d^2 \cdot [1-(dr/d)^2-(e/d)^2]} \tag{9}$$

Based on Lopina and Bergles [17], consider the centrifugal convection effect to be similar to the natural convection circulation established over a heated plate facing up under the influence of a body force field, the magnitude of the effect should be predictable in a like manner. So the centrifugal convection heat transfer may be predicted by,

$$Nu_2 = 0.114 (Gr \cdot Pr)^{1/3} \tag{10}$$

where, Gr -- Grashof number, it can be shown to be [19],

$$Gr = \frac{32G^2 \cdot \sin^2\Theta \cdot \cos^2\Theta \cdot \beta \cdot (Tw-Tb)}{\pi^2 \cdot d^2 \cdot \rho^2 \cdot \nu^2 \cdot [1-(dr/d)^2-(e/d)^2]^2}$$

$$= (\frac{4G \cdot \sin\Theta \cdot \cos\Theta}{\pi \cdot d \cdot \rho \cdot \nu \cdot [1-(dr/d)^2-(e/d)^2]})^2 \cdot 2 \cdot \beta \cdot (Tw-Tb) \tag{11}$$

Combining above equations yields the prediction for heat transfer in swirl flow,

$$Nu = 0.023 \ [1+\pi^{\circ}/(s/d)^{\circ}]^{\circ \cdot \cdot} \cdot (\frac{4G \cdot cos\Theta}{\rho \cdot \pi \cdot d \cdot \nu \cdot [1-(dr/d)^{\circ} -(e/d)^{\circ}]})^{\circ \cdot \circ} \cdot Pr^{\circ \cdot \cdot}$$

$$+0.114 \ (\frac{4G \cdot cos\Theta \cdot sin\Theta}{\rho \cdot \pi \cdot d \cdot \nu \cdot [1-(dr/d)^{\circ} -(e/d)^{\circ}]})^{\cdot \cdot} \cdot [2 \ \beta \cdot (Tw-Tb)Pr]^{\cdot \cdot} \qquad (12)$$

The part of comparisons between the predicting value of equation (12) and the experimental data are shown in table 3. The absolute value of maximum error is less than 38%. It seems that the error mainly comes from the assumption that the whole flow to be along the spiral channel. As it does, the flow described by equation (7) is only the outer part of flow, or the flow part within the near tube wall area, the main core flow is still axial flow.

It is worthy of note that the numeral calculation shows Nu1>>Nu2, that means the major heat transfer are attributed to the swirl flow caused by the cleaner -augmentor spiral wire, as a matter of course, it involves the effect of inter-rupting the development of boundary layer.

CONCLUSIONS

1. The cleaner-augmentor does enhance heat transfer in single-phase. Its heat transfer characteristics can be evaluated either by correlation (1) or by equation (12) and the drag characteristics is estimated by equation (2). The prediction of equation(12) is basically consistent with the experimental data.

2. The major heat transfer are attributed to the swirl flow caused by spiral wire of cleaner-augmentor.

3. The effects of reinforcing wire on heat transfer and drag characteristics originate from the obstruction to swirl flow.

4. Under the experimental conditions, 3800<Re<42000 and 5.5< Pr<7.9, the optimal geometric parameters of cleaner-augmentor are s/d=1.565, and e/d=0.0452.

NOMENCLATURE

Cp   specific heat of fluid at constant pressure, KJ/kg °C
d   inside diameter of tube, m
D   outside diameter of tube, m
e   diameter of cleaner-augmentor spiral wire, m
f   Fanning friction factor
G   mass flow rate, kg/s
g   gravitational acceleration, m/sec²
h   heat transfer coefficient, kw/m² °C
K   thermal conductivity of tube wall, kw/m °C
L   tube length, m
q   rate of heat transfer, kw
s   pitch of cleaner-augmentor coil, m
T   temperature, °C
ΔT   wall minus fluid temperature, Tw-Tb, °C
U   flow stream mass velocity, kg/s m²
β   volumetric coefficient of thermal expansion, 1/°C
Θ   spiral angle, (Fig.4) radian

ρ fluid density, kg/m^3
μ dynamic viscosity, $kg/s \cdot m$
ν kinematic viscosity, m^2/s
$Nu = h\,D/k$, Nusselt number
$Pr = Cp\mu/k$, Prandtl number
$Re = U\,D/\nu$, Reynolds number
$Gr = g\rho\,L^3\,\Delta T/\nu^2$, Grashof number

SUBSCRIPTS

b   bulk fluid condition
w   tube wall characteristic
m   average value or measured value
r   reinforcing wire

REFERENCES

1. Webb,R.L.,The Use of Enhanced Surface Geometries in Condensors An Overview, in << Power Condensor Heat Transfer Technology >> Edited by Marto, P. J. and Nunn, R. H. Hemisphere Publishing Corp., Washington D. C., pp. 287-324, 1981.

2. Marto, P. J. and Nunn, R. H.,The Potential of Heat Transfer Enhancement in Surface Condensors, in << Condensors, Theory and Practice >>, The Institution of Chemical Engineers Symposium Series, No.75, pp. 23-47, 1983.

3. Withers, J. C. , Heat Exchange Apparatus and Method of Controlling Fouling Therein, U. S. Patent No. 4,007,774, 1977.

4. Mehta,M. H. and Rao,M. R., Heat Transfer and Frictional Characteristics of spirally enhanced tubes for horizontal condensers, in << Advances in Enhanced Heat Transfer >> Edited by J. M.Chenoweth et al,ASME, New York, pp.11-12,1979.

5. La Rue, J. C.,Libby, P. A. and Yampolsky, T. S. , Fluid Mechanics and Heat Transfer Spirally Fluted Tubing, GA-A16541, General Atomic Company, 1981.

6. Marto, P.J., Reilly D. J. and Fenner J. H., An Experimental comparison of enhanced heat transfer condenser tubing, in << Advances in Enhanced Heat Transfer >> , Edited by J. M. Chenoweth et al, ASME, New York, pp. 1-10 1979.

7. Li Heungming, Ye Kousing and Deng Songjiu, A Reivew on Research Works and Applications of Enhanced Heat Transfer Element-Spirally Corrugated Tube, J. of Chemical Industry and Engineering , No. 4, pp. 359-367. 1982. (in Chinese)

8. Klaczak, A., Heat Transfer in Tubes With Spiral and Helical Turbulators, J. of Heat Transfer, Nov. 557-559. 1973.

9. Nagaoka, Z. and A. Watanabe, Proc. 7th Int. Cong. Refrigeration, vol.3, p.221, 1936.

10. Migai, V. K., Enhancement of modern heat exchanger efficiency, Energiya, 1980. (in Russian).

11. Cui, L. Y., Lu, Y. S. Investigation of enhancement heat transfer in single-phase flow by insert in tube, << Chemical Engineering and Universal machine >>, No. 9, 1983. (in Chinese).

12. Yang, S.R.,Zai, G. D. Experimental investigation on a cleraner-augmentor

in tube side of surface condensor, will be presented at The Scond U. K. National Heat Transfer Conference, Glasgow. 1988, 9.

13. Dittus, P. W. and L. M. K. Boelter, Univ. of Calif., Eng. Publ. 2, p.443, 1930.

14. McAdams, W. H., Heat Transmission, 3nd Ed., McGraw-Hill, 1954.

15. Drew, T. B. E. C. Koo and W. H. McAdam, Trans. ALChE, 28, pp.56-72, 1932.

16. Newson, I. H., Hodgson, T. K., The 4th International Symposium on Fresh Water from the sea, vol. 1, p.69, 1973.

17. Lopina,R. F. and Bergles, A. E., Heat Transfer and Pressure Drop in Tape--Generated Swirl Flow of Single-Phase Water, J. of Heat Transfer, vol. 91, 8, pp. 434-442, 1969.

18. J. P. Chiou, Experimental investigation of the Augmentation of Forced Convection Heat Transfer in a circular tube using spiral spring inserts, J. Heat Transfer, Trans. ASME, vol. 109, pp.300-307, 1987.

19. Zai. G. D.,Experimental Investigation on a Cleaner-Augmentor in tube-side of surface condensor, Master Thesis, Northeast China Institute of Electric Power Engineering. 1987. (in Chinese).

20. Yang, S. R., Zai, G. D., Performance evaluation criterion of the cleaner-augmentor, will be published. 1988.

Experimental Study on Corrosive Resistance of Plate Heat Exchanger in Geothermal Water

ZHONGZHENG WANG, ZHENNAN ZHAO, CHENGCHAO FANG, and YIHAN CAI
Department of Thermophysics Engineering
Tianjin University
Tianjin, PRC

SHIWEN LIU
Yin Kou Cooler Factory
Yin Kou, PRC

ABSTRACT

For the purpose of seeking such a heat exchanger materials that should be good corrosive resistance in geothermal water and excellent heat transfer performance and cheapper in price, authours of the paper did an exploratory research work in respect of various platings and coatings used in plate heat exchanger. Exposure tests and operation trial at site of geothermal well with a modified special plating and five nonmetalic coatings showed that anticorrosive materials were found. corrosive charicteristics in geothermal water are described, the data for exposure test and the situations of disassembling and checking are reported and analyzed in detail in the paper. Besides, prospect is done for the application of these materials in the future.

1. INTRODUCTION

With the increasing demands for energy in the whole world, much attention has been paid to energy conservation techniques and exploitation of so called new energy sources centred on solor, geothermal, biological and wind energies etc., since 1980s. along with the launch of four—modernizations construction in china and the rising of living standards of chinese people, energy consumption has soared rapidly, the situation of short supply of energy has generally and prominently been presented. Therefore, while energy conservation is being encouraged and the availability and reasonableness of energies available are constantly rising, speeding up the pace of development and application of various new energy sources at the same time are also a basic and urgent task for us.

Geothermal energy has its distinctive merits: clear, dense and polutionless, its important role in the energy supply is being recognized by more and more people. Based on the statistics of some materils, the world installed electric power of geothermal origin, by the end of 1985, was 4766 MW which might have almost doubled from 1982, the pace is astonishing.

As compared with the world eleetric power this figure represents only 0.2% and plays a very minor role in the world energy scene. In the developing countries, however, with an as yet limited electrical consumption but rich geothermal resources the electrical energy of geothermal origin could make quite a significant contribution to the total. For instance, it is around 18 pen cent at present in Philippines and EL Salvador. Besides, the total power for direct use of geothermal energ was even much more than that of generating, reached up to 7072 MW by 1985.

There are quite a lot geothermal resources in China which have been used in many mays such as generating, heating of buildings, agricultural greenhouse and medical treatment etc., good results and huge economic benefits have been obtained. Wherein heat exchange equipments were always neccessary and took play a crucial role. With them effects of corrosive medium would be limited within the primary loop of the heat exchanger, so as to protect all other equipments from heavy corrosive damage.

Geothermal water has generally strong corrosiveness because many detrimental chemical compositions such as H^+, Cl^-, H_2S, CO_2, NH_3, SO_4 even F^- involved. Besides, there usually are oxygen coming from atmosphere to take part in the corrosive and greatly accelerate it. According to the previous experinces of operation, these harmful compositions could cause serious damage to most of engineering materials, for example, ordinary stainless steel plate might be penetated in a short period of time. It's therefore a prominent problem for us to seek effective solutions to the corrosion of heat exchanger in geothermal application and give it both good corrosive resistance in geowater and excellent heat transfer performance, at the same time, the price can't rise too much.

In chemical engineering entorprices,   traditional shell—and tube heat exchangers which are made of nonmetalic materials such as glassfibre reinforced plastic or graphite are used for various corrosive cases. Because of the structure of shell—and—tube heat exchanger and quite lower conductivity of these nonmetalic materials, their heat transfer coefficients are far from satisfactory. We think that not only corrosive resistance and heat transfer performance, but also the national conditions of China should be fully considered, that means the price must be acceptable for most of chinese users. So it's absolutely neccessary to open a new channel. Plate heat exchanger is a kind of compact and effective heat exchange equipment which has much higher (by a factor of about three) heat transfer coefficient than that one of the shell—and—tube heat exchanger when liquid to liquid heat exchange takes place. In addition, it can flexibly be assembled and easy to mash up. The only drawback is the higher pressure drop through the plate channel. Although with a proper design it will not be excessive. For this purpose we took plate heat exchanger as a basic object of study in the past two years, corrosive tests with many materials, plating and coatings were done. Detailed reports on the results drawn from those tests in 1985 were reported in references (4), (5). Based on these tests, a further try with a modified special plating and a few nonmetalic coatings were done, this year and good results have been obtained.

2. CORROSION CHARICTERISTICS IN GEOTHERMAL WATER

It's known that corrosion of materials in geothermal water are usually in the form of local corrosion which distinquishes from general corrosion. The local corrosion is often presented in a small part of the surface while the other part may still be in good condition. Damage caused by this type of corrosion, however, is certainly enough for the exchanger's plates with whin thickness, to make whole set of heat exchanger scrapped.

The basic patterns of corrosion in geothermal water of low temperature are as follows:

(1) Pitting. Attack takes place in some active points, and forms small cavity penetrating deeply into metal at a rate much greater than general corrosion. A classic example of this corrosion is stainless steel in geothermal water wherein cl⁻ involved. Especially, if there is a lot of oxygen leaking into the system, the rate of pitting will be doubled and redoubled.

(2) Crevice corrosion. where narrow fissures exist between metals or metal and nonmetal, corrosive fluids stagnate and crevice corrocion may take place. Because of the structural requirements of plate heat exchanger, there must be a certain amount of contact points between adjacent plates so as to ensure rigidity and pressure—bearing capacity of the plates. Therefore, crevice corrosion is easy to produce because of the low velocity of fluids there.

(3) Erosion corrosion. It's such a type of corrosion that mechanical abrasion by high speed turbulent flow can combine with the corrosion when the protective oxide film was mechanically removed and fresh metal was constantly exposed. Fretting is an important sort of erosion corrosion, it occurs at the interface of two contacting surfaces subject to slight relative movement, usually produced by vibration. Although no apreciable vibration in plate heat exchangers, a slight relative movement of plates does frequently happen because of the deformation of paltes by the unstatic flow and the change of pressures on both sides. In addition, there are narrow channels, strong turbulent in plate heat exchangers and a large amount of mud and sand are always involved in geo—water, all of these factors lead to severe erosion corrosion.

Besides mentioned above, intergranular corrosion, galvanic corrosion, stress corrosion soeaking and selective leaching may also take palce in some special occasions.

3. CONTENTS OF TESTS

The whole test programme was divided into three parts:

1) At first, observations and exposure tests were done in some strong corrosive mediums with various metalic plating and nonmetalic coatings to select the better among them.

2) Exposure tests in real geothermal water. The exposure tests for first stage were done in 1985 with ordinary carbon steel, stainless steel, titanium and a special plating. In 1987, corrosive rate measurements were done again with a modified special plating and some kinds of nonmetalic coatings.

3) Operasion trial at the site of geothermal well. Two sets of plate heat exchanger, one the special plating and another titanium were put into operative trial for three months in 1985. Further operation examinations with the modified special plating and five nonmetalic coatings, herringbone and sawtooth corrugated types of plates involved, were done for about six months this year.

The water from 2° geothermal well located on the campus of Tianjin University was used for both exposure tests and operation trials, Analysis on the water quality was detailed in ref. (5)

4. RESULTS OF THE TESTS

1) The data of corrosive rate for the special plating in some strong corrodents were reported in ref. (5).

2) Some samples of nonmetalic coatings in corrodents.
= In the artificial sea water. 80°C for 8 hours, then ambient temp. for
16 hours, 116 cycles altogether.
85—A coating: Some tiny bubbles could be seen.
85—B coating: no change observed.
85—C coating: as original except fading in colour.
= 50% sulphuric acid, ambient temp. for 150 days.
85—A coating: fissuring 28 days later;
85—B and 85—C coatings: no change.
= 25% caustic soda, ambient temp. for 150 days.
no change for all samples.

3) Exposure test in geothermal water about 50°C.
The ASMT Standard and the chinese national standard were refered in the tests, some results of exposure test done in 1985 are also listed here for comparison(with = symbol)

NAME OF SAMPLES	CORROSIVE RATE (mm/year)	
	successive immersion	intermittent immersion
Stainless steel=	0.00125	0.00327
Titanium=	0.0000	0.0000
Special plating=	0.01184	0.02154
Special plating	0.00664	0.012010
Nonmetalic coatings	0.0000	0.0000

(Note: the exposure test lasted for seventy one days)

Appearance of the samples. The modified special plating. no remarkable sign of corrosion were found except a few tiny pitting corrosion in one sample. The nonmetalic coatings, a very thin layer of fouling covered samples. There was no mark of any corrosion to be seen. They became as orginal again after washed with dilute hydrochloric acid which inhibitor was added to. Weights of all samples had a little bit larger (about several thousandth gram) after exposure. It considered that this is caused by micro—absorption of water, and would recover after drying.

4) Disassembling and checking. The plate heat exchangers consisted of all anticorrosive plates mensioned above were thoroughly taken apart after operating for about half a year.

• The modified special plating (herringbone). heavy fouling was found at the side of geo—water. There were notable marks of extrusion at the contact points, i. e. crossed points of rhombic net formed by adjacent plates. pitting cavities were discovered in some plates (about 20% of the total) after washing up, other plates were basically in ggod condition. (refer to Fig. 1, 2 and 3)

• Stainless steel plate (herringbone). fouling was less than the above one, but there was remarkable pitting pressnted at the points of contact for one plate. (refer to Fig 4)

• 86—P coating. there were much fouling and the coating peeled off when the plate was being washed. (refer to Fig 5)

• 85—A, 85—B, 85—C, 85—L coatings, (both herringbone & sawtooth corrugated plates involved) All of these plates were in good condition and there were onlya thin layer of fine fouling about 10 to 20 micrometers covered on them, sings of extrusion and erosion could be seen just at the contact points, but no dam age was observed on the surfaces of all coatings. Intact coatings apperred while the fouling was washed away. (refer to Fig. 6 and 7)

5. ANALYSTS

1) In comparison to that of the first round of exposure test with the special plating in 1985, the corrosive rate in this test has been reduced. Surfaces of the samples were basically in good condition except a trifle of pitting in one or two positions of the samples. This shows that the inherent quality of the modified special plating is better than that in past. The thickness of the plating, based on the measurents, is in the range of 45 to 60 micrometers. In consideration of the general relation of the corrosive rate with time, five years of service live may be possible when the plating is used in geothermal water.

2) The results of exposure tests with the nonmetalic coatings are satisfactory. There was absolutely no sign of any type of corrosion after seventy one day's immersion. This shows all of the coatings tested have very good corrosive resistance against geothermal water.

3) Most of the special plating plates were basically in a quite good condition after 150 day's operation trial except a minor part of the plates wherein pitting presented. Some of the attack cavities even were severe (refer to photo 3). However, except for two plates, no obvious attack was found at the points of contact of the plates which is the weakest link of the plate against corrosion. Besides, there were no puncture and peeling off discovered

in this round of test. By comparison, there were both of them in the last round of test in 1985. These indicate, on one hand, overall quality of the special plating has risen, on the other hand, both the stability of electroplate technique and the quality of the base metal should neccessarily be improved.

4) Influnces of surface finish of the plating on corrosive resistance and heat transfer charicteristics. Thicknesses of the fouling in the plating plates were generally much more than that in the coating ones, even reached up to 400 to 500 micrometers in some positions. It rose heat transfer resistance by a big margin and sharply increased the pressure drop across the narrow channels because of reduction of the sectional area. Futhermore, it tended to produce unequal distribution of flow in the pararell channels. It´s believed that the main reason for thick fouling is poor surface finish of the plating layer. In comparison, the fouling thickness of nonmetalic coating plates in the same set of heat exchanger was much thinner and well–distributed. A poor surface finish is also unfavourble when the surface subjects to a corrosion. The fluids tend to be stagnated or reduce its velocity in the hollows of the surface and there may usually be some tiny chinks beneath the fouling. All of these factors provide a nice surounding to the crevice corrosion. Hence the risè of surface finish would be crucial to the improvement of heat transfer and corrosive resistance. It´s gratifying that considerable progress of the electroplating technique had been made while this round of test was being done. The surface finish of the special plating has apperently risen.

5) Comparison between the metal palting and the nonmetalic coating. Except for 86–P, all of the other nonmetalic coatings having two corrugated types of plates withstood the severe tests in geothermal water, the results were even better than that we expected. It can be seen on photo 7 that there are some small hollows at the contact points of the coating plates. It is caused by extrusion against each other in adjacent plates, not by corrosive attack. This indicates that these four nonmetalic coatings have very good resistance against crevice corrosion and strong enough adhesive force to the base metal. Neither peeling off nor bending upwards happened for all coating plates. As far as the results we have got up to now, the nonmetalic coatings are obviously better than the special plating in respect of corrosive reisitance. However, there is also another aspect of the problem to be seen, i. e. heat transfer performance. Heat tranfer coefficient of exchanger which consists of coating plates must be down because of the poor conductivity of the coating. Since we hav´t got enough test data for heat transfer performance, we can´t affirm how much reduction would bring about. While as far as long term service goes, that the fouling thickness on the coating paltes was much less than that on the plating plates is certainly advantageous to the former.

6) Comparison between the two corrugated types. It seems that the sawtooth corrugated type is more favourable than the other in point of corrosive resistance, this means the difference between the patterns of their contact points. The contact points on herringbone corrugated plate are formed oblique crossing of two convex edges, plating or coating layer tends to be injured because extruded force concentrates on these points. while the points of contact on the sawtooth plate are lines, in fact, they are small strips. so extruded force is spread out, almost no damage would take place. So far as heat transfer and pressure drop of the two types of plates are concerned, further tests and analysis are expected.

6. BRIEF CONCLUSIONS

A few heat exchangers made from more than ten sorts of materials, platings and coatings were successively tested in the past two years. We have got a group of test data and first hand informations from which conclusions are drawn as follows:

1) Some progress have been made for the modified special plating in corrosive resistance, however further works are certainly necessary in anticorrosive performance and surface finish.

2) Among five nonmetalic coatings four have satisfactory corrosive resistance with the exceptance of one failure because of poor adherence, study in detail on the heat transfer performance is required in the next stage of the work.

3) We have such a directive thinking, i.e. it's necessary to consider chinese conditions in this topic of research work. Titanium, which has certainly excellent performances, is very expensive. It has generally been selected to use in geothermal utilizations in many developed countries. We hope to find out cheap, corrosive resisting and good heat transfer materials. According to a rough estimation, an ordinary carbon steel plate with the special plating or one of the coatings are both cheapper than that of stainless steel with the same area of heat exchange. We think the successful experinces in this respect will not only promote the development of geothermal application in China but also give a lot of help to many chemical engineering enterprices. It's profitable not only to China but also to those developing countries just as China.

REFERENCES

1. J. S Gudmundsson, Direct use of geothermal energy in 1984, Proc. of Intern. Symposium on Geothermal Energy. Hawail, U.S.A, Aug 1985
2. Enrico Barbier, Geothermal energy in the world energy scenario, Proc. of the first Afro—Asian Geothermal Seminar, Thailand, Nov 1985
3. Chemical Engineering college of Beijing. Application of the heat exchanger made by graphite modified polypropylene, Corroosion $ Protection. No. 2, 1981(in chinese)
4. Z. Z. Wang, Z. N. Zhao, etc., An experimental study of plate heat exchan gers for heating building with geothermal water, Proc. of 7th Geothermal Workshop in N. Z. Nov 1985
5. Z. Z. Wang, Z. N. Zhao, etc., A preliminary exploration to corrosive resistance of heat exchangers in geothermal application. The annual meeting of Chinese Society of Engineering thermophysics. 1986. (in chinese)
6. 1982 Annual Book of ASTM Standard, part 10.

HEAT PIPE

The Calculation of Temperature and Velocity Field as Well as Heat Pipe Exchanger of Constant Gas Velocity and Varying Cross Section

SONG-ZHE LI and RONG-ZHI HE
Guangzhou Institute of Energy Conversion
Chinese Academy of Sciences
P.O. Box 1254
510070 Guangzhou, PRC

ABSTRACT

In this paper, a mathematic model is developed which may be used to calculate the temperature fields of working fluids (air and gas) at high and low temperature sides, thus to find out heat pipe temperature and heat transfer intensity. By such calculation it is possible to estimate the safety of the heat pipes and to adjust the flow fields in the heat exchanger so as to obtain a design of varying-passage heat xchanger. In this paper, also an exemplary calculation is made and the result is plotted for comparison.

INTRODUCTION

Heat pipe is a good thermal conductive element enabling enhanced heat transfer on both high and low temperature sides. It is used to from a high efficiency heat exchanger. In this paper, used in industry (especially for waste heat recovery) as a new type of energy saving device. In design calculation of conventional heat pipe heat exchanger of constant cross section, the rows of heat pipe are treated as an entirety and the average fluid temperature is taken as characteristic temperature for heat transfer calculation, allowing the calculation simplified. However, as the cool and hot fluids pass by a heat pipe, heat exchange takes place and the state parameters of the fluids will change consequently. Fluid velocity will be higher at the first row of pipe in the diretion of flow and then will decrease progressively with the rows due to decreased fluid temperature and specific volume. As a result of this uneven velocity distribution when a heat pipe heat exchanger is used for waste heat recovery of tunnel gas containing ash and unburnt particals, the front rows of pipe will be subject to quick wear under the higher gas velocity while ash is liable to deposite on the rear rows due to lower gas velocity and thus heat transfer is reduced. Hance, a detailed flow field calculation for heat exchanger is of general importance for analysis of flow flied distribution, research for heat transfer efficiency improvement, optimization of design, and improvement of operation performance, safety, relinbility and economics. Through flow field calculation, heat pipe heat exchanger of varying flow area may be designed.

In China, the section-wise method has been used to design a conventional constant-flow-area heat exchanger with stepwise cross sections, that is, a heat exchanger formed by sections of

637

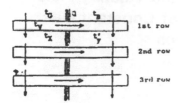

Fig.1 Co-current heat transfer   Fig.2 Countercurrent heat transfer

different cross section area. Operation has shown that the result
is good.

To better energy saving, a mathematic model is proposed in this
paper regarding detailed flow fluid calculation.

Thermodynamic calculation of flow field in a heat exchanger
begins from the heat pipe at the entrance of heat source fluid.
The temperature of heat source is taken as the heat source fluid
temperature before the heat pipe; the heat sink fluid temperature
is taken as heat sink fluid temperature at the exit (ambient
temperature) or discharged temperature, depending on co-current
or countercurrent mode. (Fig.1 and Fig.2)

Beginning from the first row of pipes, calculated are the
temperatures of heat source fluid and heat sink fluid behind the
rows as well as temperatures of working medium inside the pipes
and heat transfered by pipe rows.

Fluid temperature before a pipe row is considered equal to the
downstream temperature of the row immediately upstream.

Perform row-by-row calculation until the final fluid temperature
reaches the preset state.

1.THERMODYNAMIC CALCULATION FOR CO-CURRENT CONDITION (Fig.1)

Basing on thermodynamics and heat transfer theory, the following
equation are written.

Equation of thermal balance

$$Q_2 = \eta_t \cdot Q_1 \tag{1}$$

Thermodynamic equations

$$Q_1 = V_G \cdot \rho'_G \cdot \overline{C}_{p1} \cdot (t_G - t_X) \tag{2}$$

$$Q_2 = V_L \cdot \rho_L \cdot \overline{C}_{p2} \cdot (t_y - t_A) \tag{3}$$

where $\overline{C}_{p1} = C_{p1}(t) \, dt \tag{4}$

$\overline{C}_{p2} = C_{p2}(t) \, dt \tag{5}$

Heat transfer equations for high and low temperature sides:

$$Q_1 = K_1 \cdot F_1 \cdot \triangle T_G \tag{6}$$
$$Q_2 = K_2 \cdot F_2 \cdot \triangle T_L \tag{7}$$

Heat transfer temperature difference is taken as the arithmetic average temperature concerned. Temperature of working medium inside a heat pipe is considered the same for high and low temperature sides. Hence, equations to calculate the temperature difference for heat transfer are as follows:

$$\triangle T_G = \frac{1}{2} \cdot (t_G + t_X) - t_V \tag{8}$$

$$\triangle T_L = t_V - \frac{1}{2}(t_A + t_y) \tag{9}$$

Introduce the following parameters for calculation:
Heat flux (also called 'water equivalent') of hot and cool fluid:

$$U_1 = V_G \cdot \rho_G \cdot \overline{C}_{p1} \tag{10}$$

$$U_2 = V_L \cdot \rho_L \cdot \overline{C}_{p2} \tag{11}$$

Thermal resistance at high and low temperature sides:

$$R_1 = \frac{1}{K_1 \cdot F_1} \tag{12}$$

$$R_2 = \frac{1}{K_2 \cdot F_2} \tag{13}$$

Introduce the following equivalent calculation factors:

$$C = 2R \cdot U + 1 \tag{14}$$
$$B = 2R \cdot U - 1 \tag{15}$$

For heat source side and heat sink side:

$$C = 2R_1 \cdot U_1 + 1 \tag{16}$$
$$B = 2R_1 \cdot U_1 - 1 \tag{17}$$
$$C = 2R_2 \cdot U_2 + 1 \tag{18}$$
$$B = 2R_2 \cdot U_2 - 1 \tag{19}$$

After having complicated derivation, the following simplified formulas may be obtained from the above equations (when thermal resistances of secondary importance are neglected):
Fluid temperature after a heat pipe at heat sink side:

$$t_X = \frac{(U_1 \cdot \eta_t \cdot C_2 + U_2 \cdot B_1) \, t_G + 2U_2 \cdot t_A}{U_1 \cdot \eta_t \cdot C_2 + U_2 \cdot C_1} \tag{20}$$

Fluid temperature after a heat pipe on heat sink side:

$$t_y = \frac{(U_2 \cdot C_1 + U_1 \cdot \eta_t \cdot B_2) t_A + 2U_1 \cdot t_G}{U_1 \cdot \eta_t \cdot C_2 + U_2 \cdot C_1} \qquad (21)$$

Temperature of working medium inside a heat pipe:

$$t_v = \frac{U_1 \cdot \eta_t C_2 \cdot t_G + U_2 \cdot C_1 \cdot t_A}{U_1 \cdot \eta_t \cdot C_2 + U_2 \cdot C_1} \qquad (22)$$

Heat absorbed by a row of pipes:

$$Q_2 = \frac{2U_1 \cdot U_2 \cdot \eta_t \cdot (t_G - t_A)}{U_1 \cdot \eta_t \cdot C_2 + U_2 \cdot C_1} \qquad (23)$$

After the above thermodynamic parameters are found, velocity at a pipe row may be calculated:

$$V_1 = \frac{V_G \cdot \rho_G}{\rho_1 \cdot P_1} \qquad (24)$$

$$V_2 = \frac{V_2 \cdot \rho_2}{\rho_2 \cdot P_2} \qquad (25)$$

2. THERMODYNAMIC CALCULATION FOR COUNTERCURRENT CONDITION

In countercurrent heat transfer, flow direction of cool fluid is opposite to that of hot fluid, therefore, for cool fluid the temperature before a pipe is the outlet temperature t_B of the fluid, while the temperature t_y' to be found is the inlet temperature of the fluid. The rest part of derivation precedure is similar to that for co-current heat transfer. Then the following simplified formulas are obtained.
Fluid temperature behind a pipe in high temperature side:

$$t_x' = \frac{(U_2 \cdot B_1 + U_1 \cdot \eta_t \cdot B_2) t_G + 2U_2 \cdot t_B}{U_1 \cdot \eta_t \cdot B_2 + U_2 \cdot C_1} \qquad (26)$$

3. PRACTICAL EXAMPLE

From design calculation of a heat exchanger for the roller tunnul in a ceramics plant, the following may be obtainde:

3.1 In co-current mode. temperature of working medium inside heat pipes in different rows is fundamentally uniform (Fig.3). In counter-current mode, temperature of working medium in the first row exhibit a little higher (Fig.4) but still in the range of 250 °C for safety operation. Besides, in this example, heat recoverd by counter-current heat transfer is 20% more than that by the co-current mode. Hence, counter-current heat transfer mode is selected

3.2 In these designs, the number of elemental heat pipes is the same. For conventional heat exchanger of constant flow passage area, velocities decrease progressively with rows (Fig.5). This gives rise to the fact that the first rows will be subject to premature wear due to higher velocity and the last rows will be

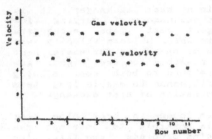

Fig.4 Temperature distribution curve of counter-current heat transfer

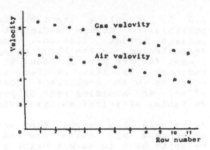

Fig.3 Temperature distribution curve of co-current heat transfer

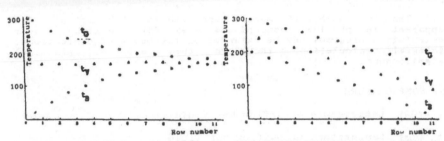

Fig.5 Variation of velovity in heat exchanger of constant flow passage area, for co-current

Fig.6 Variation of velovity in heat exchanger of constant flow passage area, for counter current

subject to ash deposit due to lower velocity. During design of heat exchanger of varying cross section, uniform fluid velocities were attained by varying the number and interval of heat pipes in a row to adjust the cross section of flow passages. For heat exchanger of convergently varying flow passage area, velocities at the rows can be adjusted and become roughly uniform (Fig.6), thus problems arising from ununiform velocity field can be eliminated.

At present, heat exchanger of constant velocity and varying flow passage area, designed through calculations of temperature and velocity fields have already been in practical operation, and the measured operation data conform quite well with the calculations.

4. CONCLUSIONS

4.1 As a new way to design and improve conventional heat exchangers of constant flow passage area, the two mathematic models suggested in this article,i.e.Eq.s 20 - 25 and Eq.s 24 -29 may be used to calculate temperature field, heat transfer intensity and velocity field in heat pipe heat exchangers of both co-current and counter-current heat transfer modes.

4.2 The flow field calculation program includes calculations of temperature and corresponding saturation pressure of working medium inside a heat pipe. the latter being the operation pressure of the heat pipe. This allows the operation safety of the heat pipe to be assessed and the upper operation limit of heat source temperature for a heat exchanger to be defined.

4.3 By adjusting the flow field in a heat exchanger, it is possible to design different heat exchangers of varying flow passage cross section. Convergent heat pipe exchangers demonstrated in the above example in which uniform velocity was sought for are more rational in view of operation for waste heat recovery. By this program it is possible to use higher gas velocity with concurrent considerations to both soot blowing effect and avoiding wear of pipe wall, hence to enable long term reliable, efficient energy-saving operation of heat exchanger.

Experience has shown that pressure losses in heat pipe heat exchangers are quite low and it is possible to increase gas velocity by 9 to 10 m/s without causing problems, and this, plus a periodic self soot blowing for either low and high temperature sides alternatively, will lead to a very well heat transfer performance.

4.4 The two mathematic models suggested in this article are apparent in physical concept and easy for iteration, their calculation programs may be extended on the basis of conventional integral calculation, therefore they are of practical significance.

5. NOMENCLATURE

t_G (^oC)   temperature of inflow hot fluid

t_X (^oC)   temperature of outflow hot fluid

t_V (^oC)   temperature of working medium inside a heat pipe

t_Y (^oC)   temperature of outflow fluid in co-current condition

t_A (^oC)   temperature of inflow cool fluid

t_B (^oC)   temperature of outflow cool fluid

t'_y (^oC)   temperature of inflow cool fluid in countercurrent condition

ΔT_L, ΔT_G (^oC)   temperature difference at high and low temperature sides

V_L, V_G (Nm^3/h) volumetric flow rate of cool and hot fluids (standard condition)

ρ_L ρ_G (kg/m^3) special gravity of cool and hot fluids (standard condition)

C_{p1}, C_{p1} ($kJ/kg \cdot {}^oC$) specific heat of cool and hot fluids at constant pressure

K_2, K_1 ($kJ/h \cdot m^2 \cdot {}^oC$) heat transfer coefficient of heat pipe at high and low temperature sides

F_2, F_1 (m^2) heat transfer area at high and low temperature sides

U_2, U_1 ($kJ/h \cdot {}^oC$) heat flow of fluids at high and low temperature sides (water equivalent)

R_2, R_1 ($m^2 \cdot h \cdot {}^{\circ}C/kJ$.) thermal resistance of heat pipes at high and low temperature sides

ρ'_1, ρ_2 (kg/m^3) specific gravity of cool and hot fluids at pipe row temperature

V_1, V_2 (m/s) velocity of cool and hot fluid at pipe row

Q_1, Q_2 (kJ/h) heat absorption and heat release at high and low temperature sides

η_t heat reservation efficiency of heat exchanger

LITERATURE (ALL IN CHINESE)

1. Hao Cheng-Ming, Designing of Heat Pipe Heat Exchanger, Heat Pipe and Heat Pipe exchanger, 1981, Foshun City Science & Technology Information Institute.

2. Lang Kui, Qiao Zhang-Xia, Shang Guo-Shou, Heat Pipe Technology and Application, 1984, Liaoning Province Science & Tecnology Publisher.

3. Li Zeng-Mi, Calculation of Heat Pipe Heat Exchanger for Foshan City Chemical Ceramics Factory (calc.1), May 1984,Guangzhou Institute of Energy Conversion, Chinese Academy of Sciences.

4. Li Song-Zhe, Row-by row Detailed Calculation of Heat Pipe Heat Exchanger for Foshan City Chemical Ceramics Factory (Cal.2) June 1984, Guangzhou Istitute of Energy Conversion, Chinese Academy of Sciences.

5. Xin Ming-Dao, Chen Yuan-Gou, Xie Huan-De, Gao Ming-Chong,Tong Ming-Wei, Shi Cheng-Ming, Experimental Research of Water-Steel Heat Pipe for Waste Heat Recovery Installations, June 1984, Journal of Chong Qing University, No.2.

6. Li Song-Zhe, On Calculation of Temperature & Velocity Fluids in Heat Pipe Heat Exchangers of Varying Flow Passage Cross Section, May 1985, Guangzhou Istitute of Energy Conversion, Chinese Academy of Sciences.

7. Southwest China Electricity Bureau Experiment & Research Istitute, Dujiang Electric Equipment Manufacture & Repair Factory, Thermal Engineering Department of Chongqing University: Configuration and Test Running of Heat Pipe Air Preheater for Furnace No.1, May 1984.

Active Fins as Heat Exchange Enhancers

M. I. HARVEY, R. J. LAVERTY, J. M. MCGIVEN, P. RENNER, C. J. RINCKES,
and A. G. WILLIAMSON
Department of Chemical and Process Engineering
University of Canterbury
Christchurch, New Zealand

ABSTRACT

This paper describes a particular form of flat plate heat pipe
and its application in two widely different devices, a solar
water heater and a refrigerant to air heat exchanger.

INTRODUCTION

Fins have long been used as a means of extending heat transfer
surfaces. This is usually done to achieve a better match
between the two sides of a heat transfer interface. For
example, in air cooling refrigeration equipment a very high
heat transfer coefficient on the refrigerant side has to be
matched to a very low heat transfer coefficient on the air side.

The usable length of a fin is determined largely by its thermal
conductivity. For aluminium fins in air the maximum practical
length is about 5 cm. Large ratios of external surface to
internal surface must therefore be achieved by having large
numbers of closely spaced fins as shown in Figure 1.

A second type of finned tube heat exchanger is that frequently
used in flat plate solar water heaters of the type shown in
Figure 2.

In these, heat from the solar energy absorbed by the fins is
conducted along the fins to the risers. In this case the
spacing of the risers depends on the thermal conductivity of
the fins. In a well designed all-copper tube-on-fin solar
collector the optimum spacing of the risers is about 150 mm
so that no point on a fin is more than about 70 mm from a
riser. In this case the spacing of the risers and hence the
cost of construction of the collectors could be reduced if the
fin had a higher thermal conductivity. In the limit of very
high conductivity the solar collector could consist of a
single fin with a heat exchanger tube attached to one edge as
shown in Figure 3.

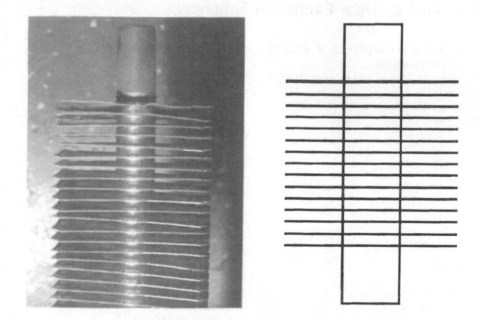

FIGURE 1.   Finned tube heat exchanger.

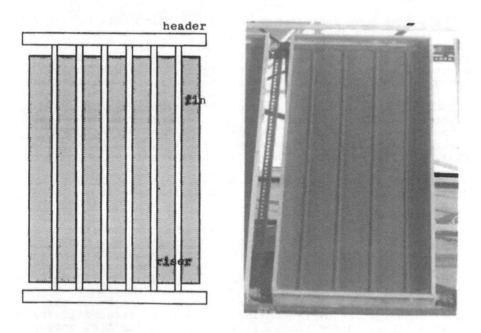

header

fin

riser

FIGURE 2.   Conventional tube-on-sheet solar collector.

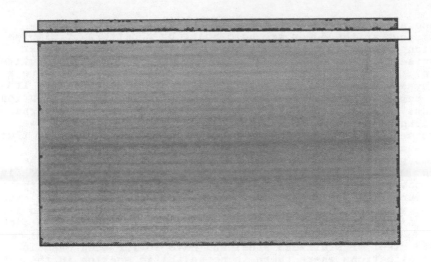

FIGURE 3. Panel with highly conductive fin.

The possibility of highly conducting fins is offered by the heat pipe or even more simply by the two phase thermosyphon as depicted in Figure 4(a). A flat plate version of the two phase thermosyphon is shown in Figure 4(b).

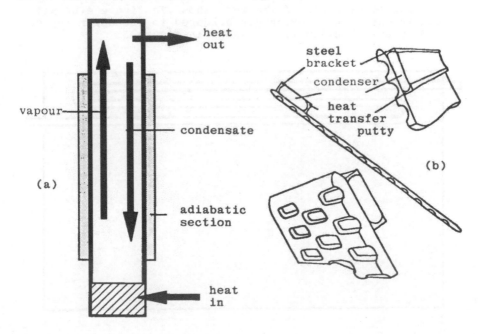

FIGURE 4. Two phase thermosyphon.

Such an arrangement, in which two sheets of metal welded together at the edges are held apart by a pattern of dimples which serve also to redistribute the liquid over the inner surfaces, has the characteristics of a very highly conductive fin. It is possible to use such a structure as if it were a fin, and because of the very high effective conductivity it is possible to make the fins very much longer than conventional fins. Because the conductivity is the result of an internal mechanism rather than a simple property of the material we choose to call these active fins. The choice of working fluid is such that the pressure in the envelope remains below atmospheric even at the highest working temperature of the fin and the only welding required is at the edges of the fin. Moreover, since the only metallic conduction is through the thickness of the metal and perhaps along a short section with poorly wetted surface the overall conductivity of the fin is not highly dependent on the choice of metal. The active fins we have constructed on this pattern have been made from mild steel or stainless steel and have so far been used in two situations where there is no adiabatic section in the heat pipe.

These applications have been to low cost solar water heating collectors and to refrigeration equipment as described below.

SOLAR COLLECTOR

This consists of a fin of the type shown in Figure 4(b) with a tubular copper heat exchanger soldered to the upper region of a mild steel active fin as shown in Figure 5.

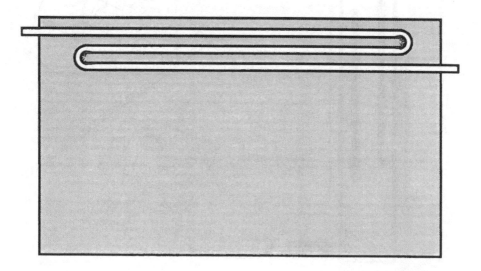

FIGURE 5. Final version of solar panel.

The configuration of the heat exchanger was selected to use low cost small bore (7 mm i.d.) copper tubing rather than a single larger bore copper tube. The working fluid is xylene with a boiling point of 144°C which is about 10-15°C above the highest stagnation temperature of the panel. The panel is flat enough that at any angle to the horizontal greater than about 15 degrees, the condensate from the condensing region is distributed evenly over the panel surface. All of the area which is not directly under the condensing region remains isothermal and the panel, although made of steel has a performance (when mounted in a suitable case) which is similar to that of a good quality all-copper tube-on-sheet collector of the type shown in Figure 2. There are now about 500 of these panels in operation in New Zealand, the oldest having run for 10 years. An installation is shown in Figure 6.

REFRIGERATION

As was indicated in the introduction the typical finned tube refrigerant-to-air heat exchanger has a large number of short closely spaced fins. A major disadvantage of this arrangement is that when ice forms on the fins it tends to build up to the point where it completely fills the space between them as shown in Figure 7 and the system is reduced to a very low efficiency.

FIGURE 6. Installation of active fin collectors.

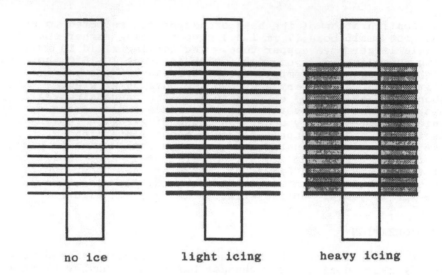

| no ice | light icing | heavy icing |

FIGURE 7.   Stages in ice formation.

In many systems the coils are de-iced with a cold water spray
and typically a complete de-icing takes several hours.   A
modification of the fin used for solar collection offers a
way out of these problems.   The refrigeration fin that we
have devised is a "double acting" flat plate heat pipe which
takes advantage of the unidirectional characteristics of the
two phase thermosyphon.   The prototype fins were made exactly
the same as the solar collectors with one modification shown
in Figure 8.

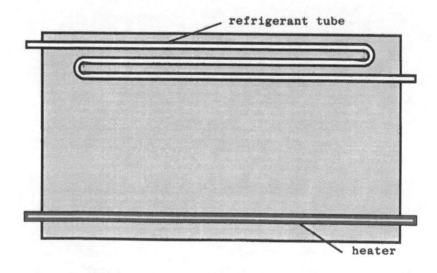

FIGURE 8.   Double acting active fin.

The refrigerant tube is attached to the top of the panel and to the bottom is attached an electrical heating element. The working fluid is either pentane or acetone. In the refrigeration mode refrigerant is passed through the top tube and heat is absorbed by the panel (fin) and conducted to the refrigerant. Because of the large size of the fins they may be set at a reasonable distance apart so that ice formation does not interfere with the air flow past the fin. In the defrost mode the refrigerant flow is stopped and the heater is turned on. Heat is conducted up the panel and ice is melted quickly over the whole surface. Complete deicing is achieved in about 20 minutes. Experiments with single panels again showed that the whole panel except the region near the refrigerant tubes is isothermal (±1°C) at all times.

The experimental heat exchanger shown in Figure 9 which contains 28 fins totalling 42 m² and is capable of about 16 kW of cooling has been tried at temperatures down to -20°C and has proved successful, operating slightly better than the design calculations suggested. The cost of heat exchangers of this type is comparable with that of conventional equipment.

FIGURE 9. Complete refrigeration heat exchanger set.

REFERENCES

1. J F Kreider and F Kreith. Solar Heating and Cooling (1977). Hempshire Publishing Corporation, Washington DC, USA.

2. P P Dunn and D A Reay. Heat Pipes (2 ed) 1978. Pergamon Press, Oxford, UK.

3. N W Foot, K L Wallace and A G Williamson. Proc. 9th Australasian Conference on Chemical Engineering, Christchurch, New Zealand.

4. K L J Webb, T S Prosad and A G Williamson. Dryland Resources and Technology, Vol 3, 211-218, 1986.

5. J M McGiven. Research Report. Department of Chemical Engineering, University of Canterbury, New Zealand, 1986.

Condensation Heat Transfer Research in Rotating Heat Pipe

JIAN CHEN and CHUANJING TU
Department of Thermoscience and Engineering
Zhejiang University
Hangzhou, PRC

ABSTRACT

The condenser heat transfer characteristics of three basic
types of rotating heat pipe (co-axial, parallel and vertical)
are investigated. The physical models of working liquid flow
corresponding to three rotating heat pipe types are creatively
proposed, the condenser heat transfer mechanisms are discussed
according to various flow models. The theoretical formulas of
the average Nu number are obtained through solving the flow
differential equations. The effect of centrifugal acceleration
on the condenser heat transfer characteristics is also theoretically and experimentally investigated. The relations between
the rotational speed and average condenser heat transfer coef.
and local condenser heat transfer coef. are shown. There are
good agreements between theoretical and experimental results.

INTRODUCTION

The rotating heat pipe was first formulated by Gray(1) in 1969
in unite states. It is a closed device to transport thermal
energy either radially or axially in rotating machinery components. It relies upon the evaporation and condensation of
a small amount of working liquid, and thus may be apporopri-
ately referred to as a rotating two-phase thermosyphon. Unlike
ordinary heat pipes, which rely upon a capillary wicking structure to return the condensate to the evaporator(or an ordinary thermosyphon which relies upon gravity), the rotating
heat pipe contains no wick and relies centrifugal forces to
return the condensate.

Figure 1 shows the three basic forms of rotating heat pipes.

1. co-axial --- heat pipe axis and axis of rotation are
                 identical,
2. parallel --- heat pipe axis and axis of rotation are
                 parallel,
3. vertical --- heat pipe axis and axis of rotation are
                 vertical,

653

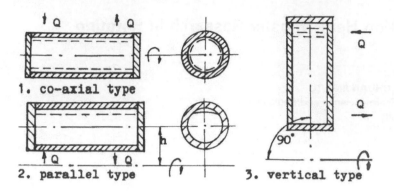

FIGURE 1.   Various rotating heat pipe types.

Various rotating heat pipes have various performance characteristics. The purpose of this study is, therefore, to analyze
the condenser heat transfer of three basic forms of cylindrical rotating heat pipe, discuss the condenser heat transfer
mechanisms according to various models, then compare the theoretical results with experimental results.

CONDENSER HEAT TRANSFER

The rotating heat pipes may be used in many important areas,
for example, it can be used to cool electrical motors, drills
which are rapidly rotating, or to recover the waste heat in a
heat exchanger. Up to now, some investigators have studied the
performence of rotating heat pipe, but unfortunately, most of
the eariler research were the experimental study, the theoretical research about the rotating heat pipe are hardly carried
out, therefore, in this study we presented three cylindrical
rotating heat pipe condensation models, theoretically analyzed
the condenser heat transfer characteristics.

The analyses are all based on the assumption similar to those
of Nusselt's basic condensation theory, otherwise, we have
some alternations.

Co-axial rotating heat pipe

For co-axial rotating heat pipe, we discussed that case-- the
axis of rotating heat pipe is inclined, here we considered the
effect of vapor shear, and also we assume the tilt angle of
heat pipe is small. Figure 2 shows the cordinate system and
geometry for laminar film condensation model. In this analysis
we assume that the complete liquid annulus has been formed
over the inner wall surface of heat pipe. We obtained the differetial equation which gives the distribution of condensate
film thickness.

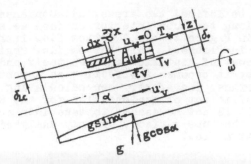

FIGURE 2.   Analytical model and geometry.

$$\left(g\sin\alpha\,\delta^3 - \frac{\tau_v}{\rho}\delta^2 - \frac{k\Delta T\delta}{h_{fg}\rho 2}\frac{u}{v}\right)\frac{d\delta}{dx} - (\omega^2 r - g\cos\alpha\,)\delta^3\left(\frac{d\delta}{dx}\right)^2$$

$$- (\omega^2 r - g\cos\alpha\,)\,\frac{\delta^4}{3}\,\frac{d^2\delta}{dx^2} = \mu\frac{k\Delta T}{h_{fg}\rho 2} \tag{1}$$

A Gill numerical intergration was used to solve the thickness
differetial equation to find the heat transfer coefficient,
Figure 3 compares the theoretical results with experimental
results, where the rotational speed is 1300 rpm, vapor temperatures are 40 °c and 50 °c. Agreement between them is within 10%.

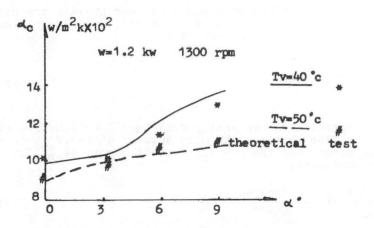

FIGURE 3.   Comparision of experiment and theoretical results.

Parallel rotating heat pipe

For condensation within a cylindrical parallel rotating heat
pipe, the presented condensation model was shown in Figure 4,

the patters will be defined two areas, a): Condensate with the film flows along the tube interior surface as shown by the broken line in Figure 4, b): A bottom flow represented by the hatched area in Figure 4, this bottom flow region, in general, blocks most of heat flow across itself and the tube surface covered by it becomes ineffective as a condensation heat transfer surface. Therefore, when calculating the overall heat transfer coefficient of condensation for the whole condenser section, it is not necessary to take into accout the bottom flow over the tube surface, we only analyzed the behavior of the bottom flow.

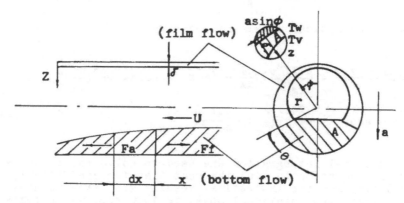

FIGURE 4. Analytical model and geometry.

In this model, the effect of vapor shear and gravity are neglected. For the film flow area, we obtained the film thickness non-dimensional differential equation,

$$Z = \frac{4}{3(\sin\phi)^\frac{1}{3}}\int_0^\phi (\sin\phi)^\frac{1}{3}d\phi \tag{2}$$

From above equation we can easily find that Z is a function of ϕ alone. We also obtained the mean Nusselt number, N_{um}, over the circumference and the overall Nusselt number, N_{uM}, for the whole condenser,

$$N_{um} = \frac{1}{\pi}\int_0^{\pi-\theta} N_u \, d\phi = \frac{1}{\pi}\left(\frac{2GaPrG}{3HZ}\right)^\frac{1}{4} d\phi \tag{3}$$

$$N_{uM} = \frac{1}{l}\int_0^1 N_{um} \, dx \tag{4}$$

The experimental results for 0 to 100 rpm are presented in Figure 5, the experimental and theoretical results show a good agreement.

From Figure 5, it can be said that the overall coefficient

of heat transfer increases definitely as the rotational speed
increases. From Figure 5 we also see that the agreement between experimental and theoretical results is better at high rotational speed than at low rotational speed.

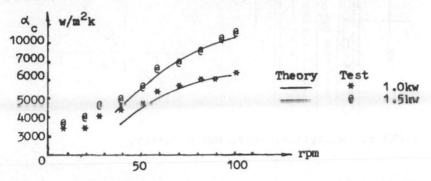

FIGURE 5.   Comparision of experiment and theoretical results.

Vertical rotating heat pipe

For the third form rotating heat pipe,vertical rotating heat
pipe,we also have a condensation model, see Figure6. In this
model the effect of vapor shear and gravity are not considered
either. In the help of condensation model we get the differential equation of film thickness distribution inside rotating
heat pipe condenser.

$$\delta = \left(\frac{3k\Delta T \mu}{\rho^2 h_{fg} \omega^2} \left(1 - \frac{1}{(1 - \frac{x\omega^2}{g\cos\theta})^{\frac{4}{3}}} \right) \right)^{\frac{1}{4}} \tag{5}$$

The local Nusselt number is,

$$N_{ux} = 0.76 \left(\frac{\omega^2 x^4}{\nu^2} \frac{Pr\, h_{fg}}{c_p \Delta T} \right)^{\frac{1}{4}} \left(1 - \frac{1}{(1 - \frac{x\omega^2}{g\cos\theta})^{\frac{4}{3}}} \right)^{-\frac{1}{4}} \tag{6}$$

From the film thickness equation we can see that with the increasing of x, the δ increases, but the increasing magnitude
become small. For the local Nusselt number, N_{ux}, the trend is
similar, it increases when the x increases .

Figure 7 is the comparision of experimental and theoretical
results of overall condenser heat transfer coefficient. We see
both results also increase with the increasing of rotational
speed. There are comparetively good agreement between theoretical and experimental results.

SUMMARY

657

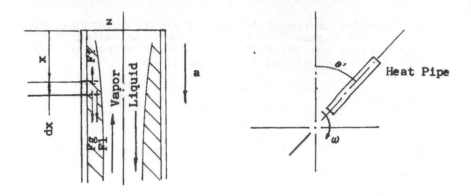

FIGURE 6.   Analytical model and geometry.

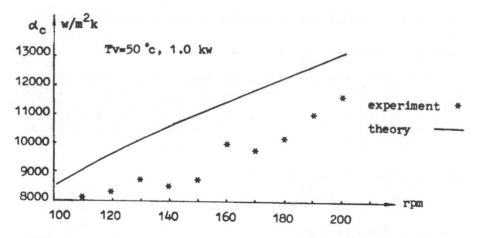

FIGURE 7.   Comparision of experiment and theoretical results.

Rotating heat pipes are very effective devices to transfer energy in rotating system, very high heat transfer coefficient is possible for rotating heat pipe, they may be used in many areas. The condensation performance of three forms rotating heat pipe mentioned above can be adequately predicted using presented analyses and correlations in this paper. Further research is needed, however, with regard to the application test study.

NOMENCLATURE

μ -- liquid dynamic viscosity,
ρ -- liquid density,
δ -- condensate layer thickness,
K   -- thermal conductivity of condensate,

c_p -- specific heat at constant pressure,
h_{fg} -- latent heat of vaporization,
α -- tilt angle of heat pipe,
α_c -- overall condenser heat transfer coef.,
ΔT -- $T_v - T_w$,
T_v -- vapor temperature,
T_w -- wall temperature,
u_v -- vapor velocity,
ω -- angular velocity,
τ_v -- vapor shear force,
ϕ -- co-ordinate in circumferential direction,

$$Z = (\frac{2GaPrG}{3H})(\frac{\delta}{d})^4 = \frac{\rho a\, h_{fg} \delta^4}{3r\nu k(T_v - T_w)},$$

a -- centrifugal acceleration,
ν -- kinematic viscosity of condensate,
θ -- angle of bottom flow level,
G -- acceleration ratio a/g,
Ga -- gd^3/ν^2,
d -- tube diameter,
r -- tube radius,
Pr -- prandtl number $\rho\nu cp/k$,
H -- $c_p(T_v-T_w)/h_{fg}$,
x -- distance along tube axis,
z -- distance normal to tube interior surface,
A -- cross sectional area of bottom flow,
l -- condenser section length,
F_f -- viscous friction force at wall,

Fa -- force due to difference in liquid level of bottom flow under centrifugal acceleration,

REFERENCES

1. Gray,V.H., The Rotating Heat Pipe - A Wickless, Hollow Shafe for Transfering High Heat Fluxes, ASME Paper, No. 69-HT-19,1969.

2. Chen Jian and Tu Chuanjing, Theoretical and Experimental Study of Condensation Heat Transfer in Parallel Rotating Heat Pipe, Int. Heat Pipe Sym., Osaka, Japan, 1986.

3. Chen Jian and Tu Chuanjing, Investigation of Condenser Heat Transfer Characteristics of Rotating Heat Pipe in Horizontal Position and Inclination, Int. Heat Pipe Sym., Osaka, Japan,1986.

4. R. Marto and H. Weigel, The Development of Economical Rotating Heat Pipe.

HEAT STORAGE

A System Investigation on Chemical Heat Storage

I. FUJII, K. TSUCHIYA, and Y. TAKAHASHI
Department of Mechanical Engineering
Meiji University at Kawasaki
Kawasaki, Kanagawa 214, Japan

M. S. MURTHY
Department of Chemical Engineering
Indian Institute of Science
Bangalore 560012, India

INTRODUCTION

When we evaluate calcium hydroxide, $Ca(OH)_2$, as the material for thermochemical heat storage or the absorbent for chemical heat pump, it is easily seen that the substance has many advantages such as nonpoison, cheapness and handling easiness caused by using water as the working fluid. Therefore, in our laboratory the substance has been investigated as promising energy storage material for more than ten years [1][2][3].

As is known, however, the substance has turning temperature of about 500°C. In other words, its dehydration under atmospheric pressure requires heating temperature of more than 500°C in practice. Hence, when we apply the substance to recovery of waste heat or to temporary storage of excess heat, such heat sources do not always have temperature level enough to dehydrate it. Naturally, such a situation enabled us to reduce reaction temperature of the reactant without any damage of the handling easiness. Usage of zinc catalyzer and reduced surrounding pressure form a part of the temperature reduction.

In the meantime, we could recently come upon a promising material having turning temperature of about 300°C in the air. Unexpectedly, the material could be obtained through chemical reaction of $Ca(OH)_2$, H_2O and an additive.

Then, to begin with, this paper shows several physical and chemical characteristics of the new material named here "CALO" for convenience' sake. In addition, we will describe a performance of heat storage system holding the new as heat storage material.

FORMATION AND GENERAL BEHAVIOR OF CALO

Visualization of water flow uses often Al-powder as visible tracer. Particle size of the powder is said to be about several micron, which enables us to mix $Ca(OH)_2$ and Al powder finely. Moreover, water addition to the mixed powder makes the following exothermic chemical reaction as

663

$$3Ca(OH)_2 + 2Al + 6H_2O \longrightarrow Ca_3Al_2(OH)_{12} + 3H_2 + 885kJ \qquad (1)$$

This is fairly vigorous reaction accompanying water vaporization in general and generation of hydrogen gas also. In fact, as the gases are continuously released from upper surface of the reactant, the surface has a look of hell-like spot.

The product $Ca_3Al_2(OH)_{12}$, namely CALO, has turning temperature of about 300°C in the air, and its dehydration induces the next endothermic reaction as

$$7Ca_3Al_2(OH)_{12}+1292kJ \longrightarrow 9Ca(OH)_2 + Ca_{12}Al_{14}O_{33} + 33H_2O \quad (2)$$

This reaction has reversibility, which naturally promises a material available for chemical heat storage. Storable energy quantity per unit mass, that is, storable energy density of CALO is 489kJ/kg which becomes about half of that of $Ca(OH)_2$. Nevertheless, CALO can be seen to be attractive because of much presence of low temperature heat sources in general.

On the other hand, subject to heating at temperature above 500°C, it is also possible to decompose $Ca(OH)_2$ appeared in the right side of Eq.(2). And, as you known, its governing reaction is

$$Ca(OH)_2 + 65.2kJ \longrightarrow CaO + H_2O \qquad (3)$$

Hence, CALO dehydration under high temperature heating makes after all

$$7Ca_3Al_2(OH)_{12} + 1879.3kJ \longrightarrow 9CaO + Ca_{12}Al_{14}O_{33} + 42H_2O \quad (4)$$

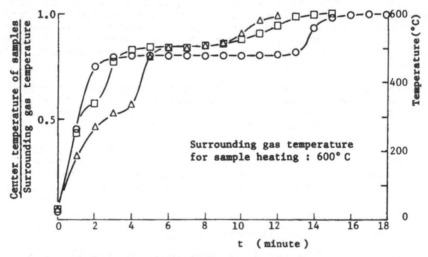

Fig.1 Temperature history of heat storage material under heat charging.(The marks O, □ and △ are respectively the case measured for the samples having Al additive of zero , 7 and 15 weight percent.)

664

Material check with X-ray micro-analyzer also confirmed this reaction.

Figure 1 shows temperature histories of several kinds of the heat storage material observed under dehydration process. All reactants formed into a cylindrical shape of about 12.6mm diameter and 22mm length were heated in an electric furnace controlled at temperature of 600°C. The illustrated temperatures are all ones detected at the center of each cylinder.

The abscissa t in the figure means the time elapsed since the heating is started.

As is easily supposed, heat absorption on the reaction process blunts temperature rising of the histories, and the signs can be seen at temperatures around 300°C and 500°C. Naturally, the first sign is recognized to be upon heat absorption by CALO, because there are no relevant sign at around 300°C for the sample with no Al additive. The second means heat absorption related to Ca(OH)$_2$ dehydration.

On the other hand, as mentioned before, mixed powder of Ca(OH)$_2$ and Al is also solidified cylindrically in the same manner as the case of the sample with no Al additive. To be concrete, after making the mixed in a paste state through water addition and later churning it we dry the formed paste without any application of artificial pressure. Hardness of the solidified, however, changes with weight percentage of Al additive, which can be seen in Fig.2. Here, the size of groove width scratched on the surface of the solids gives, so to speak, index of the hardness and its detection was made by Mantence scratch tester. As this figure shows, the highest hardness is observed in the case when the weight percentage of Al against Ca(OH)$_2$ is around 7%, which is easily confirmable if we observe transition of η defined as

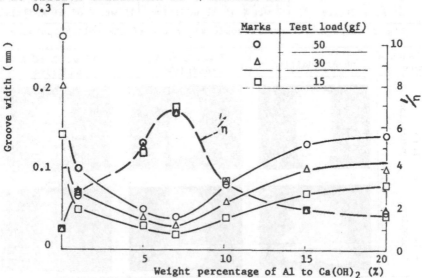

Fig.2 Hardness transition with degree of Al addition

$$\eta = \frac{\text{groove width for the sample with Al additive}}{\text{groove width for the sole } Ca(OH)_2 \text{ sample}} \qquad (5)$$

Further, mass rate of Al to $Ca(OH)_2$ necessary for realizing the chemical reaction of formula (1) is theoretically 19.6%. Hence, 7 percent of the rate leaves still large proportion of $Ca(OH)_2$ without any contribution of the reaction, even if the reaction is completed. Anyhow, the hardness depends fairly upon microscopic structure of the material. To be concrete, a fall of the hardness may be closely related to the increment of very fine lacunae existing among the structure.

A few electron micro-graphs of the lumpy material with various amount of Al additive are shown in Fig.3 together with that of the material with no Al additive.

Comparison of these pictures may back up rightness of the current discussion. Next, let us observe the looks just after their dehydrations. Typical examples are shown in the upper column of Fig.4.

In general, such a lumpy material has some voluminal expansion and shrinkage under hydration and dehydration respectively.

| no Al additive | 7 wt. % of Al additive | 15 wt. % of Al additive |

Fig. 3 Comparison of microscopic structure of the related samples

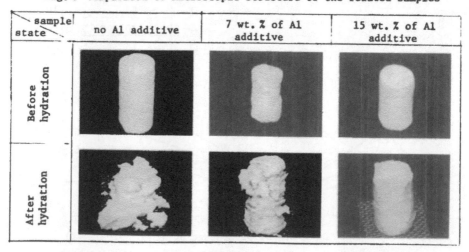

Fig. 4 Looks of various samples before and after hydration

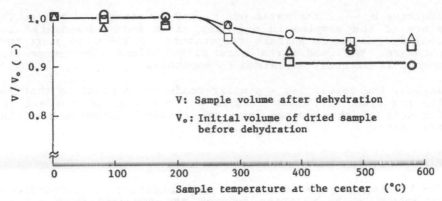

Fig. 5 Volume change observed after dehydration, The marks o, ⊓ and Δ are respectively for the case of zero, 7 and 15 wt. % of Al additive.

Degrees of the shrinkage can be seen from Fig.5, indicating the relation between voluminal change and heating temperature. The picture also shows that the material has generally voluminal shrinkage by endothermic reaction. Naturally, in addition to such a shrinkage it is also conceivable to be produced further increment of the lacunae on the other side. But this phenomena is not always inconvenient in handling, considering the subsequent hydration. Namely, we can guess that under hydration process the lacunae work as buffer for structural expansion and reduce lump breaking. The looks just after the coming hydration are also shown in the lower column of the previous picture 4.

As a matter of course, the presence of the lacunae has also much effect upon the observed thermal properties such as thermal conductivity and specific heat. Hence, if we denote thermal conductivity of lumpy $Ca(OH)_2$ by λ_o and others with Al

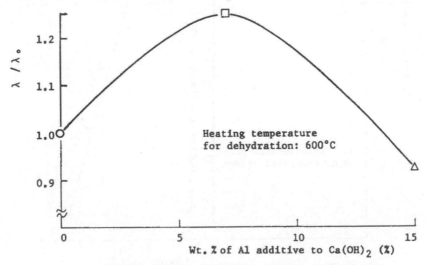

Fig. 6 Influence of Al additive on thermal conductivity.

additive by λ , the ratio of λ and λ_o becomes as Fig.6. The state of the samples is one just after being dehydrated and cooled from their uniform temperature of $200^{\circ}C$ to room temperature. The used marks are alike before. The inspection method is minutely explained in Appendix.

Anyhow, the result has a similar tendency in graph to that for the hardness η , which enables us to recognize abundant influences of the lacunae upon the physical properties of the reactants.

EXPERIMENTS ON HEAT STORAGE

The turning temperature changes with pressure of surroundings. In general, the relation between the temperature T and the pressure P is represented by the Clausius-Clapeyron's equation and we have

$$\ln P = - \Delta H / RT + \Delta S / R \qquad (5)$$

where R, ΔH and ΔS are respectively the gas constant, the reaction enthalpy and reaction entropy for the release of the substance as a gas. The relation for CALO is drawn in Fig.7 together with that for H_2O.

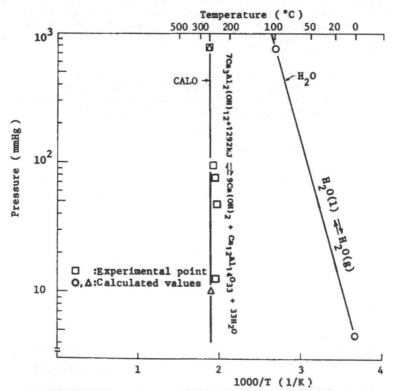

Fig. 7 Vapour pressure diagram - lnP vs. 1/T - for CALO.

As this figure shows, large value of ΔH for CALO makes its curve steep, which means that even much reduction of surrounding pressure produces slight turning temperature drop. In other words, the temperature at heat charging and discharging becomes approximately identical, regardless of surrounding pressure.

Now, taking account of the preliminary experimental results, we built a testing system with small scale. The system consists of two containers with several attachments, of which schema is shown in Fig.8. The container I with 120mm diameter and 270mm height is made of stainless steel because of avoiding corrosion. Six cylindrical energy storage elements with 26mm diameter are held in the container I with regular arrangement. They are, as seen in the figure, pipes of 13mm outside diameter with spiral fins, between which solidified CALO is packed as steam absorbent. The total amount of packed CALO is approximately 300g, which is prevented from falling off by covering the elements with stainless-wire gauze. Naturally, as the gauze has high gas permeability, efficient vapour passage is hardly held back on energy charging and discharging process. Heat transfer medium flowing through the pipe of the element is air and plays heat transmission from and to the

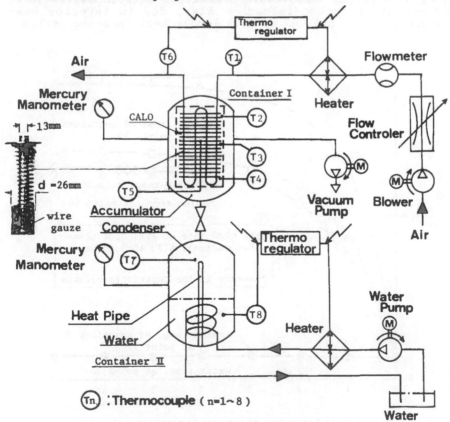

Fig. 8 Schematic diagram of the testing system.

669

elements. In short, the container I can be regarded as energy accumulator. On the other hand, the remainder Ⅱ which corresponds to both evaporator and condenser is made of acrylic resin. Consequently, the transparency enables us to see inside water level directly with the eye, which also helps to estimate reaction proceeding taking place in the container I under system operation. Further, in order to achieve active water evaporation and condensation the container Ⅱ holds heat transfer tube spirally wound at its lower inside position enough to be easily submerged in the water. The heat transfer medium flowing through the tube is also water, whose temperature is timely changed with temperature regulator, heater and pump according to demand.

In addition, to estimate system performance detection of temperature and pressure transitions was carried out at several key points of the system. Their concrete positions can be seen in the figure.

Now, Fig.9 shows an example of temperature histories observed at the key positions concerning the container I under heat charging process. Adoption of air for heat transfer medium makes here bigger temperature difference between medium inlet and outlet of the container. Also, due to inferior heat transfer, we are obliged to have longer heat charging period

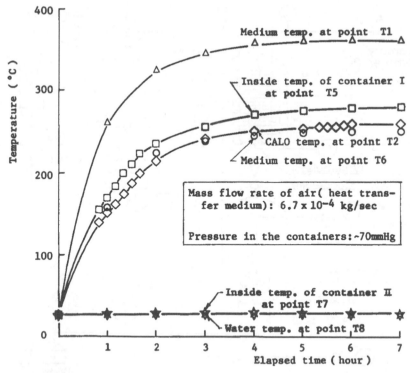

Fig. 9 Temperature histories of system key points under heat charging process

670

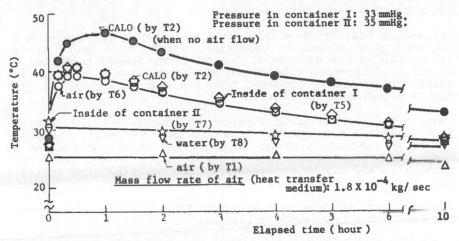

Fig. 10 Temperature histories of system key points under heat recovery.
(Water temperature in the container Ⅱ is kept to be constant.)

for CALO dehydration. Therefore, liquid medium will undertake certainly improvement of such weak points.

The following two pictures, Figs. 10 and 11, indicates behaviour of the system under heat recovery process. The former is a case when water temperature of the container Ⅱ is kept to be constant, while the latter is one when the container is thermally insulated. On a whole, hydration is inactive for both cases, which makes consequently inactive temperature rises of the absorbent and others together with long heat generation period reaching several hours. The conceivable cause of this result is, after all, low hygroscopic nature of the absorbent. Nevertheless, as Fig.11 shows, heat pump effect such as temperature drop of water in container Ⅱ can be observed.

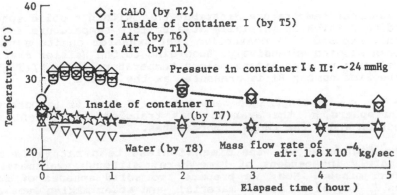

Fig. 11 Temperature histories of system key points under heat recovery.
(Outside of the container Ⅱ is thermally insulated.)

CONCLUSIONS

The system test on chemical heat pump teaches us that promising heat pump effect is unobtainable with ease for a system using CALO as the absorbent. In addition, temperature control by means of pressure regulation is also difficult in practice, which can be easily supposed from the diagram of lnP vs. 1/T for CALO indicated in Fig.7.

Judging from heat storage density and comparative low temperature level for dehydration, however, CALO has still the power not to be ignored. Although this testing result obliged us to consider several promising means to solve these technical problems, CALO may be fundamentally the material more suitable for chemical heat storage.

Anyhow, heat storage system using CALO will be also investigated in our laboratory hereafter.

REFERENCES

1. I.Fujii, Heat Energy Storage Based on Reversible Chemical Reaction (Part 1 - Principle and Basic Experiments),Trans. of SHASE of Japan, No.4, pp.21 27, June 1977.

2. I.Fujii, K.Tsuchiya, Experimental Study of Thermal Energy Storage by Use of Reversible Chemical Reactions, Alternative Energy Sources 1, Vol. 9, pp.4021 4035, Hemisphere Publishing Co., 1978.

3. I.Fujii, K.Tsuchiya, Heat Energy Storage based on Reversible Chemical Reactions (Part 2 - Influences of Reduced Pressure Surrounding for Thermal Decomposing Process), Trans. of SHASE of Japan, No.12, pp.1 6, February 1980.

4. M.P.Heisler, Trans. ASME 69, pp.227, 1947

APPENDIX

The transient temperature distribution T within a solid sphere of radius R which is initially at a uniform temperature and is suddenly exposed to a convective heat transfer on its surface has been studied extensively. According to the Heisler charts, the relation between dimensionless temperature $(T-T_{\infty})/(T_{c}-T_{\infty})$ and the Biot number Bi is presented as the figure [4].

Where T_{∞} is heating fluid temperature, T_{c} center temperature of the sphere, $\bar{h}$ the average heat transfer coefficient for convection and r the radial coordinates.

As the figure shows, the dimensionless temperature is approximately independent of time here, although exactness is somewhat missed. Now, we prepare two solid spheres of same radius R_{0} but of different material, and after making same and uniform initial temperature for both we expose them to the fluid having temperature of T_{∞} with same heat transfer coefficient $\bar{h}$. Then, if we detect two point temperatures for each,

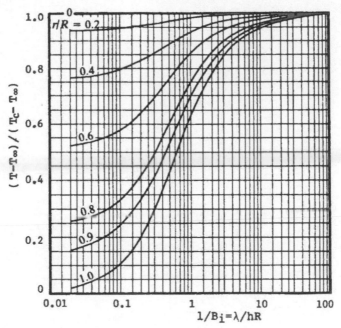

Fig. A Correction chart for a solid sphere[4]

for example, at the center and any other r-location in the body at a specified time, dimensionless temperature $(T-T_\infty)/(T_C-T_\infty)$ can be determined for each sphere. Accordingly, the value of $\lambda/\bar{h}R$ ($=1/Bi$) which corresponds to the value of $(T-T_\infty)/(T_C-T_\infty)$ can be also decided for each with the aid of Fig. A.

So, denoting these values by m_1 and m_2 respectively, we can get

$$m_1 = \frac{\lambda_1}{\bar{h}_1 R_1} \quad \text{and} \quad m_2 = \frac{\lambda_2}{\bar{h}_2 R_2} \tag{A-1}$$

And by aforesaid conditions

$$R_1 = R_2 = R_0, \quad \bar{h}_1 = \bar{h}_2 = \bar{h}$$

Hence, the ratio λ_1/λ_2 becomes as

$$\frac{\lambda_1}{\lambda_2} = \frac{m_1 \bar{h}_1 R_1}{m_2 \bar{h}_2 R_2} = \frac{m_1 \bar{h} R_0}{m_2 \bar{h} R_0} = \frac{m_1}{m_2} \tag{A-2}$$

This relation means that m_1/m_2 becomes the ratio of thermal conductivity in itself.

673

Superconductivity and Energy Storage

JAMES J. HURTAK
Technology Marketing Analysis Corporation (TMAC)
118 World Trade Center
San Francisco, California 94111, USA

INTRODUCTION

Superconducting magnetic energy storage (SMES) is currently the only method known for energy storage specifically in the form of electricity. This device stores energy directly as an electric current passing through an inductor made out of special alloy metals (or in the future ceramics). Other known storage devices must convert the energy to thermal, chemical gas, or air and then reconvert it into electricity.

This paper also looks at parallel areas of energy research which were conducted using energy generators having superconducting field windings. Current technology has been developed in the past year that takes us away from low temperature metallic superconductors to a future that looks forward to higher temperature ceramic systems.

SMES (SUPERMAGNETIC ENERGY STORAGE) DESIGN FOR THE FUTURE

One of the most promising utility applications for advanced super-conductors is supermagnetic energy storage, which allows electricity to be stored directly without conversion to other forms, as with batteries (chemical energy) or pumped-hydro storage (potential energy). The conductor coils for a magnetic storage system (1000-MW) might be half a mile in diameter and would need to be buried in solid rock to contain the extreme physical forces produced during operation.

SMES could used as an efficient energy storage device to displace expensive generation. Its history goes back to prototype work that had transpired well over a decade ago in the US and USSR. [1] The concept of energy storage in a superconducting magnet is based on the facts that in superconducting material the resistance to the flow of current is essentially zero. Therefore, there is no energy loss associated with maintaining the magnetic field. Superconductors operate within three major variables: the size of the magnetic fields (teslas [or gauss]) and current density (amps/cm^2), which are both in relationship to temperature (K).

675

THE SMES SYSTEM

The key design of the SMES system is that energy is stored in a coil which is a high current (765kA) conductor in the form of a circular geometry. In experiments with metallic superconductors, near-zero resistivity has been achieved at temperatures approaching absolute zero (0 K.), the point at which all molecular motion ceases.

In superconductors the current can circulate with practically no energy loss. According to early theories (BCS), conducting electrons are able to move through a superconductor without resistance because they form pairs that do not collide (or are not effected by collisions) with surrounding atoms. These are called Cooper pairs which are part of an atomic and crystallographic ordering such that the lattice vibrations of the atoms in the crystal structure of the materials have little interaction with the conducting electrons.

Tests that were performed on certain components of the SMES low temperature storage system in the early 1980's, showed efficiency rates as high as 90%. It's life expectancy has been predicted to be as much as 50 years. And since it has few moving parts and is kept at low temperatures, it has a very high reliability factor. Electric response has proved to be virtually instantaneous (tens of milliseconds), enabling its use for system regulation and spinning reserve.

Since there is no energy loss associated with maintaining the magnetic field, SMES, even when operating at extremely low temperatures, could be economically advantageous as an energy storage device when used during day time high load hours and recharged by inexpensive generation during night time low load hours. The preliminary plan would be for direct storage of energy in a geometric coil system which efficiently displaces expensive generation of energy during high load hours.

Designs for a high field conductor using metallic materials have been made from a large number of fine filaments, usually less than 100 micrometers in diameter, located in a matrix of high conductivity normal metal such as copper. Superconductor designs such as those researched at Westinghouse and by the Bonneville Power Administration (Tacoma substation) used the more common multi-filament construction with niobium-titanium alloy filaments embedded in a copper matrix. With the critical temperature for NbTi at 9 K or Nb_3Sn at 17 K, the only practical liquid coolant for these superconductors is the expensive and difficult-to-handle liquid helium whose atmospheric boiling point is 4.2 K.

There are some benefits to using liquid helium as a coolant. It demonstrates a low heat of vaporization (which limits the amount of heat that can be transferred); it has a low viscosity; high compressibility; and high specific heat which gives it the ability to store heat and thereby limit the temperature excursions of the superconductor.

One of the key components in the SMES, as designed by Energy Management Associates, has been the single-layer, 112-turn, solenoidal superconducting coil, operating at a nearly constant temperature of 1.8 K by a helium bath. The coil is charged and discharged through an ac-dc power conditioner (PCS). During discharge the PCS converts the dc voltage/current from the coil into ac for transmission. During charge, the ac is converted back to dc for storage. The PCS must work with a 3 phase 60 Hz utility system at a standard utility line voltage (230kV). [2]

As previously mentioned the coil design consists of 112 turns at a nominal total diameter of 1568 m (for a 5500 MWh unit/energy stored in the coil at the end of charge). This composes a single layer, series-wound solenoid (See Figure 1.), with an inductance of 67.7 Henries and an aspect ratio of 0.010 [3]. A multi-layered coil may also be feasible.

The structure is held within a coil axial support structure (CASS) made of aluminum bricks which holds the conductor in place. The conductor is one of the largest costs of the construction of the component system since it must carry 765 kA when the coil is fully charged. In the system designed by the Bechtel Group, the conductor was to contain eight super-conducting cables, embedded within high purity aluminum wedges. The cables, made of Nb50.5Ti, were within each of the wedged sectors. They were braided and made from .69mm Cu/NbTi superconducting wire. [4]

The coil and CASS assembly is enclosed within an apparatus that is con-tinually flooded with helium at 1.8 K and 1 atmosphere. The apparatus is made up of a thin aluminum shell, corrugated for stiffness against the internal pressures (2 atm).

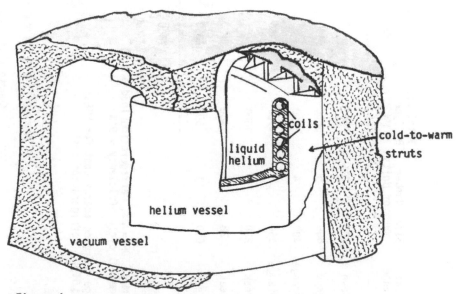

Figure 1.

Typical Underground Solenoidal SMES

Plant Type	Load Range	Size MW	Number of Units	Total Cap By Type MW	Total Cap By Range MW	% of Cap By Range	1981 Average Operating Cost $/MWH	Annual Capacity Factor %	Annual Capacity Factor w/o SMES %
COSPR	Base	800	3	2400			19.3	75	74
COSUB	Base	600	5	3000	5400	55	20.3	69	63
STMGAS	Midrange	400	4	1600			39.9	28	33
STMOIL	Midrange	150	5	70	2350	24	53.5	7	12
COMCYC	Peak	250	4	1000			59.7	2	6
CT	Peak	75	6	450			82.6	0	2
SMES	Peak	600	1	600	2050	21	24.5	29	-
				9800		100			

SMES = 6000 MWH, FOR = 2%, Refrigeration load = 5.5%
Incremental cycle efficiency = 91%

Average charging energy cost for SMES:
22.7 $/MWH + refrigeration cost

SYNTHETIC UTILITY MODEL WITH 600 MW SMES SERIES

Courtesy of Electric Power Institute

ENERGY COSTS AND SMES

For the SMES system, there are two major losses where energy is dissipated:
1) the refrigerator load and 2) the loss through the convertor of
ac - dc. The central refrigerator load is used to maintain a cool environment for three major elements in the system:

  1) to make up for heat losses from the magnet;
  2) heat generated during the charging cycle; and
  3) for maintaining an adequate temperature for the power leads.[5]

The two energy losses (refrigeration and conversion) must be seen as
fixed costs to operate the system. The cost of energy due to the refrigeration process is based on additional variables such as the decision
to charge and discharge the magnets as determined during the cyclic
operation of the system defined to be the incremental cycle efficiency.
During charging there is heat created in the magnet due to eddy currents,
as well as the energy requirement which occurs in the time span from
zero charge to full charge and back to zero.

According to EPRI findings, a typical helium coolant SMES system with
a maximum storage capacity of 6000 MWh, Forced Outage Rate = 2%, refrigeration load of 5.5% and incremental cycle efficiency of 91% would
have in 1981 dollars a total production cost of 900 million dollars
with a break even capital cost (BECC) of 695$/kW. However, as the hours
of storage capacity increase, the incremental operating savings (because
of charging and discharging) decreases.

In the past year scientists have taken seriously that there is now a
better future for SMES devices whereby the development of superconducting
materials at higher temperatures would eventually lead to superconductors
at: 1) room temperature thus reducing much of the refrigerator load;
and 2) the development of higher current densities. With a technology
of only Cu/NbTi superconductors operating at close of 1.8 K, 600 MW
SMES was determined to function best in the capacity of supplying energy
to replace the peak load generation otherwise required from other high
cost peaking units such as STMOIL (Residual) and CT (Distillate) units.
The Table that follows shows the synthetic utility model where the energy
stored in SMES during the low load hours is used to peak shave the high
loads during the day. [6] Yet, it may be concluded that it does not
pay to oversize SMES storage capacity from 6 hours to 10 hours, because
the break even capital cost has been shown to best be achieved when
the size of the SMES is very small relative to the utility size. Nevertheless, it can be economically feasible to operate at peak load hours
a SMES each day of the week.

Given the possible oil/gas escalation rate of about 10%, a coal fuel
escalation rate of 9.3%, an inflation at 8.5% and a discount rate of
12.5%, the results show that the SMES breakeven cost ranges from 1219
$/kW to 600 $/kW and that the comparable cost quickly decreases with
the increasing oil escalation rate. However, at present prices, coal
systems seem to be the best alternative which can produce the greatest
system fuel cost savings.

Final conclusions from the EPRI is that using metallic superconductors which require extremely low temperatures, the cost of the system is about twice the expense of currently available energy storage for power plants, in the form of batteries, compressed-air or pumped hydro storage devices. Even with advances in superconducting materials, we must remember that about half of the costs are related to the conductor coil material itself, its axial support structure, and its fabrication. However in large scale power-plants, improved superconducting systems that will operate closer to 273 K will allow a more competitive BECC due to the coolant/refrigeration costs, but prototypes for these systems might still be 10-15 years away.

Currently, the best system would be a 600 MW SMES with a forced outaged rate of 2%, with a 5.5% refrigeration load and a 91% incremental cycle efficiency (over the conventional pumped hydro storage) provides an annual savings of $45,699,000 and a BECC of 695 $/kW. When this is compared to futuristic systems which show a 0% refrigeration (due mainly to ceramic superconductor) but only a 70% incremental cycle efficiency (same as conventional pumped hydro storage) the BECC is 608 $/kW. This represents only a 14% increase in BECC.[7] Others have predicted only a 5 to 8 percent savings for current nitrogen cooled ceramics.

Thus, it is important that whether we begin to develop ceramic super-conductors within the coming years, additional research and development needs to be done on related support systems which could also reduce the costs by about $88/kWh (where kW + kWh x 5h = capital cost in kW.) lowering the capital cost to a competitive figure.

During the early 1980's, two important studies were performed to determine the design and function of these systems, one was by Energy Management Associates (California), and the other jointly by Bechtel Corporation and General Atomics Inc. The EMA study showed the SMES system could be operated at 91% efficiency, had 98% availability, and designed prototypes needed for the carrying of a 5.5% refrigeration load. This 600-MW, 6000-MWh SMES plant had an average economic cost of about $1200/kW (1982 $). The University of Wisconsin also evaluated the EMA lead using the single-year analysis approach (based on 1981 dollars). The design parameters for one of the cases was Discharge/Charge MW Capacity = 350/350, for a 5500 MWh unit with 100% availability and a refrigeration load 8.6 MW (3.75%) using an incremental cycle efficiency 95%, which yielded a 34 $/kWh, a similar study using a 900 MW cost capacity showed a value change of only 28 $/kWh demonstrating that oversizing the converter was not extremely cost efficient. [8]

In December 1987, the Defense Nuclear Agency of the United States awarded both Ebasco Services, Inc (New York) and Bechtel National Inc (San Francisco) contracts for preliminary designs for niobium-based materials. At the end of 1989, one of the two companies will be awarded the next phase of the project which will be to build an engineering test model. The coil itself could range from one-half to three-quarters of a mile in diameter. The reason why niobium is still being looked at is that the current densities must operate upwards of 600,000 A/cm^2 with fields from 1.6 to 5 teslas.

300-MVA SUPERCONDUCTING GENERATORS

Two studies were conducted on cost feasibility for energy generators
using metallic superconductor field windings: one by Energy Power Research
Institute (EPRI) and the other at Westinghouse Electric Corporation.
During the early 1980s, the design and certain developmental work was
begun on a 300-megavoltampere (270-MW) superconducting generator. This
included the development of a data acquisition and telemetry system
that would operate and monitor a cryogenic rotor environment by means
of multiple concentric cylinders cooled by liquid helium. The system
(300-MVA) that was built operated for more than 400 hours, including
100 hours at speeds of up to 3600 rpm, and was instrumental in establishing test parameters for rotating experiments conducted in liquid
helium temperatures of 4 K (-452 degrees Fahrenheit).

HELIUM FLUID UNDER HIGH-SPEED ROTATION

Experiments were done using both nitrogen and helium as the coolant
for the 300-MVA system. Due to the low temperatures that are necessary
for metallic superconductors, the focus of the study was the transfer
coefficients to liquid helium from the superconductor under conditions
of generator rotor rotation at approximately 3600 rpm having niobium-
titanium superconducting field windings.

A series of experiments were performed by Westinghouse engineers whereby
a test chamber was first partially filled with liquid nitrogen and the
test rig was rotated to 500 rpm to promote circulation throughout the
test chamber. When the test chamber temperature stabilized at liquid
nitrogen temperature (77 K), the rig speed was increased to 1330 rpm.
The vibration of the rig bearing support brackets was measured as 13-15
mils. The safety ring that surrounded the apparatus also experienced
some of the vibrations.

The cooldown was then continued to helium temperatures to measure dewar
performances. The first phase was with a limited rotational speed of
approximately 1000 rpm. The cooldown procedure continued, liquid nitrogen
evaporated, and the interior was purged with helium gas to remove any
residual nitrogen gas. Next, the temperature in the chamber was cooled
to the critical T_c of 4.2 K by injecting liquid helium into the test
chamber. This process took about 90 minutes and consumed 100 liters
of liquid helium. Other experiments were performed at 1330, 1500, 2900
and finally 3600 rpm. The vibrations levels were lowered from 13-15
mils down to 5-6 mils.[9] It is important to note that the vibrations
were able to be sustained at a low range except at critical thresholds
at speeds of 2300, 3300 and 3630 rpms.

The test rig was a standard dewar structure adapted for rotation.
Inconel 706 was determined to be the material which could be used for
most of the parts, but there was some weaknesses found in the tensile
strength depending on the temperature. In addition, Inconel 718 plate
and weld wire was thought to need further investigation and could provide
additional strength. [10]

The dewar was a double-walled vacuum insulated structure with a
100 K conduction cooled radiant heat shield made of copper which surrounded

the helium area, creating a cryogenic insulation system. Copper has
a higher conductivity at cryogenic temperatures. The copper shields
almost all the high-frequency components of the magnetic fields and
is spaced as a "damper." The shield is thermally anchored to the vapor
cooled inner cylinder to fix this temperature at 100 K. There is also
a support cylinder on the top unit which together with the inner cylinder
is cooled by helium vapor brought in from the central core of the test
chamber. OFHC copper was used that could withstand 45,000 psi. The cooling
channel between the two cylinders was specially designed to prevent
axial convection loops.

A machine designed at MIT however had a different shielding system,
whereby the outer shield is replaced by a damper winding operating at
cryogenic temperatures composed of highly pure copper wires. The purpose
of this design was to transfer by compression some of the strong
electromagnetic forces imposed on the outer shield into the rotor support
tubes.[11] Although final construction of the 300-MVA system was never
completed, these critical preliminary tests have helped to determine
the thermal and hydraulic design of the system. The main purpose of
the experiments was to test certain factors that would be necessary
to adapt superconductor technology for large utility equipment. Tests
performed for high speed rotation of the helium level gauge/effort were
among those not complete at the time of the termination, but the effort
was later completed under USAF funding which demonstrated that the superconducting wire gauge could be used to follow the liquid helium level
under speeds of 3600 rpm.

CONVECTION MODELS

One of the important tests was the verification of heat transfer correlations to liquid helium in superconducting rotors due to the increased
convection produced by centrifugal forces. Experiments determined whether
the secondary effects, as well as Coriolis acceleration of fluids flowing
in the rotating channels might degrade the heat transfer. E.R. Eckert
had predicted heat transfer in the thermosyphons well in excess of that
of turbulent natural convection. To test the theory a simulation was
performed using windings 9.1 cm long, containing heaters and temperature sensors which were spun in the rotating dewar. Included in the
convection coefficient was the Kapitza conductance effects. The experiments
showed that under certain conditions, Coriolis acceleration could reduce
heat transfer in the rotor cooling circuit by as much as 60%. Secondary
coolant flows also inhibited the transfer. According to Westinghouse
research: "Conventional buoyancy induced flow analysis and conventional
heat transfer correlations for supercritical flow provide a good indicator
of heat transfer coefficients in rotating thermosyphons if regimes of
strong secondary flow are avoided, but conventional heat transfer correlations do not adequately describe the thinning of the boundary layer
reduced by high rotational "g" fields and the resulting improved heat
transfer of high tip speed surfaces." [12]

NEW HARDWARE ADAPTATIONS FOR THE 300-MVA

As part of the development of the 300-MVA system, an important telemetry
system was established. The telemetry system converted the analog

temperature sensor signals to digitial form. They were then passed
on by frequency shift keying telemetry at a sampling rate of 100/sec.[13]
The data acquisition and telemetry system as developed under the 300-MVA
program has a broad range of applications to all types of rotating
apparatus and can be used with higher temperature ranges for new superconducting ceramics, as well as in conventional motor rotors, turbines
and pumps. Therefore, unlike the sensors that were specifically
designed for cryogenic temperatures, the basic data acquisition and
telemetry system could be adapted to multiple outlets and applications
through minor design changes of "on-board" signal conditioning circuits.

Some of the sensors tested in these experiments can also be used at
room temperature such as the resistance strain gauges and thermocouples.
In fact many components of the module could be adapted to room temperature
or nitrogen coolant systems, namely, the multiplexers, the processor,
the Frequency Shift Keying, and the ac-dc converters. Thus, research
done could be adapted for future applications of ceramic superconducting
generators and in other instances where rotating apparatus was needed
for the successful development of this system for world use in the
monitoring of most conventional generators and motors.

THE FUTURE FOR 300-MVA and SMES

The two systems which were built to test apparatus capable of spinning
heat transfer test modules immersed in liquid helium provided specific
information on the effects of various rotor geometries and rotation-
ally induced secondary coolant flows on heat transfer. The final tests
showed that a cryostable field winding made of niobium titanium is technically feasible in large mostly underground superconducting energy
systems that use a helium based coolant system.

New advances in superconductors have not made this research obsolete.
Although ceramic superconductors will allow for a lesser cost for the
refrigeration systems and the cryostats or thermal insulating systems,
this is only a small savings out of the original manufacturing costs
of the total system. The daily costs however will be greatly reduced
if the superconductors can operate at room temperatures. Even with
superconductors operating still within the 95 K range, the systems will
use nitrogen (77 K) instead of hydrogen or helium, which can be liquefied
in situ, since it is a component of the atmosphere. In studies that
took place in late 1987, liquid nitrogen was determined to cost about
20 cents per liter, compared to about $3 per liter for liquid helium.

Superconducting generators appear to have several inherent advantages
that might well justify development within the next few years and extensive
development effort is still needed to perfect them. Even as the system
stands today, there are still losses in energy when the systems convert
their energy to ac power. Therefore for true savings, energy transmission
wires should be developed along with transformers, computers (used extensively in all utility systems) and phase shifters, to name a few.

Currently even the US government has recognized that low temperature
superconducting generators are still in need of greater developmental
mental design. One important aspect is the current needed to bathe
the rotor in a refrigerated cooling fluid. The current rotor design
involves concentric cylinders insulated from each other by a vacuum

683

space, where the interior parts of the rotor are subjected to extreme mechanical stresses, but the location is very difficult to reach for inspection or repair.

Given adequate current carrying capacity, for example, smaller conventional generators could be developed perhaps only one half the size of the proposed systems and thus potentially be less expensive to manufacture and transport. Since they feature longer response time to power surges on a line and have very high magnetic flux levels, such generators could enhance the stability of utility networks and thus perhaps reduce needed reserve margins. [14]

CERAMIC SUPERCONDUCTORS

Current research at the University of Houston, IBM (Zurich Research Laboratory and Almaden, California), Stanford University, as well as in Japan (University of Tokyo) and China (Institute of Physics, Beijing) has already begun to make breakthroughs in superconducting technologies that will eventually make this technology not only less expensive but eventually even commonplace. At which time, superconducting materials will not only be used in storage devices, but in superconducting cables that could transmit electricity from a power plant to a distant city with essentially no energy loss. The new discoveries have critically focused on raising the temperature in superconducting materials which would allow other more inexpensive elements such as liquid nitrogen to replace the costly use of liquid helium per unit volume. In only the past few years we have jumped from 23 K to 125 K temperatures.

In 1986, the first new breakthrough took place in the development of high temperature superconductors. Georg Bednorz and Alex Muller at IBM Zurich laboratories discovered a mixed oxide material that was super- conducting at 30 K.[15] The oxide (La_2CuO_4) belonged to a family of ceramic materials called perovskites. Perovskites have now been found to be superconducting at temperatures up to approximately 100 K (-280°F). Certain materials seem to be superconducting even at room temperature, but substantial tests cannot maintain this criteria. The common feature are sheets of Copper-Oxygen bonds which play a role in both conductivity and superconductivity. Here the wave function amplitude on the oxygen (p levels) is about the same as on the copper (d levels) and thus the bond forms the conduction path.

The latest (March 1988) superconducting material has a T_c of approxi- mately 225 K. The structure is $La_2Sr_1Nb_5O_{15}$ in a thin film created by T. Oguchi. However, most researchers have achieved superconductivity at 125 K by structures such as $Bi_2Sr_{3-x}Ca_xCu_2O_{8+y}$, where x ranges from 0.4 to 0.9. The $Bi_2Sr_{3-x}Ca_xCu_2O_{8+y}$ structure contains alternating double copper-oxygen sheets and double bismuth-oxygen sheets. Calcium and strontium ions exist between the copper-oxygen sheets with Ca(Sr)-O bond lengths at approximately 2.45 and 2.55 angstroms. The copper-oxygen bond lengths within the sheets are 1.91 angstroms, where strontium cations exist just above and below these double sheets. Electron microscopy shows that this structure has platy crystals (diameters from 2-30 micro- meters) which are known to be high T_c superconductors.[16] The structural size of the cell in one configuration is a=27.0 angstroms b=5.4 angstroms and c=31.0 angstoms. One of the key benefits of this structure seems to be its lower reactivity with humid air.

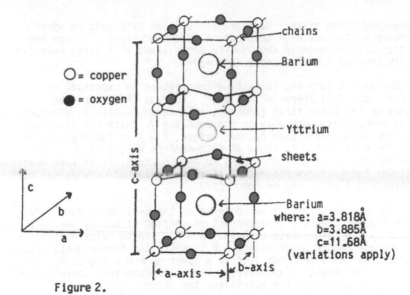

= copper

= oxygen

c-axis

chains

Barium

Yttrium

sheets

Barium

where: a=3.818Å
b=3.885Å
c=11.68Å
(variations apply)

|← a-axis →|← b-axis

Figure 2.

A BASIC CELL: Crystal structure of YBaCuO
Showing Cu-O sheets and Cu-O chains

Currently researchers have been working with three major ceramic super-conducting structures: $La_{2-x}A_xCuO_4$ where A_x is Ba, Sr, or Ca (and Cu can be substituted for Nb according to T. Oguchi); $RBa_2Cu_3O_7$ where R is a rare earth such as Y (Yttrium); and $Bi_2Sr_{3-x}Ca_xCu_2O_{8+y}$. In these instances, there are copper-oxygen sheets with copper as a square-planar coordination. Some variations appear in La_2CuO_4 where two additional oxygens are bound to copper forming a distorted octahedron. In $YBa_2Cu_3O_7$, a fifth oxygen exists which has a longer Cu-O distance that forms a type of square pyramidal coordination for copper.(See Figure 2.) According to Subramanian et al, in $Bi_2Sr_{3-x}Ca_xCu_2O_{8+y}$ there are no oxygens between the adjacent sheets; thus, the copper coordination is four or five. [17] One of the general features of all these cases is the mixed oxygen states Cu^{II}-Cu^{III}.

The ordering of occupied sites in the Cu-O planes appears to be linked directly to the occurrence of T_C. It has also been noted that even when the Cu-O sites are only one forth full (where x=0.4)in $YBa_2Cu_3O_7$, superconductivity is present as long as the occupied sites are ordered. Analyzing these three structures and their Cu-O sheets it has been noted that distorting the tetragonal sheets to orthogonal, aids in their superconductivity, making an exact Cu-O symmetry (specifically the oxygen arrangement) essential for high T_C. In $YBa_2Cu_3O_7$ the T_C increases as the sites become one-half full and ordered. However, if x is greater than 7 then the Cu-O planes are staturated and the orderings within each unit of Ba, Cu, and Y are changed which can create an insulator.

Theoretically it has been hard to explain superconducting between the copper-oxygen pairs. In 1957, Bardeen, Cooper, and Schrieffer (BCS) put forth a theory of pairs which stated that an attractive electron-electron interaction mediated by the lattice vibrations (phonons) causes the electrons to form pairs (Cooper pairs) and condense into the superconducting state. However, it has been difficult to apply this theory to the new high temperature superconductors. It has been found that the exchange in the Cu-O-Cu bonds leads to a large negative J_{dd} which creates high-frequency spin waves.

In conventional theory the transition temperature is expressed as $T_c = (\omega) \exp(-1/\lambda)$. Where (ω) is the average characteristic frequency of the boson field (phonons, spin waves, excitons, plasmons...). (ω) is reduced exponentially by the coupling λ with the electrons. Here as the exponential term becomes larger, (ω) is found to decrease so that the maximum T_c is found at intermediate values of λ. If this holds true for high temperature superconductors, it is then possible that higher temperatures can result if either chemical or physical ways to control (ω) or λ can be found.[18]

Most of the new structures seem centered on the d-electron band. Most theories suggest that the supercurrent is carried either by the Cu-O chains or by the Cu-O sheets or both. Here T_c can be related directly to the average magnon polarization (τ), where the ferromagnetic coupling between chains and sheets may lead to a lower τ and a higher T_c in both chains and sheets. Calculations have also shown that there is some oxidation that occurs in the sheets and the chains.

One theory suggested by Emergy uses the model of a magnon-pairing mechanism where he constructed a Hubbard Hamiltonian model that would have partially occupied copper dσ and oxygen pσ bands, suggesting that an attractive coupling by means of magnetic exchange with copper d spins might lead to a strong attractive coupling of the oxygen pσ electrons. This emphasizes the band of oxygen pσ holes rather than the pπ holes, using a one-electron band calculations.

According to Chen and Goddard, the GVB calculations in certain instances show that strongly coupled pairs are not stable, and found that the attractive coupling arises in the weak coupling limit (via magnetic interactions), this leads to the coupling of triplet pairs and p-wave coupling rather than the coupling of singlet pairs and s-wave coupling. In support of their theory on magnon-pairing, they worked with $Y_1Ba_2Cu_3O_7$, which should ideally have one oxygen pτ hole for every three copper atoms and use both chains and sheets for superconductivity.

In Chen and Goddard's theory, ferromagnetic pairing between the oxygen pτ holes and copper d electron couples the conducting electrons to the copper d magnons, leading to the attractive pairing that is accountable for superconductivity. The process is three-fold: as a conduction electron, the oxygen pπ-like electron travels, it tends to leave ferromagnetically paired copper spins. As another second conduction electron is scattered it tends to travel in the direction already polarized by the first electron and a final result of attractive interaction takes place. The key to this theory is the hole in the proper oxygen pπ orbitals of both the sheets and chains.[19]

A companion theory put forth by associates of this author suggest that since coupling of electron pairs implies a higher mass which keeps the pairs from being broken or dissipated, the carriers may not be the electrons but the (paired) holes themselves, since the effective mass of holes is almost always larger than the effective mass of electrons.[20]

An additional theory for superconductivity was put forth by Philip Anderson at Princeton University which claims that some forms of antiferromagnetism takes place in the new ceramic superconductors (i.e., the alternating of electron spin on rows of atoms), whereby these ceramics conduct charge not via electrons but particles such as bosons.

An overall picture of the new ceramic conductors shows that the T_c and H_c values far outweigh the benefits of other superconducting materials. However the J_c (current density) values are only between 10^2 and 10^3 A/cm^2. (YBCO was shown to have the highest current density at 1.8 MA/cm^2 at 77 K.

The difference between the new ceramic and older metallic materials are a factor of 100 in the ampere-turns thickness for the magnetic winding. The effectiveness of superconducting magnetic applications comes from a tightly packed winding usually close to where the magnetic field is needed.

A major problem with the new ceramics is the brittleness of their structure. In certain types of accelerators, it might still be necessary to support the new ceramics within a ductile metal supporting tube during the formation of conductors. While in their "green" state the new materials are flexible, once they are prepared their form cannot be changed without cracking. Special containers might have to be built to help form these special materials, whereas they would be formed in situ. In the final structure, current metal conductors will have to play and important role in current switching and stability, where the metal also serves as a heat by-pass. These metals could be deposited epitaxically on the structures. Additional research is needed in this area, since copper or aluminum can diffuse into the ceramic superconductor and ruin the material.

WHISKERS

One of the major problems with developing ceramic superconductors to energy devices is the brittleness of the structure. Small fibers anywhere from 1 to 100 microns would be necessary to create the windings. Parallel research has been underway at Los Alamos using materials called super whiskers which are single growth (beta) systems that allow a single, pure silicon carbide crystal in cubic form to grow only in one direction. Unlike--alpha and beta whiskers grown commercially, with stacking faults that result in weak spots, these are grown in a graphite reactor placed inside a quartz tube furnance. A catalyst--metal balls five to 30 microns in diameter--are placed on the graphite growth plate and exposed to vapors containing carbon and silicon at high temperature--about 1,400 degrees Centigrade.[21]

The final product are whiskers--tiny (1-3 microns) single crystal fibers-- which have a tensile strength of two million pounds per square inch (psi), more than five times the strength of the strongest steel. Their elastic modulus (a measure of stiffness) of 100 million psi is more than three times higher than steel, and they can be stretched up to two percent of their length without breaking. These remarkably strong whiskers, when mixed with ceramic powder and consolidated, offer a high potential for a structural ceramic with properties far superior to existing ceramics in high temperature strength, fracture toughness and thermal and mechanical shock resistance.

The whisker-reinforced ceramics hold promise for applications in energy conversion processes for valve parts, turbine blades, heat changers, regenerators and recuperators. Metals ordinarily used for these components lack the corrosion temperature and wear resistance of ceramics, but ceramics until now have lacked the thermal stress resistance and fracture toughness that makes metal so desirable.

A combination of these techniques shows promise in the development of new materials to be used with ceramic superconductors for energy induction.

WHAT THE FUTURE HOLDS

The first large-scale applications of superconductivity for energy storage will come through massive energy-utility complexes above and below ground. (See Figure 3.) Ultimately, these systems will be coupled with underground transmission lines which will travel for miles with extremely low losses of energy. The current-carrying capacity of the new materials will have to be increased first. In 1983, Brookhaven National Laboratory tested a conductor cable constructed with an Nb_3 Sn tape mounted on an aluminum stabilizer. It can now be suggested that a film of ceramic super-conducting material such as $YBa_2 Cu_3 O_7$ be used with more research into the 60-Hz loses of such films.

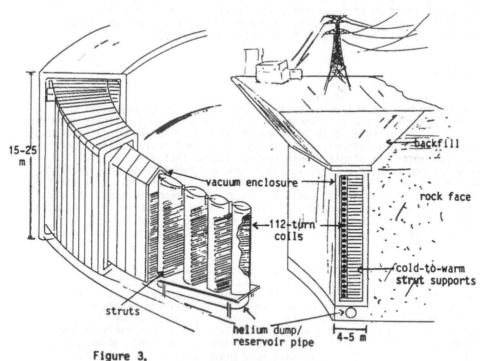

Figure 3.

MODEL OF SMES
Showing Helium Vessel Coil Assembly
Below Ground Level

In the future superconducting magnets could open the door for direct storage of electricity, while using superconducting materials in generators which could reduce their size and increase their efficiency. Long lead lines will be needed for such potential applications to be proved and scaled up to the size required for utility operations in underground structures. [22]

A summary of new ceramic superconductivity:

. Temperature versus Critical current density and Magnetic Fields. Researchers throughout the world are not only working on achieving a higher T_c from these materials, but also a higher J_c. Although the critical currents may soon be acceptable for transmission purposes, they would be only marginal for magnetic uses (according to John Hulm of Westinghouse/working with EPRI).

. Oxygen content. The amount of oxygen in the material and the way it is arranged have a lot to do with the effectiveness of superconductivity. Through neutron scattering methods, researchers have found that in the characteristic triple perovskite structure of superconducting materials, the oxygen comes out of one of the four oxygen sites. Thus the site arrangement of the atoms may be the important trigger of superconductivity.

. Stability. The new compound's stability when reacting to other substances has yet to be resolved. We do not know all the variables of the super-conducting compound reactivity to air and water which may open new avenues of use for the materials.

New Areas of Energy Related Storage:

Railway transportation. Superconducting magnetically levitated trains (MAGLEV) using linear synchronous motor propulsion, now being researched and developed by the Japan Railway Technical Research Institute. Additional research into Type-I superconductor that have only one critical magnetic field, and Type-II superconductors that have a second higher critical magnetic field (which can trap magnetic flux inside the material is still needed. (10-15+ years until commercialization of new ceramics.)

Josephson junctions. Superconducting transistors act as a switch by amplifying small voltage signals in "on" and "off" states. The junctions, made up of two layers of superconductors sandwiching a layer of insulator, currently are being manufactured by one company in the US to make high-speed oscilloscopes. Applying the new superconducting materials to the junctions would decrease the amount of energy needed to perform a switching function. This, in turn, would decrease heat output and allow them to be down-sized. One obstacle is the switching function of the new superconductors might have more thermal background "noise" because it is cooled at a higher temperature than are conventional super-conductors. Even if the new material proves to have more background noise, it might be possible to accommodate that through circuit design or by altering cooling levels.

Semiconductor-superconductor hybrids. Most superconductors are also good semiconductors above the T_c. This technology would marry the current capabilities of the semiconductor with those offered by the new super-conductors. One way to hyribize both systems would be to take an existing computer and use superconducting technology to make the interconnects with the power distribution lines that service the chips. First, critical

current density and film stability (for annealing and moisture) must
be improved. Since the superconductors have to be heated to about 900
degrees Celsius after being desposited on a wafer or chip, they undergo
temperatures that semiconductor circuits probably wouldn't be able to
withstand. (5-10 years until commercialization of ceramics)

SQUID (superconducting quantum interference devices). Simply two Josephson
junctions connected in a ring used to measure extremely faint magnetic
fields with applications in magnetic resonance imaging and biomagnetism
studies, as well as in the search for gravity waves and magnetic monopoles.
Here thin film superconductors would need to be developed. Experimental
high-temperature SQUIDS operating in liquid nitrogen are already being
designed. However, the most sensitive SQUIDS at 4.2 K can detect 10^{-6}
of a flux quantum.

Magnetic Resonance imaging (MRI). Here the key benefit for these high
resolution imaging devices for soft body tissues would be in switching
from liquid-helium to liquid-nitrogen cooling, which could save in yearly
cooling costs, although the change probably would not cut costs substantially in the million dollar magnetic systems. (5-10 years until
commercialization of ceramics)

Biomagnetism. Cooling with liquid nitrogen will enable researchers
in this discipline, to replace a bulky liquid-helium container system
with a more flexible liquid-nitrogen system. This would allow closer
placement of sensors to various parts of the body, thus increasing
sensitivity.

Tokamak superconducting magnetics. When fusion power is finally achieved,
within the configuration of the Tokamak, there are designs for 8-tesla
superconducting magnets. (10-20 years before commercialization of ceramics)

Particle accelerators. Here the superconducting Tevatron at the Fermi
National Accelerator Laboratory has demonstrated by superconducting
magnetic, the achievements of higher particle energy with much lower
electric power input. (10-20 years)

Motor generators. Los Alamos National Laboratory has been working on
a demonstration scale motor based on ceramic superconductors. This work
is expected to lead to a compact, extremely efficient motor using an
improved design strategy that eliminates both the iron cores and copper
windings characteristic of contemporary motors. The new compounds can
be used to develop a new class of wires, motors, electronic devices
and electromagnets that operate with unprecedented efficiency when
refrigerated with inexpensive liquid nitrogen. The new ceramic components
properties that include the ability to carry large amounts of current
and operate in strong magnetic fields as well as store energy. First
the brittle ceramic components must be developed into electrical machinery.
If successful, this improved technology could benefit power generation,
ship propulsion, heavy manufacturing, and perhaps transportation.
(15-20 years)

Launch vehicles. Linear motor assisted take-off system, via a superconducting platform could launch a horizontal take-off space vehicle.
(20 years plus)

CONCLUSION:

Research on material characteristics and exploration of alternative fabrication techniques for ceramic materials is being currently studied throughout the world. In some instances, in order for a material to obtain critical fields and current densities it must operate at one-half to three-quarters of the critical temperatures. So although these new materials may be superconducting at 125 K, the systems would have to be designed for 100 K to get desirable effects. However, other ceramics have, for example, a limited superconducting range between 121 K and 107 K. Progress in understanding these materials, their potential for improvements, and fabrication methods that increase stability and performance is essential before the the the suitability of ceramic super-conductors for electric power applications, such as transmission cables, magnetic coils, and other utility equipment, can be assessed and identified with end-use applications.

Our challenge is also to look for new opportunities for high-temperature superconductors, rather than simply see these materials only as substitutes for conventional uses. The greatest effect may come not by changing the way we make harness present-day utility equipment but rather by creating productive new uses for electricity and power where none existed before.

The changes in superconductivity have surpassed the changes in most other research fields or disciplines. Growth has come rapidly. We now have the potential to look positively towards high-temperature super-conductors. The importance of the electric power industry alone justifies committing significant resources, even at this early stage of technology. In addition to superconductivity, the entire structure of SMES and MVA generators must be developed to meet the growing energy needs of planet earth.

REFERENCES:

1. University of Wisconsin in cooperation with the US Energy Department designed one of the first SMES prototypes in the early 1970's. In addition, General Motors (Warren, MI); Los Alamos Scientific Laboratory also designed early prototypes. (1975-1976) Also Institute of Atomic Energy, Moscow, USSR, 1975.
2. Wood,P., AC/DC Power Conditioning and Control for Advanced Conversion and Storage Technology, final report prepared by Westinghouse Electric Corporation for EPRI under project 390-1-1, August 1975.
3. EPRI, Conceptual Design and Cost of a Superconducting Magnetic Energy Plant, EM-3454, 1984, p3-5.
4. Ibid., p.3-13.
5. EPRI, Evaluation of Superconducting Magnetic Energy Storage Systems. EM-2861, February 1983, p. 3-1.
6. Ibid., Table 4-5, p4-15.
7. Ibid, p.4-26.
8. Ibid, p. 4-8 and 4-9
9. 300-MVA Superconducting Generator Documentation: Heat Transfer to Helium Fluid under High Speed Rotation, EL-4910, Westinghouse Electric Corporation, 1986, p.2-26 and 2-27

10.Westinghouse Electric Corporation, "Metallurgical Development and Manufacture of Large Heavy Structures for Cryogenic Applications" Final Report, February, 1987, p. 349 / Inconel is made by the Inco family of companies.

11.James L. Kirtley, "Supercool Generation" IEEE Spectrum, April 1983, pp. 28-35

12.300-MVA Superconducting Generator Documentary: Heat Transfer to Helium Fluid under High Speed Rotation, EL-4910, Westinghouse Electric Corporation, 1986, p. S-2

13.Westinghouse Electric Corporation, "Development of Telemetry for High Speed Rotor Instrumentation and Monitoring" Final Report, June 1987.

14.Guanhua Chen and William Goddard III, "The Magnon Pairing Mechanism of Superconductivity in Cuprate Ceramics,Science, Feb, 19, 1988 pp. 901-902

15.Both Paul Chu (Houston) and Chinese Academy of Science in Beijing, shortly after Muller and Bednorz findings, substituted Sr^{2+} for La^{3+} and reached 40-46 K.

16.According to M.A. Subramanian (Science, Feb 26, 1988): $YBa_2Cu_3O_7$, T about 95 K has a 1-2-3 phase which is a "line" compound with separate sites for each of the Y, Ba, Cu and O atoms and with intrinsic metallic conductivity. The structure is orthorhombic with an extended c-axis of about 11.7 angstroms where the a and b axes are about equal. The structure becomes tetragonal and semiconducting when the oxygen stoichiometry is reduced from 7 to about 6.5

17.M.A. Subramanian, et al. "A New High-Temperature Superconductor: $Bi_2Sr_{3-x}Ca_xCu_2O_{8+y}$" Science. vol 239. February 26, 1988, pp.1015-1017

18.T.H. Geballe and J.K. Hulm, "Superconductivity--The State That Came in from the Cold" Science. vol 239, January 22, 1988 pp370-371.

19.Guanhua Chen and William A. Goddard III, "The Magnon Pairing Mechanism of Superconductivity in Cuprate Ceramics" Science. vol. 239, February 19, 1988 pp 899-902.

20.Discussions on the modeling of Superconductivity with Paul Ponish, (Silicon Valley) California, Feb. 1988.

21.Discussions with John Milewski, Los Alamos National Lab, New Mexico, Jan. 1988.

22.Conversations with John Hulm at EPRI February, 1988).

Techniques for Thermal Storage of Phase Change Metal

ZHIGUANG HUANG, GUANGZHONG WU, SILONG XIAO, SHAOHUA MEI,
and GUOQING XIONG
Department of Mechanical Engineering
Huazhong University of Science and Technology
Wuhan, PRC

ABSTRACT

A new device used for thermal storage and exchange of phase change metal
has been investigated in this paper. The volume of device is about 0.0072
M^3 and it can store and export heat 11613.6KJ.(3.23Kw.h). That is, the
density of storage heat is about 1608.5MJ/M^3(446.8Kw.h/M^3). The device
can be adopted as a household electric energy storage stove.
At the same time, the characteristic of thermal storage and export and the
service life of the Al-Si, Al-Mg-Si and Al-Cu-si eutectic alloy have been
measured systematically. Through the experiments and analyses, it indicated that techniques for thermal storage of phase change metal possess
promising prospects for applications.

INTRODUCTION

The utilization ratio of energy can be raised by using techniques of effective heat energy storage especially in the high temperature, Therefore
the techniques have application in these fields such as recovering the
waste heat, adjusting the load of electric power and changing the disconti-
nous energy such as solar energy into the continous energy.
The techniques for utilizing wax and salts for thermal storage by phase
change (solid=liquid) have been developed by Beghi [1], Beckman, Gilli
[2], Gary [3] and others. But metals (or alloys) as phase change materials
for thermal storage have more advantages than wax and salts, especially
when high temperature of storing and shorted time of the charge- uncharge
period is required. Birchenall[4][5]. Kauffman [6]. Mobley [7] have made
a systematic study in the field of the phase change metal (PCM) for thermal storage. But the applied techniques for thermal storage of PCM ahve
reported rather few.
Any phase change (latent heat) energy storage system should have three
components; a phase change material; a container and a heat exchange device.
A new device used for thermal energy storage and exchange of PCM has been
investigated in this paper. The storage heat of the device is about
11613.6KJ(3.23Kw.h). The device can be used as a small electric energy
storage stove.

EXPERIMENTAL METHOD AND RESULT

1. Device.

The device is as shown in Fig.1. There are three kinds of element used

693

for test(see Table 1). The outer shell of the elementes made of Fe-C alloy as a container and Al-12.6Si eutectic alloy as a core filled within the shell. During the cycle of heating and cooling between 800 C and 300 C, PCM is melted and overheated for storage of the sensible heat and latent heat, the outer shell is used as a container in this range of temperature. The shell itself, of course, can store the sensible heat too. Through the heat exchange system, the capacity of thermal storage of the device can be obtained by heating a given weight of water. The experimental results are listed in Table 1. It shows that:

a, The volume of the device is about $0.0072M^3$ ($\phi 200*230$mm in size), the element No2 can export heat 11613.6KJ(3.23Kw.h) between 800 C and 300 C. That is, the density of storage heat is about $1608.5MJ/M^3$($446.8Kw.h/M^3$). The device can heat 32Kg water from 16 C to 100 C (As well heat 3.2Kg water in lower 100 C), and that is quiet enough for household cooking.
B, The capacity of thermal storage is affected by the shape and size of the element. The overall sizes of element No2 and element No1 are identical but element No2 has a hole at its center for increasing the area of heat exchange. Therefore, the capacity of export heat of the element No2 is higher then that of element No1, although the element No.1 possesses the highest theoretical capacity of heat storage. On the other hand, the efficiencies of heat export are increased from 57.8% to 89.8% with their void ratios rise from 20.8% to 31.1%.
C, Besides the result listed in Table 1, when the range of temperature of heating-cooling cycle is from 680 C to 300 C(using the element No3) the capacity of export heat is about 7684.6KJ, but the capacity is about 9106.1KJ When the temperature range is from 800 C to 300 C, the latter is 18.5% higher than the former. Similarly, the higher the temperature, the larger the rate of heat export will be. During the range of temperature in the deive is from 800 C to 760 C, it takes 11 min. to heat 3.2kg water from 16C to 100 C. However, when the range of temperature drops down to 602C-555C, it will spend 32 min to do the same thing.

2. The capacities of thermal storage of PCM .

Four kinds of eutectic alloy such as Al-Si, Al-Mg-Si, Al-Cu-Si and Zn-Al alloy are chosen as the PCM in this experiment.
A, Measuring the parameters of capacity of thermal storage of PCM.
The latent heat of PCM is measured by the method of DSc (Differential Scanning Calorimetery, type thermalflex 8081 Japan) and the method of ICC (isothermal Cupper-Block Calorimeter, see Fig.2). The specific heat of solid and liquid is measured by ICC. The result is listed in Table 2. It shows htat the Al-Mg-Si, Al-Si, Al-Cu-Si eutectic alloys are arranged in the order to their density of heat storage and exergie. If the temperature of phase change is at 500 C-600 C, Al-Mg-Si, Al-Si and Al-Cu-si alloy will be chosen to use.
Though the specific heat and latent heat of Zn-Al alloy is lower its density of thermal storage can be equal to 67-72% that of Al-Si alloy, because specific gravity of the former is higher. Hence if the temperature of phase change is at 380-400 C, Zn-Al alloy will be chosen to use.
The latent heats have some difference between determined by ICC and DSC. But the method of ICC is similar to the service condition of the element of PCM and can be measured again and again.

B, The cycle of heating-cooling of PCM
The element No 3 was filled with Al-Si alloy, or Al-Mg-Si alloy, or Al-Cu-Si alloy as a core, it was then heated and cooled repeatedly between 800 C and 300 C. At the same time, the thermal analysis curves of those three al-

694

Table 1

Shape and Size of Elements

Parameters of thermal exchange			
Num. of element	No 1	No 2	No 3
Counts	2	2	34
Export heat in charge Q1. KJ	10984 (3.05Kwh)	11613.6 (3.23Kwh)	9106 (2.53Kwh)
Export heat discharge Q2.KJ	3898	3898	4319.7
Export heat of element Q3. KJ	7086	7715.6	4759.3
Weight of element Kg	22.17	22.08	10.29
Export heat per Kg element KJ/Kg	319.6	349.4	462.5
Theoretical Q4 storage heat KJ	12252	12001	5295.7
Percentage in weight of PCM element %	35.4	33.8	31.9
Efficiency of heat export %	57.8	64.3	89.8
Void ratio* %	20.8	22.5	31.1

*Void ratio=(Vdevice-Velement)/Vdevice

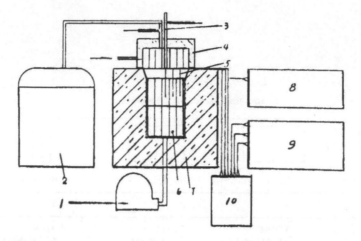

Fig.1. Heat exchange device

1. ELECTRIC FAN
2. STORAGE WATER DEVICE
3. THERMOMENTER
4. EXCHANGE HEAT DEVICE
5. THERMOCOUPLE

6. STORAGE HEAT ELEMENT
7. HEATING FURANCE
8. CONTROLLOR OF TEMPERATURE
9. ICE BOTTLE
10. MULTIPOINT RECORDER

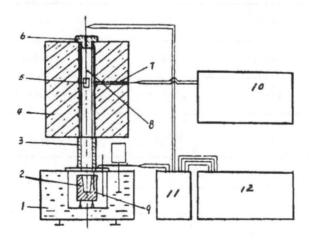

Fig.2. Isothermal cupper block calorimeter

1. ISOTHERMAL DEVICE
2. CUPPER BLOCK
3. GUIDE TUBE
4. HEATING FURNACE
5. SAMPLE
6. ISOLATION HEAT COCK

7. THERMOCOUPLE
8. THERMOCOUPLE
9. THERMOMENTER
10. CONTROLLOR OF TEMPERATURE
11. ICE BOTTLE
12. MULTIPOINT RECORDER

Table 2

Alloy	Al-12.6Si	Al-27Cu-5.25Si	Al-5Mg-13.2Si	Zn-4Al
Density g/cm³ d	2.609	3.315	2.554	6.703
Solid heat capacity Cps J/g.K	1.033	0.871	1.12	0.44
Liquid heat capacity Cpl J/g.K	1.475	1.430	1.243	0.595
Latentheat L J/g	461.3(ICC) 504.3(DSC)	360.2(ICC) 393.9(DSC)	530.6(ICC) 498.5(DSC)	135.0(ICC) 131.7(DSC)
Melting point Te K	843-846 848-853	790-793 792-797	825 827-835	654-658
Density of Storage heat MJ/m³	1203.5 1315.7	1194.1 1305.8	1355.1 1273.2	838.2 882.9
Density of exergie MJ/m³	781.0 853.9	746.3 816.1	868.6 816.1	457.6 482.0

* Ex=L*d((1-(Tr+273)/(Te+273))
 Tr: room temperature
 Te: eutectic temperature

Table 3

Condition	Heating		Cooling	
Al-Si times	24	60	23	60
alloy Te C	568-576.5	574.8-570.5	580.5	571.75-573
te min	26.5	22	78	83
ΔT C	2.5	4.3	0	1.25
Al-Cu-Si times	22	57	21	58
Te C	517.6-520	512.8-514	508.2-515.2	512-514
alloy te min	23.2	19.9	61	74
ΔT C	2.4	1.2	7	1.2
Al-Mg-Si times	22	58	21	56
Te C	558.6-560.2	570.5-540.4	550.5-554	0
alloy te min	23	12.5	75	0
ΔT C	1.6	30.15	3.5	-

loys were measured. Since those three alloy being eutectic alloy, the change of eutectic transformation time can reflect the change of chemical compositions.

Once the composition of eutectic alloy changes, the eutectic transformation time becomes shorter and the range of solidification temperature between starting and ending stage will be larger in the thermal analysis curve. The result is listed in Table 3. It states that, after heated-cooled 58 times, the eutectic transform platform disappears and the range of solidification temperature increases to 30.15 C, for the Al-Mg-Si alloy. By the metallographic test, magnesium oxide and Fe-phase can be observed. The compositions of Fe-phase are consist of Mg 4.7% Al 49.7%. Si 15.8%, Fe 29.8% measured by the electron probe test. So the service life of Al-Mg-Si alloy becomes shorter. However, after heated-cooled 60 times the range of solidification temperature of the Al-Si alloy enlarges only slightly and the sillion floating up is observed by the metal lographic test. For the Al-Cu-Si alloy, the segregation of silicon and cupper is smaller and the range of solidification temperature has a descreasing tendency . So the service life of al-Cu-si alloy will be the longest among those three eutectic alloys.

C, Service characteristics of the element.
Thermal fatigues resistance--when the elements No3 were heated and cooled in the range of 640 C to 40 C repeated 302 times, the variations in size or in weight were equal to about 1.%. And no cracks were observed on the surface of the element.
Deformation at high temperature--After the elements were kept at 640 C for 304 hours. the deformative values were less than 1%.
Mechanical properties--The tensile strength of the outer shell material is equal to 400-600 Mpa and elongation is equal to 4-8% at room temperature. However, the tensile strength is equal to 34-38 Mpa, and elongation is equal to 10-20% at 820 C.
Anti-Fe permeability--After holding the element at 640 C for 304 hours, An intermediate layer of FeAlSi was found in the surface of the core. The micro-hardness of intermadiate layer is about Hv633 and the thickness is up to 2mm. However, the diffusion of Ferro will be basically controlled by adopting a special treatment technique.

DISCUSSION

1. The application prospect of the technique of thermal storage of PCM. Both of salt categories and eutectic alloys can be used as materials of phase change for thermal storage. But the adventages used the PCM could be found according to the analysis of exergie Ex as following:

$Ex=V^*r ((Cp(Te-T_{\ell})+L)^*((1-(Tr+273)/(Te+273))$
here:
V: Volume cm; r density g/cm³
Cp: specific heat of solid J/g.C
Te: eutectic temperature C;
Tr: room temperature C
L: latent heat J/g

Comparing the experimental results of the Al-Si eutectic alloy with that of the sodium salt or calcium salt, it indicated that;
A, The storage thermal density in unit volume r.L. For Al-Si eutectic alloy it is about 1203.5MJ/m³ to 1315.7 MJ/m³. However for sodium salt or calcium salt, it is about 403 MJ/m or 558MJ/m³. The former is larger

over 2 times than the later.

B, The temperature of phase change of Al-Si alloy Te is 577 C; But for sodium or calcuim salt Te is 297 c. or 490 C. Their mass-energy ratios (1-(Tr+273)/(Te+273)) are 0.648 (Al-Si alloy), 0.473 (sodium salt), 0.607 (calcium salt) respectively.

C, According to the results of above experiments, the higher the tempera ture, the faster the rate of heat export will be. So the Al-Si alloy has better capacity of heat exchange than that of salts. Besides, the PCM have other advantages such as: the ratio of volume change is about 7% during melting; the undercooling degrees are 1-8 C; good thermal stabili- ty; low tendency to segregation; non-texic and non-corrosion.

2. The selection of the PCM.

According to the capacity of thermal storage, the Al-Mg-si electric alloy is the highest among these three eutectic alloys. But considering the service life, the Al-Cu-si eutectic alloy is the best. However, if con- sidering the capacity of thermal storage, service life and cost, the Al- Si alloy is preferable to choose.

Besides, The higher the percentage in weight of the PCM in the element, the larger the capacity of thermal storage will be. However according to the demand of service life, it hoped that the wall thickness of outer shell of element is large enough to prevent deformation of element at high tem- perature. So the percentage in weight of the PCM will be selected by con- sidering both capacity of thermal storage and service life.

CONCLUTIONS

1. A new device used for thermal storage and exchange of PCM has been ob- tained. The volume of the device is about 0.0072 M³ and store and export heat 11613.6KJ(3.23Kw.h). The dnesity of thermal storage is about 1608.5 MJ/m³(446.8kw.h/m³).

2. Considering comprehensively the capacity of thermal storage, service life and cost, the Al-Si eutectic alloy is the best selection for use of thermal storage of PCM.

3. Though the latent heats have some difference between determined by ICC and DSC, the method of ICC is similar to the service condition of the element of PCM. So it is still a very useful method.

ACKNOWLEDGEMENTS:

This research was finished in supporting of scientific fund of Huazhong (Center China) University of Science and Technology. The DSC test was performed in the test center of Wu Han Industrial University.

REFERNCES

1. Beghi, G., Thermal Energy Storage, ed. pp.1-40. D.Reidel Publishing Company Dordrecht/Boston/Lancaster 1981.

2. Beckman, G., Gilli, P.V., Thermal Energy Storage,ed.pp 40-50 Spring Verlag Wien, New York 1984.

3. Gary, H.P. Mullick, S.C., Bhargava, A.K,Solar Thermal Energy Storage, ed. pp 154-182. D. Reidel Publishing Company Dordrecht/Boston/Lan- caster 1985.

4. Birchenall, C.E., Riechman, A., Heat Storage in Eutectic Alloys, Metall. Trans. Vol.11A, pp 1414-1420, Aug. 1980.

5. Farkas, Diana. Birchenall, C.E. New Eutectic Alloys and Their Heats of Transformation, Metall. Trans, Vol 16A pp323-328 March 1985.

6. Kauffman, K.W., Lorsch, H.G, Design and Cost of High Temperature Thermal Storage Devices Using Salts or Alloys preprint 76 WA/Ht-34 ASME Dec. 1976 Flanklin Institute Research center.

7. Mobley, c.E., Rapp,R. Hypereutectic Heat Storage Shot. The Ohio State University 1983 unpublished paper.

Heat and Mass Transfer Analysis of Metal Hydride Beds

DA-WEN SUN, ZU-XIN LI, and SONG-JIU DENG
Chemical Engineering Research Institute
South China University of Technology
Guangzhou, PRC

ABSTRACT

A numerical model for two-dimensional transient heat and mass transfer with internal heat sources and variable coefficients is presented. By choosing proper dimension-control factors and shape-control factor, the model can be used for one- or two-dimensional heat and mass transfer processes in different coordinate systems. By solving differential heat balance equations with alternating direction implicit method, it is possible to predict temperature profiles, hydrogen concentration profiles and cumulative reacted hydrogen profiles. The numerical results of dehydriding reaction in MlNi4.5Mn0.5 cylindrical hydride bed are in good agreement with the experimental data.

INTRODUCTION

Metal hydrides as energy transforming materials have many prosperous usages. They can be used as heat pumps, heat transformers, refrigerators, thermal hydrogen compressors, ect. One of the technical keys in these applications is the heat transfer enhancement in metal hydride beds and the optimum design of reactors. Therefore, it is very important to analyse the heat and mass transfer characteristics of metal hydride beds.

Several models describing the heat and mass transfer process in a hydride bed have been presented by now [1-5]. Unfortunately, almost all these models considered the process as one-dimensional problem and all these models assumed the physical properties of hydrogen and metal hydrides as constants. many experiments show the fact that the values of some properties change in hydriding or dehydriding reaction and the variations of these values (e.g. the effective thermal conductivity in hydride bed) actually have severe effects on the heat and mass transfer process. Therefore, above models can not reflect the practical cases well. Considering the above facts, the authors presented a more practical one-dimensional model in 1987 [6]. The model not only includes the effect of the effective thermal conductivity in the hydride bed, but includes the effects of the physical properties of the hydride and other operating parameters.

In order to extend the use scope of the model, in present study a two-dimensional model is put forward. In this new model, the general equations for effective thermal conductivities in hydride beds, P-C-T curves and hydriding or dehydriding kinetics of metal hydrides are used . In order to

701

check the new model, a testing system was constructed to measure temperature profiles and cumulative discharge hydrogen profiles generated during reactions in hydride beds. The measurements of the dehydriding reaction in MlNi4.5Mn0.5 cylindrical hydride bed (Ml:Lanthanum-rich misch metal, developed by our University) show that the numerical results are in good agreement with the experimental data.

HEAT AND MASS TRANSFER MODEL

The following differential heat balance equation is derived to solve the two-dimensional heat transmission problem in metal hydride beds:

$$K_e(\frac{\partial^2 t}{\partial r^2}) + K_e\frac{\lambda}{r}(\frac{\partial t}{\partial r}) + K_e(\frac{\delta}{r^2} + \gamma)\frac{\partial^2 t}{\partial z^2} - \rho_e\chi_h\Delta H = \rho_e C(\frac{\partial t}{\partial \theta}) \quad (1)$$

The item $-\rho_e\chi_h\Delta H$ is greater than zero in the hydriding reaction and is less than zero in the dehydriding reaction, and the heat of reaction ΔH is always less than zero in both reactions. t is the absolute temperature. λ is the shape-control factor. γ and δ is the dimension-control factors. The combinations of different λ, γ and δ describe the following heat and mass transfer processes:
(1) $\delta = 0, \gamma = 1, \lambda = 1$ for two-dimensional process in cylindrical coordinates(r coordinate, z coordinate);
(2) $\delta = 0, \gamma = 1, \lambda = 0$ for two-dimensional process in rectangular coordinates(r for x or y coordinate, z coordinate);
(3) $\delta = 1, \gamma = 0, \lambda = 1$ for two-dimensional process in cylindrical coordinates(r coordinate, z for φ coordinate);
(4) $\delta = 0, \gamma = 0, \lambda = 0, 1$ or 2 for one-dimensional processes in rectangular, cylindrical or spherical coordinates.

For r=0 in cylindrical coordinates,the values of items $K_e\frac{\lambda}{r}\frac{\partial t}{\partial r}$ and $K_e\frac{\delta}{r}\frac{\partial^2 t}{\partial z^2}$

are indefinite. They can be estimated by using the L'Hospital law. For this special case Eq(1) is rewritten as:

$$K_e(1+\lambda)\frac{\partial^2 t}{\partial r^2} + K_e\frac{\partial^2 t}{\partial z^2} - \rho_e\chi_h\Delta H = \rho_e C(\frac{\partial t}{\partial \theta}) \quad (2)$$

In the process of hydriding or dehydriding reaction, metal hydrides crack into very small particles. The effective density of the hydride bed ρ_e is defined as:

$$\rho_e = (1-\epsilon)\rho_m \quad (3)$$

where ϵ is void fraction of bed, ρ_m is density of metal hydride.

The initial conditions for Eq(1) and (2) is given by

$$t(r,z,0) = t_I , x(r,z,0) = x_I ; r_{in} < r < r_{out}, z_{down} < z < z_{up} \quad (4)$$

The boundary conditions can be the different combinations of constant temperature boundary conditions and convection boundary conditions according to the specific configurations of hydride beds.
for convection boundary conditions:

$$K_{ei}\frac{\partial t}{\partial r}\bigg|_{r=r_{in}} = U_i[t(r_{in},s,\theta)-t_w]; \quad s_{down} \leq s \leq s_{up}, \theta > 0 \tag{5a}$$

$$K_{eo}\frac{\partial t}{\partial r}\bigg|_{r=r_{out}} = -U_e[t(r_{out},s,\theta)-t_w]; \quad s_{down} \leq s \leq s_{up}, \theta > 0 \tag{5b}$$

$$K_{ed}\frac{\partial t}{\partial s}\bigg|_{s=s_{down}} = U_d[t(r,s_{down},\theta)-t_w]; \quad r_{in} \leq r \leq r_{out}, \theta > 0 \tag{5c}$$

$$K_{eu}\frac{\partial t}{\partial s}\bigg|_{s=s_{up}} = -U_u[t(r,s_{up},\theta)-t_w]; \quad r_{in} \leq r \leq r_{out}, \theta > 0 \tag{5d}$$

for constant temperature boundary conditions:

$$t(r_{in},s,\theta) = t_{in}; \quad s_{down} \leq s \leq s_{up}, \theta > 0 \tag{6a}$$

$$t(r_{out},s,\theta) = t_{out}; \quad s_{down} \leq s \leq s_{up}, \theta > 0 \tag{6b}$$

$$t(r,s_{down},\theta) = t_{down}; \quad r_{in} \leq r \leq r_{out}, \theta > 0 \tag{6c}$$

$$t(r,s_{up},\theta) = t_{up}; \quad r_{in} \leq r \leq r_{out}, \theta > 0 \tag{6d}$$

U in Eq.(5) is the overall heat transfer coefficient. The above conditions become adiabatic boundary conditions for U=0. In general, U_d and U_u are expressed by

$$U_d = [1/h_d + (s_{down}-s_{wd})/K_{ms}]^{-1} \tag{7a}$$

$$U_u = [1/h_u + (s_{wu}-s_{up})/K_{ms}]^{-1} \tag{7b}$$

The expression of U_i and U_e are different for slab hydride beds and cylinder hydride beds.
for slab beds:

$$U_i = [1/h_i + (r_{in}-r_{wi})/K_{mr}]^{-1} \tag{7c}$$

$$U_o = [1/h_o + (r_{wo}-r_{out})/K_{mr}]^{-1} \tag{7d}$$

for cylinder beds:

$$U_i = \left[\frac{r_{in}}{r_{wi}h_i} + \frac{r_{in}\ln(r_{in}/r_{wi})}{K_{mr}}\right]^{-1} \tag{7e}$$

$$U_o = \left[\frac{r_{out}}{r_{wo}h_o} + \frac{r_{out}\ln(r_{wo}/r_{out})}{K_{mr}}\right]^{-1} \tag{7f}$$

where K_m is the conductivity of reactor vessel, h is the surface heat transfer coefficient. For spherical bed in one-dimensional problem, U_i takes the value 0, and U_i also takes 0 for r_{in} =0.

The reaction heat is removed or provided by the cooling or heating fluid. When fluid is in forced turbulence flow, h in Eq(7) is represented by

$$hd_e/k = 0.023(d_e u \rho/\mu)^{0.8}(c_p\mu/k)^{\alpha} \tag{8}$$

where d_e is the equivalent diameter of fluid channel. Exponent α takes the value 0.4 for hydrogen charging process and 0.3 for discharging process.

The effective thermal conductivity Ke in Eqs(1) and (2) is severely affected by the hydrogen pressure and hydrogen concentration. We can use the theoretical Ke model [7] developed by the same authors to describe Ke, we can also use the following general form derived from the Ke model for Ke expression:

$$K_e = A_0 + A_1 x + \frac{A_2 \, P_b}{A_3 + P_b} + \frac{A_4 P_b + A_5 x P_b}{A_6 + A_7 x + A_8 P_b + A_9 x P_b} \tag{9}$$

The coefficients A_0 – A_9 have different values for different kinds of metal hydrides. They are determined either by inserting the fundamental physical data of hydride into the Ke model or by regressing from the experimental data.

γ_h in Eqs(1) and (2) is the reaction rate of hydride bed. All the reported heat and mass transfer models used the specific forms to describe the reaction kinetics basing on the considered hydrides, and hence these forms are different one another. After analysing the reported kinetic experimental data for kinds of hydrides, the following general forms for kinetic expressions are given:
for hydriding reaction:

$$\frac{dx}{d\theta} = \eta \left(\frac{P_b - P_e}{P_e}\right)^m \left(\frac{x - x_m}{x_0 - x_m}\right)^n \, e^{-E/Rt} \tag{10a}$$

for dehydriding reaction:

$$\frac{dx}{d\theta} = \eta \left(\frac{P_e - P_b}{P_0}\right)^m \left(\frac{x - x_m}{x_0 - x_m}\right)^n \, e^{-E/Rt} \tag{10b}$$

where x is the hydrogen concentration(i.e. hydrogen to metal atom ratio) and x_m is its maximum value. p_b is the bed hydrogen pressure. The coefficients η, m,n,ΔE are determined by regressing from the experimental data. They will have different values for charging or discharging process and for different kinds of hydrides. Therefore γ_h is expressed as:

$$\gamma_h = \frac{1000N}{2M} \frac{\partial x}{\partial \theta} \tag{11}$$

where N is the number of metal atoms per molecule of hydride and M the molecular weight of hydride.

p_e in Eq (10) is the equilibrium pressure in hydriding or dehydriding reaction. All the reported heat and mass transfer models often used Van't Hoff expression to describing p_e. Unfortunately, this expression only gives relationship between p_e and reaction temperature. So the following general form is put forward to express reasonably the P-C-T relations:

$$\ln p_e = A - B/t + E \tan \frac{\pi(x - x_m/2)}{x_m} \tag{12}$$

where the coefficients A, B, E are determined by regressing the experimental data.

The above equations form the heat and mass transfer model. It is a two-dimensional transient problem with internal heat sources and variable coefficients.

EXPERIMENTAL APPARATUS AND PROCEDURE

A testing system shown in Fig. 1 was constructed to measure temperature profiles and cumulative discharge hydrogen profiles. Since the lack of the physical properties of the hydride needed in the numerical calculation, the effective thermal conductivities, the P-C-T curves and the kinetic data are required to be measured in the experiments.

The cylindrical reactor was packed with MlNi4.5Mn0.5 alloy (488.57g). After evacuating the system for several times, hydrogen gas of 99.9% purity was introduced at a pressure 30-35atm and temperature about 300°C. This activation procedure spent about 2-3h. More than ten hydriding-dehydriding cycles were repeated before data were obtained. The temperature gradient in the hydride bed were measured over a period of 60 min. using a computerized data acquisition system with an adjustable scanning speed. The cumulative hydrogen flow rates were recorded using a flowmeter. The pressure-concen-

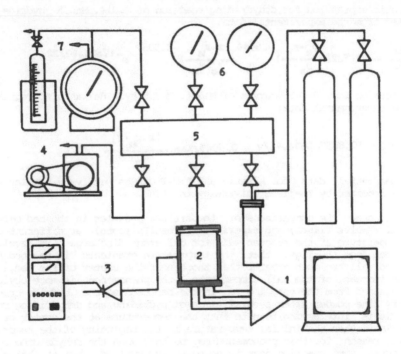

Fig. 1 A schematic diagram of the experimental apparatus

1. computerized data acquisition system for temperature; 2. reactors;
3. precision temperature control instrument; 4. vacuum pump; 5. pipes;
6. precision pressure gauge; 7. flow measurement system.

tration relation under various isothermal conditions and the kinetic data of MlNi4.5Mn0.5 hydride, and the effective thermal conductivities of the bed are also measured in the experiments.

NUMERICAL RESULTS AND EXPERIMENTAL DATA

The heat and mass transfer model can only be solved numerically. In the present study, the alternating direction implicit method was used. The model was transformed into dimensionless forms before programming. All the calculations were completed using the large capacity computer DPS8/49 and the physical properties of MlNi4.5Mn0.5 hydride used in the calculations were all taken from the experimental data.

Inserting the relative data of MlNi4.5Mn0.5 hydride bed into the theoretical Ke model[7], then the Ke expression was derived as follows:

$$K_e = 0.17 + 0.527x + \frac{0.1061p_b}{0.389 + p_b} + \frac{22.744p_b + 70.5064xp_b}{99.26 + 307.71x + 56.1p_b + 171.47xp_b}$$

The kinetic expression for dehydriding reaction of MlNi4.5Mn0.5 hydride was regressed from the experimental data:

$$\frac{dx}{d\theta} = -0.722\left(\frac{p_e - p_b}{p_0}\right)^{0.6544}\left(\frac{x - x_m}{x_0 - x_m}\right)^{-0.3882}e^{-17603.67/Rt}$$

The expression for P-C-T curves of MlNi4.5Mn0.5 hydride was also regressed from the experimental data:

$$\ln p_e = 10.8837 - 3516.46/t + 0.60432\tan\frac{(x - x_m/2)}{x_m}$$

The experimental data for kinetic and P-C-T curves were very similar with the data reported by Zhejiang University in China.

Figs. 2-4 show the comparisons of temperature profiles in the bed between numerical results (lines) and experimental results (dots) at different constant temperature at the reactor wall for different discharge time level. It can be seen from the Figs. that the temperature gradients in the bed near the heated wall are much greater than those near the center of the bed, and with the increase of the wall temperature, the phenomenon is more obvious. It is resulted from the very low Ke in the bed because the hydride crashed into very fine powder after several times of hydriding and dehydriding reactions. It is interesting to note that the temperature at the center of the bed is lower than the initial temperature in the beginning of the reaction. The main reason for this phenomenon may be that when the dehydriding reaction occurs, some reaction heat is necessarily supplied, but at that time, the heat supplied by the outside wall still has not reached the center of the bed because of the very low Ke. From the Figs, it can be seen that the numerical results are in good agreement with the experimental results.

Fig. 5 shows the hydrogen concentration profiles in the bed at a constant wall temperature 102°C for different discharge time level. If the curves shown in Fig. 5 and Fig. 4 are compared, the very important phenomenon that the heat transfer in the hydride bed is the main rate-controlling step in the reaction is observed. This means that an important technical key to quicken the reaction process and to shorten the reaction time is to enhance the heat transfer in hydride beds and to optimize the reactor designs.

Fig. 6 shows the comparisons of cumulative discharge hydrogen profiles between numerical results(lines) and experimental results(dots) at different constant wall temperatures. It is noted that with the increase of wall temperature and reaction time , the cumulative discharge hydrogen increases.

The Fig. also shows a good agreement between the numerical results and experimental results.

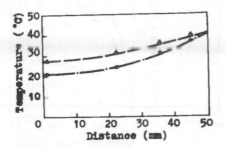

Fig. 2 Consistency of the numerical results(wall temperature 41°C, reaction time: ▲ —— 42.96 min.; • – · – 17.6 min.)

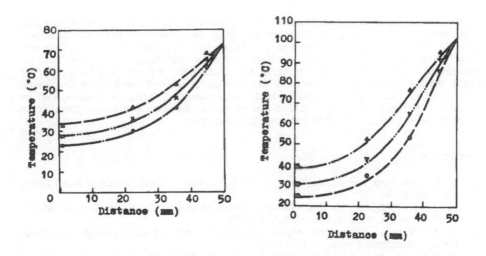

Fig. 3 Consistency of the numerical results (wall temperature 73°C, reaction time: ▲—— 49.5min.; ✕ –···– 31.5min.; • –·– 19.0 min.)
Fig. 4 Consistency of the numerical results (wall temperature 102°C, reaction time : ▲ –·– 55.4min.; ▼ –···– 34.2min.; • —— 20.2 min.)

707

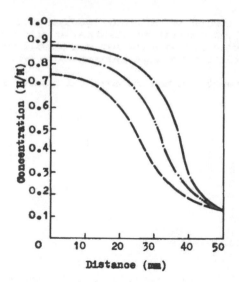 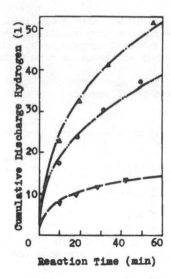

Fig. 5 Numerical results of hydrogen concentration in the bed (wall temperature 102°C, reaction time: -·- 20.2 min.; -··- 34.2 min.;   ── 55.4 min.)
Fig. 6 Consistency of the numerical results (wall temperature: ▲ -·- 102°C;
● -··- 73°C ; ▼── 41°C)

CONCLUSIONS

A two-dimensional numerical model for unsteady heat and mass transfer in
hydride beds was developed. The model can be used to describe the two-
dimensional heat and mass transfer problems in cylindrical or rectangular
coordinates and the one-dimensional problems in rectangular, cylindrical or
spherical coordinates. The calculated transient dehydriding behaviour of
MlNi4.5Mn0.5 cylindrical reaction bed (temperature and cumulative discharge
hydrogen) is in a good agreement with experimental data. The model can be
used as a helpful tool for finding the ways to enhance the heat transfer in
hydride beds and to optimize the reactor designs.

REFERENCES

1. D. L. Cummings and G. J. Powers, Ind. Engng. Chem. Process Des. Dev.,
    13, 182(1974)
2. W.S. Yu, E. Suuberg and C. Waide,"Hydrogen Energy(T. N. Veziroglu ed.)",
    Part A, p621, Plenum Press, N.Y.(1975)
3. I. A. ElOsery, Proc. 6th Miami Int. Conf. on Altern. Energy sources,
    p583, Pergamon Press, N.Y.(1985)
4. S. Suda, N. Kobayashi, E. Morishita and N. Takemoto, J.Less-Common Met.,
    89, 325(1983)
5. U. Mayer, M. Groll and W. Supper, J. Less-Common Met.,131, 235(1987)
6. Da-Wen Sun and Song-Jiu Deng, Paper Presented at 8th Miami Int. Conf.
    on Altern. Energy Sources, 14-16 Dec, 1987, Florida, USA, Proc.
    of Condensed Papers, vol. 2, p564 (1987)
7. Da-Wen Sun, Song-Jiu Deng and Zu-Xin Li, ibid, Proc. of Condensed
    Papers, vol. 2, p561(1987)

Study of a Fin-Tube Model for Latent Heat Storage

JIE HO and YAN SHONG
Department of Shipbuilding
South China University of Technology
Guangzhou, PRC

INTRODUCTION

Among the heat storage techniques, Latent Heat Storage possesses some advantages, such as higher heat storage density
and smooth output temperature during heat extraction, and is
widely used in thermal storage apparatus. Its heat conduction
process is a classical moving boundary problem, and has been
developed over many years.

In applications, due to irregular shape of container and the
temperature of cooling fluid depending on time, the difficulties in computing would be emphasized. A simplified method
applied in engineering design is the one dimensional treatment
[1],[2], but it will cause unacceptable error except for very
simple cases. Finite difference and finite-element method
for solving multidimensional problems are available. Those
numerical computations have different methods based on the
choice of dependent variables. One of the most interest is
the enthalpy method which has been investigated [3]. As stated
in [3], the interface between solid and liquid is eliminated
from consideration in calculation and the problem is made
equivalent to one of nonlinear heat conduction without change
of phases,so it can reduce the computing time. However, when
the Biot number is high where the heat extract rate is fast,
the computational effort is still hard for too complicated
calculations.

N.Shamsundar raised a similarity method, [4], for analysing
the multidimensional transient solidification problem. The
similarity rule is that the quantity $(1/Q-1)/Bi$ is a sole
function of the frozen fraction in the process of solidification, and independent of Bi and coolant temperature. It is
the base of this investigation. This rule enables the solution of some intractable problems and leads to considerable
reduction in computational time.

Althought there are many investigations on heat conduction of

709

solid-liquid phase change problems, but their objects are of cylinders[5],capsules[6] and bare tubes[7]. The intention of this investigation is applying the similarity rule to solve the whole heat transfer process in a fin-tube complex unit, and thus developing the practical usage in engineering.

ANALYSIS AND FORMULATION

An element is divided from a fin-tube, in which the coolant flows inside the tube, and outside the tube there is full of Phase Change Material (PCM). This element is composed of two half annular fins and a short tube which length equals to one spacing of fins. Along the outer radius of fins, it was assumed to be insulated. Therefore except the heat carried by the flow of coolant, there is no heat across the boundaries. The total length of the tube is much longer than that of the element. Within an element, the axial temperature gradients in the coolant and in the wall of tube compared with radial temperature gradient is small enough to be neglected, so that the temperatures at fin base and tube wall can be assumed the same.
More assumptions must be made as follows :

I.   The sensible heat of PCM is much smaller than its latent heat.
II.   The fins and the wall of tube are thin enough so that its heat capacity can be ignored.
III. There is a smooth interface between solid and liquid with no cavities in the solid and no superheat in the liquid.
IV.   The flow rate of coolant is assumed to be constant.
V.   The freezing rate is slow enough so that the temperature distributions in the fin and in the PCM can be regarded as quasi-steady.

During the heat extracting process, the solid PCM is formed on fins and the outer surface of tube, so the following analysis will be divided into two parts, fins and tube, and then combined. For fins, at a certain moment, the actual profile of the interface between solid and liquid PCM will be a curve, but following the assumption V.and taking a small enough time interval of solidification process,it is reasonable to assume that the freezing of PCM takes place in a plane parallel to the fin at any moment, thus the temperature distribution in the solid during freezing will be linear. On the tube, because of the assumption of uniform axial temperature,it is a simple one dimensional axisymmetric heat conduction problem. To connect the above two parts, a dimensionless variable is defined ϕ = Qfin/Qtub, and it is a function of time. The NTU and the Efficiency of Heat Exchanger Eff are defined as follows, they are slightly different from the usual definations.

NTU= 2πRihL/mCp

Eff = (Tfo-Tfi)/(Tsat-Tfi)

The basic equation of energy balance for an annular fin and the attached PCM:

710

$$- \frac{\partial}{\partial r}\left(2\pi \text{Kfin } wr \frac{\partial \text{Tfin}}{\partial r}\right) = 2\pi r\rho \text{Lpw} \frac{\partial \text{Ffin}}{\partial t} = \frac{\text{Ks } 2\pi r(\text{Tsat}-\text{Tfin})}{w \text{ Ffin}}$$

The dimensionless time of fin:

$$\tau_{\text{fin}} = \int_0^t \frac{\text{Ks(Tsat}-\text{Tfin)dt}}{\text{Lp } w^2}$$

Combine the above two equations, we get:

$$\tau_{\text{fin}} = (\text{Ffin})^2/2$$

and

$$\frac{\partial}{\partial r}\left[\frac{r \ \partial(\text{Ffin})^2}{2 \ \partial r}\right] = \frac{r \ \text{Ks}}{w^2 \ \text{Kfin}} \text{Ffin}$$

It is a partial differential equation and without analytic solution, but it can be calculated by munerical method to find the Ffin distribution along the radius of fin under a certain boundary conditions at any moment. At the base of fin, $\text{Ffin}^2 = (2\tau \text{fin})$. At the outer edge, two situations should be considered:one is that the freezing has not reached the outer edge of fin, so Ffin=0 at the intersection point between the freezing front and the surface of fin, the other is that the freezing over the whole fin and the boundary condition is $\partial \text{Ffin}/\partial r=0$ for an insulated edge. The total volume of solid PCM formed on one side of the fin is:

$$\text{Vs} = \int_1^{r'} \text{Ffin } 2\pi r'\text{wdr}'$$

From above equations, it is clear that both the distribution of Ffin and Vs are functions of τ fin only.
The heat conduction through a fin:

$$\text{Qfin} = 2\rho \text{Lp} \frac{d\text{Vs}}{dt}$$

Similarly, for freezing on tube, there are the following relations:

$$\text{Qtub dt} = - \frac{2\pi \text{Ks(Tsat}-\text{Tw)(wg}-2\text{wp})}{\ln(\text{Rc/Ro})} \text{dt} = \frac{2\pi\rho \text{Lp(wg}-2\text{wp)Rc}'}{\text{Ro}^2} \text{dRc}'$$

$$\tau_{\text{tub}} = \int_0^t \frac{\text{Ks(Tsat}-\text{Tw)dt}}{\text{Lp Ro}} = \frac{\text{Rc}'^2 \ln(\text{Rc}')}{2} - \frac{(\text{Rc}'^2-1)}{4}$$

From the definition of dimensionless time, it results in $\tau \text{fin(w)}^2 = \tau \text{tub(Ro)}^2$. The total heat trasfer to the coolant of the element:

$$\text{Qc} = 2\pi \text{Rih(Tw}-\text{Tf})/\text{nf} = \text{Qfin}+\text{Qtub} = (1+\phi)\text{Qtub}$$
$$= (1+\phi)2\pi\rho(\text{wg}-2\text{wp})\text{LpRo}^2 R'\text{dR}'/\text{dt} = \text{LpRo}^2\rho f(R')\text{dR}'/\text{dt}$$

711

$$= \frac{Tsat-Tf}{1/[2Ri(wg+w)h] + \ln R'/[(1+\phi)2\Pi Ks(wg-2wp)]}$$

where, $f(R')=(1+\phi)2\Pi(wg-2wp)R'$ is a function of τ fin only.
The heat balance for the total length of the tube:

$$m\ Cp\ \partial Tf/\partial x = nf\ Qc = 2\Pi RiLh(Tw-Tf)$$

The dimensionless time of a whole element:

$$\tau_{wh}= \int_0^t \frac{Ks(Tsat-Tf)}{\rho LpRo^2}dt= \int_1^{Rc'} [\frac{(1+\phi)(wg-2Ffin\ w)}{Bi\ (wg+w)} +\ln R']\ R'dR'$$

At a given time, the τ_{wh} then $f(R)$, ϕ and Ffin are calculated
in the relevant section. Thus for convenience of calculation,
the final reprensentations for a fin-tube can be rearrange as
follows (refers to [8] for the details of derivation and
computation):

$$Eff = 1- \int_1^{Rco'} f(R')dR'/ \int_1^{Rci'} f(R')dR'$$

$$-NTU = \ln(1-Eff)+ \int_{Rci'}^{Rco'} \frac{Bi\ \ln(R')\ f(R')dR'}{nf(1+\phi)(wg-2wp)\int_1^{R'} f(R')dR'}$$

$$= \ln(1-Eff)+ \int_{Rci'}^{Rco'} \frac{Bi\ \ln(R')(K3+K2)f(R')dR'}{(1+\phi)(K3-K2Ffin/Ro)\int_1^{R'} f(R')dR'}$$

Where $K1=(Rfo^2/Ro^2)-1$, $K2=2Ro/wg$, $K3=2Ro/w$, $Dk=Ks/Kfin$, which
are the dimensionless structural parameters. The result of
numerical calculations are shown by line 2 in the figures for
Bi is 0.5, 0.75, 1.0, 1.25 respectively. The line 3 was
obtained through calculation from the equations in [7] for a
bare tube under the same conditions. It is evident that
adding some fins to a bare tube can get a relatively uniform
Eff and heat output during heat extraction.

In the experiments, Naphthalene was used as PCM,its properties
were quoted from [9] . The model was made from copper, its
particulars are listed as follows:

K1=24   K2=0.4   K3=26.7   Dk=0.0041   Tsat=352.5 K
Lp=128.9 KJ/KG   Cs=1.006 KJ/KG C   Cp=1.05 KJ/KG C

Seven chromel-constantan thermocouples were set on the fin
along its radius, the other four were for measuring the temperature of tube wall and coolant at inlet and outlet. The
PCM was melted by an electric heater, then the liquid PCM was
cooled gradually down to the melting point. At this moment,
the air was blown into the tube at a constant rate to regulate
the specific value of Bi and the measurement was begun.

712

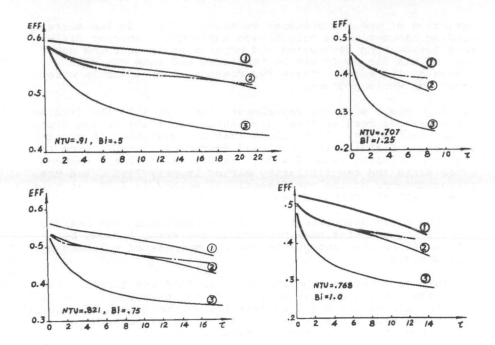

DISCUSSION

The results of experiments are shown by dash-dot line in the respective figures. It is in good agreement with the calculations for most values of τ when Bi is 0.5 or 0.75, and is larger when Bi is 1.0 or 1.25. The discrepancy could be explained as follows:

1. Superheating. At the starting point i.e.t=0, there were some nonuniform residual superheat in PCM, so it ruined the assumption III and cause errors when t is very small.

2. Bi and Ste. According to assumption I, the ste must be small enough to ensure that the sensible heat can be neglected, but in this experiment Ste is 0.70. this causes the result of calculation lower than that of experiments, furthermore, from the figures, it can be seen that the values of Eff from calculations decrease sharply when t is larger than a certain value, and meanwhile the bigger the Bi and Ste, the more evident this effect will be. This phenomenon coincides with the predictions of N.Shamsundar in [10] for bare tube. Fortunately, the moment that Eff drops sharply is near the end of solidification and it is of no practical meaning.

3. From the records of experiment, it can be seen that the temperatures of tube wall and fin base within an element are nonuniform and increased with Bi. When Bi=1.0, the maxium

deviation of the temperatures is about 1.7 C. So the corresponding assumption is not correct anylonger. Another difference between the assumption and experiment is that the outer
edge of the fin could not be insulated and some PCM will
freeze on its edge.   These two reasons cause a little error
cover the whole process.

4. To compare with the experiment, the line 1 in the figures
represent the results from calculation according to one dimensional method [1].   It is clear that the deviation is greater
than that of similarity model and decreases with τ . From the
figure of Bi=1.0, when τ = 9 , the deviation of Eff between
experiments and the similarity method is only 1.5% , and the
correspondence of the one dimensional method is 3.8%, at this
moment, 65% of PCM was frozen.

From the above discussions, it can be concluded that besides
simplifying computational procedure, the error induced by
this model is also acceptable for the purpose of engineering
applications.

The further study will be continued to find the largest value
of Bi that this method is still valid for practice .   It is
also to improve the model to make its assumptions and boundary
conditions more suitable to the conditions of experimnts,
thus reduce the errors for practical usage.

NOMENCLATURE

Bi	Biot Number, hD/K
Cp,Cs	Specific Heat of Coolant and Solid PCM
Ffin	Dimensionless Thickness of Solid PCM on fin,wp/w
h	Convective Heat Transfer Coefficient
Ks,Kfin	Thermal Conductivity of Solid PCM and Fins
L	Total Length of Tube
Lp	Latent Heat of PCM
m	Mass Flow Rate of Coolant
nf	Number of Fins per Tube
Qfin,Qtub	Heat Flux of fin and tube
Ri,Ro	Inner and Outer Radius of Tube
Rfo	Outer Radius of Fins
Rc	Outer Radius of PCM Frozen on Tube
Rco',Rci'	Rc' at Outlet and Inlet of Tube
r,R	Radius Coordinates of Frozen PCM on Fin and Tube
r',R',Rc'	Dimensionless Radius, r/Ro,R/Ro,Rc/Ro
Ste	Stefan Number, Cs((Tsat-Tf)/Lp
T	Temperature. Subscript : F-fin, w-wall, f-coolant, i- inlet, o-outlet, sat-melting point of PCM.
t	Time
Vs	Volume of Solid PCM
wg,w	Spacing and Half Thickness of fins
wp	Thickness of frozen PCM on Fins
ρ	Denity of Solid PCM

714

REFERANCES

1.   London,A.L. and Seban,S.A., "Rate for Ice Formation"
      Trans. ASME Vol.65, 1943, pp.771-778.

2.   Muehlbaner,J.C. and Sunderland,J.E ,"Heat Conduction
      with Freezing or Melting" Applied Mech. Reviews,
      Dec.1965, pp.951-959.

3.   Shamsundar,N.,Sparrow,E.M.,"Analysis of Multidimensional
      Conduction Phase Change via the Enthalpy Method" ASME,
      J.of Heat Transfer, Vol.97, 1975, pp.333-340.

4.   Shamsundar,N.,Srinivasan.R.,"A New Similarity Method for
      Analysis of multi-Dimensional Solidification"
      ASME, J.Heat Transter, Vol.101, 1979, pp.585-591.

5.   Song Von-Wang , J.of Engineering Thermophysics. China
      Vol.2, No.4, 1981, pp.359-365.

6.   Kozo Katayam et.al.,Solar Energy, Great Britain, Vol.27
      No.2, 1981, pp.91-97.

7.   Shamsundar,N.,Srinivasan.R., ASME, J. of Solar Energy
      Engineering , Vol.102, 1980, pp.263-270.

8.   Shong Yan, M.E. Thesis, South China Institute of
      Technology, 1988.

9.   Saito,A. et.al.,"On the Heat Transfer in the Latent Heat
      Energy Storage Capsule" ,Int. Heat Transfer Conf. 1982,
      EUB. ,pp.548-590.

10.   Shansundar,N., Srinivasan,R.,"Analysis of Energy Storage
      by Phase Change with an Array of Cylindrical Tube" ,
      ASME Winter Annual Meeting, San Francisco, 1987.

Performance Characteristics of a Natural Circulation Cooling/Latent Heat Storage System

KATSUMI SAKITANI
Mechanical Engineering Laboratory
Daikin Ind. Ltd.
1304, Kanaoka-cho, Sakai
Osaka 591, Japan

HIROSHI HONDA
Department of Mechanical Engineering
Okayama University
3-1-1 Tsushima-naka
Okayama 700, Japan

ABSTRACT

An experimental study has been made of the performance characteristics of a natural circulation cooling/latent heat storage system. This system was designed to cool a shelter in which electronic equipment is installed without using a power source. It consists of an outdoor condenser, an indoor evaporator, and an indoor latent heat storage unit, all of which are connected by piping. Lauric acid and refrigerant R-22 were used as the latent heat storage medium and the heat transport medium, respectively. Natural convection heat transfer performance of the outside surfaces of the condenser and the evaporator (cross-fin coils) and the transient behavior of the system during the melting and solidification periods of the latent storage medium were tested. It was found that the system can maintain the temperature within the shelter below 50°C even when the shelter is located in subtropical deserts or similar regions.

NOMENCLATURE

a = thermal diffusivity of air

717

A_1 = sum of wall and floor areas

A_2 = ceiling area

g = gravitational acceleration

h = fin height

K = overall heat transfer coefficient

Nu = Nusselt number = $\alpha s/\lambda$

Q = heat transfer rate of evaporator and condenser during steady state operation

Q_e = total heat load of cooling system

Q_1 = heat loss through outer envelope of shelter

Q_i = heat dissipation rate of telecommunication equipment

Q_o = electric power supplied to incandescent lamps

Q_w = heat gain through wall

Ra = Rayleigh number = $g\beta\Delta ts^4/\nu ah$

s = fin spacing

t_a = air temperature around test shelter

t_i = air temperature in shelter

t_1 = surface temperature of latent heat storage unit

t_m = melting temperature

t_o = outdoor temperature; also air temperature in test chamber

t_s = solidification temperature

Δt = temperature difference (= t_i-t_r for evaporator, t_r-t_o for condenser, and t_i-t_o for system)

α = heat transfer coefficient

β = volumetric expansion coefficient

λ = thermal conductivity

ν = kinematic viscosity

1. INTRODUCTION

There has been a remarkable growth of microwave communication systems able to cover wide areas in various regions of the world. For radio relay stations located in subtropical deserts or similar regions, cooling of a shelter housing electronic telecommunication equipment is a very important problem because the efficiency and reliability of the electronic equipment is inversely proportional to its operating temperature (the maximum allowable temperature for commonly used semiconductors is around 100°C). As humidity, dust, and corrosive gas are detrimental to electronic equipment, direct cooling of the shelter via the ambient air is practically impossible. Therefore, a conventional radio relay station with a shelter volume of 10 m^3 and a power dissipation of about 1 kW, requires installation of a refrigeration system with a maximum power the consumption of several kilowatts. As a result, the reliability of the telecommunication equipment is dependent on the reliability of the power source and the refrigeration system. However, due to the recent development of semiconductor devices, the heat dissipation rate of the telecommunication equipment has been reduced to 50 to 150 W. Consequently, small power sources such as a battery or a solar battery are sufficient for the telecommunication equipment. Mori and Sakitani[1] proposed a non-powered cooling system utilizing a latent heat storage unit which can be used in an area where there is a large difference between daytime and nighttime temperatures. The present study was undertaken to develop the proposed cooling system.

2. DESCRIPTION OF THE COOLING SYSTEM

A schematic diagram of the cooling system is shown in Fig. 1. Telecommunication equipment is installed in a shelter in which the walls are covered with a layer of insulating material. The cooling system consists of an outdoor condenser mounted on the top of the shelter, an indoor evaporator, and an indoor latent heat storage unit, which are placed above the telecommunication equipment. These three components are connected through piping to form a closed circuit charged with an evaporable heat transfer fluid (refrigerant). The plate-fin and tube type condenser and the evaporator exchange heat with the ambient air by natural convection. These heat exchangers are tilted so that the refrigerant circulates naturally by gravity. The latent heat storage unit, which is a horizontal double-tube heat exchanger with a high-finned inner tube, is placed lower than the outdoor condenser but higher than the indoor evaporator. The double-tube annular space is charged with a phase change material (PCM) having a melting temperature near the room temperature for cooling storage. The inner tube is filled with liquid refrigerant up to the center line level of the tube.

At night, when the outdoor temperature is low, the refrigerant vapor in the condenser is cooled by the ambient air and condenses. The condensate flows down the piping and a portion of it flows into the indoor evaporator via a header. The condensate is heated by the air in the shelter and evaporates. The rest of the condensate flows into the inner tube of the latent heat storage unit via a liquid pipe. The condensate is heated by the liquid PCM in the annular space and evaporates, while the PCM is solidified. The generated vapor flows through the piping into the outdoor condenser. The diameter of the piping must be large enough to maintain a smooth countercurrent flow of the refrigerant gas and liquid. As a result of the natural circulation of the refrigerant involving liquid/vapor phase changes, the latent heat is stored in the PCM and the air in the shelter is efficiently cooled.

During the day time, when the outdoor temperature is high, the refrigerant vapor generated in the indoor evaporator cannot be cooled by the outdoor condenser. Therefore, the vapor does not flow upward through the piping, but is introduced into the inner tube of the latent heat storage unit via the header and a vapor pipe. The vapor is then cooled by the solidified PCM in the annular space and condenses, as the PCM is melted. The condensate flows back to the indoor evaporator through the liquid pipe. As this circulation of refrigerant continues, the latent heat stored in the PCM is used to effectively cool the air in the shelter.

Radio relay stations using the proposed cooling system are assumed to be located in subtropical deserts or similar regions. The assumed conditions are as follows:

(1) Heat dissipation rate of the telecommunication equipment $Q_i = 150$ W.

(2) Hourly variation of the outdoor temperature t_o

time(h)	t_o(°C)	time(h)	t_o(°C)
7 - 10	40	18 - 20	45
10 - 12	45	20 - 23	40
12 - 18	50	23 - 7	35

(3) Air temperature in the shelter $t_i \leq 50°C$
(4) Sun-earth geometry   latitude: 27° North
                                solar declination: 20°

The heat gain through the walls of the shelter Q_w is calculated with the procedure described in ref. [2]. The total heat load of the cooling system Q_e is given by

$$Q_e = Q_i - Q_w \tag{1}$$

Table 1 shows the hourly variation of Q_e for a shelter described in section 4 for parametric values of t_i.

3. ASSESSMENT OF LAURIC ACID AS PCM

Since the operating condition of the proposed cooling system is $t_o \geq 35°C$ and $t_i \leq 50°C$, the desired melting temperature of the PCM is from 41 to 45°C. A large number of organic, inorganic, and eutectic PCMs are available. Kai [3], Kamimoto [4], and Clark [5] have reviewed PCMs and PCM units used for thermal energy storage. Taking into account practical conditions such as melting temperature, magnitude of latent heat, toxicity, availability and cost, lauric acid was selected as the PCM for this system. Table 2 shows the physical properties of pure lauric acid [6].

Commercially available lauric acid had a purity of 99.7 %. The greater part of the remaining impurities was myristic acid ($CH_3(CH_2)_{12}COOH$). Experiments were performed to assess the feasibility of using this material as the PCM. The items studied were the solidification temperature t_s, the melting temperature

721

t_m, the degree of subcooling, corrosiveness to candidate metals for manufacturing the latent heat storage unit (copper, iron, and stainless steel) and thermal degradation due to repetition of cyclic melting and solidification.

The solidification and melting temperatures and the degree of subcooling were determined by visual observation of the test fluid during gradual cooling and heating processes. The values of t_s and t_m were determined as $42.5 \pm 0.2°C$ and $43.0 \pm 0.2°C$, respectively. The degree of subcooling was estimated to be less than $0.5°C$.

Corrosiveness of lauric acid to copper, iron, and stainless steel was studied by immersing metal test pieces ($40 \times 10 \times 1$ mm^3) in lauric acid kept at $70°C$. The corrosion rate of the copper test piece after 72 hours of immersion was 68.2 $mg/dm^2/day$. The iron and stainless steel test pieces showed no indication of corrosion after 72 hours. Iron test pieces with and without a weld were further tested for long-term corrosion. Four types of welding processes (arc, tig, mig, and seam) were compared. All the test pieces rusted uniformly after 17,000 hours of immersion. The corrosion rate ranged from 4 to 77 $mg/dm^2/day$ and was langer for welded test pieces. However, it was concluded that iron can be used safely for the latent heat storage unit provided a sufficient margin for corrosion is allowed in the thickness.

Thermal degradation of lauric acid was tested using a closed natural circulation loop of R-22 consisting of a latent heat storage unit, an electric heater, and a condenser. Melting and solidification of lauric acid was repeated by cyclic heating and cooling of R-22 within the temperature range of 30 to 65°C. The period for one cycle was about 5.5 hours. It was found that the heating-cooling curves of lauric acid and R-22 were almost unchanged between the early stage and the 1,870th cycle. This indicates that thermal degradation is negligible up to as many as 1,870 cycles. This corresponds to fifteen years of continuous operation, since melting of lauric acid in the latent storage unit occurs only for four months during the summer season.

Further details of experimental apparatuses, procedures, and results will be reported in a separate paper.

4 HEAT TRANSFER PERFORMANCE OF COOLING SYSTEM

4.1 Experimental Apparatus

Figure 2 shows the experimental apparatus used for simulating operation of the natural circulation cooling/latent heat storage system. The rectangular shelter had dimensions of $2.3 \times 2.3 \times 2.7$ m^3 (width x depth x height). The wall and the roof were made of 150 mm thick hard urethane foam board sandwiched between 1 mm thick steel sheets. The outer surfaces were laminated with a white polyvinyl chloride paint. The outdoor condenser was housed in a test chamber with an electric heater to control the air temperature.

The outdoor condenser and the indoor evaporator were made of a one-row bundle of 15.9 mm O.D. copper tubes with continuous cross fins made of aluminum. The fin dimensions were: 120 mm in height, 8 mm in pitch and 0.2 mm in thickness. The number of parallel tubes, the projected surface area, and the total outside surface area of the evaporator were 24, 0.86 m^2, and 24.5 m^2, respectively. Those of the condenser were 32, 1.36 m^2, and 38.7 m^2, respectively. The latent heat storage unit, shown schematically in Fig. 3, consists of five double-tube heat exchangers connected in parallel by R-22 piping. The outer tube had an outer diameter of 127 mm and a length of 1,010 mm. The fin dimensions were: 32 mm in height, 9.5 mm in pitch, and 0.5 mm in thickness. The inner tube was filled with R-22 liquid up to the center line level. The annular space was almost completely filled with lauric acid except for a small gap allowed for thermal expansion. The total amount of lauric acid used for the PCM was 38 kg. The gap was charged with dry nitrogen gas.

The foregoing three components were connected by one or two copper tubes of 15.9 mm O.D. In order to simulate the heat load Q_e of the shelter, eight incandescent lamps were set on the inner wall. Electric power was supplied to the lamps through a voltage stabilizer and a wattmeter.

The temperatures of the indoor and outdoor air, the refrigerant piping and the latent heat storage unit were measured

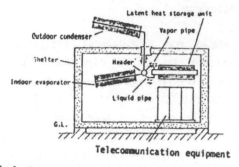

Fig.1 Schematic diagram of a natural circulation cooling/latent heat storage system

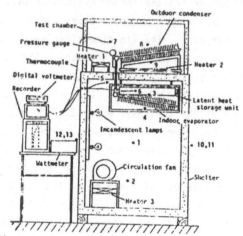

Fig.2 Experimental apparatus used for simulating operation of natural circulation cooling/latent heat storage system

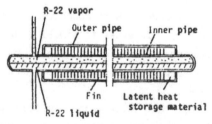

Fig.3 Structure of latent heat storage unit

with 0.65 mm O.D.T type sheathed thermocouples at thirteen points indicated in Fig. 2. Hourly variations of these temperatures were monitored by a recorder. The refrigerant pressure in the piping was measured with a precision Bourdon tube gauge (2.5 MPa F.S.). The temperature of the refrigerant piping agreed to within ± 0.2°C with the saturation temperature corresponding to the measured pressure. Using these measurements, the representative temperature of each part was defined as follows.

Air temperature in the shelter, t_i:
  average value of the measured temperatures
  at points 1, 2, 3, and 4

Air temperature in the test chamber, t_o:
  average value of the measured temperatures
  at points 7, 8, and 9

Air temperature around the shelter, t_a:
  average value of the measured temperatures
  at points 10, 11, 12, and 13

Refrigerant temperature, t_r:
  saturation temperature corresponding to the
  measured pressure

Surface temperature of the latent heat storage unit, t_1:
  the measured temperature at point 6

4.2 Air-Side Heat Transfer Performance of Evaporator and Condenser

Experiments were conducted to study the steady state heat transfer performance of the evaporator and condenser. The experiment was begun by supplying constant electric power Q_o to the incandescent lamps in the shelter. After a steady state was reached, the temperature at each measuring point was measured by digital voltmeter reading to 10 μV. The heat transfer rate of the evaporator and condenser Q was obtained from

$$Q = Q_o - Q_1 \qquad (2)$$
$$Q_1 = K \{ A_1 (t_i - t_a) + A_2 (t_i - t_o) \} \qquad (3)$$

where Q_1 is the heat loss through the outer envelope of the shelter, $K = 0.327 \text{ W/m}^2/\text{K}$, $A_1 = 26.7 \text{ m}^2$ is the sum of the wall and floor areas, $A_2 = 4.0 \text{ m}^2$ is the ceiling area. The K value was obtained from preliminary experiments without refrigerant in the cooling system.

Figure 4 shows the heat transfer performance of the evaporator, condenser, and system, where Q is plotted as a function of the temperature difference Δt (= $t_i - t_r$ for evaporator, $t_r - t_o$ for condenser and $t_i - t_o$ for system). In Fig. 4, the open symbol data represents the case of two connecting pipes, while the solid symbol data represents the case of one connecting pipe. The data agrees well with each other at small values of Δt. However, the solid symbol data deviates toward a smaller Q value than the open symbol data as Δt increases, indicating a deficiency in the circulating refrigerant as a result of the increase in the flow resistance. This tendency is not observed for the open symbol data with two connecting pipes up to Q = 1.2 kW.

Figure 5 shows the experimental data for the evaporator and condenser (excluding four points at higher Δt for a single connecting pipe) plotted on the coordinates of Nu vs. Ra, where $Nu = \alpha s/\lambda$, $Ra = g\beta \Delta t s^4/\nu a h$, α is the heat transfer coefficient based on the actual surface area, s is the fin spacing, h is the fin height, and λ, β, ν, and a are the thermal conductivity, volumetric expansion coefficient, kinematic viscosity, and thermal diffusivity of air, respectively. The experimental data for a three-row bundle with the same fin pitch and fin height as the present evaporator and condenser are also shown for comparison. The thick solid line in Fig. 5 shows the empirical equation for natural convection heat transfer from vertical parallel plates proposed by Aihara [7]:

$$Nu = Ra \{ 1 - \exp(-13.6/Ra^{3/4})\}/ 24 \qquad (4)$$

The thin solid lines show the limiting cases for strong interaction between the plates (Ra → 0) and for a single plate (Ra → ∞) expressed as (see, e.g. ref. [8]):

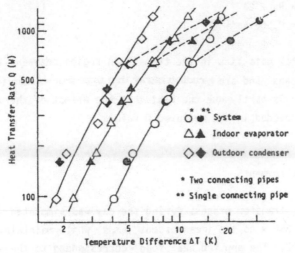

Fig.4 Heat transfer performance of evaporator, condenser and system

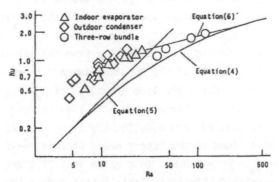

Fig.5 Air-side heat transfer performance of evaporator and condenser

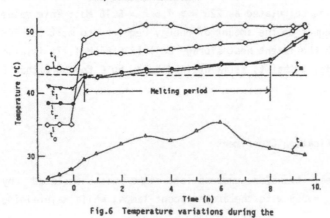

Fig.6 Temperature variations during the melting period

$$Nu = Ra / 24 \qquad (5)$$
$$Nu = 0.566 \, Ra^{1/4} \qquad (6)$$

The experimental data lies in the transition region between the two limiting cases and are generally higher than those from equation (4). This difference is partly due the fact that the effect of thermal radiation is included in the measured Q values.

4.3 Melting Experiment

The heat transfer process during the day was simulated by supplying Q_o = 350 W to the incandescent lamps, while maintaining t_o at about 50°C. The applied heat load corresponded to the sum of the calculated value of Q_e and the estimated value of Q_1. The initial values of t_r and t_1 were set a few degrees below t_m.

Figure 6 shows the hourly variations of the measured temperatures t_i, t_1, t_r, t_o, and t_a. The value of t_m is also shown by a broken line in Fig. 6. It is seen from the variation of t_1 that the melting period of the PCM is between 0.5 and 8.0 hours on the abscissa. Using the time averages of t_o, t_i, and t_a during the melting period, the average heat loss through the shelter wall is calculated from equation (3) as Q_1 = 122 W. Thus, the net heat load of the latent heat storage unit is calculated as $Q_e = Q_o - Q_1$ = 228 W. This value is 1.25 times the maximum value of daytime Q_e listed in Table 1 (i.e. Q_e = 182.6 W at t_i = 50°C and 19 hours). The total heat transferred during the period is calculated as 228 W x 7.5 h = 6.16 MJ. This value is roughly equal to the amount of energy required to melt the PCM contained in the latent heat storage unit (= 6.84 MJ). It is also seen from Fig. 6 that t_i is maintained below 50°C for more than 8 hours.

4.4 Solidification Experiment

The heat transfer process at night was simulated by supplying Q_o = 350 W to the incandescent lamps, while maintaining

t_o at about 35°C. The initial values of t_l and t_r were set several degrees above t_s. Figure 7 shows the hourly variations of the measured temperatures. The value of t_s is also shown by a broken line in Fig. 7. It is seen from the variation of t_l that the solidification period of the PCM is between 1 and 9 hours on the abscissa. Using the time averages of t_o, t_i, and t_a during the solidification period, the average heat loss through the shelter wall is calculated from equation (3) as Q_l = 130 W. Thus, it follows that Q_e = Q_o - Q_l = 220 W. This value is 1.12 times the maximum value of nighttime Q_e listed in Table 1 (i.e. Q_e = 96.7 W at t_i = 50°C and 21 hours).

4.5 Simulation Test

Simulation of continuous operation of the experimental apparatus for two days was conducted to verify the performance of the cooling system. Based on the heat load calculation summarized in Table 1, hourly variations of t_o and Q_e were set as shown in Table 3. Figure 8 shows the hourly variations of the measured temperatures for the second day. In Fig. 8, the values of t_s and t_m are also shown by broken lines. It is seen from Fig. 8 that a daily cooling cycle, involving complete solidification and melting of the PCM, is established by the cooling system, and the t_i value is maintained constantly below 50°C.

5 CONCLUSION

Experiments were conducted to develop a reliable, non-powered cooling system for a shelter accommodating telecommunication equipment located in subtropical deserts or similar regions. The cooling system consists of an outdoor condenser, an indoor evaporator and an indoor latent heat storage unit connected by piping. Refrigerant R-22 and 99.7% pure lauric acid were used as the heat transfer medium and the PCM, respectively. The results may be summarized as follows:

Table 1   Shelter heat load calculation results

Time (h)	t_o (°C)	Q_e (W) t_i (°C)				
		35	40	45	50	55
1	35	231.6	178.3	125.6	72.6	19.5
2	35	231.6	178.3	125.6	72.6	19.5
3	35	231.6	178.3	125.6	72.6	19.5
4	35	200.3	147.3	94.3	41.3	-11.7
5	35	200.3	147.3	94.3	41.3	-11.7
6	35	200.3	147.3	94.3	41.3	-11.7
7	40	200.3	147.3	94.3	41.3	-11.7
8	40	200.3	147.3	94.3	41.3	-11.7
9	40	200.3	147.3	94.3	41.3	-11.7
10	45	200.3	147.3	94.3	41.3	-11.7
11	45	252.0	199.0	145.9	92.9	39.9
12	50	296.4	243.4	190.3	137.3	84.3
13	50	291.7	238.7	185.7	132.7	79.7
14	50	280.7	227.7	174.7	121.6	68.6
15	50	307.1	254.1	201.0	148.0	95.0
16	50	298.5	245.5	192.4	139.4	86.4
17	45	319.7	266.6	213.6	160.6	107.6
18	45	334.4	281.4	228.4	175.3	122.3
19	45	341.6	288.6	235.6	182.6	129.5
20	40	347.4	294.4	241.4	188.4	135.3
21	40	354.8	301.7	248.7	196.7	142.7
22	40	352.8	299.8	246.7	193.7	140.7
23	35	312.2	259.2	206.2	153.1	100.1
24	35	263.0	210.0	157.0	104.0	50.9

Table 2   Physical properties of lauric acid

Molecular formula	$CH_3(CH_2)_{10}COOH$
Molecular weight	200.3
Heat of melting	178 kJ/kg
Melting point	316.8 K (43.6 °C)
Specific heat	2.15 kJ/kg (L*),2.14 kJ/kg (S*)
Density	869 kg/m³(L*)
Thermal conductivity	0.147 W/m/K (L*)
pH	Slightly acidic

* L: Liquid, S: Solid

Table 3   Hourly variations of t_o and Q_e

Time (h)	t_o (°C)	Q_e (W)
10 - 12	45	183
12 - 18	50	183
18 - 20	45	183
20 - 23	40	247
23 - 7	35	259
7 - 10	40	94

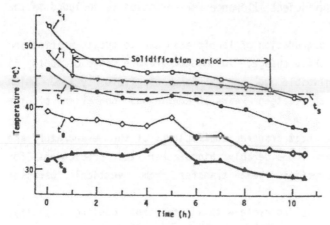

Fig.7 Temperature variations during the
solidification period

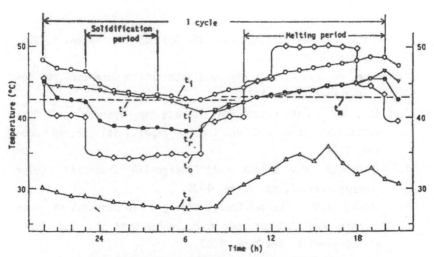

Fig.8 Temperature variations during the second day
of continuous operation

(1) Commercially available lauric acid had a melting temperature of 43.0 ± 0.2°C, a solidification temperature of 42.5 ± 0.2°C and a degree of subcooling less than 0.5°C.

(2) Iron can be used safely for the latent heat storage unit, provided a sufficient allowance for corrosion is included in the thickness.

(3) Thermal degradation of lauric acid due to cyclic heating and cooling with phase change, in the temperature range of 30 to 65 °C, was negligible up to as many as 1,870 cycles, which corresponds to fifteen years of continuous operation of the system.

(4) Air-side heat transfer coefficients of the evaporator and condenser were considerably higher than the predictions for natural convection heat transfer from vertical parallel plates.

(5) A daily cooling cycle with a sufficient cooling capacity, involving complete solidification and melting of the PCM, was established by the present cooling system.

REFERENCES

1. Mori,S. and Sakitani,K., Cooling System, U.S. Patent 4, 285, 027, August 18, 1981.

2. Inoue, U., Handbook of Air conditioning (in Japanese),pp. 45-87, Maruzen, Tokyo, 1973.

3. Kai, J., On the Latent Heat Storage (in Japanese), J. Soc. Heating Air-Cond. and San. Eng. of Japan, Vol. 52, pp. 885-890, 1978.

4. Kamimoto, M., Latent Heat Storage (in Japanese), Solar Energy, Vol. 5, pp. 37-45, 1979.

5. Clark, J.A., Thermal Energy Storage, in Handbook of Heat Transfer Applications, ed. W. M. Rohsenow et al., 2nd.,Chpt. 8, McGraw-Hill, New York, 1985.

6. Japan Oil Chem. Soc., Handbook of Oil Chemistry (in Japanese), ed. Komori,S., et al., 2nd ed., pp. 195-199, Maruzen, Tokyo, 1971.

7. Aihara, T., Heat Transfer due to Natural Convection from Parallel Vertical Plates, Trans. Japan Soc. Mech. Engrs., Vol. 29, pp. 903-909, 1963.

Coordinative Complex Compounds as Working Media in Energy Conversion and Storage Technologies

U. ROCKENFELLER
Rocky Research
P.O. Box 1086, Boulder City, Nevada 89005, USA

J. I. MILLS
EG&G Idaho, Inc.
Idaho National Engineering Laboratory
P.O. Box 1625, Idaho Falls, Idaho 83415, USA

T. R. ROOSE
Gas Research Institute
8600 West Bryn Mawr Avenue
Chicago, Illinois 60631, USA

INTRODUCTION

Complex, multicomponent fluid flows, expansion/compression processes, chemical reactions, heat and mass transfer, and additional, more specific processes and principles underpin the operation of advanced energy conversion systems and associated components. Even with efficient combustion systems that optimize release of the chemical energy of fuel as heat, the amount of energy converted to useful work ultimately depends upon the efficiency of the conversion technology and its associated processes and components. In heat engine operation, for example, there are inherent irreversibilities resulting in heat (not available for conversion to work) that must be rejected to a naturally accessible heat sink. Hence, efficient use of fuels depends not only on efficient combustion, but also on efficient energy conversion technologies.

Target opportunities that have been identified in the various end-use sectors suggest a number of opportunities that could benefit from innovative energy conversion technologies. However, critical problems continue to impede the timely development of technically feasible innovative energy conversion systems and improved heat and mass transfer systems characterized by enhanced performance, reliability, and attractive or competitive economics. The lack of complete knowledge and understanding of the fundamental processes of energy conversion has led to the development of energy conversion technologies and their associated components, including heat exchangers, that are inherently oversized and designed with overestimated safety factors. This has resulted in unnecessarily large costs for components that can represent as much as 40 to 60% of the design hardware required for conventional energy conversion technologies.

Despite improvements made to energy conversion technologies and systems over the last decade, there still exists a sizable gap between maximum theoretical and actual performance limits of devices designed and developed to convert energy to some form of work. Improved understanding of fundamental energy conversion and associated heat and mass transfer processes and subprocesses will provide the private sector with the scientific and technology knowledge base required to surmount critical pacing technology barriers, leading to the development of practical systems with closer-to-theoretical efficiencies.

In addition, to the development of innovative energy conversion technologies that result in both economic and thermodynamic improvements, leading to widespread application, an expanded energy conversion technology knowledge base can be simultaneously useful for addressing critical short-term problems that pose a threat to conventional technologies. The recent, cooperative worldwide initiatives to limit and replace the use of chlorofluorocarbons (CFCs) is an example of one response to such a threat.

This identified need to limit and replace the use of CFCs will put an increasing burden on energy conversion technologies. The heating and cooling technology industry, in particular, will be increasingly and severely affected by worldwide protocols to limit the production and use of chlorofluorocarbons, 30% of which were required for various refrigeration applications in 1985.

Solutions to the CFC issue will come in two forms: (1) the replacement of existing CFCs with less toxic (and possibly less efficient and more expensive) alternatives; and (2) innovative new energy conversion technologies that do not require the use of CFCs. The first option represents a necessary but short-term fix; the second option represents the most feasible approach for the longterm.

One of the most promising technologies that satisfies the requirements of the second option above is based upon the application of complex chemical compounds to energy conversion. The objective of this paper is to introduce the concept of complex compounds as applied to energy conversion systems, with particular emphasis, for illustrative purposes, upon temperature and heat amplifiers. The employment of complex compounds to these technologies allows for a design with no moving parts yielding efficiencies much superior to state-of-the-art liquid/vapor and inclusion compound systems, with additional benefits of potentially improved cost-effectiveness and absence of CFCs in the working media.

Background of Complex Compound Media

Complex compounds are coordination complexes of an adsorbent, often a solid metal inorganic salt, and an adsorbate, often referred to as the ligand of the complex. The adsorbate is a gas with one or several lone electron pair(s), such as H_2O, NH_3, CH_3OH. The adsorbent and adsorbate undergo a thermochemical reaction forming the so-called complex compound.

The adsorption reaction is an exothermic and the desorption reaction is an endothermic process.

$$\text{Salt} + \text{Ligand} \rightleftharpoons \text{Salt} \cdot \text{n Ligand} \ . \tag{1}$$

The vapor pressure-temperature equilibrium is commonly described with the law of mass action or the Nernst approximation formula.[1]

$$\log p = \frac{Q_o}{2.303 \ R_m \ T} + 1.75 \log T - aT + C \tag{2}$$

with:

Q_o = heat evolution (Joules)

p = partial complex pressure (bar)

T   =   absolute temperature (Kelvin)

R_m =   universal gas constant (Joules/kilogram Kelvin)

C   =   constant for ligand

a   =   parameter function of coordination number and condensation energy of ligand.

The major difference between complex compounds and liquid-vapor fluids, such as LiBr - H_2O or H_2O - NH_3, is there is the independence of the vapor pressure from the fluid concentration within wide concentration ranges. This allows for high temperature lifts and coefficients of performance (COP), because the dilution does not influence the driving differential pressures.

Figure 1 shows the vapor pressure - temperature - concentration diagram for the $AlCl_3$ - n NH_3 complexes. The molar compositions next to the vapor pressure curves denote the maximum ligand concentration represented by the respective vapor pressure equilibrium line.[2]

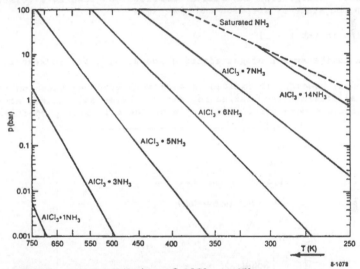

FIGURE 1. P-T Equilibrium of $AlCl_3$- n NH_3

The mere calculation of enthalpy and entropy, however, which leads to the vapor pressure equilibrium, does not reveal any information on the stability or even the existence of a complex compound in a given temperature range.

Almost all metal inorganic salts that form a coordination complex with water, ammonia, alcohol, or any other suitable ligand gas turn into solutions at high ligand concentrations. The critical concentrations that cause liquefaction of the complex can be determined with the solution diagram of the respective components. Figure 2 shows the solution diagram of NaCl - NH_3. The solubility, a definite limit for the complex stability, is a function of the temperature.[3]

During a sorption reaction, however, the system is not in the state of equilibrium and therefore the solution diagram does not apply. Especially during the adsorptions, the nonideal complex surface yields local concentration and temperature gradients, which shift the saturation point to lower ligand concentrations.

The actual application of complex compounds as a working fluid in an energy conversion technology, such as a heat or temperature amplifier, requires a finite layer thickness of the complex with heat and mass transfer in the third dimension. This causes relatively large concentration and temperature gradients which are a function of the reaction rate and the absolute temperature.

The reaction kinetics of surface sorption reactions can be described with the Elovich coverage model. The Elovich equation is a simplified approach to describe slowly activated chemisorption processes.[4]

$$d\theta/dt = a \exp(-b\theta) \qquad (3)$$

where θ is the coverage of the adsorbent area and a and b are constants. Integration and simple substitutions lead to a relation between the state of sorption and the time:

$$n = C \; 1/b \; \ln(ab \; t + 1) \qquad (4)$$

where n is the amount of gas adsorbed and a, b, c are constants.

This model, however, is limited to sorption reactions with constant differential pressure and limited bed depth. Layer thicknesses exceeding 4 mm require a more detailed analysis of the diffusion properties.

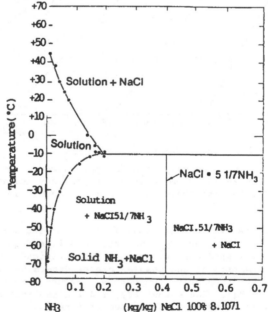

FIGURE 2. Solution Diagram for NaCl-NH$_3$Mixture

Since the above model is in excellent agreement with experimental results for thin layers over the entire concentration range, Equation (4) can be used to describe the sorption process at the surface and in thin layers. Figure 3 shows experimental results of an adsorption with a layer of 2 mm and the respective model function.

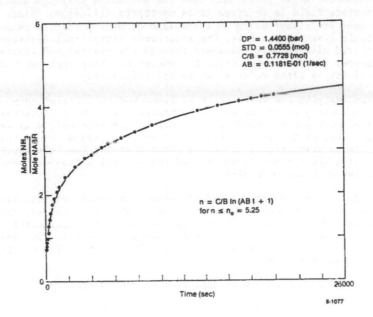

FIGURE 3. Adsorption Progres vs Time (2mm Layer Thickness)

In order to describe the sorption processes in thicker layers and to account for effects such as local saturation and surface liquefaction, the heat and mass transfer in sorbents must be investigated in more detail.

Sorption rates in porous sorbents are mostly controlled by the sorbate transport within the pore network, rather than by the kinetics of the surface sorption. The bulk flow through the pores is generally negligible so that the intraparticle transport must be considered a diffusive process. Such processes are described by Fick's first equation, which correlates the sorbate flux J with the diffusivity $D(c)$ and the concentration gradient:

$$J = -D(c) \frac{\partial c}{\partial x} \ .$$
(5)

Pore diffusion may occur via several different mechanisms depending on the pore size, the sorbate concentration, and other conditions.[5]

737

The transport in intracrystalline pores is usually referred to as micropore diffusion. Molecules diffusing in micropores never escape from the force field of the adsorbent surface and the transport requires an activated process involving jumps between sorbent sites.

Diffusion in larger pores such that the diffusing molecule escapes from the surface field is referred to as macropore diffusion. Since actual reactions are always using an assemblage of microparticles rather than on single isolated particle, the additional diffusional resistance associated with sorbate transport through the crystal bed must also be considered in a micropore diffusion analysis. This particular macropore diffusivity is often referred to as bed diffusivity.

For the analysis and description of fixed bed solid/vapor sorption reactions using an assemblage of microparticles, the diffusional resistance of macropores through the crystal bed is referred to as bed diffusion. For salt crystals that adsorb gaseous ligand and do not form large macro assemblages, it can be shown that the two relevant diffusion mechanisms are the micropore and bed (macropore) diffusion, the latter being the dominant mechanism.

To analyze such a system, a bed of uniform spherical microparticles shall be considered. Figure 4 shows a sketch of the bed. The adsorbent layer thickness or the depth of the bed is assumed to be much larger than the particle diameter. The system is considered isothermal with equilibrium between sorbate and sorbent at the particle surface. The equilibrium relationship shall be linear.

The bed diffusion can now be expressed by Fick's equation:

$$D_p \frac{\partial^2 c}{\partial x^2} = \frac{\partial c}{\partial t} + \left(\frac{1 - \epsilon}{\epsilon}\right) \frac{\partial \bar{q}}{\partial t} \tag{6}$$

with

D_p = pore diffusivity $\qquad$ [cm^2 s^{-1}]

c = local sorbate concentration within macropores $\qquad$ [mmol g^{-1}]

$\bar{q}$ = sorbate concentration averaged over crystal $\qquad$ [mmol g^{-1}]

l = bed depth $\qquad$ [mm]

ϵ = bed voidage $\qquad$ [-].

The boundary conditions are:

$$\bar{q}\,(x,\ t) = \frac{3}{r_c^2} \int_0^{r_c} q\ r^2\ dr \tag{7}$$

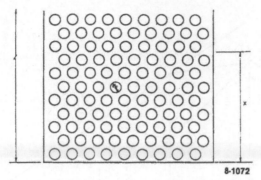

8-1072

FIGURE 4. Sketch of bed of uniform spherical microparticles

$$\frac{\partial c}{\partial x} (0, t) = 0, \ c (1, t) = C (t) \ . \tag{8}$$

If bed diffusion is dominant q is only a function of time and in equilibrium with c. Then it can be shown that the equation for the uptake curve is:

$$1 - \frac{m_t}{m_\infty} = \sum_{n=0}^{\infty} \frac{2}{\left(n + \frac{1}{2}\right)^2 \pi^2} \exp\left[-\left(n + \frac{1}{2}\right)^2 \frac{\pi^2}{1^2} \frac{D}{1^2} \frac{e^t}{1^2}\right] \tag{9}$$

Cycle Configurations and Operating Modes

In order to test the applicability of the theory of complex compounds for energy conversion, several concepts for heat pumps were developed and analyzed. These concepts comprise single and multistage temperature amplifiers and heat amplifiers.

The suggested multistage systems were designed to increase the coefficient of performance rather than the temperature lift. However, staging with respect to lift improvements can be performed as well.

Heat amplifier cycles. The single-stage heat amplifier cycle is shown in Figure 5. The process can be described as follows:

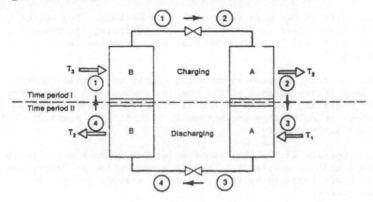

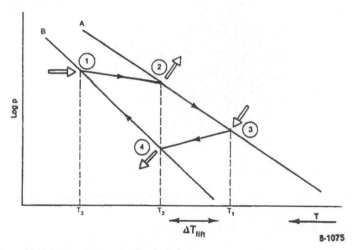

FIGURE 5. Single-state heat amplifier

1->2   Complex B is heated with prime energy or high temperature waste
       heat at the generator temperature T_3. This results in a ligand
       vapor pressure higher than the pressure of complex A at the deli-
       very temperature T_2. Therefore, complex B undergoes an endo-
       thermic desorption reaction and complex A undergoes an exothermic
       adsorption reaction which releases thermal energy at the delivery
       temperature T_2.

2->3   After the above reaction is completed, complex A is exposed to the
       waste heat stream at T_1 and complex B is exposed to the delivery
       temperature T_2. Then the vapor pressure of complex A at temper-
       ature T_1 is higher than the vapor pressure of complex B at T_2.

3->4   Complex A undergoes a desorption reaction drawing heat from the
       waste heat stream, the adsorption process in complex B produces
       thermal energy at the useful delivery temperature T_2.

4->1   complex B is heated by the driving heat source to temperature T_3.
       complex A self-heats to temperature T_2 during the following
       sorption process. Then heat is released at temperature T_2.

The cycle operates in two main sequential time periods. The engine
reaction 1->2 takes place in Period I. After switching the temperature from T_2 to T_1 and T_3 to T_2, the heat pump reaction takes
place in Period II.

Two parallel container systems allow for a continuous operation even
during the temperature changes of the complex compounds A and B. For
continuously operating temperature amplifiers and heat amplifiers as well
as single-stage or multistage arrangements, sequentially operated dual
configurations are necessary.

The cycle can operate with either two complex Compounds A and B or the
plain ligand A and one complex B. Then, the adsorption or desorption
processes in A are replaced by condensation or evaporation of the
ligand, respectively.

740

Figure 6 shows the COP-staged version of the two-stage heat amplifier, which operates with two discharging and only one charging reaction and does not require any internal heat coupling. The time sequence is controlled by the adsorption-desorption of complex B.

This two-stage configuration has the following advantages over conventionally staged cycles:

- No internal heat coupling, less heat capacity losses

- Three instead of four containers

- Three instead of four temperature levels

- Improved reaction rates for single engine reaction instead of heat coupled engine process.

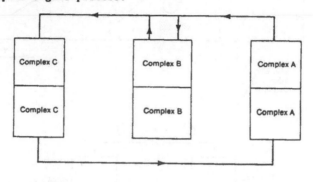

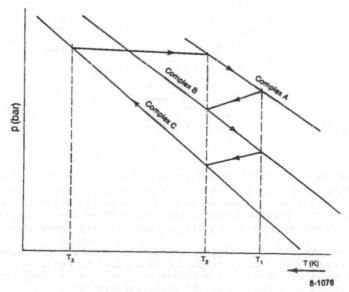

8-1076

FIGURE 6. Two-stage refrigeration cycle

741

Temperature amplifier cycles. The single-stage temperature amplifier (heat transformer) cycle is shown in Figure 7. The process can be described as follows:

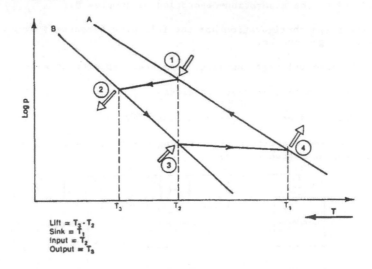

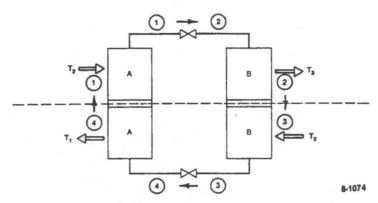

FIGURE 7. Single stage temperature amplifier

1->2   Complex A is heated by the waste heat (input energy) at temperature level T_2. This results in a ligand vapor pressure higher than the complex pressure of complex B at the delivery temperature. Therefore, complex B undergoes an exothermic adsorption reaction which delivers thermal energy at the delivery temperature.

2->3   After the above reaction is completed, complex B is exposed to the waste heat stream and complex A is exposed to the sink temperature. The vapor pressure of complex B is higher than the pressure of complex A at the sink temperature.

3->4   Complex B undergoes a desorption reaction, drawing heat from the
        waste heat stream, and complex A rejects the heat to the sink
        during the adsorption process.

4->1   Complex A is heated by the waste stream until it reaches temper-
        ature level T_2. Complex B self-heats to the delivery temper-
        ature during the following adsorption.

Again the cycle operates in two main sequential time periods as discussed
for the HA concept. Two parallel container systems allow for a continuous
operation during the temperature changes of the respective complexes.

The cycle can operate with either two complex compounds A and B or the
plain ligand A and one complex B. Then, the adsorption or desorption
processes in A are replaced by condensation or evaporation of the
ligand, respectively.

Figure 8 shows the COP staged two-stage temperature amplifier configuration. The hardware concept is very similar to the two-stage heat
amplifier concept shown in Figure 6. All components are identical. The
flow direction, however, is reversed. This cycle operates with two heat
pump and only one bottoming engine reaction and does not require any
internal heat coupling.

CONCLUSIONS

The preliminary investigation of complex compounds as working media in
heat amplifiers and temperature amplifiers leads to promising results.

The exploitation of the heat of reaction and the high temperature
stability of many complex compounds yield favorable performance data for
innovative industrial and commercial refrigeration/heat pump technologies.
The stoichiometrically determined COP and the capability to internalize
high firing temperatures in the HA concepts give these working media a
considerable advantage over liquid/vapor systems.

The high lift characteristics of solid/vapor complex compound concepts
allow for the use of single-stage heat pumps in applications where state-
of-the-art fluids require two-stage hardware configurations. This leads
to much lower first cost of complex compound heat pumps especially in
applications with high lift requirements.

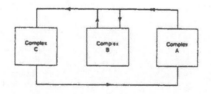

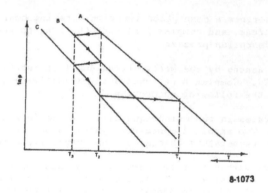

8-1073

FIGURE 8. Two-stage temperature amplifier

The disadvantages arising from the stationary solid working media can be avoided by suitable hardware configurations or the use of inert solvents.

REFERENCES

1.   Nernst, W., "The New Heat Theorem," Dover Publications, Inc., New York, NY, 1969.

2.   Rockenfeller, U., "Entwicklung eines Niedertemperatur-Spiecherwaermepumpensystems fuer Kuehlung and Heizung," Ph.D. Thesis, RWTH Aachen, December 16, 1985.

3.   Rockenfeller, U., Stojanoff, C. G., Horn, G., "Development of Dual Temperature Amines for Latent Heat Pump Storage Applications," U.S. DOE Report, Oak Ridge National Laboratories, 1986.

4.   Wilkinson, F., "Chemical Kinetics and Reaction Mechanisms," Van Nostrand Reinhold, 1980.

5.   Ruthven, Douglas M., "Principles of Adsorption and Adsorption Processes."

HEAT PUMP

Closed Cycle Zeolite Regenerative Heat Pump

DIMITER TCHERNEV and DANIEL EMERSON
The Zeopower Company
Natick, Massachusetts 01760, USA

INTRODUCTION

A closed cycle regenerative zeolite heat pump was designed, constructed and tested in performance. The system combines the sorption properties of solid zeolites/refrigerant vapor pairs with the principle of regenerative heat exchangers.

Present absorption cooling technology is characterized by low seasonal Coefficient of Performance (COP less than 1.0 for single effect systems) and high initial equipment costs and parasitic energy demand usually associated with cooling towers, large numbers of pumps, fans, etc. Double-effect or two-stage systems exhibit higher COPs but at the price of even higher initial costs and parasitic energy demands. These facts have resulted in very low commercial market penetration and a residential heat pump market completely dominated by electric vapor-compression units.

The solid-gas zeolite adsorption system reported here utilizes the extremely open frame of the zeolite crystal structure and the resulting adsorption of very large quantities of most polar gases. The interaction of the dipole or quadrupole moment of the gas molecule with internal crystal field of the zeolite frame is very strong and extremely nonlinear, thereby providing this family of materials with unique adsorption properties which permit extremely high efficiencies for adsorption heat pump cycles with air-cooled condensers. Combining the zeolite technology with the principle of energy regeneration results in a single-effect system with seasonal cooling COPs of 1.2 and heating COPs above 1.8 and initial equipment cost comparable to electric heat pumps.

THE ZEOLITE CYCLE

Zeolites form a family of materials with a unique open, cage-like crystal structure. The atoms, oxygen, silicon, aluminum, and one of the alkalene metals or earths, are arranged in such a manner that they form cavities interconnected by channels or tunnels on the atomic scale. There are more than 40 types of zeolites with cavity sizes ranging from 3 to 10+ angstroms and connecting channels of variable openings. Because of this open porous structure, zeolites are capable of adsorbing large quantities of a variety of refrigerant gases, ranging from water vapor, ammonia and carbon dioxide to the different fluoro-, chloro- and hydro-carbons in the vicinity of room temperature. The adsorption energy is caused mainly by the interaction of the polar moments of the refrigerant molecules with the strong internal crystal field of the cage-like structure. This interaction is extremely nonlinear and exhibits a saturation behavior in their pressure dependence. The nonlinearity and saturation behavior of zeolites adsorbing

747

water vapor is such that at pressures above 10mm Hg the adsorbed amount is almost indenpendent of pressure.

To describe the situation in another way, one can compare the thermal activation of the different processes invoved. The solubility of ammonia in water, of water vapor in lithium bromide and the adsorption of refrigerant gases on surface adsorbents such as silica gel, activated alumina or activated carbon all depend exponentially on H/RT where H is the energy of solution or adsorption and T is the absolute temperature (i.e. they obey the Arrhenius equation). Adsorption in zeolites, on the other hand has been shown by Dubinin to depend exponentially on at least the second, and as high as the fifth, power of H/RT. It is this extreme nonlinearity of thermal activation that makes zeolites so well suited for heat pump applications by reducing the influence of condensation pressure and temperature on the COP of the cycle.

When the zeolite is at ambient temperature it can adsorb large quantities of refrigerant vapor even at low partial pressure. On the other hand, when the zeolite is heated it desorbs most of the refrigerant vapor even at high vapor pressure, corresponding to high condenser temperatures. Thus, the shape of the adsorption isotherm and its only slight dependence on evaporator and condenser temperatures makes the zeolite system superior to conventional adsorption heat pump cycles since the nonlinear properties of zeolites eliminate the need for cooling towers and the accompanying high initial costs. Considering a number of potential refrigerants as part of the solid zeolite-gas system we have determined in the past that of all possible candidates -- water, ammonia, alcohols and freons -- in the temperature range of interest ($45°F$ to $55°F$ evaporator temperatures), water vapor delivers the highest cycle COP since zeolites adsorb the same amounts of refrigerant (about 20% by weight) and water has the largest heat of vaporization of all candidate refrigerants (1,000 Btu/lb compared to 500 Btu/lb for alcohols and ammonia, and 100 Btu/lb for freons). In addition, water vapor is the only refrigerant which is stable at $400°F$ in the presence of zeolites. Alcohols decompose at about $250°F$ while freons break down at even lower temperatures, thereby proving that zeolites are excellent catalysts.

Theoretical analysis focusing on the COP if the cycle was performed for synthetic zeolite 13X since adsorption data for it is available up to $600°F$. The analysis was performed utilizing the isosteres for the zeolite as indicated in figure 1 where the cycle is represented by the letters A-B-C-D. When heating the zeolite from point A to B no desorption takes place since the pressure in the container increase from 0.1 to 1 psia and therefore the energy entering the zeolite is used up only by the heat capacity of the zeolite with 23% water adsorbed in it (about 0.35 Btu/lb°F). Desorption takes place between points B and C where the heat of adsorption is determined by the slope of the isostere and varies between 1,200 and 2,000 Btu/lb. The heat capacity in the process is reduced continuously from 0.35 at point B to 0.2 Btu/lb°F at point C. Integrating the total energy entering the zeolite from point A to C gives us the heat input. The cooling output generated equals the heat of condensation of the water vapor (at $100°F$ it is about 1,000 Btu/lb) which was desorbed between points B and C. Dividing the cooling output by the heat input gives the COP of the cycle for this zeolite. For $100°F$ condenser and zeolite temperature above $200°F$ the COP is between 0.4 and 0.5 and varies very little with evaporator temperature -- a feature typical of zeolites and their nonlinear adsorption properties. When the condenser temperature is in-

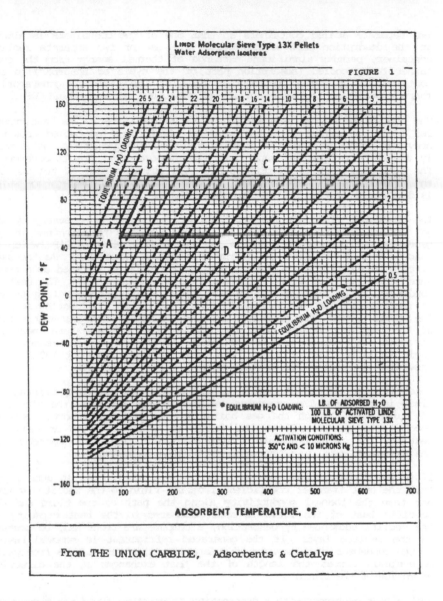

Linde Molecular Sieve Type 13X Pellets
Water Adsorption Isotheres

FIGURE 1

EQUILIBRIUM H₂O LOADING: LB. OF ADSORBED H₂O / 100 LB. OF ACTIVATED LINDE MOLECULAR SIEVE TYPE 13X

ACTIVATION CONDITIONS: 350°C AND < 10 MICRONS Hg

DEW POINT, °F

ADSORBENT TEMPERATURE, °F

From THE UNION CARBIDE, Adsorbents & Catalys

creased from 80°F to 120°F at 400°F zeolite temperature the cycle COP changes only from 0.46 to 0.44 thus indicating that the zeolite cycle can operate without a cooling tower with almost no loss in effiency.

ENERGY REGENERATION

Since the solids are difficult to move in closed system, and in order to avoid intermittent operation, the zeolite is divided into at least two separate sealed containers so that one of the containers is always in the desorption part of the cycle. The use of two separate zeolite containers permits significant exchange of thermal energy from the container being cooled (adsorption part of the cycle or adsorber) to the container being heated (desorption part of the cycle or generator), thereby resulting in the capability of obtaining extremely high COPs.

The concept of energy regeneration is shown in Figure 2. The heat transfer fluid is preheated by the adsorber which is being cooled from the previous generator mode. The adsorber slowly cools down, while continuously adsorbing refrigerant vapor from the evaporator at low temperatures. The generated heat of adsorption plus the specific heat of the zeolite and containers heat up the heat transfer fluid to a temperature which is decreasing in time but remains above 100°F.

The preheated fluid leaving the adsorber is heated as necessary by the gas boiler to about 400°F and then enters the zeolite container in the generator mode. Here the zeolite is heated from about 100°F (which it reached during the previous adsorption mode) to about 400°F. As the zeolite in this container is heated, refrigerant will be desorbed at a pressure sufficient for it to condense at a high temperature. The heat of condensation is rejected to the outside air during the cooling season or used to provide part of the heating load during the heating season. The fluid leaves the generator zeolite container considerably cooler at the beginning of the generation cycle and its temperature increases to about 400°F at the end of the cycle. During the heating season, the fluid could be further cooled in a heat exchanger to about 100°F, thus providing the rest of the heating load.

After this portion of the cycle is completed, the gear pump driving the fluid is reversed. The generator zeolite becomes adsorber, the direction of fluid flow through the gas boiler is reversed and adsorber zeolite becomes the generator with its output now being rejected or used for heating. The cycle is repeated again and again during the operation of the system.

If a zeolite/hot fluid heat exchanger is now constructed in such a way that the heat transfer coefficient from the fluid to the zeolite is larger than the thermal conductivity along the path of the fluid and the specific heat of the fluid is small compared to the heat capacity of the zeolite augmented by desorption, a temperature front will be created in the zeolite layer. If the generated refrigerant is removed faster to the condenser than the rate of generation, this temperature front will move slowly across the length of the heat exchanger in the direction of motion of the fluid.

Such a heat exchanger with a propagating temperature front can regenerate more than 80% of the thermal energy from the adsorber to the generator so that the COP of the system becomes closer to the theroetical Carnot efficiency.

We have analyzed and constructed such a heat exchanger and demonstrated regeneration efficiencies of 75% to 80%. The analysis of a breadboard prototype was performed using the first and second laws of thermodynamics to set up the coupled differential equations describing the temperature distribution in time and space within a regenerative zeolite heat exchanger and the heat transfer fluid following the process utilized by

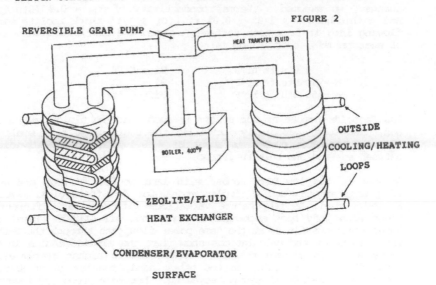

FIGURE 2

REVERSIBLE GEAR PUMP

HEAT TRANSFER FLUID

OUTSIDE
COOLING/HEATING
LOOPS

BOILER, 400°F

ZEOLITE/FLUID
HEAT EXCHANGER

CONDENSER/EVAPORATOR

SURFACE

Schmidt and Willmott[2]. The equations were then solved analytically and by a finite difference computer model using numerical solutions. The necessary data on zeolite physical and thermal properties was genereated and the rate limiting steps for heat and mass transfer determined.

We consider also the approximations involved in the model. The most important is the assumption of large thermal conductivity in the direction normal to the flow. In reality there are two principal resistances encountered in the transfer of heat from the fluid to the solid. One occurs at the surface and is inversely proportional to the heat transfer coefficient h. The other thermal resistance is associated with the transfer of heat from the surface to the interior and is inversely proportional to the bulk termal conductivity.

If the surface resistance dominates the heat flow process, the thermal gradients in the solid normal to the flow will be very small and can be neglected and the solutions given by the model will apply. If, on the other hand, the resistances are comparable or the internal resistance is greater, temperature gradients will be present in the solid and a different mathematical model has to be used.

The relative influence of the surface and internal thermal resistance may usually be associated with the value of the Biot number $Bi = hw/k_m$ where h is the surface heat transfer coefficient, w is the characteristic length associated with the solid geometry (thickness of zeolite), and k_m is the thermal conductivity of the solid. If the Biot number is small, temperature gradients in the solid will be insignificant and the data from the model can be used. If the Biot number is large, a different ial equation has to be used which is two-dimensional for the solid.

In order to determine the Biot number for the copper-zeolite heat exchanger, we assumed a thermal conductivity of $k_m = 0.9$ (Btu/hr ft°F) and a thickness $w = 1/8" = 0.01$ ft (for a 1/4" thick zeolite with heat flowing into it from both surfaces). Then, the Biot number as a function of heat transfer coefficient is:

$$h = \ \ 4 \ \text{Btu/ft hr°F} \qquad Bi = 0.04$$
$$h = 10 \ \text{Btu/ft hr°F} \qquad Bi = 0.11$$
$$h = 50 \ \text{Btu/ft hr°F} \qquad Bi = 0.55$$

The Reynolds number of our system is such that the fluid flow is laminar and, therefore, especially for oils, the coefficient h is small about 10 and the Biot number is about $Bi = 0.1$. This is the dividing line between surface and bulk resistance.

Because of the low Biot number with laminar flow of oils and any flow of gases, we can use a simpler model for our theoretical evaluation. We calculated the temperature variations with time at different positions along the heat exchanger for both hot oil (liquid fluid) and hot gases (gaseous fluid) at the same power flow rate through the exchanger. The results of the calculations show that the liquid results in sharper temperature fronts and therefore eventually in higher system efficiencies. The gaseous fluid, on the other hand, results in broader fronts and a lower degree of energy recycling, therefore producing lower system efficiency. These results were confirmed later by experimental observation of thermal fronts propagation in long insulated copper pipes when a step of hot fluids (gases or liquids) is introduced at one end.

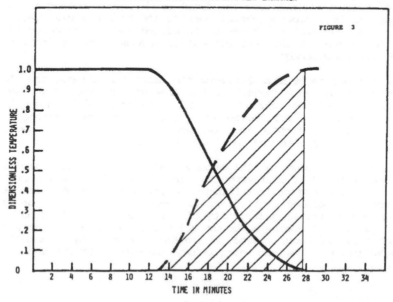

HEAT REGENERATION FOR A ZEOLITE HEAT EXCHANGER

FIGURE 3

Figure 3 represents the calculated normalized output fluid temperature for the two zeolite heat exchangers as a function of time during the first half of the cycle. The solid line represents the output temperature of the heat exchanger during the cooling cycle. This output is the input to the boiler and we can see that for the first 12 minutes we do not have to add any energy to the liquid. Only after 12 minutes does the gas burner turn on to keep the temperature of the liquid constant (at 400°F) and the amount of energy supplied by the burner increases in time until the end of the cycle at 28 minutes. Simultaneously, the output temperature of the heat exchanger being heated (desorption -- dashed line), remains at 0 (100°F) for the first 12 minutes and only afterwards increases to its maximum of 400°F at the end of the cycle. Since the fluid has to be kept cool (100°F) at this output, the heat represented by the hatched area under the dashed line is rejected to the ambient. This area is exactly the same as the area above the solid line, i.e. the energy that the gas burner has to supply as it should be expected from the conservation of energy considerations. It was determined that 75% to 80% of the cycle energy was regenerated form the adsorber to the generator so that the cycle COP becomes 3 to 4 times larger than the one without energy regeneration.

DESIGN AND CONSTRUCTION OF HEAT PUMP PROTOTYPE

The breadboard prototype was then designed, constructed and tested. It consisted of two 45-pound zeolite regenerative heat exchangers (90 pounds total) with modulation capability of 10% to 100% of its capacity of 1/2 ton (6,000 Btu/hr).

The heat pump prototype was operated over a period of 10 months and demonstrated a cooling COP above 1.2 and heating COP above 1.8. The complet performance maps for the heat pump were generated and the expected weak dependence on condenser and evaporator temperature was demonstrated. The analytical model was thereby validated. A novel serpentine-like regerative zeolite/hot fluid heat exchanger was designed, consturcted, and tested. Its efficiency was optimized while minimizing weight, volume and cost, and keeping thermal losses low. The proof of concept of the zeolite heat pump was established successfully.

Next, the modeling and design of a laboratory prototype of a 3-ton zeolite heat pump for residential and small commercial use was completed. Again, analytical and computer models were established and thereafter validated by comparison with experimental data determined previously.

complete performance maps were generated describing the operation of the laboratory prototype. The ideal control strategy matches the capacity to the load while maximizing the regenerator effectiveness. Once the

surface area & zeolite mass are fixed, the remaining controlling variables are the oil flow rate and the period. For operation around the design point the period is held constant and the mass flow modulated to meet the load. Further away form the design point the period may be modified to accurately control the system.

Figure 4 shows the COP-capacity curves for the baseline laboratory system (including combustion efficiency). These performance maps generated by the analytical model and verified with the computer simulation repersent the operating characterstics of a zeolite heat pump with set design paramters.

The performance maps were used to determine operating performance and costs in Dallas, Chicago and Boston. The predicted seasonal heating COPs for Chicago and Boston were between 1.45 and 1.6 while seasonal cooling COPs were 1.6 in Chicago, 1.8 in Boston and 1.3 in Dallas thus establishing considerable operating cost savings.The annual operating costs comparison between the high efficiency zeolite heat pump and an electric heat pump in Chicago and Dallas is shown in Figure 5. It indicates that zeolite heat pump cuts the operating cost roughly in half at both locations. This, combined with installed first cost comparable to electric heat pumps, indicates a very bright future for the regenerative zeolite system.The laboratory prototype was contructed and preliminary test results confirm the performance maps generated during the design phase while indicating manufacturing system cost only slightly higher than that of comparable electric units.

CONCLUSIONS

A high-efficiency regenerative zeolite heat pump was analyzed, designed, constructed and tested.
The system combines the nonlinear sorption properties of solid zeolites/ refrigerant vapor pairs, which reduce the system's first cost with the principle of regenerative heat exchangers with propagating temperature fronts to significantly increase the COP and achieve large operating cost reductions. The system contains no corrosive chemicals or fluorocarbons and presents no danger to the environment. Its low initial and operating costs indicate rapid penetration of the residential and small commercial HVAC markets.

1M.M. Dubinin and V.A. Astakhov, in Molecular Sieve Zeolites-II, American Chemical Society, Washington, DC 1971, pp. 69-85
2 F.W. Schmidt and A.J. Willmott, Thermal Energy Storage and Regeneration McGraw Hill and Hemisphere Publishing Company, 1981

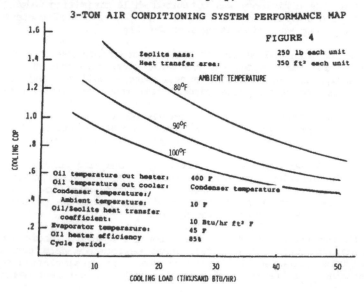

3-TON AIR CONDITIONING SYSTEM PERFORMANCE MAP

FIGURE 4

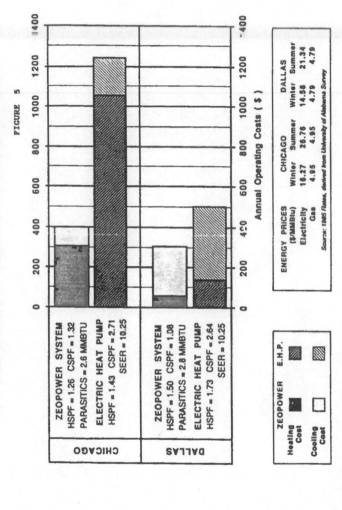

ANNUAL OPERATING COSTS FOR ZEOPOWER SYSTEM
AND ELECTRIC HEAT PUMP IN CHICAGO AND DALLAS

FIGURE 5

From Delima Associates,   062387-8701

An Evaluation Method of the Performance of Working Fluid in Heat Engine and Heat Pump for Effective Use of Low Temperature Thermal Energy

FUMIMARU OGINO
Department of Chemical Engineering
Kyoto University
Kyoto 606, Japan

ABSTRACT

An evaluation method of the performance of heat engine and heat pump has been proposed. Simple analysis of the heat engine and heat pump cycles has showed that the use of the heat pump is preferable to the heat engine for effective use of low temperature thermal energy. More detailed calculation has showed that the use of a mixture of water and organic compound as a working fluid results in the increase of the performance of the heat pump cycle.

INTRODUCTION

A hydrocarbon or a halogenated hydrocarbon of low boiling temperature is used in general as a working fluid in a heat-engine or a heat-pump cycle for recovery of energy from low temperature heat source. However, such compounds have some drawbacks that the latent heat is low, which requires a high circulation rate of the fluid, and the thermal conductivity in liquid state is small, which causes low heat transfer coefficients in condensation and boiling. These drawbacks would be eliminated by employing an azeotropic mixture of such an organic compound with water as a working fluid because water has a higher latent heat and a higher thermal conductivity in liquid.

The objective of the present investigation is to show that the use of such a mixture is effective for recovery of energy from a low temeperature heat source.

AZEOTROPIC MIXTURE

The essential feature of an immiscible system such as organic compound-water is that each liquid phase exerts its own vapor pressure regardless of the quantity of the other liquid present.

The phase characteristics of such a system are well illustrated on the temperature-composition diagram. The diagrm for a binary system of n-heptane and water is shown in Fig. 1.

Suppose initially the mixture is entirely in the vapor phase at a compositon. This would be represented by a point A in Fig. 1. Upon

757

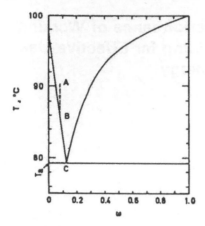

FIGURE 1. Temperature-composition diagram for binary system of n-heptane and water.

cooling at constant pressure, a point B is reached such that the vapor pressure of pure n-heptane is equal to the partial pressure of n-heptane in the vapor. With continued cooling, pure liquid n-heptane will be condensed and the vapor composition will decrease in n-heptane along the line BC. When the temperature T_a is reached, the partial pressure of the water in the vapor will have become equal to the vapor pressure of pure water, that is, the sum of the vapor pressure of the two liquids will become equal to the total pressure. At this temperature, there will be three phases in equilibrium, pure liquid n-heptane, pure liquid water, and vapor of composition C. Upon further heat removal, both components will condense from the vapor mixture and the temperature and the vapor composition will remain constant until the vapor phase disappears. The mixture at the composition C can be called azeotropic mixture although this term is used in general for miscible systems.

HEAT ENGINE CYCLE

The ideal heat engine cycle would be of the Carnot type shown in Fig. 2. The heats absorbed and rejected are given by

$$Q_1 = U_1 A_1 (T_1 - T_b) = m \Delta H_b \tag{1}$$

$$Q_2 = U_2 A_2 (T_c - T_2) = m \Delta H_c \tag{2}$$

If it is assumed that $U_1 A_1 = U_2 A_2 = UA$, the following equation can be obtained by using the relation $\Delta H_b = T_b \Delta S$ and $\Delta H_c = T_c \Delta S$,

$$(T_1 - T_b)/T_b = (T_c - T_2)/T_c = m \Delta S/UA = x \tag{3}$$

The quantity x can be considered to represent the dimensionless temperature difference.

The work done by the engine is given by

$$W = Q_1 - Q_2 = UA(T_1 x/(1 + x) - T_2 x/(1 - x)) \tag{4}$$

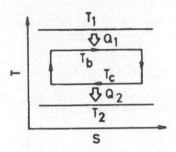 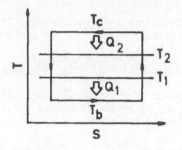

Figure 2. Ideal heat engine cycle.   Figure 3. Ideal heat pump cycle.

It is desirable that the work done by the engine is as large as possible. Since the work has a maximum value at $0 < x < 1$ as seen from eq. (4), the maximum value of W can be used as an index to evaluate the performance of a working fluid. However, from the thermodynamic point of view it is important to reduce energy wasted through irreversibility in the heat transfer process.

The lost works resulted from the irreversible transfer of heat from the temperature T_1 to T_b and T_c to T_2 are respectively given by the following equations.

$$W_{\ell 1} = T_0(1/T_b - 1/T_1)Q_1 \tag{5}$$

$$W_{\ell 2} = T_0(1/T_2 - 1/T_c)Q_2 \tag{6}$$

where T_0 denotes the temperature of surrounding. Consequently it is derirable that the value of

$$E = W - (W_{\ell 1} + W_{\ell 2})$$
$$= UA(T_1 x/(1 + x) - T_2 x/(1 - x) - 2T_0 x^2/(1 - x^2)) \tag{7}$$

is as large as possible. However, the mechanical energy and the lost energy are entirely different each other from the view point of energy utilization. Therefore, the following quantity will be useful as the performance index.

$$Y = \alpha_1 W - \alpha_2(W_{\ell 1} + W_{\ell 2})$$
$$= UA(\alpha_1 (T_1^2 x/(1 + x)^2 - T_2 x/(1 - x)) - 2\alpha_2 T_0 x^2/(1 - x^2)) \tag{8}$$

where α_1 and α_2 denote the prices of electricity and steam. Here the price of the lost energy is assumed to be equal to that of steam. The quantity Y has also a maximum value at $0 < x < 1$.

HEAT PUMP CYCLE

The ideal heat pump cycle is shown in Fig. 3. The heats rejected and absorbed and the dimensionless temperature difference are given by eqs. (1)-(3) under the same assumption as that for heat engine cycle.

The work added to system is given by

$$W = Q_2 - Q_1 = UA(T_2 x/(1 - x) - T_1 x/(1 + x)) \tag{9}$$

759

The lost works are given by eqs. (5) and (6). Therefore, the following quantity will be used as the performance index of the heat pump cycle.

$$Y = \alpha_2 Q_2 - \alpha_1 W - \alpha_2 (W_{\ell 1} + W_{\ell 2})$$
$$= UA(\bar{\alpha}_1 T_1 x)(1 + \bar{x}) - \ell_1(\alpha_1 - \alpha_2)T_2 x/(1 - x) - 2\alpha_2 T_0 x^2/(1 - x^2)) \qquad (10)$$

Comparison of eq. (10) with eq. (8) gives that the value of Y for the heat pump is larger than that for the heat engine when the following relation is satisfied.

$$T_{2 \text{ heat pump}} < (\alpha_1/(\alpha_1 - \alpha_2))T_{2 \text{ heat engine}} \qquad (11)$$

The temperatures of the heat source and of the surrounding are here assumed to be equal in the heat engine and the heat pump cycles. If we put $T_{2 \text{ heat engine}} = 30\ °C = 303\ K$, $\alpha_1 = 8.36 \times 10^{-6}$ ¥/J and $\alpha_2 = 2.63 \times 10^{-6}$ ¥/J, we can conclude that we had better adopt the heat pump than the heat engine when $T_{2 \text{ heat pump}} = 442\ K = 169\ °C$ which is in practice encountered. Hence we restrict the discussion to the heat pump cycle in the following.

From eq. (10) the quantity Y has a higher value when the temperature of the heat source T_1 and the heat transfer performance UA are higher and it does not depend on the thermodynamic properties. However, in the practical cycle with, for example, water as the working fluid, the vapor should be superheated in the condenser, resulting in larger temperature difference between the working fluid and the heat sink. This fact shows that Y depends on the thermodynamic properties.

In the present study the value of Y was calculated by using the temperature-entropy diagram of water, Freon R-11, n-octane and the azeotropic mixtures of n-octane-water and R-11-water.

The conditions of the heat source and sink were specified as follows;

heat source: water, $T_1 = 50\ °C$ (constant)
heat sink: water, $T_2 = 120\ °C$ (constant)
surface area of heat transfer: $A_1 = A_2 = 20\ m^2$

The heat transfer coefficients in boiling, condensation and forced convection were calculated by the relevant equations except that the heat transfer coefficient of the heat source was fixed at $4.5\ kW/(m^2\ K)$. In the calculation of heat transfer coefficients of the azeotropic mixture, the physical properties of pure organic compound were taken. The temperature of the surrounding was taken to be 293 K.

The expansion process of the liquid was assumed to be accomplished by an expansion valve, so that the lost work $W_{\ell 3}$ of the expansion process was calculated by assuming constant enthalpy.

The calculated result is depcited in Fig. 4. Since the temperature difference $T_1 - T_b$ was larger than 25 K and the value of Y was negative for pure n-octane, the result for n-octane is omitted in Fig. 4. From Fig. 4 the maximum value of Y for the azeotropic mixture of water with n-octane is 13.6×10^{-3} ¥/s = 49 ¥/h at $T_1 - T_b = 14\ K$ and this value is 2 - 2.5 times as large as those for pure water and pure R-11. The maximum value of Y for the mixture of water with R-11 is also larger than those for pure water and pure R-11. Therefore, the use of the azeotropic mixture as working fluid is effective for recovery of energy from the low temperature heat source.

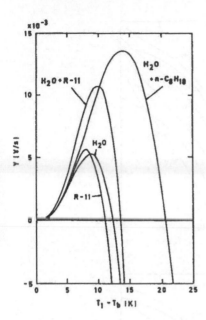

FIGURE 4. Comparison of performance of heat pump with several working fluids.

The heats rejected and absorbed, the work added and the lost works are depicted in Fig. 5. The lost work in the boiler $W_{\ell 1}$ is larger than that in the condenser $W_{\ell 2}$ for the mixture, showing that the boiling heat transfer coefficient must be increased for the mixture. In addition the throttling valve should be replaced by turbine or expansion engine in the expansion process of the mixture to reduce the lost work $W_{\ell 3}$.

CONCLUSIONS

(1) The use of the heat pump is preferable to the heat engine for the effective use of low temperature thermal energy.

(2) The use of the azeotropic mixture of organic compound and water as a working fluid in a heat-pump cycle is effective.

NOMENCLATURE

A   area of heat transfer, m^2
E   energy gain, W
m   mass flow rate of the working fluid, kg/s
Q   heat transferred, W
S   entropy, J/(kg K)
T   temperature, K
U   over-all heat transfer coefficient, $W/(m^2 K)$
W   work, W
W   lost work, W
Y   performance index, ¥/s

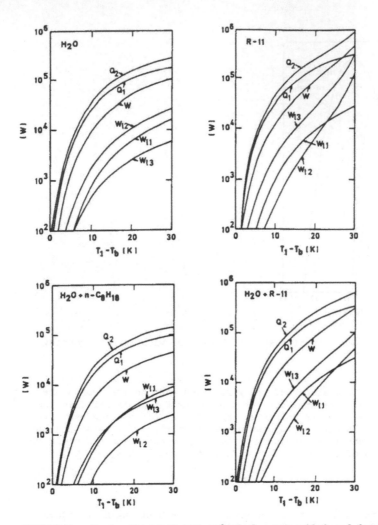

FIGURE 5. Heats absorbed and rejected, work added and lost works in heat pump cycle.

α_1	price of electricity, ¥/J
α_2	price of steam, ¥/J
ΔH	heat of vaporization, J/kg
ω	mass fraction of water

Subsctipts

a	azeotropic
b	boiling
c	condensing
0	surrounding
1	heat source
2	heat sink

Design and Operation of a Solar-Powered Absorption Air-Conditioning System in Hong Kong

MANKIT RAY YEUNG and KWOK-KEUNG MAK
Department of Mechanical Engineering
University of Hong Kong
Hong Kong

C. Y. CHU
Department of Mechanical and Marine Engineering
Hong Kong Polytechnic
Hung Hom, Hong Kong

ABSTRACT

To study the feasibility of utilizing solar power for comfort cooling in Hong Kong, a solar-powered absorption air-conditioning system has been designed and constructed on the campus of the University of Hong Kong (HKU). The system consists of a flat-plate collector system with a surface area of 38.2 m^2, a 4.7 kW cooling capacity LiBr-H_2O absorption chiller, a 2.75 m^3 hot water storage tank, a cooling tower, a fan coil unit, an electrical auxiliary heater, and the associated control systems. The system is also equipped with all types of sensors to measure the vital parameters of the system. The heart of the data acquisition system is an IBM-Compatible PC/XT personal computer designed to monitor the performance of the system. The purpose of the present paper is to introduce the design features of the HKU solar-powered air-conditioning system and some of its operational results.

INTRODUCTION

In Hong Kong, the yearly average insolation of solar energy is approximately 15 MJ per m^2 per day [1]. With a physical area of 1000 square kilometers approximately, this represents a potential energy of 5.82 x 10^{12} MJ/year. Since Hong Kong has a latitude of 22.3°N, the heating season is extremely short, the demand for heating is not as high as comfort cooling and there is considerable interest in the utilization of solar energy for cooling in the summer.

The earliest study on solar energy application in Hong Kong was an investigation of using solar energy for winter heating of swimming pools [2]. In 1980, a pilot solar heating system consisting of 60 m^2 of collector panel area was built at the Stanley Public Bath-House. Performance results indicated that the system was not exconomical because the demand for hot water at the bath-house is only between November and March [3]. Other major solar installations include those at the Maximum Security Prison at Shek Pik (450 m^2 of collector area) and the Drug Addiction Treatment Centre at Hei Ling Chau (400 m^2 of collector area). They were both completed in 1983. In 1983, a conceptual design of a solar absorption air-conditioning system was completed by Chan et al [4] at the University of Hong Kong. The design was finalized and actually constructed by Yeung and Mak in mid-1986[5]. The HKU solar-powered absorption air-conditioning system consists of twenty flat-plate solar collectors, five circulating pumps, a hot water storage tank, an

763

auxiliary heater rated at 9 kW, a LiBr absorption chiller with a nominal cooling capacity of 4.7 kW, a cooling tower, fancoil unit, piping connecting these components and the associated control system. The entire solar system is located on the roof of the Redmond Building of the University of Hong Kong and the layout of the system is given in Figure 1.

DESIGN ON THE HKU SOLAR-POWERED AIR-CONDITIONING SYSTEM

The entire solar-powered air conditioning system is made up of four flow circuits; namely, the collector panel cirucit, the generator flow circuit, the chilled water circuit and the cooling water circuit. All the flow circuits except the cooling water circuit are insulated with armaflex to reduce heat losses. Valves are installed at various locations for flowrate and flow path control.

The collector panel circuit consists of twenty Yazaki flat-plate single-glazed solar collectors. The absorber plate of the collectors is of tube-in-sheet type with an effective area of 1.91 m^2. The twenty solar collectors are arranged in two rows each of which consists of ten solar collectors tilted at an angle of 22.3° facing south as shown in Figure 2. Water is circulated through the collector panels by a circulating pump. The hot water is stored in a hot water storage tank with an effective storage capacity of 2.75 m^3. The outside surfaces of the tank are insulated by 7.6 cm thick polystyrene covered by cement plastering to minimize heat loss to the surroundings.

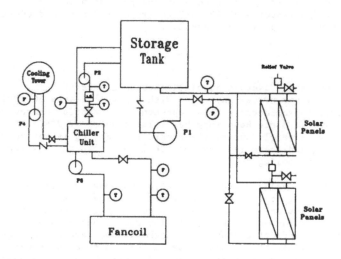

FIGURE 1. Layout of the H.K.U. Solar Air-Conditioning System

FIGURE 2.   Solar Panels of the HKU Solar Air-Conditioning System

During operation, water is circulated by the circulating pump to the
collectors where it is heated up by the solar energy and returned to the
storage tank.   On the other hand, the generator pump in the generator
flow circuit circulates hot water from the storage tank to the generator
of the absorption chiller.   After the hot water gives up its heat to the
absorption chiller in the generator, the water is returned to the storage
tank.   Chilled water is produced in the absorption chiller and is
circulated by the chilled water pump of the chilled water circuit to the
fancoil unit inside the building for comfort cooling.   The heat generated
in the condenser of the chiller is carried by the cooling water to the
cooling tower where its heat is rejected to the environment.   Figure 3
shows the arrangement of the storage tank, chiller, cooling tower and
auxiliary heater.   All important design parameters of the system are
summarized in Table 1.

DATA ACQUISITION AND PROCESSING SYSTEM

The performance of the HKU solar air-conditioning system is monitored by
a personal computer-based data acquisition system.   Three types of
sensors are used to measure the vital parameters of the system: platinum
resistance thermometers (RTDs) for temperatures, turbine flow meters for
water flow rates and a pyranometer for solar irradiance.   The heart of
the moniotring system is an IBM-Compatible PC/XT computer together with
an analog-to-digital converter card.   The sensors and the computer are
linked by this analog-to-digital card which converts the analog singals
generated by the sensors to digital signals for processing.

765

FIGURE 3. Arrangement of Major Components

TABLE 1. Major Design Parameters of the HKU Solar Air-Conditioning System

Collector area (20 panels)	38.2 m^2
Panel type	flat plate
Circulating pump power	150 W
Storage tank capacity	2.75 m^3
Tank insulation	polystyrene & cement plastering
Chiller manufacturer	Yazaki
Chiller capacity	4.7 kW
Chiller heat input	7.8 kW
Chiller generator water flow	0.27 l/s
Auxiliary heater capacity	9.0 kW
Auxiliary heater setpoint	75 °C
Cooling tower type	wet
Cooling tower capacity	12.4 kW
Cooling tower water flow	0.59 l/s
Fan-coil manufacturer	York
Fan-coil capacity	5.3 kW
Chilled water flow	0.22 l/s

The input data from the RTDs, pyranometer and flowmeter sensors are recorded every 30 seconds and stored in the IBM-Compatible PC/XT Computer. Every 5 minutes, the recorded data are averaged, analyzed, stored on a floppy disk and batched to the printer for output. The data is recorded daily from 6:30 to 19:30. The entire data acquisition system is shown in Figure 4.

RESULTS OF OPERATION

Since commissioning in the summer of 1986, the HKU solar-powered air-conditioning system has been in use, accumulating continuous operating data in both summer and winter. The data has indicated that the water temperature from the collectors is well into 80°C or higher, which is high enough to drive the LiBr absorption chiller. However, auxiliary heating is still necessary in the morning because of the large inventory of the storage tank and the heat loss during night time. On the other hand, the chilled water temperature is generally in the teens which is fairly ideal for air-conditioning. Since the operation season of 1987 was from May to September, it would not be practical to show all the operational data here. The present paper only presents the operation history of July 12, 1987 for illustration and the overall energy summary for the entire cooling season.

FIGURE 4.   Data Acqusition System

Figure 5 shows the time histories of solar power incident the panels and the actual solar gain for July 12, 1987. It can be seen that the actual solar gain for most of the day is significantly higher than 7.8 kW - the required input to the absorption chiller. Also, Figure 6 shows the time histories of the thermal power supplied to the absorption chiller and the actual cooling load delivered at the fan-coil unit. It is obvious that the thermal power supplied to the chiller is significantly lower than the solar gain from the panels. This implies that a large portion of the energy gained from by panels was never delivered to the chiller and eventually lost from the storage tank.

Table 2 is a summary for overall system energy balance of the HKU solar air-conditioning system for the 1987 cooling season. It can be seen that the solar gain is generally much less than the incident energy on the panels. In fact, the overall panel efficiency for the entire season was about 38%. For the entire season, the chiller produced about 9553 MJ of cooling effect and the overall chiller C.O.P. is approximately 0.51 indicating that the chiller performance is basically satisfactory. On the other hand, the energy supplied by the auxiliary heater is generally quite high and this is especially obvious for May. The reason for this is the absence of the cooling load control. As a result, the chiller was basically operating to meet a constant demand resulting in an over contribution from the auxiliary heater. In reality, the cooling load demand in May is generally lower than July or August but the cooling effect of the present system did not reflect this in the summer of 1987. Also, the relatively high contribution from the auxiliary heater was caused by the heat loss from the storage tank during night-time and the chiller had to rely on the auxiliary heater in the early morning hours. Naturally, this can be improved by upgrading the tank insulation and the installation of a cold storage system for the chilled water.

TABLE 2. Overall Energy Balance for HKU Solar Air-Conditioning System (1987)

	Incident Energy (MJ)	Solar Gain (MJ)	Auxiliary Heat (MJ)	Energy to Chiller (MJ)	Cooling Effect (MJ)
May	10240	2479	4843	4002	2040
June	13573	4785	3008	3512	1772
July	15328	6936	3081	4222	2290
August	18358	7779	1914	4011	2073
September	9197	3379	2857	2865	1463
Season	66696	25358	15703	18612	9553

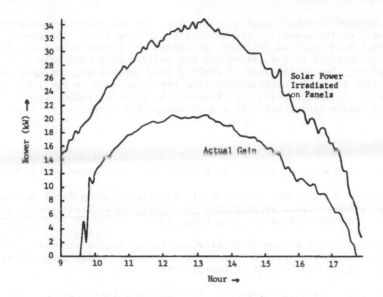

FIGURE 5. Energy Irradiated on Panels and Actual Gain on July 12, 1987

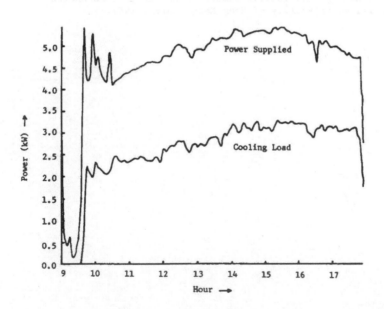

FIGURE 6. Power Supplied to Chiller and Cooling Load on July 12, 1987

SUMMARY AND CONCLUSION

A solar-powered air-conditioning system has been successfully designed
and erected on the campus of the University of Hong Kong. Continuous
operational data has been obtained for almost the entire summer of 1987
and the performance of the system and the chiller were basically
satisfactory. Due to the lack of cooling load control and the heat loss
during night time, the contribution from the solar gain is too low.
Improvements in theses areas are necessary in order to reduce in
auxiliary heater contribution to a more acceptable level.

REFERENCES

1. Royal Observatory of Hong Kong, Monthly Weather Summary, July, 1987.

2. Bruges, E.A., The Winter Heating of Open-Air Swimming Pools in Hong
    Kong, a first report prepared for the Public Works Department and
    Urban Services Department, Hong Kong, 1978.

3. Osborne, G.J., A Review of Alternative Energy Source and the
    Adminstration's Efforts in Energy Saving, Green Productivity, Hong
    Kong Productivity Council, October, 1985.

4. Chan, S.K., Design,Construction and Continuous Evaluation of a Medium
    Scale Solar Energy System, Hong Kong Engineer, pp. 43-48, January,
    1983.

5. Mak, K.K., Design of a Solar-Powered Absorption Air-Conditioning
    System, HKU-SP-001, Interim Report, Department of Mechanical
    Engineering, University of Hong Kong, March, 1987.

An Experimental Study of a Diesel Engine Driven Water-to-Water Heat Pump System

TINGJIN SHI, HUANG GU, YUANJI MA, FENGNING LIU, WEIMING LU,
and HAOREN REN
Department of Thermoscience and Engineering
Zhejiang University
Hargzhou, PRC

ABSTRACT

This paper presents an experimental study of a diesel-engine driven heat pump system (DEHP). A comparison is also made between the performance of the DEHP and the electric heat pump system (EHP). Experiments showed that the primary energy ratio (PER) for the DEHP was about 50% higher than that for the EHP, if the recovery of waste heat from the cooling water and from the exhaust gas of the diesel-engine was included. If we consider the prices of electricity and oil, the operating cost for the DEHP is only about 50% that for the EHP. Moreover, the exit temperature of the hot water for the DEHP is about 10°C higher than that from the EHP. If the same temperature of exit hot water is required, the compression ratio of the heat pump compressor may be decreased, and the coefficient of performance (COP) of the DEHP will be higher than that of the EHP.

INTRODUCTION

Heat pump technology as an energy conservation measure has been developed widely all over the world, especially, after the second "energy crisis". To save energy, it is necessary to consider the primary energy input to the whole chain of energy flow. The efficiency from primary energy to the electricity supply is about 32%, and, if the COP of a EHP is about 3.0, the PER for the EHP is near to 1. Because of the recovery of waste heat from cooling water and exhaust gas, the PER for a gas engine heat pump and a DEHP, is higher than that of an EHP, it can reach 1.45. Therefore,gas engine heat pump and DEHP are energy saving systems.

This paper presents an experimental study of a DEHP. Experimental results show that the DEHP can supply 50% more heat than that provided by the EHP. Its PER reachs 1.45 and the temperature of the hot water from the DEHP is about 10°C higher than that from the EHP.

TEST FACILITY AND INSTRUMENTATION

Fig.1 shows a schematic of the experimental system. The designed heating capacity for this system is about 12KW, and the driven power is about 4KW. The heat source was taken from a tank arrangement with electrical immersion heaters controlling the inlet water temperature of the evaporator. The heat sink was an open water system. A condenser (1.8M²), an evaporator ($2.5M^2$), a diesel engine (4KW), a compresor (displacement is 4.81E-04 M^3) and a thermostatic expansion valve were also used in this system.

Necessary primary measurements included: (1) The compressor entering temperature; (2) the entering pressure; (3) the discharge temperature; (4) the discharge pressure; (5) the heat input to the evaporator; (6) the heat rejected from the condenser; (7) the primary energy consumption. In addition, the pressure, the temperature, and the water flow rates were measured at various points in the system to understand and characterize system behavior. Fig.1 also indicates the instrumentation layout.

To measure the heat rejected from the condenser(Q_c), the water flow rate and the temperature differences across the condenser were measured. Qc was calculated from : $Q_c = G_c*(h_0(t_0) - h_1(t_1))$, where, G_c is the water flow rate, t_1, t_0 are the water temperatures entering and leaving the condenser respectively. $h(t)$ is the enthalpy entering and leaving the condenser corresponding to t.

The waste heat energy recovered from the cooling water and the exhaust, and the heat input to the evaporator were measured in a similar way.

The primary energy consumption was calculated from the consumption of oil using the Eq. $Q_{o1} = G_{o1}*Q_d/T$, where G_{o1} is the oil consumption in a time of T seconds and Q_d is the calorific value of diesel oil.

In order to compare the performance between a DEHP and an EHP, an electric motor was used in the system to replace the diesel engine and the measurements were repeated.

DATA PROCCESSING AND EXPERIMENTAL RESULTS

A test plan was arranged beforehand, to assist the performance analysis and decrease the errors in the results. The various levels of factors such consisting of as the water flow rates flowing through the condenser, evaporator and the temperature entering the evaporator, were arranged to form a test plan. The EHP performance was also tested by the same plan.

A computer program was composed to include the calculation of the thermodynamic properties of the refrigerants using the Martin-Hou equation and the water using the IFC equation. This program can be used to calculate the coefficient of performance of the cycle, the PER, etc. The computer program diagram as shown in Fig.2 .

Another computer program was used to calculate the perfomance of an EHP, to compare the performance of a DEHP with that of an EHP, it is necessary to apply the PER. However, the EHP uses electricity produced in power plants; the COP of an EHP should be converted into PER. If we consider the thermal efficiency of power plants (33%) and the efficiency of electricity transportation (92%), the PER of an EHP can be expressed as follows:

PER. = COP*92%*33% = 30%COP ......(1)

A statistical regression analysis method to process experimental data, was included in another computer program. The performance of the DEHP and of the EHP were analysed by using this program.

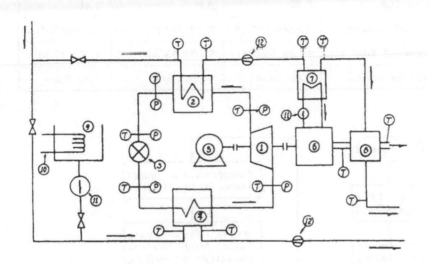

FIGURE 1. Sysetm Schematic and Instrumentation Layout

T,P   represent temperature sensors, pressure gages, respectively
1--compressor; 2--condenser; 3--thermostatic expansion valve;
4--evaporator; 5--electric motor; 6--diesel engine;
7,8--heat exchanger for recovering waste heat from cooling water
and exhaust gas, respectively; 9--water tank; 10--electric heater;
11--water pump; 12--flowmeter.

TABLE 1. Operating Cost of Different Equipments

	Boiler	EHP	DEHP
Heating capacity	10^6 KJ	10^6 KJ	10^6 KJ
Type of energy	Coal	Electricity	Diesel oil
Calorific capacity	20930 KJ/KG	3600 KJ/KW.hr	42496 KJ/KG
Price	0.08 Yuan/KG	0.16 Yuan/KW.hr	0.45 Yuan/KG
PER or COP	0.70	3.0	1.40
Energy of consump.	1.4286×10^6 KJ	0.3333×10^6 KJ	0.6896×10^6 KJ
Consump. of fuel,ele.	68.26 KG.coal	92.59 KW.hr	16.23 KG.oil
Operating cost	5.46 Yuan	14.81 Yuan	7.30 Yuan

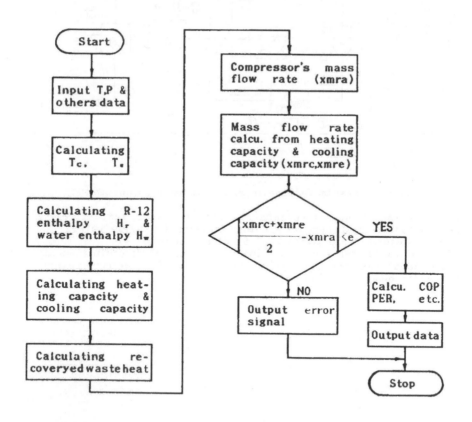

FIGURE 2. Flow Chart of Program

The experimental results are plotted in Fig.3 - Fig.8 and expressed in empirical formulae (2) - (10). Empirical formulae (2) - (5) represent the relation between the various parameters and the primary energy consumption; empirical formulae (6) - (10) show the relation between various parameters (including the temperature of exit hot water and of leaving condenser) and the condensing temperature (t_c) and the evaporating temperature (t_e). Formulae obtained are as follows:

$$PER_d = -0.444 + 0.331*Q_{ii} - 0.015*Q^2_{ii} \qquad \text{-----(2)}$$

$$Q_{tot} = -22.560 + 5.560*Q_{ii} - 0.194*Q^2_{ii} \qquad \text{-----(3)}$$

$$Q_c = -8.472 + 2.410*Q_{ii} - 0.061*Q^2_{ii} \qquad \text{-----(4)}$$

$$Q_{re} = -12.455 + 2.908*Q_{ii} - 0.124*Q^2_{ii} \qquad \text{-----(5)}$$

$$PER_d = -2.581 + 0.176*t_c - 0.02*t_c^2 + 0.002*t_e + 0.000237*t_e^2 \qquad \text{-----(6)}$$

$$Q_{tot} = -27.64 + 1.726*t_c - 0.019*t_c^2 + 0.306*t_e + 0.0075*t_e^2 \qquad \text{-----(7)}$$

$$Q_c = -15.016 + 0.998*t_c - 0.013*t_c^2 + 0.153*t_e + 0.0089*t_e^2 \qquad \text{-----(8)}$$

$$t_{co} = 29.01 - 0.140*t_c + 0.010*t_c^2 + 0.048*t_e - 0.0024*t_e^2 \qquad \text{-----(9)}$$

$$t_{exit} = 57.44 - 1.258*t_c + 0.024*t_c^2 + 0.569*t_e - 0.025*t_e^2 \qquad \text{-----(10)}$$

Here:

PER$_d$————PER of DEHP
Q_{tot}————total heating capacity of DEHP
Q_{re}————heat energy recovering from cooling water and exhaust
t_{co},t_{exit}—the temperature of exit hot water and of leaving condenser, respectively.

DISCUSSION

In this section, the performances of the DEHP will be analysed, and the performance of the DEHP and the EHP will be compared under the same conditions. In addition, a simple estimation of the operating costs will be made using the performance data from the previous section.

The Performance of the DEHP

Fig.3 shows curves of the total heating capacity, the PER, the recovery of waste heat and the heat energy rejected from the condenser versus the primary energy consumption. The heating capacity increases in direct ratio with the primary energy consumption, and then, its increment gradually becomes slower, until $Q_{ii} > 13KW$, when it is nearly a constant. Because the power produced by the diesel engine is greater than the power economical power the thermal efficiency of the diesel engine will decrease (Fig.4), and the PER becomes worse. In the whole range of variation, there is an optimum PER value, corresponding to 11KW primary energy consumption, which is about 1.45 . When the diesel engine works in the range of the optimum (RPM = 1925, power = 3.5), the specific oil consumption reachs the most economical value and the performance of the DEHP is also best within the same range. When the power produced by the

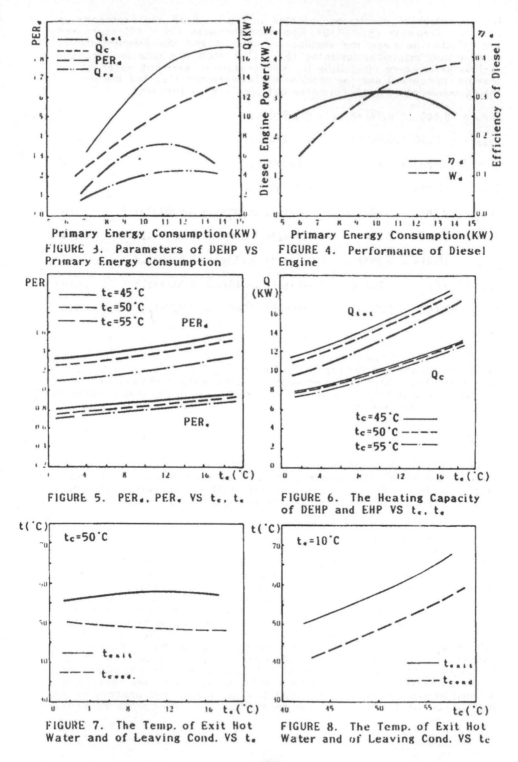

FIGURE 3. Parameters of DEHP VS Primary Energy Consumption

FIGURE 4. Performance of Diesel Engine

FIGURE 5. PER₄, PER₄ VS t_c, t_e

FIGURE 6. The Heating Capacity of DEHP and EHP VS t_c, t_e

FIGURE 7. The Temp. of Exit Hot Water and of Leaving Cond. VS t_e

FIGURE 8. The Temp. of Exit Hot Water and of Leaving Cond. VS t_c

diesel engine increases further direct ratio with the primary energy consumption , its increment becomes slower. Because of the decrease of efficiency of the diesel engine caused by overload, the performance of the DEHP deteriorates.

The performance of the DEHP depends strongly on the thermal efficiency of the diesel engine and the amount of heat energy recovered from the cooling water and the exhaust gas. Fig.3 - Fig.4 show these properties. It is important for the diesel engine to work in the optimum zone in order to save primary energy consumption.

The curves of the PER and the heating capacity versus the condensing and the evaporating temperature are showed in Fig.5 - Fig.6. The experimental curves show the same laws as the results of theoretical analysis. The heating capacity and the PER are increased with the decrease of the condensing temperature and increase of the evaporating temperature.

Further correlation curves between the temperatures of the exit hot water and of the exit condenser and the condensing and the evaporating temperature are shown in Fig.7 -Fig.8. The condensing temperature has a great influence upon the temperature of the exit hot water, but, the influenced of the evaporating temperature on the variation of the temperature of the water is smaller.

Comparison

The performance of the DEHP is better than that of the EHP, because of the recovery of waste heat energy from the cooling water and the exhaust gas. Fig.5 - Fig.6 show the curves of the PER of the EHP and the DEHP (in this scheme PER_e = 0.3*COP). The PER_e fluctation is in the range of 0.8 - 0.9. If the diesel engine operates in the zone of high efficiency by controlling the rotation speed, the PER_d can reach 1.45, be about 50% more than the PER_e , and the fluctation of PER_d is greater than that of PER_e.

The temperature of the exit hot water from the DEHP is also higher than that of the EHP, for the same reason as given above. Fig.7 - Fig.8 show correlations between the temperature of the exit hot water and the con- densing and the evaporating temperature . It is easy to see that the temperature of the exit hot water supplied by the DEHP is about 10°C higher than that supplied by the EHP. Both temperatures depend strongly on the condensing temperature rather than on the evaporating tempera- ture. The hot water coming from the condenser, flows through both heat exchangers recovering waste heat from cooling water and exhaust gas, in series. The temperature increases further from this so that hot water with a higher temperature is obtained . This may be important in an industrial satuation.

If the same temperature of exit hot water as that of the EHP is requir- ed, the compression ratio of the heat pump compressor may be decreased and the performance of the heat pump itself which is driven by the diesel engine will be higher. This is explained in Fig.8 and Fig.5. From Fig.8, if the temperature of the exit hot water required is 55°C, the condensing temperature will be decreased from 55°C to 47°C, when the evaporating temperature is equal to 10°C. The reversed cycle thermal efficiency will also be increased. From Fig.5 we may get the result that the PER_d is increased from 1.2 to 1.4 .

Operating Costs

The operating costs for the different equipment supplying heat energy were analysed in this section.

If we suppose the boiler efficiency to be 0.70, the COP of an EHP is 3.0 (from Eq.(1) and we may obtain the PER of the EHP as: $PER_e=30\%COP=0.90$) and the PER of a DEHP as 1.40 (from Fig.5). Considering the different prices of electricity (RMB:0.16Yuan/KW.hr), coal (RMB:0.08Yuan/KG.coal) and oil (RMB:0.45Yuan/KG.oil), the operating cost per 10^6 KJ heating capacity was obtained in Tab.1 .

From Tab.1, we can see that the PER of the DEHP is about 50% more than that of the EHP, and the operating cost of the DEHP was also decreased greatly from 14.81 RMB Yuan/10^6KJ of the EHP to 7.30 RMB Yuan/10^6KJ. Although, the PER of the DEHP (1.4) is hogher than that of boiler(0.70). THe saving of operating costs by the DEHP is dependent on the prices of the different fuels and of electricity.

CONCLUSION

The DEHP, because it uses oil as the energy source, uses little or no electricity. Its PER will be increased because of the recovery waste heat energy from the cooling water and tha exhaust gas.

(1)   The PER of the DEHP can reach 1.45, and is about 50% more than that of the EHP and twice as high as that of coal boiler, because of the recovery waste heat energy from the cooling water and the exhaust gas.

(2)   The temperature of the exit hot water supplied by DEHP in industrial application is about 10°C higher than that of the EHP, for the same reason as above. This temperature depends strongly on the condensing temperature.

(3)   If the same temperature of the exit hot water as that of the EHP was required, the compression ratio for a DEHP can be decreased and the performance increased. When the temperature of the exit hot water is equal to 55°C, the PER increases from 1.2 to 1.4 .

(4)   The performance of a DEHP depends strongly on the thermodynamic efficiency of the diesel engine.

(5)   Although the PER of the DEHP is nearly twice as high as that of a boiler, its operating cost is not less, because of the difference in price of oil and coal. But, in comparison with an EHP, the operating cost will be decreased from RMB 14.81 Yuan/10^6KJ to RMB 7.30 Yuan/10^6KJ.

REFERENCE

(1)   A.Patand, Ph.D, Analysis of A Gas-Engine-Driven Heat Pump System, *ASHRAE Transaction*, No3,1982.

(2)   San.V.Shelton, Ph.D, Natural Gas I.E Engine Driven Heat Pump, *6th Heat Pump Technology Conference*, 4-5, Nov. 1982.

(3)   F.W.Steimle, Ph.D, Developments in Heat-Operated Heat Pumps in West

West Germany, *ASHRAE Transaction*, No5, 1982.

(4)   J.E.Christian, A Unitary Air-to-Air Heat Pump, ORNL,DOE, 1977.

(5)   F.B. Wang, Experimental Reseach and Energy Saving Effect Analysis
for Gas-Engine Heat Pumps, (in Chinese).

Stirling Refrigerator with Gas Extraction for Energy Saving

ZHONG-QIU SUN and HONG-XING HAN
Refrigeration and Cryogenics Engineering Lab
Shanghai Institute of Mechanical Engineering
Shanghai, PRC

Stirling refrigerator has many advantages, e.g. large refrigerating capacity, wide range of refrigerating temperature, compact structure, high cooling rate and high thermodynamic efficiency etc.. But when Stirling refrigerator was used for the liquefaction of gases as usually applied in the laboratories, its thermodynamic efficiency becomes considerably low. Take the air liquefaction for example, the theoretical calculation indicated that 0.202 kw-hr is the minimum work required to produce one kilogram of liquid air. It is generally know that the coefficient of performance of an ideal Stirling refrigerator is the same as that of Carnot cycle. But even if the Stirling refrigerator were ideal, the work required for liquefying one kilogram of air would be as much as 0.329 kw -hr, because during the liquefaction process, the total refrigerating capacity was produced at 77 K, while the inlet air was at the room temperature. It is this large temperature difference in the heat exchanger that results the irreversible loss, thus increasing the power consumption and decreasing the coefficient of performance. The same loss is incurred when two-stage Stirling refrigerator is used for precooling and/or liquidizing.

To improve the coefficient of performance, a gas extraction Stirling refrigerator is introduced in this paper as illustrated in Fig. 1. The low temperature working gas $G_{\$}x$ is extracted from the expansion space E of the Stirling refrigerator into the cold side of the heat exchanger HE. The gas to be liquefied is induced into the warm side of the heat exchanger and cooled down form room temperature to its saturation temperature, then is liquefied in the cold head L of the Stirling refrigerator. Simultaneously the working gas is heated up to the room temperature and returned back to the compression space C of the Stirling machine. Obviously in the system with the gas extraction, the temperature difference in the heat transfer between the working gas and the gas to be liquefied can be reduced and the effective refrigerating capacity is enlarged. Hence the exergy efficiency of the system is increased.

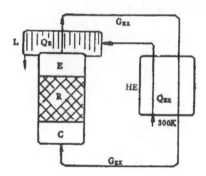

FIG.1 Schemic diagram
of the gas extraction
refrigeration

Calculation of Gas Extraction Cycle

The total effective refrigerating capacity of the gas extraction cycle can be expressed as follows:

$$Q = Q_{EX} + Q_T = \frac{G_{EX}}{G_{EXmax}} Q_{EXmax} + \frac{G_{EXmax} - G_{EX}}{G_{EXmax}} Q_{Tmax} \qquad (1)$$

we must determine Q_{Emax}, Q_{EXmax}, G_{EXmax}, G_{EX} and P, V, T, G parameters of the cycle for calculating Q. Usually the complicated analysis calculating methods were used for solving the Stirling models, in this paper a new method which is named 'Equivalent Volume Method' is proposed. For the given Stirling refrigerator, both expansion and compression volume $V(\alpha)$, $V_C(\alpha)$ versus crank angle can be found, the clearance volumes in the different parts V_{si} are fixed. The assumed gas volumes V_{EN}, V_{CN}, V_{si} are called the equivalent volumes which represent the gas volumes occupied by V_E, V_c and V_{si} respectively at a same equivalent temperature.

$$V_{EN}(\alpha) = V_E(\alpha) * \frac{T_N}{T_E(\alpha)} \qquad (2)$$

$$V_{CN}(\alpha) = V_C(\alpha) * \frac{T_N}{T_C(\alpha)} \qquad (3)$$

$$V_{SN}(\ ) = \sum V_{SiN} = \sum V_{si} * \frac{T_N}{T_{si}} \qquad (4)$$

the selection of the equivalent temperature T_N influences only the absolute values of the equivalent volumes not the relative ones. The sum of the equivalent volumes is called the total equivalent volume V_N.

$$V_N(\alpha) = V_{EN}(\alpha) + V_{CN}(\alpha) + V_{SN} \qquad (5)$$

The pressure of the Stirling refrigerator $P(\alpha)$ is linearly inverse proportion to the total equivalent volume V (α), which can be proved:

$$P(\alpha) * V_E(\alpha) = G_E(\alpha) * R * T_E(\alpha) \tag{6}$$

$$P(\alpha) * Vc(\alpha) = Gc(\alpha) * R * Tc(\alpha) \tag{7}$$

$$P(\alpha) * Vs = Gs(\alpha) * R * Tsi \tag{8}$$

$$G = G_E(\alpha) + Gc(\alpha) + Gsi(\alpha) = constant \tag{9}$$

$$P(\alpha) * V_{EN}(\alpha) = G_E(\alpha) * R * T_N \tag{10}$$

$$P(\alpha) * Vc_N(\alpha) = Gc(\alpha) * R * T_N \tag{11}$$

$$P(\alpha) * Vs_N = Gsi(\alpha) * R * T_N \tag{12}$$

$$G = G_E(\alpha) + Gc(\alpha) + Gsi(\alpha) = constant \tag{13}$$

From equations (5), (10), (11), (12), (13) we have:

$$P(\alpha) * V_N(\alpha) = G * R * T_N = constant \tag{14}$$

From the boundary condition the minimum pressure is the charging pressure and correlated to the maximum total equivalent volume:

$$P(\alpha) = P_{min} * \frac{V_{Nmax}}{V_N(\alpha)} \tag{15}$$

According to the variable mass thermodynamics, the gas flows into the expansion or compression space while the crank sweeps from an angle of α_1 to α_2, then we have:

$$\frac{P(\alpha_1) * [V_E(\alpha_1)]^k}{\dfrac{T_E^*}{T_E(\alpha_1)}} = \frac{P(\alpha_2) * [V_E(\alpha_2)]^k}{\dfrac{T_E^*}{T_E(\alpha_2)}} \tag{16}$$
$$[G_c(\alpha_1)] \qquad\qquad [G_E(\alpha_2)]$$

$$\frac{P(\alpha_1) * [Vc(\alpha_1)]^k}{\dfrac{Tc^*}{Tc(\alpha_1)}k} = \frac{P(\alpha_2) * [Vc(\alpha_2)]^k}{\dfrac{Tc^*}{Tc(\alpha_2)}k} \tag{17}$$
$$[Gc(\alpha_1)] \qquad\qquad [Gc(\alpha_2)]$$

If the gas flows out of the expansion or compression space:

$$\frac{[T_E(\alpha_1)]^{\frac{k}{k-1}}}{P(\alpha_1)} = \frac{[T_E(\alpha_2)]^{\frac{k}{k-1}}}{P(\alpha_2)} \tag{18}$$

783

$$\frac{[Tc(\alpha_1)]^{\frac{k}{k-1}}}{P(\alpha_1)} = \frac{[Tc(\alpha_2)]^{\frac{k}{k-1}}}{P(\alpha_2)} \qquad (19)$$

where : $k = \dfrac{Cp}{Cv}$; $T_E^\bullet$ ---- refrigerating temperature

The boundary conditions:
When P equals P_{min}, Tc equals Tc_{min} that is room temperature Tc_{min}. When P equals P_{max}, T_E equals T_{Emax} that is refrigerating temperature T_{Emax}, P_{min} ·is the charging pressure of the Stirling refrigerator.

From equations (2)--(9) and the boundary conditions, all the parameters of the Stirling cycle can be determined by a computer. Q_{Emax} can be calculated from P-V diagram of the expansion space.

According to the heat balance of the refrigerator (Fig. 2), the Q_{EXmax} can be determined:

$$Gc^\bullet * Cp * Tc^\bullet + G_E^\blacktriangle * Cp * T_E^\bullet = Gc * Cp * Tc + G_E * Cp * T_E \qquad (20)$$

$$G_E^\blacktriangle = G_E + G_{EXmax} \qquad (21)$$

$$Gc = Gc^\bullet + G_{EXmax} \qquad (22)$$

$$G_{EXmax} = G_E * \frac{T_E^\bullet - T_E}{Tc^\bullet - T_E} \qquad (23)$$

where G_E, $T_E^\bullet$ and $Tc^\bullet$ can be determined from G--α, T_E--α, and Tc--α curves respectively, then we have:

$$Q_{EXmax} = G_{EXmax} * Cp * (Tc' - T_E^\bullet) \qquad (24)$$

So far, the solution is completed. Table 1, Fig.3,4,5,6,7 are results cycle of the calculation for an ideal cycle of the Stirling refrigerator with the gas extraction which was used in our test-set.

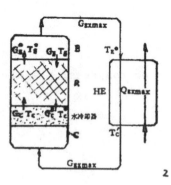

FIG.2 Heat balance of regenerator

TABLE 1. : Calculation Results

		Conventional Stirling	Gas extracted at low pressure	Gas extracted at high pressure
Refrigerating temperature	K	75-300	75-300	75-300
Refrigerating capacity	kw	2.217	4.211	3.192
Power consumption	kw	9.567	9.567	9.632
Quantity of extracted gas	kg/hr	0.	4.251	3.485
Coefficient of performance		0.2317	0.4402	0.3314
Exergy efficiency		0.1941	0.3687	0.2270

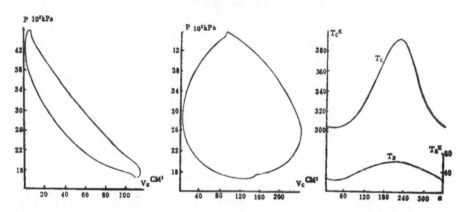

FIG.3 P-V diagram of expansion space V_E

FIG.4 P-V diagram of compression space Vc Tc

FIG.5 Temperature T_E vs.crank angle α

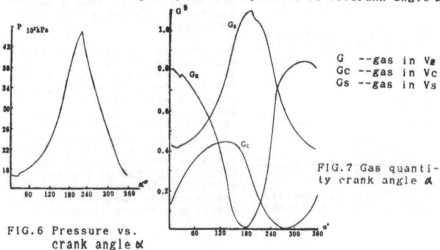

G --gas in V_E
Gc --gas in Vc
Gs --gas in Vs

FIG.7 Gas quantity crank angle α

FIG.6 Pressure vs. crank angle α

Test Arrangement and Results

The test arrangement is shown in Fig.8. The low temperature gas was extracted out by an extraction valve connected to the expansion space, and produced refrigerating capacity in the detached heat exchanger, then passed through the regulating valve, turbo-flowmeter, balance vessel, and returned back to the Stirling refrigerator. The temperature, refrigerating capacity, power consumption and flow rate were measured.

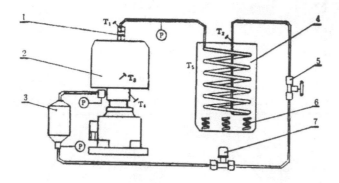

FIG.8 Test arrangement

1. Extraction valve
2. Balance vessel
3. Stirling refrigerator
4. Heat exchanger
5. Heater
6. Tubo-flowmeter
7. Regulating valve

The extraction valve is one of the main parts in the system, it must be switched on/off quickly, less resistance, easy assemble, and suited to low temperature. All these factors should be considered in the design and manufacture.

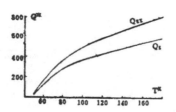

FIG.9 Cooling capacity of extraction Q_{EX} and Stirling Q_s vs. temperature

For the gas extraction Stirling machine the relationship between the refrigerating capacity and the Stirling refrigerating temperature were measured. The result is show in Fig.9, were the charging pressure is 1100 kPa. It was found that the refrigerating capacity of the gas extraction cycle was about 30% larger than that of conventional Stirling. The variation of the refrigerating capacity with the charging pressure is shown in

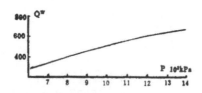

FIG.10 Cooling capacity Q_{EX} vs. charging pressure

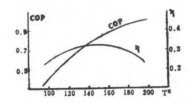

Fig.10,Fig.11 shows the COP and exergy efficiency versus the refrigerating temperature of the gas extraction Stirling temperature of the gas extraction Stirling refrigerator.

FIG.11 Coefficient of performance
COP and exergy efficiency versus
refrigerating temperature

SYMBOL NOTATION

Q ----- total refrigerating capacity
Q_{ex}---- refrigerating capacity of the extracted gas
Q_s ---- refrigerating capacity of Stirling cold head
Q_{smax}--- maximum refrigerating capacity of the cold head
G_{ex} --- extracted gas quantity
G_{exmax}-- maximum G_{ex} under the heat balance of regenerator
Q_{exmax}-- maximum Q_{ex} under the heat balance of regenerator
T_N----- equivalent temperature
$T_g(\alpha)$ - temperature of the expansion volume versus
$T_c(\alpha)$ - temperature of the compression volumes versus
T_{si} -- temperature of clearance volumes in different parts
$G_g(\alpha)$-- gas quantity in the expansion volumes
$G_c(\alpha)$-- gas quantity in the compression volumes
$G_{si}(\alpha)$- gas quantity in the clearance volumes
R ----- the gas constant
G ----- total gas quantity in the Stirling refrigerator
T_c ---- room temperature
C_p ---- specific heat of the working gas

REFERENCES

1. Hong-xing Han, Some improvements of Stirling refrigerator,
   SIME working paper, 1981

2. Hong-xing Han, Thermal calculation and parameters revise
   of Stirling, Journal of Huazhong Engineering Institute,
   No.2,1978

787

The Research on Saving Energy of Heat Pump Drying System and Its Thermodynamic Analysis

MING FANG
Petrochemical Process Development Centre
Tianjin University
Tianjin, PRC

MINGHAN LIANG
Department of Chemical Engineering
Tianjin University
Tianjin, PRC

ABSTRACT

An energy-saving system which consists of a heat pump and a dryer has been developed in this paper. The experiments of its dynamic performence have been investigated. A set of optimum conditions have been found under which specific energy consumption is minimum. The comparision of energy consumptions between the heat pump dryer and the electrical heating dryer shows that the energy consupmtion of the former is about 50% less than that of the latter. The thermodynamic analysis of the heat pump drying system has been made with the help of the experimental data in the paper. Various energy and exergy losses in the system have been estimated in order to improve it further.

1. INTRODUCTION

Drying operation is widely used in many branches of industry and agricuture in every country. Of all the unit operations, it is one of the most enery intensive operations. Thus saving energy in drying process has become an important problem. A heat pump has been put into use as an energy-saving device in the industry, agriculture and civil life in some countries. It also can be used to recover a large amount of latent heat of the exhaust gas in the drying process. The present parper deals with the heat pump drying system in two respects: the technology of saving energy and the thermodynamic analysis.

2. THE HEAT PUMP DRYING PROCESS AND ITS EXPERIMENTAL DEVICE

The heat pump drying exprimental device is shown in Fig.1. It consists of two systems: a heat pump cycle and a drying cycle. The refrigerant used in the heat pump system is R-142, and the drying medium is air. The liquified refrigerant recovers a part of the exhaust heat of the air from dryer at a lower temperature, thus evaporating itself in the evaporator. Then it is compressed in the compressor, that is to say, some energy is added to it. After that, it is condensed in the condensor, transferring some of its heat at a higher temperature to the air which is going to the dryer. The intercooler is set to complete heat exchange between the hot refrigerant and the cool one. All this can make the refrigerant vapour slightly superheated to prevent the refrigerant in the form of drops from entering the compressor. The condensed refigerant being slightly supercooled leads to increasing

789

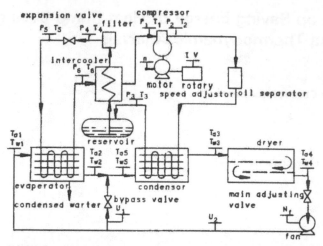

FIGURE 1. The scheme of the heat pump drying experimental device

its potentiality of recovering heat in the evaporator, thereby improving the performance of the whole system. The hot air, which has been heated in the condensor, enters the dryer to dry the moist material. The moist air from the dryer passes through the fan and then is divided into two parts. One part of it enters the evaporator and transfers some of its latent heat and sensible heat to the refrigerant in order that its exhaust heat may be made better use of. At the same time, the temperature of the moist air will drop to the dew point so that some of the water vapour contained in the moist air may be condensed into liquid water, which is then removed from the systems, or rather, the air is dehumidified to some degree. The other part of the moist air passes through the bypass and recirculates. Its flow volume can be adjusted with the help of the bypass valve so as to make the heat loads of both the evaporator and the condensor match each other under the optimum conditions, thus reducing the energy consumption of the system.

The measuring instruments mounted in the system are shown in the Fig.1. P_1-P_6 represent the pressures of the refrigerant, T_1-T_6 to the temperatures of the refrigerant, T_{a1}-T_{a6} and T_{w1}-T_{w6} to the dry and the wet bulb temperatures of the air, I and V to the electric curent and voltage supplied to the motor, U_1 and U_2 to the velocities of air flow, N_f to the power of the fan, and n to the rotary speed of the compressor.

3. DYNAMIC PERFORMENCES OF THE HEAT PUMP DRYING PROSESS

When the heat pump drying system works by the batch method, its various operating parameters and performances will vary with time. The experiments of the dynamic performance of the heat pump drying system have been made in the device as shown in Fig.1, with clay being material to be dried. The curves in Fig.2 prove that the temperature and the absolut humidity of the moist air from the dryer vary with time in the drying process. The curves in Fig.3 show the changes of various drying performances, such as the changes in drying rate Rc, the specific power consumption SPC of the system, the coneefficient of performance

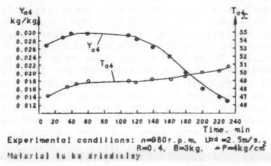

Experimental conditions: n=980r.p.m. U_ad=2.5m/s.
R=0.4, B=3kg, ~P=4kg/cm²
Material to be dried:clay

FIGURE 2 Temperature and humidity of the air from the dryer

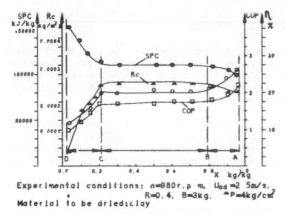

Experimental conditions: n=980r.p.m, U_ad=2.5m/s.
R=0.4, B=3kg, ~P=4kg/cm²
Material to be dried:clay

FIGURE 3. Drying performance curves of the heat pump dryer

of heat pump COP, and the efficiency of the system η with the moisture
content of the material to be dried in the whole drying period. Because
the humidity of the air is lower when the temperature is higher at
the begining of the period, which is referred to as the interval of
A-B in Fig.3, the drying rate Rc increases with the increase in
temperature of the moist material in the dryer. At the interval of
A-B, SPC increases gradually while COP and η decrease slightly. At
the constant rate interval of B-C, the input energy of the system
balances with the sum of energy losses of the system and consumption
for evaporating the water from the surface of the moist material.
Therefore, SPC, COP and η are all in the stable state. More than half
of the water in the moist materal is evaporated and the moisture content
decreases from 0.8kg/kg to 0.3kg/kg at this interval. At the interval
of C-D known as the falling rate interval, SPC increases, while COP
and η decrease in a large range. Compared with the performances at
the constant rate interval, all those of the falling rate interval
deteriorate as the moisture content decreases. Therefore, the heat
pump drying is more suitable to dry the material whose constant rate
interval is longer and whose falling rate interval is shorter.

4. OPTIMUM OPERATING CONDITIONS OF THE HEAT PUMP DRYER

791

When the environmental conditions are kept constant, there are only five operating parameters that can be adjusted individually in the experimental system shown in the Fig.1. They are n(the rotary speed of compressor), ΔP(the difference between the condensing and the evaporating saturation pressure), R(the air volume flow rate between the bypass and the main pipeline), U_{ad}(the velocity of air flow in the dryer), and B(the gross weight without the moisture content in the material to be dried. In order for the total energy consumption of the system to be minimum, it is necessary to study the paramters in various operating conditions to find a set of optimum operating conditions. The method of orthogonal experiment has some advantages, such as a smaller number of experiments, larger coverage and optimization. For this reason , a group of orthogonal experiments were carried out in the normal experiment range. The above mentioned five operating parameters were adjusted as factors each of which could be changed at five different levels. The material to be dried was clay. Its original absolute moisture content was 1.0kg/kg and its final content of the product was 0.07kg/kg.

The experimental data and results are shown in Tab.1, which takes the specific energy consumption as the performance target. Comparing the experimental values of SPC on the right in the table with each other, we can see that SPC of the eighth experiment reached minimum. It might be one of the optimum points.

In Fig.4 there are a group of curves of the orthogonal experiment analysis based on the data listed in Tab.1. The $\overline{SPC}$ is the average value of the five different values of SPC in the Tab.1 and is calculated by the orthogonal experiment analysis when one of the factors is set

TABLE 1. The orthogonal experiment data and results of the heat pump drying

experiment number	adjusted factors					performence target
	n r.p.m	Ugd m/s	R	B kg	ΔP kg/cm²	SPC kJ/kg
1	700	0.84	0.0	1.0	4.0	37890
2	700	1.4	0.2	3 0	4.5	12680
3	700	2 0	0.4	6 0	5.0	5818
4	700	2 5	0 6	7.0	5.5	5899
5	700	3 0	0 7	9 0	8 0	7372
6	770	0.84	0.2	6 0	5.5	8631
7	770	1.4	0.4	7.0	8 0	6229
8	770	2 0	0 6	9 0	4.0	5466
9	770	2 5	0.7	1.0	4.5	28750
10	770	3 0	0.0	3 0	5.0	12080
11	840	0.84	0.4	9.0	4.5	7474
12	840	1.4	0 6	1.0	5.0	33100
13	840	2 0	0 7	3.0	5.5	11670
14	840	2 5	0 0	6 0	8 0	8758
15	840	3 0	0.2	7.0	4.0	7225
16	910	0 84	0.6	3 0	6 0	14680
17	910	1.4	0.7	6 0	4.0	10230
18	910	2 0	0.0	7.0	4.5	7849
19	910	2 5	0.2	9.0	5.0	6244
20	910	3 0	0.4	1.0	5.5	35850
21	980	0.84	0 7	7.0	5.0	10190
22	980	1.4	0.0	9.0	5.5	6170
23	980	2 0	0.2	1.0	8 0	31200
24	980	2 5	0.4	3 0	4.0	11790
25	980	3 0	0.6	6.0	4.5	7239

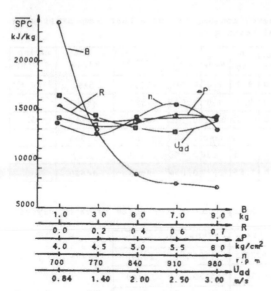

FIGURE 4. Orthogonal experiment analysis curves of the heat pump dryer

at the same level. The curves display the changing tendency of the $\overline{SPC}$ and the effect of each operating parameter on it. Among the five operating parameters, the effect of B is the largest, next to it are U_{ad} and n, while effects of ΔP and R are slight. Even though the effect of R is small, the $\overline{SPC}$ at the point (R=0.0) is higher than those at other points on the same R-$\overline{SPC}$ curve. It also shows the necessity to set air bypass. By analyzing and comparing the curves, we were able to find a group of working conditions (n=770r.p.m, U_{ad}=2.5m/s, R=0.2, B=9Kg, and ΔP=4.5Kg/cm^2) which correspond to the $\overline{SPC}$ minimum. It might be another optimum point of SPC. The experiments were conducted under the new conditions, and its SPC was equal to 5534 KJ/kg. It was slightly larger than that of the eighth experiment in Tab.1,which was equal to 5466 KJ/kg. The results of the eighth experiment, consequently, could be considered to be optimum, and its operating conditions are as follows: n=770 r.p.m, U_{ad}=2m/s, R=0.6, B=9kg, and ΔP=4kg/cm^2.

5. THE EFFECT OF SAVING ENERGY IN THE HEAT PUMP DRYER

In order to investigate the effect of saving energy in the heat pump dryer on the different materials, we chose three typical kinds of materials to be dried—clay, timber and cloth. The contrasting experiments with a heat pump dryer and an electrical heating dryer were carried out when their operating conditions were similar to each other. Tab.2 has listed their operating conditions and the results of three couples of experiments. It is obvious that the heat pump dryer is a highly effective energy-saving device. It will be even more effective if the size of the device is larger and more material is fed in.Besides the above-mentioned three kinds of materials, the device can also be used to dry many other kinds of materials, such as various chemical products, grain, food, tea, fruit, and so on.[1-3] It is especially suitable to dry materials with high humidites at

793

TABLE 2 Comparison of the energy consumption of a heat pump dryer with that of an electrical heating dryer

material type	dryer type	T_a °C	T_{a3} °C	U_{ad} m/s	B kg	X_b kg/kg	X_q kg/kg	total input energy kJ	SPC kJ/kg	saving energy
clay	heat pump dryer	31.5	50.0	2 0	9.0	1.0	0.074	45554	5466	50 6%
	electrical heating dryer	31.4	50.1	2 0	9.0	1.0	0 075	92133	11067	
timber	heat pump dryer	34.0	57.0	2.0	13.32	0.42	0.160	43243	12486	55.3%
	electrical heating dryer	34.3	56 0	2 0	13 32	0.42	0.162	97906	27948	
cloth	heat pump dryer	34.0	56 5	2 0	1.65	1.636	0.056	18302	6257	56.9%
	electrical heating dryer	33 7	56 0	2 0	1.65	1.636	0.050	38800	14530	

TABLE 3 Primary energy souce cosumptions of the three types of dryers

PEC kJ/kg	clay	timber	cloth
heat pump dryer	18030	41290	20690
steam—heating dryer	20530	51390	26100
electrical heating dryer	35350	89340	48400

lower temperatures.

Besides the electrical heating dryer, the steam-heating dryer is also used in many areas because a lot of factories have thier own boilers to produce steam. Although both electricity and steam come from the primary energy source, such as coal or oil, the efficiency of converting primary fuel into electricity is dictinct from that of converting it into steam. Their primary energy consumptions are different when the same quantity of water has been evaporated with the help of the heat pump dryer, the electrical heating dryer or the steam-heating dryer. Under the same operating conditions as shown in Tab.2, the primary energy consumptions PEC of the three types of dryers are estimated in Tab.3. With coal used as primary energy source, the efficiency of the motor is assumed to be 0.96, the efficiency of the electrical generator 0.35, the efficiency of the electrical network 0.9, the efficiency of the boiler 0.68, and the efficiency of the heat-supplying pipeline 0.95. On conditions that the same production task is completed, as shown in Tab.3, the PEC of the elctrical heating dryer is the largest, the PEC of the steam-heating dryer is next to it, and the PEC of the heat pump dryer has the smallest amount. In order to save the primary energy source of the world, it is essential to make our best to study and develop the energy-saving devices with high efficiency in the form of the heat pump dryers.

6. THE THERMODYNAMIC ANALYSIS OF HEAT PUMP DRYING SYSTEM

The drying process is not only concerned with the conversion and utilization of energy,but also accompanied with the decrease in quality of energy.It can also be said that there must be exergy loss in any practical drying process.It is necessary to analyze and improve the system by the thermodynamic method to make it more effective. In the following discussion, the energy conservation method based on the first law of thermodynamics is combined with the exergy analysis method based on the second law of thermodymanics to analyze the heat pump drying system in Fig.1.

Both energy and exergy analysises of the system and its unite apparatus were made in detail when the operating conditions were given below: $n=770$ r.p.m. $U_{ad}=2.5m/s$, $R=0.2$, $B=9kg$, and $\Delta P=4.5kg/cm^2$. Various energy losses and exergy losses and their causes were estimated as shown in Tab.4. As for both the energy and the exergy, the losses in the compressor are the largest in the system. Its improvment no doubt depends upon its design and manufacture quality. The compressor shoud be made to work as effectively as possible. The energy losses, which result from the heat loss and the air leak out of the dryer, condensor and air pipeline, account for a larger proportion in the whole system. More insulation and seal measurements should be carried out on these sections to reduce the energy consumption of the whole system. There is hardly heat loss in the evaporator and the intercooler because their wall temperatures are close to the environmental temperature. The heat losses of the expansion valve in the heat pump cycle and the confluence of both the main air pipeline and the bypass are much less than those of the other apparatuses and can also be assumed to be approximately equal to zero. Although it can be assumed that there is no energy loss in these four apparatuses, from the view-point of the second law of thermodynamics, they all cause the quality of the input energy to drop down to different extents because of the irreversibility in heat transfer and mass transfer. Their exergy losses can not be negleted in the thermodynamic analysis of the system. If the size of the apparatus is larger and more material is fed to the dryer, both the energy and exergy efficiencies of the system will be higher than those given in Tab.4.

TABLE 4   The energy and the exergy analysises of the heat pump drying system

analyzed items	energy loss KW	relative energy loss	efficiency of energy	cause of energy loss	exergy loss KW	relative exergy loss	efficiency of exergy	cause of exergy loss
compressor	0.639	44.4%	68%	leak, friction and heat loss	0.694	24.5%	65.3%	leak, friction and heat loss
condensor	0.457	31.8%	92 9%	neat loss	0.194	6 86%	80 6%	heat loss and irreversibility in heat transfer
intercooler			_100%		0.42	14.9%	48 9%	irreversibility in heat transfer
expansion valve			_100%		0.197	6 9%	72 8%	irreversibility in flowing of refrigerant
evaporator			_100%		0.539	19.1%	36.7%	irreversibility in heat transfer
confluence both main air pipeline and bypass			_100%		0.109	3 9%	94.89%	irreversibility in mixing, mass trasfer and heat transfer
dryer	0 26	18 1%	84.5	leak and heat loss	0.064	2 3%	33 3%	leak, heat loss, flowing friction, irreversibility in heat transfer and mass transfer
an	0.083	5.77%	90%	leak, friction and heat loss	0.611	21.6%	29%	leak, friction and heat loss
the whole system	1.439	100%	50%		2 828	100%	1.12%	

NOMENCLATURE

B   the gross weight of material to be dried without the moisture content (kg)

COP   the coefficient of performance of a heat pump

I   electrical current (A)

H_f the input power of fan (kw)

n   the rotary speed of compressor (r.p.m.)

P   the pressure of the refrigerant (kg/cm^2)

ΔP the difference between the condensing and the evaporating pressure (kg/cm^2)

PEC   the primary energy consupmtion to evaporate 1kg water out of the material to be dried (KJ/kg)

R   the rate of the air flow between the bypass and the main pipeline

Rc   drying rate ($kg/m^2.s$)

SPC   specific power consumption to evaporate 1kg water out of the material to be dried (KJ/kg)

$\overline{SPC}$ the average value of SPC (KJ/kg)

T   the temperature of refrigerant (°C)

T_a the dry bulb temperature of air (°C)

T_o the temperature of environment (°C)

T_w the wet bulb temperature of air (°C)

U_{ad} the velocity of air flow in the dryer (m/s)

V   electrical voltage (V)

X_o the original absolute moisture content of the material to be dried (kg/kg)

X_f the final absolute moisture content of the material to be dried (kg/kg)

η the energy efficiency of the system

REFERENCES

[1] M.Y. Cech'et al: Forest Products Journal, V.28, No.3, 1978

[2] Ernst W.Mann: "New Energy Conserv. Technol. and Commer. Proc. Int. Conf.", Berlin, 6-10, Apr. 1981, Vol.2

[3] B.Geert: "Heat Pump and Their Contribution to Energy Conservation", Proceedings of the NATO Advanced Study Institute, 1975.

Study on a High Temperature Two Stage Heat Pump (Part 1: Experimental Results)

Q. S. YUAN, J. C. BLAISE, and C. MISSIRIAN
Electricite de France
Centre des Renardières
Départment Applications de l'Electricité
77250 Moret-sur-Loing, France

1 - INTRODUCTION

Some industrial sectors such as drying, papermaking and food require steam with a temperature of 130°C-160°C. At the same time, low temperature heat source (60-90°C) is usually available. Production of the steam required from heat recovery by a high temperature heat pump has great advantages for these sectors. At the moment, there is no refrigerant heat pump that can work at this high temperature level.

In the earlier studies [1,2,3], the authors have proposed the possibility of a combination of a refrigerant heat pump and a water heat pump. This heat pump takes heat from a cooling source and increases its temperature to about 150°C. The influence of the cooling source for this heat pump system have been studied. The importance of intermediate temperature (refrigerant condensing temperature) has also been evaluated. It has been confirmed that this cascaded heat pump is not only technically but also economically feasible. A coefficient of performance (COP) of 3 can be attained for a refrigerant evaporating temperature of 75°C and a steam condensing temperature of 155°C. The present work outlines the main experimental results on a high temperature heat pump test-up constructed for this purpose. The emphasis is on the performances of two cycles and the whole system. The technical aspects for high temperature heat pumps are analysed and have general guiding importance.

2 - EXPERIMENTAL SET-UP

A schematic diagram of the experimental set-up is given in Figure 1. The heat pump system consists of mainly three circuits : R114, water and auxiliary. The R114 circuit recovers heat from the wist (steam + air) and increases its temperature to about 115°C, the water circuit takes the energy released by R114 circuit and produces steam of about 150°C, while the auxiliary circuit is for the purpose of providing steam for R114 evaporator.

The basic components which form the R114 cycle are a compressor, a condenser, an evaporator, an expansion valve and a superheater/subcooler. The evaporator is a shell-tube type with R114 liquid evaporating in tube side and wist condensing in shell side. The condenser is constructed of finned tubes with R114 condensing in shell side and water evaporating in tube side (this heat exchanger is equally used as the water evaporator). The expansion valve works as an automatic one that means the superheat at outlet of the evaporator is regulated to maintain a constant value. As for the superheater/subcooler, it is also a shell-tube construction.

The water cycle is composed of a compressor, a condenser, an evaporator (R114 condenser), an expansion valve and an auxiliary condenser. The compressor is a twin screw type with variable rotation rate for capacity regulation. The condenser is a shell tube construction and is for the purpose of simulating an industrial dryer. The auxiliary condenser is for balancing heat power. The expansion valve is based on a liquid level in the separator.

The test set-up is instrumented for temperature, pressure, mass flow rate and electric power measurement. The temperatures are measured by using calibrated platinum resistance thermometers with a nominal accuracy of 0.1°C. The strain-gauge pressure transducers are used for measurement of pressures. The different flowmeters such as electromagnetic, vortex and turbine flowmeters are used for measurement of mass flow rates in the set-up. Electrical consumption of two motors driving the compressors and pumps is measured using transducers. The measurements are effected by a HP data acquisition system and are commanded by a micro-computer.

A programmable automate commands the operation of this heat pump.

3 - DATA REDUCTION

The heat power recovered from the wist is :

$$\dot{Q}_{er} = \dot{m}_r \, (h_{er.o} - h_{er.i})$$

The heat power delivered to R114 condenser is :

$$\dot{Q}_{cr} = \dot{m}_r \, (h_{cr.i} - h_{cr.o})$$

The COP for this cycle is defined as :

$$COP_r = \dot{Q}_{cr}/\dot{W}_r$$

With the same method, we have :

$$\dot{Q}_{cw} = \dot{m}_w \, (\dot{h}_{cw.i} - h_{cw.o})$$

$$COP_w = \dot{Q}_{cw}/\dot{W}_w$$

The COP of this installation is :

$$COP = \dot{Q}_{cw}/(\dot{W}_r + \dot{W}_w)$$

The isentropic and volumetric efficiencies for R114 compressor are written as :

$$\eta_{is.r} = \dot{W}_{is.r}/\dot{W}_r$$

$$\eta_{v.r} = (\dot{V}_{measured}/\dot{V}_{swept})_r$$

The steam compressor efficiencies are :

$$\eta_{is.w} = \dot{W}_{is.w}/\dot{W}_w$$

$$\eta_{v.w} = (\dot{V}_{measured}/\dot{V}_{swept})_w$$

4 - RESULTS AND DISCUSSION

The studies carried out are focused on the working performances of the R114 cycle, the water cycle and the whole cascaded heat pump. The compressors are especially studied due to their important roles in the system [4].

The R114 evaporating temperatures vary from 70°C to 83°C and the steam condensation temperatures vary from 150°C to 160°C. As for the intermediate temperature (R114 condensing temperature), the variations are between 112°C to 117°C. The heat power is about 250 kW.

The isentropic and volumetric efficiencies as a function of pressure ratios of R114 compressor are presented in Figure 2. As one can see, the efficiencies are relatively low, $\eta_{is.r}$ is about 0.55 for a compression ratio of 2.5, one cause being the size effect. This compressor is the smallest in the series. Some parameters such as oil flow, superheat are not optimised. The second cause is the lower viscosity of the oil. In fact, the oil for lubricating, sealing and injection is an oil-R114 mixture. R114 represents about 35 % in the mixture. This results in a low viscosity of the oil. Sealing in the compressor is no longuer ensured. The third cause is partial running conditions that result in a lower performance.

As a result of lower compressor efficiencies, the COP of this cycle defined earlier varies from 4.3 to 3.5. The ratio of COP to COPc, which is the maximum value attainable at working conditions, is about 40 % for a compression ratio varying from 2.25 to 2.7 (Figure 3).

The steam compressor has an equally low efficiency. The small values results especially from a high discharge temperature that deteriorates the compressor sealing. A temperature of 150°C-160°C is obtained while the proposed value by the manufacturer is below 140°C. The COP of this cycle varies from 4.7 to 4 throughout the whole range as shown by Figure 4. Also, the ratio of COP to COPc has a value of 40 %. This is a consequence of the low efficiencies of the compressor.

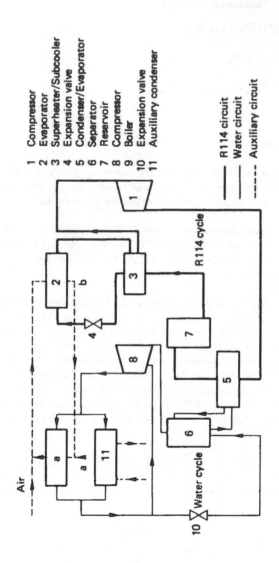

1 Compressor
2 Evaporator
3 Superheater/Subcooler
4 Expansion valve
5 Condenser/Evaporator
6 Separator
7 Reservoir
8 Compressor
9 Boiler
10 Expansion valve
11 Auxiliary condenser

——— R114 circuit
——— Water circuit
------- Auxiliary circuit

Fig. 1 – Schema of the high temperature cascaded heat pump

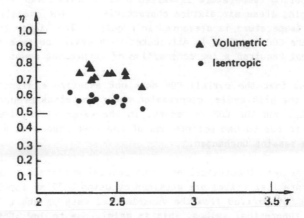

Fig. 2 — Evolution of R114 compressor efficiencies vs pressure ratio

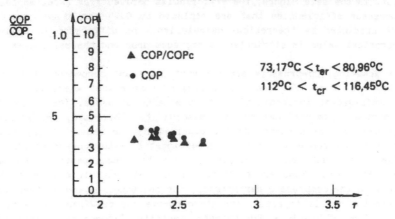

$73,17^{\circ}C < t_{er} < 80,96^{\circ}C$

$112^{\circ}C < t_{cr} < 116,45^{\circ}C$

Fig. 3 — COP of the R114 cycle vs pressure ratio

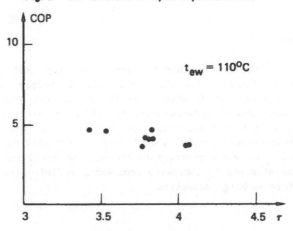

$t_{ew} = 110^{\circ}C$

Fig. 4 — COP of the water cycle vs pressure ratio

The R114 evaporing temperature is varied over a small range from about 70°C to 83°C by changing steam/air mixture characteristics. The overall COP as a function of this temperature is presented in Figure 5. The higher the value of t_{er}, the greater the COP is. This is attributed to a small compression ratio for the R114 stage that requires a low consumption of compressor power.

Figure 6 shows that the overall COP does not experience significant variation with t_{cr}. In the R114 cycle, compression power increases proportionally to the increase of t_{cr} and the COP decreases. In the water cycle, improvement of COP is not observed due to bad performance of the compressor. As a result, the COP of the system remains unchanged.

A comparison between theoretical and experimental performance is given in Figure 7. The theoretical values are attained by using 0.75 as isentropic efficiencies [2]. The COP derived from the experimental data is about 2.2 and is smaller than the theoretical values. This is mainly due to low efficiencies of the compressor. In the same Figure, the efficiencies derived from experimental data with compressor efficiencies that are replaced by 0.75 are given, the COP is near that predicted by theoretical calculation. The difference in comparison with theoretical value is attributed to pressure drop and thermal losses.

Generally speaking, compressors are the most important component of this high temperature heat pump. For the oil lubricated compressor, miscibility of oil with the refrigerant increases with the discharge pressure. The viscosity of oil is inversely proportional to this miscibility. Therefore, the sealing in the compressor can be ensured. On the other hand, the vapour formed from the refrigerant liquid contained in the oil/refrigerant mixture needs an additional consumption. The bad sealing conditions and the supplementary consumption decreases efficiency. Some modifications on the oil circuit may be envisaged for increasing the compressor performancy, for instance, having another separation before the oil is injected. The steam compressor requires a more detailed study on sealing of bearings. Furthermore, centrifugal compressors are suitable for this kind of application.

5 - CONCLUSION

The experimental studies on this cascaded heat pump test-up, using R114 and water as working fluids, show that this test-up operates without particular problems. The overall COP is about 2.26 for a R114 evaporating temperature of 75°C and a steam condensing temperature of 155°C. The difference with the theoretical value is due to lower compressor efficiencies. The influences of R114 evaporating temperature for COP is confirmed. As for R114 condensing temperature, there is no influence observed in the working conditions. The improvement for COP can be attained by improving compressor efficiencies as well as by optimizing certain working parameters.

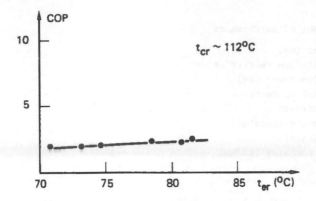

Fig. 5 — COP of the system vs R114 evaporating temperature

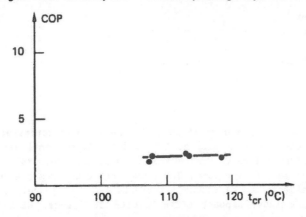

Fig. 6 — COP of the system vs R114 condensing temperature

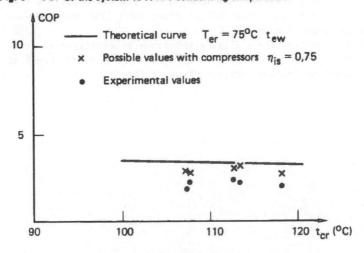

Fig. 7 — COP of the system vs R114 condensing temperature

NOMENCLATURE

COP	coefficient of performance
$\dot{Q}$	heat power (kW)
$\dot{V}$	volumetric flow rate (m³/s)
$\dot{W}$	compression power (kW)
cr	refrigerant condenser
cw	steam condenser
er	refrigerant evaporator
ew	water evaporator
h	specific enthalpy (kJ/kg)
i	inlet
is	isentropic
$\dot{m}$	mass flow rate (kg/s)
o	outlet
r	refrigerant
v	volumetric
w	water
τ	pressure ratio

REFERENCES

1. Yuan, Q.S., Blaise, J.C., and Missirian, C., Design and optimization of a high temperature cascaded heat pump at 155°C. 3rd International Symposium on the LARGE SCALE APPLICATIONS OF HEAT PUMPS, Oxford, England, 1987.
2. Blaise J.C., Missirian, C. and Yuan, Q.S., Couplage d'une pompe à chaleur et d'une compression mécanique de vapeur : présentation d'un projet de prototype et impact économique. Rapport d'Electricité de France, HE 142 W 2339, 1986.
3. Blaise, J.C., Missirian, C. and Yuan, Q.S., Pompe à chaleur à très haute température - Couplage d'une PAC et d'une CMV, 17th International Congress of Refrigeration, Vienna, 1987.
4. Yuan, Q.S., Contribution à l'étude théorique et experimental et à la modélisation d'une pompe à chaleur en cascade à très haute température, ph.D. Thesis, Institut National Polytechnique de Grenoble, 1987.

Study on a High Temperature Two Stage Heat Pump (Part 2: Modelization)

Q. S. YUAN, and J. C. BLAISE
Electricite de France
Centre des Renardières
Départment Applications de l'Electricité
77250 Moret-sur-Loing, France

The high temperature cascaded heat pump consists of a number of elements. A good operation of this high temperature cascaded heat pump needs knowledge of each element. This knowledge can be obtained not only by experimental study but also by system simulation. That is the reason for the development of a mathematical model.

The goal of the present paper is to give a description of the mathematical model developed and the programming. The calculated results are presented. A comparison between the measured and calculated values is made and some general conclusions are drawn from this comparison.

1 - DESCRIPTION OF THE MODEL

As mentioned in [1], the heat pump is constructed of different elements such as compressors, heat exchangers and expansion valves. The running of each element is determined, on the one hand, by construction of the element and, on the other hand, characteristics of fluids at the inlet. Therefore, it is possible to have a mathematical model for each element. The model calculates output parameters by using construction data and transformation function. As there are particularities for each machine, some coefficients which are derived from the experimental data, are used in cases where they are indispensable. The connection of the model for each element constitutes the model for the R114 stage and the water stage. These two models make up the model of the whole system.

The model, for a given condition of cooling source and production, calculates the thermodynamic performance of each element of the system. The input parameters are the temperature, the percentage of air and the mass flow rate of the mist, the temperature of the production, the steam compressor speed, the water evaporating temperature. The output parameters from the model are the R114 mass flow rate and its thermodynamic state at approprate points in the cycle, the characteristics of water and steam at eppropriate points, the thermal power of each heat exchanger, the power consumption of two compressors and the coefficients of performance of two stages and the whole system.

2 - MODELING ASSUMPTION

For simplicity of calculation, we assume :

- pressure drop and heat losses of pipelines are negligible,
- pressure drop in heat exchanges is negligible,
- compression is non-isentropic,
- internal leakage of the compressors follows the orifice flow with a constant coefficient.

3 - ELEMENT MODELS

The compressor models are based on their geometric calculation and physical principles. The subroutines of geometry calculate the compressor volume for each rotation angle. The mass in the compressor is estimated by deducing the assumed internal leakage, following the orifice flow. The specific volume of vapour is then determinated. On the other hand, the pressures after each elementary rotation are estimated by using isentropic and adiabatic compression with an empirical coefficient. Its role is to consider the particularities of the compressors. The specific volume and pressure allow the temperature to be obtained using the equation of state. The variation of the enthalpy between the two points corresponds to an elementary compression power. This procedure is repeated from the suction to the discharge.

The heat exchanger models are based on their geometric characteristics and material properties. According to the cases, the heat exchangers are divided into one, two or three distinct zones. In the model, the different zones are treated as separate heat exchangers connected in series. The heat transfer coefficients are calculated by using inlet parameters and heat exchanger geometries. The R114 evaporating heat transfer coefficient is calculated by using the correlations given by Dhami [2]. The condensation heat transfer is estimated by using the correlation of Kenneth [3]. As for single-phase heat transfer, the correlation proposed by [4] are used. The evaporation of water is considered by the correlation given in [5]. The steam condensation is evaluated using the correlation given by [6]. The condensation of steam in the mist is delicate to treat because the heat and mass transfer co-exist, Colburn's equation [7] is used for this application. The heat exchangers are discreted into small elementary surfaces by heat power. For each elementary surface, by applying the conservation of energy and heat transfer coefficient calculated, one can determine the necessary surface. This procedure is repeated until the calculated surface is equal to the real surface.

The model of the expansion valve is based on orifice flow. With the inlet parameters, the model calculates the pressure drop and the vapour quality at the entrance of the evaporator using an empirical parameter, which is determined by experimental data.

The heat pump model is organised in two principal sections : R114 section and water section. The R114 section is an assembly of models of the compressor, evaporator, condenser, expansion valve and superheater/subcooler in R114 cycle, while the water section is an assembly of models of the compressor and heat exchangers in this cycle.

An mentioned above, the pressure drop in the heat exchangers and the variation of viscosity of the oil as function of temperature are neglected in the model.

4 - ORGANISATION OF CALCULATION

The calculation begins with the steam compressor. Using the evaporating temperature, the condensing temperature and the compressor speed, the model calculates the motor power, isentropic efficiency and the mass flow rate. The state of steam at the entrance of the boiler is estimated from the compressor output conditions. Using mass flow rate and temperature at the entrance of water, the boiler model calculates the heat transfer rate and determines the temperature of the condensate at the outlet of the boiler.

The suction state of the R114 compressor is determined by an estimated temperature and suction pressure. Using the position of the slide valve calculated according to mass flow rate in the steam compressor, the R114 compressor subroutine calculates the motor power, isentropic efficiency, the mass flow rate and the state at the outlet of the compressor. The characteristics of R114 at the entrance of the condenser are given by the compresser model. Using the construction data of the condenser and the temperature and mass flow rate of water, the R114 condenser model calculates the heat transfer power and the condensing temperature. Iteration is required in order to have convergence by adjusting the condensation temperature. The superheat/subcooler model, using R114 properties at the outlet of the condenser and at the outlet of the evaporator, calculates the heat power and the outlet temperatures for the subcooled liquid and the superheated vapour.

Using the mass flow rate and condensation pressure of R114, the expansion valve model calculates the pressure drop and vapour quality at the entrance of the evaporator.

For the R114 evaporator model, it is necessary to give the temperature, mass flow rate and air percentage at the entrance of the mixture for calculation. The model calculates the heat power, the superheat of R114 and the characteristics of the mixture at the outlet of the heat exchanger. The evaporation pressure calculated by two successive iterations is compared. If the difference is less than a given value, the calculation is stopped. Otherwise, the calculation started from the beginning in order to have the convergence. The programme flow diagram is shown in Figure 1.

807

5 - RESULTS

The model has been validated with the experimental results carried out in 1987.

Figure 2 shows the variation of the efficiencies of the R114 compressor as a function of pressure ratio. Firstly, the same variation tendancy for experimental and calculated values is obtained. Secondly, there is a divergence between experimental and calculated values. The calculated values are always greater than the experimental ones. This difference results from the compression of the R114 vapour dissolved in the oil. In fact, there is about 30-40 % of R114 in the oil/R114 mixture. The R114 liquid evaporates when entering into the compressor and forms vapour. This vapour needs consumption for reaching the discharge conditions.

The comparison between experimental and calculated heat power of the R114 condenser is shown in Figure 3. A good accordance can be observed. The maximum error is less than 5 % over the range of conditions tested. Figure 4 shows the comparison of COP of the R114 stage. It can be seen that the calculated values differ by no more than 5 % from the corresponding experimental values, this result is considered satisfactory.

As for the steam compressor, the difference between experimental and calculated values increases when the capacity is greater (Figure 5). This tendancy is partially explained by the fact that the mass flow rate increased by higher compressor rotation rate takes a different value for two cases. The increase is less important for experimental values owing to leakage. The COP calculated for the water stage is generally in accordance to experimental values (Figure 6).

Figure 7 shows the comparison between experimental and calculated values of COP of the system. The error is less than 10 %, which is relatively satisfactory. Futhermore, the difference results from the steam mass flow rate, as mentioned above.

Table 1 shows a list of experimental and calculated values. The accordance for the condensation temperature is good, the difference remains less than 0.52°C. As for the evaporating temperature, the difference becomes greater as a result of the evaporation temperature calculated by the expansion valve.

6 - CONCLUSIONS

- The model of the R114 stage gives good results, the calculated values can be estimated with a precision of 10 %,

- The model of the water stage is less precise. A good value can be obtained in a large range of running conditions. But error becomes greater when the capacity is higher,

- The models of heat exchangers calculate heat power and global heat transfer coefficient. They are equally capable of estimating local heat transfer coefficient.

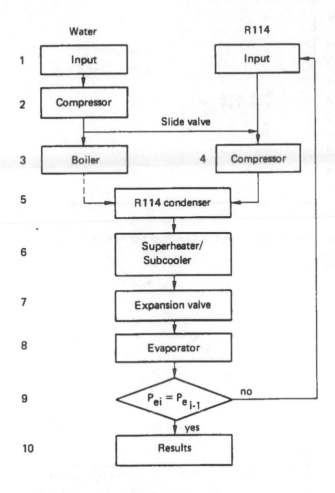

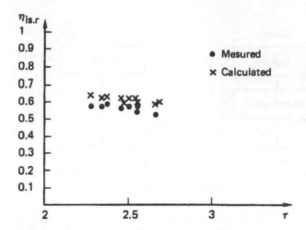

Fig. 2 – Mesured and calculated isentropic efficiencies

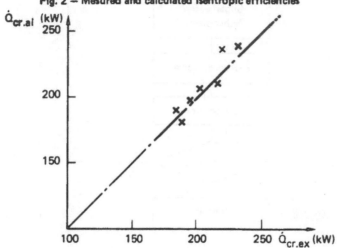

Fig. 3 – Measured and calculated heat power of the R114 condenser

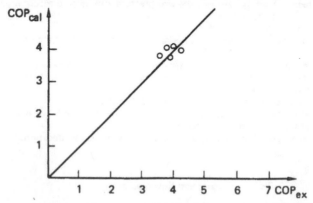

Fig. 4 – Mesured and calculated COP of the R114 stage

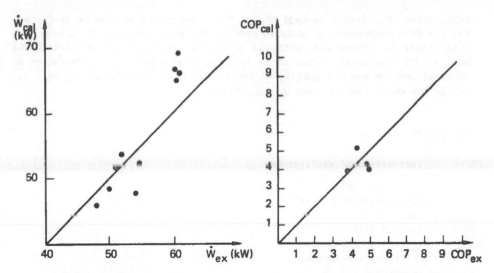

Fig. 5 — Measured and calculated compression power of the R114 compressor

Fig. 6 — Measured and calculated COP of the water stage

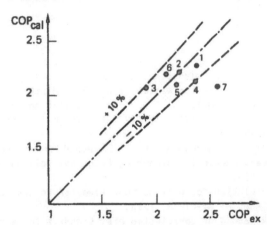

Fig. 7 — Mesured and calculated COP of the system

TABLE 1 — Comparison of mesured and calculated values

N°	T_{er} (°C)		T_{cr} (°C)		COP	
	Calculated	Mesured	Calculated	Mesured	Calculated	Mesured
1	82.9	80.74	113.76	113.35	2.27	2.37
2	78.2	74.69	112.39	112.2	2.21	2.21
3	78.59	73.17	112.92	112.5	2.07	1.91
4	78.33	76.72	113.15	112.64	2.12	2.36
5	80.64	77.94	113.35	112.89	2.09	2.19
6	81.57	79.68	113.66	113.57	2.2	2.09
7	80.26	78.72	113.18	112.72	2.08	2.57

- Calculation of internal leakage of the two compressors are to be perfected. For the R114 compressor, a variable coefficient as a function of working conditions may be introduced. Owing to a strong concentration of R114 in the oil. As for the water stage, it is necessary to obtain more information on internal leakage and the quality of water injection in order to use this model in the whole range of running conditions.

NOMENCLATURE

COP	coefficient of performance
$\dot{Q}$	heat power (kW)
$\dot{W}$	compression power (kW)
cal	calculated
ex	experimental
cr	refrigerant condenser
er	refrigerant evaporator
is	isentropic
τ	pressure ratio
η	efficiency

REFERENCE

[1] Yuan, Q.S., Blaise, J.C., Missirian, C., Study on a high temperature two stage heat pump (Part 1 : Experimental results), ISHTEEC, Guangzhou, 1988.

[2] Dembi, H.J. Dhar, D.L. and Arora, C.P., Heat transfer and pressure gradient for R22 boiling in a horizontal tube, XV International Congress of Refrigeration Vol. II, 1979.

[3] Kenneth, O., Beatty, Jr. and Katy, D.L., Condensation of vapours on outsides of finned tubes, Chemical Engineering Progress, Vol. 44, pp55-69, 1948.

[4] ASHRAE handbook, 1980.

[5] Chen, J.C., A correlation for boiling heat transfer to saturated fluids in convection flow, ASME N° 63-HT-34, 1963.

[6] Rohsenow, W.M., A method of correlation heat transfer for surface boiling liquids, Trans. ASME, Vol. 74, pp969-975, 1952.

[7] Colburn, A.P. and Hougen, O.A., Design of cooler condensers for mixtures of vapours with incondensing gases, Industrial and Engineering Chemistry, Vol. 26, pp1178-1182, 1934.

The Experiment and Research of 1744 kW Low Pressure Steam-Operated Double-Effect Lithium Bromide Absorption Refrigerating Machine

YUQUING ZHENG
Shanghai Marine Equipment Research Institute
Shanghai, PRC

1. INTRODUCTION

The successful development of steam-operated double-effect lithium bromide absorption refrigerating machine has important meaning in full utilizing the thermal energy of high pressure $(5.88 - 7.84) \times 10^5$ Pa (G) steam and increasing the energy recovery effect.

Neverthless, the pressure of steam source in many factories is about $(2.94-3.92) \times 10^5$ Pa(G)for example, in textile factories, in which the package boilers are often used, only the process steam with pressure $(2.94-3.92) \times 10^5$ Pa (G)is used, so the supply of steam with $(5.88-7.84) \times 10^5$ Pa (G) pressure is not only difficult, but also uneconomic. Besides, the discharged steam from turbogenerator has the pressure about 2.94×10^5 Pa(G).

If low pressure steam was throttled to 9.8×10^4 Pa(G) and then used in the single-effect lithium bromide refrigerating machine. the COP (refrigerating capacity/heating energy) in general is about 0.6-0.7, so we had to develop 2.45×10^5 Pa (G) low pressure steam operated double-effect lithium bromide absorption refrigerating machine.

The prototype machine was delivered to Jia Xing Silk Spinning factory in september 1986. The results of commissioning indicated that the unit run smoothly and the operation was convenient.

When steam pressure is $(0.96 - 2.25) \times 10^5$ Pa (G) the temperature of chilled water at the outlet of machine is 12.6-12.9 °C and the refrigerating capacity of unit (without octanol) is 1651-1794 kW, the specific steam consumption was 1.23-1.45 kg/kW and the coefficient of performance is higher than 1.1. In comparison with single-effect refirgerating machine, the double-effect one could save steam by as much as 45%.

2. MAIN TECHNICAL DATA

(1) Heating steam pressure: 2.45×10^5 Pa (G) superheat steam (superheat temperature no more than 50°C);
(2) Refrigerating capacity: 1744 kW;
(3) Temperature of chilled water at outlet: 13°C ;

(4) Inlet temperature of cooling water: 32°C;
(5) Specific steam consumption: 1.45 kg/kW;
(6) Power consumption of unit: 5.5 + 5.5 + 2.2 = 13.2 kW;

3. TECHNIQUE KEY POINTS AND THEIR SOLUTIONS

There are high pressure and low pressure generators in the double-effect lithium bromide absorption refrigerating machine. Therefore, how to make the refrigerant in the high and low pressure generators boil and generate required refrigerant vapour is the key point of refrigerating machine.

When external heating steam supplied with constant pressure, the boiling condition of refrigerant in high and low pressure generators depands mainly on the pressure and concentration of the solution.

1) Pressure of solution
The process of generating refrigerant vapour in generator is a mass transfer process. It's occurrence and severity is associated with mass transfer driving potential. Assuming that the average pressure of refrigerant vapour in generator is P_1 and partial pressure of vapour of solution is P_2, then the mass transfer driving potential Δ m may be expressed as follows:
$$\Delta m = P_2 - P_1 \dots\dots\dots\dots\dots\dots\dots\dots\dots\dots\dots\dots\dots(1)$$
As we could see from equation (1) that in order to increase the mass tranfer driving potential Δ m, P_2 should be increased, while P_1 should be decreased as much as possible.

P_2 is restricted by operating steam, and P_1 is relevant directly to condensing pressure i.e. to condensing temperature.

Therefore the temperature of cooling water has important effect on low pressure steam operated double-effect lithium bromide absorption refrigerating machine. Besides, to reduce as much as possible the flow resistance of refrigerant vapour and the effect of hydrostatic height of liquid on solution boiling also play an important role.

2) Concentration of solution
Solution in high and low generators is transformed from weak solution supplied from absorber and the concentration of solution depends on pressure and temperature of solution in absorber: the higher the pressure of solution and the lower it's temperature, the lower the concentration of solution.

Under ideal conditions, the pressure of solution in absorber may be taken as the average pressure of refrigerant vapour in this vessel. It is relevant to evaporating pressure in evaporator.

In order to raise absorbing pressure, when evaporating pressure (evaporating temperature) is a constant value, the flow resistance of refrigerant vapour in evaporating-absorber should reduced as much as possible, so as to reduce the concentration of solution.

4. FLOW CHART AND STRUCTURE CHARACTERISTICS

Low pressure steam-operated double-effect machine is composed
of high pressure generater, low pressure generator-condenser,
evaporator-absorber, high temperature heat exchanger, low temperature heat exchanger, condensate heat recoverer, and herme-
tic pumps, valves etc. The cycle of solution is realized by
parallel flow, as shown in Fig. 1

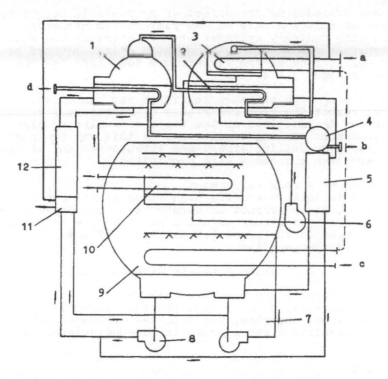

Fig.1 Flow chart of 1744 kW low pressure steam-operated
double-effect absorption refrigerating machine

1 - high pressure generator   2 - low pressure generator
3 - condenser   4 - condensate heat recoverer   5 - low
temperature heat exchanger   6 - evaporator pump   7 - absorber pump   8 - generator pump   9 - absorber   10 - evaporator   11 - precooler   12 - high temperature heat exchanger
a - cooling water   b - condensate   c - cooling water
d - operating steam

It can be seen from Fig.1 to compare with standard steam-
operated double-effect machine, the difference is that the
cooling water goes first through the condenser and then
through absorber, by this way it is able to reduce condensing
pressure, and the pressure in high and low generators.

Moreover, since the cooling water at the inlet of the absorber is higher (2°C higher than that at the inlet of the con-

815

denser), solution from the high temperature heat exchanger
is precooled by the cooling water before the solution flows
into the absorber.

The machine have some special construction characteristics
as shown below.

1) High and low pressure generators use flat bottom
structure.
For the sake of reducing the effect of hydrostatic height to
the boiling of solution here we use flat bottom structure.
To compare with ordinary cylinder shaped generator the flat
bottom structure can make the hydrostatic height reduced by
about 120 mm. It reduces the temperature head loss for the
low generator about 3 - 4 °C.

2) Eliminator in high pressure generator was cancelled.
In order to reduce the flow resistance of refrigerant vapour
the eliminator is cancelled on the basis of analysing high
pressure generator working conditions. By this way the high
pressure generator boiling effect was strenthened and at same
time the carryover phenomena were prohibited.

3) Precooling of strong solution
The precooler is integrated with high temperature heat exchanger. Thus, it is compact in construction and convenient
to installation. The strong solution outside tube and cooling
water inside tube in the precooler are counterflowed to
increase heat-exchange effect.

5. THE OPERATING TEST RESULTS AND PRIMARY ANALYSIS

The field test results are shown in table 1. The heat load
of the main apparatus, such as high pressure generator, condenser, evaporator, absorber, the relative errors of heat
balance, and specific steam consumption are calculated as
follows:

1) Heat load
High pressure generator Q_h.

$$Q_h = G_h (t_h'-t_h) C_h + G_h r_h + G_h(t_h-t_h'') \ldots\ldots\ldots(2)$$

Where G_h is operating steam quantity kg/h, t_h', t_h'' are
relatively the temperature of steam at inlet and condensate
at outlet from high pressure generator; t_h stands for the
steam saturation temperature r_h is condensing heat of steam,
C_h stands for of super heat steam specific heat, kWh/kg °C.

Condenser Q_k

$$Q_k = G_w (t_{w1} - t_w) \times 1.163 \text{ kW} \ldots\ldots\ldots\ldots\ldots\ldots(3)$$

Where G_w stands for the cooling water quantity, t/h,
obtained from vane-wheel type water meter.

t_w, t_{w1} relatively stand for the inlet and outlet temperature
from condenser, obtained by mercury thermometer with 0.1 °C

scale.

Evaporator (refrigerating capacity) Q_o

$$Q_o = G_c (t_c' - t_c'') \times 1.163 \text{ kW} \dots\dots\dots\dots(4)$$

Where G_c stands for chilled water quantity T/h, obtained by the vane-wheel type watermeter.

t_c', t_c'' relatively stand for the inlet and outlet chilled water temperature from evaporator, obtained by mercury thermometer with 0.1 °C scale.

Absorber Q_a

$$Q_a = G_w (t_{w2} - t_{w1}) \times 1.163 \text{ kW} \dots\dots\dots\dots(5)$$

Where t_{w2} stands for the outlet cooling water temp. or absorber obtained by mercury thermometer with 0.1 °C scale.

Condensate heat recoverer Q_t

$$Q_t = G_t (t_h'' - t_h''') \times 1.163 \times 10^{-3} \text{ kW} \dots\dots(6)$$

Where t_h''' stands for the outlet condensate temperature of the condensate heat recoverer. °C, obtained by the mercury thermometer.

2) Relative errors of heat balance

$$\delta = \frac{(Q_o + Q_h + Q_t) - (Q_a + Q_k)}{Q_o + Q_h + Q_t} \leqslant 5 - 7.5\% \dots\dots(7)$$

The data in table 1 give the results in accordance with the equation (7)

3) Specific steam consumption d.

$$d = \frac{G_h}{Q_o} \qquad \text{kg/kW}$$

According to the data in table 1, the curve of refrigerating capacity in function of operating steam pressure is shown in Fig. 2 for reference use. The ordinats of the figure show the percentage of design value (1744 kW).

From table 1 and Fig 2 we can see:

(1) Near the design condition, without octanol, the refrigerating capacity of the low pressure steam-operated double-effect machine achieved 95% design value, with octanol, the capacity will be more than design value. The specific steam consumption is no more than 1.38 kg/kW, better than design value. So the design and trial of prototype unit is sucessful.

(2) The relationship between the operating steam pressure and the refrigerating capacity:

From Fig. 2 we obtain that operating steam pressure have much effect on refrigerating capacity. When steam pressure

Table 1. Test run data

	test item	test number 1	2	3	4	5	6	7	8
high pressure generator	operating steam inlet pressure x10^5 Pa	1.1	1.47	1.93	2.05	2.25	2.25	2.25	2.25
	operating steam flow kg/h	1308	1677	1995	2571	2125	2123	2083	2024
	heat load kW	799	1018	1201.4	1544.5	1272.3	1271.2	1247.9	1211.8
evaporator	chilled water inlet temp. °C	16.8	19.4	18.7	18.7	18.6	20.7	19.6	19.5
	chilled water outlet temp. °C	13.7	15.2	13.1	12.0	12.1	13.7	12.8	12.8
	chilled water flow T/h	219	221	221	223	216	206	211	211
	refrigerating capacity kW	789.7	1079.3	1439.8	1737.5	1640	1694.5	1674.7	1651.5
condenser	cooling water inlet temp. °C	31.6	31.5	31.8	28.1	30.7	31.6	31.1	31.0
	cooling water outlet temp. °C	32.4	32.6	32.9	29.9	32.3	33.1	32.7	32.6
	heat load kW	446.6	605.9	612.9	994.4	886.2	832.7	893.2	885
absorber	cooling water outlet temp. °C	34.5	35.3	36.1	33.3	35.7	36.5	36.1	36.0
	cooling water flow T/h	480	474	479	475	476	477	480	476
	specific steam consumption kg/kw	1.65	1.55	1.39	1.48	1.30	1.25	1.24	1.23

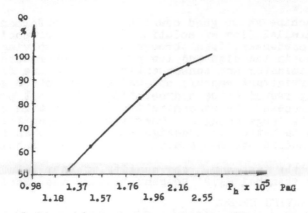

Fig. 2 The relationship between refrigerating capacity and operating steam pressure.

varies between (2.0 - 2.45) x 10^5 Pa G the refrigerating capacity will be increased or reduced by 2.5% with increasing or reducing the pressure by 9.8 x 10^5 Pa. When the steam pressure is lower than 2.0 x 10^5 Pa the changement of the refrigerating capacity will be increased (about 5% the design value) with the same amount of steam pressure variation.

(3) The relationship between the superheat of operating steam and the specific steam consumption

We obtain that the specific steam consumption is reduced with increasing the steam superheat. For example: at the similar operating conditions with the steam pressure 2.25 x 10^5 Pa G the chilled water outlet temperature 13 $\pm$ 1 $^{\circ}$C, the cooling water inlet temperature 3 1 $\pm$ 0.5 $^{\circ}$C, increase the steam inlet temperature from 158 $^{\circ}$C to 164°C, that is, increase the superheat from 22 $^{\circ}$C to 28 $^{\circ}$C, will cause the specific steam consumption to drop from 1.30 to 1.22 kg/kW. However, when the operating steam inlet temperature is near the saturation temperature, such as at working conditions with 1.47 x 10^5 PaG and 2.0 x 10^5 Pa G the specific steam consumption was increased considerablely, equals 1.55 and 1.47 kg/kW. The reason of improving the steam comsumption with superheated steam may be the less humidity contained in the superheated steam.

(4) The problem about liquid carry over from generator
Because of cancelling the eliminator in the generator, the important problem we consider in the process of the design is there any liquid carryover occured in the 1744 kW low pressure double-effect machine under operating conditions, or whether the refrigerant water is polluted or not. The test-run results show, that to cancel the eliminator in the generator for low pressure double-effect machine is feasible. Under the working condition with the steam pressure of 1.1 - 2.25 x 10^5 Pa G and the cooling water inlet temperature of 28 - 32.3 $^{\circ}$C, we obtain the refrigerant water specific gravity in the evaporator is 0.998 - 1.014, it lower than 1.02, so it is suitable for running.

(5) The machine shows good coefficient of performance because of using parallel flow of solution, the flow of cooling water first into condenser, then through absorber, solving the boiling problem in the high and low pressure generators, cancelling the eliminator and consequently reducing the flow resistance of refrigerant vapour, and using flat bottom generator, consequently reducing the hydrostatic height. At practical operating process, the concentration difference in both generators is large enough.   Therefore the specific steam consumption is better than design target, it comes to the value of standard steam-operated double-effect machine.

(6) During the test run, the quality of LiBr solution produced by Shanghai $No.2$ Reagent Factory was stable.

6. ENERGY SAVING EFFECT

In the past years, $(1.96 - 2.94) \times 10^5$ Pa G low pressure steam can only be used in single-effect LiBr absorption refrigerating machine, usually with the specific steam consumption of 2.53 - 2.58 kg/kW. In comparison with above machine, specific steam consumption of this unit is reduced to 1.39 kg/kW, it shows the 45% of energy saving effect. To the 1744 kW unit, the steam quantity per hour:

for low pressure double-effect machine   2430 kg

for single-effect machine   4410 kg

Thus, if these machines continuously run 90 days in summer of the year, then in comparison with single effect machine, the low pressure double-effect machine saves steam

$(4410 - 2430) \times 24 \times 90 = 4276.8$ (T)

If we consider that 1 ton of standard coal can produce 7 ton of steam one machine can save standard coal in a year:

$$\frac{4276.8}{7} = 611 \quad (T)$$

The low pressure double-effect machine has more remarkable energy saving effect than low pressure steam jet refrigerating machine.

Because the low pressure steam-operated double-effect machine has remarkable energy saving effect, it has lower operating costs and notable economic benifit. As above mentioned, a 1744 kW machine can save 4276.8 Ton steam a year, at the present energy costs, 20 yuan per ton of steam, we can save operating costs in a year:

$4276.8 \times 20 = 85536$ yuan

On the Heat and Mass Transfer Process in the Absorber of the Lithium Bromide Absorption Refrigerator

YONGHUI FAN, ZHIYUAN CHEN, and HEGAO ZHAO
Huazhong University of Science and Technology
Wuhan, PRC

ABSTRACT

The operating line equations of the absorber are derived by analysing the simultaneously heat-and mass-transfer process in the absorber. A more reasonable absorption and diffusion model is presented, in which the Stefan flow is introduced. According to the differential equations governing the process, the rates of heat and mass transfer, heat-and mass-transfer coefficients of the absorber are calculated theoretically. An experiment has been carried out in an apparatus with a coil absorber. A comparison of the theoretical results with the experimental data shows satisfactory agreement.

1. INTRODUCTION

The aqueous solution of LiBr is sprayed on a set of tubes in the falling liquid film absorber of a lithium bromide absorption refrigerator. The falling liquid film outside the tube absorbs steam from the evaporator. The liberated absorption heat is carried away by the cooling water inside the tube. This is a simultaneous heat-and mass-transfer process. Studies on the process were insufficient. Most of the earlier researchers used the classical models for gas absorption, such as the film theory, the penetration theory and the surface renewal theory, etc. to study the process separately [1,2]. Recently, some researchers analysed the simultaneous heat-and mass-transfer process , but they used the equimolal counterdiffusion model, in which the transfer phenomenon in the falling liquid film was treated as a mutual diffusion of water and lithium bromide, and Fick's Law was simply used to evaluate the mass flux of water[3,4,5]. Experiments carried out in water-cooled wetted-wall columns showed that all of the models had considerable differences with actual process[2,5].

The authors present new equations for heat and mass transfer to study the simultaneous process, from which the operating line equations of the absorber are obtained. In order to determine the values of heat-and mass-transfer coefficients in the equations, the authors present a more reasonable model in which the Strfan flow is introduced according to the nature of aqueous LiBr and the actual process. In the region of laminar flow of the liquid film, the simultaneous heat and mass transfer is analysed theoretically by solving the fundamental differential equations. Formulae for evaluating the rates of heat and mass transfer are obtained under some assumptions and simplifications. The heat-and mass-transfer coefficients of the absorber are calculated. Experiments have been carried out in an apparatus with a coil absorber, which indicate that the theoretical analysis is correct. The results of this paper provide a reasonable method and reference data for the study and design for the absorber of the lithium bromide absorption refrigerator.

2. OPERATING LINE EQUATIONS OF THE ABSORBER

The impetus of absorption of steam is the concentration difference between the interface and the bulk liquid of the liquid film. Because of the linear relation between the concentration and the saturated temperature at constant absorption pressure, the following equation for mass transfer (absorption) is acceptable.

$$M = \beta(t_i - t_m)\pi d_0 \tag{1}$$

The equation for heat transfer from the liquid film to the coding water is:

$$Q = K(t_m - t_c)\pi d_0 \tag{2}$$

where t_m is the logrithm mean temperature, it can be obtained by E.L. CoKoпoB's approximate equation[6]:

$$t_m = t_1 - 0.65(t_1 - t_1') \tag{3}$$

The energy conservation applies to every tube:

$$MR - Q = 2\Gamma C_p(t_1' - t_1) \tag{4}$$

let

$$\theta_m = t_m - t_c \qquad\qquad \theta_1 = t_1 - t_c$$
$$\theta_1' = t_1' - t_c \qquad\qquad \theta_i = t_i - t_c \tag{5}$$

Substituting equations (1) – (3) and (5) into (4) which yields:

$$\theta_1' = (\theta_m - \theta_1)/0.65 + \theta_1 \tag{6}$$

$$\theta_m = \frac{\theta_1 + A\theta_i}{A + B} \tag{7}$$

where

$$A = \frac{0.65\pi d_0 R\beta}{2\Gamma C_p} \tag{8}$$

$$B = 1 + \frac{0.65\pi d_0 K}{2\Gamma C_p} \tag{9}$$

Thus, the rate of water absorbed by the liquid film on the tube can be rewritten as follows:

$$M = \frac{\pi d_0 \beta B}{A + B}\, \theta_i - \frac{\pi d_0 \beta}{A + B}\, \theta_1 \tag{10}$$

The flow rate per unit wetted perimeter and concentration of the liquid falling to the next tube are:

$$\Gamma' = \Gamma + 0.5M$$
$$= \Gamma + \frac{0.5\pi d_0 \beta B}{A + B}\, \theta_i - \frac{0.5\pi d_0 \beta}{A + B}\, \theta_1 \tag{11}$$

$$C_1' = \Gamma C_1/\Gamma' \tag{12}$$

For LiBr-H_2O absorption, if the concentration variation is small, the relation between the concentration and the saturated temperature is nearly linear at constant pressure[5], i.e.

$$C = e + ft$$

then

$$\theta_i - \theta_1' = (C_1 - C_1')/f$$
$$= (1 - \Gamma/\Gamma')(e/f + t_c + \theta_i)$$

i.e.

$$\theta_1' = \theta_i - (1 - \Gamma/\Gamma')(e/f + t_c + \theta_i) \tag{13}$$

If the heat-and mass-transfer coefficients K and β are known, all the paramenters on every tube can be easily calculated with the use of the above equations. These equations are called operating line equations of the absorber. Following theoretical analysis is made to determine the values of K and β.

3. THEORETICAL ANALYSIS

First, the following assumptions are mode:

1) The pressure in the absorber is constant;

2) Thermal conduction and diffusion in the direction of flow are negligible;

3) The physical properties of the aqueous LiBr are constant and determined by the averge temperature and concentration;

4) The heat expenditure is negligible in the process;

5) The steam-liquid interface is always in equilibrium;

6) The thickness of the liquid film on a certain tube is considered constant.

3.1 Model of Flow, Heat and Mass Transfer of the Liquid Film

The aqueous LiBr solution is sprayed uniformly on the uppermost tube of the absorber and drops spontaneously in order to the lower stages. The falling liquid film outside the tube absorbs the steam from the evaporator, meanwhile, it is cooled by the coding water inside the tube. It is considered that the flow state, heat-and mass-transfer process are the same on every tube without accounting the influences of surface tensile force, collision, etc.

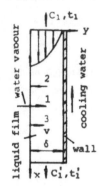

The falling liquid film is shown in Fig.1. δ is the average thickness of the liquid film and is determined by[3]:

Fig.1 1. H_2O

2. LiBr

3. Stefan flow

$$\delta = 1.35 \left(\frac{3\Gamma\nu}{g\rho} \right)^{1/3} \tag{14}$$

824

Because the steam is absorbed at the interface, the concentration of water component at the interface increases, so, water conponent will diffuse into the liquid film, meanwhile, LiBr component will diffuse to the interface. Since LiBr is a nonvolatile material, LiBr component can not be evaporated out from the liquid film, therefore, a very weak flow will appear in the opposite direction of the diffusion of LiBr component. This is so-called Stefan flow. The only flux of mass transfer across the interface is that of water, thus

$$v = \frac{D}{C} \frac{\partial C}{\partial y} \tag{15}$$

$$m = \frac{\rho D}{C} \frac{\partial C}{\partial y} \tag{16}$$

Over a small concentration range, an average concentration may be used, with this assumption, equation (16) reduces to

$$m = \frac{\rho D}{C_{av}} \frac{\partial C}{\partial y} \tag{16a}$$

The velocity of the Stefan flow can not be ignored when discussing its influence on the absorption of steam, although it is very small. When Re < 30, the flow state of the liquid film is laminar[7]. In this region, it is found that the velocity of the Stefan flow is small enough to be neglected by order of magnitude analysis when studying the equations of continuity, momentum, energy and diffusion of the liquid film. Then, we get

$$g + v \frac{\partial^2 u}{\partial y^2} = 0 \tag{17}$$

$$u \frac{\partial t}{\partial x} = a \frac{\partial^2 t}{\partial y^2} \tag{18}$$

$$u \frac{\partial C}{\partial x} = D \frac{\partial^2 C}{\partial y^2} \tag{19}$$

From equation (17) and its boundary conditions, we obtain

$$u = \frac{\rho g \delta^2}{2\mu} (1 - y^2/\delta^2) \tag{20}$$

and

$$u_{av} = \frac{\rho g \delta^2}{3\mu} \tag{21}$$

3.2 Solution for the Equations and Formulae for Evaluating the Rates of Heat and Mass Transfer

Due to the cooling water inside the tube and the absorption heat released by the steam absorbed by the liquid film at the interface, there exist temperature boundary layers both at the interface and at the outer wall of the tube, simultaneously, at the interface also exists a concentration boundary layer, They are shown in Fig.2.

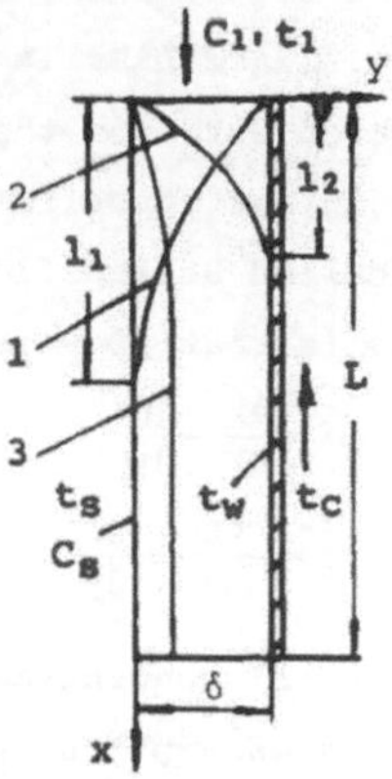

From the assumptions, the model presented, the boundary conditions and some simplifications, the equations (18) and (19) can be solved by integration to yield:

$$l_2 = 0.05625 u_{av} \delta^2/a \tag{22}$$

when $x < l_2$, the rates of heat and mass transfer are

$$Q_1 = 0.3192 Pe\lambda (t_1 - t_w) \tag{23}$$

$$M_1 = 0.3458 \frac{f}{C_{av}} \rho D (Pe Re Pr L_w)^{0.5}$$

$$\times \frac{1}{1 + \frac{Ka L_w^{-0.5}}{C_{av}}} (t_i - t_1) \tag{24}$$

where, the following dimensionless numbers are introduced:

$$Re = \frac{4\Gamma}{\mu} \qquad Pe = \frac{u_{av}\delta}{a} \qquad Pr = \frac{\nu}{a} \qquad L_w = a/D \qquad Ka = \frac{fR}{C_p}$$

when $l_2 < x < L$, the temperature profile within the liquid film is considered linear, according to the model presented and the method introduced in[3], we get

$$Q_2 = \frac{2\lambda L}{\delta^*} (t_i - t_c) [T(X_0)H_1 + AH_2] \tag{25}$$

$$M_2 = \frac{2f\rho D}{C_{av}} (\frac{3PeL_w}{2\pi\delta/L})^{0.5} (t_i - t_c)[2(1 - \sqrt{X_0}) - T(X_0)e^{F(X_0)}S_1 - AS_2] \tag{26}$$

here

$$\delta^* = \delta + \lambda[\frac{\delta_w}{\lambda_w} + \frac{d_0}{d_i}(\frac{1}{\alpha_c} + r_b)]$$

$$A = \sqrt{\frac{6}{\pi}} \frac{Ka/C_{av}}{(2 - \delta/\delta^*)\sqrt{PeL_w\delta/L}}$$

$$B = \frac{2\delta/\delta^*}{(2 - \delta/\delta^*)Pe\delta/L}$$

$$X = x/L \qquad X_0 = l_2/L$$

Fig.2. 1. δt_1
2. δt_2
3. δ_c

826

$$T(X_0) = 1 - \frac{1}{1 + \dfrac{KaL_w}{C_{av}}^{-0.5}} \cdot \frac{t_i - t_1}{t_i - t_c}$$

$$F(X) = 2A\sqrt{X} + BX$$

$$H_1 = \int_{X_0}^{1} e^{-F(X)} dX$$

$$H_2 = \int_{X_0}^{1} e^{-F(X)} \int_{X_0}^{X} \frac{1}{t} e^{F(t)} dt dX$$

$$S_1 = \int_{X_0}^{1} \frac{1}{\sqrt{X}} e^{-F(X)} dX$$

$$S_2 = \int_{X_0}^{1} \frac{1}{\sqrt{X}} e^{-F(X)} \int_{X_0}^{X} \frac{1}{\sqrt{t}} e^{F(t)} dt dX$$

The rates of heat transfer and mass transfer (absorption) on the tube are

$$Q = Q_1 + Q_2 \tag{27}$$

$$M = M_1 + M_2 \tag{28}$$

The temperature, concentration and flow rate per unit wetted perimeter of the solution leaving the tube or falling to the next tube are

$$t_1' = t_1 + \frac{MR - Q}{2\Gamma c_p} \tag{29}$$

$$C_1' = C_1 \Gamma / (\Gamma + 0.5M) \tag{30}$$

$$\Gamma' = \Gamma + 0.5M \tag{31}$$

Having had the above formulae, the rates of heat and mass transfer on every tube of the absorber can be calculated with computer. Therefore, the heat- and mass-transfer coefficients K and β can be obtained from the definitions (1) and (2) made previously in section two of this paper.

4. EXPERIMENT

The experimental apparatus is shown schematically in Fig.3. After the system had been evaculated and checked, proper amount of LiBr solution was introduced into it. The main part of the apparatus is the absorber, which is a coil of 20-stage and 112mm in diameter, made of copper tube of inner diameter 10mm and outer diameter 12mm. The coding water in the coil is from a constant temperature bath. The rich solution cooled by the cooler is sprayed on the top of the coil and absorbs the steam from the evaporator. The poor solution after the coil is enriched and lifted

to the steam-liquid separator
by the electrical heaters
which operate as thermosiphon
pump. The coolant water to
the evaporator is from the
condenser throughout a
throttling tube.

The inlet and outlet temperatures of the absorber were
measured by mercury thermometer having a resolution
of 0.1°C. The inlet and outlet concentrations of the
liquid were obtained from
the temperature and the
density of the sample
solution. The rate of flow
of liquid was measured by
enthalpy difference method,

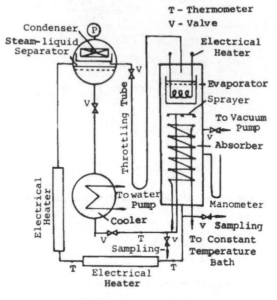

T - Thermometer
V - Valve

Fig.3 Schematic diagram of
experimental apparatus

then the flow rate of the liquid per unit wetted perimeter can be obtained
by

$$\Gamma = \frac{G}{2\pi d} \tag{32}$$

The pressure in the absorber was determined by a manometer and a mercury
barometer. The temperature of the cooling water was measured by mercury
thermometer having a resolution of 0.1°C, its flow rate was determined
by a receiver tank and the measuring time.

The experiment was carried out under such conditions: inlet temperature
32°C - 44°C, inlet concentration about 50%, rate of flow of liquid 0.0095 -
0.015Kg/s, inlet temperature of cooling water in the absorber kept at
30°C, absorption pressure 10 - 14mmHg.

After having measured the above parameters of the absorber, we can calculate the absorber by computer, using the equations presented in section
two of this paper by a trial-and-error solution until the calculated
outlet parameters are equal to the measured ones. Then, we can obtain the
values of K and β from the equations (8) and (9) by the final values of

828

A and B, thus

$$K = (B - 1)\frac{2\Gamma Cp}{0.65\pi d_0} \tag{33}$$

$$\beta = A \frac{2\Gamma Cp}{0.65\pi d_0 R} \tag{34}$$

The theoretical and experimental values of the heat-and mass-teansfer coefficients K and β are shown in Fig.4 and Fig.5, which indicate satisfactory agreement in the region of Re < 30.

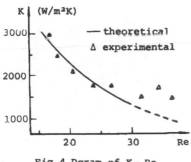

Fig.4 Dgram of K - Re

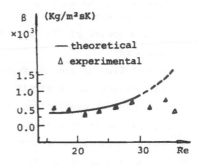

Fig.5 Diagram of β - Re

CONCLUSIONS

1) The Operating line equations obtained can be used to calculate the absorber easily, and provide convenience for investigating the relation between heat and mass transfer and utilizing energy efficiently.

2) The process in the absorber is that of simultaneous heat and mass transfer, there exists Stefan flow in the liquid film, only water is absorbed and diffuses into the liquid film across the interface.

3) When Re < 30, the flow state of the liquid film is laminar, the theoretical values of the heat-and mass-transfer coefficients agree with the experimental results, i.e. the theoretical analysis is correct. When Re > 30, the flow state of the liquid film will be turbulent.

4) The results obtained provide reference for the study and design for the absorber of the lithium bromide absorption refrigerator.

829

NOMENCLATURE

A - constant

a - thermal diffusivity of liquid, m^2/s

B - constant

C - concentration of liquid, %

C_i - saturated concentration under P and t_1, %

C_p - heat capacity of liquid, KJ/KgK

D - diffusivity of liquid, m^2/s

d - diameter of the coil, m

d_i - inner diameter of the tube, m

d_0 - outer diameter of the tube, m

e,f- constants

G - rate of flow of liquid, Kg/m^3

g - acceleration due to gravity, m/s^2

K - heat-transfer coefficient, W/m^2K

L - half perimeter of the tube, m

M - rate of mass transfer (absorption), Kg/ms

m - flux of mass transfer, Kg/m^2s

P - absorption pressure, mmHg

Q - rate of heat transfer, W/m

R - absorption heat of liquid, KJ/Kg

r_b - filth coefficient of cooling water, m^2K/W

t - temperature of liquid, °C

t_i - saturated temperature under P and C_1, °C

u - velocity of liquid film, m/s

v - velocity of Stefan flow, m/s

Greek Symbols

α - heat-transfer coefficient, W/m^2K

β - mass-transfer coefficient, Kg/m^2sK

μ - viscosity of liquid, Kg/ms

ν - kinematic viscosity of liquid, m^2/s

λ - thermal conductivity of liquid, W/m

Γ - flow rate of liquid per unit wetted perimeter, Kg/ms

δ - thickness of liquid film, m

θ – temperature difference, °C

ρ – density of liquid, Kg/m³

Subscripts

1 – inlet of a certain tube

av – average

s – cooling water

m – logrithm mean

s – at the interface

w – wall

REFERENCES

1. James R.Welty et al, Fundamentals of Momentum, Heat and Mass Transfer
   (Third Edition), John Wiley & Sons, 1984.

2. Wang Jianpin et al, Heat and Mass Transfer in Down Film Absorption of
   Lithium Bromide-Water Solution under Small Renolds Numbers (in Chinese)
   Proceedings of the 1984 Heat and Mass Transfer conference, P268–273.

3. Chen Zhiyuan et al, Theoretical Analysis on the Absorption Process in
   the Lithium Bromide Absorption Refrigerator (in Chinese), Journal of
   the Chinese Association of Refrigeration, No.2, 1983, P18–26.

4. Zou Xinxi et al, The Analysis of Heat and Mass Transfer of the LiBr-
   H_2O Solution Film Absorption over the Horizontal Tubes, Proceedings of
   the 17th International Congress of Refrigeration, Vienna, 1987,
   P1026–1031.

5. Naoyuki Inoue, Absorption Process in the Absorber of Absorption
   Machine (in Japanese), Ebera Corporation, 1985.

6. E.L.Cokoтов, Thermal Characteristics of Heat-exchanger Apparatus (in
   Russian), Thermodynamics, No.5, 1958.

7. B.M.Ramm, Gas Absorption, Chemical Industry Publishing House, 1985.

Experimental Research on Heat Transfer of LiBr Solution in Circular-Finned Tubes

WEN-HUI XIA, WEI-BIN MA, WEI-KANG CHEN, LI-YING LIN,
and YUAN-QING XIA
Guangzhou Institute of Energy Conversion
Chinese Academy of Sciences
P.O. Box 1254
Guangzhou, PRC

ABSTRACT

Although low-finned tube heat exchanger are widdely used in LiBr absorption refrigerations to enhance heat transfer at concentrated solution side, calculation formulas for the purpose have not yet been found reported. Through experiments of LiBr two stage absorption system, 300 data groups of heat transfer in heat exchanger were collected and multi-dimensional regression was used to derive the Nusselt criterion formula.

Experiments showed that solution velocity should be greater than 0.4 m/s if low-finned tubes were used, otherwise enhancing effect would be relatively little.

1. FEATURES OF SOLUTION HEAT EXCHANGER

In a LiBr absorption refrigerator, solution heat exchanger is an important component which can promote the system's performance coefficients. For hot water type LiBr absorption refrigerators, it is even more important because their heat source is of low grade and heat transfer in generator is carried out under small temperature difference. Apparently it is possible to lower the load on a generator by raising the temperature of solution entering it by use of a heat exchanger. Heat exchange in a solution exchanger is the heat exchange carried out between the in-tube low temperature thin solution coming from absorber and ex-tube high temperature concentrated solution. The flowing of concentrated LiBr solution is causedd by the difference in water head and the pressure difference (40-60 mmHg) in the system. Hence, design velocity is only about 0.5 m/s and can not be higher. At this velocity, the flow is in laminar or transitional state and heat transfer coefficient is small. On the contrary, thin solution is delivered by a pump and its velocity may be higher. Therefore, at present in China, low external-finned tubes are used in steam type LiBr absorption refrigerators to enhance heat transfer at concentrated

833

solution side. Unfortunately, neither universal formula for this purpose has yet been found reported, nor special research work. Consequently, in recent years we made experiments on a hot water type LiBr two-stage absorption refrigerator using annular-low-finned tube heat exchanger and got 300 groups of data for different concentration and velocities in the range of practical application. In this paper, these data were collated, analysed and synthesized into a criterion formula for future use in design.

2. APPARATUS AND EXPERIMENT PROCEDURE

2.1 The heat exchanger used in experiment was of double wall tube, Fig 1. The outer tube was carbon steel pipe of 32*3.5 while the inner tube was low external-finned copper pipe, see Fig.2 .

2.2 Experimental System

Experiments were made on a two-stage LiBr absorption system. Concentrated solution comes out from the generator, flows into the annular gap outside the low-finned tube (Fig.3) where it is cooled down, then flows into the absorber. At the same time, the refrigerant water from the condenser also flows into the absorber. These two are mixed into thin solution in the absorber and after cooled by cooling water, the latter is delivered by a screened pump into the heat exchanger. When thin solution flows in the copper tubes, it is heated by concentrated solution and then flow through the rotor flowmeter into the generator where it is heated and concentrated again. The cycle is then completed.

Concentrated solution inlet

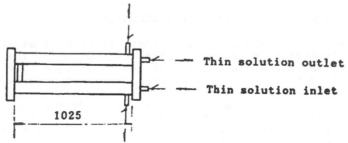

— Thin solution outlet

— Thin solution inlet

1025

Concentrated solution outlet

Fig.1 Schematic configuration of solution heat exchanger

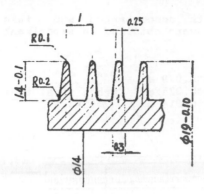

Fig.2 Low external-finned copper tube

2.3 Experimention Parameters

(1) Concentration (in percentage of weight)
    Thin solution : from 46% to 55% in 1% increment.
    Concentrated solution : from 48% to 56% .
(2) Velocity range
    Thin solution : from 0.24 to 0.58 m/s .
    Concentrated solution : from 0.17 to 0.42 m/s .
(3) Intel temperature range
    Thin solution : from 33 to 40 'C .
    Concentrated solution : from 65 to 75 'C .
(4) Instruments
    Temperature : Primary instrument , Cu-Const. thermo-
        coupple. Secondary instrument: PF15 Digital or FLUKE
        2240c Digital meter, for direct measurement.
    Flow rate : Primary instrument, rotor flowmeter;
        secondary instrument, PP11a Frequence meter.
    Concentration : Precision specific gravity meter.

2.4 Nomenclature of Fig.3

Concentrated solution	Thin solution	
XH --- Concentration	XL --- Concentration % (in weight)	
THi --- Temperature	TLi --- Temperature	'C
MH --- Flow rate	ML --- Flow rate	Kg/h
hHi --- Enthalpy	hLi --- Enthalpy	Kcal/Kg

Effective length of heat transfer tube L = 2.05 m

Finned factor ϕ = 2.5 , supplied by factory.

Thin solution flows inside the tube with constant
concentration and increasing temperature. Concentrated
solution flows inside annular gap, with constant
concentration and decreasing temperature. Inlet and outlet

835

temperatures of both concentrated and thin solution, THi,THo,TLi and TLo, were obtained in experiment.

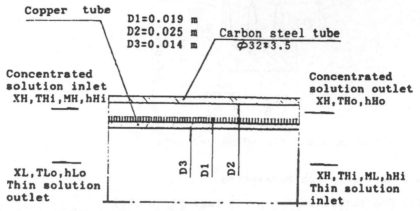

Fig.3   Configuration of double-wall tube heat exchanger and direction of solution flow.

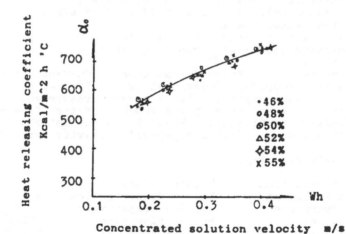

Fig.4   The relationship between concentration,velocity and heat transfer coefficient of low-finned tube side.

3. SYNTHESIZING THE EXPERIMENTAL DATA

3.1 Heat Transfer from Thin Solution inside the Tudes

Heat transfer from thin solution inside the tubes may be calculated by universal formula, i.e.

$$Nu = 0.15\ Re^{\wedge}0.33\ Pr^{\wedge}0.43\ Gr^{\wedge}0.1\ (Pr/Prw)^{\wedge}0.25 \tag{1}$$

Since at small temperature difference, $Pr/Prw = 1$, it is allowable to take.

$$Nu = 0.15\ Re^{\wedge}0.33\ Pr^{\wedge}0.43\ Gr^{\wedge}0.1 \qquad \text{for calculation,}$$

where $Re = Wi\ D3\ /\mathcal{V}$, $Wi = 4\ ML\ /(3600\rho\pi D3^{\wedge}2)$
$Pr = \mathcal{V}/a$, $Gr = g\ D3^{\wedge}3\ \beta\ \Delta t\ /\mathcal{V}^2$, $Nu = d_i\ D3\ /K$
$a = K/(\rho\ Cp)$, Temperature conduction coefficient, m^2/h
$\rho = 1000\ d$, Kg/m^3
$d = d(TL,XL)$, Factor of specific gravity, Kg/l
g is acceleration of gravity, $9.81\ m/s^{\wedge}2$
$\Delta t=((THi+THo)/2-(TLi+TLO)/2)/2$, temperature difference
$\mathcal{V} = \mathcal{V}(TL,XL)$, Factor of kinematic viscosity, $m^2\ /s$
$Cp = Cp(TL,XL)$, Factor of specific heat, $Kcal/Kg.'C$
$K = K(TL,XL)$, Heat conductivity, $Kcal/m.h.'C$
$\beta = \beta(TL,XL)$, Volumatric expansion factor, $1/'C$

Characteristic temperature for calculation of material properties $TL = (TLi+TLo)/2$.

3.2 Heat Transfer in Annular Jacket

Since one of the tube surfaces is treated for enhanced heat transfer, the flow is practically a longitudinal flow past solid surface for which no formula is available. (Here the problem is the handled by a way of from the whole to the locality) :

First, calculate Kt from inlet and outlet temperatures, concentration and flow rate of the heat exchanger by use of approximate formula.

$$Kt = QH/(F2(THi-TLo-0.35(TLo-TLi)-0.65(THi-THo)))$$

Where Qh is the heat released by the concentrated solution.

$$QH = MH(hHi-hHo) \tag{2}$$

Since $Kt = 1/(1/d_i+1/d_o+r+\delta/\lambda)$ \hfill (3)

Where $r = 0.0001$ --- Factor of dirt, in $(m^{\wedge}2\ h\ 'C)/Kcal$
$\delta = (D1-D3)/2$ ---   Thickness of wall across which
$= (0.019-0.014)/2$ heat transfer takes place.
$=0.0025\ m$
and
$F2 = \pi D1\ L\ \phi$
$= 3.1416*0.019*2.05*2.5$
$=0.3059\ m^{\wedge}2$ ------ Heat transfer area

The following parameters may by calculated :

$$d_o= 1/(1/Kt-(1/d_i+r+\delta/\lambda)) \tag{4}$$

Nuo = d_o(D2-D1)/Ko   Nusselt Number   (5)

Reo = Wo(D2-D1)/ν_o Reynolds Number

Wo = MH /($\rho_o \pi$ (D2^2 - D1^2)3600/4)

Pro = ν_o/(Ko/ρ_oCpo) = ν_o/a$_o$ Prandlt Number

Gro = g(D2-D1)^3 ρ_o Δt /ν_o^2 Grachof Number

In the above calculation of parameters, the characteristic temperature is take as :

TH = (THi+THo)/2

Δt = ((THi+THo)/2-(TLi+TLo)/2)/2

The relationship between concentration, velocity and transfer coefficient of low-finned tube side was plotted in Fig.4.

By multi-dimensional linear regression, the following criterion formula is derived :

Nuoc = Nuo(Reo,Pro,Gro)

or in a more general form :

Nuoc = a Reo^A Pro^B Gro^C

Take the logarithmic form :

Ln Nuoc = Ln a + A Ln Reo + B Ln Pro + C Ln Gro

This equation is similar to

Y = ao + a1 X1 + a2 X2 + ... + an Xn   (n=3)

Carry out regression for the 290 data groups and a criterion formula is obtained :

Nuoc = 1.28E-6 Reo^1.6123 Pro^-0.245 Gro^0.5   (6)

After rejecting the data groups of large error, for the rest 218 data groups,the error of the above formula is less then 10% .

4. ANALYZING THE EXPERIMENTAL DATA

In the range of concentration (48-56%) used in the experiments, the concentration of solution on low-finned tube side has little effect on heat transfer coefficient d_o (See Fig.4), and d_o increases with solution velocity.

The total heat transfer coefficient Kg is calculated by

Table : Velocity at low-finned tube side in function of Kt,Kg

WH \ K	0.16-0.18	0.24-0.25	0.28-0.30	0.36-0.38	0.42
Kt Kg	145.0 233.1	187.7 258.7	218.2 274.1	259.4 293.4	308.9 310.1
Kt Kg	200.3 242.0	237.6 263.7	254.6 271.6	338.1 303.8	406.9 319.1
Kt Kg	167.7 244.9	245.2 263.1	257.1 304.0	326.4 304.8	429.0 316.1
Kt Kg	175.0 238.0	199.2 263.8	265.3 284.4	252.9 295.0	387.7 310.8
Kt Kg	182.8 242.6	239.4 265.3	286.6 287.5	371.7 300.8	410.5 316.0
Kt Kg	198.8 239.6	258.4 265.3	321.7 287.3	336.6 288.3	456.0 317.0
Kt Kg	124.5 242.4	183.8 267.7	206.5 186.4	291.5 303.7	405.0 319.9
Kt Kg	130.8 246.0	163.9 264.0	209.8 288.7	277.3 304.7	328.6 321.2
Kt Kg				291.7 307.3	358.6 320.6

WH --- Solution velocity   m/s

K, Kt, Kg --- Heat transfer coefficient   Kcal/m^2 h 'C

equation (3) after the in-tube and ex-tube heat transfer coefficients are obtained by universal heat transfer crierion formula :

$Nu = 0.15\ Re^{0.33}\ Pr^{0.43}\ Gr^{0.1}$

Kt is the total heat transfer coefficient for low-finned tube, which is calculated by the approximate formula (Eq.3) through meassured solution inlet and uotlet temperatures and a heat transfer area taken as the product of smooth tube surface area and the finned factor. Kt = Kg means the expected enhancing is attained, namely finned factor reaches 2.5 . In our experiments, at solution velocity of 0.42 m/s, we have Kt > Kg . At velocity smaller than 0.42 m/s, we have Kt < Kg for most cases; but in the worst case, Kt is still not smaller than 0.5 Kg, showing an enhancing effect, but not great, which reduces with decreasing velocity.

Nuoc calculated by criterion relationship (6) derived by multi-dimensional regression is in an error smaller than 20% and averaged 10% when compared with Nuo calculated by approximete formula. This means that formula (6) can serve engineering purpose.

5. CONCLUSIONS

If low-finned tubes are used as heat transfer tubes in a solution heat exchanger, solution velocity should be greater than 0.4 m/s , otherwise there will be little enhancing effect.

The criterion fromula derived here:

$Nuoc = 1.28E\text{-}6\ Reo^{1.6123}\ Pro^{-0.245}\ Gro^{0.5}$

can serve design purpose, in cases of similarity.

REFERENCE   (All in Chinese)

1.   Yang Shi-ming, Chen Dai-xie, Heat transfer, revised edition, Chinese Industry Publisher, Beijing, October 1961.

2.   Chen Weikang, Xia Wenhui, An Expression of LiBr Water Solution Properties, Refrigeration Magazine, Volume 3, September, 1987.

3.   Fudan University, Atlas of Physical Properties of Water Solution of China-made LiBr, P.O. Box 5022, Shanghai, 1976.

Simulative Experiment of an Industrial Application Using a Diesel Engine Driven Heat Pump

FENNING LIU, MEILI HU, and GUANGMING LUO
Department of Thermoscience and Engineering
Zhejiang University
Hargzhou, PRC

ABSTRACT

For simulation of a proposed heat pump plant utilising the warm waste-water in a silk scoring and dyeing workshop, a dual powered water/water compression experimental installation is set up and tested. The performance of the installation is discussed and a comparison of diesel and electric drives is made, which is of some reference value for the application of heat pumps.

1. INTRODUCTION

A heat pump is effective to extract heat at a lower temperature and make it available at a suitable higher temperature. With the development of industry and agriculature in China, heat pump technology sees broad prospects in energy conservation while its inherent advantages becomes more salient under the current shortages of conventional energy sources. The heat pump has already found wide commercial applications in the field of air-conditioning, heating etc.

In the silk industry, especially in its scoring and dyeing workshops, a large continuous supply of hot water at about 70°C is needed, while the waste water at about 40°C is simply drained away. A simulative experimental installation was set up and tests made with an attempt to explore the field of heat pump applications before starting work with a commercial plant. The results of experiments consist of performance of both motor and diesel engine driven plant, which is compared and analysed. With its academic and practical merits, the diesel engine driven heat pump is proved to be a promising unit to justify application and popularisation in China.

2. A BRIEF DESCRIPTION OF THE EXPERIMENTAL HEAT PUMP INSTALLATION (FIGURE 1.)

A compression water/water heating plant of 11.6 kW heating capacity is adopted using a conventional reciprocating refrigeration compressor and RIZ (freon-12) working medium, both of which are widely used and inexpensive in China. The evaporator and condenser are tubular in design. An electric motor and a diesel engine are

841

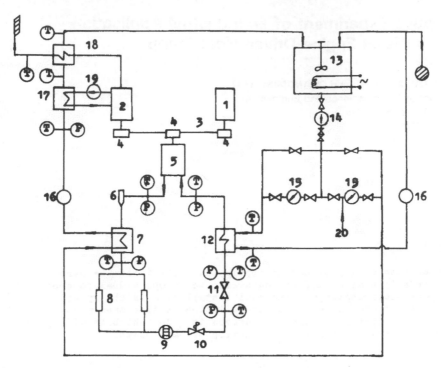

FIGURE 1. The experimental heat pump plant system

1. Electric motor	11. Expansion valve
2. Diesel engine	12. Evaporator
3. Belt	13. Hot water tank
4. Pulley	14. Water pump
5. Compressor	15. Integrating flowmeter
6. Oil-trap	16. Rotameter
7. Condenser	17. Water/water heat exchanger
8. Liquid collector	18. Exhaust gas/water heat exchanger
9. Strainer	19. Water jacket circulating pump
10. Solenoid valve	20. Supply water valve

provided for a shift of driving power by means of changeover belting.

For the convenience of monitoring the parameters at various points of the cycle, instruments such as pressure gauges, thermometers, flowmeters etc are installed, as shown in the figure.

3. EXPERIMENTS, RESULTS, AND DISCUSSION

3.1 The Experiment Under Design Conditions
With electric drive, fresh water can be heated from 26°C to 59°C when the temperature of waste water entering the evaporator is 35°C. When the drive is shifted to the diesel engine, and all other conditions held unchanged, the fresh water, after first being heated to 59°C, is further heated in the water/water and gas/water heat

exchangers, utilizing the heat energy in the cooling water and exhaust gas of the diesel engine. Thus, the fresh water is brought to 71°C, which fulfils the requirements of silk scoring and dyeing on hot water supply.

Under these operating conditions, the vaporisation and condensation temperatures of the heat pump cycle are 13°C and 62.5°C respectively. The energy distribution diagram is shown in FIGURE 2.

Condenser output	11.60kW
Evaporator input	8.70kW
Heat dissipation	0.78 kW
Compressor input	3.72 kW
Heat input of fuel	12.00 kW
Input from water jacket	2.54 kW
Input from exhaust gases	1.56 kW

Coefficient of performance of motor driven heat pump, $COP_e = 3.1$
Primary energy ratio of motor driven heat pump, $PER_e = COP_e \, \eta_t^* = 0.96$
Primary energy ratio of diesel engine driven heat pump, $PER_d = 1.3$

As is generally the case, small diesels have a larger fraction of heat dissipation than large engines. Our simulative plant has a small engine, the recovery from its cooling water and exhaust gases was less than that attainable in large engines. Hence its primary energy ratio (PER_d) reached merely 1.3.

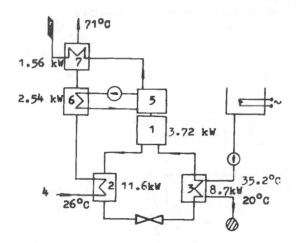

FIGURE 2   The energy distribution diagram

1. compressor   2. condenser   3. evaporator   4. water supply
5. diesel engine 6. water/water heat exchanger
7. exhaust gas/water heat exchanger

η_t^* is the overall efficiency of electricity supply, taken as 0.31

3.3 Energy flow diagrams.

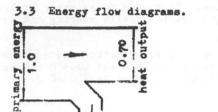

<u>Hot water boiler</u>

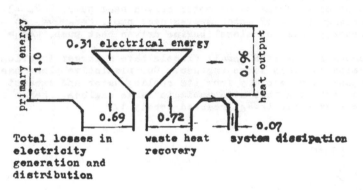

<u>Electrical motor driven heat pump</u>

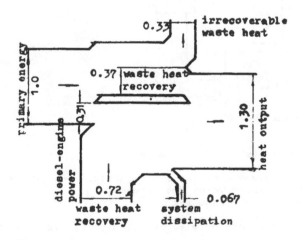

<u>Diesel-engine driven heat pump</u>

FIGURE 3. Energy Flow Diagrams

3.3 Comparison of the Three Heating Schemes

TABLE 1. Comparison of the Three Heating Schemes

	COP	PER	Water temp. °C	Supply of fuel	Pollu- tion	Price of fuel RMB	Operating cost (incl. pumping) RMB
Heat pump, electric drive	3.1	0.96	59	in shor- tage (electri- city	none	0.16/kWh	0.755/h
Heat pump, diesel drive	3.1	1.3	71	fairly good (oil)	mode- rate	0.44/kg	0.695/h
Hot water boiler		0.7	high	fairly good (coal)	heavy	0.08/kg	0.388/h

3.4 Analysis of experimental results under varying working conditions

In order to explore the variation of performance of the heat pump plant under working conditions other than the design condition, we made experiments with varied waste water temperature and varied water flow through the evaporator and the condenser, and, eventually, ob- tained the relationship between such data as the hot water tempera- ture, vaporization and condensation temperatures, PER_s etc: For details see the following figures.

FIGURE 4. Shows the relationship between the primary energy ratios PER_d, PER_e and the vaporization and condensation temperatures t_e and t_c. It is concluded from the curves that (1) for the same t_e and t_c, $PER_d > PER_e$; (2) when t_c is held unchanged, the PER_s varies as t_e with either electric or diesel drive. Therefore, it can be stat- ed with regard to large industrial heat pump installations that the primary energy ratio of diesel drive units is higher than that of electric drive units.

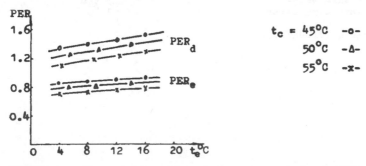

FIGURE 4. The relationship of the curves between PER_s and t_e, t_c

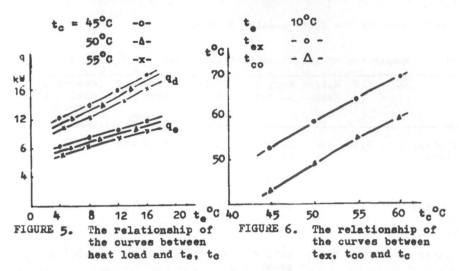

FIGURE 5. The relationship of
the curves between
heat load and t_e, t_c

FIGURE 6. The relationship of
the curves between
t_{ex}, t_{co} and t_c

FIGURE 5. Displays the relationship between the heat load of the
heat pump and t_e and t_c, here, it is evidenced that, with t_c held
constant, both the heat output q of the diesel driven heat pump
system and the heat loading of the condenser q_c vary as t_e.

In FIGURE 6, it is shown that, while t_e is held constant, the hot
water temperature of both the motor driven and diesel engine driven
heat pump installations varies as the condensation temperature
t_c. Its maximum value, however, is limited by the maximum permis-
sible pressure of the compressor.

4. PROPOSED SCHEME OF THE WATER/WATER COMPRESSION HEAT PUMP PLANT
IN HANGZHOU SILK DYEING & PRINTING MILL'S SCORING AND DYEING WORK-
SHOP

Based on the fore-mentioned experimental data and the conditions
of the said workshop's waste warm water, we have worked out a scheme
for the proposed heat pump plant as described in the following;

Original data:
Quantity of waste warm water   5 t/h
Temperature of waste warm water   42 °C

The heat pump system (FIGURE 7):

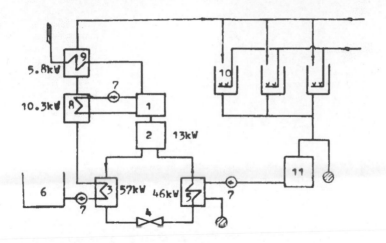

FIGURE 7. Proposed scheme

1. Diesel engine
2. Compressor
3. Condenser
4. Expansion valve
5. Evaporator
6. Water supply tank

7. Water pump
8. Water/water heat exchanger
9. Gas/water heat exchanger
10. Hot water tank
11. Waste water tank

Energy ratings of the components, see FIGURE 7.

Coefficient of performance, pump work excluded $COP' = 4.38$

Coefficient of performance, pump work included $COP = 3.35$

Primary energy ratio, pump work included $PER = 1.38$

To conclude from the above-mentioned experiments and proposed design, it is worthy of our notice that the recovery of waste heat with heat pumps is a good means of energy conservation and the use of diesel drive further increases the primary energy ratio of the installation. This is especially true for those industries suffering from power shortage. Therefore, it can be said that the use of diesel drive in large industrial heat pump plants is favorable.

REFERENCES

1. D.A. Reay and D.B.A. Macmichael, "Heat Pumps Design and Application," 1979.
2. Hans Ludwig von Cube and Fritz Steimle, "Heat Pump Technology", 1981.
3. D.A. Reay and V.A. Eustace, "Industrial Application of High Temperature Heat Pump Driven by Gas Engine", New Ways to Save Energy, 1979.
4. Harry J. Sauer, Jr. and Ronald H. Howell, "Heat Pump Systems",1983.

Figure 3. Proposed scheme.

Modeling of a Water-to-Water Heat Pump

WEIMING LU, YUANJI MA, HUANG GU, MEILI HU, TINGJIN SHI,
and HAOREN REN
Department of Thermoscience and Engineering
Zhejiang University
Hargzhou, PRC

ABSTRACT

In order to improve the performance and components of a heat pump system and make a optimum design of system, a computer program by which the performance of a water-to-water heat pump system can be simulated was composed. It included power consumption of the compressor, thermodynamic states for the refrigerant and water, working fluid mass rates and models of the condenser and evaporator. This program is suitable for a system which consists of an open-compressor, condenser, evaporator and other supplement.

The computer simulation data has been contrasted with the experimental performance of the heat pump system. It showed that the calculated results could reflected the performance laws as well-know, and the performance predicted by the computer model is in reasonable agreement with the performance data observed in the tests. According to the theoretical analysis and experimental data, finally, some conclusions were reached.

INTRODUCTION

In recent year, the heat pump technique, an efficient measure of energy conservation, is widely investigated and rapidly developing. It is necessary to build experimental devices in laborators for deeper study.

Because the working process of a heat pump is considerably complicated, it is not enough yet only through experimental study. If a computation model, by which the work process of a heat pump is modeled, is composed, we can investigate it by a computer rapidly cheaply and comprehensively.

This paper presents a brief overview of the structure of a computation model, including a discussion of the fuction of each of its major program, which is designed for a water-to-water compressive heat pump system. Then the performance of the heat pump system which is used to simulate a system to be applicated in a factory were calculated by the program.

In order to check the computer model,the simulation data has been contrasted with the experimental performance of the heat pump system.The comparison indicates that the calculated results qualititively tallied with experimental data.

THE EXPERIMENTAL SYSTEM AND HEAT PUMP CYCLE

Fig.1 is a schematic diagram of the experimental heat pump system. The detail of the system can be found in reference [5].

The theoretical cycle of a heat pump is showen in Fig.2-(a), a somewhat destorted pressure VS enthalpy (p-h) diagram. And Fig.2-(b) is the simplified real thermodynamic process being modeled. The subscripts in Fig.2 refer to the state points indicated in Fig.1.

In this model, the mass flow rate of refrigerant is computed from volumetric efficiency as follows:

$$m_{ref} = \eta_{vol} * \rho * N * V_h \tag{1}$$

where:

η_{vol} -----volumetric efficiency
ρ -----refrigerant density at suction port
V_h -----compressor displacement
N -----compressor rotation speed

Comsumption of motor power is calculated from:

$$POW = m_{ref} * (H_{2'} - H_1)/(\eta_{mot} * \eta_{ise}) \tag{2}$$

Where:

$H_1, H_{2'}$ -----refrigerant enthalpy
η_{mot} -----compressor motor efficiency
η_{ise} -----isentropic efficiency of compressor($=(H_{2'} - H_1)/(H_2 - H_1)$)

When the characteristics of heat exchangers (condenser and evaporator) are computed, some basic relationship would be used as follows:

$$Q_{ref} = Q_w = K_{co} * A_o * \Delta t \tag{3}$$

Where:

Q_{ref} -----the amount of heat exchange($=m*(h_i - h_o)$)
A_o -----the area of heat transfer
Δt -----the logarithemic mean temperature difference
$\quad$ ($=(t_o - t_i)/\ln((t - t_i)/(t - t_o))$)
K_{co} -----the overall coefficient of heat transfer

it can be gained by eq.(4)

$$K_{co} = [\frac{1}{\alpha_o} + \frac{\delta_p}{\lambda_p} \frac{A_o}{A_o} + (R_f + \frac{1}{\alpha_i})\frac{A_o}{A_i}]^{-1} \tag{4}$$

Where:

α_o, α_i -----the coefficients of heat transfer of outside or inside
δ_p -----the thickness of tube
λ_p -----the thermal conductivity
A -----the area of heat exchanger (inside, outside or middle)

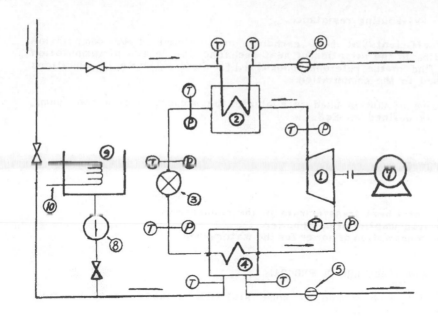

FIGURE 1.   Sysetm Schematic and Instrumentation Layout

T,P   represent temperature sensors, pressure gages, respectively
1--compressor; 2--condenser; 3--thermostatic expansion valve;
4--evaporator; 5,6--flowmeter; 7--electric motor; 8--water pump;
9--water tank; 10--electrical heater

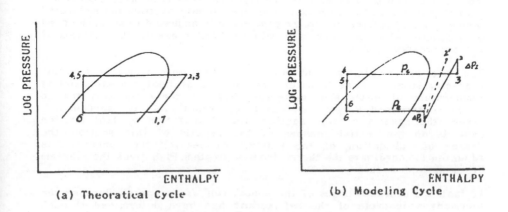

(a) Theoratical Cycle

(b) Modeling Cycle

FIGUTRE 2.   Pressure VS Enthalpy Diagram
for the Heat Pump Cycle

R_f ----Fouling resistance

The coefficients of heat exchange are determined by computation according to the situation of heat exchange. The details of computation could find in the reference books[3],[4]. Some experimental equations are used in the computation.

The value of COP is used to evaluate the properties of a heat pump, which is defined as Eq5.

$$COP = \frac{Q_c}{POW + P_w} \qquad (5)$$

Where:

Q_c ----total heat transfer rate in the condenser
POW----consumption of motor power
P_w-----consumption of power for the water pump

MODEL AND STRUCTURE OF PROGRAM

The model is composed by some siplified methods, based on some hypotheses as follows:

(a) The heat pump works at the steady-state condition.
(b) Heat energy lost to surroundings could be neglected. It is reasonable if some measures of heat preservation are taken.
(c) The refrigerant is unsaturated liquid at the outlet of the condenser ,in other word, there is some subcooling.
(d) The working fluid flow in single-phase region is turbulent flow due to the slender tube and great flow vilocity.

The total program flow diagram shown in Fig.3 displays the sequence of calculations, the decision points and main subroutines. Input prarmeters for all of the submodels, including structure parameters of components, thermodynamic state parameters, initial guesses, some constants and so on, are read in the subroutine DATAIN. Then compressor and flow balance models are used to determine the refrigerant flow rate($m_{r,t}$), consumption of motor power and pressure balance. In this section, the saturated temperature of refrigerant in the condenser is adjusted to meet the need of pressure balance,or flow rate balance. Fig.4 presents the diagram of the compressor model.

Next section is condenser subroutine, in which such parameters as the characteristics of heat transfer, the amount of heat transfer, outlet temperature of water and refrigerant leaving the condenser, and so on, are determined. Because the calculated results of last section will affect the results of this section, and computation of last section depends on the initial guesses of the results of this section, the degree of subcoolilng of the refrigerant leaving the condenser is adjusted to coordinate the thermodynamic system. Fig5 gives the diagram of condenser model.

At last, the characteristics of the evaporator are computed. Because the thermodynamic cycle of the refrigerant has been determined through computation above, the input temperature of cooling water entering the evaporator is adjusted to fit for the heat balance and heat transfer. In

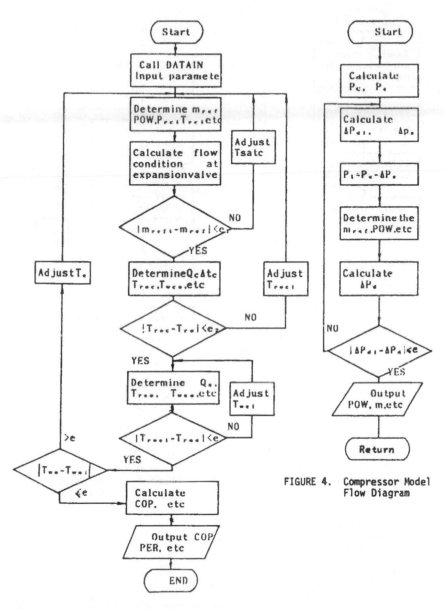

FIGURE 3. Computer Program Flow Diagram

FIGURE 4. Compressor Model Flow Diagram

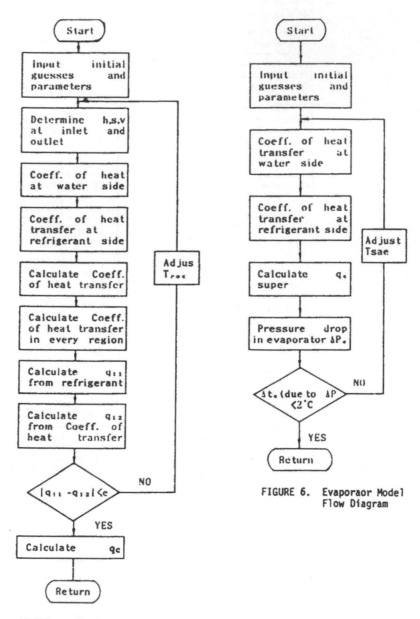

FIGURE 5.   Condenser Model
            Flow Diagram

FIGURE 6.   Evaporaor Model
            Flow Diagram

other word , the calculated results in this section will not affect the results computed above. The computation process presents in Fig.6.

The thermodynamic properties of refrigerants using the Martin-Hou equation and water using IFC equation are also computed in subroutines.

MODEL RESULTS AND COMPARISON

Runing the program, which is writing in FORTRAN IV, a series of relations between the parameters, such as COP, heating capacity (Q_c), water flow rates of the heat exchangers, refrigerant saturated temperatures in the heat exchangers and so on are attained Priliminary computation has been finished, and model results shown in Fig.7-Fig.10, simultaneously experimental curves presented for the comparison in these charts. The details about experiments are discussed in our other paper[5].

Fig.7 through 8 show the influence on performance of the temperature in the evaporator, and condenser. Each graph shows one quantity (COP or Q_c) while each curve shows the result of varying the temperature of condenser (T_c). It is proved that, with the increase of the temperature of evaporator, the quantities (COP and Q_c) also increased alike, and with the increase of the temperature of condenser, the quantities will decrease as expected. The laws of the performances by model computation are near to the experimental laws. But the absolute value of the quantities from calculation is better than from the experimants at the same conditions, because of simplication and idealization.

Fig.9 give the relation of the $W_{\cdot}vs.T_{\cdot}$ or T_c. It is found that with the increase of $T_{\cdot}$ or decrease of T_c, the power consumputation will increase. Within the range of the motor power, from 2.4 to 6.0 kw,the change of the performance (COP) takes place only a few (from 2.5 to 3.0). And heating capacity (q_c) change from 7 to 15 kw.

CONCLUSION

(1)   According to the comparison and analysis, it could be said that the model quantitively reflect the working mechanism of a water to water heat pump system.

(2)   Used the program, the performances of a heat pump can be investigation. The laws of computed data is in reasonable agreement with the real laws as well-known.

(3)   Because the thermodynamic properties of refrigerants and water are computed by subroutines in this program, the program is good at the general   characteristic.

(4)   The relative error of the computed data is small for the quantitive analysis. And further improvement should be developed in order to perfect the program.

REFERENCE

[1]   R.D.Ellision,F.D.Creswick, A Computer Simulation of Steady-state Performance of Air-to-Air Heat Pump . ORNL,1978.

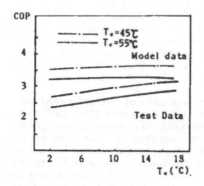

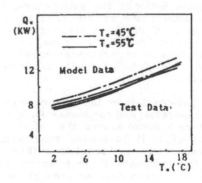

FIGURE 7. COP as a Function of T_e at Different T_c

FIGURE 8. Heating Capacity as a Function of T_e at Different T_c

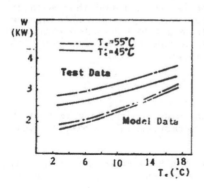

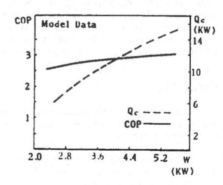

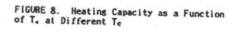

FIGURE 9. Power Consumption VS T_e and T_c FIGURE 10. COP and Q_c VS Power Consumption

|2   V.W.Goldschmidt. G.H.Hart. Heat Pump System Performance Experi-
      mental and Theoretical Results .*ASHRAE Transactions* 1982.

|3]   J.P.Holman,*Heat Transfer*, Mc Graw Hill International Book Company
      1981.

|4]   Q.S.Yan, Refrigerating Technology Using in Air Conditioning ,(In
      Chinese).

|5| T.J.Shi,W.M.Lu, An Experimental Study of A Diesel-Engine Driven
      Water-to-Water Heat Pump System . (to be published).

FURNACE AND REACTORS

An Investigation on Heat Transfer in Catalytic Fixed Bed Reactor: The Effects of Bed Length and Wall Temperature on Heat Transfer Parameters

CHUNJIE ZHANG, FUXIN DING*, DEHUA LIU, ZHENG LIU, and NAIJU YUAN
Department of Chemical Engineering
Tsinghua University
Beijing, PRC

ABSTRACT

This paper presents the experimental study of the effective thermal conductivity and wall heat transfer coefficient for low d_t/d_p ratio fixed bed reactor which is 29.8 mm in inner diameter and 1.2 m in length. The reactor is packed with different kinds of industrial catalyst pellets. Based on the two-dimensional pseudo-homogeneous model and the experimental measurements of the radial temperature distribution the heat transfer parameters have been determined and correlated with the Reynolds Number. A linear relationship between these parameters and Re is held over the experiment range. It is clear that the bed length and wall temperature are shown to have significant effects on the temperature distribution as well as the heat transfer parameters. The behavior of heat transfer in a fixed bed reactor is found to bear close relation with the catalyst geometry.

---

*Author to whom correspondence should be addressed.

INTRODUCTION

The fixed bed reactors which are packed with various kinds of catalyst have been widely used in chemical and petroleum processing industry. Many of the catalytic reactions carried out in fixed bed reactor are strongly exothermic and the energy released during the reaction must be removed by cooling at the wall. The behavior of the heat transfer in this reactor plays an important role in limiting the reactor productivity and reducing the catalyst life. Therefore, the knowledge of the heat transfer and enhancement mechanism in the reactor forms essential aspects of the reactor design and operation. These research areas have drown more attention in these years due to their theoretical and industrial importance.[1-6] The multitubular fixed bed reactor with low tube-to-particle diameter ratio(d_t/d_p) is particularly important because many industrial catalytic processes are accomplished in this kind of reactors, and numerious investigations on the heat transfer in such reactor systems have been reported in recent years.[7-9,17]

The heat transfer in fixed bed reactor is commonly described by two-dimensional pseudohomogeneous model which leads the following equation by introducing two parameters namely the effective heat transfer conductivity, Ke, and the wall heat transfer coefficient, Hw:

$$-GC_p \frac{\partial T}{\partial Z} + K_{er}(\frac{1}{r}\frac{\partial T}{\partial r} + \frac{\partial T}{\partial r}) - r_p \rho_B(-\Delta H) = 0 \qquad (1)$$

$$Z=0, \ 0<r<R, \ T=T_0 \ ; \qquad r=0 \ \frac{\partial T}{\partial r} = 0$$

$$r=R, \ K_{er}\frac{\partial T}{\partial r} = H_w(T_R - T_w)$$

The theorectical development and the application of the model are given elsewhere.[10-15] In this work a 1.2 m long tubular reactor is used. The reactor is packed with several kinds of catalyst pellets which are different from each other in the sizes and geometric shapes. The radial temperature distributions are measured and analysed besed on the two-dimensional pseudohomogeneous model and the heat transfer coeficients are calculated by using numerical approach. Both the catalytic reactor and the catalysts used in this work are widely employed in the industry and are typical of industrial operations. It could be seen that the experimental results and the calculations of this study are useful for the design and analysis of the industrial fixed bed reactor.

EXPERIMENTAL

The experimental apparatus used in this study is diagrammatically shown in Figure 1. It consists essentially of a 29.8 mm inner diameter copper tubular reactor which is 1.2 m long with 2.6 mm thick wall. The reactor is heated and controlled electrically with six precision temperature controllers. Inside the reactor there is a specially designed frame which could move in the axial direction and on which six thermocouples were fixed in different radial locations to monitor the radial temperature distribution. The fluid, air, is provided by two low noise blowers and enters the reactor after passing through the pressurizer, regulator, pressure gauge, flowmeters and the preheater.

862

The radial temperature profiles are measured in different wall temperatures and various bed lengths while varifying the air flow rate as well as the inlet air temperature. All the experiments have been conducted at wall temperature range up to 200 °C which is more close to the industrial operation condition.

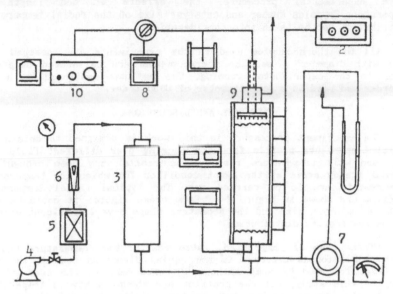

Fig. 1 Experimental set-up:
1. reactor   2. temperature control system   3. preheater
4. blower   5. presurizer   6. flowmeter
7. pressure transmitter   8. XWT recorder   9. thermocouple frame
10. potentiometer

The industrial catalysts are provided by Fushun Petro-chemical Corporation and some of their physical properties are listed in table 1.

Table 1 Physical properties of the catalysts

Catalyst	Geometric shape	particle size (mm)	Bulk density (g/ml)	Equivalent diameter (mm)	Heat conductivity of static bed (100°C) (w/m k)
Ag/Al$_2$O$_3$ (s-140)	spherer	d_p=4.8	1.05	4.80	0.325
Model 3822	trifoliate	l=4.14	0.82	1.01	0.141
Model 3823	long cylindrical	ϕ1.53 x7.0	0.80	2.07	0.163
Model 3812	cylindrical	ϕ4.0 x4.0		4.0	0.150

The experiments consist of measuring the radial temperature profile in the fluid just above the packing for each kind of catalyst, such measurement is done of several lengths. On each bed length level a wide range of the fluid velocities is used while the wall is also maintained at different temperatures ranging from 80 to 200°C. By carrying out the above experimental procedure, the effects of bed length, wall temperature, Reynolds Number and catalyst size on the radial temperature distribution can be taken into consideration.

All the thermocouples used in the experiments are sheathed EA-2 type with diameter 1 mm. The thermocouples are calibrated carefully prior to the experimental running. The emf was monitored on a XWT recorder and read on a potentiometer of UJ35 model.

RESULTS AND DISCUSSION

The experiment processure in this work is designed to obtain the radial temperature profile for each packing over different fluid flow rate and wall temperature ranges. The reactor has been brought to thermal steady state at the exact condition for which the temperature measurements are to be carried out. The typical radial temperature profiles are shown in Figure 2-4. From these figures it could be seen that the catalyst size and the geometric shape have significant effects on the temperature distribution.

In the case of large d_t/d_p ratio reactor, the temperature profile exhibits monotonous decrease in the radial direction towards the tube center, as reported by some authors.[18] However, in the case of low d_t/d_p of this study, all the profiles are shown to have a temperature "hump" after their falling down in the neighbourhood of the reactor wall, and then decrease. The similar phenomena were also reported.[8] Such kind of the radial temperature profile for low d_t/d_p bed may be attributed to the complicated heat and mass transfer mechanism as well as the specific radial porosity distribution. The influence of the bed length on the temperature profile is given in Figure 5 from which it should be clear that with increasing the bed length, the temperature distribution tends to be smoothed.

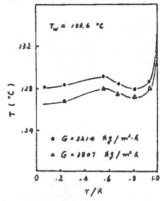

Fig. 2 Radial temperature profile: Ag/Al$_2$O$_3$ catalyst T_w =/326°C, d_t/d_p = 6.2

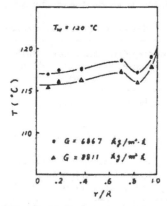

Fig. 3 Radial temperature profile: hydrocracking catalyst (Model 3823, T_w=/20°C, d_t/d_p=14.4)

864

On the basis of the two-dimensional pseudohomogeneous model and applying numerical methods, Hw and Ke are turned out and shown to be linearly related with Reynolds Number, as seen in Figure 6 and 7. This linear correlation was also reported for other fixed beds.[11,16,19] However, these two parameters are falling down to reach asympototical levels with increasing bed length when other factors are kept constant (Figure 8 and 9). The comparison between these two figures shows that the bed length has similar effects on the two parameters. Several studies also pointed out the length effect on Ke and Hw,[8,19,20] but it was found with ceramic spheric pellets at Re=430 that the length effect was less pronounced for Hw.[19] The bed length corresponding to the asympototical point is usually called entrance-length. The Ke-L and Hw-L curves show that the entrance-length has close relation with Reynolds Number: the larger the Reynolds Number, the shorter the entrance-length.

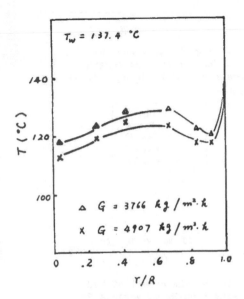

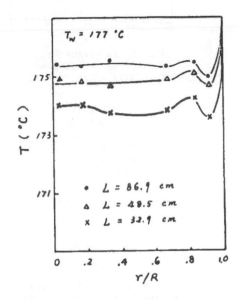

Fig. 4 Radial temperature
profile: hydrocracking cat.
(trifoliate, Model 3822)
T_w = 137.4 °C, d_t/d_p = 29.5

Fig. 5 The effect of bed length on
radial temperature distribution
(cylindrical catalyst 3812)
T_w=177 °C, d_t/d_p=7.45, Re=218

The Ke and Hw are also ploted versus Re at different wall temperatures, as shown in figuer 10 and 11. By examining these figures one could easily find the effect of wall temperature Tw on the heat transfer coefficients. Obviously, besides Tw and bed length there could be many factors which would exert their influence on the magnitute of Hw and Ke, to say a few, the particle size and its geometrical shape, the radial porosity distribution of the bed, etc. This may give a reason for the wide spread data of the Hw and Ke reported in the literature.

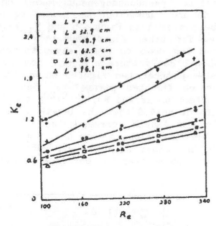

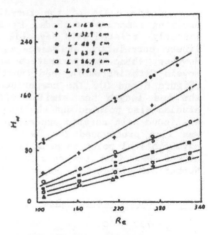

Fig. 6 Ke versus Re
for various bed length
(catalyst 3812, d_t/d_p = 7.45)

Fig. 7 Hw versus Re
for various bed length
(catalyst 3812, d_t/d_p = 7.45)

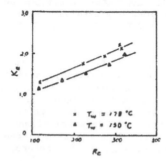

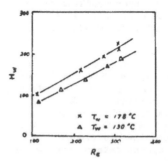

Fig. 8 The effect of bed
length on Ke at various Re
(catalyst 3812, d_t/d_p = 7.45)

Fig. 9 The effect of bed
length on Hw at various Re
(catalyst 3812, d_t/d_p = 7.45)

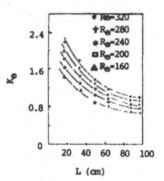

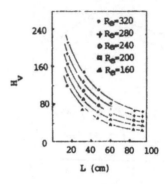

Fig. 10 Ke versus Re at
different wall temperatures
(catalyst 3812)

Fig. 11 Hw versus Re at
different wall temperatures
(catalyst 3812)

CONCLUSION

An experimental investigation of heat transfer behavior in the low d_t/d_p fixed bed reactor has been presented. The radial temperature profiles are determined experimentally and shown to be quite different from those reported in the literature with higher d_t/d_p. Because the resistance to the heat transfer near the reactor wall is quite substantial, the radial temperature profiles are inevitably to have a drop in this region, however, a "hump" is noticed to follow this temperature drop.

The two-dimensional pseudohomogeneous model has been applied to calculate the heat transfer coefficients, and a linear relationship is found when ploting Hw and Ke versus Reynolds Number.

The effects of bed length and wall temperature on the heat transfer behavior are examined. The experimental data reveal that both Hw and Ke are decreasing quickly with increasing bed length and reaching a asympototical levels which depend on wall temperature and Reynolds Number.

The results of this work are useful for the design and modeling of low d_t/d_p fixed bed reactor.

Acknoledgement

The authors would like to thank Fushun Petro-Chemical Corporation for supplying the catalysts.

NOMENCLATURE

d_t	inner diameter of reactor tube (m)
d_p	diameter of catalyst pellet (m)
G	superficial mass flow rate (kg/m^{2}h)
Hw	wall heat transfer coefficient (kcal/m^{2}h °c)
Ke	effective thermal conductivity (kcal/m·h °c)
L	length of packed bed (m)
r	radial coordinate (m)
R	radius of bed (m)
Re	Reynolds Number, $Re = Gd_p/\mu$
t	temperature (°c)
Tw	temperature of reactor wall (°c)
To	inlet temperature (°c)
x	dimensionless radial coordinate, $x = r/R$
z	axial coordinate in bed (m)

Greek symbols

μ	viscosity of the fluid (kg/m.s)
ρ	density (kg/m^3)

REFERENCE

1. DIXON, G. A. and Gresswell, D. L., *AIChE. J.*, 25,676-676(1979).

2. Dixon, G. A. etal, *Int. J. Heat Mass Transfer*, 27, 1701-1712(1984).

3. Leron, J. J. and Froment, G. F., *C.E.S.*, 27, 567-576(1977).

4. Wellauer, T., etal, *ACS Symp. Series*, 196, 527,(1982).

5. Gunn, D. J. and Ahmad, M. M., *First UK National Heat Transfer Conference, Leeds*, 3-5 July(1984).

6. Rao, S. M. and Toor, H. L., *IEC. Fund*, 23, 294(1984).

7. Melanson, M. M. and Dixon, G. A., *Int. J. Heat Mass Transfer*, 28, 383-393(1985).

8. Dixon, G. A. and Paterson, W. R., *ACS. Symp. Series*, 65, 238(1978).

9. Lerou,J. J. and Froment, G. F., *Chem. Eng. Sci.*, 32, 853(1977).

10. Zhang Chunjie etal, *J. of Chem. Reaction Eng. and Tech.*, in press.

11. Froment, G. F. and Bischoff, K. B., *Chemical Reactor Analysis and Design*, John Wiley and sons, 1979.

12. Froment, G. F., *Chem. Eng. Adv. Chem. Ser.*, 109. 1-34(1972).

13. Coberly, C. A. and Marshall, W. R., *Chem. Eng. Prog.*, 47, 141(1951).

14. Yagi, S. and Wakao, N., *AIChE. J.*, 5, 79(1959).

15. Lu Dewei etal, *Proc. CIESC-AIChE Meeting*, 1, 278(1982)

16. De Wasch, A. P. and Froment, G. F., *Chem. Eng. Sci.*, 26, 629(1971).

17. Marin, M. M. and Dixon, G. A., *Int. J. Heat Mass Transfer*, 28, No 2, 383(1985).

18. De Wasch, A. P. and Froment, G. F., *Chem. Eng. Sci.*, 27,567(1972).

19. Dixon, G. A., *Chem. Eng. Journal*, 31, 163-173(1985).

20. Paterson, W. R., *Ph. D Thesis, University of Edinburgh* (1975).

Experiment Research of the Ash Cutting in Boilers

JIAQING SUN, JINQI WANG, and XIANGYONG LU
Shanghai Institute of Mechanical Engineering
516 Jun Gong Road
Shanghai 200093, PRC

INTRODUCTION

A large number of low grade coal with high ash content have been used in the power boilers in China in order to use the fuel source rationally. Therefore the problems of ash cutting on the convection heating surface of boilers have become more and more importance to ensure the economy and safety operation of boilers. Such as the reheater of a boiler with capacity 400 T/H occured the tube failure caused by ash cutting after 5000 hours operated. Thus it is very important to carry out the study of erosion for controlling occurrence of accident by ash cutting.

Since 1962, numerous studies, concerning the rule and the pro-tection measures of ash cutting, have been conducted in China [1,2,3,4,5,6,7].However the studies on the basic behavior could not satisfy the requirement to solve the technical problems in practice about ash cutting,such as; the researches on main factors influencing ash cutting and the calculation of the erosion rate and safety gas velocity. Thus, we performed the experiment researches of the property of ash cutting using AL_2O_3 grains and five type coal-ash, and obtained the technical information for the design and economy, safety operation of coal fired boilers in power station.

TEST APPARATUS AND MEASUREMENT METHOD

Tests were conducted in a specific rig (Fig. 1) using compressed air as gas source.

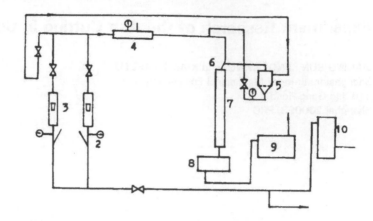

FIGURE 1. Schematic drawing of the test system

The gas coming from gas tank 10 flows through the path with
bypass, in which the temperature, pressure and discharge of the
gas are measured by thermometer 1, the pressure gauge 2 and rotameter
3. Then it divides into two pursuing two ways. One of them enters
the ash funnel 5 as complement gas, while the other enters jet 6
as main efflux, there it introduces grains from the ash funnel forming
the phase flow, through the acceleration section 7 then impacts
specimen fixed on the test section. After impacting the specimen,
it passes through the grains remover 8. Finally, it passes through
the micro-grains remover 9 into the atmosphere.

The test section and the specimen see Fig. 2.

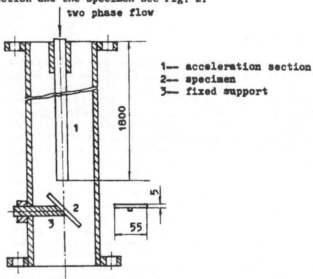

1— acceleration section
2— specimen
3— fixed support

FIGURE 2. Test section specimen

The weight of ash used in test was measured by a standard
scale with 1/5000 precision. The loss of weight of the specimen
was determined by a balance with 1/10000 gram precision.

To measure the accurate velocity of the grains and stream impacting specimen surface a rotameter was used to calculate the total
flow of stream, and again convert it into the velocity of flow by
measuring the circulation section. As the grains introduced by
the jet pass through a long enough constant area of acceleration
before impacting the specimen, the velocity of the grains may be
calculated. To ensure the precision of the results, we performed
an adjustment of the test before the test, until the flow condition
of the specimen surface obtained by practice measurement was satisfactory.

THE MATHAMATIC MODEL OF EROSION RATE AND AVERAGE SIZE OF GRAIN

The following mathematic model of erosion rate was adopted
on the basis of the theory study and analysis for some test information [7,8,9,10,11,12,13].

$$\Delta I = a \cdot d_s^m \cdot W_s^n \cdot f(\theta)$$
(1)

where: ΔI ---- erosion rate, mg/g
         a ---- erosion behavior coefficient of grain, $S^2 \cdot mg/g \cdot \mu m \cdot m^2$
         d_s ---- average size of grain, μm
         W_s ---- velocity of grain, m/s
         $f(\theta)$ ---- a function of impact angle, τ
         m,n ---- exponential of size and velocity of grain, respectively

It is convenient to use the mathematic modle above for data
processing and determination of the formula calculated erosion rate
in practice.

The formula of calculating average size may be given by

$$d_S = [\sum_{i=1}^{n}(x_i \cdot d_{si}^m)]^{\frac{1}{m}}$$
(2)

where: x_i ---- percentage of the grain whose size is d_{si} %
        d_{si} ---- size of i kind grain, μm

THE TEST RESUITS OF AL_2O_3 GRAINS

A series of tests have been done to seek the rule between size
of grain and impact erosion. The carbon steel specimen with an impact
angle of 40° was impacted by using six grains of Al_2O_3 wigh different
size (38.83, 41.74, 59.75, 68.73, 117.13, 145.24 μm respectively),
at a certain velocity ranging under fifteen working conditions.

Fig. 3 shows the experiment results

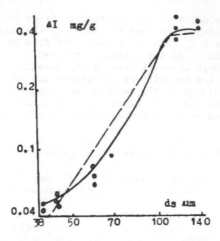

FIGURE 3. Relationship between erosion rate and size of grain

We may obtain by regression analysis for test data that size exponential of grain m = 1.876 under the test condition above. It was shown by the test that the erosion rate ΔI will increases with increasing of size of grain. And there is a critical size in that process; ΔI will not increase continually with its size and maintain a constant of value while reaching critical size. The critical size of grain obtained by test is near 110 μm.

The tests about abrasion resistance of different materials were done by using the carbon steel and copper specimens.

The Fig. 4 gives the test results of them at a velocity ranging from 40 to 88 m/s. Wich impact angle of 45°, using Al_2O_3 grain with size of 40.42 μm.

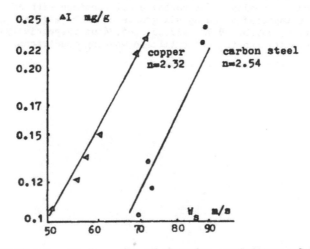

FIGURE 4. Test result of abrasion resistance of two materials

The test showed that the abrasion resistance of both is different. The velocity exponential of carbon is 2.54; copper is 2.32.

TEST RESULTS OF COAL-ASH

The tests under fifty-four working conditions have been conducted by using five type ash from different power plant, at a velocity ranging from 38 to 115 m/s, with impact angle of 45°.

Fig. 5 gives the property of ash cutting of five type ash.

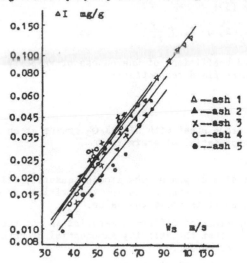

FIGURE 5. Test curve of the property of ash cutting

MATHEMATICAL MODEL OF EROSION RATE AND SAFETY GAS VELOCITY

The erosion rate was used to express the degree of erosion in experiment study on ash cutting, that is

$$\Delta I = \frac{\Delta g}{\Delta G} = \frac{f \delta\, r}{F W_s\, \mu \tau} \qquad\qquad mg/g \qquad (3)$$

and

$$\Delta I = a\, d_s^m\, W_s^{n'} \qquad\qquad mg/g \qquad (4)$$

where:
ΔI ---- erosion rate,   mg/g
Δg ---- loss weight of metal by erosion, mg
ΔG ---- weight of grains (ash),   g
f ---- erosion area, cm^2
δ ---- erosion thickness,   mm
r ---- specific density of material cutted, mg/cm^3
F ---- jet area, cm^2
W ---- velocity of grain,   m/s
μ ---- density of grain g/m^3
τ ---- time,   s
a ---- erosion behavior coefficient determined by test,

$$\frac{S^2 \cdot mg}{g \cdot \mu m \cdot m^2}$$

d_s ---- average size of grain, μm
m, n' ---- exponential of size and velocity of grain respectively.

873

The erosion thickness may be obtained by formula (3) and (4), that is:

$$\delta = 10^{-3} \cdot a \cdot d_s^m \cdot W_s^n \cdot \mu \cdot \tau \cdot \frac{F}{rf} \qquad \text{mm} \qquad (5)$$

where: $n = n' + 1$

It is should be noticed that the unevenness of velocity and desity field have a great influence on ash cutting in practice. Then the applied model of calculating erosion thickness may be given by modifying the formula (5):

$$\delta_{max} = 10^{-3} \cdot a \cdot d_s^m \cdot W_s^n \cdot K_w^n \cdot \mu \cdot K_u \cdot \tau \frac{F}{rf} \qquad \text{mm} \qquad (6)$$

where: $K_w K_u$ —— modified coefficient of unevenness of velocity and desity field respectively.

CONCLUSION

1. When impacting carbon steel with six Al_2O_3 grains at an impact angle of 45°, the size exponential of grain is

$$m = 1.876$$

There is a critical size of grain, the erosion quantity will increase with increasing of size when it is low the critical value; that will not increase when it is above the value.

2. The abrasion resistance of different material is not the same. It is shown by test that the velocity exponential of carbon steel $n = 2.54$; $n = 2.32$ for copper.

3. The average size may be calculated by

$$d_s = \left[\sum_{i=1}^{n} (x_i \cdot d_{si}^m) \right]^{\frac{1}{m}} \qquad \mu m$$

4. Establishing the mathematical modle of ash cutting is a basic to study erosion rule and measures of reducing the erosion failure. That is:

$$\Delta I = a \cdot d_s^m \cdot W_s^n \cdot f(\theta) \qquad \text{mg/g}$$

and the modle of calculating erosion rate and safety gas velocity is

$$\delta_{max} = 10^{-3} \cdot a \cdot d_s^m \cdot W_s^n \cdot K_w^n \cdot \mu \cdot K_u \cdot \tau \frac{F}{rf} \qquad \text{mm}$$

REFERENCES

1. Chen, X., Xi'an Jiao-Tong University, "Study of Gas Velocity in the Convection Heating Surface of the Boiler", Journal of Mechanical Engineering, Vel. 10, No. 5, June 1962.

2. Xian Thermal Engineering Research Institute, "Wear of the Boiler heating Surface", Boiler Technique, No. 9, 1972.

3. Qing-Hua University, "Research of Preventing Erosion in the Fluid Bed Boiler", Boiler Technique, No. 4, 1973.

4. Power Industry Agency of Jiang Xi, "Test Research Summary of the Erosion of the Economizer", Oct. 1979.

5. Chou, A., "Test Study of the Coal Erosion Behavior", Heating Power, No.. 5, 1979.

6. Sun, J., C.   Shanghai Institute of Mechanical Engineering "Study of Ash Erosion on Power Boiler Convection Heating Surface", Boiler Technique, No. 7, 1980.

7. Sun, J., C. "Erosion Calculation of the Convection Heating Surface of the Boiler", Power Machinery, No. 2, 1980.

8. Finnie, I., "An Experiment Study of Erosion Proceedings of the Society for Experimental Stress Analysis, Vel. XVII, No. 2, 1960.

9. Finnie, I., "Erosion of Metals by Solid Particles", Journal of Materials, Vol. 2, No. 3, 1967.

10. Raask, E., "Tube Erosion by Ash Impaction, Wear 13", 1969.

11. Translated by Beijing Boiler Plant, "Standard of Method of Boiler Thermal Calculation", Mechanical Industry Publishing House, Nov., 1976.

12. Tabakoff, W., Kotwal, R. and Hamed, A., "Erosion Study of Different Materials Affected by Coal Ash Particles, Wear 52", 1979.

13. Sun Jiaqing and Wang Jinqi "Two-Phase flow and heat transfer" China-U.S. Progress. P.577 - 585. Hemisphere Publishing Corporation, 1984.

14. Wang Jinqi and Sun Jiaqing. "An Experiment Study of the Property of Ash Cutting", JOURNAL OF SHANGHAI INSTITUTE OF MECHANICAL ENGINEER-ING. Vol.9 NO.2. June 1987.

Mathematical Model for Radiation Heat Transfer in Combustion Chamber

WENDA YAO, ZHIYONG BAO, and JIAWU YAO
Graduate School
North China Institute of Electric Power
Qinghe, Peking, PRC

INTRODUCTION

As early as 1958, H. C. Hottel put forward the Zone Method, which provides an integral mathematical description for radiative heat interchange in the close system of real gases and gray surface. This is the sound theoretical basis of the three dimensional calculation of radiative heat transfer in the combustion chamber of a boiler. Later it was applied to solve the space distribution of the gas temperature in furnace and the heat flux distribution over the wall. In fact, it takes too much memory capacity of computer and too many calculation administrative level. For modern large boiler, if the combustion chamber is crudely divided into one thousand zones(including gas and surface zones), the Zone Method need to take 1000 k memory units. Therefore it is difficult for the Zone Method to calculate in detail the combustion chamber. In 1964, J. R. Howell applied the Monte Carlo modelling method to the problem of radiative heat exchange between gases and surfaces. Howell's method takes less memory capacity of computer and less calculation administrative levels. Thus Howell's method has put to practical use because it has made some advancement on the radiative heat exchange of a boiler. It still has a shortcoming. It is difficult to analyse and estimate calculation errors. Its convergence speed is $O(1/\sqrt{n})$. This is one of the method of the numerical calculation with slower convergence.

This paper gives the mathematical model, which retains the merit of two methods mentioned above and at the same time overcomes their shortcoming.

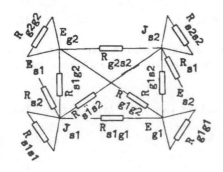

FIGURE 1. Radiation network

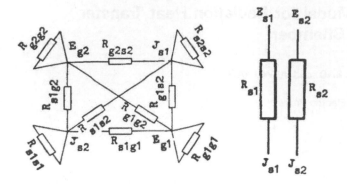

FIGURE 2. Simple network

THE MATHEMATICAL MODEL FOR RADIATION HEAT TRANSFER

For above purpose, the following measures are taken: The close system of gray gas and gray surface can be analysed by the network method that Oppenhein set up for the first time. For the sake of simplity and convenience, first of all the system consisting of only two gas zones and surface zones is analysed, as shown Fig.1. This network belongs to a complicated pattern. It is difficult to seek direct solution of this network by mathematical method. After taking the nodes J_{s1}, J_{s2}, J_{s3}, the above complicated network is resolved into the two simple network shown in Fig.2. The solution of the two simple networks becomes easy.

According to refences (Holman, 1976; Hottel, 1967), by inference, it is easy to obtain the space resistance of the radiation network in Fig.2. to be:

$$R_{s_i s_j} = 1 / s_i s_j \; ; \quad R_{s_i g_j} = 1 / s_i g_j \; ; \quad R_{g_i g_j} = 1 / g_i g_j \tag{1}$$

The radiasity of the S_1 wall zone to be (see Fig.2):

$$J_{s1} = E_{s1} + (Q_{ns1} - Q_{cs1})(1 - \varepsilon_{s1}) / (\varepsilon_{s1} \cdot \delta A_{s1}) \tag{2}$$

If the combustion chamber of boiler is divided into M gas zones and N surface zones, from Fig.2 the energy balance equation of each zone can be written as follow:

$$\sum_{j=1}^{n} s_j g_i \cdot (\sigma T_{sj}^4 + (Q_{nsj} - Q_{csj})(1 - \varepsilon_{sj}) / (\varepsilon_{sj} \cdot \delta A_{sj})) + \sum_{j=1}^{m} g_j g_i \cdot \sigma T_{sj}^4$$

$$-4K_{gi} \cdot \sigma T_{gi} \cdot \delta V_{gi} + Q_{ngi} - Q_{cgi} = \Delta H_{gi} + Q_k$$

$$(i = 1, 2, 3, \cdots \cdots, m) \tag{3}$$

878

$$\sum_{j=1}^{n} s_j s_i \cdot \{\sigma T_{s_j}^4 + (Q_{n,s_j} - Q_{c,s_j})(1 - \varepsilon_{s_j})/(\varepsilon_{s_j} \cdot \delta A_{s_j})\} + \sum_{j=1}^{n} g_j s_i \cdot T_{s_j}^4$$

$$- \{\sigma T_{s_i}^4 + (Q_{n,s_i} - Q_{c,s_i})(1 - \varepsilon_{s_i})/(\varepsilon_{s_i} \cdot \delta A_{s_i})\} \cdot \delta A_{s_i} + Q_{c,s_i} = Q_{n,s_i}$$

$$(i = 1, 2, 3, \cdots \cdots, n)$$

Where :

$$Q_{c,s_i} = a_{s_i}(T_{g x} - T_{s_i}) \cdot \delta A_{s_i}$$

$$\tag{4}$$

$$Q_{n,s_i} = (\lambda / \delta)_{s_i} \cdot (T_{s_i} - T_s) \cdot \delta A_{s_i}$$

Equation(3) is the three dimension mathematical model for radiative heat transfer in combustion chamber. It is applicable for the close system of the gray gas and the surface.

The above model is based on the fundament of gray gas. If the model is applied to the close system of real gas and surface, the model need to revise.

H. C. hottel supposed that real gas consisted of one part of clear gas and the several parts of the gray gas, thus:

$$\varepsilon_g = \sum_n a_{g,n} \{1 - \exp(-\int_0^{r} K_{g,n} \cdot dr)\}$$

$$\tag{5}$$

$$a_g = \sum_n a_{g,n} \{1 - \exp(-\int_0^{r} K_{g,n} \cdot dr)\}$$

By inference the directed flux area from one zone to other zone can be obtained by:

$$\overline{s_i s_j} = \sum_n a_{g,n}'(T_{s_j}) \cdot (s_j s_i)_n$$

$$\overline{s_j g_i} = \sum_n a_{g,n}'(T_{s_j}) \cdot (s_j g_i)_n \tag{6}$$

$$\overline{g_i g_j} = \sum_n a_{g,n}'(T_{s_j}) \cdot (g_j g_i)_n$$

If only the direct interchange areas can be replaced by the direct flux areas in equation(3), the three dimensional mathematical model of the radiative heat exchange in combustion chamber of a boiler will be the same to real gas as equation(3).

THE DIRECT INTERCHANGE AREAS

In cartesian coordinates the direct interchange areas are multi-dimensional integral. For the direct interchange areas the two method are applicable. The two method are the Monte Carlo integral method and the Monte Carlo modelling method.

About the Monte Carlo integral method: The multi-dimensional integral is regarded as an expected values of some stochastic variables. The point column is chosen from the random sampling. The summation of the integrand at the point column approaches to the multi-integral mentioned. The probability density function is taken as 0-1 uniform distribution,has:

1. Between two parallel planes zones:

$$s_i s_j = (1/n) \cdot \sum_{m}^{n} (1/\pi)(\Delta Z_m^2/r_m^4) \cdot \exp(-\int_0^{r_m} K_e \cdot dr) \cdot \delta A_{s i} \cdot \delta A_{s j}$$

2. Between two perpendicular planes zones:

$$s_i s_j = (1/n) \cdot \sum_{m}^{n} (1/\pi)(\Delta Z_m \cdot \Delta Y_m/r_m^4) \cdot \exp(-\int_0^{r_m} K_e \cdot dr) \cdot \delta A_{s i} \cdot \delta A_{s j}$$

3. Between plane zone and gas zone:

$$s_i g_j = (1/n) \cdot \sum_{m}^{n} (K_{e i}/\pi)(\Delta Z_m/r_m^3) \cdot \exp(-\int_0^{r_m} K_e \cdot dr) \cdot \delta A_{s i} \cdot \delta V_{e j}$$

4. Between two gas zones:

$$g_i g_j = (1/n) \cdot \sum_{m}^{n} (1/\pi)(k_{e i} \cdot K_{e j}/r_m^2) \cdot \exp(-\int_0^{r_m} K_e \cdot dr) \cdot \delta V_{e i} \cdot \delta V_{e j}$$

The random points $(X_i^{(m)}, Y_i^{(m)}, Z_i^{(m)})$ and $(X_j^{(m)}, Y_j^{(m)} Z_j^{(m)})$ are chosn in i zone and j zone by uniform distribution.
where:

$$r_m = \sqrt{(X_i^{(m)} - X_j^{(m)})^2 + (Y_i^{(m)} - Y_j^{(m)})^2 + (Z_i^{(m)} - Z_j^{(m)})^2} \tag{7}$$

$$\Delta X_m = | X_i^{(m)} - X_j^{(m)} | , \quad \Delta Y_m = | Y_i^{(m)} - Y_j^{(m)} | , \quad \Delta Z_m = | Z_i^{(m)} - Z_j^{(m)} | \tag{8}$$

The linking line r_m runs though some gas zones. If the parameters of the gas zone are uniform. We has:

$$\int_0^{r_m} K_e \cdot dr = \{ \sum_i K_{e,n} \cdot \Delta r_m |_{n \in r_m} \} \tag{9}$$

About the Monte Carlo modelling method: The whole of the system is regarded as close system of gray gases and black surface. The modelling test is made with the probable model of the radiative heat transfer process. The frequency approaches to the probability. The statistical estimation is taken for the solution of the radiative heat transfer process at the system. Thus the direct interchange areas can be obtained. These two method are compared through practical calculation. The CPU time of the two method is near.
For the Monte Carlo integral method,the completeness error is less than the Monte Carlo modelling method.What is called the completeness error,is:

$$R = \sum_{1}^{n} \left(1 - \left(\sum_{j}^{n} s_i s_j + \sum_{j}^{\cdot} s_i g_j \right) / \delta A_{\bullet i} \right)^2 +$$

$$\sum_{1}^{\cdot} \left(1 - \left(\sum_{j}^{n} g_i s_j + \sum_{j}^{\cdot} g_i g_j \right) / 4K_{\bullet i} \cdot \delta V_{\bullet i} \right)^2 \tag{10}$$

In short for the calculation of the direct interchange areas, the Monte Carlo integral method is a good method.

CALCULATION EXAMPLE

The combustion chamber of the boiler produced in the Harbin Boiler Factory is taken as the calculation on object. The boiler parameter are: pressure 9.9 MPa, temperature 540°c, flux 410000 kg/h. The size of the combustion chamber are: width 13.585 m, depth 7.54 m, volume 2204 m³.
The test of the heat transfer in the combustion chamber has been made, and the heat flux of incident upon the wall and the gas temperature at the exit of the chamber are measured.
In the calculation, the combustion chamber is treated as the close system with gray gases and gray surface.
The results of the calculation of the gas temperature distribution on the vertical section is shown in Fig.3. The results of the calculation and the measurement of the distribution of the heat flux along the central line on both side walls shown in Fig.3.
The results of the theoretical calculation and the practical measurement are compared. The error of the gas temperature at the exit equals 31 k . The error of the heat flux is less 10% .

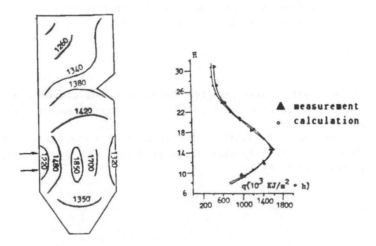

FIGURE. 3   Distribution of gas temperature K   and heat flux.

CONCLUSION

The three dimensional numerical   method for radiative heat transfer in
furnace that the paper gives has many advantages.
The advantages that above method has over the Zone Method are listed now
as follows: It is not necessary for the mathematical model that the paper
gives to calculate the total interchange areas, so the calculation of the
(n+m) group linear equation will be simplied and the less memory capacity
of computer will be taken. After the solution method of the equation (3)
has adopted the conjuate gradient method, the model may save memory
capacity in great quantity.
It is recomeneded to adopt the Monte Carlo integral method for the
calculation of the direct interchange areas. The calculation error can be
estimated and analysed by the strict formuiation. If the Monte Carlo
modelling method is adopted for the calculation of the radiative heat
transfer in the combustion chamber, it will be difficult to astimate and
analyse the calculation error(Wenda, 1984).

NOMENCLATURE

$a_{v,n}$	weighting factor of the each gray gas zone;
ΔH	enthalpy decrease of gas flowing from feed to exit, KW;
$gg, gs, ss, sg,$	direct interchange area M^2;
J	radiasity KW/h;
K_a	absorption coefficient M^{-1};
Q_c, Q_λ	energy due to convective and turbulent heat exchange KW/h;
Q_a	heat value of the combustion in gas zone KW/h;
T_a, T_{ux}	saturated temperature, gas temperature k;
$\delta A, \delta V$	area and volume of zone M^2 or M^3.

REFERENCES

Holman. J. P. (1976). Haet Transfer. McGaw-Hill Book Company. New Youk.
Hottel. H. C. and Sarofim. A. F. (1967). Radiative Transfer. McGaw-Hill
Book company. New York.
Wenda. Y. and Yusen. B. (1984). The Problem of Monte Carlo solution of
Radiat Heat Transfer in Furnace of Large Boiler. Journal of Engineering
Thermophysics. Vol.5. No.3

On the Study of the Thermophysical Properties of Inorganic Materials Fired in High Efficiency Kiln

QINGREN WU, JIANQING WU, JIANTING LO, and YAOJUN YAN
Department of Inorganic Materials Science and Technology
South China University of Technology
Guangzhou, PRC

ABSTRACT

In order to reduce the firing heat consumption of inroganic materials, high efficiency kiln with optimum firing schedule should be designed to make the product fired under uniform temperature and stress. But now no standard data of the thermal physical properties of the inorganic material remains, which can be used during the firing process and can make optimum firing schedule. Therefore, the relationship between the therolphysical properties, and chemical composition, microstructure of this material under different temperatures must be studied. Then, the internal condition of the body in firing process can be actually controlled. In this paper, the variation of the thermal conductivity of several china ceramics under different temperatures and porosities is reported.

1. INTRODUCTION

Heat transfer within ceramic body and product is mainly by conduction. However, no exact conductivity values remain, which can be used to calculate the temperature and stress distribution and make the optimum firing schedule. Now the conductivities of four china ceramic products and a ceramic body under different temperature and porosities are studied with the laser flash method.

2. EXPERIMENTAL METHOD

The thermal diffusivity was measured with the laser flash thermal constant instrument. The phusical model of this method is a peripherally insulated thin circular specimen facing a bundle of one dimensional heat flow, the temperature distribution at the back of this specimen was measured. If the thickness of the circular specimen is L, and the initial temperature distribution at any position of the specimen is $T(x,o)$, then the temperature distribution $T(x,t)$ at an arbitrary instant, t, is[1]:

$$T(x,t) = \frac{1}{L}\int_0^L T(x,o)dx + \frac{2}{L}\sum_{n=1}^{\infty}\exp(\frac{-n^2\pi^2\alpha t}{L^2})\cos(\frac{n\pi t}{L})\int_0^L T(x,o)\cos\frac{n\pi x}{L}dx \qquad (1)$$

Project supported by The Science Fund of the Chinese Academy of Sciences

When considering the special condition at the back of the specimen, (x=L), substituting the boundary conditions into Equation(1), and defining two non-dimensional parameters the thermal diffusivity, α, can be derived from the Equation(1).

$$\alpha = \frac{0.139L^2}{t_{1/2}} \tag{2}$$

Where, $t_{1/2}$ represents one-half of the time needed to reach maximum temperature at the rear face of the specimen.

After the diffusivity α is found, the value of the thermal conductivity, λ, can be calculated by using the following equation:

$$\lambda = \alpha \rho C_p \tag{3}$$

where, (ρC_p) is heat capacity per unit volume. In our experiment we easily obtain (ρC_p) using the following equation:

$$\rho C_p = \frac{Q}{L.T_m} \tag{4}$$

where, Q is the amount of pulsed laser radiation absorbed by the specimen and T_m is the maximum temperature increase at the rear face of the specimen after absorbing the laser radiation. Q and T_m were measured simuteneously by the laser thermal constant instrument.

The specimens were measured under vacuum conditions from the room temperature to 2037K. The accuracy for the thermal conductivity data is ±10% based on accuracy of ±5% for the thermal diffusivity. The size of the specimen is: diameter ϕ = 0.9-1.0 cm and thickness L = 0.1-0.15 cm. The porosity of ceramic body and the porosity of the ceramic product were measured with oil-filled method and water-filled method, respectively.

3. EXPERIMENTAL RESULTS

The specimens were supplied by the ceramic plant at Jingdezhen and Shiwan in China. Their chemical compents and porosities are shown in Table 1. Fig.1 shows the experimental data and λ -T curves.

TABLE 1. The chemical components and porosities for four specimens

Specimens	Chemical components (%)						Firing temp. (K)	POrosity (%)	Plant
	SiO_2	Al_2O_3	Fe_2O_3	CaO	MgO	K.NaO			
Porelain dish	69.89	19.88	0.42	0.31	0.21	3.58	1573	9.55	Shiwan
Lustre brick	72.07	4.67	16.67	0.49	0.89	5.21	1433	18.15	Shiwan
Pottery ware	71.68	18.07	0.84	0.38	0.63	4.08	1173	29.38	Shiwan
Porous body	68.05	16.16	0.47	3.65	2.26	1.54	1483	37.68	Jingdezhen

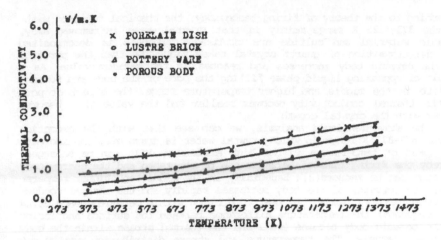

Figure 1.Effect of temperature on thermal conductivity
for four ceramics

Figure 2 shows how the thermal conductivity and porosity of each ceramic body very with temperature increase. When the temperature raises at the rate of 5 K/min. the thermal conductivity will increase slowly, the porosity of the body reach to the maximum and the decrease in the range of 373-1123 K. The value of λ increases quickly and the value of porosity decreases continuously within the 1123-1223 K range; the increasing speed of λ from 1123 K to firing temperature is larger than that from 373 K to 1123 K, but the porosity value decreases rapidly within the higher temperature range.

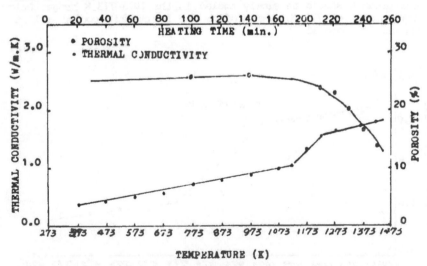

Figure 2. Thermal conductivity and porosity curves for the body of
lustre brick

According to the theory of firing technology, the chemical reaction within in the 373-1123 K range mainly is that crystal water is removed away, organic material and sulfide are oxidized, carbonate is decomposited and the transition of quartz crystal takes place. So that the porosity of the ceramic body increases and reaches to the maximum value. As a result of appearing liquid-phase filling the pore of the body and forming mullite in the middle and higher temperature range, the effect of pore on the thermal conductivity becomes smaller and the value of   becomes larger with the crystal growth.

From the above-mentioned analysis, we can see that with the exception of the 673-873 K range in which crystal water is drawn off, the increasing speed of temperature in the high efficiency kiln can be enlarged. Thereby the firing time of ceramics can be shortened and the energy consumption can be reduced[2]. Especially in the middle and higher temperature, the porosity of the body decreases rapidly and the thermal conductivity increases quickly, but heat conduction within ceramic speeds up. As a consequence, the temperature difference between the surface and inside of the ceramic body becomes small and the thermal stress within the body decreases largely. The temperature and stress distribution easily get to uniform. Heat consumption can be decreased by speeding up the temperature increase and shortening the firing time.

Figure 3 shows that how the thermal conductivity of the lustre brick varies with the decreasing temperature. The thermal conductivity values for the lustre brick from the firing temperature cooling to 1073K decreases slowly, but values of λ in the 1073-773 K range decreases rapidly. It is to due to the fact that the liquid-phase in the product is in the plastic state and the value of λ is larger. In order to keep way from devitrification of the liquid-phase and the crystal growth, it is good to make the product cooling rapidly. In contract, when the liquid-phase becomes solid state and the quartz crystal transition takes place, the volume of the product changes. So that the values of λ rapidly decreases and the product should be slowly cooled in the 1073-773 K range. Below the temperature of 773 K, the product can be quickly cooled and its quality can be improved.

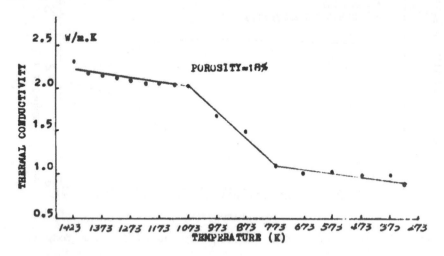

Fig.3. Thermal conductivity curve for lustre brick in cooling process

4. DISCUSSION AND CONCLUSIONS

A. Thermolphysical property measurement method can be used as a experimental method to study the physical and chemical changes of the ceramic body fired in the high efficiency kiln. It not only can be used to determine the temperature which the changes within the body in the firing process are taking place, according to the transition point of the curve for the thermal conductivity, but also can be used to give the information of the thermal behavior in the different firing processes. It is helpful to show the natural relationship betwwen the microscopic heat conduction and the optimum firing schedule to make the ceramic quality better and the heat consumption lower. Also, by measuring a lot of inorganic materials, we can bring together the data of thermolphysical property and build up the data system of computer storage in order that we can finish the computerized control schemes for the ceramic kiln and promote the thermal efficiency of kiln. This is an important basic research work.

B. By using the thermal constant instrument with the laser flash technique to measure the thermal diffusivity and conductivity of the body and study the temperature and stress distribution within the body in the firing process, and find the optimal firing schedule, this method has the following advantages:
(a) Measuring temperature range is wide, from room temperature to 2073 K.

(b) The size of the specials is small, diameter ϕ = 0.9 - 1.0 cm, thickness L = 0.10 - 0.15 cm.

(c) Measuring speed is high, the interval between the laser emission and the display of the thermal diffusivity result is approxinately 5 seconds.

C. The experimental results show that the effect of porosity on the thermal conductivity for the body with the same composition is mainly depends on the volume percent of the pore. The values of λ are inversely propotional to the values of porosity. For ceramics, however, the effect of pore on the thermal conductivity is complex. It not only relates to the values of porosity, but also the shape, size and direction of the pore[3].

ACKNOWLEDGE

Finally, the authors grately acnowledge the direction and help of Professor Liu Zhenqun.

REFERENCES

1. Carslaw,H.S., Jaeger,J.C., Conduction of Heat in Solids, 2nd., pp.101. Oxford University Press, New York, 1956.
2. Liu Zhenqun, Calculating the Heat-consumption Equation of a Tunnel Kiln, Ceramic Proceedings, pp. 1023-1025, 1983.
3. Wu Qingren, The Microscopic and microscopic Analysis Method of Heat Conduction for Inorganic Materials, Guangdong Cement, No.4, pp.11-14, 1986.(in Chinese)

Heat Consumption Analysis and Structure Improvement of Roller Hearth Kiln Wall

LINGKE ZENG, XIAO MO, XIAOFAN HU, and LI ZENG
Department of Inorganic Non-Materials Science and Technology
South China University of Technology
Guangzhou, PRC

ABSTRACT

The Roller hearth kiln is one of new type kilns in ceramic industry.
It has a lot of advantages, available to realization of mechanization
and automation of the firing process, and suitable for fast firing of
the products. Owing to the limitation of the roller length, the thickness of the kiln wall has to be built thinner, therefore its heat loss
is quite high. In this paper, the wall temperatures have been measured
in oil-burning roller hearth kiln, the experimental data have been processed with computer, the reason of the increase of heat loss through
the wall is analysed and the measurement method is presented. Furthermore, the improvement of the wall structure in the kiln is proposed.

1. INTRODUCTION

The Roller hearth kiln is a new type of kiln used in ceramic industry,
used for clay products, brick and tile, sanitary ware, with the advantages of small cross-section and without needing kiln car. The kiln
is commonly used with different types all over the country, such as
oil-burning single or double firing roller hearth kilns in Foshan, coal-
burning or electric-heating roller hearth kilns in Shan-dong province,
electric-heating roller hearth kilns in North East of China, and several
imported gas-burning roller hearth kilns throughout the country. The
roller hearth kiln became into use in china only in recent time, so
its structure isn't fully satisfactory, the rollers can't be long enough
(2.1 m home-made, 2.4 m by import), resulting in small thickness of the
kiln wall to ensure the mass production. These make the external kiln
face temperature as high as $300°c$. The temperature of the roller ends,
sticking out of the wall 0.1 m or so, can even get to $400°c$, for the
supporting of the rollers are filled with loosen referactory filler.
Nowadays, no standard of assesing these heat emission waste remains
because the sort of kiln is quite new. In this paper this problem is
tackled with our experimental data.

2. MEASUREMENT METHOD AND DATA PROCESS

The temperatures of external kiln face are measured by DP type digit-
direct-read surface temperature meter. The total length of external
kiln walls are divided into 102 sections. Each section has 7 temperature measured points(3 points above the central roller, 3 points below
the central roller and 1 point for the ceramic fiber). They are shown
in Fig.1.

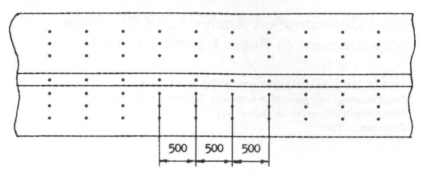

Fig. 1 Temperature measuring point distribution

Because the temperature on the surface of the external kiln wall is not homogeneous, the heat consumption in each section is calculated by using measured point temperature and the element area near the point. The sum of the heat consumption can be calculated as follows:

$$Q_q = \sum A_q * F_q (t_q - t_k)$$

Where:

Q_q — Heat consumption of the wall (KJ/h)

t_q — Measured temperature of the external kiln face (°c)

F_q — Element area of the external wall face (m')

t_k — Environmental temperature out of the wall (°c)

A_q — The convectional and radiational coefficient between the element and the environment

$$A_q = 9.20(t_q - t_k)^{0.25} + \frac{16.72[(\frac{t_q + 273}{100})^4 - (\frac{t_k + 273}{100})^4]}{t_q - t_k}$$

$$F_q = B_q * H_q$$

B_q — Length of the element area (0.5 m)

H_q — Height of the element area (m)

Because there are so many points and elements, it is impossible to calculate the heat loss without the help of computer. These data have been processed with microcomputer. The wasted heat of two external kiln wall faces are as follows:

Q_A = 361238 kJ/h

Q_B = 371777 kJ/h

Q_{sum} = 733015 kJ/h

The heat consumption through the roller is calculated with the measured temperature of ceramic fiber near the roller, the same as the roller measured by the same method for the wall face. The amount of loss heat are calculated as follows:

Q_{Ar} = 42739 kJ/h

Q_{Br} = 42504 kJ/h

Q_{sumr} = Q_{Ar} + Q_{Br} = 85243 kJ/h

Q = Q_{sum} + Q_{sumr} = 818258 kJ/h

The results of the calculation are realizable in respect to engineering, because the divided sections are short(0.5 meter) and the number of sections are many enough to be approximate to the practical heat loss.

3. MEASUREMENT METHOD ANALYSIS

The methods to measure the temperature and to calculate the heat loss make the number of the points increase and the calculation work be very tedious. Can the number be decreased and how many points are enough for the calculation? The method used in Foshan is that the kiln wall was divided as 3 groups of pre-heating zone, firing zone and cooling zone. The average temperature of each zone is calculated the locating temperature, the area is the whole area of that zone. For the same kiln the measurement and calculation are as follows:

Q_a = 245584 kJ/h

Q_b = 263889 kJ/h

Q_s = 509473 kJ/h

The ΔQ = 308785 kJ/h, relative error is 37.74%, compared with the method we used. So the method applied in Foshan is not available. If we divide the kiln wall face as 1m, 2m, 3m for one section, the same points as previous case of 3 zones (top, bottom, middle part of each zone). The average temperature and heat loss are calculated as Table 1.

Table 1. The heat loss for dividing different length of kiln wall face

length of each section (m)		1	2	3
Q_a	(kJ/h)	300846	299758	250915
Q_b	(kJ/h)	297130	296216	226823
Q_{fiber}	(kJ/h)	71929	68570	65689
Q_{roller}	(kJ/h)	84207	86436	86906
Q_{total}	(kJ/h)	754112	750980	630334
ΔQ	(kJ/h)	64145	67277	187924
Relative error (%)		7.84	8.22	23

891

It is clear that the longer the section is, the more deviation it has. As for the deviation, the smaller the distance between the measuring and calculation points is, the better it is. From Table 1,the difference of the relative error is little within the length of one meter and two meters. The number of the points decrease to 1/2 compared with the 1 meter case, it increases quite a lot with the 3 meters case. So the best length of the section is 2 meters. This is an accurate and suitable measurement and calculation.

4. IMPROVEMENT OF THE KILN WALL STRUCTURE

From the calculation above, the heat loss on both two external surfaces of the roller hearth kiln wall is 818258 kJ/h, which is about 24% of the total heat waste of the kiln. The reason for the heat waste is kiln wall structure exempting from the wall thin. The structure of the wall in Foshan has been shown in Figure 2. The first two inner linings are made up of light refractory brick. Thickness of one lining is 115 mm. The third lining made up of ceramic fiber with 68mm of thickness, the external surface made up of steel with 2mm of thickness. If we try to make the wall thicker to decrease the heat consumption. With the help of calculation, it has proved that this arrangement is of bad economy. Is it possible to make the wall structure with the same thickness to save the energy? The way is to external reduce the ceramic fiber and add the ceramic fiber between the two light refractory bricks remaining the same thickness. The heat consumption is shown in Table 2.

Table 2 The heat loss for changing the thickness of ceramic fiber between the two light refractory bricks

Thickness (mm)	10	20	30	40	50	60	68
Heat loss (kJ/h)	366014	352147	348634	327643	312461	310006	302877

Therefore, it is efficient to add ceramic fiber lining to step down the heat loss through the wall. In consideration that the thinner the ceramic fiber is, the more stable the structure is. From Table 2. it is apparent that the heat loss in the case of having a thickness of 10 mm ceramic fiber lining is about the same as that for 68 mm ceramic fiber lining. So that it is proved to add a thickness of 10 mm ceramic fiber can decrease the heat consumption by 50%. But it is not widely used as with tow-row light refactary bricks lining, the lining should be cross built as shown in Figure 3. The optimum structure is shown in Figure 4, and based on both experiments and theretical calculations the ceramic fiber can withstand the temperature of the lining under consideration. The heat loss throughout the wall shown in Figure 4 is 542016 KJ/h, which has been decreased by 27%, and 5.5% to the total. A factory having three roller hearth kilns may save 100000 yuan with this improvement. It is also recommended to use the structure in Figure 5. More energy can be saved with the same investment. The method to improve the wall structure is necessary and this could improve the environment condition as well.

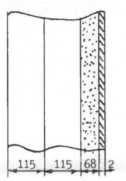

Fig.2. Kiln wall structure

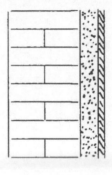

Fig.3. Building figure of
        Kiln wall structure

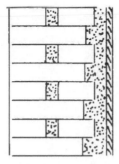

Fig.4. Improvement of kiln
        wall structure

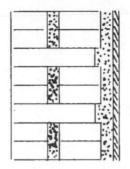

Fig.5. Another improvement of kiln
        wall structure

5. CONCLUSIONS

A. Calculation of the heat consumption used in the roller hearth kiln should not be the same as that in tunnel kiln with temperature measurment, because of the special structure of roller hearth kiln. It is necessary to think of not only the heat loss through the roller, the ceramic fiber for the insulation of the roller, but also the distribution of the temperature measurment points.

B. In order to descrease the heat loss through the wall, it is necessary to add a layer of 10 mm ceramic fiber between refractary bricks.

C. The method of improving the kiln wall structure not only can be applied to oil-burning kiln but also could be used to the gas-burning kiln and electric-heating kiln etc..

REFERENCES

1. Lin Zhen-qun at al. Ceramic industry kilns. Construction Industry Press. China. 1978.
2. Zeng Lingke, The Software Package for the Calculation of Heat Balance and Heat Efficiency of Tunnel Kiln, Proceedings of EPMESC vol.2 336.1987.

ANALYSIS OF INDUSTRIAL PROCESS

The Exergy-Economic Evaluation and Optimization of Shell-and-Tube Heat Exchanger and Its Heat Transfer Enhancement

BEN HUA and TIAN-HUA XU
Chemical Engineering Research Institute
South China University of Technology
Guangzhou, PRC

ABSTRACT

Based on The Second Law analysis and exergy-economic evaluation of shell-and-tube heat exchanger, using τ (heat transfer economy factor, Y RMB/(W/K)) and λ (power consume factor, W/(W/K)) as the suggested performance evaluation criteria, a heat exchanger cost equation related to some comprehensive parameters of ΔT, τ, λ and the unit cost of exergy C_t and C_f has been presented. Taking the equation as the objective function, the simultaneously optimization of both sides of shell-and-tube heat exchanger has been carried out, and a mode-opting computer program has been developed, which can be used for practical design. As the feedback information for modifying the ΔT given by network integration, an equation of ΔT_{opt} related to the τ and λ has been derived. The effect of some technical and economic factors on the result of optimization has been discussed.

NOMENCLATURE

a	coefficient in constraint equation	Q	heat transfer duty
C	cost	q	heat flux per unit area
D	diameter	Re	Reynolds number
D_k	exergy dissipation	S	entropy
F	heat transfer area	T	temperature
h	film heat transfer coefficient	V	volume flow rate
I	investment of heat exchanger		
K	overall heat transfer coefficient	α	ratio of $C_{\Delta T}$ to $(C_{\Delta P} + C_{eq})$
L	Lagrange function	β	depreciation rate
N	operation hours per year	Λ	Lagrange multiplier
n	coefficient in equation (9)	λ	power consume factor
Nu	Nusselt number	τ	heat transfer economy factor
P	pressure	ϕ	constraint equation

Subscripts

c	cold	i	inside	ΔP	due to pressure drop
eq	equipment	m	maintenance	ΔT	due to temperature drop
f	flow exergy	o	outside	0	reference, fundamental
h	hot	opt	optimized		

INTRODUCTION

The equipment for heat exchange is of significance in chemical processes.

The design level of both heat exchangers and its network has great influence on the benefit and capital investment of the process system. Along with the enhancement of energy price and the development of systematic optimization technique, the R & D has been deepening along two aspects. One is to develop a lot of efficient enhanced heat transfer techniques and new equipment, in order to get higher heat transfer coefficient K but paying less pumping power [1,2,3]. Another is to develop the method for the optimum match between the multiple hot and cold streams in order to get better trade-off between ΔT and q (or energy cost and capital cost) and to minimize the total annualized cost of the network [4,5]. Each focuses on one aspect and evades the other for the reason of simplification and easiness to deal with.

Recently, many researchers such as A.Bejan [6] and B.Linnhoff [7] et al. made great contribution to the subject by using The Second Law analysis. But so far, the study is either too theorized far from practice or too conceptualized without quantification, and still lacks the consideration of the relationship between the two aspects.

In this paper, starting with The Second Law analysis but taking account of engineering-economic factors as well, and aiming at the practical design optimization, we will present the new performance evaluation criteria and a new optimization strategy for practical design.

THE SECOND LAW ANALYSIS OF HEAT EXCHANGERS

A.Bejan took the entropy generation ΔS, being the sum of $\Delta S_{\Delta T}$ (due to the irreversible heat transfer) and $\Delta S_{\Delta P}$ (due to fluid friction), as the objective function and the criterion for comparing the different enhancing techniques [6]. Based on this, W.H.Huang et al. presented another criterion — the ratio of Nu to Re [8]. Both of them respected $\Delta S_{\Delta T}$ as a function of Nu through the heat transfer correlation. Meanwhile, by assuming the constant heat transfer per unit tube length q' and changing Re, they had made trade-off between $\Delta S_{\Delta T}$ and $\Delta S_{\Delta P}$ to obtain the ΔS_{min}. However, when changing Re or using enhanced elements, q' can hardly keep constant.

As a matter of fact, the entropy generation or exergy dissipation due to irreversible heat transfer of a heat exchanger is a pure thermodynamic quantity. It only depends upon the temperature of both streams, and is independent of h and Nu. In the case of industrial heat exchanger design, the trade-off between $\Delta S_{\Delta T}$ and $\Delta S_{\Delta P}$ is not so simple and direct but much concerning with equipment cost. If Cp=const, we can have:

$$D_{k,\Delta T} = T_0 \, S_{\Delta T}/Q = T_0(1/T_{o,1} - 1/T_{h,1}) \qquad \qquad J/J \qquad (1)$$

where $T_{o,1}$ and $T_{h,1}$ are the logarithmic mean value of inlet and outlet temperature for cold and hot streams respectively. T_0 is the reference temperature.

As to the entropy generation or exergy dissipation due to fluid friction, it is the price of flow exergy paying for the certain heat transfer coefficient K. For a heat exchanger, there is:

$$D_{k,\Delta P} = T_0 \, S_{\Delta P}/Q = \Sigma(V_i \Delta P_i)/ \, K \cdot F \cdot \Delta T \qquad \qquad J/J \qquad (2)$$

where, V_i, ΔP_i —— the volume flow rate (m^3/s) and the pressure drop (Pa) of two streams respectively;

Q —— the heat transfer duty (W);
K —— the heat transfer coefficient (W/m²K);
F —— the heat transfer area (m²);
ΔT —— the heat transfer temperature difference (K)

In the strict sense, there is another term of exergy dissipation due to heat radiation to the ambient through the outside surface of the heat exchanger $D_{k,D}$. But we leave it out of account because of its less quantity and being of no concern with design optimization.

THE EXERGY-ECONOMIC ANALYSIS

Since, practically, the cost of flow exergy is many times higher then that of thermal exergy [9], it is meaningless to sum up $D_{k,\Delta T}$ with $D_{k,\Delta P}$ (as well as $S_{\Delta T}$ with $S_{\Delta P}$). From the view point of exergy-economics, under the condition of neglecting the cost of heat radiation and assuming the crossflow temperature difference correcting factor F_t equals 1, the total cost of unit transfered heat C should consist of following three parts:

1. The cost of exergy dissipation due to heat transfer:

$$C_{\Delta T}=C_t D_{k,\Delta T}=C_t T_0 \left(\frac{1}{T_{c,1}} - \frac{1}{T_{h,1}}\right)=C_t T_0 \frac{\Delta T}{T_c (T_0+\Delta T)} \qquad \text{Y RMB/J} \qquad (3)$$

2. The cost of exergy dissipation due to fluid friction:

$$C_{\Delta P}=C_f D_{k,\Delta P}=C_f \cdot \Sigma(V_i \Delta P_i)/KF\Delta T \qquad \text{Y RMB/J} \qquad (4)$$

3. The cost of equipment and maintenance:

$$C_{eq}=\frac{(\beta_0+\beta_m) I}{3600 \cdot N \cdot Q} = \frac{(\beta_0+\beta_m) I}{3600 \cdot N \cdot KF\Delta T} \qquad \text{Y RMB/J} \qquad (5)$$

Then, we have the total cost equation as follows:

$$C = C_{\Delta T}+ C_{\Delta P}+ C_{eq}$$

or:
$$C = C_t T_0 \frac{\Delta T}{T_c (T_c+\Delta T)} + C_f \frac{\Sigma(V_i \Delta P_i)}{KF\Delta T} + \frac{(\beta_0+\beta_m)I}{3600 \cdot N \cdot KF\Delta T} \qquad \text{Y RMB/J} \qquad (6)$$

where, T_c, T_h —— the arithmetic mean temperature of cold or hot stream, K;
C_t —— the unit cost of thermal exergy, depending on the fuel price and the efficiency of the furnace, Y RMB/J;
C_f —— the unit cost of flow exergy, depending on the power price and the efficiency of the pumps (compressors), Y RMB/J;
β_0 —— the fundamental depreciation rate, a^{-1};
β_m —— the annual rate of maintenance, repairs and operation costs to the primary investment, a^{-1};
I —— the cost of investment of the heat exchanger, Y RMB;
N —— the operating time per year, h/a.

From the view point of heat exchanger design, we can see that there could have two kinds of trade-off. One is the trade-off between C_{eq} and $C_{\Delta P}$, i.e., under a certain ΔT and Q, the trade-off between ΔP (concerned with K) and I (concerned with F). The other is that of between $C_{\Delta T}$ and $C_{eq} + C_{\Delta P}$, i.e., trade-off between ΔT (concerned with thermal exergy cost) and C'(see eq.(11)). Because ΔT depends upon the process condition and the energy use in the total process system, the latter trade-off belongs to the problems of system or network optimization, while the former, between $C_{\Delta P}$ and C_{eq} at certain ΔT and Q, belongs to the single heat exchanger optimization. Nevertheless, they must be concerned each other; and this is the harmony question between the optimization of the total system and that of the subsystems.

THE PERFORMANCE EVALUATION CRITERIA

A.Bergles and R.Webb had presented many performance evaluation criteria for the comparison of enhanced heat transfer surface with a smooth one [10,11]. But they all omitted an important factor, that is the cost paid for the manufacture of enhancing elements. Besides, most enhanced techniques only enhance heat transfer at one side. So, in order to get the overall effect of a heat exchanger, it has to have the coordination of the other side. Take account of these, based on the suggestion in reference [12], here presented two factors as the performance evaluation criteria.

The Heat Transfer Economy Factor: $\qquad \tau = I/K \cdot F \qquad$ Y RMB/(W/K) $\qquad$ (7)

it means the primary investment paid for unit heat transfer rate under the temperature difference of 1K.

The Power consume Factor: $\qquad \lambda = \Sigma(V_1 \Delta P_1) \ /K \cdot F \qquad$ W/(W/K) $\qquad$ (8)

it means the flow work (exergy) consuming rate for unit heat transfer rate under the temperature difference of 1K.

Obviously, a better enhancing technique must have lower τ and/or λ values.

DESIGN OPTIMIZATION OF SINGLE HEAT EXCHANGER

Substituting equation (7) and (8) into (6), and let:

$$n = (\beta_0 + \beta_m)/3600 \cdot N \qquad\qquad \text{1/s} \qquad (9)$$

means the equipment cost rate per second, relatively, it is a constant being independent of the concerned parameters. Then we have:

$$C = C_t T_0 \frac{\Delta T}{T_0(T_0 + \Delta T)} + C_f \frac{\lambda}{\Delta T} + n \frac{\tau}{\Delta T} \qquad \text{Y RMB/J} \qquad (10)$$

That is the objective function for optimization of heat exchangers. We can see that except for the system parameter ΔT, economic condition parameters C_t , C_f and the process parameter T_0, the parameters remained for reflecting the performance of the heat exchanger are only τ and λ , instead of K, F, V_1, ΔP_1, I, β_0, β_m as in equation (6).

Under certain ΔT and T_0, the design optimization of single heat exchanger includes two problems: (1) how to design a heat exchanger with both shell side and tube side optimized simultaneously; (2) how to choose a proper enhancing heat transfer technique at given process condition. In order to solve the first problem, we use the simplified objective function:

$$C' = C_{\ell}\lambda + n\tau \qquad\qquad Y\ RMB/(J/K) \qquad (11)$$

In the case of forced convective heat transfer with no phase changed in both side at certain T_h and T_o, Listing all the concerned equations and variables, we can get a set of 12 variables and 10 constraint equations. After substituting and eliminating the variables τ, λ, F, h_i, h_o, ΔP_i, ΔP_o, f_i, f_o (friction factor), there are only three variables Re_i, Re_o, K and one constraint equation remained. Using the Lagrange undetermined multiplier method and taking Re_i and Re_o as decision variables, the problem of simultaneous optimization can be solved. That is through solving the following equations:

$$L = C_{\ell}\lambda + n\tau + \Lambda\phi \qquad (12)$$

$$\phi = \frac{1}{a_1 Re_i{}^{a_2}} + \frac{1}{r} + \frac{1}{a_3 Re_o{}^{a_4}} - \frac{1}{K} = 0 \qquad (13)$$

$$\partial L\,/\,\partial Re_i = 0 \qquad (14)$$

$$\partial L\,/\,\partial Re_o = 0 \qquad (15)$$

$$\partial L\,/\,\partial K = 0 \qquad (16)$$

Thus the optimal $(Re_i)_{opt}$, $(Re_o)_{opt}$ can be found out and the optimal geometric parameters of the heat exchanger can be obtained. The result shows that $(Re_i)_{opt}$ and $(Re_o)_{opt}$ only depend upon the properties of the fluids in both sides, the heat transfer and friction performance of the elements, and the price ratio of the equipment to the flow exergy, but are independent of heat duty, flow rate and ΔT. According to the definition then, the corresponding τ_{opt} and λ_{opt} can be calculated; obviously, they are influenced by the same factors that affect $(Re_i)_{opt}$ and $(Re_o)_{opt}$. In as much as τ_{opt} and λ_{opt} only depend upon the technique-economic performance of the heat exchanger (with or without certain enhancing elements), it is verified that they are suitable as the performance criteria for enhanced heat exchangers.

In order to meet the practical design demand, a computer software is developed, which evaluates each heat exchanger listed in the current National Standards of Series of Shell-and-Tube Heat Exchangers (NSSSTHE) and choose the optimal one in accordance with the above objective function. It has been verified under a wide range of all parameters that the results of $(Re_i)_{opt}$ and $(Re_o)_{opt}$ obtained from Lagrange undetermined multiplier method are consistent with the results given by the mode-opting program, see Fig.1.

As to the second problem, we have taken into account three enhancing elements, which formed four enhancing schemes shown in Table 2.. Using the previous one-by-one evaluating procedure for each enhancing scheme, the computer program estimates every possible coordination with the modes in NSSSTHE and chooses the optimum one.

THE OPTIMUM HEAT TRANSFER TEMPERATURE DIFFERENCE

In the design procedure, ΔT is given by the primary network optimization before the mode-option of each single equipment. In this stage it is

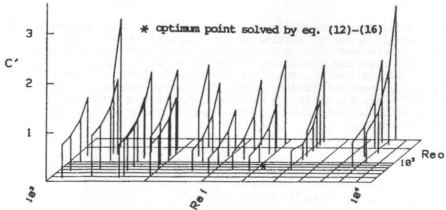

Fig.1 The effect of Re optimization on the cost C'

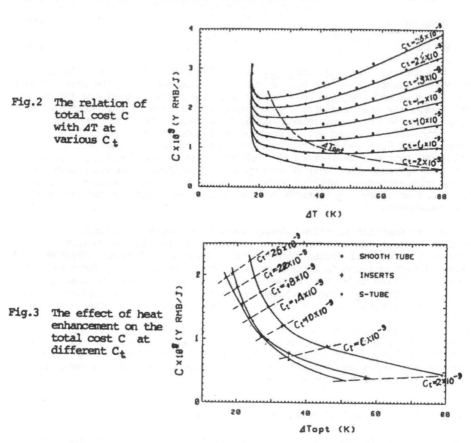

Fig.2 The relation of total cost C with ΔT at various C_t

Fig.3 The effect of heat enhancement on the total cost C at different C_t

general and roughly estimated, for it could not suit the distinction of K of each single heat exchanger. But, from the cost equation, it can be seen that for each single heat exchanger with specific τ and λ, there must have corresponding distinct ΔT_{opt}. This is an important feedback information for the network synthesis or optimization.

Let the first partial derivative of C with respect to ΔT equals zero. According to equation (10), there is:

$$\frac{\partial C}{\partial \Delta T} = C_t \frac{T_0}{T_c} \left(\frac{1}{T_e + \Delta T} - \frac{\Delta T}{(T_0 + \Delta T)^2} \right) - \frac{C\lambda + n\tau}{\Delta T^2} = 0 \tag{17}$$

Rearranging it, we have:

$$\Delta T_{opt} = \frac{T_c}{\sqrt{C_t T_0/(C_f \lambda + n\tau)} - 1} = \frac{T_c}{\sqrt{C_t T_0/C'} - 1} \tag{18}$$

or: $$\Delta T_{opt} = T_h \sqrt{\frac{C_f \lambda + n\tau}{C_t T_0}} = T_h \sqrt{\frac{C'}{C_t T_0}} \tag{19}$$

We can see from the above equation that:

1. ΔT_{opt} would decrease along with the reduction of τ and λ. However, which may differ due to following reasons: a) different physical properties of the fluids; b) diversity of the pressure, temperature and corrosiveness then the material and thickness of the equipment; c) variation of price of the material and manufacture; d) distinction of the performance of the heat exchanger. Of course, there is big gap between the optimum ΔT and a lot of inferiors which are being taken in practical design.

2. The thermal exergy (or fuel) cost has significant influence on ΔT_{opt}. While the other parameters are fixed, $\Delta T_{opt} \propto C^{\frac{1}{2}}$ approximately.

3. ΔT_{opt} is not in proportion to T_c, for which has great effect to τ and λ through changing the fluid properties, especially viscosity. By the way, it is interesting to find out that while assuming a parameter α defined as:

$$\alpha = C_{\Delta T}/(C_{\Delta P} + C_{eq}) \tag{20}$$

then, from eq.(18) or (19), the following formula can be derived:

$$\alpha_{opt} = T_h / T_c \tag{21}$$

Notice that it is a general relationship, and is independent of any other parameters. So, it can be as an explicit criterion for judging if the given ΔT is far from ΔT_{opt}.

DISCUSSION OF THE INFLUENCES

1. For the optimization of single heat exchanger, the optimizing of Re_1 and Re_0 (i.e. the mode-opting) has great effect on the total cost as shown in Fig.1, and some of practical examples are shown in Table 1. In many instances there is great difference between the optimum result and that given by experimental method applied in industrial design.

Table 1. Comparison of the optimal design with the experiential
on the mode-opting of shell-and-tube heat exchangers

term	example 1		example 2		example 3	
	exp.	opt.	exp.	opt.	exp.	opt.
basic model	F_B500-55-4 B300*2	F_B900-200-6 B450	F_B800-165-4 B200*3	F_B900-215-4 B300*3	F_B600-85-4 B150 LF	F_B800-165-4 B300 LF
K, W/ m^2K	241	144	244	179	183	102
F, m^2	109	201	497	644	87	166
τ, Y/ W/K	1.4	1.8	1.1	1.4	1.6	2.7
λ, W/ W/K	0.25	0.03	0.10	0.02	0.45	0.13
C, Y/ GJ	1.994	1.434	1.800	1.664	3.687	3.52
$C_{\Delta p}+C_{eq}$,kY/a	19	11	48	33	16	11

2. At present state-of-the-art and the economic condition, some of the
enhancing techniques have significant benefit. Table 2. shows the optimum
results of some enhanced heat exchangers under the same design condition.

Table 2. Comparison of the performance of some enhanced heat exchangers
with non-enhanced one

term	non-enhanced	insert	S-tube	low-finned tube	insert+low-finned
basic model	F_B800-180-4 B450*2	F_B1000-285-4 B300	F_B900-225-6 B300	F_B700-135-6 B480*2	F_B900-225-2 B300
K, W/ m^2K	143	225	266	215	259
F, m^2	362	283	237	271	241
τ, Y/W/K	1.8	1.2	0.9	1.4	1.1
λ, W/W/K	0.06	0.04	0.03	0.06	0.03
C, Y/GJ	2.341	2.29	2.229	2.328	2.261
$C_{\Delta p}+C_{eq}$,kY/a	19	16	12	18	14

3. The optimization of heat transfer temperature difference has more effect
along with the increase of the thermal exergy cost. As shown in Fig.2, the
higher the C_t the bigger the diversity of C along with the variation of ΔT.

4. Fig. 2 also shows the great effect of the thermal exergy cost. It can
be seen that along with the increase of the C_t, ΔT_{opt} tends to decrease
and the total cost C increase.

5. Along with the increase of thermal exergy cost and the decrease of
optimum ΔT_{opt}, the effect of heat transfer enhancement are getting bigger
and bigger, as shown in Fig.3. This indicates the significance of heat
transfer enhancement at the energy-intensive future.

REFERENCES

1. Bergles, A. E., Appl. Mech. Rev. No.26, pp.675, 1973.
2. Li, X. M., et al., Proc. 7th Int. Heat Transfer Conf., pp.75-80, 1982.
3. Xu, Tian-hua, et al , Chem.& Petro. Processing Equipment, Vol.13,
   No.6, pp.18-24, 1984, (in Chinese).
4. Nishida, N., et al., AIChE J., No.23, pp.77-93, 1977.
5. Linnhoff, B. and Flower, J., AIChE J., No.24, pp.633-654, 1978.
6. Bejan, A., ENERGY, No.5, pp.721-732, 1980.
7. Linnhoff, B. et al., Chem. Engr., Vol.88, No.22, pp.56-70, 1981.
8. Huang, W. H. et al., Proc. 2nd ASME-JSME Thermo-Engr. Joint Conf., 1987.

9. Hua, Ben, Analysis and Synthesis of Energy Use in Process, Hydrocarbon Processing Press, Beijing, 1988, (in Chinese).
10. Bergles, A. E. et al., Proc. 5th Int. Heat Transfer Conf., 1974.
11. Webb, R.L., Int. J. Heat Mass Transfer, Vol.24, pp.715-726,1981.
12. Hua, Ben. Petrochemical Equipment, No.1, pp.11-18, 1985, (in Chinese).

Energy Degradation Analysis of a Benzene Hydrogenation System

LIEQIANG CHEN and FUCHANG RUAN
Chemical Engineering Research Institute
South China University of Technology
Guangzhou, PRC

ABSTRACT

In view of the rising costs of fuels and shortage of energy sources, lost energy from production process to the surrounding or the extent of available energy degradation due to the irreversibilities of the process need be reduced to a minimum. In this paper a systematically thermodynamic analysis of a benzene hydrogenation system has been made based on the thermodynamic first and second laws.

The results of this energetic analysis shows that the major energy and available energy degradation are present in the hydrogenization reactor and the process can be significantly improved by changing the reaction temperature.

INTRODUCTION

The process shown in Fig.1 is a benzene hydrogenation system. The primary effort of this work has two objecties: a) to define and locate the energy loss and available energy degradation existing in each subsystem of the benzene hydrogenation system based on thermodynamic criterion; b) to investigate temperature effect on the thermodynamic energy properties of the reactor.

Our analysis is based on thermodynamic basic and rigorious laws: a) the energy conservation and energy transformation, thus resulting in the energy loss and the process and incremental efficiencies upon the thermodynamic first-law; b) an application of the combination of the first- and second laws of thermodynamics, thus attaining the energy degradation and second-law process and incremental efficiencies.

THERMODYNAMIC BACKGROUND

The procedures employed are based on the first and second laws of thermodynamics, which result in energy and available energy balances. The expressions of energy and available energy balances, energy loss and available energy degradation and various thermodynamic efficiencies

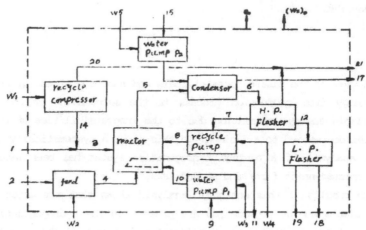

Fig.1. The Process Sheet of a Benzene Hydrogenation System.

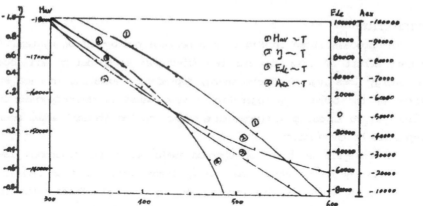

Fig.2. The Dependence of H_{av}, η, E_{de}, and A_{ex} on the Reaction Temperature

are outlined below.

Energy Balance

Application of the first law of thermodynamics to the system gives rise to

$$[(H_{av})_i] = [(H_{av})_{e,u} + (H_{av})_{e,d}] + [(H_{av})_c] \qquad (1)$$

Where $(H_{av})_i$ is the sum of the energies associated with the input streams, $(H_{av})_{e,u}$ is the sum of the energies associated with useful output $(H_{av})_{e,d}$ is the sum of the energies associated with the discarded output streams, and $(H_{av})_c$ is the energy consumed by the process, including the heat loss from the walls of the system to the environment and the work loss to the environment except that due to the expansion of the system boundary.

Available Energy Balance

Application of the second law of thermodynamic to system obtains entropy balance. Combining the energy balance and entropy balance results in the available energy balance.

$$[(A_{ex})_i] = [(A_{ex})_{e,u} + (A_{ex})_{e,d}] + [(A_{ex})_{dis}] \qquad (2)$$

where the first four terms are identical to those in the energy balance, Equation(1), with energy or enthalpy (H_{av}) being replaced by available energy or exergy (A_{ex}) and the last term is the available energy degraded by the system due to all kinds of irreversibilities, such as heat transfer irreversibility, mass transfer irreversibility, and so on.

Energy loss

In Equation(1), the sum of $(H_{av})_{e,d}$ and $(H_{av})_c$ is the energy loss, which can be expressed as follows

$$(H_{av})_{loss} = (H_{av})_{e,d} + Q_. + (W_x)_. \qquad (3)$$

or $(H_{av})_{loss} = (H_{av})_i - (H_{av})_{e,u}$ $\qquad (4)$

where $Q_.$ denotes the sum of heat loss to the surroundings excluding that due to the expansion of the boundaries of the system under consideration. It is obviously found that combination of $Q_.$ and $(W_x)_.$ comes into $(H_{av})_{loss}$.

Available Energy Degradation

From Equation(2), available energy loss can be defined as follows considering entropy generation during the process

$$(A_{ex})_{loss} = [Q_o(1-\frac{T_a}{T_m})] + (W_x)_o + (T_o)\qquad (5)$$

Because the leakage streams are included in energy gaining and/or losing streams which do not lose ability to do work, therefore comparing Equation(5) with Equation(2) gives rise to

$$(A_{ex})_{dis} = [Q_o(1-\frac{T_a}{T_m})] + (W_x)_o + (T_o)\qquad (6)$$

According to the real content of available energy depradation which means the decrease of ability to do work, it is interchangeably to define.

$$E_{de} = (A_{ex})_{dis}\qquad (7)$$

For many organic material containing streams, available energy degradation can be expressed as follows:

$$(E_{de})_k = Q_o[1-\frac{T_a}{T_m}] + [(W_x)_o] + (T_o)\qquad (8)$$

where subscript K means component K contained in the inlet stream, K=1, 2, 3,...n.

The speed of available energy depradation is a function of concentration, tempreature, pressure and other factors, that is

$$E_{de} = E_{de}(c) + E_{de}(T) + E_{de}(P) + ...\qquad (9)$$

here $E_{de}(C)$ may be named chemical available energy degradation, the sum of $E_{de}(T)$ and $E_{de}(P)$ may be named physical available energy degradation. For further details, $E_{de}(T)$ may be named thermal available energy degradation and $E_{de}(P)$ may be named pressure available energy degradation. The other kinds of available energy degradation may be named according to their chemical and/or physical meaning.

An new important concept for available energy degradation can be created and it is available energy degradation rate, which measures the speed of ability lossing to do work per unit of time.

$$E'_{de}(t) = E_{de}(C) + E_{de}(T) + E_{de}(P) + ...\qquad (10)$$

or the average degradation rate is often used

$$\bar{E}_{de} = \frac{E_{de}}{t} = \frac{\varepsilon_i - (\varepsilon_{eu} + \varepsilon_e)}{t}\qquad (11)$$

where ε_i is the specific available energy of the inlet stream, $\varepsilon_{e,u}$ is the specific available energy of useful output stream, and ε_e is the specific available energy of the leakage stream, t is the duration of the process.

Thermodynamic Efficiencies

Based on the energy and available energy balances, various thermodynamic efficiencies can be meaningfully defined.

The first law process efficiency which indicates the extent of energy conservation, is defined as follows

Table 1. The Process Conditions, Energy and Available Energy of Each Stream.

No.	Stream	temperature T(K)	pressure P(atm)	Flow Rate (kg/hr)	(J/kg)×10⁶	(J/kg)×10⁶	H_{av}(J/Hr)×10⁹	A_{ex} (J/Hr)×10⁹
1	Gaseous Make-up Hydrogen	322.05	22.80	269.55	136.56	115.83	36.810	31.222
2	Liquid Benzene to Feed Pump	310.95	1.02	3264.40	43.453	43.679	141.85	142.60
3	Gaseous Mixture	331.79	22.80	473.38	115.48	98.964	54.666	46.85
4	Liquid Benzene to Reactor	350.00	22.8C	3264.40	43.458	43.682	141.864	142.60
5	Gaseous Reactor Effluent	473.15	21.44	5892	48.767	47.42	287.33	279.39
6	Gas-Liquid Effluent from Condenser	473.15	21.44	5892	48.451	47.312	285.47	278.76
7	Recycle from High Pressure Flasher	322.05	20.41	2154	47.023	47.045	101.29	101.34
8	Recycle from Recycle Pump	322.45	22.45	2154	47.025	47.048	101.291	101.3405
9	Water to Pump P1	298.15	1.0	2007	0	0	0	0
10	Water to Reactor	298.15	4.67	2007	0.0004	0.0004	0.0004	0.0004
11	Steam	423.67	4.76	2007	2.64	0.71	5.2985	1.4250
12	Gas-Liquid Effluent to Low Pressure Flasher	322.05	20.41	3516	77.054	75.871	270.90	266.74
13	Gaseous Mixture from High Pressure Flasher	322.05	20.41	222	55.086	48.620	12.232	10.796
14	Gaseous Recycle from Compressor	335.95	22.45	204	55.144	48.658	11.240	9.9180

(Cont.Table 1.)

No.	Description							
15	Inlet Cooling Water to Pump P2	298.15	1.0	19487	0	0	0	0
16	Cooling Water from Pump P2	298.15	1.53	19487	0.0002	0.0002	0.0039	0.0039
17	Outlet Cooling Water	343.95	1.40	19487	0.0649	0.0127	1.2650	0.2475
18	Vent	321.95	1.02	30	46.373	44.854	1.3842	1.3389
19	Cyclohexane Product	321.95	1.02	3489	77.077	75.665	268.68	263.76
20	Gaseous Recycle from High Pressure Flasher	322.05	20.41	204	55.086	48.620	11.228	9.9102
21	Purge	335.95	22.41	18	55.086	48.620	1.0037	0.8859

Table 2 . Energy Loss, Available Energy Degradation and Thermodynamic Efficiencies of Each Subsystem

Subsystems	$(H_{av})_{loss}$ (J/Hr)X10	$\dfrac{(H_{av})_{loss}}{\Sigma(H_{av})_{loss}}$X10	E_{de} (J/Hr)X10	$\dfrac{E_{de}}{\Sigma E_{de}}$	$(\eta)_{s\,P}$	$(\eta)_{s\,D}$	$(\eta)_{s\,P}$	$(\eta)_{s\,D}$
Liquid Benzene Feed Pump	0.0026	0.029	0.0066	0.050	99.99	84.34	99.99	62.50
Water Pump P1	0.0003	0.003	0.0003	0.002	72.73	72.73	72.73	72.73
Reactor	5.1925	0.500	9.9850	0.595	98.25		96.57	
Recycle Compressor	0.0024	0.027	0.0066	0.050	99.98	83.33	99.93	54.17
Recycle Pump	0.0002	0.002	0.0007	0.005	99.99	83.33	99.99	41.67
Water Pump P2	0.0016	0.018	0.0016	0.136	70.91	70.91	70.91	70.91
Condensor	0.5989	0.067	0.3846	0.029	99.35	67.80	99.77	38.67
High Pressure Flaher	2.3360	0.261	1.2230	0.009	99.18		99.56	
Low Pressure Flaher	0.8325	0.093	1.6411	0.124	99.69		99.40	

$$(\eta)_p = \frac{(H_{av})_{e,u}}{(H_{av})_i} \quad \text{or} \quad (\eta)_p = 1 - \frac{(H_{av})_{loss}}{(H_{av})_i} \tag{12}$$

and first law incremental efficiency can be written as follows by re-arranging Equation(1)

$$(\eta)_D = \frac{[(H_{av})_{e,u}]_D}{[(H_{av})_i]_D} \quad \text{or} \quad (\eta)_D = 1 - \frac{(H_{av})_{loss}}{[(H_{av})_i]_D} \tag{13}$$

where [(Hav)] denotes the sum of the energies transfered to the system from the energy sources, including the heat sources, work sources and process streams which lose energy contents during their passage through the system; [(Hav)e]D, the sum of the energies transfered from the system to the energy sinks, including the heat sinks, work sinks and process streams which gain energy contents during their passage through the system.

Similarly, from Equation(2) and by some necessary rearrangement, the second law process and incremental efficiencies can be defined as

$$(\eta)_p = \frac{[(A_{ex})_{e,u}]}{[(A_{ex})_i]} \quad \text{or} \quad (\eta)_p = 1 - \frac{E_{de}}{(A_{ex})_i} \tag{14}$$

and

$$(\eta)_D = \frac{[(A_{ex})_e]_D}{[(A_{ex})_i]_D} \quad \text{or} \quad (\eta)_D = 1 - \frac{E_{de}}{[(A_{ex})_i]_D} \tag{15}$$

RESULTS AND DISCUSSION

Conditions, flow rate, energy(enthalpy) and available energy(exergy) contents of each stream are tabulated in Table 1. Table 2 gives energy losses, available energy depradation of each sunsystem and its constributions to those of the entire system through using Equations 1 to 15.

Energy Loss(Hav)loss and Available Energy Degradation(Ede)

The energy loss and available energy degradation of each subsystem listed in Tabla 2 indicate that the reactor consume 5.1925×10^9 J/Hr of energy, prssessing 50% of the total consumed quantity. Comparing this to the total energy entering the system, it is only about 3% of the amount. However, the reactor dissipates a remarkable available energy, 9.9950×10^9 J/Hr, which sheres 59.5% of the total values degraded, 13.2513×10^9 J/Hr.

The energy and available energy associated with the discarded materials of stream 18 are 1.3842×10^9 J/Hr and 1.3398×10^8 J/Hr, which have comparatively small effect on the efficiencies, which obviously depend on the energy loss $(H_{av})_{loss}$ and available energy degradation (E_{de}).

No incremental efficiencies are presented for the subsystems, reactor and high and low pressure flashers because of no energy lossing and increasing streams exist respectively in these subsystems. It is necessary to point out that a thermodynamic efficiency, in some cases does not mean everything at all. For example, the process thermodynamic efficiencies $(\eta)_b$ and $(\eta)_p$ are high for the reactor, but it contributes 50% of total energy losses and 59.5% of total available energy degradation mainly caused by the heat diffusion and the recreating conditions.

Temperature Effect on the Available Energy Degradation in the Benzene Hydrogenation Process

It can be seen in Table 2 that the avaiable energy degradation in the reactor is 59.5% of total quantity. It is necessary to make the available energy degradation in the reactor minimize.

The thermodynamic properties of $C_6H_6(g)$; $H_2(g)$; $C_6H_{12}(g)$ and $N_2(g)$ are tabulated in Table 3. The latent heat of vaporation of benzene is 30.761×10^3 J/gmole and specific heat of liquid benzene can be expressed as follows:

$$C_p = 62.551 + 0.2335T \text{ J/gmole} \tag{16}$$

For the following reaction:

$$C_6H_6 + 3H_2 - C_6H_{12} \tag{17}$$

energy and avaiable energy can be calculated as follows at any tepreature:

$$H_{av}(T) = -1.9146 \times 10^5 - 54.505T + 0.3054T^2 - 8.405656 \times 10^{-5}T^3$$
$$+ 3.301176 \times 10^{-9}T^4 \text{ J/gmole} \tag{18}$$

$$A_{ex}(T) = -2.2270 \times 10^5 + 193.41T + 54.505T\ln T - 0.3054T^2 + 4.202828 \times 10^{-5}T^3$$
$$-1.100392 \times 10^{-9}T^4 \text{ J/gmole} \tag{19}$$

Suppose that environmental temperature be 298.15K, then the available energy degradation in the reactor can be expressed as

$$E_{de} = (-A_{ex}) - (-H_{av}) \times (T-298.15)/T$$
$$= \frac{5.7082 \times 10^7}{T} + 4.7491 \times 10^4 - 338.98T - 54.505T\ln T + 0.63593T^2$$
$$- 1.2707 \times 10^{-4}T^3 + 4.4016 \times 10^{-9}T^4 \text{ J/gmole} \tag{20}$$

The proportion of available energy degradation is

$$= \frac{E_{de}}{-A_{ex}}$$

$$= (5.7082 \times 10^7 + 4.7491 \times 10^4 T - 338.98 T^2 - 54.505 T^2 \ln T + 0.63593 T^3 - 1.2707$$
$$\times 10^{-4} T^4 + 4.4016 \times 10^{-9} T^5) \ / \ T(2.2270 \times 10^5 - 193.41 T - 54.505 T \ln T + 1.7584$$
$$\times 10^{-4} T^3 + 0.3054 T^2 - 1.926 \times 10^{-8} T^4)$$

$$(21)$$

Form Fig.2, if this process is feasible, $A_{ex} = 0$; that is $T = 594.45K$.

According to the expressions above, H_{av} and A_{ex} increse with temperature $T(T = 72.95$ to $2500K)$. In order to make $C_6 H_6$ and H_2 change into $C_6 H_{12}$ as completely as possible, the value of A_{ex} should be as small as possible. But when doing so, it can be found that the proportion of available energy degradation would be very large. The most suitable temperature minimizing the available energy degradation can be obtained as $T = 430K$, this time $= 0$, $A_{ex} = -5.0582 \times 10^4$ J/gmole, $H_{av} = 1.6499 \times 10^5$ J/gmole.

References

1. Denbigh, K.G. " The Principles of Chemical Equilitrium ", 3rd edition, Cambridge University Press, Cambridge(1971).
2. Fan, L.T. and J.J. H. Shieh, " Thermodynamically Based Analysis and Synthesis of Chemical Process Systems " Energy, 5, 995- 966(1980).
3. Gaggiloli, R.A. and P. J. Petit, " USE the Second Law First ", Chemtech, 497, August(1975).
4. Keenan, Joseph H. " Thermodynamics ", The M. I. T. Press, Cambridge, Massa Chusettes(1970).

Computer-Aided Comparison of Fluids for ORC Systems

M. LIU, Z. D. SHEN, and Y. M. XIANG
Division of Thermal Turbomachinary
Department of Energy and Power Engineering
Xi'an Jiaotong University
Xi'an, Shaanxi Province, PRC

ABSTRACT

As a prerequisite to the preliminary design of high efficiency organic Rankine cycle systems, a thermodynamic model describing the steady-state performances is presented in this paper. Four similarity parameters correlating the ORC engine efficiency are defined and then used as independant variables to optimize the engine efficiency. Heat transfer characteristics of both the evaporator and the condensor are compared, respectively, in terms of the required heat transfer surfaces per unit mass flow rate of the waste streams. A general computer code used to evaluate the turbine work, the pump work and the cycle exergy effectiveness of six potential working fluids is developed. The optimum evaporation temperatures are computed. Trade-offs which must be considered are discussed. Conclusions from this investigation are also presented.

1. INTRODUCTION

Low-grade heat streams are aboundant in industrial processes and geothermal fields. The most practical way of utilizing such low-grade energy is by the so called "flashing " systems. But the effectiveness of this scheme is thought to be relatively poor compared with the "binary systems". So, attentions have been partially shifted to the investigation of organic Rankine cycle systems. Quit a lot of fluids and systems have been suggested. The suitabilities of many potential working fluids to a given situation have been compared to some extent. It is found that the obtainable performances of a specific ORC system depends on the performances of its major components and the selected thermodynamic parameters at the cycle state points for a given application. So the selection and optimization of a particular fluid become very important, even at the research stages, for optimum design of high efficiency low-grade heat recovery systems. The early papers available dealt with this problem in such a way that either the thermodynamic parameters at the cycle's state points are fixed or the engine efficiency operating with the unconventional fluids is supposed to be constant .

In order to compare the relative performance of the potential

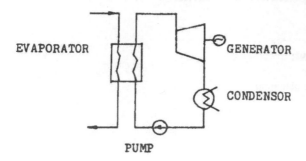

FIGURE 1. Schematic of the proposed ORC.

working fluids, a thermodynamic model is firstly presented in this paper. The dependence of ORC engine efficiency on thermodynamic properties of a particular fluid is taken into account. Heat transfer performances of both the evaporator and the condensor are evaluated. The practical optimum evaporation temperatures corresponding to the maximum net work output for different fluids and different source temperatures are calculated. Suggestions concerning the selection of working fluids are put forward.

2. THERMODYNAMIC AND ECONOMIC ANALYSES

2.1 Thermodynamic Analyses

It has been noticed that, for a given initial temperature and flow rate of the variable-temperature waste heat stream, the higher the evaporation temperature is, the less the amount of the organic fluid evaporated, nevertheless, the larger the turbine isentropic enthalpy drop at a fixed "pinch point" temperature difference. So, a optimum evaporation temperature with respect to the maximum useful net power output may exists. This can only be obtained by thermodynamic modelling. To simplify both the configuration of the system and the derivation of the model, the turbine inlet temperature is supposed to be saturation temperature corresponding to the saturated pressure at the evaporator. Figure 1 shows the major components considered in the derivation of this model.

The net power output of the cycle may be written as :

$$P_n = \frac{G_o \cdot C_p \cdot (T_s - T_n)}{H_1 - H_5} \left\{ \eta_{oi} \cdot (H_1 - H_2') - \frac{1}{\eta_p} (H_5' - H_4) \right\} \tag{1}$$

The equivalent exergy effectiveness may be defined as follows:

$$E_e = \left\{ \eta_{oi} \cdot (H_1 - H_2') - (H_5' - H_4)/\eta_p \right\} / \left\{ (1 - T_a/T_0) \cdot (H_1 - H_5) \right\} \tag{2}$$

It is clear from equation (1) and equation (2) that the ORC
performances in terms of net power output and exergy effectiveness depend largely on the selected thermodynamic parameters
at the cycle state points and the ORC engine efficiency. In this
way of reasoning, the engine efficiency must be predicated
quantitatively before the thermodynamic performances of the ORC
can be compared practively.

2.2 Engine Efficiency Predications

Because of the relatively low turbine isentropic enthalpy drop,
ORC engine efficiency exerts a more important impact on the
total obtainable thermodynamic performance than on conventional
Rankine cycles. The method adopted presently to evaluate engine
efficiency needs to perform a lot of complicated calculations
on the gas dynamics. The calculations must be repeated times to
find the optimum. So a simple predication method may be useful
for this purpose. The methodology adopted here is to use a set
of similarity parameters defined as the specific speed, the size
factor, the compresibility factor and the Mach number. Why the
Mach number is introduced to this set is that supersonic flow
may be encounted in small turbine stages, for the low sonic
speed and the high molecular weight of the fluids. To evaluate
ORC engine efficiency, efficiency equations for single-stage
axial-flow turbines are developed, which are based on the information presented in references 1 and 2. Equations and the
optimized results, together with the special FORTRAN subroutine
are available from reference 3.

2.3 Heat Transfer Performances

The criteria for comparing the relative performance of heat
transfer equipments are the required heat transfer surfaces per
unit mass flow rate of the heat stream. Both the evaporator and
the condensor are supposed to be of shell-and-tube type with
the working fluids on the shell side.

Performance of the evaporator. First of all, nucleate boiling
is supposed to be the heat transfer process in the evaporator.
The transfer coefficients of organic fluids can be predicated
by the following equation (see reference 4):

$$A_e = 1.132.10^{-3} \, P_e^{3.933} \, P_r^{0.916} \, (D/0.01588)^{-1.03}$$

$$(V_g/V_1)^{3.664} \, (D(T_w - T_e)/(U_1 \cdot H_e))^{1.32} \tag{3}$$

The required evaporation heat transfer surface is given by:

$$S_e = Q_e / (A_e \cdot G_o \cdot (T_w - T_e)) \tag{4}$$

Performance of the condensor. Suppose the principle transfer process in the condensor is by film cooling and the Nusselt relation is used to evaluate the heat transfer coefficient:

$$A_c = 0.725 \; \frac{(1/V_1 - 1/V_g) g \lambda_1^3 H_c'}{V_1 \cdot D \cdot \mu_1 \cdot (T_c - T_w)} \tag{5}$$

The required condensation heat transfer surface is:

$$S_c = Q_c / (A_c \cdot G_o \cdot (T_c - T_w)) \tag{6}$$

Where H_c' is the revised latent heat of vaporization given by:

$$H_c' = H_c + 0.375 \, C_{pl} (T_c - T_w) \tag{7}$$

3. ANALYSES OF RESULTS

3.1 Net Power Output

The net power output of a given ORC system is determined by the heat source temperature, the amount of working fluid which can be evaporated, the selected thermodynamic parameters at the turbine inlet and outlet, the ORC engine efficiency and the saturated evaporation and condensation pressures, i.e, the pump work. Figure 2 shows the net power output versus evaporation temperature of the six potential working fluids at the heat source temperature of 150°C. The relative performance of R114 and R600 a. are superior to that of others. R717 shows a relatively poor performance, it is excluded from the following analyses.

3.2 Pump Work Requirements

Pump work requirements is an very important factor to the optimum selection of working fluid for a specific cycle, for it determines the selection of the pump, which, in turn, determines the economic performance of the cycle. Only those fluids with both a high net power output and a lower pump work requirements may be faourable economically. Figure 3 represents the effects of the evaporation temperature on the pump work requirement of the potential working fluids. Pump work requirements may be determined by the working fluid flow rate , the saturated evaporation pressure and the condensation pressure saturated.While both R114 and R6ooa may offer a higher net power output, R114 is more superior to R6ooa for its relatively lower pump work. It is also clear from figure 3 that the specific selection of working fluid affects greatly the pump work of a given system.

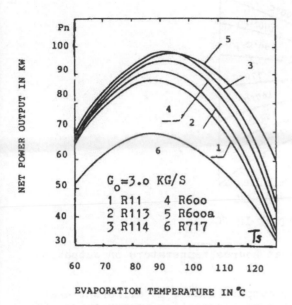

FIGURE 2.   Net power output versus evaporation temperature.

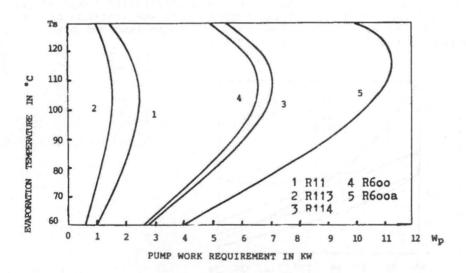

FIGURE 3.   The effects of evaporation temperature on the pump
work requirements of five potential working fluids.

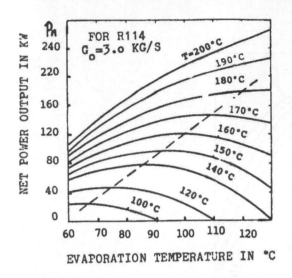

FIGURE 4. Effects of the heat source temperature on output.

Figure 4 shows the net power output of R114 at different heat source temperatures and different evaporation temperatures, for its high relative net power output and low pump work required. The net power output of ORC systems depends largely on the heat source temperature. The output increases sharply with the rise of source temperature. And the optimum evaporation temperatures for the maximum power output increase with the increase of heat source temperatures too. The trace of the maximum net power is shown by the dotted-line in figure 4.

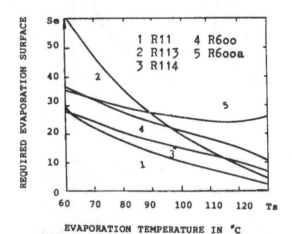

EVAPORATION TEMPERATURE IN °C

FIGURE 5. Effects of evaporation temperature on the required evaporation heat transfer surface.

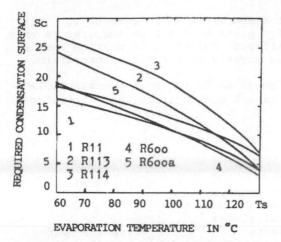

FIGURE 6. Effects of evaporation temperature on the required condensation heat transfer surface.

3.3 Required Heat Transfer Surfaces

In order to provide a basis for the comparison of ORC systems from the economic point of view, performances of heat transfer equipments are compared in terms of the required evaporation and the condensation heat transfer surfaces per unit mass flow rate of the heat streams. The heat transfer coefficients needed to perform the computations are evaluated from equations 3 and 5 ,respectively, for the evaporation and condensation processes. Figure 5 shows the effects of the evaporation temperatures on the required evaporation surface , while figure 6 shows the condensation surface required of the five potential working fluids.

As shown in figure 5, the evaporation surface of working fluid R113 is far larger than that of others at low evaporation temperatures and decreases sharply with the increase of evaporation temperature.R11 presents a comparatively lower heat transfer surface requirement, with the working fluid R114 in the middle.

The required condensation heat transfer surfaces of fluid R11 are superior to that of others. While R114 shows a relatively poor performance, the difference among the rest are negligible within the whole temperature range, as shown in figure 6.

4. CONCLUSIONS

The conclusions of this study are as follows:

1. Trade-offs concerning the ORC net power output and heat transfer performances of the evaporator and the condensor must be taken into consideration before the selection of a suitable fluid for a given application, since the two factors are almost always contradictory with each other.

2. The optimum evaporation temperature with respect to the maximum power output at a given heat source temperature vary very little with the different selection of working fluids. This temperature changes greatly with source temperature.

3. The working fluid R114 seems to be the most recommendable for its high net power output and moderate heat transfer surface requirements.

NOMENCLATURE

A =heat transfer coefficient
C_p =heat capacity at constant pressure
D^p =characteristic diameter of the heat tube
G =mass flow rate of the heat stream
H^o =enthalpy P =pressure P_r =Prandtle number
T temperature Q =heat transfer rate V^r =specific volume

Subscripts

1 =turbine inlet   2 =turbine outlet
2 =turbine isentropic outlet   4 =condensor outlet
5 =pump outlet   5'=pump isentropic outlet
a =atmospheric   w =wall
g =gas   l =liquid
e =evaporator   c =condensor

REFERENCES

1. E.Macchi and A.Perdichizzi, Efficiency Predication for Axial-flow Turbines Operating with Nonconventional Fluids, Trans. of the ASME, Vol. 1o3,1981.

2. V.Maizza, The Use of Unconventional Fluids for Single Stage Supersonic Turbines of Low Power Output, IECEC,7692o3.

3. M.Liu,etc, Computer-aided Efficiency Predication for Axial-flow Turbines Operating with Organic Fluids, to appear.

4. S.Chongrungreong and H.J.Sauer, Nucleate Boiling Performance of Refrigerants and Refrigerant-oil Mixtures, ASME, J. of Heat Transfer,198o.

Second Law Analysis of Standard and Supercharging Heat Pump Cycles and Systems

HUANG GU
Department of Thermoscience and Engineering
Zhejiang University
Hangzhou, PRC

J. SAUNDERS
Department of Applied Science
Brookhaven National Laboratory
Upton, New York 11973, USA

ABSTRACT

Calculations are presented for exergy and irreversibility of the components in a standard vapor compression heat pump and in an advanced concept, the supercharging cycle. This cycle, recently developed at Brookhaven National Laboratory, has demonstrated substantial increases in capacity and COP compared to the standard cycle.

NOMENCLATURE

Q_{rev}	heat transfer in the reversible process from the system point of view (Btu)
Q_{surr}	heat transfer from the surrounding point of view (Btu)
H	enthalpy (Btu)
S'	entropy (Btu/°R)
m	mass (lb)
h	enthalpy per unit mass (Btu/lb)
s	entropy per unit mass (Btu/lb-°R)
q	heat transfer per unit mass (Btu/lb)
T_H	absolute temperature of heat source
T_o	absolute surrounding (ambient) temperature
$T_{evap\ surr}$	absolute evaporator surrounding temperature
T_R	absolute room temperature
ds_z	entropy differential per unit mass due to irreversibility
W_t	work done by system in the reversible process (Btu/lb)
W_t'	work done by system in arbitrary process (Btu/lb)
W_l	work loss due to irreversibility (Btu/lb)
ϕ	exergy (Btu/lb)
ϕ_z	exergy of heat flux (Btu/lb)
COP	coefficient of performance

Subscripts:

cond	condenser
evap	evaporator

INTRODUCTION

In heat pump design, the second law of thermodynamics provides a method for determining the importance of component performance in system efficiency . This method has become quite popular in recent years [1,2,3], since the first law does not completely reveal ways to improve efficiency. For this purpose the second law introduces the concepts of irreversibility, which is defined as the difference between the ideal and actual work of a cycle or process. Since the sum of the component irreversibilities equals the system irreversibility, the role of each component can be assessed in determining the overall work requirement for a given capacity.

These techniques are often called exergy method, where for a steady flow open system, exergy (or available energy) is defined as the maximum work that can be obtained from the combination of the system and its surroundings as the system goes from a given state to the state which is in equilibrium with the surroundings.

In this paper, this concept is used to analyze and compare the irreversibilities associated with heat pumps using the standard vapor compression cycle and those using a new cycle known as the supercharging cycle [4].This cycle, which has been recently developed at Brookhaven National Laboratory, under sponsorship of the Electric Power Research Institute, is documented in detail in [4]. In this effort we modified a commercially available heat pump compressor and tested it in a laboratory version of the supercharged cycle. Significant improvements in capacity and COP were seen; for instance at an evaporating temperature of -15°F and a condensing temperature of 110°F, the evaporating capacity is 1.8 times the standard compressor capacity, while the COP is 1.25 times higher. Based upon the laboratory data, increases in heating season performance factors of 19 to 21% were predicted for three northern cities, with lesser savings for more moderate climates. The payback period (compared to the sandard heat pump) was conservatively estimated to be about 4 years or less--it may be significantly lower with optimal sizing of components.

In this paper we illustrate the procedure for availability calculations for both heat pump cycles and analyze some specific cases. Additional data should be analyzed in future work, particularly a case where the supercharged cycle COP is substantially greater than the standard cycle COP.

THE SUPERCHARGING CYCLE

The standard and supercharging heat pump cycle are shown in Figures 1 and 2. Three primary modifications are used in the supercharging cycle: a special heat pump exchanger is employed, a second expansion value is used, and special ports are introduced in the comprssor cylinder. Unlike the standard cycle, the refrigerant flow leaving the condenser in the supercharging cycle is split into two streams. One stream is throttled to a temperature somewhat above the evaporating temperature and used to subcool the other stream. Consequently, the temperature of the liquid entering the evaporator expansion valve will be near the evaporator temperature,so that little flash gas will be generated during expansion. Hence, there will be a greater ratio of liquid to vapor entering the evaporator than the standard cycle, increasing the capacity. After subcooling, the intermediate pressure stream (which is now mainly gas) cools the motor and compressor, and is injected into the cylinder throuhgh ports that are briefly uncovered while the piston is near bottom dead center. In this process, the cylinder pressure is raised

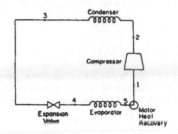

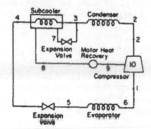

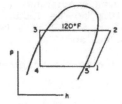

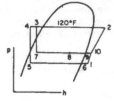

FIGURE 1. Schematic and Pressure-
Enthalpy Diagram for Standard Cycle

FIGURE 2. Schematic and Pressure-
Enthalpy Diagram for Supercharged Cycle

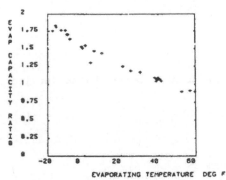

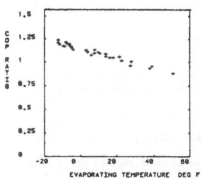

FIGURE 3. Ratio of Evaporating Capacity of Super-
charged Machine to Standard Machine Versus
Evaporating Temperature at 110°F Condensing
Temperature

FIGURE 4. Ratio of COP of Supercharged
Machine to Standard Machine Versus Evap-
orating Temperature at 120°F Condensing
Temperature

from the suction pressure to nearly the intermediate pressure. Note that the evaporator vapor can be ducted directly to the suction valve, since the intermediate stream cools the motor and compressor.

Figure 3 plots the increase in evaporating capacity of the supercharged compressor compared to the conventional compressor, while Figure 4 showns the increase in COP. This data was obtained in our earlier work with R-22 [4]. It can be seen that the increase is higher as the evaporating temperature drop.

THERMODYNAMICAL PRINCIPLES OF SECOND LAW ANALYSIS

We know from the definition above that the available energy of a given state is the maximum work that can be obtained from a system and its surroundings.If the process is irreversible the work will be less than the maximum work available.

In the following analysis we will derive the exergy and irreversibility of a steady flow, open system under surrounding conditions T_o, P_o. An excellent introduction to availability accounting can be found in [5]. The gas parameters at a given state are P, T, H, S' and at a state in equilibrium with the surroundings are P_o, T_o, H_o, and S'_o.

Neglecting the potential and kinetic energies, the energy equation for a reversible process will be:

$$dQ_{rev} = dH + dW_{max} \qquad (1)$$

Where, as usual, we adopt the sign convention of work performed by the system as positive and heat added to the system is positive. The entropy equation for the open system in the process is:

$$dS' = \frac{dQ_{rev}}{T}$$

Because the process is reversible, the total entropy change for the system and surroundings is zero.

$$dS' + dS'_{surr} = 0 \qquad (2)$$

$$dS' = -dS'_{surr}.$$

The entropy change for the surroundings is:

$$dS'_{surr} = \frac{dQ_{surr}}{T_o} = \frac{-dQ_{rev}}{T_o}. \qquad (3)$$

Combining Equations (2) and (3), we get:

$$dQ_{rev} = T_o dS'. \qquad (4)$$

Substituting Equation (4) in Equation (1), we get:

$$dW_{max} = -dH + T_o dS'.$$

After integration

$W_{max} = (H - T_oS') - (H_o - T_oS'_o)$

Using the specific parameter

$$\frac{W_{max}}{m} = \frac{(H - T_oS') - (H_o - T_oS'_o)}{m} = (h - T_os) - (h_o - T_os_o) = \emptyset$$

where $\emptyset$ is called the specific available work done in going from the given state to the surrounding state, $P_o T_o$. This is also called the specific exergy for the given state P,T. Note that for a refrigeration or heat pump cycle dW_{max} is negative, so the actual work must be larger in absolute value than dW_{max}. Consequently, W_{max} represents the ideal work for systems where work is done on the system as well as by the system.

For a steady flow system which passes from state 1 to state 2, and receives heat Q_H, from a reservoir at temperature T_H, it can be similarly shown [5] that

$$\frac{W_{max}}{m} = \emptyset_1 - \emptyset_2 + \frac{Q_H}{m} (1 - \frac{T_o}{T_H}) \tag{5}$$

To compute the ireeversibility, we write the energy equation for both reversible and irreversible processes, as follows:

$$dq = dh + dW'_t \tag{6}$$

For the reversible process the work increment dw'_t is equal to the reversible work increment dW_t, otherwise $dW'_t < dW_t$. The entropy equation for the process is:

$$\frac{dq}{T_H} = ds - ds_s \tag{7}$$

where dS is the system entropy change and dS_s is the system entropy change due to irreversibilities within the system.

Combining Equations (6) and (7), we get:

$$(1 - \frac{T_o}{T_H}) dq = dh - T_ods + dW'_t + T_o ds_s \tag{8}$$

Because $d\emptyset = dh - T_ods$, Equation (8) becomes:

$$d\emptyset = (1 - \frac{T_o}{T_H}) dq - dW'_t - T_ods_s$$

or

$$dW'_t = -d\emptyset + (1 - \frac{T_o}{T_H}) dq - T_ods_s$$

But the first two terms on the right hand side are the differential forms of the ideal work shown in equation (5), so $dW'_t = dW_{max} - dW_1$. where ϕ = exergy, = availability, ϕ_q = exergy of heat flux.

Thus, the irreversibility, or the difference between the actual work and minimum work will be

$$W_1 = \phi_1 - \phi_2 + \phi_q - W'_t \tag{9}$$

Equation (9) is the expression of the first and second laws of thermodynamics using exergy.

Similarly we can write the equations for each component process of cycle, and summing for the cycle we have:

$$\Xi W_1 = \Xi \phi_q - \Xi W'_t \tag{10}$$

Specifically for the heat pump, we have:

$$\Xi W_1 = -\phi_{q,cond} + \phi_{q,evap} - W'_t \tag{10'}$$

If $\Xi W_1 = 0$, it would be a reversible cycle. The minimum cycle input work needed to remove q_{evap} from $T_{evap, surr}$ to T_R will be:

$$W'_t = W_t = \phi_{q,cond} + \phi_{q,evap} \tag{11}$$

For comparing two cycle the loss change equation according to Equation (10') will be:

$$(\Xi W)_1 - (\Xi W_1)_2 = (-\phi_{q,cond})_1 - (-\phi_{q,cond})_2$$
$$+ (\phi_{q,evap})_1 - (-\phi_{q,evap})_2 - (W'_t)_1 + (W'_t)_2 \tag{12}$$

For heat pump systems, the COP for heating can be determined as follows:

$$COP = \frac{q_{cond}}{W_{motor}}$$

RESULTS AND DISCUSSION

Energy flow diagrams for both systems are shown in Figure 5.

To calculate the irreversibility, catalog data was used for the standard system, while test data from [4] was used for the supercharged system. Table 1 summarizes the data and conditions. Exergy at various points in the cycle are tabulated in Table 2, while irreversibility for a 10,000 Btu/hr heating capacity is shown in Table 3 and Figure 6.

From the calculated data in Tables 1-3 we can conclude the following:

The heating capacity of supercharging cycle is 9.5 -17.4% higher than that in the standard cycle for the cases we studied.

The system loss, about 60 to 70%, is caused by three main components: the condenser, compressor and motor.

For both systems, the condenser loss is high due to the temperature dif-

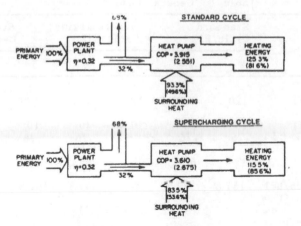

FIGURE 5. Energy Blance

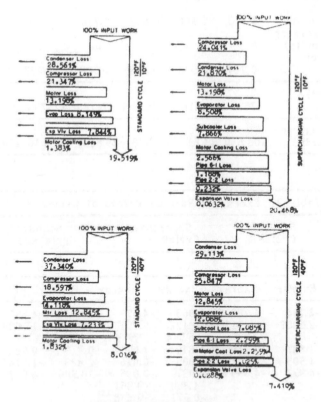

FIGURE 6. Exergy Balance

Table 1. General Data of Cycles

	Standard Cycle R22		Supercharging Cycle R22		Standard Cycle R22	
Cycle Number	1	2	3	4	5	6
Condensing temperature (°F)	120	120	120	120	120	110
Evaporating temperature (°F)	10	40	10	40	10	10
Room temperature(°F)	76	76	76	76	76	76
Surrounding temperature (°F)	35	65	35	65	65	35
Massfloworcondenser mass flow(lb/hr)	161.8	369.5	230.5	449.5	161.8	174.9
$\dfrac{\text{Evaporator mass flow}}{\text{Condenser mass flow}}$			.6585	.7334		
$\dfrac{\text{Superchargemassflow}}{\text{Evaporator mass flow}}$			.518	.3634		
Motor power(watts)	2019	2576	2260.5	3058.5	2019	1827
Motor efficiency (%)	85	85	85	85	85	85
Exergy of condenser heat flux (Btu/lb)	8.310	1.911	6.848	1.720	2.230	8.302
Exergy of evaporator heat flux (Btu/lb)	0	0	0	0	0	0
Exergy of motor heat (Btu/lb)	0.767	0.513	0.603	0.500	0.427	0.642
Heating capacity $Q_{....}$ (Btu/hr)	17579.0	34416.5	20639.4	37678.4	17579.0	18980.0
Specific heating capacity $q_{....}$(Btu/lb)	108.640	93.136	89.538	83.828	108.640	108.532

Table 2. Exergy at Various Points of Cycles

	Standard CycleR22		Supercharging CycleR22		Standard Cycle R22	
Cycle Number	1	2	3	4	5	6
State Point	Exergy	EXergy	Exergy	Exergy	Exergy	Exergy
1	-4.794	-4.282	-4.963	-4.579	-10.704	-4.857
2	22.307	11.512	16.122	9.791	15.845	17.009
2'			16.044	9.553		
3	1.837	0.719	1.879	1.073	0.719	0.982
4	-1.503	-1.001	-0.005	-0.016	-2.823	-1.114
5	-4.972	-4.359	-0.037	-0.025	-10.506	-4.972
6			-4.360	-3.851		
7			-0.671	-0.538		
8			-2.197	-2.103		
9			2.945	-2.193		

TABLE 3. Availability loss, or Irreversibility, and Cycle Balance (Heating Capacity = 10,000 Btu/hr)

	Standard Cycle - R22				Supercharging Cycle - R22				Standar Cycle - R22			
	Cycle 1 Avail. loss		Cycle 2 Avail. loss		Cycle 3 Avail. loss		Cycle 4 Avail. loss		Cycle 5 Avail. loss		Cycle 6 Avail. loss	
	Btu/hr	%	Btu/hr	%	Btu/hr	%	Btu/hr	%	Btu/hr	%	Btu/hr	%
Compressor	836.568	21.347	474.943	18.597	898.432	24.041	715.903	25.847	887.269	22.641	777.062	23.659
Motor	517.202	13.198	328.035	12.845	493.213	13.198	355.760	12.845	548.548	13.998	433.471	13.198
Pipe (2 -2')					8.666	0.232	28.392	1.025				
Condenser	1,119.264	28.561	953.628	37.340	817.297	21.870	806.364	29.113	1,187.098	30.292	711.815	21.672
Subcooler (3-4, 3-7-8)					293.969	7.866	196.248	7.085				
Expansion valve 4-5 (supercharging cycle) 3-4 (standard cycle)	307.391	7.844	184.710	7.233	2.361	0.0632	0.797	0.0288	326.021	8.319	193.038	5.878
Evaporator	319.350	8.149	360.563	14.118	317.962	8.508	334.797	12.088	707.256	18.047	355.525	10.824
Pipe (6-1)					44.413	1.188	63.675	2.299				
Motor cooling	54.196	1.383	46.777	1.832	95.890	2.566	62.575	2.259	57.480	1.467	48.618	1.480
Σ loss Btu/hr %	3,153.971	80.481	2,348.656	91.964	2,972.203	79.532	2,564.511	92.590	3,713.672	94.763	2,519.529	76.711
Minimum work Btu/hr %	764.925	19.519	205.224	8.036	764.925	20.468	205.224	7.410	205.224	5.237	764.925	23.289
Σ work percentage Btu/hr %	3,918.896	100.00	2,553.880	100.00	3,737.128	100.00	2,769.735	100.00	3,918.896	100.00	3,284.455	100.00
COP	2.551		3.915		2.675		3.610		2.552		3.045	

933

ference between the refrigerant and room surrounding. The cooling of the superheated vapor to the saturation temperature accounts for a large portion of the irreversibility (58% in the standard cycle - cycle 1 and 33% in the supercharged - cycle 3). The lower discharge temperature in the supercharged system accounts for the lower irreversibility.

For the standard cycle, the expansion valve loss is higher. In the supercharging cycle, the expansion valve loss is lower because of the deeply subcooled liquid expansion, however, the subcooler loss is high. The expansion valve loss plus the subcooler loss for the supercharging cycle approximately equals the expansion valve loss for the standard cycle. Consequently, the benefits of subcooling in these cases are in increasing capacity. It is not clear whether this conclusion is true over the entire performance spectrum.

The motor cooling irreversibility is higher for the supercharged cycle than the standard, although the motor temperature in the data seleted for study, 102.6°F, is lower than usual. Experimentally, the power seemed to be the lowest at the highest allowable motor temperatures, although some exceptions have been noted.

Compared with the standard cycle the supercharging cycle can demonstrate a higher COP for high temperature lifts and in contrast, the standard cycle has a higher COP during the low temperature lifts. The main reason is due to the performance characteristics of the compressors.

Compared with the cycle 5 , under the same room and ambient temperature, the cycle 2 has a 53.4% higher COP. The reason is that in the cycle 2 the evaporative surface is about 2.2 times that in the cycle 5. Therefore, under the same evaporating capacity the evaprating temperature in cycle 2 will be higher (40°F) and this will reduce the temperature lifts from 120°F/10°F to 120°F/40°F, and the temperature difference of heat transfer from 55°F to 25°F. Thus, increasing the evaporative surface reduces the cycle irreversibility and improves COP.

From the comparison between the cycle 6 and cycle 1 we can find under the same room and ambient temperature 76°F/35°F, increasing the condenser surface, is also an effective means for improving COP (even if not so effective as for the evaporator). Table 3 shows that doubling the condenser area decreases the refrigerant ── room temperature difference from 44°F to 24°F and improves the COP 19.3%.

CONCLUSION

From the exergy analysis above, we can demonstrate that most losses in the heat pump systems are in the condenser, compressor and motor indicating that opportunities for improvement should be concentrated in these components.

Additional data should be analyzed for irreversibility, particularly a case where the COP of the supercharged cycle is substantially greater than that of the standard cycle.

BIBLIOGRAPHY

1. *ASHRAE Handbook and Product Directory*, 1977 Fundamentals.

2. Anand, D.K., K.W. Lindler, S.Schweitzer and W.J. Kennish, Second Law Analysis of Solar Powered Absorption Cooling Cycles and Sysyems. *Journal of Solar Energy Engineering*, 106, Aug. 1984.

3. Ahern, John E. *The Exergy Method of Energy Systems Analysis*, John Wiley, 1980.

4. Saunders, J.H., M. Catan, P.LeDoux, J. Sanchez and D.Shaw, An Advanced Residential Heat Pump Using the Supercharging Cycle. EPRI Final Report, (in press).

5. Jones, James B. and George A. Hawkins. *Engineering Thermodynamics*, John Wiley, 1960.

Waste Heat Utilization System of Power Plant Boiler Stack Gas

YONGKANG XU, GUANGDUO LIU, and WANCHAO LIN
Department of Energy and Power Engineering
Xi'an Jiaotong University
Xi'an, PRC

INTRODUCTION

There is a very large quantity of waste heat from the boiler stack gas of a power plant. If a low pressure energy-saving device is installed in a gas flue in the rear of the boiler air heater and connected to the thermal system, the waste heat from the boiler stack gas can be used to heat the condensate of the thermal system so as to save part of the extraction steam for the low pressure feed water heater. The amount of extraction steam saved, in turn, can be used to do extra work or generate extra electric energy in the turbine. Therefore, more electric energy can be obtained by utilizing the waste heat from the boiler stack gas. If the generator is kept on the rated output, energy will be saved. This idea has been put in to practice on a 100 mw electric generating unit in Longkou Power Plant. This paper is aimed at discussing the features of the waste heat utilization system of the boiler stack gas in Longkou Power Plant, the thermal efficiency of the waste heat from the boiler stack gas, and the conditions under which the waste heat utilization system can bring economic benefit. It also presents information on the installation, design parameter, operating characteristics and safety of the low pressure energy-saving device as well as the investment and the repayment period of such system.

THE WASTE HEAT UTILIZATION SYSTEM OF BOILER STACK GAS

Figure 1 illustrates the waste heat utilization system of the boiler stack gas in Longkou Power Plant. The feed water of the low pressure energy-saving device 5 is fed from the exit condensate of the low pressure feed water No.2 or No.3. The condensate, after being heated by the waste heat from the boiler stack gas, goes to the entrance of the deaerator 4, where it meets the exit condensate of the low pressure feed water heater No.4. In this way, part of the extraction steam of the low pressure feed water heaters No.3, No.4 and the deaerator can be saved. As a result, extra work can be done or extra electric energy can be generated in the turbine by the saved extraction steam. If the generator is kept on the rated output, energy will be saved. As is shown in Figure 1, the characteristics of such a system are as follows.

1. If only the low pressure energy-saving device, connecting pipework and control valves are added to the original electric generating unit, electric energy can be generated by using the waste heat from the boiler stack gas.

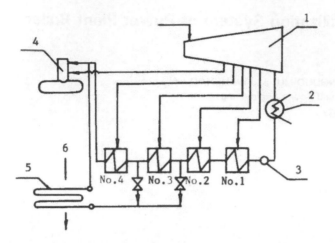

1 — turbine; 2 — condenser; 3 — condensate pump; 4 — deaerator;
5 — low pressure energy-saving device; 6 — waste heat of boiler
stack gas; No. 1 to No.4 — low pressure feed water heaters.

FIGURE 1. The waste heat utilization system of boiler stack gas

Such a system is cheap in cost and easy in application.
2. The feed water of the low pressure energy-saving device is fed from
the exit condensate of the low pressure feed water heater No.2 or No.3
and the flow of feed water can be controlled by an electrical drive valve.
This may affect the heat-exchange surface metal temperature of the low
pressure energy-saving device and the temperature of the boiler gas after
the device. The low pressure energy-saving device has a multi-channel
supply of feed water, which enables the system to adapt itself to the fuel
change in operation, and also provides the safe and economical operation of
the low pressure energy-saving device during their service life.
3. Because the low pressure energy-saving device is in parallel connection
with the low pressure feed water heaters of the thermal system, and the
low pressure energy-saving device bypasses more than one low pressure feed
water heater, the reduction of the water flow resistance is enough to offset the increase of flow resistance by the low pressure energy-saving
device and its connecting pipework. Therefore, there is no need to replace
the condensate pump. This is of great advantage to the innovation of operating power plants.

THE THERMAL EFFICIENCY OF WASTE HEAT FROM BOILER STACK GAS

When the condensate passes through the low pressure energy-saving device,
it carries with its waste heat from the boiler stack gas into the thermal
system. Then how much waster heat can actually be changed into work? This
problem concerns the economic benefit of the utilization of waste heat.

As is illustrated in Figure 1, if the feed water of the low pressure energy-
saving device is fed from the exit condensate of the low pressure feed
water heater No.2, the thermal efficiency n of the waste heat from the
boiler stack gas would be

$$n = \frac{(i_3 - i_2)\, n_3 + (i_4 - i_3)\, n_4 + (i'' - i_4)\, n_5}{i'' - i'} \qquad (1)$$

If the feed water is fed from the exit condensate of the low pressure feed water heater No.3,

$$n = \frac{(i_4 - i_3)\, n_4 + (i'' - i_4)\, n_5}{i'' - i'} \qquad (2)$$

where i_2, i_3 and i_4 are the condensate enthalpies of the low pressure feed water heaters No.2, No.3 and No.4 respectively; n_3, n_4 and n_5 are the extraction steam efficiencies for pushing away the extraction steam of the low pressure feed water heaters No.2, No.3 and the deaerator, respectively; i' and i'' are the entrance and exit condensate enthalpies of the low pressure energy-saving device, respectively.

The numerator term in both formulae above represents the work done by the waste heat from the boiler stack gas when pushing away the multi-stage extraction steam; the denominator term represents the quantity of the waste heat. For engineering unit system, the condensate enthalpy is almost equal to the temperature of the water. Because the values of i_2, i_3, i_4, n_3, n_4 and n_5 in the formulae are determined by the thermal system, when the value of the entrance condensate enthalpy i' of the device is kept constant, the thermal efficiency of the waste heat depends on the value of the exit condensate i'' of the device, i.e., on the exit condensate temperature t''. If the feed water of the low pressure energy-saving device is fed from the exit condensate of the low pressure feed water heater No.2, the thermal efficiency of the waste heat from the boiler stack gas is shown in Figure 2. If the value of the exit condensate enthalpy i'' of the low pressure energy-saving device is smaller than that of the condensate enthalpy at meeting point, the $(i'' - i_4)\, n_5$ term in the numerator term of formulae (1) and (2) will be a negative value. This indicates that although the extraction steam of the low pressure feed water heater No.3 and No.4 is reduced through the utilization of the waste heat from the boiler stack gas, the higher energy level extraction steam of the deaerator is increased. The total work of the numerator term in the formulae should be reduced. The value of the numerator term in the formulae might appear less than zero which means that the economic benefit of the utilization of the waste heat is negative. In order for the utilization of the waste heat from the boiler stack gas to yield economic benefit, the numerator terms of formulae (1) and (2) must be greater than zero. In the case of the system in Figure 2, the exit condensate temperature t'' of the low pressure energy-saving device should be greater than 107.4 C.

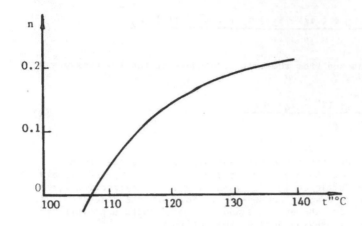

FIGURE 2. The thermal efficiency of the waste heat from the boiler stack gas

THE LOW PRESSURE ENERGY-SAVING DEVICE

The Installation and Parameters of the Low Pressure Energy-Saving Device

The low pressure energy-saving device is installed in a vertical gas flue after the boiler heater, the design parameters are listed in Table 1 below.

TABLE 1. Design parameter of the low pressure energy-saving device

term	unit	value
pressure	MPa	1.225
heat-exchange surface	m^2	2045
temperature of entrance/exit water	°C	(98, 113)/138
temperature of entrance/exit gas	°C	169.9/(146,149)
heat load	GJ	17.48, 14.87
velocity of gas	m/s	6.32
velocity of water	m/s	0.55, 0.75
flow resistance of gas	Pa	240
flow resistance of water	MPa	0.031, 0.057
diameter of tube	mm	32
total weight	t	47.4

The Optimum Quantity of the Entrance Water of the Low Pressure Energy-Saving Device

As the quantity of entrance water of the low pressure energy-saving device increases, the temperature of the exit condensate of the device will drop, resulting in a reduction in the thermal efficiency of the waste heat from the boiler stack gas. But when the entrance gas temperature of the device is held constant, the quantity of heat absorbed by the device Q can be increased due to an increase in the temperature difference of heat transfer.

The amount of work to be done by the waste heat is the value of Q·n which will be maximium on condition that the quantity of the feed water of the low pressure energy-saving device is a certain optimum value (see Figure 3). For the system in Figure 2, the optimum quantity of entrance water is 84038 kg/h.

Safety of the Low Pressure Energy-Saving Device

1. If the low pressure energy-saving device breaks down the feed water supply may be stopped. The low pressure energy-saving device can bear heating without feed water in it. The normal operation of the original electric generating unit will not be influenced.
2. The feed water of the low pressure energy-saving device is fed from the exit condensate of the low pressure feed water heater No.2 or No.3 and the quantity of the feed water can be controlled by electrical drive valve, thus the temperature of the boiler stack gas and the metal tempe-rature of the low pressure energy-saving device could be controlled.
The waste heat utilization system of the LongKou Power Plant boiler stack gas has been operating for three years. While the boiler is under regular standardised inspection, there has been no evidence of corrosion at low temperature and ash fouling in the low pressure energy-saving device.

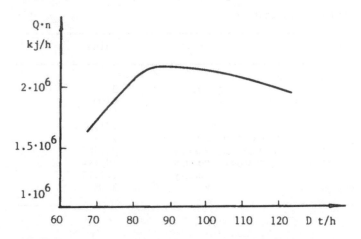

FIGURE 3. The optimum quantity of entrance water

ENERGY-SAVING BENEFIT OF THE WASTE HEAT UTILIZATION SYSTEM OF POWER PLANT BOILER STACK GAS

When the low pressure energy-saving device is in operation, with the tem-perature of boiler stack gas being 150°C and the heat-exchange surface of the low pressure energy-saving device being 2045 m^3, the test result shows that the boiler stack gas temperature is reduced by 16-20°C, the specific steam consumption is decreased by 0.02625-0.05359 kg/kw.h, the specific heat consumption is reduced by 65.61-119.24 kj/kw.h, the coal consumption of the generating unit is decreased by 2.58-4.64 g/kw.h and the quantity of standard coal saved annually is 1806-3248 t.

941

THE TECHNICAL ECONOMY OF THE WASTE HEAT UTILIZATION SYSTEM OF BOILER STACK GAS

The investment and the repayment period of the waste heat utilization system of boiler stack gas of the LongKou Power Plant is shown in tables 2 and 3, provided that the cost of electricity is 0.047 yuan/kw.h, the cost of standard coal is 81.44 yuan/t, the operation time of the system is 7000 h/y, the depreciation charge of equipment is 5% in one year.

TABLE 2. Investment of waste heat utilization system

term	yuan
the low pressure energy-saving device (including the device support)	357,882
pipework (including pipework for heat preservation)	25,000
total investment	382,882

TABLE 3. The period of repayment

term	unit	value
reduction of coal consumption of the generating unit	g/kw.h	3
annual reduction of standard coal consumption	t	2100
reduction of cost of coal	yuan	171,024
electricity consumption of exhaust fan	kw	75.1
annual cost of electricity consumption	yuan/y	24,708
annual deprectation charge of equipment	yuan/y	19,144
net income	yuan/y	127,172
the period of repayment	y	3.01

CONCLUSIONS

1. Because the low pressure energy-saving device has a multichannel feed water supply, and the quantity of feed water can be controlled, the temperature of the boiler stack gas and the metal temperature of the low pressure energy-saving device can therefore be controlled, thus ensuring safety and economical operation of the low pressure energy-saving device within its service life.
2. It is an effective energy-saving measure to adopt the waste heat utilization system of power plant boiler stack gas. The test result shows that the boiler stack gas temperature is decreased by 16–20°C, the specific heat consumption is reduced by 65.61–119.24 kj/kw.h, the coal consumption of the generating unit is decreased by 2.58–4.64 g/kw.h under the condition that the temperature of the boiler stack gas is 150°C and the heat-exchange surface of the low pressure energy-saving device is 2045 m^2.

REFERENCES

1. Xu Yongkang Lin Wanchao Liu Guangduo, "Thermal Efficiency of Waste Heat in Steam Regenerative Power Cyole", Journal of Engineering Thermophysics. Vol.7, no.1, pp. 48–50, 1986. (in Chinese).

2. Lin Wanchao Xu Yongkang Liu Guangduo, "On the Utilization of Stack Gas Heat in a Power Cycle", Journal of Xi'an Jiaotong University. Vol.19, no.6, pp.43–49, 1985. (in Chinese).

MISCELLANEOUS

Gas-Steam Combined Cycle and Heat Recovery Steam Generator (HRSG)

YUAN-XIN WU
Hangzhou Boiler Works
Hargzhou, PRC

THE ELEMENTLY EXPLANATION OF THERMODYNAMIC PRINCIPLE OF GAS–STEAM COMBINED CYCLE [1][2]

It is well known that the thermal efficiency of an ideal heat engine working in the light of Carnot Cycle is given by

$$\eta_{ideal} = 1 - \frac{T_2}{T_1}$$

where, T_1 —— temperature of heat source,

T_2 —— temperature of cold source.

The significance of the above formula is in such, that it can evaluate and point out the right way to increase the thermal efficiency of any heat engines.

In the steam power cycle, the steam temperature results from heat transfer through burning fuel in the boiler combustion chamber. The flue gas temperature from burning fuel is more than 1400°c or higher, whereas the steam temperature is hardly over 600°c. This heat absorption process of

947

steam in conventional steam power cycle (Rankine cycle) makes the energy stored in the fuel less available. However, on the other hand, the discharged steam temperature can be reduced to very low level in the cycle, say 45°c or so, the thermal efficiency of steam cycle is less than 40%. Now, the gas turbine is developing with the use of high temperature materials and cooling techniques, the initial temperature (t_1) of gas turbine has reached 1200°c, the thermal cycle of gas turbine unit (Braton cycle) shows the exhaust gas temperature behind gas turbine could be even more than 550°c, usually would be at the range of 450—550°c, the thermal efficiency of gas turbine is less than 35%.

The gas steam combined cycle makes use of the characteristics of high gas inlet temperature of gas turbine and lower discharged steam temperature of steam turbine to increase its thermal efficiency.

If the gas turbine unit uses fuels such as light oil and natural gas, it consists of three main components:compressor, combustion chamber and gas turbine itself. The ambient air is drawn into the compressor and compressed to a certain pressure then enters into combustion chamber mixed with fuel. Through burning, the high temperature flue gases with high pressure pass through the gas turbine and work is done. After this, the exhaust gases temperature and pressure has been reduced. The total mechanical work produced by the gas turbine has been consumed about two thirds for the compressor, only one third can be used for producing electricity or for driving other machine directly. Of course, the higher the temperature of inlet gases of gas turbine, the higher the thermal efficiency is. The thermal efficiency of gas-steam combined cycle can be reached as high as 47%.

The flow scheme of gas-steam combined cycle and its thermodynamical temperature-entropy diagram are shown in FIGURE 1. The gas turbine thermal cycle is defined by line 1-2-3-4-1 and the steam thermal cycle is defined by line a-b-c-d-e-f-a.

The thermal efficiency of gas turbine could be expressed as:

$$\eta_g = \frac{Q_{23} - Q_{41}}{Q_{23}} \tag{1}$$

The efficiency of heat recovery is

$$\eta_b^\circ = \frac{Q_{45}}{Q_{41}} \tag{2}$$

The thermal efficiency of steam turbine can be written as

$$\eta_s = \frac{he - hf}{he - ha} \tag{3}$$

where, he —— initial heat content (enthalpy) of steam entering the steam turbine,
hf —— final heat content of steam discharged from steam turbine,
ha —— heat content of feedwater.

The thermal efficiency of gas-steam combined cycle could be

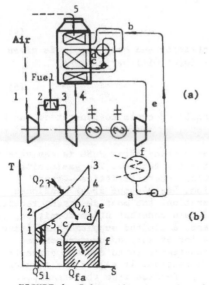

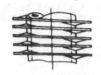

1. Ambient air entering the compressor
2. Compressed air leaving the compressor
3. Gases leaving the combustion chamber
4. Exhaust gas from the gas turbine
5. Exit gas of HRSG entering the stack
a. Feedwater entering the economizer of HRSG
b. Heated feedwater leaving the economizer of HRSG
c. Saturated water entering the evaporator of HRSG
d. Steam-water mixture entering the drum of HRSG
e. Superheated steam entering steam turbine
f. Exhaust steam entering the condenser

FIGURE 1. Schematic process and its T--s diagram of gas-steam combined cycle

FIGURE 2. Spiral finned tube

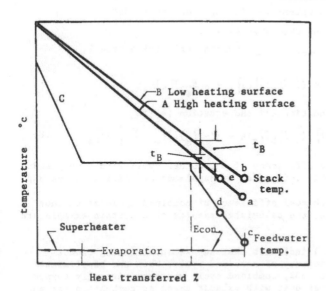

FIGURE 3   Heat transferred-temperature profile

$$\eta_{gs} = \frac{Q_{23}-Q_{41}}{Q_{23}} + \frac{Q_{41}}{Q_{23}} \cdot \frac{Q_{41}-Q_{51}}{Q_{41}} \cdot \eta_s$$
$$= \eta_g + (1-\eta_g)\eta_b^{\circ}\eta_s \tag{4}$$

If gas turbine is used in producing electricity and heat supply is given by HRSG, the effectiveness of total heat imput will be

$$\eta_{co} = \frac{Q_{23}-Q_{51}}{Q_{23}} \tag{5}$$

The thermal effectiveness of gas steam combined cycle co-generation can be reached as high as 85%.

In the combined cycle, the power of steam turbine after HRSG is equivalent to about 45% of the power generated by the gas turbine. The gasturbine-generator unit has several intrinsic dynamic characteristics——rapid start-up and shut-down, flexible operation. Besides the above mentioned requirement for matching gas turbine operation, the most important require-ment for the steam power cycle is to maintain constant steam condition including steam temperature. In some cases, a limited supplemental firing in HRSG should be carried out. Because a lot of surplus oxygen content about 16-18% by volume existed in the high-temperatured exhaust gases from gas turbine, it can be used as air for combustion. If all of the surplus oxygen could be used up. the electricity production from the steam power cycle could be raised to 70% of total power of combined cycle. In this case, the HRSG could be designed as a conventional utility boiler.

The thermal efficiency of a combined cycle with a supplementary fired HRSG can be written as

$$\eta_{gs-f} = \frac{Q_{23}}{Q_{23}+Q_b}[\eta_g + (1-\eta_g)\eta_b\eta_s + \eta_b\eta_s\frac{Q_b}{Q_{23}}] \tag{6}$$

where, Q_b —— heat from supplemental firing added to a HRSG,

$\quad$ b —— efficiency of HRSG when firing,

let $\quad \frac{Q_b}{Q_{23}+Q_b} = s$; where s —— supplemental firing rate %,

then, $\eta_{gs-f} = (1-s)[\eta_g + (1-\eta_g)\eta_b\eta_s + \frac{s}{1-s}\eta_b\eta_s] \tag{7}$

The difference between equation (7) and equation (4) is

$$\Delta\eta = \eta_{gs-f} - \eta_{gs} = -\eta_g(1-\eta_b\eta_s)s + (1-\eta_g)(\eta_b - \eta_b^{\circ})\eta_s \tag{8}$$

From this, we can see the efficiency of a combined cycle with a supplemen-tary firing is unfavourably influenced by supplementary firing in the HRSG.

In ordwe to explain the thermal efficiency of combined cycle at various supplemental firing rates, the calculated results of a certain example are exemplified in TABLE 1.

The calculation in TABLE 1 indicates that supplemental firing cannot increase the thermal efficiency of combined cycle. When the supplemental firing rate is larger then 50%, combined cycle has been virtually turned into a combined cycle power unit with exhaust gases as combustion air and with steam power station as a main part.

TABLE 1.   Thermal efficiency of combined cycle after firing

1. Oxygen content in gas turbine exhaust gases		16%		
2. The percentage of consumed oxygen in the oxygen content of gas-turbine exhaust gases %	0	30	70	100
3. Supplemental firing rate %	0	35.7	56.4	64.9
4. Exhaust gas temperature after firing °c	542	930	1487	1800
5. Stack temperature after firing °c		185		
6. Efficiency of heat recovery %	66.1	80.8	87.4	89.9
7. Thermal efficiency of steam power cycle %		35		
8. Thermal efficiency of gas turbine %		30.8		
9. Thermal efficiency of combined cycle %	46.8	42.5	39.9	38.9
10. The percentage of power from steam turbine in the combined cycle %	29.5	48.3	61.7	68.0

THE PERFORMANCE CHARACTERISTICS OF EXHAUST GASES FROM GAS TURBINE AND THE DESIGNING OF HRSG

Quantity and temperature of gas turbine exhaust gases.

The quantity of ambient air entering the combustion chamber of gas turbine requires approximately five times as much as that for a nomal boiler at the same rated capacity. It makes the exhaust gases from the gas turbine a large volume and heat content. Each kilogram exhaust gases can produce about 0.12—0.14 kilogram steam. The temperature of gases is usually at the range of 450—550°c. Therefore, the mean temperature difference between gases side and heated medium (steam and water) is quite small, the coefficient of heat transfer is low, too. This can be seen from the comparison of the HRSG with the normal boiler in TABLE 2.

Under the same capacity, the heating surface of HRSG is 5—10 times as much as the normal boiler, so the tube with extended surface as spiral-finned tube (FIGURE 2) has to be used [3]

The dimensions of spiral finned tubes commonly used in HRSG are given in TABLE 3.

TABLE 2   The comparison between HRSG and normal boiler

Item		HRSG	Normal boiler
Range of temperature difference	°c	40-160	200-900
Gas velocity in heating surface m/sec		13-20	9
Coefficient of heat transfer kcal/m²hr°c		28-35	45-55
Percentage of heating surface	%	500-1000	100

TABLE 3   Dimension of spirally finned tubes

Outside diameter of tube mm	Height of fin mm	Thickness of fin mm	Numbers of fin per meter
29-51	13-20	1.2-1.5	150-250

It must be pointed out that increasing every 100mm watergauge of back pressure at the gas turbine outlet would lose 1 percent of its capacity, so the heating surface must be arranged suitablly.

Startability of gas turbine.

Gas turbine has rapid startability and flexibility in operation. It can come to full load in 15-20 minutes from a cold start. During this short time, a lot of exhaust gases with temperature of 500°c or so enter into HRSG. The HRSG after gas turbine should be able to withstand the thermal shock from the opening of a damper to allow 500°c gases to enter the HRSG. The whole water circulating system must be helpful in the rapid start-up.

Temperature profile of HRSG [4].

The typical temperature profile of single pressure HRSG is shown in FIGURE 3. Line A and line B indicate how the exhaust gas temperature drops in the HRSG at different heating surfaces. The broken line C demonstrates the situation of water and steam rising temperatures. The heat transfer between A and B, the C is carried out in counter flow manner, It is always suitable to HRSG design. Δt_A and Δt_B represent temperature differences between gas temperature leaving the evaporator and the steam temperature at saturated pressued respectively. They are also called pinch points.

Studying FIGURE. 3, the value of pinch point Δt_B is greater than Δt_A thus, the heating surface operating at line B may be reduced, but its stack-temperature point b would be higher than point a. Although the heat recovery efficiency is lower, the capital cost of HRSG could fall down. If we increase the feedwater temperature from point c to point d, according to line A, the stack-temperature will be raised from a to e; and the heat recovery efficiency will be reduced.

From the same FIGURE, the minimum temperature differences are Δt_A and

Δt_B, which take place in the evaporators where flue gas leaves.

The calculation of a medium pressure HRSG (3.82MPa) indicates that reducing pinch point would substantially increase the quantity of heating surface. But the increase of steam output is slight. The results of calculation are shown in TABLE 4.

TABLE 4. Calculating Results

Pinch point	°c	10	14	18.2	24.4
Stack-temperature	°c	184	186.5	189	192
The percentage of evaporator heating surface	%	140	126	113	100
The percentage of evaporation capacity	%	102.96	102.18	101.4	100

Thus, it can be seen that, if pinch point is reduced from 24°c to 10°c, the heating surface has to increase 40% and the evaporation capacity would merely increase 3%. For medium steam condition, the value of the pinch point may be chosen for 15°c or so.

In order to further improve the thermal efficiency of steam side of the combined cycle, in respect of the reduction of the stack temperature and the improvment of the heat transfer, the dual-pressure or multipressure HRSG may be used in combined cycle. More waste heat could be recovered from the exhaust gases. HRSG can not only produce the high pressure steam or medium pressure steam, but can heat the lower temperature feedwater up to higher level, so that the temperature of steam expelled from the steam turbine will be reduced. Then the steam cycle power output and thermal efficiency will increase.

Meanwhile, more attention must be paid to the arrangement of heat transfer surface in such a condition that it can make heat transfer more effective under large temperature difference. Even though the thermal system is rather complicated, it can get the maximum thermal efficiency.

REFERENCES

1. Luming Pang, Mengle Wang and Haixian Feng, Engineering Thermodynamics 1st ed., pp. 399-402, People's Education Publishing House, Beijiang, China, 1981.
2. Ir.B.van den Hoogen, Steamproduction be Heat Recovery from Gasturbine Exhaust Gases, HCG-NEM Boiler and Process Equipment Division, Legden-Holand.
3. Yuanxin Wu, Waste heat Recovery Boiler pp.100-101 Machine Building Industry Purlishing House, Beijiang, China 1980.
4. Leroy.O. Tomlinson and Richard S Rose, Combined Cycle Repowering for Steam Plant Efficiency Improvement, pp.9-10. General Electric Co. United States 1984.

The Theoretical and Experimental Researches of the Pressure Wave Supercharger

WEN WU and XIAO HUANG
Guangzhou Institute of Energy Conversion
Guangzhou, PRC

ABSTRACT

Both the theoretical and experimental research results of the pressure wave supercharger (PWS) were descripted. The PWS was designed for the diesel engines rating about 120 HP to boost their output 60%. The methods of the theoretical solution and the calculated results were related briefly. The experimental results of PWS at rated speed and at lower speeds were both given. The performances of a diesel engine equipped with the PWS at different speeds were also shown. The improvements needed for further studies were finally mentioned.

INTRODUCTION

The pressure wave supercharger (PWS) is a new device to utilize the engine exhaust gas energy to compress the atmospheric air before entering the engine to achieve the output increase purpose. The PWS was first developed by the Brown Boveri company of Switzerland [1]. The PWS works on an entirely different principle from the conventional turbo-supercharger. In the PWS, the engine exhaust gas contacts and transfers the energy to the air directly through a series of compression and expansion waves. It has the merits of simple in construction, low rotating speed, and no surge. When the diesel engine is equipped with the PWS, the engine will have higher torque at low speed, higher accelerationg rate for starting, fast respond to load change and lower fuel consumption at partial load. Now the PWS has been used in varions rating diesel engines for tractors, trucks and cars in Finland, Western Germany and Switzerland[2].

THEORETICAL RESEARCH

Two methods, both based on one dimensional unsteady flow have developed to design the PWS at rated speed.

(a) Graphical method--based on isentropic one dimensional flow. To cross-plot the flow regions on the X/L_0 & $a_f^0 t/L_0$

physical plane and region flow properties on the W/a&
(p/p)$^{\frac{\gamma-1}{\gamma}}$ state plane. The method is easy to master and can be
used as a preliminary design.

(b) Computer method [3]--to treat one dimensional unsteady
flow with friction and heat transfer. The basic nondimensional equations to be treated are:

1. The continuity equation,

$$\frac{\partial \rho}{\partial t} + \frac{\partial (\rho w)}{\partial x} = 0 \tag{1}$$

2. The momentum equation,

$$\frac{\partial (\rho w)}{\partial t} + \frac{\partial}{\partial x}(\rho w^2 + P) = -\rho F_\mu \tag{2}$$

3. The energy equation,

$$\frac{\partial}{\partial t}\left[\rho(e + \frac{w^2}{2})\right] + \frac{\partial}{\partial x}\left[\rho w(h + \frac{w^2}{2})\right] = \rho \mathfrak{g} \tag{3}$$

4. The equation of state,

$$\gamma_i P = R \rho T \tag{4}$$

The above first three equations can be written in vector form
as:

$$\frac{\partial}{\partial t}(\vec{D}) + \frac{\partial}{\partial x}[\vec{F}(\vec{D})] = \vec{B} \tag{5}$$

Where

$$\vec{D} = \begin{pmatrix} D_1 \\ D_2 \\ D_3 \end{pmatrix} = \begin{pmatrix} \rho \\ \rho w \\ \frac{P}{\gamma-1} + \frac{\rho w^2}{2} \end{pmatrix} \tag{6}$$

$$\vec{F} = \begin{pmatrix} F_1 \\ F_2 \\ F_3 \end{pmatrix} = \begin{pmatrix} \rho w \\ (\frac{3-\gamma}{2})\rho w^2 + (\gamma-1)(\frac{P}{\gamma-1} + \frac{\rho w^2}{2}) \\ \gamma w(\frac{P}{\gamma-1} + \frac{\rho w^2}{2}) - (\frac{\gamma-1}{2})\rho w^2 \end{pmatrix} \tag{7}$$

$$\vec{B} = \begin{pmatrix} B_1 \\ B_2 \\ B_3 \end{pmatrix} = \begin{pmatrix} 0 \\ -\rho F_\mu \\ \rho \mathfrak{g} \end{pmatrix} \tag{8}$$

F_μ and q are the frictional effect term and heat transfer
effect term respectively.

With the initial values and the boundary values given, the calculation can be proceeded in a medium speed computer. The calculated results are shown in figure 1. The high pressure gas inlet relative velocity Wk is quite regular while the high pressure air exit relative velocity Wg is also regular except a small portion of reverse flow near the opening end of the exit. The low pressure gas exit relative velocity Wd and the low pressure air inlet relative velocity Wi are both very irregular. The interface trace first goes into the rotor up-to about 60% of the rotor axial length, then retreats back and exits at the low pressure gas port.

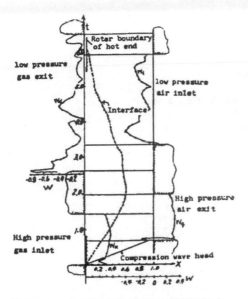

FIG. 1   A CALCULATION EXAMPLE

EXPERIMENTAL INVESTIGATION

To use the data obtained by the graphical solution as the design guide, several experimental PWS were constructed and tested. One series of the tested results were shown in figure 2 & figure 3. In figure 2, it can be seen that the air compression ratio Pg/Pi reaches 1.80 at speed of 8500 rpm with gas inlet temperature of 620 C, while the maxium value of Pg/Pi is 1.86 at speed of 7000 rpm. The atmos-pheric air inlet flow rate Gi is always higher than that of compressed air exit flow rate Gg. This means that portion of the inlet atmospheric air will exit at the low pressure port together with the gas. In figure 3, it can be seen clearly that the compression ratio Pg/Pi, the density ratio and the air flow rate Gg & Gi all increase with the increase of the gas inlet temperature Tk while the speed keeps constant.

957

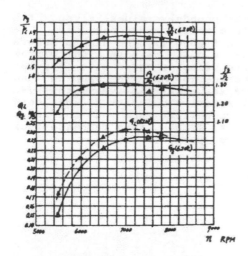

Fig. 2   PERFORMANCE CURVE OF PWS MODEL II-2

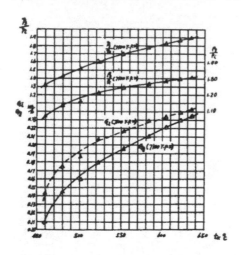

FIG. 3   PERFORMANCE CURVE OF PWS MODEL II-2

THE DIESEL ENGINE PERFORMANCE EQUIPPED WITH THE PWS

The figure 4 shows the performance of a diesel engine
equipped with PWS [4]. It can be seen that the engine output
is increased fropm 120 HP of no PWS to 190 HP with PWS, an
increase of 58%, at the rated speed of 1500 rpm. The fuel
consumption was also reduced to 174 g/HP.hr, a reduction of
about 3%.

It was found that the engine noise level of the PWS engine
is somewhat higher than that without the PWS. Another demerit

is that the size and the weight of the PWS are a little
higher than that of the turbo-supercharger of the simillar
rating.

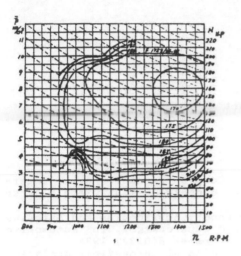

FIG. 4 PERFORMANCE MAP OF A DIESEL ENGINE WITH PWS

CONCLUSION

The performance of the PWS investigated through both theoretical and experimental researches were found satisfactory to
supercharge the diesel engine. The engine output could be
increased by nearly 60% at rated speed at inlet gas
temperature of 620 °C.

The PWS construction is simple, low rotating speed and no
surge. Diesel engine equipped with PWS will be quick in
acceleration for starting, fast respond to the load change
and low fuel consumption at part load.

NOMENCLATURE

a--sound speed
e--internal energy
G--mass flow rate
h--enthalpy
L--characteristic length of PWS (rotor)
N--engine power
n--rotation speed
P--pressure
R--gas constant
T--temperature
t--time variable
W--relative velocity
X--coordinate variable

959

γ--specific heat ratio
ρ--density

superscript

o--stagnation state

subscript

d--low pressure gas exit port
g--high pressure air exit port
i--low pressure air inlet port
k--high pressure gas inlet port
o--rotor length

REFERENCE

1. G.Zehnder and A.Mayer, Comprex pressure wave super-
    charging for automotive Diesels -- SAE Tech. paper
    840132, 1984.
2. G.M.Schruf and T.A.Kollbrunner, Application and Matching
    of Comprex pressure wave supercharger to automotive
    diesel engine -- SAE Tech. paper 840133, 1984.
3. W.Wu and M.chen, Some results of theoretical analysis,
    calculation and experimental research of pressure wave
    supercharger -- Journal of Engineering Thermophysics -
    V.3 n.1, p30, Feb.1982.
4. W.Wu etc, The experimental investigation of 6135ZG
    diesel engine equipped with the W100 pressure wave
    supercharger -- Research report of Guangzhou Institute
    of Energy Conversion, December, 1984.

Possibilities of Taking the Advantage of Steam Power Facility to Increase the Fuel Utilization Efficiency of a Gas Turbine Unit

QING-ZHAO WANG, ZHI-PING SONG, and LAN LIN
Graduate School
North China Institute of Electric Power
Qinghe, Beijing, PRC

ABSTRACT

It is emphasized the importance of taking the advantage of steam power facility to increase the utilization of fuel fed to gas turbine units. A generalized method of determining the waste heat efficiency is proposed. Three sample alternatives are investigated in detail and are compared with conventional arrangements.

INTRODUCTION

In China, gas turbine units are not so widely used in large-scale power production for their relatively low thermal efficiency and for the need of a fluid (rather than solid) fuel supply. However, gas turbine units have a number of attractions, such as rapid start up, low capital cost and adaptability of relatively low-grade fluid fuel. In many applications they are installed within coal-fired power plants as stand-by units. The dominant part of power is provided from coal, the smaller gas turbine units are operated only when it is necessary. Sometimes the addition of gas turbine units serves as a measure to be taken to modernize the existing plants increasing their thermal efficiency and/or their output capacity. In these applications we are much concerned about how efficiently is used the fuel fed to the combustion chamber of the gas turbine units. Generally speaking, this depends mainly on the way the gas turbine waste heat is recovered under specified conditions.

The solutions of waste heat recovery may be numerous. However, it is sometimes neglected the possibilities of taking the advantage of steam power facility to reach a solution favourable both in thermal efficiency and in unit capital cost.

In Fig.1 possible alternatives of waste heat utilization are shown. Figure 1 (A) is a gas turbine unit without waste heat utilization which suffers from low efficiency. Alternatives to Fig.1 (A) may be realized either by equiping a regenerator like Fig.1 (B), or by installing a subsidiary steam cycle driven by an unfired waste heat boiler, as shown in Fig.1 (C).

If the gas turbine unit is sited within a large coal-fired

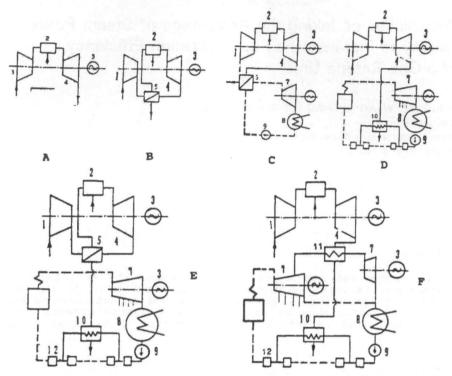

1.compressor   2.combustion   chamber   3.electric   generator
4.gas turbine   5.regenerator   6.waste heat boiler   7.steam
turbine   8.condensor   9.pump   10.gas cooler   11.steam
reheater   12.regenerative heater of feed water

FIGURE 1.   Possible Alternatives of Waste Heat Utilization

steam power plant it is preferably recommended to take the
advantage of the existing steam power facilities for waste
heat recovery, as shown in Fig.1 (D) through Fig.1 (F). In
Fig.1 (D) the waste heat is utilized in a gas cooler instead
of waste heat boiler. Figure 1 (E) is similiar to a normal
regenerative gas cycle except for an additionali gas cooler
mounted   among the regenerative heaters of the main steam
cycle. In Fig.1 (F), the exhaust gas from the turbine is led
into a steam reheater installed prior to the gas cooler.

In order to quantitatively evaluate the mentioned optional
systems under a variety of steam and gas conditions it is
imperative to work out a suitable methos for determining the
waste heat efficiency.

A GENERALIZED METHOD FOR DETERMINING THE EFFICIENCY OF WASTE
HEAT UTILIZATION

In this section it is proposed a generalized method for
determining the efficiency of waste heat utilization. Unlike

962

methods found in technical literatures this methos is simply enough and, in addition, the proposed method takes the main irreversibilities encountered in practice into consideration.

To simplify the analysis we start with an assumption that the drain water of the regenerative heaters of the main steam cycle is delivered to condenser via heaters of lower pressure in a cascade way without neither drain pumps nor drain coolers. This assumption will naturally influence the real efficency of the main steam cycle. It is, however, compensated by assuming an enlarged number of regenerative stages. Further more, we assume that the terminal temperature difference between the bled steam and feedwater as well as the feed water enthalpy rise for each regenerative heater are the same.

Under these assumptions the fractional amount of r-th bled steam with respect to unit mass flow rate of admission steam will be

$$\alpha_r = \Delta h (1 - \sum_{p=1}^{r-1} \alpha_p) / (h_r - h^L_r) \tag{1}$$

where Δh is feed water enthalpy rise in one regenerative heater, h_r ahd h^L_r are enthalpies of r-th bled steam and its condensate respectively. Let

$$\phi_r \equiv (h_1 - h^L_1) / (h_r - h^L_r)$$

which indicates the proportion of condensation heat between the 1-st (highest pressure) and r-th bleeding, then Eq.(1) becomes:

$$\alpha_r = \phi_r \alpha_1 (1 - \sum_{p=1}^{r-1} \alpha_p) \tag{2}$$

where ϕ_r is well close to unity. With the aid of equation (2) one can easily determine the value α_r for each regenerative stage starting with the first one.

The work output per kg of the main steam is

$$W_g = H (1 - \sum_{r=1}^{m} \alpha_r \mu_r) \eta_m \eta_g \psi \tag{3}$$

where

$$H \equiv h_o - h_g \tag{4}$$
$$\mu_r \equiv (h_r - h_g) / (h_o - h_g) \tag{5}$$

h_o, h_g, h_r are specific enthalpies at steam turbine inlet, at condenser inlet and at r-th extraction point respectively.

The work output of steam turbine will be increased by inserting a steam reheater and/or a gas cooler into its cycle.

The work increment due to replacing a number of regenerative heaters by gas cooler can be deduced on basis of energy balance:

$$\Delta W_{gc} = \sum_{r=a}^{b} \alpha_r (h_r - h_g) \tag{6}$$

where b and a denote respectively the highest and lowest order numbers of regenerative heaters being replaced. To facilitate its further use, Eq.(6) is modified to the form

963

$$\Delta W_{gc} = (h^L_a - h^L_{b+1}) - (h^L_a - h^L_{b \cdot 1})(h_g - h^L_b)/(q_s + h^L_s - h^L_b) \qquad (7)$$

where

$$q_s \equiv \sum_{r=a}^{b} a_r (h_r - h^L_r)/\sum_{r=a}^{b} a_r \qquad (8)$$

$$h^L_s \equiv \sum_{r=a}^{b} a_r h^L_r /\sum_{r=a}^{b} a_r \qquad (9)$$

represent the mean condensation heat of the replaced bled steam and mean specific enthalpy of their drains respectively. From Eq.(7) we obtain the final expression for the work increment by virtue of waste heat Q_{gc} recovered in the gas cooler:

$$\Delta W_{gc} = Q_{gc}[1 - (h_g - h^L_b)/(q_s + h^L_s - h^L_b)] \qquad (10)$$

The expression within the parenthesis may be conceived as the "thermal efficiency" of the recovered heat:

$$\eta_{gc} = 1 - (h_g - h^L_b)/(q_s + h^L_s - h^L_b) \qquad (11)$$

Similarly, the work increment by virtue of waste heat recovered in the steam reheater can be given by

$$\Delta W_{rk} = Q_{rk} \eta_{rk} \qquad (12)$$

where

$$\eta_{rk} = 1 - (h_k' - h_k)/(h_{rk}'' - h_{rk}') \qquad (13)$$

h_{rk}', h_{rk}'' are specific enthalpies of the steam at inlet and outlet of steam reheater respectively; h_k' is specific enthalpy of reheated steam at inlet of condenser.

EVALUATION AND ANALYSIS

By the aid of the derived equations it is not difficult to calculate the efficiencies of waste heat η_{gc}, η_{rk} and the fuel utilization efficiency of gas turbine unit η. It is obvious that

$$\eta = w/q \qquad (14)$$

$$w = w_t - w_c + \Delta w_{gc} + \Delta w_{rk} \qquad (15)$$

and the efficiency of the steam-gas combined cycle

$$\eta_{comb} = (W + m_s w_s)/(q + m_s q_s) \qquad (16)$$

where m_s is the mass ratio between steam and gas, w_s and q_s are respectively the specific work and specific fed heat of the steam cycle. The gas turbine work w_t and compressor work w_c can be computed with normal procedure. In the following we shall restrict ourselves to the calculation of η and w for alternatives A through F shown in Fig. 1.

The assumption regarding the calculation of gas cycle are as follows:

compressor intake temperature	$t_1 = 15\,°C$
gas turbine inlet temperature	$t_3 = 1100\,°C$
	$900\,°C$
compressor efficiency	0.85
gas turbine efficiency	0.86
relative pressure drop	
for combustion chamber	0.08
for regenerator	0.08
for steam reheater	0.08
for gas cooler	$0.04 \sim 0.08$
for waste heat boiler	0.12

terminal temperature difference in waste heat exchangers
$$\Delta t = 50\,°C$$

the temperature of gas at gas cooler outlet	$t_6 = 120°C$
fuel data heating value	LHV = 43124 kJ/kg
composition	$H_2 = 13.92\%$
	$C = 86.08\%$

The prerequisite for the calculation of waste heat efficiencies η_{gc} and η_{rh} is specifying the real expansion line of the steam cycle in Mollier Chart. Figure 3 represents the results of calculation for 100MW steam generating unit of national product. The throttle conditions are 88.2 bar 535°C and the condenser is under a pressure of 0.048 bar. Turbine efficiency is 0.87. Feedwater is heated by extraction steam up to 227 °C. The original cycle is without steam reheat.

It is understood that the value of efficency η_e depends greatly upon the interval in which the regenerative heaters are replaced by the gas cooler. In order to achieve a better efficency, it is desirable to arrange the cycle so that the feed water is heated in gas cooler to a temperature as high as possible, provided it is within the range below the rated value of feed water temperature. This is usually achieved by adjusting the flow rate of feedwater passing through the gas cooler to make a more homogeneuos temperature difference between the gas and the feedwater.

Figure 3 shows that the efficency of waste heat recovered in the steam reheater varies with the pressure and outlet temperature of steam reheater. For the sake of higher value of η_{rh} it is necessary to choose an adequate reheat pressure and adjust the flow rate of steam in a similar way as with gas cooler. It is in principle possible to let only a portion of the total amount of steam to pass through the reheater and

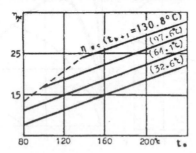

FIGURE 2   Thermal Efficiency of Waste Heat Recovered in Gas Cooler

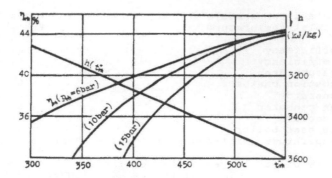

FIGURE 3. Thermal Efficiency of Waste Heat Recovered in Steam
Reheater

then expand to the back pressure in a separate turbine casing.

The results of calculation for alternatives A through F are
presented in Fig. 4 and Fig. 5. Some descriptive figures are
tabulated in Table 1.

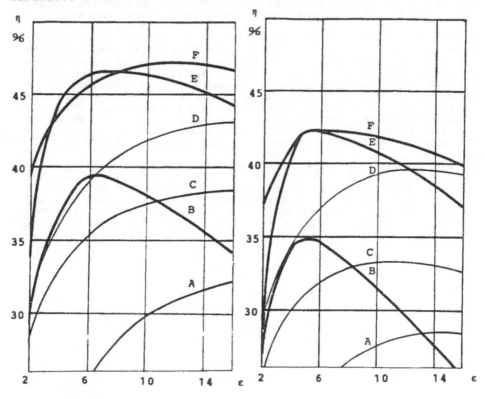

FIGURE 4   The Fuel Utilization Efficiency of Gas Turbine Unit
Shown in Figure 1

TABLE 1. Selected Figures From Calculation

		ε	t_4	t_5	t_6	q	q_{uc}	w	w_t	w_{uc}	η_a	η
A	$t_3=$	900	14	389			589		372	540			0.258
	$t_3=$	1100	16	486			807		397	657			0.322
B	$t_3=$	900	6	537	299	481	486		222	385			0.335
	$t_3=$	1100	6	681	328	598	602		222	452			0.384
C	$t_3=$	900	10	458	243		657		312	468			0.334
	$t_3=$	1100	16	506	261		807		404	635			0.384
D	$t_3=$	900	12	423			621	310	347	504	79.3	0.256	0.389
	$t_3=$	1100	16	499			807	388	404	643	99.4	0.256	0.428
	t3=	900	6	545	300	487	480	183	225	378	46.8	0.256	0.422
	$t_3=$	1100	8	630	355	566	638	239	272	505	61.3	0.256	0.467
F*	$t_3=$	900	6	545	277		748	159	225	378	40.8	0.256	0.423
	$t_3=$	1100	10	588	277		904	159	312	550	40.8	0.256	0.473

* The following figures for alternative F may also be of interest:

t_3	t_{rh}^*	w_{rh}	η_{rh}
900	487	119.1	0.430
1100	526	142.6	0.443

It is seen from the obtained figures that installation of a gas cooler among the regenerative heaters of the main steam cycle brings substantial increase both in thermal efficiency and specific work output. Alternatives E and F are compared each other in Fig.5.

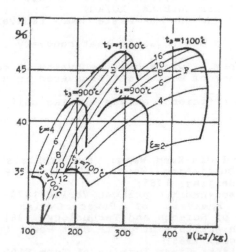

FIGURE 5. Comparison of η and w Between Alternatives E and F

CONCLUSIONS

1. The waste heat of a gas turbine unit can be utilized in a gas cooler installed among the regenerative feedwater heaters of a main steam cycle(see Figs. 1-D, 1-E). It can be utilized in a steam reheater as well (Fig. 1-F.)

2. In comparison with conventional ones (Fig. 1-A through C) these arrangements results in substantial improvments in thermal efficiency of the fuel fed to gas turbine unit by 11% to 26% under conditions encountered in current practice of China. The relative improvment in specific work output of unit gas flow rate passing through the turbine amounts up to 50%

3. The thermal efficiency of waste heat utilization in a steam cycle depends on the pressure, inlet and outlet temperature of the heated medium (see Fig.2 and Fig.3). To achieve a better efficiency it is desirable to adjust the flow rate of steam and/or feedwater passing through the waste heat exchangers. It is also important to have a steam power unit to match, suitable both in capacity, cycle arrangement and in throttle conditions.

NOMENCLATURE

t_3 gas turbine inlet temperature,$^{\circ}$C;
t_4 gas turbine outlet temperature,$^{\circ}$C;
t_5 gas temperature at outlet of waste heat exchangers(except for gas cooler)$^{\circ}$C;
t_6 air temperature at outlet of gas cycle regenerator$^{\circ}$C;
ε pressure ratio of compressor
q specific amount of heat fed to gas turbine unit, kJ/kg;
q_o specific waste heat recovered in gas cooler, kJ/kg;
$q_{r,h}$ specific waste heat recovered in steam reheater kJ/kg;
w_c specific work absorbed by compressor, kJ/kg;
w_r specific work of gas turbine, kJ/kg;
$w_{o,c}$ specific work gain due to waste heat recovery in gas cooler, kJ/kg;
$w_{r,h}$ specific work gain due to waste heat recovery in steam reheater, kJ/kg;
$\eta_{o,c}$ thermal efficiency of waste heat recovered in gas coller
$\eta_{r,h}$ thermal efficiency of waste heat recovered in steam reheater
η fuel utilization efficiency of gas turbine unit

REFERENCES

1.   Zhi-Ping Song and Jia-Xuan Wang: Fundamentals of Energy Conservation,   Publishing   House   of   Water   Resouces   & Electric Power, Beijing, 1985;
2.   Baehr, H.D, Thermodynamik, Springer-Verlag,1978;
3.   Da-Xie Chen: Analysis of Power Cycles,   Shanghai Publishing House of Science and Technology, 1981;
4.   Zhi-Ping Song: An Analysis of Waste Heat Power Cycle, J. of Electr. Engrg, 1985;
5.   Qingshan Power Plant: Steam Turbine of Type N100 - 90/535, 1979;
6.   Rui-Xian Cai: Thermodynamic Analysis of Gas-Steam Combined Cycle with Waste Heat Boiler, J. Engrg. Therm Phys.1981.

Characteristics Research of Spray Type Generator

WEN-HUI XIA, WEI-BIN MA, WEI-KANG CHEN, LI-YING LIN, and YUAN-QING XIA
Guangzhou Institute of Energy Conversion
Chinese Academy of Sciences
P.O. Box 1254
Guangzhou, PRC

ABSTRACT

In this paper, the influences of spraying density and concentration on heat and mass transfer in a spraying generator are studied. Heat transfer coefficient on solution side increases with spraying density, however, refrigerant liquid quantity does not increase in a linear way but with a point of inflexion. Heat transfer coefficient decreases with concentratoin. Owing to the physical properties of LiBr, a linear relation exhibits under 50% concentration while a non-linear relation exhibits over 50% concentration.

1. INTRODUCTION

To use waste heat for refrigeration is an important aspect in waste heat utilization. In Guangzhou Institute of Engergy Conversion, LiBr absorption refrigeration technology is used for recovery of heat from waste hot water delow 100 °C.

From the viewpoint of heat transfer, the main difference lies in the generation process in generator, in which heat transfer takes place at small temperature difference and small flow rate for hot water type chiller. In the case where 9 °C chilled water is required and a hot water temperature of 86 °C and a cooling water of 33 °C in summer are available, two stage absorption refrigeration has to be used. A two stage cycle involves generation process with two different concentrations and an intermediate pressure which connects the two stages. If heat and mass transfer could not reach a required level, the cycle would not be successful. Therefore, we carried out researches into heat and mass transfer in generation process.

In an absorption refrigeration cycle of hot water type, the key problem is the generator and generation process. At present, there are two basic configurations of generator, i.e. immerging type and spraying type. For generator of small heat transfer temperature difference of hot water type, the latter is preferred. Spraying types are not much researched and there is still no practical case of application. Scholars in Soviet Union carried out some researches in spraying type generator. But the used spraying density of g = 1000 - 3000 Kg/m.hr is difficult to attain in practical installations.

And they concluded only the influence of spraying density on heat transfer coefficient at liquid side outside the pipes. In fact, physical properties do have effects on heat and mass transfer (Ref.1), especially in heat transfer of small temperature difference. The change in liquid concentrations has to be taken into account in cases where such change almost reflects itself in various thermal properties of materials.

To solve design problems, the following research work is proposed.

a. The effect of concentration on heat transfer coeffibcient;

b. The effect of spraying density on heat transfer coefficient.

2. EXPERIMENTAL INSTALLATION (FIG.1)

2.1 System Flow

Thin solution is pumped by solution pump (2) via heat exchanger (4), where it is preheated to a higher temperature, to generattor (2). The concentrated solution after generation flow via heat exchanger where it is cooled, into absorber (3). On the other hand, steam generated in generator is condensed into chilled water in the condenser, flows to graduated cylinder (5) where it is measured, finally it is into chilled water tank (6) and is drawn by vacuum into the absorber. The circulation caused by solution pump (8) makes the solution and the chilled water mixed well and properly cooled bafore they enter the generator. Hot water which is heated by a electric heater, is put into a water trough and then pumped into the system. The whole system is operated under vacuum which is provided by a vacuum pump that operates intermittently to remove uncondensable gas and the hydrogen generated by corrision. Inlet temperature of solution to the generator is controlled by an electric heater mounted before the generator.

2.2 Principal Equipment

The generator is formed 12 by copper tubes of 1.2m long, ϕ16×1, arranged in 4 tiers of triangular pattern.

2.3 Measuring Instruments

Temperature: Cu-Const. thermocoupple, calibrated by

the province's Metrological Bureau.

Secondary meter: Readings were displayed and printed by PF-15 digital voltage meter, and also directly displayed and printed by 2240C digital meter of FLUKE, of the United States.

Flow rate :   Rotor type flowmeter. Accuracy was ensured by the possibility provided on the system to be calibrated at any time. PP11a Frequence Meter was used as secondary meter.

Pressure :   U shape mercury pressure difference guage.

Concentration :   Precission specific gravity meter.

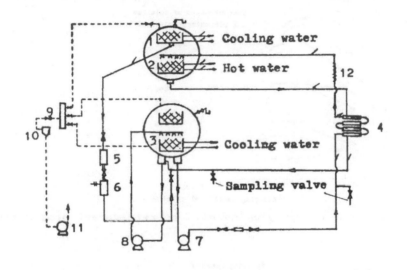

⊔   Mercury pressure difference guage

◄◄⊂⊃►► Flow meter   ╱ Cu-Const. thermocoupple

1. Condenser 2. Generator 3. Absorber 4. Heat exchanger
5. Measuring   cylinder   6.   Chilled   water   tank
7.8. Solutionpump   9. Vacuum drawing tube   10. Filter
11. Vacuum   pump   12.   Electric   heater
Fig.1 Experimental Installation

2.4 Experimental Parameters

Fixed parameters

```
            Inlet solution temperature :              49-51 °C
            Condensation pressure :                    55-60 mmHg

Variable parameters
            Inlet hot water temperature :              84 .90 °C
(hot water flow inside pipes was fixed,    unit heat flux    was
constant.)
            Concentration of solution :                46.4-55.4 %
                in weight, divided in 10 groups.
            Spraying density :                          50-120 Kg/m.hr,
                divided in 5 groups.
```

3. EXPERIMENTAL RESULT AND ANALYSIS

3.1 Measured Experimental Results are Shown in Fig.2,3 and 4.

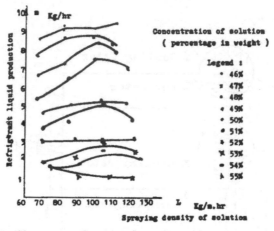

Fig.2 Influence of spraying density on refrigerant production

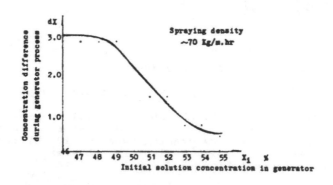

Fig.3 Relation between initial concentration in generator and
concentration difference, at spraying density 70 Kg/m.hr.

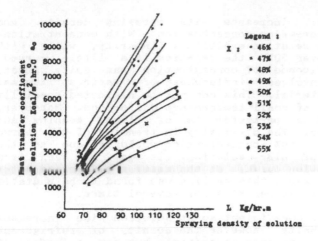

Legend :
X : • 46%
    • 47%
    • 48%
    • 49%
    • 50%
    • 51%
    ▲ 52%
    ¤ 53%
    • 54%
    ✦ 55%

L Kg/hr.m
Spraying density of solution

Fig.4 Relation between solution concentration, spraying density and heat transfer coefficient.

From Fig.4, it can be seen that the relation between spraying density, concentration and heat transfer coefficient at solution side is of some regularity and may be used to find an expression through multi-dimensional linear regression.

The expression thus found is :

$$ao = \sum_{n=0}^{7} An(2L)^n + (50-X)\sum_{n=0}^{7} Bn(2L)^n + (50-X)^2 \sum_{n=0}^{7} Cn(2L)^n$$

Where X -- concentration of solution (in percentage of weight)
       L -- spraying density of solution, Kg/m.hr
       ao -- heat transfer coefficient of solution, Kcal/m^2.hr. °C

The coefficients An, Bn and Cn are as follow:

Ao=-4714.675	Bo=6361.19	Co=771.1089
A1=161.9295	B1=227.0261	C1=24.01567
A2=2.467706	B2=2.170261	C2=9.740677E-02
A3=-1.974231E-02	B3=9.158921E-03	C3=6.722467E-03
A4=1.321519E-04	B4=2.099102E-04	C4=2.284698E-06
A5=1.312309E-06	B5=1.257971E-06	C5=9.444027E-07
A6=-1.593032E-08	B6=2.094749E-08	C6=-5.168091E-09
A7=2.138366E-11	B7=-1.551882E-10	C7=1.096349E-12

3.2 Analysis of Experimental Result

The relation between solution concentration, spraying density and heat transfer coefficient ao at solution side falls in some regularity: increases with spraying density and decreases with increasing concentration. With concentrations below 50%, the relation exhibits a linearity, while with concentrations over 50%, the relation is a little curved. However the influence of concentration was quite great. Concentration involves influences of various thermal properties of material. This can not be neglected specially for heat transfer of small temperature difference. Moreover, concentration as a parameter is of direct meaning and convenient for use. The diverisity in trending of the curve before and after 50% concentration would be due to the characteristics of LiBr solution. It may be seen from enthapy-concentration curve, at the section before and after 50%. During experiment, this section was found to be unstable and the experiment was repeated for several times.

The heat transfer coefficient ao at solution side increases with spraying density, however the quanity of refrigerant liquid does not increase accordingly but has a point of inflexion, most points of inflexion appear in the range of spraying density between 100 to 110 Kg/m.hr. On the other hand, the refrigerant liquid production decreases with increasing concentration. The principal causes are: (a) The boiling point increases with concentration. With the same heat source temperature, smaller temperature difference will be unfavourable to generation. (b) With increasing concentration, thermal properties will vary. Particularly, the increased viscosity will increase the resistance to heat and mass transfer.

A suitable increase in spraying density will enhence heat and mass transfer and this reflects itself in Fig.4. However when the spraying density is increased to certain level, the liquid film thickness thus thermal resistance and resistance to diffussion will also increase, being unfavourable to heat and mass transfer. Besides, an increase in spraying density means an increase in solution quantity, resulting in an increase in heating capacity required. Considering both the above favourable and unfavourable factor, there must be an optimun value (see Fig.2).

The relation between initial concentration and concentration difference during generation process: Concentration difference of solution in generator is influenced in a relatively large degree by initial concentration (Fig.3). The higher concentration, the stronger effect on mass transfer. This is what referred to in Chemical Engineering as time of contact. It is necessary to calculate the approximate number of required tube rows, in order to attain couppling between heat transfer and mass transfer. We had used this principle to determine the number of tube rows when a new installation was designed. After preliminary adjusting and test, the result was considered satisfactory.

4. CONCLUSIONS

Judging from the experimental results, it is neccessary to properly select spraying density rather than to possible higher value as usually considered it better to do so. Because once the refrigeration capacity and parameter are determined, the circulation quantity of solution is settled. To increase spraying density, one has to increase circulation ratio, resulting in reduced thermal coefficient; or to use self-recycling in generator, resulting in a higher concentration of circulation solution which will be unfavourable to heat and mass transfer. Therefore, proper selection is imperative.

Regarding the arrangment of tubes in spraying type generator, sufficient time of contact between solution and tube must be allowed for so that couppling of heat transfer and mass transfer is ensured. Besides, flowingg directions of hot water and solution should be opposite to each other as long as possilbe so that the temperature difference is fully exploited and the entering hot water can contact firstly with solution of higher concentration thus to maintain given motive force.

In cases where it is difficult to select sprayer because of too small refrigeration capacity, perforated tubes may be used. Relatively good result could also be got if the arrangement of perforated tubes is proper, the heat transfer tubes are well acid-pickled and inactivated, and good dampness is maintained.

Hot water type refrigerator is preferably to be used in making of industral cooling water (15-18°C). In this case, solution may be circulated in lower concentration, which is favourable to heat and mass transfer, and better result would be got.

Since generators are featured by heat transfer of small temperature difference, stress should be laid on desiging of solution heat exchanger. Temperature of thin solution entering the generator should be as high as possible ao as to minimize the load of generator.

REFERENCES

1. Dai Yong-Ching, Zheng Yu-Qing, LiBr Absorption Refrigerator, National Defence Industry Publisher, Beijing, June 1980. (in Chinese)

2. B.M. Pamm, Operation of Absorption in Chemical Industry, High Education Publisher, Beijing, 1957. (in Chinese)

Optimization of Heat Transfer in the Rectangular Fins with Variable Thermal Parameters

XIANG X. YANG
Hua Chiao University
Quanzhou, Fujian, PRC

C. L. D. HUANG
Kansas State University
Manhattan, Kansas 66506, USA

ABSTRACT

Circular fins are used extensively in heat exchanger devices to increase the heat transfer rate. For economic purposes, the optimization of heat transfer in a fin is needed to be studied. The optimization problems of fins can be classified either as the least volume fin for a given amount of heat dissipation or as the maximum heat dissipation fin for a given volume. In this study, the latter case is considered. First, the circular fins of several different profile with variable thermal parameters is investigated. The mathematical modes of heat transfer is determined for each of the following shapes of the fins: rectangular, trapezoidal, triangular, parabolic and hyperbolic. Second, the optimum dimensions of the rectangular fin with a constant volume, which yields the maximum heat dissipation, are determined by the method of invariant imbedding(a numerical schema of dynamic programming). The predominant modes of heat transfer considered in this study are conduction and convection. Linear variation of thermal conductivity and a power law for variation of the heat transfer coefficient are assumed. The results will yield a design guideline for engineers.

Nomenclature

b:	semi-thickness at the base of the fin
D_b:	dimensionless parameter, $D_b = r_b/(r_0 - r_b)$
D_0:	dimensionless parameter, $D_0 = r_b/(r_0 - r_b)$
f,g:	functionals of the derivatives of the independent variables X_1, X_2 and ξ
$h(r)$:	heat transfer coefficient
h_a:	average heat transfer coefficient
h_{nf}:	heat transfer coefficient at the wall in the absence of the fin
$H(r,m)$:	dimensionless heat transfer coefficient, $H(r,m) = h(r)/h_a$
$k(T)$:	thermal conductivity, $k(T) = k_a(1+\alpha\theta)$
k_a:	reference thermal conductivity
$K(L,m)$:	heat transfer coefficient variation parameter
L:	dimensionless length parameter, $L = r_0/r_b$
m:	index of the heat transfer coefficient variation
N:	dimensionless fin parameter, $N = \left(\dfrac{h_a(r_0-r_b)^2}{k_a \cdot b}\right)^{\frac{1}{2}}$
N_1:	dimensionless parameter, $N_1 = \delta/(r_0 - r_b)$
N_2:	dimensionless parameter, $N_2 = b/(r_0 - r_b)$

N_3:　　　dimensionless parameter, $N_3 = N_1+(N_2-N_1)(1-R)$

q:　　　heat transferred by the fin

q_{nf}:　　heat transferred in the absence of the fin

Q_h:　　dimensionless heat dissipation from the fin

r:　　　radial distance

r_b:　　inner(base) radius of the fin

r_o:　　outer (tip) radius of the fin

R:　　　dimensionless radius, $R=(r-r_b)/(r_o-r_b)$

t:　　　temperature of the fin

t_b:　　temperature at the base of the fin

t_ω:　　temperature of the surroundings

T:　　　relative temperature, $T=t-t_\omega$

T_b:　　relative temperature, $T_b=t_b-t_\omega$

T_{nf}:　　temperature on the wall surface in the absence of a fin

U:　　　dimensionless volume of the fin

V:　　　volume of the fin

w:　　　dimensionless parameter related to the thickness of the fin

X_1,X_2:　independent variable

y(r):　　y coordinate(semi-thickness) of the fin

α:　　　thermal conductivity variation parameter

δ:　　　semi-thickness at the tip of the fin

θ·　　dimensionless temperature, $\theta = T/T_b$

ξ:　　　dimensionless independent variable, $\xi = r/r_b$

Introduction

When the overall heat transfer rate is limited by a low rate of heat transfer between a solid surface and a surrounding fluid, extended surface, say fins, may often be used to improve the overall heat transfer rate. Circular fins have been used extensively in various heat exchange devices to increase the heat transfer rate. The fins are used to increase the area available for heat transfer with surroundings. However, the overall heat transfer rate is not a simple function of the area when fins are used. Therefore, an optimization problem for fins needs to be studied. In other words, it is to find the shape of which either minimizes the volume of the fin for a given amount of heat to be dissipated, or miximizes the heat dissipation for a given fin volume. In this study, only the latter case will be considered.

The fin optimization has been investigated by many researchers in the last decase[1-5]. More recently, Razelos and Imre[6] have obtained optimum dimensions for a circular fin of trapezoidal profile, with variable thermal conductivity and heat transfer coefficient. A Quasi-Newton algorithm is used to solve the set of non-linear differential equations. It is known that a set of non-linear differential equations of this type is not very stable. The numerical integration scheme must be executed very cautiously. Recently, Netrakanti, Huang and Yang[7,8] studied the same problem using the invariant imbedding principle, by which the inherent instability in a numerical integration technique can be avoided.

In consideration of the circular rectangular fins are most simple shape, and it is used extensively in practical engineering. Therefore, in this paper, the circular rectangular fin is studied with both the effect of variable thermal conductivity, as well as the effect of a variable heat transfer coefficient. First, circular fins of several different profile, such as rectangular, trapezoidal, triangular, parabolic and hyperbolic, with variable thermal parameters are investigated. Their mathematic modes of heat transfer are determined. Second, the optimum dimensions of the fin of rectangular profile are then determined. The predominant modes of heat transfer considered are conduction and convection. The effect of the radiation mode is assumed to be negligible. An invariant imbedding technique has been used in this study. The results obtained from this study will yield a design guideline for engineers.

The Mathematical Modes of Heat Transfer

Consider a circular fin of arbitrary profile, depicted schematically in Fig.1. The governing differential equation can be expressed as follows

$$\frac{d}{dr}\left(k(T)ry(r)\frac{dT}{dr}\right) - h(r)rT = 0 \tag{1}$$

The bounadry conditions are

$$T\Big|_{r=r_b} = \text{constant} = T_b \tag{2a}$$

$$\frac{dT}{dr}\Big|_{r=r_o} = 0, \text{ if } y(r_o) = \delta \neq 0 \tag{2b}$$

$$T\Big|_{r=r_o} = 0, \text{ if } y(r_o) = 0 \tag{2c}$$

Now, the mathematical modes of heat transfer for several fins of different shapes, such as fins of rectangular, trapezoidal, triangular, parabolic and hyperbolic profile, are listed below.

A. For a Fin of Rectangular Prifile
In this case, $y(r) = b = \delta$

$$\frac{d^2\theta}{dR^2} = \left(N^2 H(r,m)\theta - \alpha\left(\frac{d\theta}{dR}\right)^2 - \frac{(1+\alpha\theta)}{(R+D_b)}\frac{d\theta}{dR}\right)/(1+\alpha\theta) \tag{3}$$

B. For a Fin of Trapezoidal Profile
In this case, $y(r) = \delta + \left((b-\delta)(r_o-r)\right)/(r_o-r_b)$.

$$\frac{d^2\theta}{dR^2} = N^2 H(r,m)\theta N_2/\left(N_3(1+\alpha\theta)\right) + \left((N_2-N_1)/N_3\right.$$
$$\left. - 1/(R+D_b)\right)\frac{d\theta}{dR} - \alpha\left(\frac{d\theta}{dR}\right)^2/(1+\alpha\theta) \tag{4}$$

C. For a Fin of Triangular Profile
In this case, $y(r) = b(r_o-r)/(r_o-r_b)$.

979

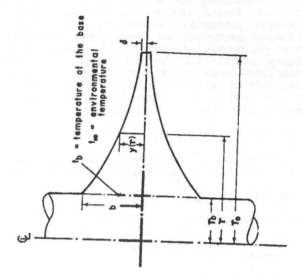

t_b = temperature at the base

t_∞ = environmental temperature

Fig. 1

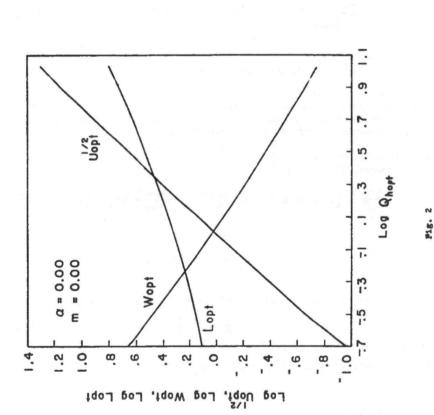

$\alpha = 0.00$

$m = 0.00$

$U_{opt}^{1/2}$

W_{opt}

L_{opt}

Log Q_{hopt}

Log $U_{opt}^{1/2}$, Log W_{opt}, Log L_{opt}

Fig. 2

$$\frac{d^2\theta}{dR^2} = \{ N^2 H(r,m)\theta/(1-R) - \alpha(\frac{d\theta}{dR})^2\}/(1+\alpha\theta)$$

$$+ \{1/(1-R) - 1/(R+D_b)\} \frac{d\theta}{dR} \qquad (5)$$

D. For a Fin of Parabolic Profile

In this case, $\quad y(r) = (b-\delta)(\frac{r_0 - r}{r_0 - r_b})^2 + \delta$.

$$\frac{d^2\theta}{dR^2} = \left\{ N^2 H(r,m)\theta N_2/[(N_2'-N_1)(1-R)^2 + N_1] - \alpha(\frac{d\theta}{dR})^2\right\}$$

$$/(1+\alpha\theta) + \left\{2(N_2-N_1)(1-R)/\right.$$

$$\left.[(N_2-N_1)(1-R)^2 + N_1] - 1/(R+D_b)\right\} \frac{d\theta}{dR} \qquad (6)$$

E. For a Fin of Hyperbolic Profile

In this case, $\quad y(r) = b \cdot r_b/r$

$$\frac{d^2\theta}{dR^2} = \left(N^2 H(r,m)\theta(R+D_b)/D_b - \alpha(\frac{d\theta}{dR})^2\right)/(1+\alpha\theta) \qquad (7)$$

Formulation of the Optimization Problem

The rectangular circular fin is the most practical one used in the heat exchange devices. Therefore, in this chapter, only the optimum dimensions of the circular rectangular fin will be determined. The optimum condition requires that following heat quantity must be maximum

$$q = -4\pi r_b \ b \ k(T) \ \frac{dT}{dr}\Big|_{r=r_b} \qquad (8)$$

The constraint condition requires that

$$q > q_{nf} \qquad (9)$$

$$q_{nf} = 4\pi r_b \ b \ h_{nf} \ T_{nf} \qquad (10)$$

The thermal conductivity is assumed to depend on temperature linearly, and the heat transfer coefficient is assumed to vary according to a power of a function of the radious of the fin.
They are

$$k(T) = k_a(1+\alpha\theta) \qquad (11)$$

$$h(r) = h_a H(r,m) \qquad (12)$$

respectively.

Now the governing differential equation(1) becomes as follows

$$\frac{d}{d\xi}\left((1+\alpha\theta)\xi \frac{d\theta}{d\xi}\right) - w^2 H(r,m)\xi\theta = 0 \qquad (13)$$

where w denotes a nondimensional parametric function as

981

$$w = r_b \left(\frac{h_a}{k_a \cdot b}\right)^{\frac{1}{2}} = w(b) \tag{14}$$

and $H(r,m)$ may be assumed in the following form (6)

$$H(r,m) = K(L,m)\left[(\xi-1)/(L-1)\right]^m \tag{15}$$

where K has been assumed as

$$K(L,m) = \frac{(L+1)(m+1)(m+2)}{2\{(m+1)L+1\}} \tag{16}$$

The boundary conditions (2) can be rewritten in the nondimensional forms

$$r = r_b, \quad \xi = 1, \quad \theta = 1 \tag{17a}$$

$$r = r_o, \quad \xi = L, \quad \frac{d\theta}{d\xi} = 0 \tag{17b}$$

The nondimensional volume U is defined as

$$U = \frac{k_a V}{4\pi r_b^4 h_a} = \frac{L^2-1}{2w^2} \tag{18}$$

where V is the given volume of the fin,

$$V = 2\pi b(r_o^2 - r_b^2) .$$

The nondimensional heat dissipation Q_h is defined as

$$Q_h = \frac{q}{4\pi r_b^2 h_a T_b} = \frac{-(1+\alpha)\theta'(1)}{w^2} \tag{19}$$

where $\theta'(1) = \left.\frac{d\theta}{d\xi}\right|_{\xi=1}$.

Method of Solution

The governing differential equation (13) is a second order nonlinear differential equation, which can be converted into two first order differential equations as follows

$$\frac{dX_1}{d\xi} = \frac{X_2}{\xi(1+\alpha X_1)} = f(X_1, X_2, \xi) \tag{20a}$$

982

$$\frac{dX_2}{d\xi} = w^2 H(r,m)\xi X_1 = g(X_1,X_2,\xi) \tag{20b}$$

where $X_1 = \theta$ and $X_2 = (1+\alpha X_1)\xi \frac{d\theta}{d\xi}$.

The boundary conditions are

$$X_1\Big|_{\xi=1} = 1 \quad \text{and} \quad X_2\Big|_{\xi=L} = 0 \tag{21}$$

Now applying an invariant imbedding technique, the governing differential equations(20) are solved by a computer. The functions f and g for this case are given as follows

$$f(X_1,X_2,\xi) = \frac{X_2}{\xi(1+\alpha X_1)} \tag{22a}$$

$$g(X_1,X_2,\xi) = \frac{(L^2-1)}{2U}\left\{\frac{(L+1)(m+1)(m+2)}{2(m+1)L + 2}\right\}(\frac{\xi-1}{L-1})^m \xi \cdot X_1 \tag{22b}$$

The boundary conditions (21) are converted to a generic form

$$X_1(a) = c \quad \text{and} \quad X_2(L) = 0 \tag{23}$$

where a denotes the starting value of the independent variable ξ $a \leq \xi \geq L$ By changing the starting value of a, the duration of the process is changed . Also c denotes the initial state of the process. The missing initial state of the process $X_2(a)$ is not only a function of the starting value of the process a, but also a function of the starting state c. Therefore, it can be expressed as

$$X_2(a) = r_m(c,a) \tag{24}$$

If a assumes different discreate values, say a =1,..., L-2Δ, L-Δ, L, where Δ is an incremental value, thus the original problem forms a set of similar problems with different durations. The equation, governing the invariant imbedding process for the missing initial condition r_m, can be shown as [9]

$$r_m(c,a) = r_m(c + \frac{r_m(c,a+\Delta)\Delta}{(1+\alpha c) a}, a +\Delta)$$

$$- \frac{(L^2-1)}{2U}\left\{\frac{(L+1)(m+1)(m+2)}{2(m+1)L + 2}\right\}(\frac{a-1}{L-1})^m \, a \, c \, \Delta \tag{25}$$

Equation(25) can be solved in a backward recursive fashion, by implementing the boundary condition(23)

$$r_m(c,a)\Big|_{a=L} = X_2(L) = 0 \tag{26}$$

The first step is to obtain the missing initial condition of the neighboring problem $r_m(c,L-\Delta)$, when a = L-Δ. In this study, the values of c used are 0, 0.1, 0.2,...,1.0, and the incremental value used isΔ =0.01. All the values of r for different values of c with a=L-Δ are obtained and stored in the computer memory as tabular form. These values are used in the next step of the neighboring process. This process is carried out in a backward fashion by reducing the value of a byΔ in each step, and will continue until a=1.0. The missing initial condition $r_m(1,1) = X_2(1)$ is of interest in the final result.

In the following computation, a value of the dimensionless volume U is selected first. The optimization process starts by assuming a value of the dimensionless length parameter L. Values of w^2 are calculated from equation(14). By an invariant imbedding technique, $X_2(1)$, representing the temperature gradient at the fin base, is obtained. The value $r_m(1,1)$ is used also in computing the total heat dissipation Q_h. Next, the value of L is incremented by Δ and Q_h is calculated again. The process is repeated until the maximum value of Q_{hopt} is reached. The corresponding optimum values of L and w are denoted as L_{opt} and w_{opt}, respectively.

Results and Conclusions

The computer program for this computation is used not only to obtain the optimum dimensions of the circular rectangular fin, but also to study the effects of the two pertinent parameters α and m on the optimum dimensions.

The maximum Q_{hopt} is obtained for values of U ranging between 0.01 and 400. The final results obtained for $U_{opt}^{\frac{1}{2}}$, w_{opt} and L_{opt} againest Q_{hopt} are poltted in Fig.2. It shows that as Q_{hopt} increases, the volume required $U_{opt}^{\frac{1}{2}}$ increases. Also w_{opt} decrease as Q_{hopt} increases. A decrease in w_{opt} implies an increase in optimum fin thickness b. The length parameter of the fin L_{opt} also increases as Q_{hopt} increase. However, for a given change in the heat dissipation Q_{hopt}, the magnitude change in w_{opt} is much more significant than the change in L_{opt}. Hence, the variation in the thickness of the fin plays an important role in the heat transfer process.

Fig.3 illustrates the effects of the conductivity parameter α on the parameter of the optimum fin length L_{opt} and the parameter of the optimum fin thickness B_{opt}, where U = 1.0 and m=0.0. It shows that as α increases, the parameter of the optimum fin length L_{opt} increases. However, the parameter of the optimum fin thickness B_{opt} decrease.

Fig.4 shows the effects of the index of the heat transfer coefficient variation m on the parameter of the optimum thickness B_{opt} while α =0.4. For equal value of heat dissipation, the optimum fin thickness increases, as m increases. Also, for the same value of the index m, the optimum fin thickness increases, as Q_{hopt} increases.

The results are expressed in terms of dimensionless parameters, and presented in the Fig. 2-4. These figures can be used for practical engineering design.

References

Brown, A., Optimum Dimensions of Uniform Annular Fins, International Journal of Heat and Mass Transfer, Vol.8, pp.655-662, 1965.

Maday,C.J., The Minimum Weight one-Dimensional Straight Cooling Fin, ASME Journal of Engineering for Industry, Vol.96, No.1,pp.161-165, 1974.

Guceri, S., and Maday,C.J., A Least Weight Circular Cooling Fin, ASME Journal of Engineering for Industry, Vol.97,No.1, pp.1190-1193, 1975.

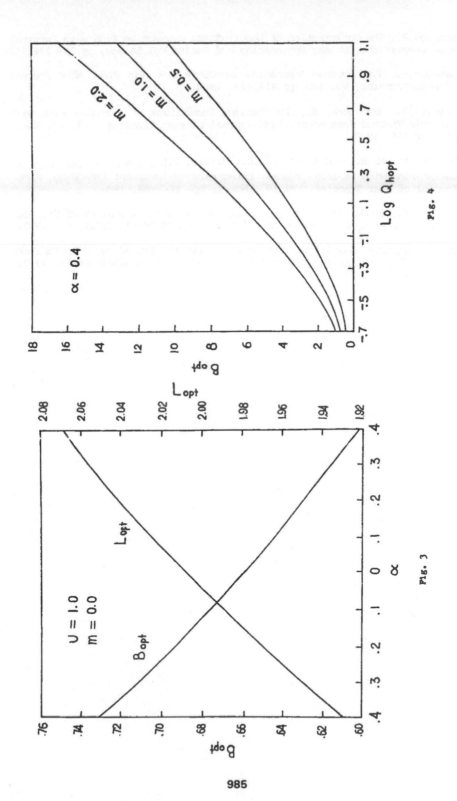

Fig. 4

Fig. 3

Razelos,P., The Optimization of Longitudinal Convective Fins with Internal Heat Generation, Nuclear Engineering and Design,Vol.54,No2, pp.289-299,1979.

Razelos,P., The Optimum Dimensions of Convective Pin Fins, ASME Journal of Heat Transfer, Vol.105, pp.411-413, 1983.

Razelos,P., and Imre, K., The Optimum Dimensions of Circular Fins with Variable Thermal Parameters, ASME Journal of Heat Transfer, Vol.102, No.3, pp. 420-425, 1980.

Netrakanti, M. N., and Huang, C.L.D., Miller, P.L., The Optimization of Annualar Fins with Variable Thermal Parameters by Invariant Imbedding, ASME Journal of Heat Transfer, Vol.107, No. 4, pp. 966-968, 1985.

Yang, X.X., Huang,C.L.D., Miller,P.L., The Optimum Dimensions of Circular Trapezoidal Fins, Proceedings of the Eleventh CANCAM'87. Vol.2, D-4.1987.

Lee, E. Stanley, Quasilinearization and Invariant Imbedding, With Applications to Chemical Engineering and Adoptive Control, Academic Press. 1968.

The Heat Transfer Analysis on Pneumatic Conveying Drying Process

CHUMING TANG and YOUSHENG YUAN
Research Institute of Thermoenergy Engineering
Southeast University
Nanjing, PRC

ABSTRACT

In this paper, a mathematical model including gas-solid two-phase flow and heat transfer is set up for pneumatic drying process in a pipe with a constant diameter based on practical operation and engineering design requirements, and numerical calculations are carried out. The computation results are consistent with the practical experiments and the parameter distributions obtained along the flow path can discribe the fundamental properties of drying process. Therefore, the authors consider this method is a more reasonable way for the design of drying pipes, and it can be taken as design bases for the improvement of drying performances and for the better utilization of thermal energy.

I. INTRODUCTION

Pneumatic conveying drying process is an effective method in drying particulate materials. Because the drying time is extremely short, the drying medium with high temperature will not cause the material overheated or burnt. Drying devices in this way can be manufactured conveniently and operated continuously, so this kind of dryer is used widely in power, chemical and foodstuff industries.

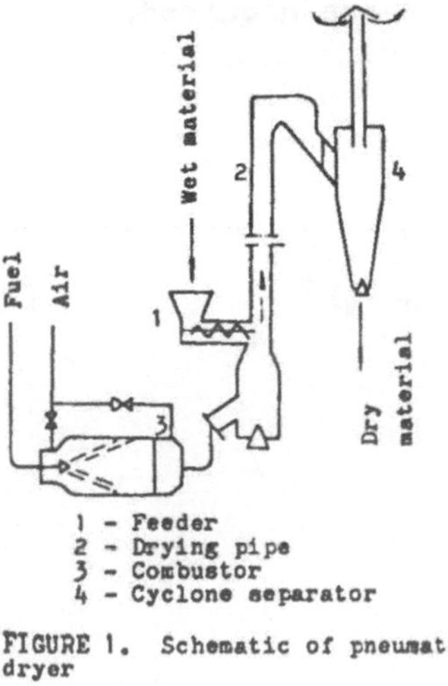

Figure 1 shows a pneumatic dryer schematically. High temperature gas is introduced into the bottom of pipe from the exit of a combustor. Heat needed for vaporization of moisture in material is provided by gas. The aqueous vapor diffuses into gas and is carried away with gas. Dried material is carried to a cyclone together with gas, then separated and collected there.

1 - Feeder
2 - Drying pipe
3 - Combustor
4 - Cyclone separator

FIGURE 1. Schematic of pneumatic dryer

The objectives of design and calculation of dryer are to determine the diameter and height of drying pipe according to moving processes of particulate material and the heat and mass transfer processes between particles and gas in the drying pipe. According to the equations of gas–solid two-phase flow and heat transfer, a mathematical model of pneumatic drying process in a constant diameter pipe has been established and the computed results have been analyzed in this paper.

II. EQUATIONS OF GAS–SOLID TWO–PHASE FLOW AND HEAT TRANSFER AND COMPUTED RESULTS

1. Coupled Equations

The particles are accelerated as soon as they are fed into pneumatic dryer. Because of higher relative velocity, large surface of heat transfer and large temperature differences between gas and particles, heat and mass transfer are intense and the drying process is mainly performed in this region. The velocity of particle increased with height of drying pipe. When relative velocity between particle and gas flow equals the terminal velocity of particle, the particles will move in uniform velocity. In uniform motion region, the relative velocity between gas and particle remains minimum. The temperature differences become smaller and smaller. And reduction of concentration of particles results in decrease of heat transfer surface. Therefore, the rate of heat and mass transfer become lower gradually.

The parameters of the gas and particle at every point along the stream (T_g, T_s, P, ρ_g, μ_g, U_g, U_s, F_s) are varying. All the above parameters obviously affect the heat transfer. For this reason, the energy equation, momentum equation, continuity equation of gas flow, motion equation of particles and heat transfer equation are involved and solved simultaneously in this paper. α, U_g, U_s and T_g at every point along the height of pipe are obtained. Finally the total height of pipe and the residence time of particles are acquired.

The coupled equations are as follows:

$$\frac{Ad(\rho_g U_g \varepsilon)}{dX} = 0 \tag{1}$$

$$\rho_g U_g \frac{dU_g}{dX} = -\frac{dP}{dX} - \frac{C_f}{2D}\rho_g U_g^2 - \frac{N}{8}\pi d_s^2 C_D \rho_g (U_g - U_s)^2 \tag{2}$$

$$\frac{dQ}{dX} = A\rho_g U_g \frac{d}{dX}(I + \frac{1}{2}U_g^2) \tag{3}$$

$$\rho_g = \frac{P}{R T_g} \tag{4}$$

$$\frac{dU_s}{d\tau} = \frac{3C_D \rho_g (U_g - U_s)^2}{4d_s \rho_s} - g - \frac{C_s}{2D}U_s^2 \tag{5}$$

$$dQ = \alpha F_s(T_s - T_g) + K_i \pi D \, dX(T_E - T_g) \tag{6}$$

The boundary conditions are expressed as follows:

When $X = 0$,

then $Q = 0$

         $U_g = U_{go} = \dfrac{4 m_g}{\pi D^2 \rho_g}$

         $T_g = T_{go}$

         $P = P_o$

         $U_s = U_{so}$

         $\tau = 0$

         $\dfrac{dU_s}{d\tau} = a_0 = \dfrac{3 C_D \rho_g (U_{go} - U_{so})^2}{4 d_s \rho_s} - g - \dfrac{C_s}{2D} U_{so}^2$

When $X = H$,

then $\left| \alpha \dfrac{6 m_g}{\rho_s d_s U_s} (T_s - T_g) \right| \geqslant Q_r$

To consider the influence of particles concentration, give $U_{so} = 3$ m/s. [1,3]

Because Nu changes with Re_r, α at every point along the length of accelerating region is variable. According to the results of experimental investigation given by Toei Liuzou [3], it can be calculated as follows:

$$N_u = b \, Re_r^k$$

where $k = \dfrac{\lg N_{uo} - \lg Nu_t}{\lg Re_o - \lg Re_t}$

         $b = 10^{(\lg N_{ut} - k \lg Re_t)}$

         $Re_r = d_s(U_g - U_s) \rho_g / \mu_g$

         $N_{ut} = 2 + 0.54 \, Re_t^{0.5}$

         $N_{uo} = \begin{cases} 0.76 \, Re_o^{0.65} & (30 < Re_o < 400) \\ 0.000095 \, Re_o^{2.15} & (400 < Re_o < 1300) \end{cases}$

         $Re_o = d_s(U_{go} - U_{so}) \rho_g / \mu_g$

         $Re_t = d_s U_t \rho_g / \mu_g$

The variation of particle temperature is shown in Figure 2. The temperature of particles rises from θ_0 to Tw in region I, remains constant at Tw in region II and rises from Tw to θe in region III. It can be calculated as follows:

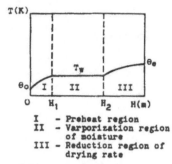

$$Q \leqslant qI \qquad T_s = \theta_0 + \frac{T_w - \theta_0}{H_1^{1/n}} X^{1/n}$$

$$qI < Q \leqslant (qI + qII) \qquad T_s = T_w$$

$$(qI + qII) < Q \leqslant Q_r \qquad T_s = T_w + \frac{\theta_e - T_w}{(H - H_2)^{1/n}} (X - H_2)^{1/n}$$

I   – Preheat region
II – Varporization region
      of moisture
III – Reduction region of
      drying rate

FIGURE 2. Variation of material
temperature with height of pipe

qI, qII and $qIII$ are heat required for rigion I, II and III. They are found by calculation of heat balance respectively.

The differential equations are solved by fourth order Runge-Kutta method, after the coupled equations are manipulated.

2. Results and Comparisons

(A) Computed results. As an example, a computation for crushed coal being dried in a pneumatic conveying dryer was carried out.

Knowns: material——crushed coal

G_s = 1200		C_m = 1.13	
C_1 = 27.10		θ_0 = 273	
C_2 = 14.5		θ_e = 353	
C_0 = 18		T_{go} = 673	
d_s = 0.002		T_{ge} = 393	
d_{max} = 0.006		d_0 = 0.025	
ρ_s = 1500			

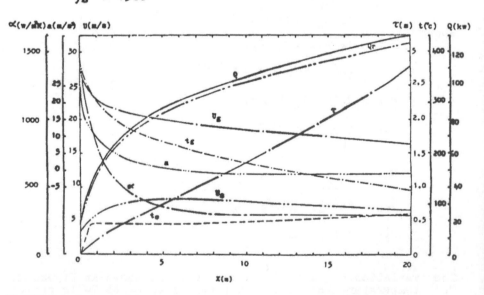

FIGURE 3. The parameter variation along the flow path

The results of computation are shown in Figure 3, indicating that, although total length of drying pipe is 19.9 meters, however, in the acceleration section which is 6.2 meters long, the heat and mass transfer is strong enough, with its α reaching from 1500 W/(m^{2}K) to 316.6 W/(m^{2}K), and its heat absorbed by the particle amounting for 71.6% of the whole heat, especially in the first three meters from the inlet of pipe, α is very high up to 454-1500 W/(m^{2}K), the heat absorbed by particles is about 58% of the total heat.

(B) Comparison this method with other results of calculation.

(a) Comparison with calculated results of Toei Liuzou:

Knowns:
G_s	= 3000		C_m	= 1.256
C_1	= 25		θ_o	= 293
C_2	= 0.3		θ_e	= 353
C_o	= 2		T_{go}	= 673
d_s	= 0.0002		T_{ge}	= 368
d_{max}	= 0.0005		d_o	= 0.025
ρ_s	= 2000			

Results:

	Toei Liuzou [3]	this method
H(m)	13.31	12.06
τ(s)	0.74	0.78

(b) Comparison with the formulas obtained from experiences and unlimited series development method respectively: The dried material and computative conditions are the same as (A). The results are:

	empirical formula [4]	unlimited series development method formula [2]	this method
H(m)	19	21.55	19.9
τ(s)		2.9	2.72

(c) Comparison with the measured data from a practical drying pipe: Based on measured data of drying pipe used in practice, the verifying computation for this existing drying pipe is carried out by means of this method. The computed height of pipe is in accordance with the height of the existing pipe approximately.

Knowns: material——crushed coal
G_s	= 500		C_m	= 1.13
C_1	= 7.5		θ_o	= 273
C_2	= 1.0		θ_e	= 353
C_o	= 2.0		T_{go}	= 463
d_s	= 0.00116		T_{ge}	= 373
d_{max}	= 0.005		d_o	= 0.025
ρ_s	= 1500			

Results:

	used drying pipe	this method
H(m)	14	13.4
$T(s)$		0.98

3. Analyses of the Main Factors Affecting the Drying Process

(A) Diameter of particles, dg. Figure 4 shows that α reduces strongly with increase in particle size and the specific surface area of coarse particles is less than that of fine particles. Thus, the rate of heat transfer along stream dQ/dx will become small (see Figure 5). Although the reduction of velocity of particles results in extension of its residence time, the pipe length is still increasing.

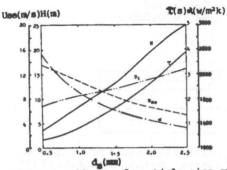

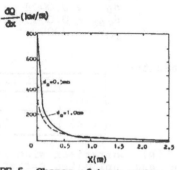

FIGURE 4. Effects of particle size on drying process

FIGURE 5. Change of heat exchanger rate along the flow path

(B) Initial velocity of gas, Ugo. Figure 6 shows the effects of initial velocity U_{go} of gas on parameters α, U_{se}, T, H, etc. With the increase of U_{go}, the increase of α is limited, while the increase in particle velocity U_s will lead to the reduction of heat transfer surface per unit volume in the drying pipe so that dQ/dx reduces rapidly, thus H will increase. From Figure 6, it can be seen that the pressure drop ΔP in drying pipe will increase with U_{go}. When U_{go} is

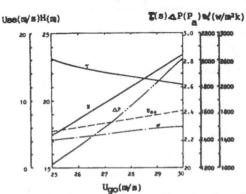

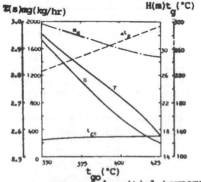

FIGURE 6. Effects of initial velocity of gas on drying process

FIGURE 7. Effects of initial temperature of gas on drying process

raised from 25m/s to 30m/s, H has to increase by 6 meters and ΔP increases by 80 mm H_2O.   For this reason, in the range of $U_{go} > U_t$ (U_t is the terminal velocity of the maximum   size particle) U_{go} should be adjusted as low as possible to minimize the power consumption.

(C) Initial temperature of gas, t_{go}. Increase in initial temperature of gas causes the increase in (T_g-T_s). Figure 7 shows that H and τ reduce with the increase in t_{go}. When t_{go} is increased, the flow rate of gas should be decreased correspondingly to ensure the gas temperature at the exit of pipe is not too high, otherwise, heat is wasted and the operation is not safe. It can be seen from Figure 7 that increase in t_{go} from 350°C to 400°C makes decrease in H from 31m to 20m.

III. CONCLUSIONS:

1. Since the computation results agree with the data obtained through practical drying operation and experience, the authors consider that this method can be used for the design of drying pipes in practice.   At the same time, the computed parameter distributions along the flow path are reasonable, thus providing bases for the selection of the structure sizes and operation parameters.

2. To reduce the height of drying pipes, the following measures can be adopted.

(A) Initial temperature of gas, t_{go} should be enhanced and gas flow rate reduced correspondingly, to keep the   exit temperature of gas constant.

(B) The proper velocity of gas, U_{go} should be chosen as low as possible on the premise of U_{go} greater than terminal velocity of maximum size particle.

NOMENCLATURE

A	gas stream cross-sectional area, m^2
a	acceleration of particle, m/s^2
C_D	drag coefficient,
C_f	friction coefficient between gas and wall of pipe,
C_s	friction coefficient between particles and wall of pipe,
C_p	isopiestic specific heat of gas, $J/(kg.K)$
C_o	critical moisture content in material, %
C_1	initial moisture of material, (dry basis) %
C_2	final moisture of material, (dry basis) %
C_m	specific heat of material, $kJ/(kg.K)$
D	diamter of drying pipe, m
d_{max}	maximum size of particle, m
d_s	mean diameter of particles, m
d_o	moisture of gas at pipe inlet, kg/kg
F_s	surface of particles contained in an infinitesimal volume $(\pi/4)D^2 \cdot dX$, m^2
G_s	mass flow of particles, kg/hr
H	total height of drying pipe, m
I	enthalpy of gas, J/kg

K_1 overall heat transfer coefficient between gas flow in pipe and environment, $W/(m^2K)$
m_s mass flow of particles, kg/s
m_g mass flow of gas, kg/s
N numbers of particle per unit volume in drying pipe, $1/m^3$
Nu_o Nusselt number at the entrance of pipe,
Nu_t Nusselt number at the end of the accelerating region,
n constant relating to drying characteristic of material, usually n=1—3,
P gas pressure, P_a
Q heat released by gas, W
Q_r total heat absorbed by material, W
Re_r Reynolds number based on (U_g-U_s),
Re_o Reynolds number at the inlet of the pipe,
Re_t Reynolds number depending on the terminal velocity of particles,
T_E ambient temperature, K
T_g temperature of gas, K
T_s surface temperature of particles, K
T_w wet bulb temperature of particles in region II. K
t_g temperature of gas, $^\circ C$
U_g gas velocity, m/s
U_s velocity of particles, m/s
U_t terminal velocity of particle, m/s
X an ordinate along the axis line of drying pipe, m
ΔP pressure drop between inlet and outlet of the drying pipe, Pa

GREEK SYMBOLS

α coefficient of convective heat transfer between gas and particles, $W/(m^2K)$
θ particle temperature, K
μ_g dynamic viscosity of gas, $kg/(m \cdot s)$
ρ_g density of gas, kg/m^3
ρ_s density of particle, kg/m^3
τ particle residence time in pipe, s
ε voidage,

SUBSCRIPTS

o   the parameters at the inlet of pipe
e   the parameters at the exit of pipe

REFERENCES

1. Shanghai Chemical Engineering Institute, Development of Drying Technology, Shanghai Scientific and Technical Information Institute, Volume 3, Shanghai, (1976).
2. Xia Chenyi, Exploration for Effect and Calculation Method of Acceleration Motion Stage of Particles in Pneumatic Dryer, Chemistry World, P. 373, Shanghai, (1965).
3. Toei LiuZou, Pneumatic Conveying Dryer, The Lecture of New Chemistry Engineering 10, Niikan Kougyo Shin-Ben-Sha. Tokyo, (1956).
4. Institute of Steel and Iron, Research Section of Coal Tar Chemistry, Coal Tar Chemical Symposium, Metallurgy Industry Publishing House, Volume 4, Beijing, (1959).

Shenzhen Solar Cooling and Hot Water Supply System

ZHI-CHENG HUANG, ZHEN-HONG ZHENG, and HAN-HAO HUANG
Guangzhou Institute of Energy Conversion
Chinese Academy of Sciences
Guangzhou, PRC

H. S. WARD, WAI-CHUNG WONG, CHUN-YING CHU, and T. C. HASSETT
Hong Kong City Polytechnics
Hong Kong

ABSTRACT

An integrated solar powered air conditioning and hot water supply system for a hotel in Shenzhen, China, is described. Some of the preliminary operational results obtained during a winter and summer are presented. Performance of the three types of medium temperature solar collector--evacuated tubes, vacuum heat pipes and flat plate collectors with a corrugated insulating film are compared. The energy saving effect of the solar system is discussed.

KEYWORDS

High performance collectors, Solar air conditioning, Solar system.

INTRODUCTION

There are a number of options available for the use of solar radiation to meet the needs of energy consumption of buildings. Because of the near coincidence of peak cooling loads with the available solar power, air conditioning is a particularly attractive application for solar energy. Furthermore, a combination of solar cooling and hot water supply systems make it possible to have a more desirable, year round operating system. An integrated solar powered air conditioning and hot water supply system has been designed and constructed to investigate the performance of this type of system in subtropical regions. This has involved a collaborative research project between China and Hong Kong. Construction was completed in November 1986 at Shenzhen, a Special Economic Zone of China (long. $113°17'$E, lat. $22°23'$N). The solar system has been operating in both its winter (hot water supply) and summer (cooling) modes. This paper outlines the main features of the system and presents some of the preliminary operational results.

DESIGN CONSIDERATION

Solar energy is used to supply cooling, heating or hot water in a small hotel. The system provides air conditioning to guest rooms, with a total area of 80 m^2, or can deliver 10 tons of hot water each day, for use in the five storey buiding. Medium temperature solar collector arrays are used with a total aperture of 116 m? Two 2-refrigeration ton LiBr absorption chillers and 5 m³storage tanks are other major components. A microcomputer, together with A/D-D/A converters and a range of instruments, control and monitor the performance of the system. An outline of the system is indicated in Fig. 1, and a view of the solar installation is shown in Fig. 2.

1. Solar Collectors and Solar Array

There are three types of medium temperature solar collector each developed as part of the project.

a. Evacuated glass tube collector

This consists of a finned copper tube, through which water passes, surrounded by a glass envelope maintained under a high vacuum (less than 10^{-4} torr). The surface of the absorber plate has a selective black nickel coat (absorptance α)=0.9, emittance ε <=0.15). Tests of prototypes of this design demonstrated a most satisfactory performance at temperatures just below 100 C.

The equation of the measured instantaneous efficiency curve of this collector is (on absorber area):

$$\eta = 0.94 - 2.76(T_f - T_a)/I - 14.73((T_f - T_a)/I_t)^2$$

b. Heat pipe vacuum collector

The second type of vacuum collector consists of finned heat pipe absorbers which heat water in the manifold of the collector. Manufacturing costs are reduced in comparision with the evacuated tube because there is a need for only one glass-to-metal seal. Both the evacuated tube and heat pipe collectors were manufactured by the Shenyang Lamp Factory in China to the design of the research team.

c. Flat plate collector with a corrugated insulating film

There are advantages to be gained by using a v-shaped corrugated transparent film beneath the glass cover of a tube-in-sheet flat plate collector. Through the proper choice of the film's properties, it is possible to suppress the convection between the cover and the absorber plate, thus reducing the collector's heat loss. As a consequence it is possible to provide a satisfactory performance of the collector at temperatures that are 50-60°C above ambient.

The equation of instantaneous efficiency curve of this collector is:

$$\eta = 0.71 - 5.18(T_f - T_a)/I_t$$

Where:

η --Instantaneous efficiency
T_f --Mean temperature of heat medium,°C
T_a --Ambient temperature,°C
I --Global solar radiation, W/m²

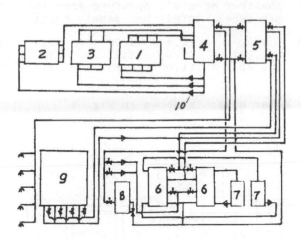

Fig. 1   Schematic of Shenzhen Solar System

1. Evacuated collector; 2. Heat pipe collector;
3. V-Flat plate; 4. hot storage; 5. Cold storage;
6. Absorption chiller; 7. Cooling tower;
8. Aux. boiler; 9; A/C room; 10. Pumps.

Fig. 2   A view of the Shenzhen solar installation

d. The solar array

Each of the three types of collector forms an independent
loop with the storage tank via a pump. This means that each

loop can be independently controlled and monitored. Essential data associated with the collectors are listed in Table 1.

TABLE 1.   Solar Array Data

Collector type	No.	Absorber area(m^2) per panel	total	Apreture area (m^2) per panel	total	Tilt
Evacuated tube	20	1.32	26.4	1.90	38	10°
Heat pipe type	21	1.287	27.0	1.81	38	15°
V-flate plate	22	1.82	41.0	1.82	41	10°

The layout of the solar array is shown in Fig. 3

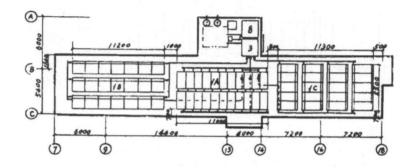

Fig. 3   Layout of solar array

1A. V-flat plate collector; 1B. Heat pipe collector;
1C. Evacuated tube collector; 3. Hot storage tank;
 8. Cold storage tank.

2. Cooling Components

Two 2-refrigeration ton Yasaki LiBr absoption chillers, model WFC-2, are used in the system. They are driven by the hot water from the solar collectors, providing this is within a range of 75-100°C. Depending on the temperature in the hot storage tank, the system is controlled   so that either one or both of the chillers are in operation. Chilled water is delivered from the cold storage tank to fan-coil units in four air conditioned rooms. Hot water can be circulated in winter to heat these rooms.

3. Energy Storage and Auxiliary Operation

A hot and a cold storage tank each with a volume of 5 m^3 allow for the continuous operation of the chillers and fan-coil units. A Yasaki boiler, model HH-2500s, which uses gas, has been installed to meet the cooling and heating demand when the solar heated water cannot operate the system.

4. Control and Data Acquisition

A 6502 microprocessor has been adapted to provide automatic
control and monitoring of the system. The control depends on
the required mode of operation and the status of the energy
in the system. Temperatures, flow rates and environmental
conditions can be sampled at any rate down to once every two
minutes. Software has been developed which allows on-line
manipulation and analysis of the measured data.

PRELIMINARY EVALUATION OF SYSTEM PERFORMANCE

The system has been available to the research team since
November 1986. In the period up to the end of 1987 it has
been possible to assess the performance of the system in
suppling hot water in the winter and cooling in the summer.
Typically measurements have been taken every 5 minutes during
the day and every hour at night. Global solar radiation on a
horizontal surface has been measured and these readings have
been integrated to give an estimate of the available energy
in any period of time. The first indications are that the
collectors and the system are behaving in a manner quite
close to that assumed in the design.

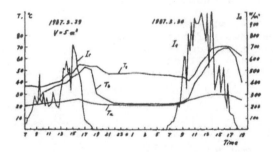

Fig. 4 Daily operation on a hazy and a clear day in winter

I_t -Solar radiation; T_t -Top temp. of storage;
T_b -Bottom temp.; T_a -Ambient temp.

1. Hot Water Supply Mode

Fig.4 provides a representation of the system's performance
on a hazy winter day. During the winter the daily average
efficiency of the collector system, based on aperture area,
is of the order of 40-60% depending on the radiation and
ambient temperature conditions. The solar threshold depends
on the type of collector and the required increase in
temperture above ambient level. If the final increase is of
the order $30°C$ the threshold value for the flat plate

collectors is around 175-250 w/m², whereas that of the evacuated tube is of the order of 100-175 w/m². This means that even on hazy days the system can provide hot water which is suitable for bathing for all of the hotel.

2. Cooling Mode

Fig. 5 and Fig. 6 represent the typical cooling performance of the system during the period of July and August, 1987. Additional relevent data for particulary days are listed in table 2.

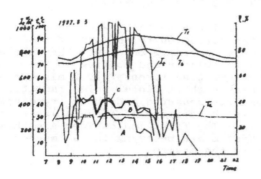

Fig.5   Daily performance of cooling operation

I$_t$ --Solar radiation; T$_a$--Ambient temp.
T$_t$,T$_b$ --top and bottom temp. of hot storage tank
A--Efficiency of flate plate collector
B--Efficiency of heat pipe collector
C--Efficiency of evacuated collector

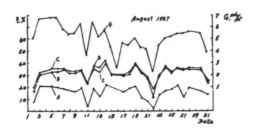

Fig. 6   Monthly performance of collectors
G-Daily global irradiance, kwh/m²·d
A,B,C-Daily efficiency of V-flat plate, heat
pipe and vacuum tube collectors

1000

Table 2   Global Energy conditions

Source	Design value	15 July	6 August	7 August
Daily global irradience wh/m^2	5980	6502	6641	5595
Daily heat collection(MJ)	478.7	777.5	878.3	705.5

The range of the efficiencies measured so far on fine days, when the temperature increase has been 50-60° C above ambient, are set out in Table 3.

Table 3   Daily collector efficiencies in the cooling mode

Criterion	V-flate plate	Heat pipe type	Evacuated tube
On aperture area	18-22%	33-42%	33-42%
On absorber area	18-22%	46-59%	47-60%

The two vacuum types of collector have a similar performance and have the advantage of being able to operate with a high efficiency at low solar threshold (250-300 w/m), but they are also expensive. The V-flat plate collector is good for the hot water supply mode, and when its far lower cost is taken into account this compensates for its poorer performance in the cooling mode.

For typical conditions of Shenzhen City (annual total global irradiance 5130 MJ/m^2, duration of bright sunshine equal to 2031 hrs), where people use electrical water heaters and air conditioners, the solar system has led to some significant savings in energy demand of around 55000 kwh/annum. One of the main objectives of design is to combine a high efficiency at temperatures around 90°C, with a minimum cost. This is likly to be achieved through a proper combination of heat pipe vacuum collectors and V-flat plate collectors.

CONCLUSION

1. The system has been operating in a satisfactory manner, close to that predicted by the design process. All three types of prototype collectors are suitable for providing water temperatures from 80-90°C, which can drive absorption chillers.

2. The greatest cost of the system is the price of the solar collectors, and one of the major initiatives that needs to be taken, before this type of system could be widely used, is to see if advanced manufacturing techniques might lower this cost.

3. There is scope for improving the overall efficiency of the system through the fine tuning of the control conditions and this will be the subject of future experimental work.

Thermal Performance Improvement of Flat Plate Solar Collector

ZHEN-HONG ZHENG and ZHI-CHENG HUANG
Guangzhou Institute of Energy Conversion
Chinese Academy of Sciences
P.O. Box 1254
Guangzhou, PRC

ABSTRACT

The flat plate solar collector with v-corrugated insulator (VFP) is a solar collector of good techno-economy. The results of performance investigation of VFP collector are presented in this paper. The results show that this type of solar collector can suppress convection heat loss, increase thermal efficiency. VFP collector has its own features, such as simple structure, high efficiency and low cost etc. So it has practical and spreading values.

KEYWORDS

Solar collector; Flat plate collector; V-corrugated insulator; Solar transmittance; convection heat loss; V-shaped film.

INTRODUCTION

A flat plate solar collector was improved by adding an internal V-shaped film between the glass cover and absorber. The V-shaped film can effectively suppress the convection heat transfer in the small gap between the glass cover and absorber , the convective heat loss of the collector is significantly reduced. At the same time, part of the radiation incident upon the V-shaped wall is reflected again to the absorbing surface, radiation loss is also reduced, as shown in Fig.1. Therefore, VFP collector can have higher thermal performance than the conventional flat plate collector (FP), especially at higher temperature.

The VFP collector has advantages of a higher solar transmittance, more effective convection suppressing ability and easy mass production. A thermoplastic forming technique is available for fabricating V-insulator, resulting the low cost of its fabrication. Polycarbonate and polyester were selected as the material of the V-film.

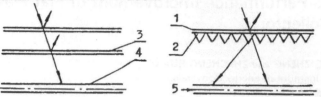

Fig. 1 Comparision of VFP and double glazing collector
        1- cover; 2-V-shaped film; 3-flat film;
        4-absorber; 5-working fluid.

The VFP collector consists of a glass cover, a internal anti-convection film and a tube-in-sheet type absorber, coated with a black paint (absorptance=0.92;emittance=0.65), as seen in Fig.2. As a type of high performance solar collector, VFP collector can meet specific requirements of heat supplying for homes, schools, hotels, hospitals and industrial applications.

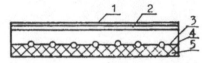

Fig. 2 Construction of VFP collector
        1-glass cover;2-V-shaped film;
        3-absorber; 4-housing; 5-insulation.

SOLAR TRANSMITTANCE

Solar transmittance measurements for single glass, single V-shaped film and glass/V-shaped film system were made under outdoor conditions. Table 1 shows the experimental values for above measurements and Fig.3 is a plot of effective solar transmittance versus angle of incidence of rays. For different conditions, effective solar transmittance can be given by:

tg=0.929 exp(- 0.099/cosθ)   (1)
tv=0.93 exp(-0.093/cosθ)   (2)
tg/v=0.804exp(-0.131/cosθ)   (3)

where tg--transmittance of single glass;
      tv--transmitance of single V-shaped film;
      tg/v--transmittance of glass/V-shaped film system;
      θ--angle of incidence.

Solar transmittance for the glass/V-film system is superior to the double glazing system. Futher studies are planed to provide the optimum dimensions of the V-shaped film. Now we use 10 mm as the width of the V-shaped and 53 °C as the opening angle.

TABLE 1   The Experimental Values of Transmittance.

Angle of inci. Item	0	7.5	10	15	22.5	30	40	50	67	71	90
Glass(3mm)	.84	.84	.84	.83	.83	.83	.82	.81	.73	.67	0
V-film(0.2mm)	.85	.85	.84	.84	.84	.83	.83	.825	.75	.69	0
Glass/V-film	.71	.71	.71	.705	.70	.69	.68	.64	.55	.51	0

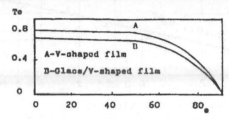

Fig. 3   Effective solar transmittance vs.
angle of incidence

CONVECTIVE HEAT TRANSFER

The heat transfer processes occurring in the conventional
flat plate solar collector are of particular importance to
solar engineers. In the present study, the free convective
heat transfer across an air layer bounded by a V-corrugated
plate and a flat plate was measured. Measurements
were performed over a range of Rayleigh number $10^4 <$ Ra $< 10^7$,
where Ra is based on the mean plate spacing L. Fig. 4 shows
the experimental results.Guided by the form of the equation
of inclined plane layer [4], the following equation was found
to be suitable over the range of 2500<Ra $\cos\theta < 10^6$

$$Nu = 0.157 (Ra \cos\theta)^{0.285} \tag{4}$$

The experimental points near curve 1 was from the present
study. In the case of the flat plate solar collector with V-
corrugated insulator with open angle 53°, experimental results can be given by

$$Nu = 0.1 Ra^{0.35} \tag{5}$$

Where   Nu=hL/k-Nusselt number;
        Ra=$L^3 \rho^2 \beta$ gCp(Th-Tc)/μk-Rayleigh number;
        h=coefficient of heat transfer across the air layer by
           convection and conduction;
        k=thermal conductivity of air;
        L=mean plate spacing;
        ρ=density of air;
        μ=dynamic viscosity of air;
        Th=temperature of hot plate;
        Tc=temperature of cold plate;
        β=coefficient of thermal expansion of air;
        Cp=specific heat at constant pressure;
        g=gravitational acceleration.

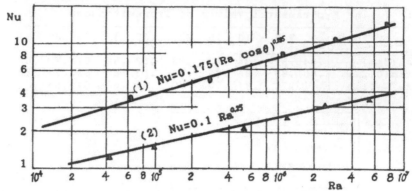

Fig. 4   Measurements of Nu as a function of Ra

The results from the present study showed that the V-shaped film system is superior to a two cover system in respect of convection suppression effect.

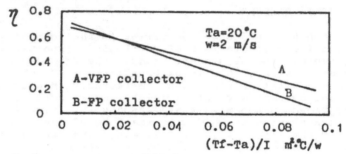

Fig. 5   Instantaneous efficiency curves

INSTANTANEOUS EFFICIENCY

In order to determine the respective collector performance, instantaneous efficiency tests were conducted by using the water loop. The solar radiation incidence was normal to the glass cover of the collector, the ambient temperature and wind velocity were fixed at 20 °C and 2.0 m/s, respectively, during the tests. Results of the performance tests are shown in Fig.5. The instantaneous efficiency equation of the collector were expressed as follows:

$$\eta_f = 0.74 - 6.74(Tf-Ta)/I \tag{6}$$
$$\eta_v = 0.71 - 5.18(Tf-Ta)/I \tag{7}$$

where η_f - instantaneous efficiency of FP collector;
        η_v - instantaneous efficiency of VFP collector;
        Tf- average temperature of heat medium (°C);
        Ta- ambient temperature (°C);
        I- solar radiation (w/m²).

The collector efficiency was mearsured to be 40%, when the global insolation, the ambient temperature and the water temperature was 700 w/m², 30°C and 80°C respectively during the tests. The efficiency of VFP collector was approximately 10% higher than that of the conventional flat plate collector.

TOP HEAT LOSS COEFFICIENTS

The top heat loss coefficients of the collector were measured. The ambient temperature and wind velocity were fixed at 20°C and 2.0 m/s, respectively, during the tests. Measured top heat loss coefficients of FP collector and VFP collector were shown in Fig.6. The top heat loss coefficients of the collectors were expressed as follows:

Uf=4.91+0.03 (Tf-Ta)   (8)
Uv=4.3 +0.01 (Tf-Ta)   (9)

where Uf- Top heat loss coefficient of FP collector (w/m².°c);
      Uv- Top heat loss coefficient of VFP collector (w/m².°c).

From Fig.6 we can see the use of the V-shaped film reduces the heat loss coefficient from 6.3 to 4.8 w/m².°c when the ambient temperature and the water temperature was 30°C and 80 °C, respectively. Therefore, the V-shaped film system is superior to a double glazing one in respect of convection suppresion effect.

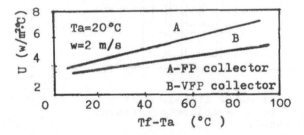

Fig. 6 Top heat loss coefficient

APPLICATION

VFP collector is particularly suitable to supply hot water of 70-100°C for solar air conditioning, boiling water supply and industrial process heat application. It may also be used to supply hot water of 40-60°C for large building, schools, hotels, dormitories, hospitals and restaurants. Up to now ,1000 m² heat collecting area of this kind of collector has been built and put into use in south China. Fig.7 shows a solar hot water system using VFP collectors. It is installed at Baiteng Lake Tourist Hotel near Zhuhai city, south China. It has been operating continuously since August 1986. It consists of 300 m² of VFP collectors arranged in 6 loops. Each row is connected with 5 collectors in series. All collectors were installed on the roof of the hotel. A hot water storage tank with electrical heaters, a circulating pump and a individual control system were connected to the collector array. The pump of the collector array starts automatically when there is a 3°C temperature difference between the inlet and outlet temperature of VFP collector array. When the temperature difference drops to 0.5°C, the circulating pump stops.

CONCLUSIONS

The flat plate solar collector with V-corrugated insulator can suppress the convection heat loss, increasing heat collection efficiency. At normal conditions, the thermal efficiency with a V-shaped film was measured to be 10% higher than that of the conventional flat plate collector. So it has practical and spreading values.

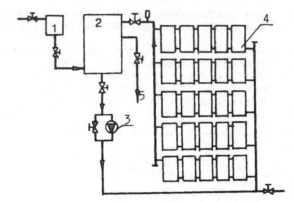

Fig. 7 Solar hot water system
1-cistern; 2-hot water storage tank;
2-circulating pump; 4-collector arrays;
5-hot water supply.

REFERENCES

1. Toshihiro Ishibashi and Masaharu Ishida, Improved flat plate solar collector with V-corrugated transparent insulator, Solar World Forum,p.198-202, 1981.
2. B.E.Sibbitt and K.G.T.Hollands, Radiant transmittance of V-corrugated transparents sheets with applications to solar collectors, ASME paper No.76-WA/SOL-1, American Society of Mechanical Engineers, New York, 1976.
3. J.G.Symons, Calculation of the transmittance-absorptance product for flat plate collectors with convectoin suppression devices, Solar Energy, Vol.33,No.6, p.637-640,1984.
4. K.G.T. Hollands, T.E. Unny, Heat Transfer, Vol. 98, P. 189-193, 1976.

Theoretical Derivation and Experiment of Gas-Solid Ejector for Energy Saving

KE-HUI GUO, DE-LIN MIAO, and WEN WU
Guangzhou Institute of Energy Conversion
Chinese Academy of Sciences
Guangzhou, PRC

1. INTRODUCTION

One of the principal shortcomings of fluidized bed boiler is the high carbon content in the ash at the exit of it. An effective measure which is used to improve efficiency of this type of boiler is to collect the exit ash and send it back to the boiler for reburning. As a transporting device, gas-solid ejector has many advantages: (1) no moving part, (2) vacuum transportation may be carrid out in a low or in a restricted space, thus facilitating the adding of the reburning system to the boiler, (3) ash may be delivered in a continuous mode without disturbing the boiler's operation.

Since the theoretical research of the gas-solid particles two phase flow, especially two phase flow inside a gas-solid particles ejector had not be well done in the previous stage, empirical formulas have to be used on this type of ejectors. Although in reference [1], a series of equations had been derived for ejector using gas to eject solid particles, the equations obtained may only be used in the circumstance where the pressure change is not great because incompressible flow relations were used for the ejected fluid and mixing fluid in the derivation. Moreover, in the formulas in ref.[1], the volumatric factor of solid particles was neglected and a heat equilibrium was assumed, thus reverse effect would at no time be incurred when they are used on ejection of hot ash from fluidized bed boiler.

In this paper, equations for ejection of solid particles by gas and the dimensions of device were derived basing directly on the basic two phase flow theory and some numerical solution results. In this derivation, the previous shortcomings were overcome. Experiment proved that the equations obtained were reliable. On this base, a universal computer program was compiled for ejection system reburning ash from fluidized bed boiler.

2. THEORETICAL DERIVATION AND NUMERICAL SOLUTION

2.1 Attainable Ejection Coefficient

Apply momentum equation at inlet and outlet of the mixing pipe, i.e. section 2-2 and 3-3 as shown in Fig.1, following equation may be got

$$\phi_2(G_p U_{p2} + G_a U_{a2}) - (G_a + G_p)U_3 = A_{p2}(P_3 - P_{a2}) + A_{p2}(P_3 - P_{p2}) \qquad (1)$$

where ϕ_2 is the velocity coefficient of mixing pipe. The subscripts indicate

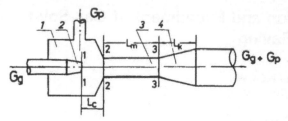

1. Mixing chamber, 2. Nozzle, 3. Mixing pipe, 4. Diffuser pipe.

FIGURE 1. Gas-solid ejector

parameters at different sections, with the first subscripts indicating the phase and second one indicating the number of section. It is assumed that before the mixing pipe i.e. in the length between section 1-1 which coincides the nozzle exit plane and the inlet plane of mixing pipe, $A_{a2}=A_{a1}$, $U_{a2}=U_{a1}$, under the design condition of ejector, $P_{a2}=P_{a1}=P_a$. Since the velocities at the inlet and outlet of ejector are low in comparison with velocity inside the mixing pipe, they are neglected. When velocity coefficients are used to correct the entropic inequality processes of different sections, we have:

$$U_{a2} = U_{a1} = \phi_1 a_{a1} \lambda_{a1} \tag{2}$$

$$U_{p2} = \phi_4 a_{p_a} \lambda_{p2} \tag{3}$$

$$U_3 = a_a \lambda_3 / \phi_3 \tag{4}$$

where a_a is the critical velocity, $\phi = U/a_a$ is the isentropic velocity, and ϕ_1, ϕ_2, ϕ_3 are the velocity coefficient of nozzle, inlet of mixing pipe and diffuser pipe respectively. According to equation of continuity, the cross section area occupied by fluid may be expressed as:

$$A_{a2} = G_a a_a / (k_a \Pi_a P_a Q_{a1}) \tag{5}$$

where Π is the ratio of pressure at a given plane to that at the stagnation plane, Q is the ratio of cross area of critical plane to that of a given plane, it is called equavalent mass velocity. Similarly, for the mixed fluid in diffuser pipe, the following equation may be obtained:

$$A_3 = G_a (1 + N) a_a (1 - Z) / (\Gamma \Pi_a P_c Q_3) \tag{6}$$

$N = G_p / G_a$ is called the ejection coefficient of ejector, Z is the volumatic coefficient of solid particles in two phase flow, Γ is the process index of two phase flow [2], when Z=0, equation (6) will reduce back to equation [5], i.e. to the single phase flow.

Combine and solve equation (1) to (6), taking consideration of $A_{p2}=A_3-A_{a2}$, ejection coefficient equation is obtained:

$$N = \frac{G_p}{G_a} = \frac{a_a \lambda_3 / a_{a_a} - K_1 \lambda_{a1} + K_3 + K_4}{K_2 \lambda_{a2} a_{p_a} / a_{a_a} - a_a \lambda_3 / a_{a_a} - K_4} \tag{7}$$

where

$$K_1 = \phi_1 \phi_2 \phi_3$$

$$K_2 = \phi_2 \phi_3 \phi_4$$

$$K_3 = P_p(\Pi_{p2}-1)\phi_3/(P_s k_s \Pi_{s4} Q_{s1})$$

$$K_4 = \frac{(\Pi_3 - \Pi_{p2}P_p/P_c)\phi_{3}a_4(1 - Z)}{\Gamma \Pi_4 Q_3 a_{s4}}$$

When equation [7] is used for calculation, since the parameters λ_{s2}, K_4, etc, are the functions of ejection coefficient N, it is necessary to calculation repeatedly before the attainable ejection coefficient is obtained. Fig. 2 shows the curve of N — $(P_c-P_p)/P_p$ calcucated by computer. It can be seen from the figure.2 that the ejection coefficient N decrease corresponding to lower outlet pressure ratio P_s/P_p.

2.2 Characteristic Equation of Ejector

Equation governing ejection coefficient of an ejector of given dimensions and the external parameters of the interacting fluids is called characteristic equation, these are equations (2) to (5) plus equation

$$P_{s2} = \Pi_{s2}P_s, \quad P_{p2} = \Pi_{p2}P_p, \quad P_3 = \Pi_3 P_c$$

solve these six equations, we have

$$N = \frac{G_p}{G_s} = \frac{K_5 - K_1\lambda_{s1} + \lambda_3 a_s/a_{s4}}{K_2\lambda_{p2}a_{p4}/a_{s4} - \lambda_3 a_s/a_{s4}} \tag{8}$$

where

$$K_5 = \frac{P_c\Pi_3/P_p - \Pi_{s2}P_s A_{s2}/(A_3 - P_p) - \Pi_{p2}A_{p2}/A_3}{k_s\Pi_{s4}A_{s4}P_s/(\phi_3 A_3 P_p)} \tag{9}$$

Fig. 3 is the characteristic curves obtained by numerical solution.

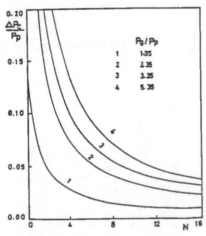

FIGURE 2. Attainable ejection coefficient curves

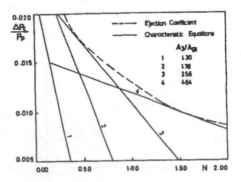

FIGURE 3. Characteristic equation curves

2.3 Factors Influencing Ejection Coefficient

<u>Temperature.</u> Fig.4 represents the influence of working fluid temperature T_a on ejector coefficient. From the figure it can be seen that N increases with the working fluid temperature increasing, this is because that when the energy level of working fluid increases, the same effect on ejected fluid may be achieved by fewer working fluid flow rate G_a, this is to say, higher energy level results in higher ejection coefficient N. Similarly, to have higher N, it is helpful to reduce the temperature of ejected fluid.

<u>Volumatric coefficient Z of solid particles.</u> As is mentioned in ref.[4], the influence of volumatric coefficient Z on parameters of mixed fluid will become stronger with the increase of ejection coefficient, i.e. loading ratio N. In Fig.5 it can be seen that with the increase of N, the mass concentration of solid particles also increases, and Z increases, the difference in N between is increased from $\triangle N=0.1$ at beginning to $\triangle N=0.8$ at the point where N is 12 whether Z is taken into account or not.

3. EXPERIMENTAL RESEARCH

3.1 Experiment Installation

Fig. 6 is the scheme of the experimental installation. the driving gas coming from a blower flows through adjusting valve (2) where it is adjusted in suitable flow rate which is measured by flowmeter (3). The gas from the nozzle (7) where the potential energy of it is changed into kinetic energy, flows into mixing chamber (4) where it draws the solid particles coming from loading funnel (6) into the mixing pipe and thus momentum exchange is carried out. At exit of mixing pipe the two phase fluids reach approximately dynamic equilibrium and then flow through diffuser pipe (9) to attain outlet pressure P_c. After a two stage separation in settling type separator (11) and cyclone type separator (12), the solid particles flow out from the flag gate, and the gas is drawn out by a draught fan. Then a cycle is completed.

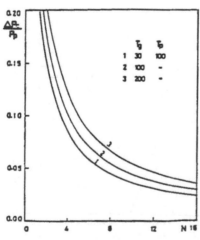

FIGURE 4. Temperature influence on ejection coefficient

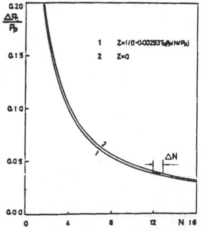

FIGURE 5. Volumatric coefficient Z influence on ejection coefficient

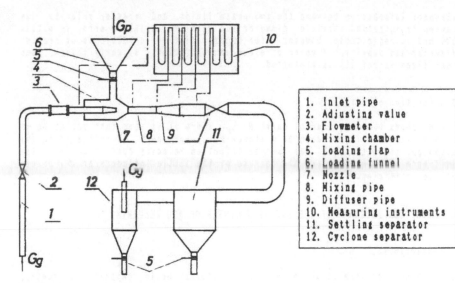

	1. Inlet pipe
	2. Adjusting value
	3. Flowmeter
	4. Mixing chamber
	5. Loading flap
	6. Loading funnel
	7. Nozzle
	8. Mixing pipe
	9. Diffuser pipe
	10. Measuring instruments
	11. Settling separator
	12. Cyclone separator

FIGURE 6. Experiment installation

Stop-watch is used to measure the flow rate of solid particles G_p in a given time interval. After mass flow of gas rate G_a is read out on the flowmeter, ejection coefficient $N=G_p/G_a$ is calculated. change the outlet pressure P_c by adjusting exit valve (2) of diffuser pipe and measure ejection coefficients at different P_c to obtain the characteristic curve of ejector. Ejectors of different dimension and their charateristic curves are obtained by changing nozzle (7) thus changing the ratio of cross section area A_{a1}/A_3. Gas velocity and pressure are measured by pitot tubes. Nozzle (7) can be turned back and forth so that the distance L_c between nozzle exit plane and mixing pipe inlet plane can be changed to obtain ejection coefficient for different L_c.

Boiler ash and sand were used in experiments as solid particles fluid. The average particle diameter was calculated by summing formula:

$$d_p = \sum_{i=1} X_i d_i$$

where X_i is the percentage in weight of solid particles with diameter d_i. Standard Tilot sieve was used to grade the solid particles.

3.2 The Checking of the Characteristic Equation

Fig.7 shows the characteristic equation curves for gas-solid particles ejectors with three different cross section ratios. The dash line in the figure was obtained by equation (8). The fact the experimental points fell around the calculation curve indicates that equation (8) is reliable when are used for calculation of ejectort. The trend of variation for experimental points is similar to that of calculated curve for given velocity coefficient, which indicates that velocity coefficients for different pipe section of ejector has little variation even in off-design conditions. With the increase of diameter and density of solid particles, the value of ejection coefficients obtained by experiment decreases a little, as indicated by points '×' and '△'. This is caused by the somewhat decrease in velocity coefficient due to stronger interaction between the two phase fluids and greater velocity lag caused by increased particle diameter and density. Relative error is within 5% and is neglectable. However, for transporting solid particles of greater diameter and density, a value of ejection coefficient slightly lower than that given in ref.[1] is preferred.

3.3 Position of Nozzle

Fig.8 shows the experimental curve $N-L_c$, where it is seen that variation of N in a relative large range of distance L_c is not great. As indicated in [5] althought the distance L_c has strong effect on velocity coefficient ϕ_4, the influence of L_c on N is not big because of the little influence of ϕ_4 on ejection coefficient.

4. COMPUTER PROGRAM FOR THE EJECTION SYSTEM OF ASH REBURNING

4.1 Description

The computer program is universal. For different boiler capacity, parameter, working conditions and the conditions for ash reburning, a single-stage or multi-stage system may be used. In single-stage system, equations given in ref.[5] are used. In multi-stage system, subsonic or supersonic ejection may be used depending on the condition of driving fluid. In the case of supersonic ejection, the losses due to shock wave is corrected by velocity coefficient. For any a boiler, after the known original data are input into computer, the computer will give out the dimensions of ejector for given condition and the list of ejector's performance value at off-design conditions.

4.2 Example of Calculation

Table 1 is the calculation results of second stage ejector in a tandem and a parallel system under cold condition. "Tandem" here means that the driving fluid in the first stage is still used as driving fluid in the second stage,

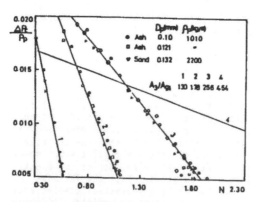

FIGURE 7. Experiment curves of characteristic equation

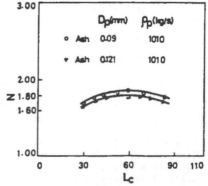

FIGURE 8. L_c influence on ejection coefficient

"Parallel" means that the driving fluid in the first stage turns into the driven fluid in the second. As is shown in the Table 1, ejection coefficient N_a or N of the ejector attainable in design condision is firstly calculated, then the dimensions of ejector to realize this ejection coefficient is found and finaly, the list of off-design performance data for this ejector is given. D in the Table respresents diameter of pipe section, $PY=(P_c-P_p)/P_p$, L_c L_m and L_k are the distance between the nozzle and mixing pipe, and the length of mixing and diffuser pipe respectively.

5   CONCLUSIONS

(1) The calculation formulas of gas solid ejector obtained in this paper are of advantages for use in high pressure ratio and non-heat equilibrium condition in comparison with those obtained in ref.[1], and are convenient for numerical solution by computer.

(2) To obtain higher ejection coefficient N, it is helpful to increase driving fluid temperature or decrease driven fluid temperature. In the case of large ejection coefficient, the neglect of volumatric percentage of solid particles will lead to large error in ejection coefficient.

(3) Experiment shows that characteristic equation is valid. For ejectors of given dimensions, velocity coefficient for different section may be taken as constant. The distance between nozzle exit plane to mixing pipe inlet plane, i.e. L_c does not have strong effect on ejection coefficient N.

(4) The computer program of ejection system for reburning boiler ash proposed in this paper can be used in multi-stage ejection thus is of universal use for different type of boiler.

NOMENCLATURE

A   Surface area   (mm)
a   Speed of sound   (m/s)
D   Pipe diameter   (mm)
d_p Solid particle diameter (mm)
Γ Process index of mixture fluid
k   Adiabatic index
L_c Ditance from nozzle exit to mixing pipe inlet   (mm)

TABLE 1.   Numerical solution of second stage gas-solid ejector

Tandem system	N_a=1.79, D_1=0.01, D_3=0.04, L_c=0.06, L_m=0.45, L_k=0.31 (m)								
	N_a	2.39	2.29	2.18	2.06	1.94	1.79	1.68	1.39	1.24
	G_2	278	264	251	237	224	248	193	160	148
	PY	0.02	0.03	0.04	0.05	0.06	0.07	0.08	0.09	0.10
Parallel system	N=2.14, D_1=0.01, D_3=0.04, L_c=0.06, L_m=0.42, L_k=0.289 (m)								
	N	4.35	4.14	3.92	3.33	3.20	2.97	2.73	2.44	2.14
	N_p	2.89	2.76	2.61	2.22	2.13	1.98	1.83	1.64	1.43
	N_a	1.46	1.38	1.31	1.11	1.07	0.99	0.90	0.80	0.71
	PY	0.02	0.03	0.04	0.05	0.06	0.07	0.08	0.09	0.10

L_m   Length of mixing pipe (mm)
L_k   Length of diffuser pipe (mm)
λ   Equivalent isoentropic velocity
N   Ejection coefficient
Z   Volumatric coefficient of solid particles
Π   Pressure ratio
φ   Velocity coefficient

SUBSCRIPTS

g   Gas
p   Solid particle
*   Stagnation
c   Ejector exit

REFERENCES

1.   A.Y.Socolov,etc.,Ejector,1977.

2.   Shih-I Pai, Two-Phase Flow, Frider.Viewes, 1977.

3.   Ke-Hui Guo, Wen Wu, De-Lin Miao, Numerical solusion of equations of gas
     and solid particles mixture for one-dimentional and stable flow, Rept.
     GIEC China, 1985.10.

4.   George Rudinger, Some effects of finite particle volume on dynamics of
     gas-solid mixture, AIAA Jour, Vol.3. 1965, PP 1212-1222.

5.   Ke-Hui Guo, Wen wu, De-Lin Miao, Isoentropic retative equations of gas
     and solid particles mixture, 5th.Conf.Eng.thermophysics.China (in Chinese),
     1985.10

ABSTRACTS

Abstracts of the following papers are published due to the length limitation of the proceedings:

113. OPTIMIZATION OF AMMONIA ABSORPTION REFRIGERATION PROCESS ACCOUNTING FOR SEASONAL AMBIENT TEMPERATURE FLUCTUATIONS

AN JINMING

CHINA TIANJIN CHEMICAL ENGINEERING CORP. CHINA

ABSTRACT

Absorption Refrigeration is a means for energy conservation via recovery of lowlevel heat energy. With the growing emphases on energy conservations, process optimization of ammonia absorption refrigeration system is becoming attractive. But very few publications dealing with this subject have appeared recently. The advanced mathematical model suitable for complex optimization with computer program was given by Prof. Schulz (1972). Unfortunately, this model and method proposed for optimization was worked out and proved only for the thermodynamical criterion of optimization without care for consumption and costs of materials.

In this study the ammonia absorption refregeration process optimization is carried out using flow sheet simulation program with realistic economic criteria. Forthermore, on the basis of the program, the optimization program accounting for seasonal ambient temperature fluctuations was established.

114. A NEW CLOSEDCYCLE THERMOCHEMICAL PROCESS FOR PRODUCTION OF HYDROGEN AND OXYGEN FROM WATER

ZHANG LONG, LI ZHIYAN AND PENG CHENGLING

JILIN INSTITUTE OF TECHNOLOGY, CHANGCHUN,CHINA

ABSTRACT

Thermochemical closedcycle production of hydrogen and oxygen from water is a new technique for energy exploitation developed in the recent 20 years in the world. The main points of this technique are to used some suitable reagents in water decomposion reactions to form a closedcycle system, thus, water can be decomposed into hydrogen by external heat supply at moderate temperature, That is, heat energy is turned into chemical emergy without resorting to any other energy exchange, and it is an efficient method for hydrogen production. Many scientists have done much research work to develop thermochemical process for production of hydrogen, but none of these processes has attained state of industrial practice. Based on some primary research work done in this field, a new fourstep thermochemical cycle for production of hydrogen from water is suggested in this paper.

115. MULTIPLE SOURCE SCREW HEAT PUMP

ZHENG ZUY I
WUHAN REFRIGERATOR WORKS, CHINA

ABSTRACT

The multiple source screw heat pump described in this report is a new type of heat pump which is developed according to the single direction compressing principle of screw refrigerating compressor. It first time has been pointed out the singlestepcompressing, two times throttling, double temperature of heat extraction, twostep suction of the multiple source screw heat pump circulating. According to requirements, it can not only select airsource, but also drawback dual source energy of the surroundings airsource and water source simultaneously. Wherever, no matter how season, it can make the largest work capability of compressor.

116. THE COMPUTER-AIDED DESIGN FOR THE HEAT PIPE HEAT EXCHANGER

CHEN XINGXIANG
UNIVERSITY OF HUNAN
CHANGSHA, CHINA

ABSTRACT

The heat pipe heat excganger is a new kind of exchanger, which is excellent device to recover industrial waste heat. This paper introduces the computeraided design for a heat pipe exchanger. It can be used in the design of gasgas type or gasfluid type of heat pipe heat exchanger.

The heat pipe exchanger consists of heat pipe, an outer shell and a splitter plate. It has the advantages of high efficiency, simple construction, compactness of size, low pressure drops, safe operation, no complementary motive power, etc. Therefore, it is an ideal device to recover industrial waste heat. An introduction of the computer aided design for heat pipe heat exchanger is made in this paper and illustrations of calculation are given as well.

117. RESEARCH ON CALCULATING METHODS FOR DISTILLATION COLUMN WITH DIRECT
VAPOR RECOMPRESSING HEAT PUMP

O'YANG FUCHENG AND GAO WEIPING
JILIN INSTITUTE OF CHEMICAL TECHNOLOGY, CHINA

ABSTRACT

It is effective to employ heat pumps to save energy in distillation. In particular , the direct vapor recompressing heat pump(DVRHP)
has become more attractive in some developed countries in recent year
In this paper, mathematic models are given, and a complete calculating method based on the objective function of minimum loss in available energy is also presented. The results of three technical problems have confirmed the above concept.

118. STUDY OF HEAT TRANSFER OF CASING PIPE UNDERGROUND SOIL STROGE

SHAN XIANFENG, SONG ZEPU AND CUI FENGRU
NORTHEAST CHINA INSTITUTE OF ELECTRIC POWER ENGINEERING
JILIN CITY, CHINA

ABSTRACT

Casing pipe underground soil thermal stroge system is an important
form of soil thermal stroge. The process of charge of the system
is that of unsteady heat conduction. Because of its boundary condition difficult to deal with, it is almost impossible to obtain th
exact solution of the temperature distribution. This paper gives the
the analytical solution of the temperature distribution in soil of
the system during thermal charge by making use of the superposition
of the real casing pipe and an imaginary casing pipe and converting
the casing pipes into sphere chains. It is acceptable to compare the
approximate analytical solution with the experiment data.

119. HEAT RECOVER FROM FLUE GAS WITH HEAT-PIPE HEAT EXCHANGERS

FANG BIN AND NING ZHENPING
HARBIN INSTITUTE OF TECHNOLOGY, CHINA

ABSTRACT

A series of effective heat-pipe heat exchangers(HPHE) have been
developed to recover waste heat from flue gases and to improve the
 burning conditions of combustion devices. The gravity-assisted heat
pipes cost less and have higher capacity. The finned surfaces enhance
the heat transfer with flue gas and make the heat exchangers very
compact. The measures to prevent fouling, wearing and corrosion make

the HPHE practically applicable. The techniques to eliminate noncondensable gas enable the HPHE to be made with cheap components, the steel-water type heat pipes. Comparison show many advantages of the HPHE over conventional heat exchangers.

Applied to boilers in middle and small size, the HPHE increased the thermal efficiency by 5% and enable the boilers to burn well even with low quality coal. Applied to steam locomotives, the HPHE increased the power output by 15% and enable the trains to be on time and fully loaded in cold winter.

The design skeme and poeration performance, the key problems and their solutions, and the economy and future of the HPHE are descreibed and discussed in detail in this paper, which may serve as a reference for similar work on energy conservation and heat transfer enhancement.

120. FLUIDIZED-BED GASIFIER/GASOLINE ENGINE SYSTEM USING RICE HULLS

VIRGIL J. FLANIGAN
UNIVERSITY OF MISSOURI-ROLLA, ROLLA, MO 65401 USA
LUIS C. BAJA, MSME
HAMLIN-ETHIER LTD., SUITE 203, SCOTTS BLDG., 717 PINE ST. ROLLA, MO 65401, USA
JOACQUIN A. TORMO, MS ENG'G MGT.
PASIG AGRICULTURAL DEVELOPMENT AND INDUSTRIAL CORPORATION, 114 PLAZA RIZAL, PASIG, METRO-MANILA, PHILIPPINES

ABSTRACT

The operation of a fluidized-bed gasifier using rice hulls and a speak-ignition, internal combustion(IC) engine on product gas at the optimum reactor conditions and on gasoline produced generator loads of 6.7 KWe and 10.5 KWe at average speeds of 1740 RPM and 1785 RPM and thermal efficiencies of 12.7% and 13.8%, respectively. The gasoline engine could be run on 100% product gas.

The engine was capable of higher generator load outputs at reduced speeds. The results of the study demonstrate that a fluidized-bed gasifier system could be an effective energy conversion system for loose, high ash and high temperature sensitive fuel like rice hulls. For long-term engine operation on product gas, the results suggest that

a pressurized reactor and fuel feeding system would offer a number
of advantages over the current reactor designs. Some improvements
on product gas and air mixing could be achieved by an appropriate
linkage and control system which respond automatically to the generator load. Additional power output may result in turbocharging the
air prior to mixing with the product gas.

121. INVESTIGATION OF THE HEAT TRANSFER LIMIT OF A TWO-PHASE CLOSED
THERMOSIPHON

HUANG LIANMIN, SUN ZENGREN, ZHANG YOUHENG
NANJING POWER COLLEGE OF CHEMICAL TECHNOLOGY
NANJING INSTITUTE OF CHEMICAL TECHNOLOGY

ABSTRACT

The heat transfer limit of a two-phase closed thermosiphon is
investigated experimentally with the methods improved in the experiment under different kinds of working conditions. A great amount
of experimental data are obtained. In the light of the mechanism
that the heat transfer is limited by the flooding, the flooding criterion for two-phase flow in a thermosiphon is developed by taking
account of the particular working conditions in it, which is practically suitable for large range. The correlation of the heat transfer limit of a thermosiphon is developed from the flooding criterion.
The very simple correlations respectively applying to some working
fluids are given..

122. THE STUDY AND APPLICATION OF HEAT TRANSFER ENHANCEMENT IN HEAT EX-
CHANGERS BY MEANS OF TURBULATORS

SUN YUANJIN, ZHANG YONGXING, CHEN YANCHUN, JIANG JINBO
FUSHUN PETROLEUM INSTITUTE, CHINA

ABSTRACT

Inserting a turbulator in the tube of a heat exchanger causes
a spiral motion of the liquid in two half circular tubes. The secondary flow caused by spiral flow destoryed the boundary layer of fluid
inside the tube, accelerated heat transfer and increased pressure
drop. Heat transfer and resistance tests were made by comparing single
tube, 25x2.5x6000 mm, with tubes containing turbulators of 1.1,
1.75 and 3.0 meters. The warm-diesel(tube pass) and vapor-water(shell
pass) were used in the tests. The f-Re and Nu-Re correlations were
obtained with Re from 2000 to 10000. The film heat transfer coeffici-

ents (h) of oil inside the tube can be increased 1.5-3.3 times and the Fanning friction factors(f) be increased 1.9-2.9 times by inserting turbulators in the tube while Re were same. The results of industrial tests showed that h were increased 1.7-1.9 times and f were increased 1.5-1.7 times

123. EXPERIMENTAL RESEARCH OF THE FORCED CONVECTIVE HEAT TRANSFER FOR TUBES WITH OVAL SHAPED CROSS-SECTION

JI ZHONG, NAN XINGSHUANG, REN JIANXUN, QIN FUKUI, QI PINGSHENG & XIE ENLI
QINGDAO INSTITUTE OF CHEMICAL TECHNOLOGY, CHINA

ABSTRACT

The heat-mass transfer anology, with the naphthalene sublimation technique, was used to investigate average heat transfer characteristics of the five kinds of oval shaped cross-section tubes in the Reynolds number range of 2800 to 23800. The results of the experiments were arranged and presented in the form of empirical equations. The results showed that the heat transfer capacity of oval tubes are related to geometrical dimensions. With Reyonlds number over 3500, the Sherwood numbers of the oval tubes are all higher than those of circular tubes, and, in a general way, higher than those of elliptic tubes for Reynolds number greater than 10000. And some types of oval tubes are specially advantageous.

124. A THERMO-HYDRODYNAMIC METHOD DESCRIBING THE TWO-PHASE FLOW IN SUBCOOLED CONVECTIVE BOILING

BARTSCH G., WANG R.
INSTITUT FUR KERNTECHNIK, TECHNISCHE UNIVERSITAT BERLIN
ZHEJIANG UNIVERSITY, CHINA

ABSTRACT

In this paper, a new method is presented to describe the behavior of a two-phase flow in the subcooled boiling region using a heated annular channel. On the basis of microscopic and macroscopic models and by means of a specitif statistical method, the average velocity of a bubble population across the cross section of the channel can be calculated. A complete set of balance equations for mass, momentum and energy is developed. In order to close this equation system, three momentum coefficients are introduced based on nonequilibrium thermodynamics. The ratio between each coefficient and the product of the dynamic viscosity of the liquid and the corresponding temperature of the liquid provides the significant dimensionless number which indicates the influence of the existance of the two-phase flow on the dynamic viscosity. Using these momentum coefficients, the volume void fraction and the slip ratio S can be evaluated. Theoretical and experimental results correspond well.

125. INVESTIGATION OF MECHANISM OF HEAT AND MASS TRANSFER IN ABSORPTION OF LOW PRESSURE WATER VAPOR BY LITHIUM BROMIDE SOLUTION IN FALLING FILM FLOW

RUAN FU CHANG, CHEN LIEQIANG, LUO YUNLU, DENG SONGJIU
SOUTH CHINA UNIVERSITY OF TECHNOLOGY, GUANGZHOU, CHINA

ABSTRACT

In order to find out the mechanism of heat and mass transfer in low pressure water vapor absorption by LiBr aqueous solution in falling film flow a mathematical model describing this absorption process is put out in this paper. It is shown that heat and mass transfer coefficients always increase with the spraying temperature and the spraying density of LiBr aqueous solution, and the calculation results of the mathematical model are content well with these of the experiments.

126. FLOW ANALYSIS OF FALLING FILM ABSORPTION PROCESS ENHANCED BY NEW-TYPE FINNED TUBES IN LiBr ABSORPTION REFRIGERATOR

MA SIPING, CHEN LIEQIANG, LUO YUNLU, DENG SONGJIU
SOUTH CHINA UNIVERSITY OF TECHNOLOGY, GUANGZHOU, CHINA

ABSTRACT

According to the mechansim of heat and mass transfer of noniso-thermal falling liquid film absorption of steam in aqueous lithium bromide(LiBr), with the help of the flow pattern experiment and high-speed photograhy, the cause is explained that low finned tube, saw-teeth-shape-finned tube and pineapple tube are terrible to this absorption process. Then, three new-type tubes: straight-fin-tilted-trough tube, rhombus-fin tube and net-fin tube are developed. By comparing the mixing and renewing, the distribution and the areas of heat and mass transfer of the film flowing down the outer surface of the above horizontal tubes qualitatively, the net-fin tube is considered to be the best for enhancement. The method is approached of decreasing or eliminating the capillary effect between the fins, and the formula for calculating the capillary retention angle of finned tube derived by Wu Peiyi is revised analytically and the revised formula is compared with the experiment result, with satisfactory agreement. The concept of critical retention angle is proposed to characterize the capillary retention.

Index